# Local and Global Aspects of Quasilinear Degenerate Elliptic Equations

## Quasilinear Elliptic Singular Problems

# Local and Global Aspects of Quasilinear Degenerate Elliptic Equations

## Quasilinear Elliptic Singular Problems

Laurent Véron
Université François Rabelais, France

World Scientific

NEW JERSEY • LONDON • SINGAPORE • BEIJING • SHANGHAI • HONG KONG • TAIPEI • CHENNAI • TOKYO

*Published by*

World Scientific Publishing Co. Pte. Ltd.
5 Toh Tuck Link, Singapore 596224
*USA office:* 27 Warren Street, Suite 401-402, Hackensack, NJ 07601
*UK office:* 57 Shelton Street, Covent Garden, London WC2H 9HE

**Library of Congress Cataloging-in-Publication Data**
Names: Véron, Laurent.
Title: Local and global aspects of quasilinear degenerate elliptic equations : quasilinear elliptic singular problems / by Laurent Véron (Université François Rabelais, France).
Description: New Jersey : World Scientific, 2017. |
Includes bibliographical references and index.
Identifiers: LCCN 2017007723 | ISBN 9789814730327 (hardcover : alk. paper)
Subjects: LCSH: Differential equations, Elliptic. | Differential equations, Parabolic. |
Quasilinearization. | Eigenvalues.
Classification: LCC QA377 .V44 2017 | DDC 515/.3533--dc23
LC record available at https://lccn.loc.gov/2017007723

**British Library Cataloguing-in-Publication Data**
A catalogue record for this book is available from the British Library.



Printed in Singapore

To the memory of Stanislav Ivanovich Pohozaev (1935–2014)
and James Burton Serrin (1926–2012), two outstanding
mathematicians, great minds and sincere friends.

# Preface

The study of local properties of solutions of quasilinear elliptic equations

$$-divA(x, u, \nabla u) + g(x, u, \nabla u) = 0 \tag{0.0.1}$$

has been initiated by James Serrin fifty years ago when he introduced genuinely nonlinear methods which did not belong to the world of classical harmonic analysis. However, many of his results were of a form that had made them considered as extensions of those already obtained in the case of linear equations. The two seminal articles he published in 1964 and 1965 contain in particular the proof of Harnack inequality and the Hölder continuity of the solutions of (0.0.1) via the Moser's method, the description of positive solutions with an isolated singularity and the removability of compact sets with zero Sobolev $W^{1,p}$ capacity. They had a tremendous influence on the development of the local theory of solutions of a large range of quasilinear equations. A few years after, a new proof of Harnack inequality for a wider class of equations was published by Trudinger. Later on the study was focused on more specific types of operators $A$, the model case of which being the $p$-Laplacian defined by

$$\Delta_p u = div(|\nabla u|^{p-2} \nabla u) \tag{0.0.2}$$

where $p > 1$. The $C^{1,\alpha}$ regularity was proved by several authors with more and more general assumptions and the *a priori* estimates obtained in these regularity results have been the key-stone in the development, in the mid eighties, of the fine description of isolated singularities of larger and larger classes of solutions of various types of perturbations of the $p$-Laplace equation in a punctured domain, the main examples being the $p$-Laplace equation which deals with solutions of

$$-\Delta_p u = 0, \tag{0.0.3}$$

the $p$-Laplace equation with a zero-order reaction with strong growth,

$$-\Delta_p u + \epsilon |u|^{q-1} u = 0, \tag{0.0.4}$$

or of Hamilton-Jacobi (or Riccati) type

$$-\Delta_p u + \epsilon |\nabla u|^q = 0, \tag{0.0.5}$$

where $q > p - 1$ and $\epsilon = 1$ (the absorption case) or $\epsilon = -1$ (the source case). The study of the behavior of singular solutions depends on the value of $q$ with respect to some critical values of $q$ and these threshold are typical of problems with strong reaction. The precise description of this behavior is linked to the construction of *separable solutions* of (0.0.3), (0.0.4) and (0.0.5), that are solutions under the form $u(x) = |x|^\beta \omega(|x|^{-1}x)$. The spherical $p$-harmonic functions correspond to the separable $p$-harmonic functions. They satisfy a quasilinear equation on the unit sphere $S^{N-1}$ in $\mathbb{R}^N$

$$-div\left(\left(\beta^2\omega^2 + |\nabla'\omega|^2\right)^{\frac{p-2}{2}} \nabla'\omega\right) - \beta\Lambda_p(\beta)\left(\beta^2\omega^2 + |\nabla'\omega|^2\right)^{\frac{p-2}{2}} \omega = 0, \tag{0.0.6}$$

where $\Lambda_p(\beta) = \beta(\beta(p-1) + N - p))$. When $p = 2$ the solutions of (0.0.6) are the spherical harmonics. When $N = 2$ this equation reduces to a separable ordinary differential equation on $S^1$ and all the solutions are known. When $N > 2$ the problem became much harder but infinitely many solution couples $(\beta, \omega)$ have already been obtained associated to regular tesselations of $S^{N-1}$. Two different types of positive separable solutions in any cone vanishing on its boundary exist and they play an important role in the characterization of positive $p$-harmonic functions therein. In some range of values of $p \neq q+1$, there exist separable solutions of (0.0.4). Then $\beta := -\beta_q = -\frac{p}{q+1-p}$ and $\omega$ satisfies

$$\begin{aligned} -div\left(\left(\beta_q^2\omega^2 + |\nabla'\omega|^2\right)^{\frac{p-2}{2}} \nabla'\omega\right) \\ - \beta_q\Lambda_p(\beta_q)\left(\beta_q^2\omega^2 + |\nabla'\omega|^2\right)^{\frac{p-2}{2}} \omega + \epsilon |\omega|^{q-1}\omega = 0. \end{aligned} \tag{0.0.7}$$

There exist also separable solutions of (0.0.5) with $\beta = -\gamma_q = -\frac{p-q}{q+1-p}$ and $\omega$ satisfies

$$\begin{aligned} &-div\left(\left(\gamma_q^2\omega^2 + |\nabla'\omega|^2\right)^{\frac{p-2}{2}}\nabla'\omega\right) \\ &\quad -\gamma_q\Lambda_p(\gamma_q)\left(\gamma_q^2\omega^2 + |\nabla'\omega|^2\right)^{\frac{p-2}{2}}\omega + \epsilon\left(\gamma_q^2\omega^2 + |\nabla'\omega|^2\right)^{\frac{q}{2}} = 0. \end{aligned} \tag{0.0.8}$$

These solutions turned out to have a key role in the classification of solutions with isolated boundary singularity.

Besides the problem of singularities, a second direction of research was open in the same period dealing with quasilinear equations with measure data, using, as a starting point, Stampacchia's truncation method. The first studies concerned the problem

$$\begin{aligned} -\Delta_p u &= \mu && \text{in } \Omega \\ u &= 0 && \text{on } \partial\Omega, \end{aligned} \tag{0.0.9}$$

where $\Omega \subset \mathbb{R}^N$ is a bounded smooth domain and $\mu$ a bounded Radon measure in $\Omega$. Initially the solutions where understood in the sense of distributions, but neither uniqueness nor any kind of stability could be considered. An important progress is made with the introduction of renormalized solutions, in the framework of which the strong convergence of the gradient is obtained and then the stability. Simultaneously, another approach inherited from the classical potential theory was developed in particular by the Finish school of mathematics. The key role was assigned to the $p$-superharmonic functions and the associated $p$-harmonic measures. It was proved later on that these two approaches are in fact equivalent. These results allows to treat problems such as

$$\begin{aligned} -\Delta_p u + \epsilon\,|u|^{q-1}\,u &= \mu && \text{in } \Omega \\ u &= 0 && \text{on } \partial\Omega \end{aligned} \tag{0.0.10}$$

in the *subcritical range* of $q > p-1$, which means that all bounded measures can be considered. These different approaches apply also to more general equations such as

$$\begin{aligned} -\Delta_p u + g(x, u, \nabla u) &= \mu && \text{in } \Omega \\ u &= 0 && \text{on } \partial\Omega. \end{aligned} \tag{0.0.11}$$

The supercritical range was considered much later, using the Wolff potential $\mathbf{W}_{1,p}$ as a key tool. The associated operator appeared to play a role similar to the one of the Newtonian potential in order to provide an explicit representation of $p$-superharmonic functions instead of mere harmonic functions. Then, necessary and sufficient conditions for the existence of renormalized solutions to the problem

$$\begin{aligned} -\Delta_p u &= u^q + \mu \qquad &&\text{in } \Omega \\ u &= 0 &&\text{on } \partial\Omega, \end{aligned} \tag{0.0.12}$$

were obtained under a Lipschitz continuity condition of $\mu$ with respect to the Bessel capacities $c_{p,\frac{q}{q+1-p}}$. In the last five years progresses were obtained in the study of the regularity properties of the Wolff potential of a Radon measure belonging to some Lorentz space. These results are expressed in terms of Lorentz-Bessel spaces. Combined with estimates of Riesz or Bessel potential, they allowed to provide sufficient conditions for existence of renormalized solutions to

$$\begin{aligned} -\Delta_p u + g(u, \nabla u) &= \mu \qquad &&\text{in } \Omega \\ u &= 0 &&\text{on } \partial\Omega, \end{aligned} \tag{0.0.13}$$

with $g$ nondecreasing satisfying some integrability condition, or

$$\begin{aligned} -\Delta_p u + |u|^{q-1} u &= \mu \qquad &&\text{in } \Omega \\ u &= 0 &&\text{on } \partial\Omega, \end{aligned} \tag{0.0.14}$$

with $q$ in the super supercritical range. Problems with an absorption term such as $g(u) = e^{a|u|^\lambda}\text{sgn}(u)$ can also be treated, always in the framework of nonlinear potential estimates. As a by product of these techniques, necessary or sufficient conditions of existence of *large* solutions of quasilinear equations with strong absorption in a domain in connexion with the regularity of its boundary expressed via capacitary Wiener tests, have been proved.

The study of quasilinear equations with measure data, in the new framework of nonlinear potential theory has become a mature subject of research. Many fascinating problems remain unsolved, and many new ones even more fascinating will appear, the solution of which will require new innovating techniques. We hope this book will incite new generations to participate to the progress of this branch of mathematics.

*Laurent Veron*

# Contents

# Chapter 1

# Regularity

## 1.1 Notations

### 1.1.1 *Domains and flow coordinates*

We denote by $\mathbb{R}^N$ the $N$-dimensional real vector space with Euclidean structure $\langle,\rangle$ (or more simply $.$) and associated norm $|.|$. An open ball of center $x_0$ and radius $r > 0$ is denoted by $B_r(x_0)$ (or $B_r$ if there is no ambiguity on its center). Its canonical basis is $\mathcal{B}_{can} := \{\mathbf{e}_1, ..., \mathbf{e}_N\}$ and the coordinates of $x$, $(x_1, ..., x_N)$. The positive half space $\mathbb{R}^N_+$ is the set of $x := (x', x_N)$ with $x' = (x_1, ..., x_{N-1}) \in \mathbb{R}^{N-1}$ and $x_N > 0$ and its boundary $\partial\mathbb{R}^N_+$ can be identified with $\mathbb{R}^{N-1}$. A domain, *i.e.* a connected open set, $\Omega \subset \mathbb{R}^N$, is of class $C^k$ ($k \in \mathbb{N}^*$) if its boundary $\partial\Omega$ is an imbedded submanifold of dimension $N-1$. If $a \in \partial\Omega$ we denote by $T_a(\partial\Omega)$ the tangent hyperplane to $\partial\Omega$ at $a$ and by $\mathbf{n}_a$ the outward normal unit vector to $\Omega$ at $a$.

If $x \in \Omega$, $\rho(x)$ is the distance function from $x$ to $\partial\Omega$. The signed distance function $\dot\rho(x)$ is equal to $\rho(x)$ if $x \in \Omega$ and $-\rho(x)$ if $x \in \Omega^c$. By the triangle inequality, $\dot\rho$ is a contraction mapping from $\mathbb{R}^N$ to $\mathbb{R}$. For $\beta > 0$ we put

$$\begin{aligned} \Omega_\beta &= \{x \in \Omega : \rho(x) < \beta\} \\ \Sigma_\beta &= \partial\Omega_\beta = \{x \in \Omega : \rho(x) = \beta\} \\ \Omega'_\beta &= \{x \in \Omega : \rho(x) > \beta\}. \end{aligned} \tag{1.1.1}$$

Then $\Sigma_0 = \partial\Omega$ and $\Omega'_0 = \Omega$. If $\Omega$ is a bounded $C^k$ domain with $k \geq 2$, there exists $\beta_0 > 0$, depending on the curvature of $\partial\Omega$ (see next section), such that for any $x \in \Omega_{\beta_0}$ there exists a unique $\sigma(x) \in \partial\Omega$ verifying $\rho(x) = |x - \sigma(x)|$. Furthermore the function $\rho$ is of class $C^{k-1}$. All the points $x$ in $\Omega_\beta$ can be represented in a unique way by the couple $(\rho(x), \sigma(x)) \in (0, \beta_0) \times \partial\Omega$, and the mapping $x \mapsto \Pi(x) := (\rho(x), \sigma(x))$ is a $C^{k-1}$ diffeomorphism from

$\Omega_{\beta_0}$ into $(0, \beta_0) \times \partial\Omega$ with inverse $\Pi^{-1}((\rho, \sigma)) = \sigma - \rho \mathbf{n}_\sigma$. This change of variables, called *flow coordinates*, plays an important role in the description of solutions of equations near the boundary.

### 1.1.2 *Riemannian manifolds*

A Riemannian manifold $(M^d, g)$ is a d-dimensional differentiable manifold (always connected) endowed with a symmetric definite positive 2-tensor $g = (g_{ij})$ with inverse matrix $g^{-1} = (g^{ij})$. The tangent space to $M$ at $x$ is denoted by $T_xM$, and the tangent bundle $TM$, *i.e.* the union of the tangent spaces, can be viewed as the set of pairs $(x, v) : x \in M, v \in T_xM$. It is endowed with a natural structure of a 2 d-differentiable manifold of class $C^{k-1}$. If $X$ and $Y$ belong to $T_xM$, we denote

$$\langle X, Y\rangle_g = g(X, Y) = \sum_{i,j} g_{ij} X^i Y^j \quad \text{and } |X|_g = \sqrt{\langle X, X\rangle_g}. \tag{1.1.2}$$

A point $x$ on $M$ can be represented in any chart by a system of local coordinates on $M$, say $(x^j)$, $j$=1,...$d$. The length $L$ of a $C^1$ curve $\gamma$ with parametric representation $\{\gamma(t) = (x^j(t)) : t \in [a, b], j = 1, ..., d\}$ is expressed by

$$L = \int_a^b |\gamma(t)|_g dt := \int_a^b \sqrt{\sum_{ij} g_{ij} \dot{x}^i \dot{x}^j}\, dt. \tag{1.1.3}$$

A curve $\gamma$ with minimal length between two points $x$ and $y$ is a *geodesic*, and this length is the *geodesic distance* between $x$ and $y$ denoted by $d_g(x, y)$; it is a metric on $M$. If $M$ endowed with this metric is a complete metric space it is called a *complete manifold.* The Euler equation corresponding to the minimization of $L$ in (1.1.3) is a second order differential equation. To any $x$ in $M$ and $X \in T_xM$ there exists a curve $\gamma$ defined on some interval $(-a, b)$ $(a, b > 0)$ such that $\gamma(0) = x$, $\gamma'(0) = X$ which has the property that for any $t, t' \in (-a, b)$, it minimizes the distance between $\gamma(t)$ and $\gamma(t')$ provided $|t - t'|$ is small enough. It is called the *exponential map* and denoted by $\gamma(t) = \exp_x(tX)$. The set of points $y = \exp_x(tX) \in M$ such that the geodesic starting from $x$ in some direction $X$ and reaching $y$ ceases to be minimizing on $[0, t + \epsilon)$ for any $\epsilon > 0$ is called the *cut locus* of $x$. The distance function $y \mapsto d_g(x, y)$ is smooth except on the cut locus. The least distance from $x$ to its cut locus is the *injectivity radius* at $x$. A subset $\mathcal{O}$ of $M$ is *convex* if for any $x, y \in \mathcal{O}$ there is a unique geodesic curve joining $x$ to $y$. The *convexity radius* $r_M(a)$ of some $a \in M$ is the supremum of all the $r > 0$ such that the ball $B_r(a)$ is convex.

The exponential map $X \mapsto \exp_x(X))$ may or may not be defined for all $x \in M$ and $X \in T_xM$, however since the differential $D\exp_x(0)$ is the identity in $T_xM$, the exponential map is a local diffeomorphism from a neighborhood of $x$ into $T_xM$. An important result in Riemannian geometry is the Hopf-Rinow theorem.

**Theorem 1.1.1** *Let $(M^d, g)$ be a Riemannian manifold. Then following statements are equivalent:*

*(i) $M$ is a complete metric space.*

*(ii) Every non-empty bounded closed subset of $M$ is compact.*

*(iii) $M$ is geodesically complete, i.e. for every $x \in M$, $X \mapsto \exp_x(X)$ is defined for all $X \in T_xM$.*

An important class of manifolds are the ones which possess a pole.

**Definition 1.1** A Riemannian manifold $(M^d, g)$ *possesses a pole $o$* if $\exp_o$ is a diffeomorphism from $T_oM$ to $M$.

We will see later that the distance function $x \mapsto r(x)$ from the pole $o$ to the point $x$ possesses important properties associated to the curvature of the manifold.

Set $|g| = \det(g_{ij})$, then $dv_g = \sqrt{|g|}dx_1dx_2...dx_d$ is the volume element on $M$. The *curvature Riemann tensor* is a (1-3)-tensor defined in the local coordinates $(x^j)$ by

$$Riem_g = (R^i_{jkl}) = \frac{\partial \Gamma^i_{lj}}{\partial x^k} - \frac{\partial \Gamma^i_{kj}}{\partial x^l} + \sum_n \left(\Gamma^i_{kn}\Gamma^n_{lj} - \Gamma^i_{ln}\Gamma^n_{kj}\right), \tag{1.1.4}$$

where the Christoffel symbols $\Gamma^i_{lj}$ are linear combinations of the derivative of the $g_{ij}$. If $\frac{\partial}{\partial x^j}$, j=1,...,d, denotes the vector field in $T_xM$ associated with the $x^j$ coordinates, then

$$Riem_g\left(\frac{\partial}{\partial x^k}, \frac{\partial}{\partial x^l}\right)\frac{\partial}{\partial x^j} = \sum_i R^i_{jkl}\frac{\partial}{\partial x^i}. \tag{1.1.5}$$

If $X, Y$ and $Z$ are vector fields on $M$, their components in $T_xM$ are expressed by linear combination of the $\frac{\partial}{\partial x^j}$; this allows to define $Riem_g(X,Y)Z$ as an element of $T_xM$. The Riemann curvature tensor describes completely the curvature of $g$, however many computations linked

to analysis of partial equations equations on manifolds, are based upon specific linear combination of the $R^i_{jkl}$, such as the *sectional curvature*, the *scalar curvature*, the *Gaussian curvature* and the *Ricci curvature.*
The sectional curvature is the natural extension to manifold of the Gaussian curvature of surface which is the product of the principal curvature of this surface. If $X$ and $Y$ are independent vectors in $T_xM$, they span a 2-plane $\pi$ tangent to $M$ at $x$. Then the sectional curvature $sec_g(\pi)$ is expressed by

$$sec_g(\pi) = \frac{\langle Riem_g\,(X,Y)\,Y,X\rangle_g}{|X|_g|Y|_g - \langle X,Y\rangle_g^2}. \tag{1.1.6}$$

Similarly to the Gaussian curvature, the sectional curvature $sec_g(\pi)$ is well-defined, independent of the choice of basis $\{X,Y\}$ for $\pi$.
The Ricci curvature $Ricc_g$ is a $(0,2)$-tensor computed on any couple $X,Y$ in $T_xM$. It is the trace of the endomorphism

$$Z \mapsto Riem_g\,(X,Z)\,Y. \tag{1.1.7}$$

If $X$ and $Y$ are expressed in an orthonormal basis $\{\mathbf{e}_j\}$ of $T_xM$, then

$$Ricc_g(X,Y) = \sum_j \langle Riem_g\,(X,\mathbf{e}_j)\,Y,\mathbf{e}_j\rangle_g. \tag{1.1.8}$$

Finally, the scalar curvature $Scal_g$ is the trace of the Ricci curvature. It is expressed by

$$Scal_g = \sum_{i\neq j} Ricc_g(\mathbf{e}_j,\mathbf{e}_i) = \sum_{i\neq j}\langle Riem_g\,(\mathbf{e}_i,\mathbf{e}_j)\,\mathbf{e}_j,\mathbf{e}_i\rangle_g. \tag{1.1.9}$$

If $u$ is a $C^1$ function on $M$, its gradient, quoted by $\nabla u$, is the vector field with covariant components $\nabla_j u = \dfrac{\partial u}{\partial x^j}$ and contravariant components $\nabla^j u = \sum_k g^{j,k}\dfrac{\partial u}{\partial x^k}$. Therefore

$$\|\nabla u\|^2 = \langle \nabla u, \nabla u\rangle_g = \sum_{j,k}\frac{\partial u}{\partial x^j}\frac{\partial u}{\partial x^k}.$$

The divergence of a vector field $\mathbf{X} = (X^j)$ is defined by

$$div_g(\mathbf{X}) = \frac{1}{\sqrt{|g|}}\sum_k \frac{\partial}{\partial x^k}\left(\sqrt{|g|}X^k\right).$$

The Laplacian $\Delta_g$ of a $C^2$ function is the divergence of the covariant gradient of $u$. It is expressed locally by

$$\Delta_g u = \frac{1}{\sqrt{|g|}} \sum_{j,k} \frac{\partial}{\partial x^j}\left(\sqrt{|g|} g^{j,k} \frac{\partial u}{\partial x^k}\right). \tag{1.1.10}$$

The following formulas hold for any $C^2$ real valued function $f$ defined on $\mathbb{R}$ and $u$ and $v$ on $M$,

$$(i) \qquad \Delta_g(uv) = u\Delta_g v + v\Delta_g u + 2\langle \nabla u, \nabla v\rangle_g$$

$$(ii) \qquad \Delta_g f \circ u = f' \circ u \Delta_g u + f'' \circ u \left\|\nabla u\right\|^2.$$

As a special case of (ii), if $w$ is a $C^2$ function, radial with respect to some point $a \in M$, which means $w(x) = \tilde{w}(r(x))$ where $r(x) := d_g(x, a)$, then

$$\Delta_g w = \tilde{w}'' + \tilde{w}' \Delta_g r, \tag{1.1.11}$$

on the set of $x$ where $r$ is regular. It holds in particular if $r(x)$ is smaller than the injectivity radius of $a$.

If $(M^d, g)$ is a compact Riemannian manifold without boundary (resp. with a nonempty boundary), then $-\Delta_g$ has a discrete spectrum in the Sobolev space $W^{1,2}(M)$ (resp. $W_0^{1,2}(M)$) denoted by $\sigma(\Delta_g; W^{1,2}(M))$ (resp. $\sigma(\Delta_g; W_0^{1,2}(M))$). Then

$$(i) \qquad \sigma(\Delta_g; W^{1,2}(M)) = \{\lambda_0 < \lambda_1 \le \lambda_2 \le ... \le \lambda_k \le ...\} \quad \text{with } \lambda_0 = 0$$

$$(ii) \qquad \sigma(\Delta_g; W_0^{1,2}(M)) = \{\lambda_1 \le \lambda_2 \le ... \le \lambda_k \le ...\} \quad \text{with } \lambda_1 > 0.$$

$H_k = N(-\Delta_g - \lambda_k I)$ is a finite dimensional subspace of $W^{1,2}(M)$ (resp. $W_0^{1,2}(M)$) and $W^{1,2}(M) = \oplus_k H_k$ (resp. $W_0^{1,2}(M) = \oplus_k H_k$.)

The second covariant derivatives of $u \in C^2(M)$ are the following

$$\nabla_{jk} u = \frac{\partial^2 u}{\partial x^j \partial x^k} - \sum_{\ell} \Gamma_{jk}^{\ell} \frac{\partial u}{\partial x^{\ell}}. \tag{1.1.12}$$

The Hessian of $u$, denoted $D^2u$ is the covariant symmetric 2-tensor with components $(D^2u)_{jk} = \nabla_{jk}u$ and its norm is $\left\|D^2u\right\|^2 = \sum_{jk}(\nabla_{jk}u)^2$. The following formula is valid for a $C^2$ function $w$ radial with respect to $a$, with $w(x) = \tilde{w}(r(x))$ and with the same restriction on $r(x)$ as above,

$$D^2 w = \tilde{w}'' dr \otimes dr + \tilde{w}' D^2 r. \tag{1.1.13}$$

The Ricci curvature is involved in many estimates of functions on manifolds via the Weinstenböck formula,

**Theorem 1.1.2** *Let $(M^d, g)$ be a Riemannian manifold, then for any $u \in C^2(M)$ there holds*

$$\frac{1}{2}\Delta |\nabla u|^2 = \left\|D^2u\right\|^2 + \langle \nabla u, \nabla \Delta u\rangle_g + Ricc_g(\nabla u, \nabla u). \tag{1.1.14}$$

An important tool for performing analysis on Riemannian manifold are Nash imbedding theorems [Nash (1954)], [Nash (1956)].

**Theorem 1.1.3** *Any closed d-dimensional Riemannian manifold $(M^d, g)$ has a $C^1$ isometric imbedding $i$ into $\mathbb{R}^{2d+1}$. It has a $C^k$ isometric imbedding $i$ into $\mathbb{R}^D$ with $D \leq \frac{1}{2}d(d+1)(3d+11)$ provided $3 \leq k \leq \infty$. If $M$ is compact $D$ turns out to be less or equal to $\frac{1}{2}d(3d+11)$.*

The interest of this result lies in the fact that a Riemannian manifold can be seen as a submanifold of the Euclidean space $\mathbb{R}^D$, the tangent space $T_xM$ is a d-dimensional linear subspace of $\mathbb{R}^D$ and for any $X, Y \in T_xM$, $\langle X, Y\rangle_g = i(X).i(Y)$ where . denotes the Euclidean scalar product in $\mathbb{R}^D$. If $f : M \mapsto \mathbb{R}$ is a $C^1$ function with differential $df_x$ identified with the vector $\delta_x f$ thanks to Riesz theorem via the scalar product $\langle .,.\rangle_g$, it is convenient to consider $\delta_x f$ as a tangent vector to $i(M)$ in $\mathbb{R}^D$.

## 1.2 Distributions, Sobolev and Besov spaces

### 1.2.1 $L^p$ and Lorentz spaces

Let $dx$ be the *Lebesgue measure* in the Euclidean N-space $\mathbb{R}^N$. If $E \subset \mathbb{R}^N$ is a measurable set, we set

$$|E| = \int \chi_E dx, \tag{1.2.1}$$

where $\chi_E$ is the characteristic function of $E$. The space $L^1(E)$ is the space of measurable functions $f : E \mapsto \mathbb{R}$ such that

$$\|f\|_{L^1(E)} := \int |f|\chi_E dx < \infty. \tag{1.2.2}$$

It is actually a quotient space of classes with respect to the equivalence property between two function of equality except for a set of zero Lebesgue measure. Then the mapping $f \mapsto \|f\|_{L^1(E)}$ is a norm on $L^1(E)$. For $1 <$

$p \leq \infty$ we denote by $L^p(E)$ the space of (classes of) measurable functions $f : E \mapsto \mathbb{R}$ such that $|f|^p \in L^1(E)$, if $1 < p < \infty$ or $\operatorname{ess\,sup} |f| < \infty$ if $p = \infty$. The followings are norms on $L^p(E)$

$$\|f\|_{L^p(E)} = \begin{cases} \left( \int |f|^p \chi_E \, dx \right)^{\frac{1}{p}} & \text{if } 1 \leq p < \infty \\ \operatorname{ess\,sup}_E |f| & \text{if } p = \infty. \end{cases} \tag{1.2.3}$$

We denote by $L^p_{loc}(E)$ the space of measurable functions $f : E \mapsto \mathbb{R}$ such that $f\chi_K \in L^p(E)$ for any compact set $K \subset E$. The conjugate of $p$ is $p' = \frac{p}{p-1}$ with the convention $1' = \infty$. If $m : E \mapsto \mathbb{R}_+$ is a nonnegative measurable function, $L^p_m(E)$ denotes the space of measurable functions such that $fm \in L^p(E)$. Then $\|f\|_{L^p_m(E)} := \|mf\|_{L^p(E)}$ is a norm on $L^p_m(E)$. Furthermore the spaces $L^p_m(E)$ are complete.

For $t > 0$, we set $S_f(t) = \{x \in E : |f(x)| > t\}$ and $\lambda_f(t) = |S_f(t)|_m$ where

$$|E|_m = \int m \chi_E \, dx. \tag{1.2.4}$$

The decreasing rearrangement $f^*$ of $f$ is defined by

$$f^*(t) : \inf\{s > 0 : \lambda_f(s) \leq t\}. \tag{1.2.5}$$

If $\Phi : \mathbb{R}_+ \mapsto \mathbb{R}_+$ is a continuous nondecreasing function it is classical that $\Phi(f^*) = (\Phi(f))^*$. We define $f^{**}$ by

$$f^{**}(t) = \frac{1}{t} \int_0^t f^*(\tau) d\tau, \tag{1.2.6}$$

and, for $1 \leq s < \infty$ and $1 < p \leq \infty$, we set

$$\|f\|_{L^{p,s}(E)} = \begin{cases} \left( \int_0^\infty \left( t^{\frac{1}{p}} f^{**}(t) \right)^s \frac{dt}{t} \right)^{\frac{1}{s}} & \text{if } s < \infty \\ \operatorname{ess\,sup}_{t>0} t^{\frac{1}{p}} f^{**}(t) & \text{if } s = \infty. \end{cases} \tag{1.2.7}$$

Then $L^{p,s}(E)$ is the space of measurable functions such that $\|f\|_{L^{p,s}(E)}$ is finite. It is complete for the norm $f \mapsto \|f\|_{L^{p,s}(E)}$. The following relation links the norm in $L^{p,s}(E)$ and the decreasing rearrangement (see [Grafakos (2010)])

$$\left\| t^{\frac{1}{p}} f^* \right\|_{L^s(\mathbb{R}_+, \frac{dt}{t})} \leq \|f\|_{L^{p,s}(E)} \leq \frac{p}{p-1} \left\| t^{\frac{1}{p}} f^* \right\|_{L^s(\mathbb{R}_+, \frac{dt}{t})}. \tag{1.2.8}$$

Furthermore, if $\Omega \subset \mathbb{R}^N$ is a domain, $1 \leq s < \infty$ and $1 \leq p < \infty$, any function in $L^{p,s}(\Omega)$ is the limit of a sequence of $C^\infty$ functions with compact support in $\Omega$. The space $L^{p,\infty}(E)$ is often called Marcinkiewicz space or weak $L^p$ space. The space $L^{p,s}$ can be defined by the real interpolation method [Lions, Peetre (1964)] between $L^1$ and $L^\infty$. The next result is due to Hardy and Littlewood,

**Theorem 1.2.1** *For any measurable functions $f, g$ in $E \subset \mathbb{R}^N$ vanishing at infinity, there holds*

$$\int_E |f(x)g(x)|\, dx \leq \int_0^\infty f^*(t)g^*(t)dt. \tag{1.2.9}$$

This inequality allows to characterize the dual of a given Lorentz space [Grafakos (2010)].

**Theorem 1.2.2** *For any $1 < p, s < \infty$*

$$(L^{p,s}(E))' = L^{p',s'}(E), \tag{1.2.10}$$

*where as usually $p' = \frac{p}{p-1}$, $s' = \frac{s}{s-1}$.*

If $f \in L^{p,s}(E)$ we can write

$$\|f\|^s_{L^{p,s}} := \int_0^\infty \left(t^{\frac{1}{p}} f^*(t)\right)^s \frac{dt}{t} \sim \sum_{j\in\mathbb{Z}} 2^{\frac{js}{p-1}} \int_{2^j}^{2^{j+1}} (f^*(t))^s\, dt, \tag{1.2.11}$$

where $\sim$ stands for equality up to two universal positive multiplicative constants, and

$$\int_{2^j}^{2^{j+1}} (f^*(t))^s\, dt = \int_{\Omega_j} |f(x)|^s\, dx,$$

where $\Omega_j := \{x \in E : f^*(2^{j+1} < |f(x)| < f^*(2^j\}$. Thus $f = \sum_{j\in\mathbb{Z}} f\chi_{\Omega_j}$. Since $|\{x : |f(x)| > t\}| = |\{s : f^*(s) > t\}|$, then

$$\left|\{x : f^*(2^{j+1}) < |f(x)| < f^*(2^j)\}\right| = \left|\{s : f^*(2^{j+1}) < f^*(s) < f^*(2^j)\}\right| = 2^j.$$

This implies $|\Omega_j| = 2^j$ (when non-empty otherwhile $f^*(2^j) = 0$). Therefore

$$\sum_{j\in\mathbb{Z}} 2^{\frac{sj}{p}} (f^*(2^{j+1}))^s \leq \|f\|^s_{L^{p,s}} \leq \sum_{j\in\mathbb{Z}} 2^{\frac{sj}{p}} (f^*(2^j))^s. \tag{1.2.12}$$

There leads to an alternative definition of Lorentz spaces based upon diadic decomposition. For $s > 0$, we denote

$$\ell^s_{\mathbb{Z}} = \{\{c_j\}_{j\in\mathbb{Z}} : \sum_j |c|_j^{\ s} < \infty\},$$

with norm

$$\|\{c_j\}\|_{\ell^s_{\mathbb{Z}}} = \left(\sum_{j\in\mathbb{Z}} |c|_j^{\ s}\right)^{\frac{1}{s}},$$

and with the usual modification if $s = \infty$. The next result can be found in [Grafakos (2010)].

**Theorem 1.2.3** *Let $1 \leq p, s < \infty$. A function $f : E \mapsto \mathbb{R}$ belongs to $L^{p,s}(E)$ if and only if it can be decomposed as*

$$f = \sum_{j\in\mathbb{N}} c_j 2^{\frac{j}{p}} f_j, \tag{1.2.13}$$

*where $\{c_j\} \in \ell^s_{\mathbb{N}}$ and the functions $f_j : E \mapsto \mathbb{R}$ are measurable, have their support mutually disjoint,* $\sup |f_j| \leq 1$, $|\mathrm{supp}(\mathrm{f_j})| \leq 2^j$.

### 1.2.2 *Hausdorff measure*

Let $0 \leq s < \infty$ and $A \subset \mathbb{R}^N$. For $0 < \delta \leq \infty$ we denote by $\mathcal{H}^\delta_s(A)$ the $(s, \delta)$-*dimensional Hausdorff capacity* of $A$ defined by

$$\mathcal{H}^\delta_s(A) = \inf\left\{\sum_j r_j^s : A \subset \bigcup_j B_{r_j}(a_j), r_j \leq \delta\right\}. \tag{1.2.14}$$

The function $\delta \mapsto \mathcal{H}^\delta_s(A)$ is decreasing and

$$\mathcal{H}_s(A) = \lim_{\delta\to 0} \mathcal{H}^\delta_s(A) = \sup_{\delta>0} \mathcal{H}^\delta_s(A) \tag{1.2.15}$$

is the $s$-dimensional *Hausdorff measure* of $A$. Then

(i) $\mathcal{H}^\delta_s$ is an outer measure (or a capacity), $\mathcal{H}_s$ is a Borel measure on $\mathbb{R}^N$.

(ii) For every set $A \subset \mathbb{R}^N$,

$$\mathcal{H}_s(A) = \inf\{\mathcal{H}_s(G) : G \supset A,\ G \text{ open}\}.$$

(iii) For every open set $G \subset \mathbb{R}^N$,

$$\mathcal{H}_s(G) = \sup\{\mathcal{H}_s(K) : K \subset A,\ K \text{ compact}\}.$$

(iv) For every measurable set $A \subset \mathbb{R}^N$ such that $\mathcal{H}_s(A) < \infty$ and every $\epsilon > 0$, there exists a compact set $K \subset A$ such that $\mathcal{H}_s(A \setminus K) < \epsilon$.

(v) If $s > N$, then $\mathcal{H}_s(A) = 0$ for every set $A \subset \mathbb{R}^N$. Furthermore, if $\mathcal{H}_s(A) < \infty$ for some $s \in [0, \infty)$, then $\mathcal{H}_t(A) = 0$ for every $t > s$. On the opposite, if $\mathcal{H}_s(A) > 0$ for some $s \in (0, \infty]$, then $\mathcal{H}_t(A) = \infty$ for every $t > s$. This allows to define the *Hausdorff dimension* of a set $A \subset \mathbb{R}^N$ by

$$dim_{\mathcal{H}}(A) = \inf\{s > 0 : \mathcal{H}_s(A) = 0\}. \tag{1.2.16}$$

**Proposition 1.2.4** *(i) $\mathcal{H}_0$ is the counting measure in $\mathbb{R}^N$*

*(ii) If $N = 1$, $\mathcal{H}_1$ is the Lebesgue measure on $\mathbb{R}$.*

*(iii) If $N \geq 2$, the normalized $N$-dimensional Hausdorff measure defined by $\mathcal{H}^n_N = \omega_N \mathcal{H}_N$ coincides with the $N$-dimensional Lebesgue measure, where we have denoted by $\omega_N$ the volume of the unit ball in $\mathbb{R}^N$ for the $N$-dimensional Lebesgue measure.*

If $X$ is a locally compact topological space, we denote by $\mathfrak{M}(X)$ (resp. $\mathfrak{M}_b(X)$) the set of Radon measures on $X$ (resp. the set of bounded Radon measures on $X$). Its positive cone is $\mathfrak{M}^+(X)$ (resp. $\mathfrak{M}^+_b(X)$). If $\mu \geq 0$, the Borel measure extension of $\mu$ is denoted in the same way. If $K \subset X$ is some closed set, $\mathfrak{M}(K)$ denotes the measures $\mu$ with support in $K$ and $\mathfrak{M}^+(K) = \mathfrak{M}(K) \cap \mathfrak{M}^+(X)$.

### 1.2.3 *Weak derivatives and distributions*

#### 1.2.3.1 *Distributions*

If $\Omega \subset \mathbb{R}^N$ is a domain and $k \in \mathbb{N}$, we denote by $C^k(\Omega)$ the set of functions $\phi : \Omega \mapsto \mathbb{R}$ which are $k$-times continuously differentiable in $\Omega$. If $k = 0$, $C^0(\Omega) = C(\Omega)$ is the space of continuous functions. By extension, $C^\infty(\Omega)$ is the set of functions which are indefinitely continuously differentiable. We also denote by $C^k_0(\Omega)$, $C_0(\Omega)$ and $C^\infty_0(\Omega)$ the subspaces of $C^k(\Omega)$, $C(\Omega)$ and $C^\infty(\Omega)$ of functions with compact support in $\Omega$. If $K$ is a compact subset of $\Omega$, $C^k_0(K)$, $C_0(K)$ and $C^\infty_0(K)$ denotes the respective subspaces of $C^k_0(\Omega)$, $C_0(\Omega)$ and $C^\infty_0(\Omega)$ of functions with support in $K$. An element $\alpha = (\alpha_1, ..., \alpha_N) \in \mathbb{N}^N$ is a multi-indice, $|\alpha| = \sum_j \alpha_j$, $\alpha! = \Pi_j \alpha_j!$ and we denote

$$D^\alpha \phi = \frac{\partial^{|\alpha|}\phi}{\partial x_1^{\alpha_1}...\partial x_N^{\alpha_N}},$$

and

$$\nabla\phi = \sum_j \phi_{x_j}\mathbf{e}_j = \sum_j \frac{\partial\phi}{\partial x_j}\mathbf{e}_j.$$

If $\alpha \in \mathbb{N}^N$ and $\phi \in C_0^{|\alpha|}(\Omega)$ we set

$$\|\phi\|_\alpha = \sup_{x\in\Omega} |D^\alpha \phi(x)|.$$

It is a norm on $C_0^{|\alpha|}(\Omega)$.

**Definition 1.2** A *distribution* $T$ in $\Omega$ is a linear form $\phi \mapsto T(\phi)$ on the space $C_0^\infty(\Omega)$ which possesses the following property: for any compact set $K \subset \Omega$, there exist a constant $C_K$ and a finite subset $I_K$ of $\mathbb{N}^N$ such that for any $\phi \in C_0^\infty(K)$ there holds

$$|T(\phi)| \leq C_K \sum_{j\in I_K} \|\phi\|_{\alpha_j}. \tag{1.2.17}$$

The distributions form a linear space denoted by $\mathcal{D}'(\Omega)$. For any $\alpha \in \mathbb{N}$, the $D^\alpha$ derivative of a distribution $T$ is defined by

$$D^\alpha T(\phi) = (-1)^{|\alpha|}T(D^\alpha\phi) \qquad \forall\phi \in C_0^\infty(\Omega). \tag{1.2.18}$$

It is a distribution. The multiplication of a distribution $T$ by a $C^\infty(\Omega)$ function $\zeta$ is the distribution $\zeta T$ defined by

$$\zeta T(\phi) = T(\zeta\phi) \qquad \forall\phi \in C_0^\infty(\Omega). \tag{1.2.19}$$

The classical Leibnitz rule of differentiation applies to $\zeta T$ at any order.

A sequence $\{T_n\} \subset \mathcal{D}'(\Omega)$ converges weakly to a distribution $T$ if

$$\lim_{n\to\infty} T_n(\phi) = T(\phi) \qquad \forall\phi \in C_0^\infty(\Omega). \tag{1.2.20}$$

Although there exists a strong topology on $\mathcal{D}'(\Omega)$, this weak convergence is sufficient for treating most of the problems encountered in the study of partial differential equations, inasmuch the mappings $T \mapsto D^\alpha T$ and $T \mapsto \zeta T$ are stable for this sequential topology.

If $\Omega = \mathbb{R}^N$ the Fourier transform of $\phi \in C_0^\infty(\mathbb{R}^N)$ is defined by

$$\mathcal{F}(\phi)(x) = \hat{\phi}(x) = \int_{\mathbb{R}^N} e^{-2i\pi\langle x,y\rangle}\phi(y)dy \qquad \forall x \in \mathbb{R}^N. \tag{1.2.21}$$

Then $\hat{\phi} \in C^\infty(\mathbb{R}^N)$ satisfies

$$\lim_{|x|\to\infty} P(x)D^\alpha\hat{\phi}(x) = 0 \qquad \forall P \in \mathbb{R}[x_1,...,x_N],\ \forall\alpha \in \mathbb{N}^N, \tag{1.2.22}$$

where $\mathbb{R}[x_1,...,x_N]$ denotes the set of real polynomial in N real variables. The space $\mathcal{S}(\mathbb{R}^N)$ is the space of $C^\infty$ functions $\phi : \mathbb{R}^N \mapsto \mathbb{R}$ which satisfy $\lim_{|x|\to\infty} P(x)D^\alpha\phi(x) = 0$, for all $P \in \mathbb{R}[x_1,...,x_N]$ and $\alpha \in \mathbb{N}^N$. It is a real algebra stable under differentiation, and the expression

$$\|\phi\|_{P,\alpha} := \sup_{x\in\mathbb{R}^N} |P(x)D^\alpha\phi(x)|, \tag{1.2.23}$$

is a norm on $\mathcal{S}(\mathbb{R}^N)$. Noticing that the Fourier transform has actually complex value and can be extended on the complexified space $\mathcal{S}^\mathbb{C}(\mathbb{R}^N)$

$$\mathcal{S}^\mathbb{C}(\mathbb{R}^N) = \{\phi + i\psi : \phi,\psi \in \mathcal{S}(\mathbb{R}^N)\},$$

by setting $\mathcal{F}(\phi+i\psi) = \mathcal{F}(\phi)+i\mathcal{F}(\psi)$, it defines an isomorphism of $\mathcal{S}^\mathbb{C}(\mathbb{R}^N)$ having for fixed point the function $x \mapsto e^{-\pi|x|^2}$. The inverse of $\mathcal{F}$ is the mapping $\phi \mapsto \mathcal{F}^{-1}(\phi) = \mathcal{F}(\tilde{\phi})$ where $\tilde{.}$ is the symmetry with respect to 0, $\tilde{\phi}(x) = \phi(-x)$. Furthermore this isomorphism is continuous for the topology defined by the norms $\|.\|_{P,\alpha}$.

The space of *tempered distributions* $\mathcal{S}'(\mathbb{R}^N)$ is the subspace of distributions $T$ in $\mathbb{R}^N$ such that there exist $C > 0$, a finite subset $I \subset \mathbb{N}^N$ and a finite set of polynomials $\{P_j\}_{j\in I}$ satisfying

$$|T(\phi)| \leq C\sum_{j\in I} \|\phi\|_{P_j,\alpha_j} \qquad \forall\phi \in C_0^\infty(\mathbb{R}^N). \tag{1.2.24}$$

If $T$ is tempered it can be naturally extended as a linear form on $\mathcal{S}(\mathbb{R}^N)$, still denoted by $T$, which satisfies the same inequality (1.2.24), but for all $\phi \in \mathcal{S}(\mathbb{R}^N)$. The Fourier transform $\mathcal{F}$ is defined on $\mathcal{S}'(\mathbb{R}^N)$ by

$$\mathcal{F}(T) = T(\mathcal{F}(\phi)) \qquad \forall\phi \in C_0^\infty(\mathbb{R}^N). \tag{1.2.25}$$

The mapping $\mathcal{F}$ is an isomorphism from $\mathcal{S}'(\mathbb{R}^N)$ into itself, which inverse is the operator $\mathcal{F}^{-1}$ defined by $\mathcal{F}^{-1}(T)(\phi) = T(\mathcal{F}^{-1}(\phi))$. This isomorphism is continuous with respect to the topology defined by duality with $\mathcal{S}(\mathbb{R}^N)$

endowed with the norms $\|.\|_{P,\alpha}$.

We recall below some standard results:

(i) If $\phi \in L^1(\mathbb{R}^N)$, $\mathcal{F}(\phi)$ is continuous and bounded, $\|\mathcal{F}(\phi)\|_{L^\infty(\mathbb{R}^N)} \leq \|\phi\|_{L^1(\mathbb{R}^N)}$ and $\lim_{|x|\to\infty} \mathcal{F}(\phi(x)) = 0$.

(ii) $\mathcal{F}$ can be naturally extended as an isometric isomorphism from $L^2(\mathbb{R}^N)$ into itself.

(iii) If for $\alpha \in \mathbb{N}$, we set $M^\alpha(x) = x_1^{\alpha_1}...x_N^{\alpha_N}$, then for all $\phi \in \mathcal{S}(\mathbb{R}^N)$ there holds

$$\mathcal{F}(D^\alpha\phi) = (2i\pi)^{|\alpha|}M^\alpha\mathcal{F}(\phi) \quad \text{and} \quad D^\alpha\mathcal{F}(\phi) = (2i\pi^{-|\alpha|})\mathcal{F}(M^\alpha\phi). \quad (1.2.26)$$

(iv) Identities (1.2.26) are valid if $\phi$ is replaced by a tempered distribution.

1.2.3.2 *Weak derivatives*

Let $g \in L^1_{loc}(\Omega)$, we say that $g_j \in L^1_{loc}(\Omega)$ is the weak derivative of $g$ following $x_j$ if

$$\int_\Omega g\frac{\partial\phi}{\partial x_j} = -\int_\Omega g_j\phi dx \qquad \forall\phi \in C_0^\infty(\Omega). \qquad (1.2.27)$$

The function $g_j$ is uniquely determined. More generally, if $\alpha \in \mathbb{N}^N$, a function $g_\alpha \in L^1_{loc}(\Omega)$ is the weak derivative of order $\alpha$ of $g$ if

$$\int_\Omega gD^\alpha\phi = (-1)^{|\alpha|}\int_\Omega g_\alpha\phi dx \qquad \forall\phi \in C_0^\infty(\Omega). \qquad (1.2.28)$$

The weak derivative $g_\alpha$ is, as usually, denoted $D^\alpha g$.

1.2.3.3 *Sobolev spaces*

**Definition 1.3** Let $\Omega \subset \mathbb{R}^N$ be a domain, $1 \leq \infty$ and $k \in \mathbb{N}^*$. The *Sobolev space* $W^{k,p}(\Omega)$ is the space of functions $g \in L^p(\Omega)$ such that $D^\alpha g \in L^p(\Omega)$ for any $\alpha \in \mathbb{N}^N$, with $|\alpha| \leq k$. It is endowed with a structure of complete normed space with the norm

$$\|g\|_{W^{k,p}(\Omega)} = \begin{cases} \left(\sum_{|\alpha|\leq k} \|D^\alpha g\|^p_{L^p(\Omega)}\right)^{\frac{1}{p}} & \text{if } p < \infty \\ \sup_{|\alpha|\leq k} \|D^\alpha g\|_{L^\infty(\Omega)} & \text{if } p = \infty. \end{cases} \qquad (1.2.29)$$

The closure of $C_0^\infty(\Omega)$ in $W^{k,p}(\Omega)$ is denoted by $W_0^{k,p}(\Omega)$. If $\Omega$ is bounded, *Poincaré inequality* asserts

$$\|g\|_{L^p(\Omega)} \leq (\mathrm{diam}\Omega)^{\frac{1}{p'}} \|\nabla g\|_{L^p(\Omega)} \qquad \forall g \in C_0^\infty(\Omega). \tag{1.2.30}$$

As a consequence, an equivalent norm on $W_0^{k,p}(\Omega)$ is

$$\|g\|_{W_0^{k,p}(\Omega)} = \begin{cases} \left(\sum_{|\alpha|=k} \|D^\alpha g\|_{L^p(\Omega)}^p\right)^{\frac{1}{p}} & \text{if } p < \infty \\ \sup_{|\alpha|=k} \|D^\alpha g\|_{L^\infty(\Omega)} & \text{if } p = \infty. \end{cases} \tag{1.2.31}$$

Below are listed some of the main properties of Sobolev spaces

(i) *Gagliardo-Nirenberg inequality.* Assume $1 \leq p < N$. For any $\phi \in C_0^\infty(\mathbb{R}^N)$ there holds

$$\|\phi\|_{L^{p^*}(\mathbb{R}^N)} \leq \frac{p(N-1)}{N-p} \|\nabla\phi\|_{L^p(\mathbb{R}^N)}, \tag{1.2.32}$$

where $p^* = \frac{Np}{N-p}$. The Sobolev constant $S_{N,p}$ is defined by

$$S_{N,p} = \inf\left\{\|\nabla\phi\|_{L^p(\mathbb{R}^N)} : \phi \in C^\infty(\mathbb{R}^N),\ \|\phi\|_{L^{p^*}(\mathbb{R}^N)} = 1\right\}. \tag{1.2.33}$$

(ii) *Gagliardo-Nirenberg interpolation inequality.* Let $1 \leq p, q, r \leq \infty$ and $m \in \mathbb{N}$. If there exist $k \in \mathbb{N}$ and $\theta \in [0,1]$ such that

$$\frac{k}{m} \leq \theta \leq 1 \text{ and } \frac{1}{r} = \frac{k}{N} + \left(\frac{1}{p} - \frac{m}{N}\right) + \frac{1-\theta}{q},$$

then one can find a constant $c = c(N,p,q,m,j,\theta) > 0$ with the property that for any $\phi \in C_0^\infty(\mathbb{R}^N)$ there holds.

$$\left\|D^j\phi\right\|_{L^r(\mathbb{R}^N)} \leq c\,\|D^m\phi\|_{L^p(\mathbb{R}^N)}^\theta \|\phi\|_{L^q(\mathbb{R}^N)}^{1-\theta}, \tag{1.2.34}$$

with the exception that if $m - j - \frac{N}{r} \in \mathbb{N}$, then $\theta < 1$.

(iii) *Morrey inequality.* Assume $N < p \leq \infty$, then there exists $c = c(N,p) > 0$ such that

$$\sup_{x\in\mathbb{R}^N} |\phi(x)| + \sup_{x\neq y\in\mathbb{R}^N} \frac{|\phi(x)-\phi(y)|}{|x-y|^\alpha} \leq c\,\|\nabla\phi\|_{L^p\mathbb{R}^N}, \tag{1.2.35}$$

for any $\phi \in C^\infty(\mathbb{R}^N)$, where $\alpha = 1 - \frac{N}{p}$.

(iv) *Meyers-Serrin approximation.* Assume $\Omega \subset \mathbb{R}^N$ is a domain, $k \in \mathbb{N}^*$ and $1 \leq p < \infty$. If $g \in W^{k,p}(\Omega)$ there exists a sequence of $C^\infty(\Omega)$ functions

$\{g_n\}$ which converges to $g$ in $W^{k,p}(\Omega)$. Moreover, if $\Omega$ is bounded and $\partial\Omega$ Lipschitz, there exists a sequence $\{g_n\} \subset C^\infty(\overline{\Omega})$ which converges to $g$ in $W^{k,p}(\Omega)$

(v) *Extension.* Assume $\Omega \subset \mathbb{R}^N$ is a $C^k$ domain with a compact boundary and $1 \leq p \leq \infty$. Then there exists a continuous linear operator $P$ from $W^{k,p}(\Omega)$ to $W^{k,p}(\mathbb{R}^N)$ such that $P(u)\lfloor_\Omega = u$. Furthermore for any domain $\Omega'$ such that $\overline{\Omega} \subset \Omega'$, $P$ can be constructed in order $P(u) \equiv 0$ outside $\Omega'$.

(vi) *Rellich-Kondrachov theorem.* Let $\Omega \subset \mathbb{R}^N$ be a bounded domain with a $C^1$ boundary and $1 \leq p < N$. Then for any $p \leq q < p^*$ the imbedding of $W^{1,p}(\Omega)$ into $L^q(\Omega)$ is a compact mapping.

A consequence of this compactness theorem is another *Poincaré inequality.* Assume $\Omega \subset \mathbb{R}^N$ is a bounded domain with a $C^1$ boundary and $1 \leq p \leq \infty$. Then there exists $c = c(p, \Omega) > 0$ such that

$$\|\phi - \phi_\Omega\|_{L^p(\mathbb{R}^N)} \leq c \|\nabla\phi\|_{L^p(\mathbb{R}^N)}, \tag{1.2.36}$$

where

$$\phi_\Omega = \frac{1}{|\Omega|}\int_\Omega \phi dx.$$

(vii) *Gagliardo-Nirenberg-Poincaré-Wirtinger inequality.* It is a variant of (1.2.36) and (1.2.32). Assume $\Omega \subset \mathbb{R}^N$ is a bounded domain with a $C^1$ boundary, $1 \leq p < N$, $q \geq 1$ and $0 \leq \theta \leq 1$. If $\phi \in W^{1,p}(\Omega) \cap L^q(\Omega)$, then

$$\|\phi - \phi_\Omega\|_{L^r(\Omega)} \leq c \|\nabla\phi\|^\theta_{L^p(\Omega)} \|\phi\|^{1-\theta}_{L^q(\Omega)}, \tag{1.2.37}$$

where $c = c(\Omega, p, q, \theta) > 0$ and

$$\frac{1}{r} = \frac{1-\theta}{q} + \theta\left(\frac{1}{p} - \frac{1}{N}\right) = \frac{1-\theta}{q} + \frac{\theta}{p^*}, \tag{1.2.38}$$

with the convention that $p^* = \infty$ if $p = N$.

(viii) *Boundary trace.* Let $\Omega \subset \mathbb{R}^N$ be a bounded domain with a $C^1$ boundary and $1 \leq p < \infty$. Then there exists a continuous linear map $\gamma_0$ from $W^{1,p}(\Omega)$ into $L^p(\partial\Omega)$ such that $\gamma_0(g) = g\lfloor_{\partial\Omega}$ for any $g \in W^{1,p}(\Omega) \cap C(\overline{\Omega})$. Furthermore $ker(\gamma_0) = W^{1,p}_0(\Omega)$.

(ix) *Co-area formula* Let $u \in C_c^1(\mathbb{R}^N)$, $w \in L^1(\mathbb{R}^N)$ and $E \subset \mathbb{R}^N$ be a measurable set. Then

$$\int_E w|\nabla u|dx = \int_0^\infty \left( \int_{E\cap s_{|u|}(t)} wd\mathcal{H}^{N-1} \right) dt, \tag{1.2.39}$$

where $s_u(t) = \{x \in \mathbb{R}^N : |u(x)| = t\}$ (see e.g. [Federer (1969), p. 258]). Note that (1.2.39) defines a local change of coordinates in any region where $\nabla u$ does not vanish. In particular

$$\int_{\mathbb{R}^N} |\nabla u|dx = \int_0^\infty Per(S_u(t))dt \tag{1.2.40}$$

where $S_u(t) = \{x \in \mathbb{R}^N : |u(x)| > t\}$ (note that $s_u(t) = \partial S_u(t)$ with the notations of Section 1.2.1) and $Per(S_u)$ denotes the perimeter of $S_u(t)$; it is the (N-1)-dimensional surface area if $\nabla u$ never vanish on $s_u(t)$. Formula (1.2.39) extends to $u \in W^{1,p}(\mathbb{R}^N)$.

### 1.2.3.4 *Fractional Sobolev spaces*

Let $\Omega \subset \mathbb{R}^N$ be a domain and $1 \le p < \infty$. If $0 < s < 1$, the *Aronszajn-Gagliardo-Slobodeckij* norm on $C_0^\infty(\Omega)$ is defined by

$$\|\phi\|_{\dot{W}^{s,p}(\Omega)} = \left( \iint_{\Omega\times\Omega} \frac{|\phi(x) - \phi(y)|^p}{|x-y|^{N+sp}} dxdy \right)^{\frac{1}{p}}. \tag{1.2.41}$$

The space $W^{s,p}(\Omega)$ is the subset of functions $\phi \in L^p(\Omega)$ such that $\|\phi\|_{\dot{W}^{s,p}(\Omega)} < \infty$. It is a complete normed space for the norm

$$\|\phi\|_{W^{s,p}(\Omega)} = \left( \|\phi\|^p_{L^p(\Omega)} + \|\phi\|^p_{\dot{W}^{s,p}(\Omega)} \right)^{\frac{1}{p}},$$

and is called a *fractional Sobolev space.* A limit case when $p = \infty$ yields the Hölder space $C^{0,s}(\Omega)$ with norm

$$\|\phi\|_{C^{0,s}(\Omega)} = \sup_{x\in\Omega} |\phi(x)| + \sup_{x\neq y\in\Omega} \frac{|\phi(x) - \phi(y)|}{|x-y|^s}.$$

We list below some useful results concerning these spaces:

(i) If $\Omega$ is bounded, the $\dot{W}^{s,p}$-norm satisfies the Poincaré inequality

$$\|\phi\|_{L^p(\Omega)} \le c(\mathrm{diam}(\Omega))^{\mathrm{s}} \|\phi\|_{\dot{\mathrm{W}}^{\mathrm{s,p}}(\Omega)} \qquad \forall \phi \in \mathrm{C}_0^\infty(\Omega). \tag{1.2.42}$$

(ii) When $sp < N$, the Gagliardo-Nirenberg inequality holds with $p^*(s) = \frac{Np}{N-sp}$ and $c = c(N,p,s) > 0$ under the form

$$\|\phi\|^p_{L^{p^*(s)}(\mathbb{R}^N)} \leq c\,\|\phi\|^p_{\dot{W}^{s,p}(\mathbb{R}^N)} \qquad \forall\phi \in C_0^\infty(\Omega). \tag{1.2.43}$$

(iii) If $sp > N$ and $\Omega$ has a compact Lipschitz boundary, the Morrey inequality is valid with $\alpha = s - \frac{N}{p}$,

$$\|\phi\|_{C^{0,s}(\Omega)} \leq c\,\|\phi\|^p_{W^{s,p}(\mathbb{R}^N)} \qquad \forall\phi \in W^{s,p}(\Omega). \tag{1.2.44}$$

(iv) If $\Omega$ is bounded and $q < p^*(s)$ the imbedding of $W^{s,p}(\Omega)$ into $L^q(\Omega)$ is compact.

If $1 \leq p < \infty$, $k \in \mathbb{N}^*$ and $0 < s < 1$ the fractional Sobolev spaces $W^{k+s,p}(\Omega)$ is the subset of $\phi \in W^{k,p}(\Omega)$ such that $D^\alpha\phi \in W^{s,p}(\Omega)$ for any $\alpha \in \mathbb{N}^N$ with $|\alpha| = k$. It is a complete space if equipped with the norm

$$\|\phi\|_{W^{k+ss,p}(\Omega)} = \left(\|\phi\|^p_{W^{k,p}(\Omega)} + \sum_{|\alpha|=k}\|D^\alpha\phi\|^p_{\dot{W}^{s,p}(\Omega)}\right)^{\frac{1}{p}}.$$

### 1.2.4 *Bessel potentials*

If $\alpha$ is a real number, the Bessel kernel of order $\alpha$ is defined by

$$G_\alpha = \mathcal{F}^{-1}\left(\left(1+|\xi|^2\right)^{-\frac{\alpha}{2}}\right), \tag{1.2.45}$$

where $\mathcal{F}^{-1}$ is the inverse Fourier transform defined in $\mathcal{S}'(\mathbb{R}^N)$ (see Section 1.2.3). Put

$$\mathbf{G}_\alpha = (I \quad c_\alpha\Delta)^{-\frac{\alpha}{2}}, \tag{1.2.46}$$

for some normalizing $c_\alpha > 0$, then there holds

$$f = \mathbf{G}_\alpha[g] = G_\alpha * g \Longleftrightarrow g = \mathbf{G}_{-\alpha}[f] = G_{-\alpha} * f \qquad \forall f,g \in \mathcal{S}'(\mathbb{R}^N). \tag{1.2.47}$$

Moreover the group property holds

$$\mathbf{G}_\alpha\mathbf{G}_\beta = \mathbf{G}_{\alpha+\beta}, \quad (\mathbf{G}_\alpha)^{-1} = \mathbf{G}_{-\alpha} \qquad \forall\alpha,\beta \in \mathbb{R}. \tag{1.2.48}$$

Finally, for any $R > 0$ the Bessel kernel satisfies a two-sided pointwise estimate (see e.g. [Stein (1971), V-3-1])

$$c^{-1}\frac{\chi_{B_{2R}}(x)}{|x|^{N-\alpha}} \leq G_\alpha(x) \leq c\frac{\chi_{B_R}(x)}{|x|^{N-\alpha}} + ce^{-\frac{|x|}{2}} \qquad \forall x \in \mathbb{R}^N \setminus \{0\}, \tag{1.2.49}$$

where $c = c(N, \alpha, R) > 0$.

#### 1.2.4.1 *Bessel potential spaces*

The classical Bessel potential spaces are defined for $p \geq 1$ and $\alpha \in \mathbb{R}$ by

$$L^{\alpha,p} := L^{\alpha,p}(\mathbb{R}^N) = \left\{f = G_\alpha * g,\, g \in L^p(\mathbb{R}^N)\right\}, \tag{1.2.50}$$

with norm

$$\|f\|_{L^{\alpha,p}} = \|G_{-\alpha} * f\|_{L^p}. \tag{1.2.51}$$

A classical result due to Calderon asserts that $W^{\alpha,p}(\mathbb{R}^N) = L^{\alpha,p}(\mathbb{R}^N)$ when $\alpha$ is an integer and $1 < p < \infty$ (see e.g. [Adams, Hedberg (1999)]). Actually, the $L^{\alpha,p}(\mathbb{R}^N)$ spaces share the same imbedding properties as the ones of Sobolev spaces when $\alpha > 0$ and $1 \leq p < \infty$:

(i) if $\alpha p < N$ and $p \leq q \leq p^{*\alpha}$ where $\frac{1}{p^{*\alpha}} = \frac{1}{p} - \frac{\alpha}{N}$, then $L^{\alpha,p}(\mathbb{R}^N) \subset L^q(\mathbb{R}^N)$;

(ii) if $\alpha p = N$ and $p \leq q < \infty$, then $L^{\alpha,p}(\mathbb{R}^N) \subset L^q_{loc}(\mathbb{R}^N)$;

(iii) if $\alpha p > N$, then $L^{\alpha,p}(\mathbb{R}^N) \subset C(\mathbb{R}^N)$.

#### 1.2.4.2 *Bessel-Lorentz potential spaces*

The scale of Bessel spaces can be refined if we replace the $L^p$ spaces by Lorentz spaces. If $1 \leq s \leq \infty$, $1 < p < \infty$ and $\alpha \in \mathbb{R}$

$$L^{\alpha,p,s} := L^{\alpha,p,s}(\mathbb{R}^N) = \left\{f = G_\alpha * g,\, g \in L^{p,s}(\mathbb{R}^N)\right\}, \tag{1.2.52}$$

with norm

$$\|f\|_{L^{\alpha,p,s}} = \|G_{-\alpha} * f\|_{L^{p,s}}. \tag{1.2.53}$$

The space $L^{\alpha,p,p}(\mathbb{R}^N)$ is identified with $L^{\alpha,p}(\mathbb{R}^N)$. When $1 \leq s, p < \infty$ the space $C_0^\infty(\mathbb{R}^N)$ is dense in $L^{\alpha,p,p}(\mathbb{R}^N)$. The following duality holds

$$\left(L^{\alpha,p,s}(\mathbb{R}^N)\right)' = L^{-\alpha,p',s'}(\mathbb{R}^N). \tag{1.2.54}$$

The Bessel potential $\mathbf{G}_\alpha$ is an isomorphism between Bessel-Lorentz potential spaces.

**Theorem 1.2.5** *Assume $\alpha > 0$, $1 < p < \infty$, $1 < s < \infty$. Then $\mathbf{G}_\alpha$ is an isomorphism between $L^{-\alpha,p,s}(\mathbb{R}^N)$ and $L^{p,s}(\mathbb{R}^N)$.*

### 1.2.4.3 *Bessel-Lorentz capacities*

The Bessel capacity of a compact set $K$ relative to $\mathbb{R}^N$ is given by

$$c^{\mathbb{R}^N}_{\alpha,p,s}(K) = \inf\left\{\|\phi\|^p_{L^{\alpha,p,s}} : \phi \in \omega_K\right\}, \tag{1.2.55}$$

where

$$\omega_K = \{\phi \in \mathcal{S}(\mathbb{R}^N) : \phi \geq 1 \text{ on } K\}. \tag{1.2.56}$$

In this definition, it is equivalent to take test functions in $C_0^\infty(\mathbb{R}^N)$. If $\alpha > 0$ and $1 < s, p < \infty$ we denote by $\overline{\omega}_K^{\alpha,p,s}$ the closure of $\omega_K$ in $L^{\alpha,p,s}(\mathbb{R}^N)$. Then there exists a unique $\Phi_K \in \overline{\omega}_K^{\alpha,p,s}$, called the *capacitary potential* of $K$, such that

$$c^{\mathbb{R}^N}_{\alpha,p,s}(K) = \|\Phi_K\|^p_{L^{\alpha,p,s}}. \tag{1.2.57}$$

This set function is extended to open sets by

$$c^{\mathbb{R}^N}_{\alpha,p,s}(\mathcal{O}) = \sup\left\{c^{\mathbb{R}^N}_{\alpha,p,s}(K), K \subset \mathcal{O},\ K \text{ compact}\right\}, \tag{1.2.58}$$

and to general sets by

$$c^{\mathbb{R}^N}_{\alpha,p,s}(E) = \inf\left\{c^{\mathbb{R}^N}_{\alpha,p,s}(\mathcal{O}), E \subset \mathcal{O},\ \mathcal{O} \text{ open}\right\}. \tag{1.2.59}$$

When $s = p$ and $\alpha$ is an integer, $c^{\mathbb{R}^N}_{\alpha,p,s}$ is a *Sobolev capacity* while if $s = p$, $c^{\mathbb{R}^N}_{\alpha,p,s}$ it is a *Bessel capacity.* Thanks to Calderon's theorem, the Sobolev capacity of a compact set can be directly computed by minimization of the energy

$$c^{\mathbb{R}^N}_{\alpha,p,p}(E) = \inf\left\{\int_{\mathbb{R}^N}\left(\sum_{|\gamma|\leq\alpha}|D^\gamma\phi|^p\right)dx : \phi \in C_0^\infty(\mathbb{R}^N) \cap \omega_K\right\}. \tag{1.2.60}$$

The *dual definition* of a capacity is

$$c^{\mathbb{R}^N}_{\alpha,p,s}(K) = \sup\left\{(\mu(K))^p : \mu \in \mathfrak{M}^+(K), \|G_\alpha * \mu\|_{L^{p',s'}} \leq 1\right\}. \tag{1.2.61}$$

The following result is proved in [Adams, Hedberg (1999), Th 2.2.7] in the $L^p$ framework, but the proofs are similar with $L^{p,s}$.

**Proposition 1.2.6** *Let $\alpha > 0$, $1 < s, p < \infty$, $K \subset \mathbb{R}^N$ be compact and $\Phi_K \in \overline{\omega}_K^{\alpha,p,s}$ the capacitary potential of $K$ defined by (1.2.57). Then the* *capacitary function* $\phi_K$ *of* $K$ *is defined by* $\Phi_K = G_\alpha * \phi_K$. *It belongs to* $L^{p,s}(\mathbb{R}^N)$ *and by (1.2.53),*

$$c^{\mathbb{R}^N}_{\alpha,p,s}(K) = \|\Phi_K\|^p_{L^{\alpha,p,s}} = \|\phi_K\|^p_{L^{p,s}}. \tag{1.2.62}$$

*There exists a nonnegative measure* $\mu_K$, *called the capacitary measure of* $K$, *such that* $\mu_K(K^c) = 0$ *and*

$$\phi_K = (G_\alpha * \mu_k)^{p'-1} \Longrightarrow \Phi_K = G_\alpha * (G_\alpha * \mu_k)^{p'-1}. \tag{1.2.63}$$

*Furthermore* $\mu_K \in L^{-\alpha,p',s'}(\mathbb{R}^N)$ *and*

$$c^{\mathbb{R}^N}_{\alpha,p,s}(K) = \int_{\mathbb{R}^N} (G_\alpha * \mu_k)^{p'} dx = \int_K \Phi_K d\mu_K = \mu_K(K). \tag{1.2.64}$$

The following very practical results are proved by parts in [Grun-Rehomme (1977)], [Adams, Hedberg (1999)], [Feyel, de la Pradelle (1977)] in the case of Sobolev spaces $W^{\alpha,p}(\mathbb{R}^N)$ ($\alpha \in \mathbb{N}$). The proof of (iii) in Bessel spaces is given in [Véron (2004)] and in the case of Bessel-Lorentz spaces in [Bidaut-Véron, Nguyen, Véron (2014), Th 2.5]; therein the convergence in the narrow topology, which requires some new arguments, is also given.

**Proposition 1.2.7** *Let* $\alpha > 0$ *and* $1 < s, p < \infty$.

*(i) If* $K$ *is a compact set such that* $c^{\mathbb{R}^N}_{\alpha,p,s}(K) = 0$ *and* $\mathcal{O}$ *an open set containing* $K$, *there exists a sequence* $\{\xi_n\} \subset C_0^\infty(\mathcal{O})$ *of functions such that* $0 \le \xi_n \le 1$, $\xi_n = 1$ *in a neighborhood ok* $K$ *such that* $\xi_n \to 0$ *in* $L^{\alpha,p,s}(\mathbb{R}^N)$.

*(ii) If* $\lambda \in L^{-\alpha,p',s'}(\mathbb{R}^N) \cap \mathfrak{M}(\mathbb{R}^N)$, *then* $|\lambda|(E) = 0$ *for any Borel set such that* $c^{\mathbb{R}^N}_{\alpha,p,s}(E) = 0$. *We say that* $\lambda$ *is absolutely continuous with respect to the* $c^{\mathbb{R}^N}_{\alpha,p,s}$*-capacity, or that* $\lambda$ *does not charge subsets with zero* $c^{\mathbb{R}^N}_{\alpha,p,s}$*-capacity.*

*(iii) If* $\lambda \in \mathfrak{M}^+(\mathbb{R}^N)$ *is absolutely continuous with respect to the* $c^{\mathbb{R}^N}_{\alpha,p,s}$*-capacity, there exists a nondecreasing sequence of nonnegative measures with compact supports* $\{\lambda_n\} \subset L^{-\alpha,p',s'}(\mathbb{R}^N)$ *which converges to* $\lambda$ *in the weak sense of measures. Furthermore, if* $\lambda \in \mathfrak{M}_b^+(\mathbb{R}^N)$, *then the convergence of* $\{\lambda_n\}$ *towards* $\lambda$ *holds in the narrow topology.*

The following terminology will be useful in the sequel.

**Definition 1.4** Let $\alpha > 0$ and $1 < s, p < \infty$.

(i) A property is said to hold $c_{\alpha,p,s}$-quasieverywhere if it holds everywhere but on a set with zero $c_{\alpha,p,s}$-capacity.

(ii) A function $f$ defined in $\mathbb{R}^N$ is $c_{\alpha,p,s}$-quasicontinuous if for any $\epsilon > 0$ there exists an open set $\mathcal{O} \subset \mathbb{R}^N$ such that $c^{\mathbb{R}^N}_{\alpha,p,s}(E) < \epsilon$ and $f$ is continuous on $\mathcal{O}^c$.

(iii) A set $\mathcal{O} \subset \mathbb{R}^N$ is $c^{\mathbb{R}^N}_{\alpha,p,s}$-quasiopen (quasiopen if there is no ambiguity) if there exists a set $E \subset \mathcal{O}$ such that $c^{\mathbb{R}^N}_{\alpha,p,s}(E) = 0$ and $\mathcal{O} \setminus E$ is open. We define similarly $c^{\mathbb{R}^N}_{\alpha,p,s}$-quasiclosed (quasiclosed) sets.

1.2.4.4 *Riesz and Sobolev capacities*

In some cases it is convenient to use other capacities than the Lorentz-Bessel ones. For $0 < \alpha < N$ we define the Riesz kernel of order $\alpha$, $I_\alpha$ by

$$I_\alpha(x) = c(N,\alpha)\,|x|^{\alpha-N} \qquad \forall x \in \mathbb{R}^N \setminus \{0\}, \tag{1.2.65}$$

for some normalizing positive constant $c(N,\alpha)$ and the associated Riesz potential $\mathbf{I}_\alpha[\mu]$ of a positive measure $\mu$ by

$$\mathbf{I}_\alpha[\mu](x) = I\alpha * \mu(y) = \int_{\mathbb{R}^N} I_\alpha(x-y)d\mu(y). \tag{1.2.66}$$

It can be finite or infinite. If $\mu$ is no longer positive $\mathbf{I}_\alpha[\mu](x)$ is well defined if $\mathbf{I}_\alpha[|\mu|](x) < \infty$. The Riesz kernel corresponds to the inverse operator of $(-\Delta)^{-\frac{\alpha}{2}}$.

For $1 \leq p < \infty$, the Riesz capacity $\dot{c}^{\mathbb{R}^N}_{\alpha,p}$ of a compact set $K \subset \mathbb{R}^N$ is defined by

$$\dot{c}^{\mathbb{R}^N}_{\alpha,p}(K) = \inf\left\{\|\mathbf{I}_\alpha[\phi]\|^p_{L^p(\mathbb{R}^N)} : \phi \in \omega_K\right\}. \tag{1.2.67}$$

This set function is extended to general sets by the same rules as for the Bessel capacity $c^{\mathbb{R}^N}_{\alpha,p}$ and there holds

$$c^{\mathbb{R}^N}_{\alpha,p}(K) \leq \dot{c}^{\mathbb{R}^N}_{\alpha,p}(K). \tag{1.2.68}$$

The main advantage of the Riesz capacity is its homogeneity invariance under dilations: for any $\ell > 0$ there holds

$$\dot{c}^{\mathbb{R}^N}_{\alpha,p}(\ell E) = \ell^{N-\alpha p}\dot{c}^{\mathbb{R}^N}_{\alpha,p}(E). \tag{1.2.69}$$

Similarly to Definition 1.4 it is possible to express that a property holds $\dot{c}_{\alpha,p}$-quasieverywhere, that a function is $\dot{c}_{\alpha,p}$-quasicontinuous and that a set is $\dot{c}_{\alpha,p}$-quasiopen or quasiclosed.

It is also useful to define the fractional Sobolev capacity $C^{\Omega}_{\alpha,s}$ of a compact set $K \subset \Omega$ relative to a domain $\Omega \subset \mathbb{R}^N$ by

$$C^{\Omega}_{\alpha,p}(K) = \inf\{\|\phi\|^p_{W^{\alpha,p}} : \phi \in \omega_{K,\Omega}\} \tag{1.2.70}$$

where

$$\omega_{K,\Omega} = \{\phi \in C_0^\infty(\Omega), \phi \geq 0\,,\ \phi \geq 1 \text{ on } K\}.$$

If $\alpha$ is an integer $C^{\mathbb{R}^N}_{\alpha,p} = c^{\mathbb{R}^N}_{\alpha,p}$ by Calderon's theorem.

## 1.3 General quasilinear elliptic equations

The purpose of this section is to present the historical framework of general quasilinear elliptic equations in a domain $\Omega \subset \mathbb{R}^N$. Consider $A : (x, r, \xi) \in \Omega\times\mathbb{R}\times\mathbb{R}^N \mapsto A(x, r, \xi) \in \mathbb{R}^N$ and $B : (x, r, \xi) \in \Omega\times\mathbb{R}\times\mathbb{R}^N \mapsto B(x, r, \xi) \in \mathbb{R}$ two functions which are Lebesgue measurable with respect to $x$ and continuous in $r$ and $\xi$ ($A$ and $B$ are called *Caratheodory functions*). It is assumed that there exist constants $p > 1$ and $c, c' > 0$ and measurable functions $c_j$, (j=1,..., 7) such that the following inequalities hold for all $(x, r, \xi) \in \Omega \times \mathbb{R} \times \mathbb{R}^N$,

$$\begin{array}{ll} (i) & \langle A(x, r, \xi), \xi\rangle \geq c|\xi|^p - c_1|r|^p - c_2 \\ (ii) & |A(x, r, \xi)| \leq c'|\xi|^{p-1} + c_3|r|^{p-1} + c_4 \\ (iii) & |B(x, r, \xi)| \leq c_5|\xi|^{p-1} + c_6|r|^{p-1} + c_7. \end{array} \tag{1.3.1}$$

A *weak solution* in $\Omega$ of

$$-div\, A(x, u, \nabla u) + B(x, u, \nabla u) = 0, \tag{1.3.2}$$

is a function $u \in W^{1,p}(\Omega)$ which satisfies

$$\int_\Omega (\langle A(x, u, \nabla u), \nabla\phi\rangle + B(x, u, \nabla u)\phi)\, dx = 0 \qquad \forall\phi \in C_0^\infty(\Omega). \tag{1.3.3}$$

By extension, a weak *subsolution* (resp. *supersolution*) of (1.3.2) is a function $u \in W^{1,p}(\Omega)$ which satisfies

$$\int_\Omega (\langle A(x, u, \nabla u), \nabla\phi\rangle + B(x, u, \nabla u)\phi)\, dx \leq 0 \qquad \forall\phi \in C_0^\infty(\Omega), \phi \geq 0, \tag{1.3.4}$$

respectively

$$\int_\Omega (\langle A(x,u,\nabla u), \nabla\phi\rangle + B(x,u,\nabla u)\phi)\,dx \geq 0 \qquad \forall\phi \in C_0^\infty(\Omega), \phi \geq 0. \tag{1.3.5}$$

The key result dealing with positive weak solutions is the fact that they satisfy Harnack inequalities as harmonic functions do. The constants in these inequalities depend in a crucial way of the assumptions on the Lebesgue spaces to which the $c_j$ belong. A classical application of Harnack inequalities lies in the proof of the Hölder continuity of the solutions, but they have also many applications for deriving local or global estimates of solutions.

### 1.3.1 *The Serrin's results*

The assumptions on the $c_j$ made by Serrin [Serrin (1964)], [Serrin (1965)] are as follows:

$$\begin{aligned}
&(i)\ \ c_3, c_4 \in L^{\frac{N}{p-1}}(\Omega),\ c_5 \in L^{N+\epsilon}(\Omega),\ c_1, c_2, c_6, c_7 \in L^{\frac{N}{p}+\epsilon}(\Omega) && \text{if } p < N\\
&(ii)\ c_3, c_4 \in L^{\frac{N}{N-1}+\epsilon}(\Omega),\ c_5 \in L^{N+\epsilon}(\Omega),\ c_1, c_2, c_6, c_7 \in L^{1+\epsilon}(\Omega) && \text{if } p = N\\
&(iii)\ c_3, c_4 \in L^{\frac{N}{p-1}}(\Omega),\ c_5 \in L^{p}(\Omega),\ c_1, c_2, c_6, c_7 \in L^{1}(\Omega) && \text{if } p > N,
\end{aligned} \tag{1.3.6}$$

for some $\epsilon > 0$. However we will concentrate on the case $1 < p \leq N$, since the weak solutions are automatically Hölder continuous if $p > N$. The local upper estimate for subsolutions is the following,

**Theorem 1.3.1** *Assume $1 < p < N$ and (1.3.6) holds. If $u$ is a weak subsolution of (1.3.2) in a ball $B_{3R} := B_{3R}(x_0)$, it satisfies*

$$R^{\frac{N}{p}} \left\|u^+\right\|_{L^\infty(B_R)} + R\left\|\nabla u^+\right\|_{L^p(B_R)} \leq m\left(\left\|u^+\right\|_{L^p(B_{2R})} + kR^{\frac{N}{p}}\right), \tag{1.3.7}$$

*where*

$$m = m\left(c, c', p, N, \epsilon, \left\|c_3\right\|_{L^{\frac{N}{p-1}}(B_{2R})}, R^\epsilon \left\|c_5\right\|_{L^{\frac{N}{1-\epsilon}}(B_{2R})}, R^\epsilon \left\|c_6\right\|_{L^{\frac{N}{p-\epsilon}}(B_{2R})}\right),$$

*and*

$$k = \left(\left\|c_4\right\|_{L^{\frac{N}{p-1}}(B_{2R})} + R^\epsilon \left\|c_7\right\|_{L^{\frac{N}{p-\epsilon}}(B_{2R})}\right)^{\frac{1}{p-1}} + R^\epsilon \left(\left\|c_2\right\|_{L^{\frac{N}{p-\epsilon}}(B_{2R})}\right)^{\frac{1}{p}},$$

*if* $1 < p < N$, *and where*

$$m = m\left(c, c', N, \epsilon, \|c_3\|_{L^{\frac{N}{N-1-\epsilon}}(B_{2R})}, R^\epsilon \|c_5\|_{L^{\frac{N}{1-\epsilon}}(B_{2R})}, R^\epsilon \|c_6\|_{L^{\frac{N}{N-\epsilon}}(B_{2R})}\right),$$

*and*

$$k = \left(\|c_4\|_{L^{\frac{N}{N-1-\epsilon}}(B_{2R})} + R^\epsilon \|c_7\|_{L^{\frac{N}{N-\epsilon}}(B_{2R})}\right)^{\frac{1}{N-1}} + \left(R^\epsilon \|c_2\|_{L^{\frac{N}{N-\epsilon}}(B_{2R})}\right)^{\frac{1}{N}},$$

*if* $p = N$.

*Remark.* By changing $u$ into $-u$ it is easy to see that the above inequality still holds for supersolution $u$ as well, provided $u^+$ is replaced by $u^-$.

The following weak Harnack inequality is fundamental for proving regularity of weak solutions.

**Theorem 1.3.2** *Assume* $1 < p \leq N$ *and (1.3.6) holds. If* $u$ *is a positive weak supersolution of (1.3.2) in* $B_{3R}$ *and* $p - 1 \leq q < \frac{N(p-1)}{N-p}$, *it satisfies*

$$R^{-\frac{N}{q}} \left\|u^{-1}\right\|^{-1}_{L^q(B_{2R})} \leq m_1 \left(\operatorname*{ess\,inf}_{B_R} u + k_1\right), \tag{1.3.8}$$

*where* $m_1$ *and* $k_1$ *depend on the same quantities as* $m$ *and* $k$ *in Theorem 1.3.1.*

Both Theorem 1.3.1 and Theorem 1.3.2 are proved by Moser's iterative scheme which consists in estimating $\|u\|_{L^{\tau a}(B_{r'})}$ in term of $\|u\|_{L^a(B_r)}$ for some $\tau > 1$ and $r' < r$ by a judicious choice of cut-off functions. Combining estimates (1.3.6) and (1.3.8) thanks to John and Nirenberg theorem (see [Serrin (1964), Lemmas 6, 7]) we infer the Harnack inequality.

**Theorem 1.3.3** *Assume* $1 < p \leq N$ *and (1.3.6) holds. If* $u$ *is a positive weak solution of (1.3.2) in* $B_{3R}$, *it satisfies*

$$\operatorname*{ess\,sup}_{B_R} u \leq m_2 \left(\operatorname*{ess\,inf}_{B_R} u + k_2\right), \tag{1.3.9}$$

*where* $m_2$ *and* $k_2$ *depend on the same quantities as* $m$ *and* $k$ *in Theorem 1.3.1.*

It is important to notice that if the terms $c_2$, $c_4$ and $c_7$ in (1.3.6) are identically 0, then all the constants $k$ are 0 and (1.3.9) inherits the following form,

$$\operatorname*{ess\,sup}_{B_R} u \leq m_2 \operatorname*{ess\,inf}_{B_R} u. \tag{1.3.10}$$

When $p > N$ the Harnack inequality takes another form. The requirements on the coefficients $c_j$ are the following

$$c_3, c_4 \in L^{p'}(\Omega)\,,\ c_5 \in L^p(\Omega)\,,\ c_1, c_2, c_6, c_7 \in L^1(\Omega). \tag{1.3.11}$$

**Theorem 1.3.4** *Assume $p \geq N$, (1.3.1) and (1.3.11) hold. If $u$ is a positive weak solution of (1.3.2) in $B_{3R}$, it satisfies*

$$u(x) \leq (u(y) + k)e^{c\left(\frac{|x-y|}{R}\right)^{1-\frac{N}{p}}} \qquad \forall (x,y) \in B_R, \tag{1.3.12}$$

*where*

$$c = c\left(p, N, c, c', R^{\frac{p-N}{p'}} \|c_3\|_{L^{p'}}, R^{\frac{p-N}{p}} \|c_5\|_{L^{p'}}, R^{p-N}(\|c_1\|_{L^1} + \|c_6\|_{L^1})\right),$$

*and*

$$k = \left(\left(R^{\frac{p-N}{p'}} \|c_4\|_{L^{p'}} + R^{p-N} \|c_7\|_{L^1}\right)\right)^{\frac{1}{p-1}} + \left(R^{p-N} \|c_2\|_{L^1}\right)^{\frac{1}{p}}.$$

As a consequence it can be proved, [Serrin (1964), Th 8], that the solutions are Hölder continuous.

**Theorem 1.3.5** *Assume $1 < p \leq N$ and (1.3.6) holds. If $u$ is a weak solution of (1.3.2) in a domain $\Omega$, then it is essentially Hölder continuous in $\Omega$. More precisely, for any bounded subdomains $D, D'$ such that $D \subset \overline{D} \subset \overline{D}' \subset \Omega$ and $|u| \leq M$ in $D'$, there exists $\alpha \in (0,1)$ subject to the same structural assumptions as the $m_j$ such that*

$$|u(x) - u(y)| \leq C\,|x-y|^{\alpha} \qquad \forall (x,y) \in D \times D, \tag{1.3.13}$$

*where $C = C(m_2, k_2, M, D, D')$.*

*Proof.* Let $z \in \Omega$ and $r_0 > 0$ such that $\overline{B}_{2r_0}(z) \subset \Omega$. For $0 < r \leq r_0$ we set

$$\overline{m}(r) = \operatorname*{ess\,sup}_{B_r(z)} u \ \text{ and } \ \underline{m}(r) = \operatorname*{ess\,inf}_{B_r(z)} u,$$

and $\overline{u} = \overline{m}(r) - u$, $\underline{u} = u - \underline{m}(r)$. Then $\overline{u}$ and $\underline{u}$ are nonnegative in $B_r(z)$. The function $\overline{u}$ is a weak solution of an equation of the form

$$-div\,\overline{A}(x, \overline{u}, \nabla\overline{u}) + \overline{B}(x, \overline{u}, \nabla\overline{u}) = 0, \tag{1.3.14}$$

in $B_r(z)$ where the functions $\overline{A}$ and $\overline{B}$ verify inequalities (1.3.1) in which $c$ and $c'$ are unchanged, the $c_j$ for $j = 1,3,5,6$, are replaced by fixed multiples $\bar{c}_j$ independent of $r$ (and thus satisfying uniform bounds in their corresponding Lebesgue spaces), while for $j = 2,4,7$ they are multiplied by positive power of $r$. Applying Theorem 1.3.3 in $B_r(z)$ to $\overline{u}$ and $\underline{u}$, we obtain

$$\begin{aligned}\overline{m}(r) - \underline{m}(\tfrac{r}{3}) = \operatorname*{ess\,sup}_{B_{\frac{r}{3}}(z)} \overline{u} &\leq \overline{m}_2 \left( \operatorname*{ess\,inf}_{B_{\frac{r}{3}}(z)} \overline{u} + \overline{k}_2 \right) \\ &\leq \overline{m}_2 \left( \overline{m}(r) - \overline{m}(\tfrac{r}{3}) + \overline{k}_2 \right).\end{aligned} \tag{1.3.15}$$

Proceeding similarly with $\underline{u}$, we obtain

$$\overline{m}(\tfrac{r}{3}) - \underline{m}(r) \leq \underline{m}_2 \left( \underline{m}(\tfrac{r}{3}) - \underline{m}(r) + \underline{k}_2 \right). \tag{1.3.16}$$

Notice that $\overline{m}_2$, $\underline{m}_2$ are independent of $r$ while there exists $a > 0$ depending on the norms of the $c_j$ such that $\max\{\overline{k}_2, \underline{k}_2\} \leq a r^{\frac{\epsilon}{p-1}}$ by Hölder's inequality. By addition

$$\overline{m}(\tfrac{r}{3}) - \underline{m}(\tfrac{r}{3}) \leq \frac{m_2 - 1}{m_2 + 1} \left( \overline{m}(r) - \underline{m}(r) \right) + \frac{m_2}{m_2 + 1} \left( \overline{k}_2 + \underline{k}_2 \right). \tag{1.3.17}$$

Put $\theta = \frac{m_2 - 1}{m_2 + 1} < 1$, $\tau = \frac{2am_2}{m_2 - 1}$ and denote by $\omega(r)$ the oscillation of $u$ in $B_r(z)$. Then (1.3.17) becomes

$$\omega(\tfrac{r}{3}) \leq \theta \left( \omega(r) + \tau r^{\frac{\epsilon}{p-1}} \right) \qquad \forall 0 < r \leq r_0.$$

Since $\omega$ is nondecreasing, for any $s \geq 3$ there holds

$$\omega(\tfrac{r}{s}) \leq \theta \left( \omega(r) + \tau r^{\frac{\epsilon}{p-1}} \right) \qquad \forall 0 < r \leq r_0.$$

Starting with $r = r_0$ and iterating this inequality, we obtain

$$\omega(\tfrac{r_0}{s^n}) \leq \theta \left( \omega(\tfrac{r_0}{s^{n-1}}) + \tau \left( \tfrac{r_0}{s^{n-1}} \right)^{\frac{\epsilon}{p-1}} \right) \qquad \forall n \in \mathbb{N}^*.$$

An easy induction yields

$$\omega(\tfrac{r_0}{s^n}) \leq \theta^n \omega(r_0) + m_3 \max\{\theta^n, s^{\frac{-n\epsilon}{p-1}}\},$$

where $m_3$ depends in the structural constants of $A$ and $B$. Choosing $s$ such that $\theta \geq s^{\frac{-\epsilon}{p-1}}$ we derive $\omega(\frac{r_0}{s^n}) \leq \theta^n(\omega(r_0) + m_3)$. For $\rho \in (0, s^{-1})$ we can find $n \in \mathbb{N}$ such that $r_0 s^{-n-1} \leq \rho < r_0 s^{-n}$, thus

$$\omega(\rho) \leq \omega(\tfrac{r_0}{s^n}) \leq \theta^n(\omega(r_0) + m_3) \leq \rho^\alpha \left( \tfrac{s}{r_0} \right)^\alpha (\omega(r_0) + m_3),$$

if we choose $\alpha \in (0,1)$ such that $\theta \leq s^{-\alpha}$. This implies the claim. $\square$

A more general form of Harnack inequality is due to [Trudinger (1967)]. The class of operators to which it applies is often called quasilinear equations *with natural growth in the gradient.* The growth assumption on $B$ is

$$|B(x,r,\xi)| \leq b_0 |\xi|^p + b_1(x) |\xi|^{p-1} + b_2(x) |r|^{p-1}. \tag{1.3.18}$$

Furthermore, the integrability requirements on the $b_j$ and the $c_j$ are lowered provided the solution satisfies some local bound. If $\rho \mapsto \epsilon(\rho)$ is a positive decreasing function on $(0,\infty)$ tending to 0 when $\rho \to 0$ we set for $k \geq 1$,

$$L^k_\epsilon(\Omega) = \{\phi \in L^k(\Omega), \|\phi\|_{L^k_\epsilon} < \infty\} \text{ where } \|\phi\|_{L^k_\epsilon} = \sup_{z\in\Omega,\rho>0} \frac{\|\phi\|_{L^k(\Omega\cap B_\rho(z))}}{\epsilon(\rho)}.$$

The assumptions on the $b_j, c_j$ in (1.3.6)-(i)-(ii) and (1.3.18) are the following:

$$c_1, b_2 \in L^{\frac{N}{p}}_{\rho^\delta}(\Omega), \; c_3 \in L^{\frac{N}{p-1}}(\Omega), \; b_1 \in L^N_{\rho^\delta}(\Omega), \tag{1.3.19}$$

for some $\delta > 0$, assuming for simplicity $1 < p \leq N$.

**Theorem 1.3.6** *Let $1 < p \leq N$ and $\Omega \subset \mathbb{R}^N$ be a domain. Assume that $A$ and $B$ satisfy (1.3.1) (i)-(ii) with $c_2 = c_7 = 0$, and (1.3.18) respectively, and that the integrability conditions (1.3.19) hold. If $R > 0$ is such that $B_{3R}(x_0) = B_{3R} \subset \Omega$, let us define $\mu = \mu(R)$ by*

$$\mu = R^{-1} \|c_3\|^{\frac{1}{p-1}}_{L^{\frac{N}{p-1}}(B_{2R})} + R^{\delta-1} \|c_1 + b_3 + b_1^p\|^{\frac{1}{p}}_{L^{\frac{N}{p}}_{\rho^\delta}(B_{2R})},$$

*if $1 < p < N$, and*

$$\mu = R^{\delta-1} \left( \|c_3\|^{\frac{1}{N-1}}_{L^{\frac{N}{N-1}}(B_{2R})} + \|c_1 + b_3 + b_1^N\|^{\frac{1}{N}}_{L^1_{\rho^\delta}(B_{2R})} \right),$$

*if $p = N$. Then, if $u$ is a weak solution of (1.3.2) in $B_{3R}$ which satisfies $|u| \leq M$ in $B_{2R}$, there exists $M_2 > 0$ depending on $c$, $c'$, $N$, $p$, $\mu(R)$ and $M$ such that*

$$\operatorname*{ess\,sup}_{B_R} \leq M_2 \operatorname*{ess\,inf}_{B_R} u. \tag{1.3.20}$$

This inequality implies also the Hölder continuity of the solutions of (1.3.2). The next result characterizes the behaviour of solutions of weak subsolution of (1.3.2) with an isolated singularity.

**Theorem 1.3.7** *Assume $1 < p \leq N$ and (1.3.6) holds. If $u$ is a weak solution of (1.3.2) in $\Omega \setminus \{0\}$ which is bounded from below near 0, then either $u$ can be extended to whole $\Omega$ as weak solution of (1.3.2) in $\Omega$, or there exist $\epsilon_0 > 0$ and $c > 0$ such that the following estimates hold for $0 < |x| < \epsilon_0 < 1$,*

$$\begin{array}{lll} (i) & c|x|^{-\frac{N-p}{p-1}} \leq u(x) \leq c^{-1}|x|^{-\frac{N-p}{p-1}} & \text{if } 1 < p < N \\ (ii) & c \ln |x|^{-1} \leq u(x) \leq c^{-1} \ln |x|^{-1} & \text{if } p = N. \end{array} \tag{1.3.21}$$

The next result due to [Reshetniak (1966)] extends Liouville theorem to weak solutions of

$$-div\ A(x, u, \nabla u) = 0 \qquad \text{in } \mathbb{R}^N, \tag{1.3.22}$$

under the assumption that $A : \mathbb{R}^N \times \mathbb{R} \times \mathbb{R} \mapsto \mathbb{R}^N$ is a Caratheodory vector field such that there exist two positive constants such that for all $(x, r, \xi) \in \mathbb{R}^N \times \mathbb{R} \times \mathbb{R}$,

$$\begin{array}{ll} (i) & \langle A(x, r, \xi), \xi \rangle \geq c|\xi|^p \\ (ii) & |A(x, r, \xi)| \leq c'|\xi|^{p-1}. \end{array} \tag{1.3.23}$$

**Theorem 1.3.8** *Assume $p > 1$. If $u$ is a bounded weak solution of (1.3.22) in $\mathbb{R}^N$, it is a constant.*

*Proof.* We give the proof in the case $1 < p \leq N$, the case $p > N$ being similar thanks to Theorem 1.3.4. For $R > 0$ we set $\overline{m}(R) = \max\{u(x) : |x| \leq R\}$ and $\underline{m}(R) = \min\{u(x) : |x| \leq R\}$. Since $\underline{k}_2 = \overline{k}_2 = 0$, inequality (1.3.17) becomes

$$\overline{m}(R) - \underline{m}(R) \leq \frac{m_2 - 1}{m_2 + 1} \overline{m}(3R) - \underline{m}(3R), \tag{1.3.24}$$

where $m_2 > 1$ is a constant independent of the solution as it is shown in Theorem 1.3.3. Iterating this inequality, we obtain

$$\overline{m}(R) - \underline{m}(R) \leq \left( \frac{m_2 - 1}{m_2 + 1} \right)^k \overline{m}(3^{k+1}R) - \underline{m}(3^{k+1}R) \qquad \forall k \in \mathbb{N}. \tag{1.3.25}$$

Letting $k \to \infty$ and using the boundedness of $u$, we infer $\overline{m}(R) = \underline{m}(R)$ for all $R$. □

*Remark.* The above proof shows actually that there exists a constant $\beta > 0$ such that any weak solution $u$ of (1.3.22) in $\mathbb{R}^N$ satisfying

$$\lim_{|x|\to\infty} \frac{u(x)}{|x|^\beta} = 0, \tag{1.3.26}$$

is a constant. If we set $\theta = \frac{m_2-1}{m_2+1}$, the exponent $\beta$ is given by

$$\beta = -\frac{\ln\theta}{\ln 3}. \tag{1.3.27}$$

### 1.3.2 *Regularity*

#### 1.3.2.1 *Local regularity*

Let $\Omega$ be a domain in $\mathbb{R}^N$, and $A : \Omega\times\mathbb{R}\times\mathbb{R}^N \mapsto \mathbb{R}^N$ and $B : \Omega\times\mathbb{R}\times\mathbb{R}^N \mapsto \mathbb{R}$ be two Caratheodory vector fields which satisfy (1.3.1) (i)-(ii) for $A$ and (1.3.18) for $B$, subject to the integrability conditions (1.3.6) and/or (1.3.19). These assumptions imply the Hölder continuity of the weak solutions of

$$-div\, A(x,u,\nabla u) + B(x,u,\nabla u) = 0 \qquad \text{in } \Omega. \tag{1.3.28}$$

In order to reach an higher regularity, it is assumed that $A$ is differentiable. If $g_0 > 0$, we introduce the class of positive $C^1$ functions on $\mathbb{R}$ subject to the inequality

$$0 \le \frac{tg'(t)}{g(t)} \le g_0 \qquad \forall t \in \mathbb{R}. \tag{1.3.29}$$

We set

$$G(t) = \int_0^t g(s)ds \qquad \forall t \in \mathbb{R}.$$

It is easy to verify that $G$ is a $C^2$ convex function and that there holds

$$\begin{array}{llll} (i) & \dfrac{tg(t)}{1+g_0} \le G(t) \le tg(t) & \forall t \ge 0 & \\ (ii) & \dfrac{G(a)}{G(b)} \le \dfrac{a}{b} & \forall b \ge a \ge 0 & (1.3.30) \\ (iii) & g(t) \le g(2t) \le 2^{g_0} g(t) & \forall t \ge 0 & \\ (iv) & ag(b) \le ag(a) + bg(a) & \forall a,b \ge 0. & \end{array}$$

Concerning $A$ and $B$, we assume that the following inequalities hold

$$
\begin{aligned}
&(i) \quad |D_\xi A| \leq \Lambda(|r|)\frac{g(|\xi|)}{|\xi|} \\
&(ii) \quad \textstyle\sum_{i,j} \dfrac{\partial A_i}{\partial \xi_j}(x,r,\xi)\eta_i\eta_j \geq \dfrac{g(|\xi|)}{|\xi|}|\eta|^2 \quad \forall \eta \in \mathbb{R}^N \\
&(iii) \quad |A(x,r,\xi) - A(y,t,\xi)| \leq \Lambda_1(|r|)(1+g(|\xi|))(|x-y|^\gamma + |r-t|^\gamma) \\
&(iv) \quad |B(x,r,\xi)| \leq \Lambda_1(|r|)(|r|)(1+g(\xi)),
\end{aligned}
\tag{1.3.31}
$$

where $\alpha \in (0,1]$, $\Lambda$ and $\Lambda_1$ are positive increasing functions and $g$ satisfies (1.3.29). We denote by $W^{1,G}(\Omega)$ the class of functions $u \in W^{1,1}_{loc}(\Omega)$ such that $G(u) \in L^1(\Omega)$. If we assume that $A$ and $B$ satisfy (1.3.31), then for any $u \in W^{1,1}_{loc}(\Omega) \cap L^\infty(\Omega)$ the functions $A(.,u,\nabla u)$ and $B(.,u,\nabla u)$ are integrable in $\Omega$, thus this class is the natural framework for studying bounded weak solutions of (1.3.28). The following local regularity theorem is proved in [Lieberman (1991), Th 1.7].

**Theorem 1.3.9** *Assume $p > 1$ and $A$ and $B$ satisfy (1.3.31). Then for any $M > 0$, there exists $\alpha \in (0,1)$, depending on $\gamma$, $N$, $g_0$, $\Lambda_1(M)$ and $\Lambda(M)$ such that any weak solution $u \in W^{1,G}(\Omega) \cap L^\infty(\Omega)$ of (1.3.28) verifying $\|u\|_{L^\infty} \leq M$, satisfies*

$$\|u\|_{C^{1,\gamma}(\overline{\Omega}')} \leq c, \tag{1.3.32}$$

*for any bounded domain $\Omega' \subset \overline{\Omega}' \subset \Omega$, where the constant $c > 0$ depends on $\gamma$, $N$, $g_0$, $g(1)$, $\Lambda_1(M)$, $\Lambda(M)$ and* $\operatorname{dist}(\Omega', \Omega^c)$.

**Applications** The conditions (1.3.31) are verified if $B$ satisfies (1.3.18) and

$$
\begin{aligned}
&(i) \quad |D_\xi A|\,(\kappa + |\xi|)^2 + |D_r A|\,(\kappa + |\xi|) + |D_x A| \leq \Lambda(|r|)(\kappa + |\xi|)^p \\
&(ii) \quad \sum_{i,j} \frac{\partial A_i}{\partial \xi_j}(x,r,\xi)\eta_i\eta_j \geq \lambda(|r|)(\kappa + |\xi|)^{p-2}|\eta|^2 \quad \forall \eta \in \mathbb{R}^N,
\end{aligned}
\tag{1.3.33}
$$

for some $p > 1$, where $\Lambda$ and $\lambda$ are positive functions defined on $\mathbb{R}^+$, respectively increasing and decreasing, and where $\kappa = 1$ or $\kappa = 0$.

### 1.3.2.2 *Boundary regularity*

In this section we consider solutions of the homogeneous Dirichlet problem,

$$\begin{aligned} -div\, A(x,u,\nabla u) + B(x,u,\nabla u) &= 0 \qquad \text{in } \Omega \\ u &= 0 \qquad \text{in } \partial\Omega. \end{aligned} \tag{1.3.34}$$

We assume that there exist positive constants $\lambda$, $\gamma$ and $\gamma \in (0,1]$ and three continuous functions $\mu : \Omega \times \mathbb{R} \mapsto \mathbb{R}$ and $\Lambda : \mathbb{R}^+ \to \mathbb{R}^+$, $C$ increasing, such that the mappings $(x,r,\xi) \mapsto A(x,r,\xi) = (A_1(x,r,\xi), ..., A_N(x,r,\xi))$ and $(x,r,\xi) \mapsto B(x,r,\xi)$ satisfy a.e. in $\Omega$ and for all $r,t \in \mathbb{R}$ and $\xi, \eta \in \mathbb{R}^N$,

$$\begin{aligned}
&(i) \quad \sum_{i,j} \frac{\partial A_i}{\partial \xi_j}(x,r,\xi)\eta_i\eta_j \geq \lambda \left(\mu^2(x,r) + |\xi|^2\right)^{\frac{p-2}{2}} |\eta|^2 \\
&(ii) \quad \left|\frac{\partial A_i}{\partial \xi_j}(x,r,\xi)\right| \leq \Lambda(|r|) \left(\mu^2(x,r) + |\xi|^2\right)^{\frac{p-2}{2}} \\
&(iii) \quad |A(x,r,\xi) - A(y,t,\xi)| \leq \beta(1 + |\xi|^{p-2} + |\xi|^{p-1})(|x-y|^\gamma + |r-t|^\gamma) \\
&(iv) \quad |B(x,r,\xi)| \leq C(|r|)(1 + |\xi|^p).
\end{aligned} \tag{1.3.35}$$

The following result is proved in [Lieberman (1988), Th 1] with $\mu(x) \equiv \kappa$ where $\kappa$ is either 0 or 1 and extended under the form stated here in [Porretta, Véron (2009), Th A1].

**Theorem 1.3.10** *Let $\Omega$ be a bounded domain in $\mathbb{R}^N$ with a $C^{1,\delta}$ boundary for some $\delta \in (0,1]$. Assume $p > 1$ and assumptions (1.3.35) hold. Then for any $M > 0$ there exist a positive constant $c = c(\lambda, \Lambda, \beta, \Omega, p, M)$ and $\alpha \in (0,1]$ such that if $u$ is a bounded solution of (1.3.34) such that $\|u\|_{L^\infty} \leq M$, then $u \in C^{1,\gamma}(\overline{\Omega})$ and more precisely*

$$\|u\|_{C^{1,\alpha}(\overline{\Omega})} \leq c. \tag{1.3.36}$$

### 1.3.3 *Maximum principle*

Two important tools in the study of second order elliptic equations are the strong maximum and the Hopf boundary lemma; they deal with the linear inequality in a domain $\Omega \subset \mathbb{R}^N$.

$$\mathcal{L}u = -\sum_{i,j=1}^{N} a_{ij}(x)\frac{\partial^2 u}{\partial x_i \partial x_j} + \sum_{i=1}^{N} b_i(x)\frac{\partial u}{\partial x_i} + c(x)u \geq 0. \tag{1.3.37}$$

We assume that the functions $a_{ij}$, $b_j$ and $c$ are measurable and locally bounded and that $\mathcal{L}$ is uniformly elliptic in the sense that there exists $\alpha > 0$ such that for almost all $x \in \Omega$,

$$\sum_{i,j=1}^{N} a_{ij}(x)\xi_i\xi_j \geq \alpha\,|\xi|^2 \qquad \forall \xi = (\xi_1, ..., \xi_N) \in \mathbb{R}^N. \tag{1.3.38}$$

The strong maximum principle is well known in the case $c(x) \geq 0$, see [Protter, Weinberger (1967), Th 7], but it is noticed in [Serrin (1971)] that it also holds with no sign assumption on $c(x)$ (see also [Gidas, Ni, Nirenberg (1979), p. 212]).

**Proposition 1.3.11** *Assume $u \in C^2(\Omega)$ is a nonnegative solution of (1.3.38). If there exists $x_0 \in \Omega$ such that $u(x_0) = 0$, then $u \equiv 0$ in $\Omega$.*

Hopf boundary lemma is the following,

**Proposition 1.3.12** *Assume $u \in C^2(\Omega)$ is a nonnegative nontrivial solution of (1.3.38). Let $x_0 \in \partial\Omega$ be such that $u \in C(\Omega \cup \{x_0\})$ and $u(x_0) = 0$, then if there exists a ball $B_r(a) \subset \Omega$ such that $x_0 \in \partial B_r(a)$, then for any unit vector $\mathbf{n}$ at $x_0$, pointing outward $\Omega$, there holds*

$$\limsup_{t \to 0_+} \frac{u(x_0 + t\mathbf{n}) - \mathbf{u}(\mathbf{x_0})}{t} < 0. \tag{1.3.39}$$

Since Proposition 1.3.11 and Proposition 1.3.12 are of a local nature, they also hold in the equations are set in a subdomain $S$ of a Riemannian manifold $(M, g)$ and $\mathbf{n}$ is a unit vector in $T_{x_0}M$ pointing outward $S$. These two results have been extended in the nonlinear framework [Vazquez (1984), Th 5]. The proof is an adaptation of the one in the linear case.

**Proposition 1.3.13** *Let $\Omega \subset \mathbb{R}^N$ be a bounded domain. If $A : \Omega \times \mathbb{R} \times \mathbb{R}^N \mapsto \mathbb{R}^N$ and $B : \Omega \times \mathbb{R} \times \mathbb{R}^N \mapsto \mathbb{R}$ are Caratheodory functions which satisfy (1.3.35) for some $p > 1$. Assume also that the function $C$ in (1.3.35)-(iv) satisfies $C(0) = 0$ and either $C(s) = 0$ for $0 < s \leq s_0$ for some $s_0 > 0$, or $C(s) > 0$ for $s > 0$ and*

$$\int_0^1 \frac{ds}{\sqrt[p]{J(s)}} = \infty, \tag{1.3.40}$$

*where $J(s) = \displaystyle\int_0^s C(\tau)d\tau$. If $u \in C^1(\Omega) \cap W^{1,\infty}(\Omega)$ is a nonnegative solution of (1.3.28) such that $A(., u, \nabla u) \in L^2(\Omega)$ which does not vanish identically in $\Omega$, it is positive everywhere in $\Omega$. If we assume moreover that $u \in$*

*$C^1(\Omega \cup \{x_0\})$ for some $x_0 \in \partial\Omega$ and that $\partial\Omega$ satisfies the interior sphere condition at $x_0$, then*

$$\frac{\partial u}{\partial \mathbf{n}}(x_0) < 0, \tag{1.3.41}$$

*where $\mathbf{n}$ is any outward unit vector at $x_0$ to a sphere $B \subset \Omega$ such that $x_0 \in \partial B$.*

## 1.4 Existence result

In this section we present some useful existence results for solutions of quasilinear equations associated to monotone operators between Banach spaces and Leray-Lions operators.

### 1.4.1 *Method of monotonicity*

We assume here that $V$ is a real separable and reflexive Banach space with norm $\|.\|_V$ and topological dual $V'$ with associated norm $\|.\|_{V'}$. A mapping $\mathcal{A} : V \mapsto V'$ is bounded if it transforms bounded sets into bounded sets.

**Theorem 1.4.1** *Let $v \mapsto \mathcal{A}(v)$ be a bounded operator from $V$ to $V'$, continuous from every finite dimensional subspace of $V$ into $V'$ endowed with the weak topology. Assume $A$ is coercive i.e.*

$$\lim_{\|v\|_V \to \infty} \frac{\mathcal{A}(v)(v)}{\|v\|_V} = \infty, \tag{1.4.1}$$

*and monotone i.e.*

$$(\mathcal{A}(v) - \mathcal{A}(v'))(v - v') \geq 0 \qquad \forall (v, v') \in V \times V. \tag{1.4.2}$$

*Then $\mathcal{A}$ is onto.*

*Proof. Step 1.* Assume $V$ is finite dimensional. Then, up to changing the norms, we can assume that $V = V'$ is endowed with an Euclidean structure and (1.4.1) remains valid. Thanks to (1.4.1) for any $k > 0$ there exists $R = R_k > 0$ such that

$$\mathcal{A}(v)(v) = \langle \mathcal{A}(v), v \rangle \geq kR \qquad \forall v \in \partial B_R. \tag{1.4.3}$$

Let $\theta \in [0, 1]$. If $v \in \partial B_R$, there holds

$$\left| \langle \theta \mathcal{A}(v) + (1 - \theta)\frac{k}{R}v, v \rangle \right| \geq kR \Longrightarrow \left\| \theta \mathcal{A}(v) + (1 - \theta)\frac{k}{R}v \right\|_V \geq k, \tag{1.4.4}$$

which implies that there is no solution of $\theta\mathcal{A}(v)+(1-\theta)\frac{k}{R}v=f$ on $\partial B_R$ if $\|f\|_V<k$. Because of homotopy invariance, the degree of the restriction of $\theta\mathcal{A}+(1-\theta)\dfrac{k}{R}I_d$ to $B_R$ relative to $f\in B_k$ is the same for all $\theta\in[0,1]$ and it is one. If $\theta=1$, it implies that there exists $v\in B_R$ such that $\mathcal{A}(v)=f$.

*Step 2.* Approximate solutions. Since $V$ is separable there exists a sequence $\{w_m\}_{m\in\mathbb{N}_*}\subset V$ of linearly independent vectors with the following property: if $V_m$ is the subspace spanned by $w_1,...,w_m$, then $\bigcup_m V_m$ is dense in $V$. Let $f\in V'$. We claim that for any $m\in\mathbb{N}_*$ there exists $u_m\in V_m$ such that

$$\mathcal{A}(u_m)(w)=f(w)\qquad\forall w\in V_m. \tag{1.4.5}$$

Indeed, we introduce $m$ elements $\theta_1,...,\theta_m$ of $V'$ such that $\theta_j(w_i)=\delta_i^j$, denote by $V'_m$ the space generated by the $\theta_j$ for $j=1,...,m$ and define the linear operator $P_m$ from $V'$ to $V'$ by

$$P_m(v')=\sum_{j=1}^{m}v'(w_j)\theta_j\qquad\forall v'\in V'.$$

If we set

$$B_m(v)=P_m(\mathcal{A}(v)),$$

then (1.4.5) is equivalent to

$$B_m(u_m)=P_m(f).$$

Since $B_m(v)(v)=\mathcal{A}(v)(v)$ the coercivity of $B_m$ in $V_m$ is ensured, therefore there exists $u_m$ satisfying (1.4.5).

*Step 3.* Convergence. By (1.4.5)

$$|\mathcal{A}(u_m)(u_m)|=|f(u_m)|\le\|u_m\|_V\,\|f\|_{V'},$$

which implies that $\|u_m\|_V\le c$ for some $c>0$ by the coercivity assumption. Since $V$ is reflexive and $\mathcal{A}$ is bounded, there exist $u\in V$, $\phi\in V'$ and a subsequence $\{m_k\}$ such that

$$u_{m_k}\rightharpoonup u\quad\text{and}\quad\mathcal{A}(u_{m_k})\rightharpoonup\phi\quad\text{weakly.}$$

Using (1.4.5) we obtain $f(v)=\phi(v)$ for all $v\in V_m$ and all $m\in\mathbb{N}_*$. Since the sequence of $V_m$ is increasing and $\bigcup V_m$ is dense in $V$ we infer $f=\phi$. Let $v\in V_{m_{k_0}}$, then for $m_k\ge m_{k_0}$, we have from (1.4.2)

$$(\mathcal{A}(u_{m_k})-\mathcal{A}(v))(u_{m_k}-v)=\mathcal{A}(u_{m_k})(u_{m_k}-v)-\mathcal{A}(v)(u_{m_k}-v)\ge 0.$$

Hence $f(u_{m_k} - v) - \mathcal{A}(v)(u_{m_k} - v) \geq 0$ by (1.4.5), therefore

$$f(u - v) - \mathcal{A}(v)(u - v) = (f - \mathcal{A}(v)(u - v) \geq 0 \qquad \forall v \in V_{m_{k_0}}. \quad (1.4.6)$$

Since $k_0$ is arbitrary, using again the density of $\bigcup V_m$, we obtain that (1.4.6) is valid for any $v \in V$. Now we fix $w \in V$, $\lambda > 0$ and take $v = u - \lambda w$, then

$$(f - \mathcal{A}(u - \lambda w)(w) \geq 0.$$

Since $\mathcal{A}(u - \lambda w) \rightharpoonup \mathcal{A}(u)$ weakly in $V'$ as $\lambda \to 0$, we derive $(f - \mathcal{A}(u)(w) \geq 0$ for all $w \in V$, which implies $f = \mathcal{A}(u)$. □

The next variant has several applications to quasilinear equations.

**Theorem 1.4.2** *Let $v \mapsto \mathcal{A}(v)$ be a bounded operator from $V$ to $V'$, continuous from every finite dimensional subspace of $V$ into $V'$ endowed with the weak topology and coercive in the sense of (1.4.1). Assume also that there exists a bounded mapping $\tilde{A} : V \times V \mapsto V'$ such that $\tilde{A}(u, u) = \mathcal{A}(u)$ with the following properties:*

*(i) For all $(u, v_0) \in V \times V$ the mapping $t \mapsto \tilde{A}(u, tv_0)$ is continuous from $\mathbb{R}$ into $V'$ endowed with the weak topology, and $\tilde{A}$ is monotone in its second argument, i.e.*

$$(\tilde{A}(u, u) - \tilde{A}(u, v))(u - v) \geq 0 \qquad \forall (u, v) \in V \times V. \quad (1.4.7)$$

*(ii) If $\{u_n\} \subset V$ converges weakly to $u$ and $(\tilde{A}(u_n, u_n) - \tilde{A}(u_n, u))(u_n - u) \to 0$, then $\tilde{A}(u_n, v) \rightharpoonup \tilde{A}(u, v)$ weakly in $V'$ for all $v \in V$.*

*(iii) If $\{u_n\} \subset V$ converges weakly to $u$ and $\tilde{A}(u_n, v) \rightharpoonup v'$ weakly in $V'$, then*

$$\tilde{A}(u_n, v)(u_n) \to v'(u).$$

*Then $\mathcal{A}$ is onto.*

*Remark.* A mapping which is continuous on any straight line is called *hemicontinuous* . Therefore for any $u \in V$, $v \mapsto \tilde{A}(u, v)$ is hemicontinuous from $V$ to $V'$-weak.

**Application.** Let $1 < p < \infty$, $\Omega$ be a bounded domain in $\mathbb{R}^N$ with a Lipschitz continuous boundary and $V$ a subspace of $W^{1,p}(\Omega)$ closed for the norm $\|.\|_{W^{1,p}}$, which contains $W_0^{1,p}(\Omega)$. Let $B : \Omega \times \mathbb{R} \to \mathbb{R}$ and

$A_j : \Omega \times \mathbb{R} \times \mathbb{R}^N \to \mathbb{R}$, $j = 1, ..., N$, be Caratheodory functions. We denote $A = (A_1, ..., A_N)$ and we consider the following assumptions

$$\begin{aligned} &(i)\ \ |B(x,r)| \leq c_0 \left(|r|^{p-1} + k(x)\right) && \forall (x,r) \in \Omega \times \mathbb{R} \\ &(ii)\ |A(x,r,\xi)| \leq c' \left(|r|^{p-1} + |\xi|^{p-1} + k(x)\right) && \forall (x,r,\xi) \in \Omega \times \mathbb{R} \times \mathbb{R}^N, \end{aligned} \tag{1.4.8}$$

where $c' > 0$, $c_0 \geq 0$ and $k \in L^{p'}(\Omega)$. Thanks to these assumptions, for any $v \in V$ the functions $A_j(x, v, \nabla v)$ belong to $L^{p'}(\Omega)$. Then for any $(u, v) \in V \times V$ the integral

$$a(u,v) = \int_\Omega \left(\langle A(x,u,\nabla u), \nabla v\rangle + B(x,u)v\right) dx, \tag{1.4.9}$$

is well defined. Furthermore $v \mapsto a(u, v)$ is a continuous linear mapping over $V$, therefore

$$a(u,v) = \mathcal{A}(u)(v) \qquad \forall v \in V, \ \text{ where } \mathcal{A}(u) \in V'. \tag{1.4.10}$$

For $f \in V'$, the following equivalence holds:

$$a(u,v) = f(v) \quad \forall v \in V \Longleftrightarrow \mathcal{A}(u) = f. \tag{1.4.11}$$

The next result is easy to prove by using Theorem 1.4.1 and Theorem 1.4.2 (see e.g. [Leray, Lions (1965), Th. 2]).

**Corollary 1.4.3** *Assume (1.4.8) holds. If A satisfies for almost all $x \in \Omega$ and all $r \in \mathbb{R}$*

$$\langle A(x,r,\xi) - A(x,r,\xi'), \xi - \xi'\rangle > 0 \quad \forall (\xi,\xi') \in \mathbb{R}^N \times \mathbb{R}^N,\ \xi \neq \xi', \tag{1.4.12}$$

*and*

$$\langle A(x,r,\xi), \xi\rangle \geq c\,|\xi|^p - c_2 \quad \forall \xi \in \mathbb{R}^N, \tag{1.4.13}$$

*where $c > 0$ and $c_2$ are independent of $x$ and $r$, then there exists a real number $\lambda_0$ such that for any $\lambda > \lambda_0$ and any $f \in V'$ one can find $u \in V$ such that*

$$a(u,v) + \lambda \int_\Omega |u|^{p-2}\, uv = f(v) \qquad \forall v \in V. \tag{1.4.14}$$

An operator $\mathcal{A}$ such that (1.4.8)-(ii), (1.4.11) and (1.4.13) are satisfied is called a second order *Leray-Lions operator*. The number $\lambda_0$ depends both on $c_0$, $p$ and $\Omega$.

### 1.4.2 *Method of super and sub solutions*

In this section it is assumed that $p > 1$ and $\Omega$ is a domain in $\mathbb{R}^N$ or on a Riemannian manifold, not necessarily bounded. We give existence results with precise two-sided estimates for some solution of equation (1.3.2) in presence of a supersolution and a subsolution.

Concerning $B$, we assume the existence of an increasing function $C$ from $\mathbb{R}^+$ to $\mathbb{R}^+$ such that for almost all $x$ in $\Omega$ and all $r \in \mathbb{R}$ and $\xi \in \mathbb{R}^N$, there holds

$$|B(x, r, \xi)| \leq C(|r|)(1 + |\xi|^p). \tag{1.4.15}$$

By a solution of (1.3.1) we mean a function $u \in W_0^{1,p}(\Omega) \cap L^\infty(\Omega)$ which satisfies

$$-div\, A(., u, \nabla u) + B(., u, \nabla u) = 0 \qquad \text{in } \mathcal{D}'(\Omega). \tag{1.4.16}$$

**Definition 1.5** A function $\zeta \in W^{1,p}(\Omega) \cap L^\infty(\Omega)$ verifying $\zeta \leq 0$ in a neighborhood of $\partial\Omega$ is a subsolution of problem (1.4.16) above if

$$-div\, A(., \zeta, \nabla \zeta) + B(., \zeta, \nabla \zeta) \leq 0 \qquad \text{in } \mathcal{D}'(\Omega). \tag{1.4.17}$$

Similarly $\zeta \in W^{1,p}(\Omega) \cap L^\infty(\Omega)$ verifying $\zeta \geq 0$ in a neighborhood of $\partial\Omega$ is a supersolution of problem (1.4.16) above if

$$-div\, A(., \zeta, \nabla \zeta) + B(., \zeta, \nabla \zeta) \geq 0 \qquad \text{in } \mathcal{D}'(\Omega). \tag{1.4.18}$$

The next result has many applications [Boccardo, Murat, Puel (1984)].

**Theorem 1.4.4** *Assume $p > 1$, $\mathcal{A}$ is a second order Leray-Lions operator, i.e. (1.4.8)-(ii), (1.4.12) and (1.4.13) hold, and $B$ satisfies (1.4.15). If there exist a supersolution $\phi$ and a subsolution $\psi$ of (1.4.16), belonging to $W^{1,\infty}(\Omega)$ and such that $\psi \leq \phi$, then there one can find a solution $u$ of of (1.4.16) satisfying $\psi \leq u \leq \phi$.*

*Proof.* Let us define the operator $\mathcal{A}$ from $W_0^{1,p}(\Omega)$ into $W^{-1,p'}(\Omega)$ by

$$\mathcal{A}(v)(\zeta) = \int_\Omega \langle A(x, v, \nabla v).\nabla \zeta \rangle dx \qquad \forall (v, \zeta) \in W_0^{1,p}(\Omega) \times W_0^{1,p}(\Omega). \tag{1.4.19}$$

In [Boccardo, Murat, Puel (1984)], in (1.3.2)-(i) it is assumed that $c_1 = c_2 = 0$, however this assumption is unnecessary in order to ensure the coercivity of $\mathcal{A}$ *i.e.*

$$\lim_{\|v\|_{W^{1,p}}\to\infty} \frac{\mathcal{A}(v)}{\|v\|_{W^{1,p}}} = \infty. \tag{1.4.20}$$

Actually, we can even assume that $c_2 \in L^1(\Omega)$. The proof is almost the same as the one of [Boccardo, Murat, Puel (1984), Th. 2.1], therefore we present a sketch of it and emphasize the modifications needed due to the fact that $c_2 \neq 0$, and we simplify some points.

*Step 1.* The truncated problem. The key idea is to truncate $A_j$ and $B$ by $\psi$ and $\phi$ and define for almost all $x \in \Omega$ and all $(r,\xi) \in \mathbb{R} \times \mathbb{R}^N$

$$H_j(x,r,\xi) = \begin{cases} A_j(x,r,\xi) & \text{if } \psi(x) < r < \phi(x) \\ A_j(x,\psi(x),\xi) & \text{if } r \leq \psi(x) \\ A_j(x,\psi(x),\xi) & \text{if } \phi(x) \leq r, \end{cases} \tag{1.4.21}$$

and

$$F(x,r,\xi) = \begin{cases} B(x,r,\xi) & \text{if } \psi(x) < r < \phi(x) \\ B(x,\psi(x),\nabla\psi(x)) & \text{if } r \leq \psi(x) \\ B(x,\phi(x),\nabla\phi(x)) & \text{if } \phi(x) \leq r. \end{cases} \tag{1.4.22}$$

Notice that $F$ is no longer a Caratheodory function. If $v \in W_0^{1,p}(\Omega)$ we set $H_j(v,\nabla v)(x) = H_j(x,v(x),\nabla v(x))$ and $F(v,\nabla v)(x) = F(x,v(x),\nabla v(x))$, for $j = 1,...,N$ and for almost all $x \in \Omega$. We put $H = (H_1,...,H_N)$ and consider the *truncated problem* of finding $u \in W_0^{1,p}(\Omega)$ solution of

$$-div\, H(u,\nabla u) + F(u,\nabla u) = 0 \qquad \text{in } \mathcal{D}'(\Omega). \tag{1.4.23}$$

It is clear that the $H_j$ are Caratheodory functions and (1.4.8)-(ii), (1.4.12) and (1.4.13) holds, thus the operator $\mathcal{H}$ defined by $\mathcal{H}(u) = -div\, H(u,\nabla u)$ is a Leray-Lions operator. The operator $\mathcal{F}$ defined by $\mathcal{F}(u) = F(u,\nabla u)$ is defined for all $u \in W^{1,p}(\Omega)$, it is continuous from $W^{1,p}(\Omega)$ into $L^1(\Omega)$, furthermore there exists $C_0 > 0$ such that

$$|F(u,\nabla u)(x)| \leq C_0(1 + |\nabla u(x)|^p) \quad \text{a.e. in } \Omega, \tag{1.4.24}$$

(see [Boccardo, Murat, Puel (1984), Lemmas 3.1,3.2]).

*Step 2.* Two-sided estimates of solutions of (1.4.23). If $u$ is a solution of (1.4.23), the function $(\psi - u)_+$ belongs to $W_0^{1,p}(\Omega)$, thus for any $m >$

$(u-\psi)_{+,m} = \inf\{m, (\psi - u)_+\}$ it is an element of $W_0^{1,p}(\Omega) \cap L^\infty(\Omega)$. Since $\psi$ is a subsolution of (1.4.16) and $A(., \psi, \nabla\psi) = H(., \psi, \nabla\psi)$, we have

$$\begin{aligned}\int_\Omega (\langle H(\psi, \nabla\psi) - H(u, \nabla u), \nabla(\psi - u)_{+,m}\rangle \\ +(B(\psi, \nabla\psi) - F(u, \nabla u))(\psi - u)_{+,m})\, dx \leq 0.\end{aligned} \tag{1.4.25}$$

Defining

$$\omega_0 = \{x \in \Omega : \psi(x) - u(x) \geq 0\},$$

and

$$\omega_0^m = \{x \in \Omega : 0 \leq \psi(x) - u(x) \leq m\} \subset \omega_0,$$

then

$$(\psi - u)_{+,m} = 0 \quad \text{a.e. on } \Omega \setminus \omega_0,$$

and

$$\nabla(\psi - u)_{+,m} = 0 \quad \text{a.e. on } \Omega \setminus \omega_0^m.$$

By the definition of $H$ and $F$, there holds a.e. on $\omega_0$ (and thus on $\omega_0^m$)

*(i)* $\qquad A(x, u(x), \nabla u(x)) = A(x, \psi(x), \nabla u(x))$

*(ii)* $\qquad A(x, \psi(x), \nabla\psi(x)) = A(x, \psi(x), \nabla\psi(x))$

*(iii)* $\qquad F(x, u(x), \nabla u(x)) = B(x, \psi(x), \nabla\psi(x)).$

Therefore

$$\begin{aligned}&\int_\Omega B(\psi, \nabla\psi) - F(u, \nabla u))(\psi - u)_{+,m} dx \\ &\qquad = \int_{\omega_0} B(\psi, \nabla\psi) - F(u, \nabla u))(\psi - u)_{+,m} dx = 0,\end{aligned}$$

and (1.4.25) becomes

$$\int_{\omega_0^m} \langle H(\psi, \nabla\psi) - H(u, \nabla u), \nabla(\psi - u)_{+,m}\rangle dx \leq 0.$$

By (1.4.12) $\nabla(\psi - u)_{+,m} = 0$ a.e. on $\omega_0^m$, which implies $(\psi - u)_{+,m} = 0$ for all $m > 0$ and finally $\psi \leq u$ a.e. in $\Omega$. Similarly $u \leq \phi$. Moreover it is classical that $\nabla u = \nabla\psi$ (resp. $\nabla u = \nabla\phi$) a.e. on the set where $u$ and $\psi$ (rest. $u$ and $\phi$) coincide. This implies that $u$ is a solution of (1.4.16).

*Step 3.* Approximate solutions. For $k > 0$ and $(x, r, \xi) \in \Omega \times \mathbb{R} \times \mathbb{R}^N$, we set

$$F_k(x,r,\xi) = \begin{cases} F(x,r,\xi) & \text{if } |F(x,r,\xi)| \le k \\ k\,\mathrm{sign}(F(x,r,\xi)) & \text{if } |F(x,r,\xi)| > k, \end{cases} \tag{1.4.26}$$

and we consider the problem of finding $v = v_k \in W_0^{1,p}(\Omega)$

$$-div\, H(v, \nabla v) + F_k(v, \nabla v) = 0 \qquad \text{in } \mathcal{D}'(\Omega). \tag{1.4.27}$$

We introduce the operator $\tilde{\mathcal{H}}$ on $W_0^{1,p}(\Omega) \times W_0^{1,p}(\Omega)$ by

$$\tilde{\mathcal{H}}(v, w) = -div\, H(v, \nabla w) + F_k(v, \nabla v).$$

Then $\tilde{\mathcal{H}}(v, w) \in W^{-1,p'}(\Omega)$ and $\tilde{\mathcal{H}}(v, v) = -div\, H(v, \nabla v) + F_k(v, \nabla v)$. The mapping $\tilde{\mathcal{H}}$ is bounded. It is monotone and hemicontinuous in its second variable. If a sequence $\{u_n\} \subset W_0^{1,p}(\Omega)$ converges weakly to $u$ and satisfies

$$\begin{aligned}(\tilde{\mathcal{H}}(u_n, u_n) - \tilde{\mathcal{H}}(u_n, u))(u_n - u)& \\ &= \int_\Omega \langle H(u_n, \nabla u_n) - H(u_n, \nabla u), \nabla(u_n, u)\rangle dx \to 0,\end{aligned}$$

then for all $w \in W_0^{1,p}(\Omega)$

$$\tilde{\mathcal{H}}(u_n, v)(w) = \int_\Omega \langle H(u_n, \nabla v), \nabla w\rangle dx \to \int_\Omega \langle H(u, \nabla v), \nabla w\rangle dx = \tilde{\mathcal{H}}(u, v)(w).$$

Finally, if a sequence $\{u_n\} \subset W_0^{1,p}(\Omega)$ converges weakly to $u$ and $\tilde{\mathcal{H}}(u_n, v) \rightharpoonup v'$ weakly in $W^{-1,p'}(\Omega)$, equivalently

$$\int_\Omega \langle H(u_n, \nabla v), \nabla w\rangle dx \to v'(w) = -\int_\Omega \langle V', \nabla w\rangle dx,$$

where $V' = (v'_1, ..., v'_N) \in (L^{p'}(\Omega))^N$, then $H(u_n, \nabla v) \to H(u, \nabla v)$ in $L^{p'}(\Omega)$ because of (1.4.8)-(ii) and Rellich-Kondrachov's compactness theorem. Therefore

$$\tilde{\mathcal{H}}(u_n, v)(u_n) = \int_\Omega \langle H(u_n, \nabla v), \nabla u_n\rangle dx \to -\int_\Omega \langle V', \nabla u\rangle dx = v'(u).$$

Thus $\tilde{\mathcal{H}}$ satisfies the assumptions of Theorem 1.4.2 and there exists $v_k \in W_0^{1,p}(\Omega)$ solution of (1.4.27).

*Step 4.* Two-sided estimates of approximate solutions. We assume $k > k_0$ where

$$k_0 = \sup\left\{\|B(.,\phi,\nabla\phi)\|_{L^\infty}, \|B(.,\psi,\nabla\psi)\|_{L^\infty}\right\}. \tag{1.4.28}$$

This is precisely here that where we use the assumption that $\phi$ and $\psi$ belong to $W^{1,\infty}(\Omega)$. Then

$$F_k(.,\phi,\nabla\phi) = F(.,\phi,\nabla\phi) = B(.,\phi,\nabla\phi),$$

and

$$F_k(.,\psi,\nabla\psi) = F(.,\psi,\nabla\psi) = B(.,\psi,\nabla\psi).$$

If $v_k \in W_0^{1,p}(\Omega)$ satisfies (1.4.27), then

$$-div\, H(.,\psi,\nabla\psi) + div\, H(.,v_k,\nabla v_k) + F(.,\psi,\nabla\psi) - F_k(.,v_k,\nabla v_k) \leq 0.$$

As in Step 2 we multiply by $(\psi - v_k)_+^m = \inf\{m, (\psi - v_k)_+\}$. Since $F(.,v_k,\nabla v_k) = F(.,\psi,\nabla\psi)$ a.e. on the set of $x \in \Omega$ where $v_k(x) \leq \psi(x)$ and since $k > k_0$, we get

$$F(.,v_k,\nabla v_k) = F(.,\psi,\nabla\psi) = B(.,\psi,\nabla\psi).$$

The proof follows by the strict monotonicity. Letting $m \to 0$ yields $\psi \leq v_k$. Similarly $v_k \leq \phi$.

*Step 5.* $W_0^{1,p}$-bounds on approximate solutions. For $k \geq k_0$ (see (1.4.28)) and $t > 0$ we set

$$\eta_k = v_k e^{tv_k^2}. \tag{1.4.29}$$

Since $\psi \leq v_k \leq \phi$, $\eta_k \in W_0^{1,p}(\Omega)$. Therefore

$$\int_\Omega \langle H(v_k,\nabla v_k), \nabla v_k\rangle e^{tv_k^2} dx + 2t\int_\Omega \langle H(v_k,\nabla v_k), \nabla v_k\rangle v_k^2 e^{tv_k^2} dx$$
$$+ \int_\Omega F_k(v,\nabla v) v_k e^{tv_k^2} dx = 0.$$

Using (1.4.13) and (1.4.15) we get, with $C^* = \max\{\|\psi\|_{L^\infty}, \|\phi\|_{L^\infty}\}$,

$$c\int_\Omega |\nabla v_k|^p e^{tv_k^2} dx + 2ct\int_\Omega |\nabla v_k|^p v_k^2 e^{tv_k^2} dx - C^*\int_\Omega |\nabla v_k|^p v_k e^{tv_k^2} dx$$
$$\leq \int_\Omega (1 + 2tv_k^2) c_2 e^{tv_k^2} dx + C^*\int_\Omega v_k e^{tv_k^2} dx.$$

The right-hand side of the above inequality is bounded if $t$ is so. Because

$$c + 2ctv_k^2 - C^* v_k \geq c - \frac{C^{*\,2}}{8ct} \geq \frac{c}{2}$$

if we choose $t = \frac{C^{*\,2}}{4c^2}$, we obtain that the norm of $v_k$ in $W_0^{1,p}(\Omega)$ is bounded independently of $k$.

*Step 6.* Convergence when $k \to \infty$. By Step 4 and Step 5, there exist a subsequence of $\{v_k\}$, still denoted by $\{v_k\}$ and a function $u \in W_0^{1,p}(\Omega) \cap L^\infty(\Omega)$ such that

$$\begin{array}{lll} (i) & v_k \to u & \text{in } L^q(\Omega) \quad \forall 1 \leq q < \infty, \\ (ii) & \psi \leq u \leq \phi & \text{a.e. in } \Omega \\ (iii) & \nabla v_k \rightharpoonup \nabla u & \text{in } L^p(\Omega) - \text{weak}. \end{array} \tag{1.4.30}$$

We have from (1.4.23),

$$\int_\Omega \langle H(v_k, \nabla v_k), \nabla(v_k - u)\rangle dx + \int_\Omega F_k(v_k, \nabla v_k)(v_k - u)dx = 0. \tag{1.4.31}$$

This can be re-written as

$$\int_\Omega \langle H(v_k, \nabla v_k) - H(v_k, \nabla u), \nabla(v_k - u)\rangle dx + \int_\Omega \langle H(v_k, \nabla u), \nabla(v_k - u)\rangle dx \\ + \int_\Omega F_k(v_k, \nabla v_k)(v_k - u)dx = 0. \tag{1.4.32}$$

By (1.4.30) and (1.4.15)

$$\int_\Omega \langle H(v_k, \nabla u), \nabla(v_k - u)\rangle dx \to 0,$$

and

$$\int_\Omega F_k(v_k, \nabla v_k)(v_k - u)dx \to 0,$$

when $k \to \infty$. This implies

$$\int_\Omega \langle H(v_k, \nabla v_k) - H(v_k, \nabla u), \nabla(v_k - u)\rangle dx \to 0.$$

Because $\langle H(v_k, \nabla v_k) - H(v_k, \nabla u), \nabla(v_k - u)\rangle \geq 0$, there exists a subsequence $\{k_\ell\}$ such that

$$\langle H(x, v_{k_\ell}(x), \nabla v_{k_\ell}(x)) - H(x, v_{k_\ell}(x), \nabla u(x)), \nabla(v_{k_\ell}(x) - u(x))\rangle \to 0, \tag{1.4.33}$$

as $k_\ell \to \infty$, for all $x \in \Omega \setminus E$ where $E \subset \Omega$ has zero Lebesgue measure. We can also assume that $v_{k_\ell}(x) \to u(x)$ for all $x \in \Omega \setminus E$. Let $x \in \Omega \setminus E$, $\xi^*$ be any cluster point of $\{\nabla v_{k_\ell}(x)\}$ and set $\xi = \nabla u(x)$. Notice that because of (1.4.8)-(ii), (1.4.13) there holds

$$\begin{aligned} c\,|\nabla v_{k_\ell}(x)|^p \le\ & c_2(x) + \epsilon_{k_\ell}(x) + \langle F(x, v_{k_\ell}(x), \nabla u(x)), \nabla u(x)\rangle \\ & - \langle F(x, v_{k_\ell}(x), \nabla u(x)), \nabla v_{k_\ell}(x)\rangle - \langle F(x, v_{k_\ell}(x), \nabla v_{k_\ell}(x)), \nabla u(x)\rangle \\ & \le c_2(x) + \epsilon_{k_\ell}(x) + 3c'\left(|v_{k_\ell}(x)|^{p-1} + k(x)\right)(|\nabla v_{k_\ell}(x)| + |\nabla u(x)|) \\ & + c'\left(|\nabla u(x)|^p + |\nabla u(x)|^{p-1}\,|\nabla v_{k_\ell}(x)| + |\nabla u(x)|\,|\nabla v_{k_\ell}(x)|^{p-1}\right), \end{aligned} \tag{1.4.34}$$

where $\epsilon_{k_\ell}(x)$ denotes the left-hand side of (1.4.33). This yields

$$\begin{aligned} c\,|\xi^*|^p \le c_2(x) + 3c'&\left((\sup\{|\phi(x)|, |\psi(x)|\})^{p-1} + k(x)\right)(|\xi^*| + |\xi|) \\ & + c'\left(|\xi|^p + |\xi|^{p-1}\,|\xi^*| + |\xi|\,|\xi^*|^{p-1}\right). \end{aligned} \tag{1.4.35}$$

Therefore $|\xi^*| < \infty$. Then, from (1.4.33)

$$\langle H(x, u(x), \xi^*) - H(x, u(x), \xi), \xi^* - \xi)\rangle = 0. \tag{1.4.36}$$

This implies that $\xi^* = \xi$, thus $\nabla v_k(x) \to \nabla u(x)$ for almost all $x$ in $\Omega$. Using Young's inequality we derive from (1.4.34)

$$\begin{aligned} M_1\,|\nabla v_{k_\ell}(x)|^p \le c_2(x) + M_2&\left((\sup\{|\phi(x)|, |\psi(x)|\})^p + (k(x))^{p'}\right) \\ & + M_3\,|\nabla u(x)|^p + \epsilon_k(x). \end{aligned} \tag{1.4.37}$$

Since $\epsilon_k \to 0$ in $L^1(\Omega)$, there exist $\Phi \in L^1(\Omega)$, $\Phi \ge 0$ and a subsequence $\{\epsilon_{k_\ell}\}$ such that $|\epsilon_{k_\ell}| = \epsilon_{k_\ell} \le \Phi$. This implies that the sequence $\{|\nabla v_{k_\ell}(x)|^p\}$ is uniformly integrable and thus

$$\nabla v_{k_\ell} \to \nabla u \quad \text{in } L^p(\Omega), \tag{1.4.38}$$

by the Vitali convergence theorem. Therefore $H(v_{k_\ell}, \nabla v_{k_\ell}) \to H(u, \nabla u)$ in $L^{p'}(\Omega)$ and $F(v_{k_\ell}, \nabla v_{k_\ell}) \to F(u, \nabla u)$ in $L^1(\Omega)$. By Step 2, $u$ is a solution of (1.4.16). □

*Remark.* It is possible to prove (see [Boccardo, Murat, Puel (1984), Prop. 3.8]) that $\nabla v_k$, and thus $\nabla u$ remains bounded is some $L^q_{loc}(\Omega)$ for some $q > p$.

The previous theorem can be extended to non-homogeneous boundary value problems.

**Corollary 1.4.5** *Assume $p > 1$, $\Omega \subset \mathbb{R}^N$ is a bounded domain, and the functions $A$ and $B$ satisfy the assumptions of Theorem 1.4.4. If there exist a supersolution $\phi$ and a subsolution $\psi$ of (1.4.16), belonging to $W^{1,\infty}(\Omega)$ and such that $\psi \leq \phi$, then for any $\chi \in W^{1,\infty}(\Omega)$ such that $\psi \leq \chi \leq \phi$ there exists a function $u \in W^{1,p}(\Omega)$ of (1.4.16) satisfying $\psi \leq u \leq \phi$ and $\tilde{u} = u - \chi \in W_0^{1,p}(\Omega)$.*

*Proof.* We put

$$\tilde{A}(x,r,\xi) = A(x, r + \chi(x), \xi + \nabla\chi(x)),$$

and

$$\tilde{B}(x,r,\xi) = B(x, r + \chi(x), \xi + \nabla\chi(x)).$$

Then

$$\begin{aligned}|B(x, r+\chi(x), \xi + \nabla\chi(x))| &\leq C(|r + \chi(x)|)(1 + |\xi + \nabla\chi(x)|^p) \\ &\leq 2^p(1 + \|\nabla\chi\|^p_{L^\infty})C\left(|r| + \|\chi\|_{L^\infty}\right)(1 + |\xi|^p),\end{aligned}$$

hence (1.4.15) holds with $C$ replaced by

$$\tilde{C}(r) = 2^p(1 + \|\nabla\chi\|^p_{L^\infty})C\left(|r| + \|\chi\|_{L^\infty}\right).$$

Next

$$\begin{aligned}&\langle \tilde{A}(x,r,\xi) - \tilde{A}(x,r,\xi'), \xi - \xi'\rangle \\ &\quad = \langle A(x, r+\chi(x), \xi + \nabla\chi(x) - A(x, r+\chi(x), \xi' + \nabla\chi(x), \xi - \xi'\rangle \\ &\quad > 0 \qquad \text{if } \xi \neq \xi'.\end{aligned}$$

Furthermore

$$\begin{aligned}\left|\tilde{A}(x,r,\xi)\right| &\leq c'\left(|r + \chi(x)|^{p-1} + |\xi + \nabla\chi(x)|^{p-1} + k(x)\right) \\ &\leq 2^p c'\left(|r|^{p-1} + |\xi|^{p-1} + k(x) + |\chi(x)|^{p-1} + |\nabla\chi(x)|^{p-1}\right).\end{aligned}$$

Finally

$$\begin{aligned}\langle \tilde{A}(x,r,\xi), \xi\rangle &\geq c\,|\xi + \nabla\chi(x)|^p - c_2(x) \\ &\geq 2^{-p}c\,|\xi|^p - (c\,|\nabla\chi(x)|^p + c_2(x)).\end{aligned}$$

Set $\tilde{\psi} = \psi - \chi$ and $\tilde{\phi} = \phi - \chi$. By the definition of $\tilde{A}$ and $\tilde{B}$, there holds

$$-div\,\tilde{A}(\tilde{\psi}, \nabla\tilde{\psi}) + \tilde{B}(\tilde{\psi}, \nabla\tilde{\psi}) \leq 0 \qquad \text{in } \mathcal{D}'(\Omega), \tag{1.4.39}$$

and

$$-div\,\tilde{A}(\tilde{\phi}, \nabla\tilde{\phi}) + \tilde{B}(\tilde{\phi}, \nabla\tilde{\phi}) \geq 0 \qquad \text{in } \mathcal{D}'(\Omega). \tag{1.4.40}$$

Since $\psi \leq u \leq \phi$ is equivalent to $\tilde{\psi} \leq \tilde{u} \leq \tilde{\phi}$, there exists $\tilde{u} \in W_0^{1,p}(\Omega)$ solution of

$$-div\,\tilde{A}(\tilde{u}, \nabla\tilde{u}) + \tilde{B}(\tilde{u}, \nabla\tilde{u}) = 0 \qquad \text{in } \mathcal{D}'(\Omega), \tag{1.4.41}$$

which satisfies $\tilde{\psi} \leq \tilde{u} \leq \tilde{\phi}$. Hence $u = \tilde{u} + \chi$ fulfills the requirements of Theorem 1.4.4. □

*Remark.* If the function $\chi$ is only given on $\partial\Omega$, then the function

$$X(x) = \max\{\chi(z) - \|\nabla\chi\|_{L^\infty(\partial\Omega)}\,|x - z| : z \in \partial\Omega\} \qquad \forall x \in \mathbb{R}^N, \tag{1.4.42}$$

is a Lipschitz continuous extension of $\chi$ to $\mathbb{R}^N$ which satisfies

$$\|\nabla X\|_{L^\infty(\mathbb{R}^N)} = \|\nabla\chi\|_{L^\infty(\partial\Omega)}\,.$$

**Theorem 1.4.6** *Let $p > 1$, $\Omega \subset \mathbb{R}^N$ be a domain non necessarily bounded and $A : \Omega \times \mathbb{R} \times \mathbb{R}^N \mapsto \mathbb{R}^N$ and $B : \Omega \times \mathbb{R} \times \mathbb{R}^N \mapsto \mathbb{R}$ be Caratheodory functions. We assume that $A$ is strictly monotone, i.e. (1.4.12) holds, and for any bounded subdomain $G \subset \overline{G} \subset \Omega$ there exist positive constants $c_G$, $c'_G$, an increasing function $\Gamma_G : \mathbb{R}_+ \mapsto \mathbb{R}_+$ and $c_{2,G} \in L^1(G)$ and $k_G \in L^{p'}(G)$ such that*

$$\begin{array}{lll} (i) & |A(x,r,\xi)| \leq c'_G\left(|r|^{p-1} + |\xi|^{p-1} + k_G(x)\right) & \forall (x,r,\xi) \in G \times \mathbb{R} \times \mathbb{R}^N \\ (ii) & \langle A(x,r,\xi), \xi\rangle \geq c_G|\xi|^p - c_{2,G}(x) & \forall (x,r,\xi) \in G \times \mathbb{R} \times \mathbb{R}^N \\ (iii) & |B(x,r,\xi)| \leq \Gamma_G(|r|)(1 + |\xi|^p) & \forall (x,r,\xi) \in G \times \mathbb{R} \times \mathbb{R}^N. \end{array} \tag{1.4.43}$$

*If there exist a supersolution $\phi$ and a subsolution $\psi$ of (1.4.16), belonging to $W^{1,\infty}_{loc}(\Omega)$ and such that $\psi \leq \phi$, then there exists a function $u \in W^{1,p}_{loc}(\Omega)$ satisfying $\psi \leq u \leq \phi$ and solution of (1.4.16).*

*Proof.* Let $\{\Omega_n\}$ be an increasing sequence of bounded smooth subdomains of $\Omega$ such that $\overline{\Omega}_n \subset \Omega_{n+1}$. Set $\chi = \frac{1}{2}(\phi + \psi)$. It follows from Corollary 1.4.5 that for each $n \in \mathbb{N}_*$, there exists a solution $u_n \in W^{1,p}(\Omega_n) \cap L^\infty(\Omega_n)$ to

$$-div\,A(., u_n, \nabla u_n) + B(., u_n, \nabla u_n) = 0 \qquad \text{in } \mathcal{D}'(\Omega_n), \tag{1.4.44}$$

satisfying $\psi \leq u_n \leq \phi$ in $\Omega_n$ and such that $u_n - \chi\lfloor_{\Omega_n} \in W_0^{1,p}(\Omega_n)$.

*Step 1. A priori* estimates. Let $G \subset \overline{G} \subset \Omega$ be a bounded open domain and $n_G \in \mathbb{N}$ such that $\overline{G} \subset \Omega_n$ for all $n \geq n_G$. Let $\theta \in C_0^\infty(\mathbb{R}^N)$, $0 \leq \theta \leq 1$, and $\theta = 1$ on $\omega \subset G$. Set $\eta_n = u_n e^{tu_n^2}$, $t > 0$.

$$\int_G \langle A(u_n, \nabla u_n), \nabla(\theta \eta_n)\rangle dx + \int_G B(u_n, \nabla u_n)\theta \eta_n dx = 0.$$

Then

$$\int_G \langle A(u_n, \nabla u_n), \nabla u_n\rangle e^{tnv_n^2}\theta^p dx + 2t\int_G \langle A(u_n, \nabla u_n)), \nabla u_n\rangle u_n^2 e^{tn_n^2}\theta^p dx$$
$$+ p\int_G \langle A(u_n, \nabla u_n), \nabla\theta\rangle u_n e^{tnv_n^2}\theta^{p-1}dx + \int_G B(u_n, \nabla u_n)u_n e^{tu_n^2}\theta^p dx = 0.$$

By (1.4.43)-(ii) the sum of first two terms on the left is larger than

$$c_G\int_G (1 + 2tu_n^2)\, |\nabla u_n|^p\, \theta^p e^{tu_n^2} dx - c_G\int_G (1 + 2tu_n^2)c_{2,G}\theta^p e^{tu_n^2} dx.$$

If $m_G = \max\{\|\phi\|_{L^\infty(G)}, \|\psi\|_{L^\infty(G)}\}$, there holds

$$\left|\int_G B(u_n, \nabla u_n)u_n e^{tu_n^2}\theta^p dx\right| \leq \Gamma_G(m_G)\int_G (1 + |\nabla u_n|^p)\, u_n e^{tu_n^2}\theta^p dx.$$

By Young's inequality, with $\epsilon > 0$,

$$\begin{aligned} p\left|\int_G \langle A(u_n, \nabla u_n), \nabla\theta\rangle u_n e^{tnv_n^2}\theta^{p-1}dx\right| &\leq \frac{1}{\epsilon^{p'}}\int_G |\nabla\theta|^{p'}\, u_n e^{tu_n^2} dx \\ &\quad + (p-1)\epsilon^{p'}\int_G |A(u_n, \nabla u_n)|^{p'}\, u_n e^{tu_n^2}\theta^p dx \\ &\leq \frac{1}{\epsilon^{p'}}\int_G |\nabla\theta|^{p'}\, u_n e^{tu_n^2} dx \\ &\quad + pp' c_G'^{p'}\epsilon^{p'}\int_G \left(|\nabla u_n|^p + |u_n|^p + k_G^{p'}\right) u_n e^{tu_n^2}\theta^p dx. \end{aligned}$$

Combining the above inequalities, we obtain

$$\begin{aligned} &\int_G \left(c_G\left(1 + 2tu_n^2\right) - \left(pp' c_G'^{p'}\epsilon^{p'} + \Gamma_G(m_G)\right) u_n\right) |\nabla u_n|^p\, \theta^p e^{tu_n^2} dx \\ &\qquad \leq \int_G \left(\frac{1}{\epsilon^{p'}}\, |\nabla\theta|^{p'} + pp' c_G'^{p'}\epsilon^{p'}\, |u_n|^p + k_G^{p'}\right) e^{tu_n^2} dx \qquad (1.4.45) \\ &\qquad\quad + c_G\int_G \left(1 + 2tu_n^2\right) \theta^p c_{2,G} e^{tu_n^2} dx, \end{aligned}$$

which yields, using the fact that $|u_n(x)| \le m_G$ in $G$ and $0 \le \theta \le 1$,

$$\int_G \left(c_G\left(1+2tu_n^2\right) - \left(pp' c_G'^{p'} \epsilon^{p'} + \Gamma_G(m_G)\right) u_n\right) |\nabla u_n|^p \, \theta^p e^{tu_n^2} dx$$

$$\le \left(\frac{\|\nabla\theta\|_{L^\infty}}{\epsilon^{p'}} + pp' c_G'^{p'} \epsilon^{p'} m_G^p\right) e^{tm_G^2}|G| \qquad (1.4.46)$$

$$+ e^{tm_G^2} \int_G \left(k_G^{p'}(x) + c_G(1+tm_G^2)c_{2,G}(x)\right) dx.$$

Hence the right-hand side of (1.4.46) is dominated by a constant $R_G(t)$ which depends on $t$, $G$, $m_G$ (*i.e.* $\phi\lfloor_G$ and $\psi\lfloor_G$), $\theta$ and $\left\|k_G^{p'} + c_{2,G}\right\|_{L^1(G)}$. If we fix

$$t := t_0 = \frac{\left(pp' c_G'^{p'} \epsilon^{p'} + \Gamma_G(m_G)\right)^2}{4c_G^2},$$

we obtain that

$$2tc_G u_n^2 - \left(pp' c_G'^{p'} \epsilon^{p'} + \Gamma_G(m_G)\right) u_n + c_G \ge c_G - \frac{\left(pp' c_G'^{p'} \epsilon^{p'} + \Gamma_G(m_G)\right)^2}{8c_G t_0} \ge \frac{c_G}{2}.$$

This implies

$$c_G \int_G |\nabla u_n|^p \, \theta^p dx \le R_G(t_0) \qquad \forall n \ge n_G. \qquad (1.4.47)$$

*Step 2.* Convergence. Using Cantor's diagonal sequence construction, we infer from Step 1 that there exist and increasing function $\tau : \mathbb{N} \mapsto \mathbb{N}$ and $u \in L^\infty_{loc}(\Omega) \cap W^{1,p}_{loc}(\Omega)$ such that

$$\begin{array}{llll} (i) & u_{\tau(n)} \to u & \text{in } L^q_{loc}(\Omega) \quad \forall 1 \le q < \infty & \\ (ii) & \psi \le u \le \phi & \text{a.e. in } \Omega & \\ (iii) & u_{\tau(n)} \to u & \text{a.e. in } \Omega & \\ (iv) & \nabla u_{\tau(n)} \rightharpoonup \nabla u & \text{in } L^p_{loc}(\Omega)\text{-weak.} & \end{array} \qquad (1.4.48)$$

Let $G \subset \overline{G} \subset \Omega$ and $\theta \in C_0^\infty(\Omega)$ be as in Step 2. Then, as in the proof of Theorem 1.4.4-Step 6,

$$\int_\Omega B(u_{\tau(n)}, \nabla u_{\tau(n)})\theta(u_{\tau(n)} - u)dx + \int_\Omega \langle A(u_{\tau(n)}, \nabla u_{\tau(n)}), \nabla(u_{\tau(n)} - u)\rangle\theta dx$$

$$+ \int_\Omega \langle A(u_{\tau(n)}, \nabla u_{\tau(n)}), \nabla\theta\rangle(u_{\tau(n)} - u)dx = 0,$$

which can be re-written as follows:

$$\int_\Omega \langle A(u_{\tau(n)}, \nabla u_{\tau(n)}) - A(u_{\tau(n)}, \nabla u), \nabla(u_{\tau(n)} - u)\rangle \theta dx$$
$$+ \int_\Omega \langle A(u_{\tau(n)}, \nabla u), \nabla(u_{\tau(n)} - u)\rangle \theta dx$$
$$+ \int_\Omega \langle A(u_{\tau(n)}, \nabla u_{\tau(n)}), \nabla\theta\rangle (u_{\tau(n)} - u) dx$$
$$+ \int_\Omega B(u_{\tau(n)}, \nabla u_{\tau(n)})(u_{\tau(n)} - u)\theta dx = 0.$$

Let us denote by $a_1(n)$, $a_2(n)$, $a_3(n)$ and $a_4(n)$ the four terms, in this order, of the above inequality. Using (1.4.48), we see that $a_2(n)$, $a_3(n)$ and $a_4(n)$ tend to 0 when $n \to \infty$. Therefore, if we set

$$\epsilon_{1,n}(.) = \langle A(., u_{\tau(n)}, \nabla u_{\tau(n)}) - A(., u_{\tau(n)}, \nabla u), \nabla(u_{\tau(n)} - u)\rangle \theta(.),$$

then $\epsilon_{1,n}(x)) > 0$ on the set of $x$ such that $\nabla(u_{\tau(n)} - u)(x)\rangle\theta(x) \neq 0$. Estimate (1.4.34) takes the form

$$\begin{aligned} c_G \left|\nabla u_{\tau(n)}\right|^p &\leq c_{2,G}(x) + \epsilon_{1,n}(x) \\ &+ 3c'_G \left(\left|u_{\tau(n)}(x)\right|^{p-1} + k_G(x)\right) \left(\left|\nabla u_{\tau(n)}(x)\right| + |\nabla u(x)|\right) \\ &+ c'_G \left(|\nabla u(x)|^p + |\nabla u(x)|^{p-1} \left|\nabla u_{\tau(n)}(x)\right| + |\nabla u(x)| \left|\nabla u_{\tau(n)}(x)\right|^{p-1}\right). \end{aligned} \tag{1.4.49}$$

If $x$ is a regular point for the convergence a.e. everywhere, $\nabla u(x) = \xi$ and $\xi^*$ is any cluster point of the sequence $\{\nabla u_{\tau(n)}\}$, we obtain again that $\xi^*$ is finite and

$$\langle A(x, u(x), \nabla \xi^*) - A(x, u(x), \xi), \xi^* - \xi)\rangle \theta(x) = 0.$$

This implies that $\xi^* = \xi$ on the subset $G'$ of $x \in G$ where $\theta(x) > 0$. Since $\theta$ is arbitrary with support in $G$, we conclude that $\nabla u_{\tau(n)}(x) \to \nabla u(x)$ a.e. in $G$ and hence a.e. in $\Omega$. Finally, we use Young's inequality and derive from (1.4.49) the upper estimate of $\left|\nabla u_{\tau(n)}\right|^p$, namely

$$\begin{aligned} M_{1,G} \left|\nabla u_{\tau(n)}(x)\right|^p \leq c_{2,G}(x) &+ M_{2,G} \left((\sup\{|\phi(x)|, |\psi(x)|\})^p + (k_G(x))^{p'}\right) \\ &+ M_{3,G} |\nabla u(x)|^p + \epsilon_{1,n}(x). \end{aligned} \tag{1.4.50}$$

Since there exist a function $\Phi_G \in L^1(G)$ and a subsequence $\{n_\ell\}$ such that $|\epsilon_{1,n_\ell}| \leq \Phi_G$ a.e. in $G$, we obtain that $\{\left|\nabla u_{\tau(n_\ell)}(x)\right|^p \theta\}$ is uniformly integrable in $G$ and consequently

$$\int_G \left|\nabla u_{\tau(n_\ell)} - \nabla u\right|^p \theta dx \to 0 \quad \text{as } n_\ell \to \infty. \tag{1.4.51}$$

This convergence holds for any function $\theta$ and the subsequence $\{n_\ell\}$ may depend on $\theta$. However, there exists a sequence $\{\theta_n\} \subset C_0^\infty(\mathbb{R}^N)$ such that

$$0 \le \theta_n \le 1\ ,\ \operatorname{supp}\theta_n \subset \overline{\Omega}_n \text{ and } \theta \equiv 1 \text{ on } \overline{\Omega}_{n-1}.$$

Finally we can extract a diagonal sequence denoted by $\{\tilde{\tau}(n)\}$ from $\{\tau(n)\}$ such that $\{u_{\tilde{\tau}(n)}\}$ possesses the following properties

$$\begin{array}{llll} (i) & u_{\tilde{\tau}(n)} \to u & \text{in } L^q_{loc}(\Omega) & \forall 1 \le q < \infty, \\ (ii) & \psi \le u \le \phi & \text{a.e. in } \Omega & \\ (iii) & u_{\tilde{\tau}(n)} \to u & \text{a.e. in } \Omega & \\ (iv) & \nabla u_{\tilde{\tau}(n)} \to \nabla u & \text{in } L^p_{loc}(\Omega). & \end{array} \tag{1.4.52}$$

This implies that $u$ is a solution of (1.4.16) and it satisfies $\psi \le u \le \phi$.
□

The framework of second order Leray-Lions operators in $\mathbb{R}^N$ can be adapted easily to a Riemannian manifold $(M^d, g)$. We assume that $A : (x,r,\xi) \in M \times \mathbb{R} \times TM \mapsto A(x,r,\xi) \in T_xM$ and $B : (x,r,\xi) \in M \times \mathbb{R} \times TM \mapsto B(x,r,\xi) \in \mathbb{R}$ are Caratheodory vector fields and they satisfy

$$\begin{array}{ll} (i)\ \ \langle A(x,r,\xi),\xi\rangle_g \ge c|\xi|^p - c_2(x) & \forall (x,r,\xi) \in M \times \mathbb{R} \times TM \\ (ii)\ \ |A(x,r,\xi)| \le c'\left(|\xi|^{p-1} + |r|^{p-1} + k(x)\right) & \forall (x,r,\xi) \in M \times \mathbb{R} \times TM \\ (iii)\ |B(x,r,\xi)| \le \Gamma(|r|)(1+|\xi|^p) & \forall (x,r,\xi) \in M \times \mathbb{R} \times TM \\ (iv)\ \ \langle A(x,r,\xi) - A(x,r,\hat{\xi}), \xi - \hat{\xi}\rangle_g > 0 & \\ & \forall (x,r,\xi,\hat{\xi}) \in M \times \mathbb{R} \times TM \times TM \text{ s.t. } \xi \ne \hat{\xi}, \end{array} \tag{1.4.53}$$

where $c$, $c'$ are positive constants, $k \in L^{p'}(M)$, $c_2 \in L^1(M)$ and $\Gamma : \mathbb{R}_+ \mapsto \mathbb{R}_+$ is an increasing function. The equation under consideration is the following

$$-div_g A(.,u,\nabla u) + B(.,u,\nabla u) = 0 \qquad \text{in } \mathcal{D}'(M). \tag{1.4.54}$$

The next result is the exact analogue of Theorem 1.4.4 in a geometric setting.

**Theorem 1.4.7** *Assume $p > 1$, $(M^d, g)$ is a complete Riemannian manifold and $S \subset M$ is a bounded domain with a nonempty boundary $\partial S$ and*

*A and B satisfy (1.4.53). If there exist two functions* $\psi$ *and* $\phi$ *in* $W^{1,\infty}(S))$ *such that* $\psi \leq \phi$ *and*

$$\begin{aligned} &(i) \quad -\, div_g A(.,\psi,\nabla\psi) + B(.,\psi,\nabla\psi) \leq 0 \;\; in\; \mathcal{D}'(S),\; \psi \leq 0 \;\; on\; \partial S \\ &(ii) \quad -\, div_g A(.,\phi,\nabla\phi) + B(.,\phi,\nabla\phi) \geq 0 \;\; in\; \mathcal{D}'(S),\; \phi \geq 0 \;\; on\; \partial S, \end{aligned} \tag{1.4.55}$$

*then one can find* $u \in W_0^{1,p}(S) \cap L^\infty(S)$ *such that* $\psi \leq u \leq \phi$ *in* $S$*, solution of (1.4.54) in* $\mathcal{D}'(S)$ *(instead of* $\mathcal{D}'(M)$*).*

Another useful result is the following transposition of Theorem 1.4.6.

**Theorem 1.4.8** *Assume* $p > 1$, $(M^d, g)$ *is a complete Riemannian manifold and* $S \subset M$ *is a subdomain. Let* $A : M \times \mathbb{R} \times TM \mapsto TM$ *and* $B : M \times \mathbb{R} \times TM \mapsto \mathbb{R}$ *be Caratheordory functions. We assume that A is strictly monotone, i.e. (1.4.53)-(iv) holds, and for any bounded subdomain* $G \subset \overline{G} \subset S$ *there exist positive constants* $c_G$, $c'_G$*, an increasing function* $\Gamma_G : \mathbb{R}_+ \mapsto \mathbb{R}_+$ *and* $c_{2,G} \in L^1(G)$ *and* $k_G \in L^{p'}(G)$ *such that*

$$\begin{aligned} &(i) \quad |A(x,r,\xi)| \leq c'_G \left(|r|^{p-1} + |\xi|^{p-1} + k_G(x)\right) \quad \forall (x,r,\xi) \in G \times \mathbb{R} \times TM \\ &(ii) \quad \langle A(x,r,\xi),\xi\rangle \geq c_G|\xi|^p - c_{2,G}(x) \quad \forall (x,r,\xi) \in G \times \mathbb{R} \times TM \\ &(iii) \quad |B(x,r,\xi)| \leq \Gamma_G(|r|)(1+|\xi|^p) \quad \forall (x,r,\xi) \in G \times \mathbb{R} \times TM. \end{aligned} \tag{1.4.56}$$

*If there exist two functions* $\psi$ *and* $\phi$ *in* $W^{1,\infty}_{loc}(S))$ *such that* $\psi \leq \phi$ *and*

$$\begin{aligned} &(i) \quad -div_g A(.,\psi,\nabla\psi) + B(.,\psi,\nabla\psi) \geq 0 \;\; in\; \mathcal{D}'(S) \\ &(ii) \quad -div_g A(.,\phi,\nabla\phi) + B(.,\phi,\nabla\phi) \geq 0 \;\; in\; \mathcal{D}'(S), \end{aligned} \tag{1.4.57}$$

*then one can find* $u \in W^{1,p}_{loc}(S) \cap L^\infty_{loc}(S)$ *such that* $\psi \leq u \leq \phi$ *in* $S$*, solution of (1.4.54) in* $\mathcal{D}'(S)$*.*

If $(M^d, g)$ is a compact complete Riemannian manifold, then the dual of $W^{1,p}(M)$ is $W^{-1,p'}(M)$. The next statement is a variation of Theorem 1.4.4.

**Corollary 1.4.9** *Assume* $p > 1$, $(M^d, g)$ *is a compact complete Riemannian manifold and the functions A and B satisfy (1.4.53). If there exist two functions* $\psi$ *and* $\phi$ *in* $W^{1,\infty}(M)$ *such that* $\psi \leq \phi$ *and*

$$\begin{aligned} &(i) \quad -div_g A(.,\psi,\nabla\psi) + B(.,\psi,\nabla\psi) + \delta\psi_+^{p-1} \leq 0 \quad in\; \mathcal{D}'(M) \\ &(ii) \quad -div_g A(.,\phi,\nabla\phi) + B(.,\phi,\nabla\phi) - \delta\phi_-^{p-1} \geq 0 \quad in\; \mathcal{D}'(M), \end{aligned} \tag{1.4.58}$$

*for some $\delta > 0$, then one can find $u \in W^{1,p}(M) \cap L^\infty(M)$ solution of (1.4.54) such that $\psi \leq u \leq \phi$.*

*Proof.* Since $\|\nabla u\|_{L^p}$ is just a seminorm on $W^{1,p}(M)$, we indicate the modifications needed in order to follow the steps of the proof of Theorem 1.4.4. For $0 < \sigma \leq \delta$, we consider the coercive problem

$$-div_g A(.,u,\nabla u) + B(.,u,\nabla u) + \sigma |u|^{p-2} u = 0 \qquad \text{in } \mathcal{D}'(M). \tag{1.4.59}$$

and define $\mathcal{A}_\sigma$ from $W^{1,p}(M)$ into $W^{-1,p'}(M)$ by

$$\mathcal{A}_\sigma(v)(\zeta) = \int_M \left( \langle A(.,v,\nabla v), \nabla\zeta\rangle_g + \sigma |u|^{p-2} u\zeta \right) dv_g, \tag{1.4.60}$$

for $(v,\zeta) \in W^{1,p}(M) \times W^{1,p}(M)$. The operator $\mathcal{A}_\sigma$ is coercive in the sense that it satisfies (1.4.20). The truncations of $A$ and $B$ by $H$ and $F$ are the same as in the proof of Theorem 1.4.4-Step 1 and the associated truncated problem is to find $u \in W^{1,p}(M)$ such that

$$-div_g H(.,u,\nabla u) + F(.,u,\nabla u) + \sigma |u|^{p-2} u = 0 \qquad \text{in } \mathcal{D}'(M). \tag{1.4.61}$$

Assuming that $u$ is a solution of (1.4.60), we obtain, instead of (1.4.32),

$$\begin{aligned} &\int_M \Big( \langle H(\psi,\nabla\psi) - H(u,\nabla u), \nabla(\psi-u)_{+,m}\rangle_g \\ &\qquad + \sigma(\psi_+^{p-1} - |u|^{p-2} u)(\psi-u)_{+,m} \Big) dv_g \\ &\qquad\qquad + \int_M B(\psi,\nabla\psi) - F(u,\nabla u)(\psi-u)_{+,m} dv_g \leq 0, \end{aligned} \tag{1.4.62}$$

where $(\psi-u)_{+,m} = \inf\{m, (\psi-u)_+\}$. Then

$$\int_M B(\psi,\nabla\psi) - F(u,\nabla u)(\psi-u)_{+,m} dv_g = 0.$$

Denoting again $\omega_0 = \{x \in M : \psi(x) - u(x) \geq 0\}$ (resp. $\omega_0^m = \{x \in M : m \geq \psi(x) - u(x) \geq 0\}$, we obtain

$$\int_{\omega_0^m} \langle H(\psi,\nabla\psi) - H(u,\nabla u), \nabla(\psi-u)_{+,m}\rangle_g dv_g + \sigma \int_{\omega_0} (\psi_+^{p-1} - |u|^{p-2} u)(\psi-u)_{+,m} dv_g \leq 0,$$

which yields $(\psi-u)_{+,m} = 0$ and thus $\psi \leq u$. In the same way $u \leq \phi$ and finally $u$ is a solution of (1.4.59). Defining $F_k$ as in (1.4.26) for $k > 0$, we find a solution $v = v_{k,\sigma} \in W^{1,p}(M)$ of

$$-div\, H(v, \nabla v) + F_k(v, \nabla v) + \sigma|v|^{p-2}v = 0 \qquad \text{in } \mathcal{D}'(M). \tag{1.4.63}$$

Defining $k_0$ by (1.4.28) we infer that $\psi \leq v_{k,\sigma} \leq \phi$.
In the same way as in Step 5 we derive that $\|\nabla v_{k,\sigma}\|_{L^p}$ is bounded independently of $k$ and $\sigma$. Then, up to a subsequence, $v_{k_\ell,\sigma}$ converges in any $L^q(M)$ for $1 \leq q < \infty$ and almost everywhere in $M$ to some function $u_\sigma$ and $\nabla v_{k_\ell,\sigma}$ converges weakly to $\nabla u_\sigma$ in $L^p(M)$. Since (1.4.33) is valid with $v_{k_\ell}$ replaced by $v_{k_\ell,\sigma}$, we deduce that $\|\nabla(v_{k_\ell,\sigma} - u_\sigma)\|_{L^p}$ converges to 0. This implies that $u_\sigma$ is a solution of

$$-div\, A(u_\sigma, \nabla u_\sigma) + B(u_\sigma, \nabla u_\sigma) + \sigma|u_\sigma|^{p-2}u_\sigma = 0 \quad \text{in } \mathcal{D}'(M), \tag{1.4.64}$$

and it satisfies $\psi \leq u_\sigma \leq \phi$. Furthermore $\|\nabla u_\sigma\|_{L^p}$ remains bounded independently of $\sigma$. Since $u_\sigma$ remains bounded, the same convergence method as the one used for letting $k_\ell \to \infty$ applies. Therefore, up to a subsequence $\{\sigma_\ell\}$ converging to 0, there exists $u \in L^\infty(M)$ such that $\nabla u \in L^p(M)$, $\psi \leq u \leq \phi$ with the property that $u_{\sigma_\ell} \to u$ in any $L^q(M)$ $(1 \leq q < \infty)$ and a.e. in $M$, $\nabla u_{\sigma_\ell} \to \nabla u$ in $L^p(M)$. At end $u$ is a solution of (1.4.54). □

## 1.5 The $p$-Laplace operator

Let $\Omega$ be a domain in $\mathbb{R}^N$. The $p$-Laplace is defined by

$$-\Delta_p u = -div\, |\nabla u|^{p-2}\, \nabla u, \tag{1.5.1}$$

where $1 < p < \infty$. If $f \in L^1_{loc}(\Omega)$, solutions of

$$-\Delta_p u = f \qquad \text{in } \Omega, \tag{1.5.2}$$

are usually understood in the *weak sense*, *i.e.* $u \in W^{1,p}_{loc}(\Omega)$ and (1.5.2) holds in the sense of distribution in $\Omega$. If $p \geq 2$, $u \in W^{1,p-1}_{loc}(\Omega)$ is sufficient to give sense to this formulation in the sense of distributions. This is not the case if $1 < p < 2$ and other definition will be given in Chapter 4.

An important property of the $N$-Laplace equation is its conformal invariance.

**Proposition 1.5.1** *Assume $\Omega \subset \mathbb{R}^N$ is a domain which does not contain*

*0 and $u \in C^1(\Omega)$ satisfies*

$$-\Delta_N u = f \qquad \text{in } \Omega, \tag{1.5.3}$$

*for some $f$, then $v : x \mapsto v(x) = u\left(\frac{x}{|x|^2}\right)$ defined in $\tilde{\Omega} = \{x : |x|^{-2}x \in \Omega\}$ satisfies*

$$-\Delta_N v = |x|^{-2N} f\left(\frac{x}{|x|^2}\right) \qquad \text{in } \tilde{\Omega}. \tag{1.5.4}$$

This property will be very helpful to study quasilinear equations with $N$-Laplacian in non-flat domains and in particular to construct local supersolutions or subsolutions (see Chapter 3).

### 1.5.1 *Comparison principles*

In this particular case the results of Section I-3 are as follows:

**Definition 1.6** A function $u \in W^{1,p}_{loc}(\Omega)$ is a weak supersolution (resp. subsolution) of (1.5.2) if

$$\int_\Omega |\nabla u|^{p-2} \nabla u.\nabla \zeta dx \geq \int_\Omega f \zeta dx \quad \forall \zeta \in C_c^\infty(\Omega),\ \zeta \geq 0, \tag{1.5.5}$$

resp.

$$\int_\Omega |\nabla u|^{p-2} \nabla u.\nabla \zeta dx \leq \int_\Omega f \zeta dx \quad \forall \zeta \in C_c^\infty(\Omega),\ \zeta \geq 0. \tag{1.5.6}$$

**Definition 1.7** A function $u \in W^{1,p}_{loc}(\Omega) \cap C(\Omega)$ is called $p$-harmonic if it is a weak solution of

$$-\Delta_p u = 0 \qquad \text{in } \Omega. \tag{1.5.7}$$

The requirement of continuity is not a restriction since any weak solution can be modified on a set with zero $c^{\mathbb{R}^N}_{1,p}$-capacity in order to become $C^{1,\alpha}$.

**Definition 1.8** A function $u \in W^{1,p}_{loc}(\Omega) \cap C(\Omega)$ is called weakly $p$-superharmonic or a supersolution of (1.5.7) if it satisfies

$$-\Delta_p u \geq 0 \qquad \text{in } \mathcal{D}'(\Omega). \tag{1.5.8}$$

It is weakly $p$-subharmonic, or a subsolution of (1.5.7), if $-u$ is weakly $p$-superharmonic.

An important property of the vector field associated to the $p$-Laplace operator is *the strict monotonicity inequality*

$$(|X|^{p-2}X - |Y|^{p-2}Y).(X-Y) \geq \begin{cases} 2^{2-p}|X-Y|^p & \text{if } p \geq 2 \\ (p-1)|X-Y|^2(1+|X|^2+|Y|^2)^{\frac{p-2}{2}}, \end{cases} \tag{1.5.9}$$

for all $X, Y \in \mathbb{R}^N$ (see [Lindqvist (2006)]).

The following strict comparison principle will be used many times.

**Proposition 1.5.2** *Let $\Omega \subset \mathbb{R}^N$ be a domain, $p > 1$ and $c \in L^\infty_{loc}(\Omega)$. Assume $u$ and $v$ belong to $C^1(\Omega)$, satisfy*

$$-\Delta_p u + cu \leq 0 \quad \text{and} \quad -\Delta_p v + cv \geq 0 \qquad \text{in } \mathcal{D}'(\Omega), \tag{1.5.10}$$

*and $\nabla v$ never vanishes in $\Omega$. If $u \leq v$ in $\Omega$ and if there exists $x_0 \in \Omega$ such that $u(x_0) = v(x_0)$, then $u \equiv v$ in $\Omega$.*

*Proof.* By the mean value theorem

$$|\nabla u|^{p-2}\frac{\partial u}{\partial x_i} - |\nabla v|^{p-2}\frac{\partial v}{\partial x_i} = \sum_j \alpha_{ij}\frac{\partial(u-v)}{\partial x_i},$$

on the set of $x \in B_\delta(x_0)$ where $\nabla u(x) \neq 0$, with

$$\alpha_{ij} = |t_i\nabla u + (1-t_i)\nabla v|^{p-4}\Big(\delta_{ij}|t_i\nabla u + (1-t_i)\nabla v|^2$$
$$+(p-2)\left(t_i\frac{\partial u}{\partial x_i} + (1-t_i)\frac{\partial v}{\partial x_i}\right)\left(t_i\frac{\partial u}{\partial x_j} + (1-t_i)\frac{\partial v}{\partial x_j}\right)\Big).$$

At the tangency point of the graphs of $u$ and $v$, $\nabla u(x_0) = \nabla v(x_0) \neq 0$,

$$\alpha_{ij}(x_0) = |\nabla v(x_0)|^{p-4}\left(\delta_{ij}|\nabla v(x_0)|^2 + (p-2)\frac{\partial v(x_0)}{\partial x_i}\frac{\partial v(x_0)}{\partial x_j}\right).$$

The matrix $\mathbf{A} = (\alpha_{ij}(x_0))$ is symmetric and the infimum of its spectrum is $\inf \sigma(\mathbf{A}) = \min\{1, p-2\}|\nabla v(x_0)|^{p-2}$. It remains positive definite in $B_\delta(x_0)$ for $\delta$ small enough. If $w = v - u$, there holds

$$-\sum_{i,j}\frac{\partial w}{\partial x_j}\left(\alpha_{ij}\frac{\partial w}{\partial x_i}\right) + c_+w \geq 0 \quad \text{in } B_\delta(x_0).$$

Thus $w \equiv 0$ in $B_\delta(x_0)$. The completion of the proof follows by connectedness. □

A variant of this strong comparison principle is the following:

**Proposition 1.5.3** *Let $\Omega \subset \mathbb{R}^N$ be a $C^1$ domain, $p > 1$ and $c \in L^\infty_{loc}(\Omega)$ Assume $u$ and $v$ belong to $C^1(\overline{\Omega})$, satisfy (1.5.10), $u \leq v$ and $\nabla v$ never vanishes in $\overline{\Omega}$ and there exists $x_1 \in \partial\Omega$ such that $u(x_1) = v(x_1)$. Then*

*(i) either $u \equiv v$ in $\overline{\Omega}$,*

*(ii) or $u < v$ in $\Omega$ and $\frac{\partial v}{\partial n}(x_1) < \frac{\partial v}{\partial n}(x_1)$.*

*Proof.* By Proposition 1.5.2, if there exists $x_0 \in \Omega$ such that $u(x_0) = v(x_0)$, then $u$ and $v$ coincide in $\overline{\Omega}$. If $u < v$ in $\Omega$, since the matrix $\mathbf{A}$ is uniformly positive definite in a neighborhood of $x_1$ and $v(x_1) - u(x_1) = 0$, it follows by Hopf Lemma that

$$\frac{\partial v}{\partial n}(x_1) - \frac{\partial v}{\partial n}(x_1) < 0.$$

□

Another approach of the comparison principle is to study the coincidence set between a supersolution and a subsolution [Guedda, Véron (1988), Prop 2.1, 2.2].

**Proposition 1.5.4** *Let $\Omega \subset \mathbb{R}^N$ be a domain and $p > 1$. Assume $u$ and $v$ belong to $C(\overline{\Omega}) \cap W^{1,p}(\Omega)$ and satisfy*

$$-\Delta_p u = f \quad \text{and} \quad -\Delta_p v = g \qquad \text{in } \mathcal{D}'(\Omega) \tag{1.5.11}$$

*where $f, g \in L^{p'}(\Omega)$ satisfy $f \leq g$. If $u \leq v$ in $\Omega$, then either $u < v$ in $\Omega$ or any connected component of the coincidence set $\mathcal{C} = \{x \in \Omega : u(x) = v(x)\}$ has a nonempty intersection with $\partial\Omega$.*

*Proof.* Assume $K \subset \Omega$ is a nonempty compact connected component of $\mathcal{C}$ and let $G$ be a bounded domain such that $K \subset G \subset \overline{G} \subset \Omega$ and $G \cap \mathcal{C} = \emptyset$. Then $v > u$ in $\bar{G} \setminus K$ and in particular $v > u_\theta := u + \theta$ for some $\theta > 0$ on $\partial G$. Since $-\Delta_p u_\theta + \Delta_p v \leq f - g$ and $(u_\theta - v)_+$ has compact support in $G$, we derive

$$\int_{G \cap u_\theta > v} \langle |\nabla u_\theta|^{p-2} \nabla u_\theta - |\nabla v|^{p-2} \nabla v, \nabla(u_\theta - v)\rangle dx \leq 0,$$

which implies $u_\theta \leq v$ in $G$, thus $K = \emptyset$. □

**Proposition 1.5.5** *Let $\Omega \subset \mathbb{R}^N$ be a bounded domain with a $C^2$ boundary and $p > 1$. Assume $u$ and $v$ belong to $C^1(\overline{\Omega})$, vanish on $\partial\Omega$ and satisfy (1.5.11) where $f, g \in L^{p'}(\Omega)$ satisfy $0 \leq f \leq g$ a.e. in $\Omega$. If the set $\mathcal{F} = \{x \in \Omega : f(x) = g(x) \text{ a.e.}\}$ has zero measure, then we have the following:*

$$
\begin{array}{llll}
(i) & v(x) > u(x) \geq 0 & \forall x \in \Omega & \\
(ii) & \dfrac{\partial v}{\partial \mathbf{n}}(x) < \dfrac{\partial u}{\partial \mathbf{n}}(x) < 0 & \forall x \in \partial\Omega. &
\end{array} \tag{1.5.12}
$$

*Proof.* It follows from the assumptions that $g$ is not zero a.e. zero thus $v$ is not identically zero. By Vazquez maximum principle Proposition 1.3.13 there holds

$$
(i) \quad v(x) > 0 \ \forall x \in \Omega \quad \text{and} \quad (ii) \quad \frac{\partial v}{\partial \mathbf{n}}(x) < 0 \ \forall x \in \partial\Omega. \tag{1.5.13}
$$

Moreover $v \geq u$ in $\Omega$ and the coincidence set $\mathcal{C}$ of $u$ and $v$ cannot have a connected component $K \subset \Omega$. If we assume that $\mathcal{C}$ is not empty, there exists $\{x_n\} \subset K$ and $x_0 \in \partial\Omega$ such that $x_n \to x_0$. Since $u$ and $v$ are $C^1$ in $\overline{\Omega}$, it follows from (1.5.13) (ii) that

$$
\frac{\partial v}{\partial \mathbf{n}}(x_0) = \frac{\partial u}{\partial \mathbf{n}}(x_0) = -a^2 < 0.
$$

Set $w = v - u$, it satisfies,

$$
-\mathcal{L}w = g - f \geq 0
$$

where $\mathcal{L}$ is the linear strictly elliptic operator defined in the proof of Proposition 1.5.3. Since $w \geq 0$ is not identically zero in a neighborhood of $x_0$,

$$
\frac{\partial w}{\partial \mathbf{n}}(x_0) = \frac{\partial v}{\partial \mathbf{n}}(x_0) - \frac{\partial u}{\partial \mathbf{n}}(x_0) < 0,
$$

contradiction. Therefore $\mathcal{C}$ is empty. □

### 1.5.2 *Isotropic singularities of p-harmonic functions*

The fundamental solution $\mu_p$ of $-\Delta_p$ is defined in $\mathbb{R}^N \setminus \{0\}$ by

$$
\mu_p(x) = \begin{cases} \gamma_{N,p}|x|^{-\frac{N-p}{p-1}} & \text{if } p \neq N \\ \gamma_N \ln |x|^{-1} & \text{if } p = N, \end{cases} \tag{1.5.14}
$$

where the constants $\gamma_{N,p}$, $\gamma_N$ are chosen so that

$$
-\Delta_p \mu_p = \delta_0 \qquad \text{in } \mathcal{D}'(\mathbb{R}^N). \tag{1.5.15}
$$

Since $\mu_p$ is rotationally invariant with respect to 0, we set $\mu_p(x) = \mu_p(|x|)$.

Let $\Omega$ be a domain in $\mathbb{R}^N$ containing 0, $\Omega^* = \Omega \setminus \{0\}$. The next result due to [Kichenassamy and Véron (1988)] makes more precise the two-sided estimates in Theorem 1.3.7.

**Theorem 1.5.6** *Let $1 < p \leq N$ and $u$ is a $p$-harmonic function in $\Omega^*$ such that $|u(x)| \leq c\mu_p(x)$ near 0. Then there exists a real number $\gamma \in [-c, c]$ such that*

$$u - \gamma\mu_p \in L^\infty_{loc}(\Omega). \tag{1.5.16}$$

*Moreover, if $\gamma \neq 0$ the following relation holds*

$$\lim_{x\to 0} |x|^{\frac{N-p}{p-1}+|\beta|} D^\beta(u - \gamma\mu_p)(x) = 0, \tag{1.5.17}$$

*for all $\beta \in \mathbb{N}^N \setminus \{0\}$. Furthermore $|\nabla u| \in L^1_{loc}(\Omega)$ and there holds*

$$-\Delta_p u = |\gamma|^{p-2}\gamma\delta_0 \qquad \text{in } \mathcal{D}'(\Omega). \tag{1.5.18}$$

Scaling techniques are among the standard methods to obtain pointwise estimates on the derivatives from similar estimates on functions itself.

**Lemma 1.5.7** *Let $1 < p \leq N$ and $u$ is a $p$-harmonic function in $B_2^*$ where it satisfies $|u(x)| \leq A\mu_p(x) + B$. Then there exist constants $\alpha \in (0,1)$ and $c > 0$ depending on $N$ and $p$ such that*

$$\begin{aligned} &(i) && |\nabla u(x)| \leq Ac|x|^{-\frac{N-1}{p-1}} \\ &(ii) && |\nabla u(x) - \nabla u(x')| \leq Ac|x|^{-\frac{N-1}{p-1}-\alpha}|x - x'|^\alpha, \end{aligned} \tag{1.5.19}$$

*for all $x, x' \in B_1^*$ satisfying $|x| \leq |x'|$.*

*Proof.* If $0 < a < \frac{1}{2}$ and $0 < |x| \leq 1$, we set $y = (1 + a^{-1})x$ and

$$v(y) = \frac{1}{\mu_p(a)} u\left(\frac{ay}{1+a}\right).$$

The function $v$ is $p$-harmonic in $\{y \in \mathbb{R}^N : \frac{1}{2} \leq |y| \leq 3\}$ and bounded by $c(A + B)$. By Theorem 1.3.9 its gradient is bounded by $c'(A + B)$ in $\{y \in \mathbb{R}^N : \frac{2}{3} \leq |y| \leq 2\}$ in which set there also holds,

$$|\nabla v(y) - \nabla v(y')| \leq c'' |y - y'|^\alpha (A + B),$$

for $\alpha = \alpha(N, p) \in (0, 1]$. Hence

$$|\nabla u(x)| \leq 2c(A + B)\frac{\mu_p(a)}{a} \leq c'(A + B)\frac{\mu_p(x)}{|x|}.$$

Similarly

$$|\nabla u(x) - \nabla u(x')| \leq c(A+B)\,|x-x'|^{\alpha}\,\frac{\mu_p(x)}{|x|} \qquad \text{for } 0<|x|\leq|x'|\leq 1.$$

□

It is easy to derive from the above theorem existence and uniqueness for the singular Dirichlet problem with Dirac measure.

**Corollary 1.5.8** *Let $\Omega \subset \mathbb{R}^N$ be a bounded domain with a $C^1$ boundary and $1<p\leq N$. Then for any $\gamma \in \mathbb{R}$ and $g \in L^\infty \cap W^{1-\frac{1}{p}}$, there exists a function $u \in C^{1,\alpha}(\Omega^*)$ such that $|\nabla u|^{p-1} \in L^1(\Omega)$, $\nabla u \in W^{1,p}(\Omega \setminus B_\epsilon$ for any $\epsilon > 0$, solution of*

$$\begin{aligned} -\Delta_p u &= |\gamma|^{p-2}\,\gamma\delta_0 && \text{in } \mathcal{D}'(\Omega)\\ u &= g && \text{on } \partial\Omega. \end{aligned} \tag{1.5.20}$$

*Furthermore $u$ is unique in the class of functions such that $|u(x)| \leq c\mu_p(x)$ near 0.*

Another consequence of Theorem 1.5.6 is the following singular rigidity result.

**Corollary 1.5.9** *Let $1<p\leq N$. If $u$ is a $p$-harmonic function in $\mathbb{R}^N\setminus\{0\}$ such that $|u(x)| \leq a|\mu_p(x)|+b$ for some nonnegative constants $a,b$, there exist two constants $\gamma$ and $\lambda$ such that $|\gamma|\leq a$, $|\lambda|\leq b$ with the property that $u(x) = \gamma\mu_p(x)+\lambda$ for all $x \in \mathbb{R}^N\setminus\{0\}$.*

*Proof. The case $p=N$.* We use the invariance of the equation under the conformal transformation $x \mapsto \mathcal{I}_0(x) = \frac{x}{|x|^2}$. By Theorem 1.5.6 we see that there exist constants $\gamma$ and $\gamma'$ such that $u-\gamma\mu_N \in L^\infty(B_1)$ and $u-\gamma'\mu_N \in L^\infty(B_1^c)$. If $|\gamma|+|\gamma'|=0$, $u$ is bounded and therefore it is constant by Theorem 1.3.8. Let us assume that $|\gamma|+|\gamma'|>0$, for example $\gamma>0$, and put, for $\epsilon>0$ and $R>1$,

$$v_\epsilon = (1-\epsilon)\gamma\mu_N + (\gamma'-(1-\epsilon)\gamma)\mu_N(R) - K,$$

where

$$K = \max\left\{\|u-\gamma\mu_N\|_{L^\infty(B_1)}, \|u-\gamma'\mu_N\|_{L^\infty(B_1^c)}\right\}.$$

By the comparison principle, $u \geq v_\epsilon$ in $B_R$. Letting $\epsilon \to 0$ yields

$$u \geq \gamma\mu_N + (\gamma'-\gamma)\mu_N(R) - K \qquad \text{in } B_R.$$

In the same way

$$u \leq \gamma\mu_N + (\gamma' - \gamma)\mu_N(R) + K \qquad \text{in } B_R.$$

Therefore the following two-sided inequality holds in $B_R$,

$$\gamma\mu_N + (\gamma' - \gamma)\mu_N(R) - K \leq u \leq \gamma\mu_N + (\gamma' - \gamma)\mu_N(R) + K. \quad (1.5.21)$$

Since $\mu_N(R) \to -\infty$ when $R \to \infty$, we infer that $\gamma = \gamma'$. By Theorem 1.5.6 $u - \gamma\mu_N$ admits a limit both at 0 and at infinity; we set $\lambda = \lim_{|x|\to\infty}(u(x) - \gamma\mu_N(x))$. For any $\epsilon > 0$ there exists $R(\epsilon) > 0$ such that

$$\gamma\mu_N(x) + \lambda - \epsilon \leq u(x) \leq \gamma\mu_N(x) + \lambda + \epsilon \qquad \forall x \text{ s.t. } |x| \geq R(\epsilon). \quad (1.5.22)$$

The three functions $u$, $\gamma\mu_N(x) + \lambda + \epsilon$ and $\gamma\mu_N(x) + \lambda - \epsilon$ are $N$-harmonic in $B_{R(\epsilon)} \setminus \{0\}$, have the same singularity at $x = 0$ and are ordered on $\partial B_{R(\epsilon)}$ because of (1.5.22). Therefore (1.5.22) holds in whole $\mathbb{R}^N$. Letting $\epsilon \to 0$ yields $u = \gamma\mu_N + \lambda$.

*The case* $1 < p < N$. There exists a real number $\gamma$ such that

$$\lim_{x\to 0} \frac{u(x)}{\mu_p(x)} = \gamma \quad \text{and} \quad u - \gamma\mu_p \in L^\infty(\mathbb{R}^N). \quad (1.5.23)$$

The following dichotomy occurs:
(i) either the maximum of $u - \gamma\mu_p$ is achieved at some point $x_0 \in \mathbb{R}^N \setminus \{0\}$,
(ii) either the maximum of $u - \gamma\mu_p$ is achieved at infinity,
(iii) or the maximum of $u - \gamma\mu_p$ is achieved at 0.
In case (i), it follows from Proposition 1.5.2 that $u = \gamma\mu_N + (u - \gamma\mu_p)(x_0)$. In case (ii), we set $\lambda(r) = \max\{(u - \gamma\mu_p)(x) : |x| \leq r\}$. Then the function $\lambda$ is increasing, again by Proposition 1.5.2, and there exists some $y_r$ with norm 1 such that $\lambda(r) = (u - \gamma\mu_p)(ry_r)$ and

$$\lim_{r\to\infty} \lambda(r) = \max\{(u - \gamma\mu_p)(x) : 0 < |x| < \infty\} := \lambda^*.$$

Set

$$v_r(y) = u(ry) - \gamma\mu_p(r).$$

Then $v_r$ is $p$-harmonic in $\mathbb{R}^N \setminus \{0\}$ and

$$\begin{array}{lll} (i) & |\nabla v_r(y)| = r\,|\nabla u(ry| \leq c(a+b)|y|^{-1} & \forall |y| \geq r^{-1} \\ (ii) & |\nabla v_r(y) - \nabla v_r(y')| \leq c(a+b)|y|^{-1-\alpha}|y - y'|^\alpha & \forall |y'| \geq |y| \geq r^{-1}. \end{array}$$

Hence there exist a sequence $\{r_n\}$ tending to $\infty$, a $p$-harmonic function $v \in C^1(\mathbb{R}^N \setminus \{0\})$ and $y_0$ with norm 1 such that $\{v_{r_n}\}$ converges to $v$ in the $C^1_{loc}$-topology of $\mathbb{R}^N \setminus \{0\}$ and $\{y_{r_n}\}$ converges to $y_0$. Then

$$v \leq v(y_0) = \max\{v(y) : y \in \mathbb{R}^N \setminus \{0\}\}.$$

This implies that $v \equiv v(y_0) = \lambda^*$ is a constant function. Therefore, for any $\epsilon > 0$, there holds $\lambda^* - \epsilon \leq u(r_n y) - \gamma\mu_p(r_n) \leq \lambda^* + \epsilon$ for $r_n$ large enough and $|y| = 1$. This implies

$$\lim_{|x|\to\infty} (u - \gamma\mu_p)(x) = \lambda^*,$$

by the maximum principle. Finally, comparing $u$ and $\gamma\mu_p)(x) + \lambda^*$ in $B_R \setminus \{0\}$ for any $R > 0$ as in the previous case yields $u = \gamma\mu_p)(x) + \lambda^*$. The proof in case (iii) is similar. □

### 1.5.3 *Rigidity theorems for p-harmonic functions*

The following result (see [Kilpeläinen, Shahgholian, Zhong (2007)]) extends Theorem 1.3.8 and the remark hereafter to $p$-harmonic functions. It is noticeable that it proofs relies only on the Hölder regularity estimate of the gradient.

**Theorem 1.5.10** *Let $1 < p < \infty$. Then there exists a number $\alpha \in (0,1)$ such that any p-harmonic function $u$ in $\mathbb{R}^N$ which satisfies*

$$|u(x)| = o(|x|^{1+\alpha}) \qquad \text{as } |x| \to \infty, \tag{1.5.24}$$

*is of the form $u(x) = \sum_j a_j x_j + b$ for some real numbers $a_j$, $b$ and $j = 1, ..., N$.*

*Proof.* Let $\{R_k\}$ be a sequence tending to infinity and

$$S_k = \max\{|u(x)| : |x| \leq R_k\}.$$

The functions $u_k$ defined by $u_k(x) = S_k^{-1} u(R_k x)$ are $p$-harmonic in $\mathbb{R}^N$ and $|u_k(x)| \leq 1$ in $B_1$. By Theorem 1.3.9 the set of functions $\{u_k\}$ is uniformly bounded in $C^{1,\alpha}(B_r)$ for some $\alpha \in (0,1)$ and any $r < 1$. Therefore the functions

$$c_k(x) = \frac{|Du_k(x) - Du_k(0)|}{|x|^\alpha} = \frac{R_k^{1+\alpha}}{S_k} \frac{|Du(R_k x) - Du_k(0)|}{|R_k x|^\alpha},$$

are uniformly bounded in $B_r$. The growth condition implies

$$\lim_{k\to\infty} \frac{R_k^{1+\alpha}}{S_k} = 0,$$

and therefore

$$\sup_{x\in B_r} \frac{|Du_k(x) - Du_k(0)|}{|x|^\alpha} = \sup_{y\in B_{R_k r}} \frac{|Du(y) - Du(0)|}{|y|^\alpha} \to 0 \quad \text{as } R_k \to \infty.$$

Thus $Du(y) = Du(0)$ for all $y \in \mathbb{R}^N$, hence $u$ is a linear functions of $(x_1, ..., x_N)$. Conversely any linear function is $p$-harmonic. □

*Remark.* It is interesting to notice that the exponent $\alpha$ is the one which gives the Hölder continuity of the gradient of $p$-harmonic functions.

The next result is optimal since linear function are $p$-harmonic.

**Corollary 1.5.11** *Let $1 < p < \infty$. Then any $p$-harmonic function $u$ in $\mathbb{R}^N$ which satisfies*

$$|u(x)| = o(|x|) \qquad as\ |x| \to \infty, \tag{1.5.25}$$

*is a constant.*

*Proof.* By Theorem 1.5.10 we should have $u(x) = \sum_j a_j x_j + b$. But all the $a_j$ are zero because of the sublinear estimate. □

The following theorem characterizes nonnegative $p$-harmonic functions in a half-space which vanishes on the boundary. Its proof is based upon a simple form of Carleson estimate. More elaborate statements will be stated in Chapter 2.

**Proposition 1.5.12** *Assume $1 < p < \infty$ and $u$ is a nonnegative $p$-harmonic function in $\mathbb{R}^N_+ = \{x := (x', x_N) : x_N > 0\}$ which vanishes on $\partial\mathbb{R}^N_+$. Then there exists a positive constant $c_1 = c_1(N, p)$ such that*

$$\frac{u(x', x_N)}{x_N} \leq c_1 \frac{u(0, R)}{R} \qquad \forall R > 0\ and\ (x', x_N) \in B^+_{2R}. \tag{1.5.26}$$

*Proof.* By the strong comparison principle we can assume that $u > 0$ in $\mathbb{R}^N_+$. If $R > 0$, we set

$$m_R^p = \int_{B_{3R}} |u|^p \, dx,$$

and define $u_R$ by $u_R(x) = R^{\frac{N}{p}} m_R^{-1} u(Rx)$. Then

$$\int_{B_3} |u_R|^p \, dx = 1.$$

If we assume that the result does not hold there exists a sequence of points $\{a_k\} = \{(a'_k, \alpha_k)\} \subset B_2^+$ such that

$$\frac{u_R(a'_k, \alpha_k)}{\alpha_k} \geq k u_R(0,1). \tag{1.5.27}$$

Since $\|u_R\|_{L^p(B_3)} = 1$, $\|\nabla u_R\|_{L^\infty(B_2)} \leq m$, where $m = m(N,p) > 0$, therefore

$$|u_R(a'_k, \alpha_k) - u_R(a'_k, 0)| = u_R(a'_k, \alpha_k) \leq m\alpha_k,$$

which implies $u_R(0,1) \leq mk^{-1}$ and $u_R(0,1) = 0$ which is a contradiction. Hence there exists $c > 0$ depending only on $N$ and $p$ such that

$$\frac{u_R(y', y_N)}{y_N} \leq c \frac{u_R(0,1))}{1} \qquad \forall y = (y', y_N) \in B_2^+.$$

By rescaling we obtain (1.5.26). □

**Theorem 1.5.13** *Let $1 < p < \infty$. Then any nonnegative $p$-harmonic function in a half-space $\mathbb{R}_+^N$ which vanishes on its boundary $\partial\mathbb{R}_+^N$ is a linear function of the $x_N$ variable.*

*Proof.* We denote by $\mathbf{e}_N$ the vector with coordinates $(0, 0, .., 0, 1)$.

*Step 1.* We claim that there exists $c_2 = c_2(N,p)$ such that

$$u(R\mathbf{e}_N) \leq c_2 R u(\mathbf{e}_N) \qquad \forall R \geq 2. \tag{1.5.28}$$

For $r > 1$ and $x_0 = 2r\mathbf{e}_N = R\mathbf{e}_N$, we define the positive $p$-harmonic function $v$ in $B_{2r}(x_0) \setminus B_r(x_0)$ by

$$v(x) = \frac{\displaystyle\int_{|x-x_0|}^{2r} t^{\frac{1-N}{p-1}} \, dt}{\displaystyle\int_r^{2r} t^{\frac{1-N}{p-1}} \, dt}.$$

This function is $p$-harmonic in $B_{2r}(x_0) \setminus B_r(x_0)$, it vanishes on $\partial B_{2r}(x_0)$ and has value 1 on $\partial B_r(x_0)$. Since by Harnack inequality (see e.g. Theorem 1.3.3) there exists $m_2 = m_2(N,p) > 0$ such that

$$u(x_0) \leq m_2 u(x) \qquad \forall x \in B_{\frac{|x_0|}{2}}(x_0),$$

we obtain by comparison,

$$m_2 u(x) \geq u(x_0) v(x) \qquad \forall x \in B_{2r}(x_0) \setminus B_r(x_0).$$

In particular $m_2 u(\mathbf{e}_N) \geq u(x_0) v(\mathbf{e}_N)$ and a straightforward computation shows that $v(\mathbf{e}_N) \geq (c_2 r)^{-1}$ for some $c_3 = c_3(N, p) > 0$. This implies $u(x_0) \leq m_2 c_3 r u(\mathbf{e}_N)$ and we get (1.5.28).

*Step 2.* To complete the proof we use boundary Harnack inequality Proposition 1.5.12 which yields

$$\frac{u(x)}{x_N} \leq c_1 \frac{u(R\mathbf{e}_N)}{R} \qquad \forall x \in B^+_{2R} \quad \forall R > 0. \tag{1.5.29}$$

Combining (1.5.28) and (1.5.29)

$$u(x) \leq c_1 x_N \frac{u(R\mathbf{e}_N)}{R} \leq c_1 c_2 x_N R u(\mathbf{e}_N) \qquad \forall x \in B^+_{2R} \quad \forall R \geq 2. \tag{1.5.30}$$

This implies that $u$ has a growth at most linear. We define now a function $\tilde{u}$ in $\mathbb{R}^N$ by

$$\tilde{u}(x', x_N) = \begin{cases} u(x', x_N) & \text{if } x_N \geq 0 \\ -u(x', -x_N) & \text{if } x_N < 0. \end{cases} \tag{1.5.31}$$

Clearly $\tilde{u}$ is $p$-harmonic in $\mathbb{R}^N$ with an at most linear growth at infinity. Thus it is a linear function by Theorem 1.5.10, which ends the proof. □

*Remark.* It is a consequence of Theorem 1.5.13 that the constant $c$ which appears in Proposition 1.5.12 and could depend on the function $u$ is actually equal to 1. The fact that $c$ is independent of $u$ is the real Carleson inequality and it is a key point in the proof of boundary Harnack inequality.

### 1.5.4 *Isolated singularities of p-superharmonic functions*

A remarkable property of positive weakly $p$-superharmonic functions with an isolated singularity lies in the presence of a Dirac mass at the singular point. This result which has many applications to equations with source reaction is proved in [Bidaut-Véron (1989)].

**Theorem 1.5.14** *Let $1 < p \leq N$ and $u \in W^{1,p}_{loc}(B_1 \setminus \{0\}) \cap C(\overline{B}_1 \setminus \{0\})$ such that $\Delta_p u \in L^1_{loc}(B_1 \setminus \{0\})$ in the sense of distributions in $B_1 \setminus \{0\}$ satisfies*

$$-\Delta_p u \geq 0 \qquad \text{in } \mathcal{D}'(B_1 \setminus \{0\}). \tag{1.5.32}$$

*Then*

*(i) if* $1 < p < N$, $u^{p-1} \in L_{loc}^{\frac{N}{N-p},\infty}(B_1)$, $|\nabla u|^{p-1} \in L_{loc}^{\frac{N}{N-1},\infty}(B_1)$ *and there exist* $g \in L^1_{loc}(B_1)$ *and* $\beta \geq 0$ *such that*

$$-\Delta_p u = g + \beta\delta_0 \qquad \text{in } \mathcal{D}'(B_1). \tag{1.5.33}$$

*(ii) if* $p = N$, $u^r \in L^1_{loc}(B_1)$ *for all* $r > 0$, $|\nabla u|^r \in L^1_{loc}(B_1)$ *for all* $r \in (0, N)$. *Furthermore, if* $\lim_{x\to 0} u(x) = \infty$ *the conclusions of (i) holds.*

In the above theorem the function $g$ is equal to $\Delta_p u$ a.e. The difficulty is to prove that $g \in L^1_{loc}(B_1)$.

## 1.6 Notes and open problems

*1.6.1.* The Hölder regularity of weak solutions of degenerate quasilinear equations associated to variational integral was initialy proved by Ladyzhenskaya and Ural'tseva. The $C^{1,\alpha}$ regularity was obtained for $p$-harmonic type equations in [Uhlenbeck (1977)], [Evans (1982)] and [Lewis (1983)]. The extension to more general operators is due to [Tolksdorff (1984)].

*1.6.2.* It was conjectured by Serrin that the exponent $\alpha$ in (1.5.24) is equal to $\frac{1}{p-1}$. There will be a discussion about optimality of this exponent at the end of Chapter 2.

*1.6.3.* In Theorem 1.5.14-(ii) the question is still open wether the assumption $\lim_{x\to 0} u(x) = \infty$ is necessary in order $-\Delta_N u \in L^1_{loc}(B_1)$ and (1.5.33) holds. A step in this direction would be the fact that $\nabla u \in L_{loc}^{N,\infty}(B_1)$.

# Chapter 2

# Separable solutions

## 2.1 Eigenvalue and eigenfunctions

### 2.1.1 *Dirichlet problem*

In this section $\Omega$ is a bounded domain of $\mathbb{R}^N$ and $p > 1$. The Rayleigh quotient $R_p[v]$ is defined for any $v \in W_0^{1,p}(\Omega)$, $v \neq 0$, by

$$R_p[v] = \frac{\displaystyle\int_\Omega |\nabla v|^p dx}{\displaystyle\int_\Omega |v|^p dx}. \tag{2.1.1}$$

If we assume that $\Omega \subset \{x \in \mathbb{R}^N : 0 < x_i < a_1\}$, then

$$R_p[v] \geq \frac{p}{a_j^p}. \tag{2.1.2}$$

This is consequence of the fact that there exists a sequence $\{v_n\} \subset C_c^\infty(\Omega)$ which converges to $v$ in $W^{1,p}(\Omega)$ and therefore $R_p[v] = \lim_{n\to\infty} R_p[v_n]$. By integration

$$\begin{aligned} v_n^p(x_1, x_2, ..., x_N) &= \left(\int_0^{x_1} \frac{\partial v_n}{\partial x_1}(t, x_2, ..., x_N))dt\right)^p \\ &\leq x_1^{\frac{p}{p'}} \int_0^{x_1} \left|\frac{\partial v_n}{\partial x_1}(t, x_2, ..., x_N)\right|^p dt. \end{aligned}$$

Since $R_p[v]$ is invariant by orthogonal transformations in $\mathbb{R}^N$, and if $R > 0$ denotes the radius of the largest ball contained in the smallest rectangular parallelepiped box containing $\Omega$, we obtain after a rotation,

$$R_p[v] \geq \frac{p}{R^p}. \tag{2.1.3}$$

Since $|\Omega| \leq \omega_N R^N$, where $\omega_N$ is the volume of the Euclidean unit ball in $\mathbb{R}^N$, it follows

$$R_p[v] \geq \frac{p\omega_N^{\frac{p}{N}}}{|\Omega|^{\frac{p}{N}}}. \tag{2.1.4}$$

**Proposition 2.1.1** *There exists $\phi_1 \in W_0^{1,p}(\Omega) \setminus \{0\}$ such that*

$$R_p[\phi_1] = \inf\left\{R_p[v] : v \in W_0^{1,p}(\Omega) \setminus \{0\}\right\} := \lambda_1. \tag{2.1.5}$$

*Equivalently*

$$\lambda_1 = \inf\left\{\|\nabla v\|_{L^p}^p : v \in W_0^{1,p}(\Omega), \|v\|_{L^p}^p = 1\right\}. \tag{2.1.6}$$

*Furthermore $\phi_1$ is $C^{1,\alpha}$ for some $\alpha \in (0,1)$, it keeps a constant sign and it is unique up to an homothety,*

*Proof.* The existence of a minimizer, say $\phi_1$, follows from the compactness of the imbedding of $W_0^{1,p}(\Omega)$ into $L^p(\Omega)$. Because $R_p[v] = R_p[|v|]$, we can suppose that $\phi_1 \geq 0$. Since the mapping $v \mapsto R_p[v]$ is $C^1$ on $W_0^{1,p}(\Omega) \setminus \{0\}$, we can compute $\frac{d}{dt}R_p[\phi_1 + t\eta]\lfloor_{t=0}$ for any $\eta \in W_0^{1,p}(\Omega)$, with $\|\eta\|_{W^{1,p}} < \|\phi_1\|_{W^{1,p}}$ and get

$$\int_\Omega |\nabla\phi_1|^{p-2} \nabla\phi_1.\nabla\eta dx = \lambda_1 \int_\Omega |\phi_1|^{p-2} \phi_1 \eta dx. \tag{2.1.7}$$

Thus $\phi_1$ is a nonnegative weak solution of

$$-\Delta_p\phi = \lambda_1 |\phi|^{p-2} \phi. \tag{2.1.8}$$

We say that $\phi_1$ is an eigenfunction of $-\Delta_p$ with corresponding eigenvalue $\lambda_1$. By regularity $\phi_1$ is bounded and continuous in $\Omega$ up to a possible modification on a set with zero measure, and by Harnack inequality, $\phi_1 > 0$. By Theorem 1.3.9, $\nabla\phi_1$ is $\alpha$-Hölder continuous for some $\alpha \in (0,1)$. Actually any $\phi \in W_0^{1,p}(\Omega)$ which satisfies (2.1.8) in the weak sense (even with $\lambda_1$ replaced by some $\lambda > 0$ if it exists) shares the same $C^{1,\alpha}$-regularity. Assume that $\psi_1$ is another minimizer of $R_p$. Since $|\psi_1|$ satisfies (2.1.8), it satisfies Harnack inequality and thus it never vanishes in $\Omega$. Therefore we can infer that $\psi_1 > 0$ and we put $w = (\frac{1}{2}\psi_1^p + \frac{1}{2}\phi_1^p)^{\frac{1}{p}}$. Then $\|w\|_{L^p} = 1$ and

$$|\nabla w|^p = \frac{1}{2}(\psi_1^p + \phi_1^p)^{1-p}\left|\psi_1^{p-1}\nabla\psi_1 + \phi_1^{p-1}\nabla\phi_1\right|^p$$
$$= \left|\frac{1}{2^{\frac{1}{p}}}\frac{\psi_1^{p-1}}{(\psi_1^p + \phi_1^p)^{1-\frac{1}{p}}}\nabla\psi_1 + \frac{1}{2^{\frac{1}{p}}}\frac{\phi_1^{p-1}}{(\psi_1^p + \phi_1^p)^{1-\frac{1}{p}}}\nabla\phi_1\right|^p.$$

Since $X \mapsto |X|^p$ is convex and $\frac{\psi_1^{p-1}}{2^{\frac{1}{p}}(\psi_1^p+\phi_1^p)^{1-\frac{1}{p}}} + \frac{\phi_1^{p-1}}{2^{\frac{1}{p}}(\psi_1^p+\phi_1^p)^{1-\frac{1}{p}}} < 1$ on the set $\mathcal{C} := \{x \in \Omega : \psi_1(x) \neq \phi_1(x)\}$, it follows

$$|\nabla w|^p < \frac{1}{2}|\nabla\psi_1|^p + \frac{1}{2}|\nabla\phi_1|^p \qquad \text{in } \mathcal{C},$$

and

$$|\nabla w|^p \leq \frac{1}{2}|\nabla\psi_1|^p + \frac{1}{2}|\nabla\phi_1|^p \qquad \text{in } \Omega\setminus\mathcal{C}.$$

If we assume that $|\mathcal{C}| > 0$, we obtain

$$\int_\Omega |\nabla w|^p\,dx < \frac{1}{2}\int_\Omega |\nabla\psi_1|^p dx + \frac{1}{2}\int_\Omega |\nabla\phi_1|^p dx = \lambda_1.$$

Because $\|w\|_{L^p} = 1$, $\|\nabla w\|_{L^p}^p \geq \lambda_1$. Therefore $|\mathcal{C}| = 0$ and $\psi_1(x) = \phi_1(x)$ a.e. in $\Omega$. □

**Definition 2.1.2** We say that a function $\phi \in W_0^{1,p}(\Omega)$ is an eigenfunction of $-\Delta_p$ in $W_0^{1,p}(\Omega)$ if there exists $\lambda$, necessarily positive, such that

$$-\Delta_p\phi = \lambda\,|\phi|^{p-2}\,\phi \qquad \text{in } \Omega. \tag{2.1.9}$$

The set of eigenvalues is denoted by $\sigma[-\Delta_p; W_0^{1,p}(\Omega)]$.

*Remark.* It is easy to see that $\sigma[-\Delta_p; W_0^{1,p}(\Omega)]$ is a closed subset of $\mathbb{R}_+$. Furthermore, if $\phi$ is an eigenfunction with eigenvalue $\lambda$ and $G$ is a connected component of $\{x \in \Omega : \phi(x) > 0\}$, $\phi\lfloor_G$ is the first eigenfunction of $-\Delta_p$ in $W_0^{1,p}(G)$ with corresponding eigenvalue $\lambda$. As a consequence of (2.1.4)

$$\lambda|G|^{\frac{p}{N}} \geq p\omega_N^{\frac{p}{N}}. \tag{2.1.10}$$

**Proposition 2.1.3** *The eigenvalue $\lambda_1$ is the smallest element of $\sigma[-\Delta_p; W_0^{1,p}(\Omega)]$. Furthermore $\phi_1$ is the only eigenfunction with constant sign and $\lambda_1$ is an isolated eigenvalue.*

*Proof.* As already noticed in the proof of Proposition 2.1.1, any eigenfunction is bounded and, up to a modification on a set with zero measure, it belongs to $C^{1,\alpha}$ for some $\alpha \in (0,1)$. The fact that $\lambda_1 = \min\{\lambda : \lambda \in \sigma[-\Delta_p; W_0^{1,p}(\Omega)]\}$ is clear from Proposition 2.1.1. If we assume that there exists $\lambda \in \sigma[-\Delta_p; W_0^{1,p}(\Omega)]$, $\lambda > \lambda_1$, with a corresponding nonnegative eigenfunction $\phi$, then $\phi > 0$ by Harnack inequality. For $\epsilon > 0$ we set $\phi_{1,\epsilon} = \phi_1 + \epsilon$, $\phi_\epsilon = \phi + \epsilon$ and

$$\eta_1 = \frac{\phi_{1,\epsilon}^p - \phi_\epsilon^p}{\phi_{1,\epsilon}^{p-1}} \quad \text{and} \quad \eta = \frac{\phi_\epsilon^p - \phi_{1,\epsilon}^p}{\phi_\epsilon^{p-1}}.$$

Then

$$\nabla\eta_1 = \left(1 + (p-1)\left(\frac{\phi_\epsilon}{\phi_{1,\epsilon}}\right)^p\right)\nabla\phi_1 - p\left(\frac{\phi_\epsilon}{\phi_{1,\epsilon}}\right)^{p-1}\nabla\phi,$$

similarly for $\nabla\eta$ by exchanging $\phi_1$ and $\phi$, thus $\eta$ and $\eta_1$ belong to $W_0^{1,p}(\Omega)$,

$$\begin{aligned} \int_\Omega |\nabla\phi_1|^{p-2}\,\nabla\phi_1.\nabla\eta_1 dx &= \lambda_1 \int_\Omega \phi_1^{p-1}\eta_1 dx \\ \int_\Omega |\nabla\phi|^{p-2}\,\nabla\phi.\nabla\eta dx &= \lambda \int_\Omega \phi^{p-1}\eta dx. \end{aligned} \tag{2.1.11}$$

After some computations we obtain

$$\begin{aligned} &\int_\Omega \left(\lambda_1\frac{\phi_1^{p-1}}{\phi_{1,\epsilon}^{p-1}} - \lambda\frac{\phi^{p-1}}{\phi_\epsilon^{p-1}}\right)(\phi_{1,\epsilon}^p - \phi_\epsilon^p)\,dx \\ &\qquad = \int_\Omega (\phi_{1,\epsilon}^p - \phi_\epsilon^p)\left(|\nabla \ln\phi_{1,\epsilon}|^p - |\nabla\ln\phi_\epsilon|^p\right)dx \\ &\qquad - p\int_\Omega \phi_{1,\epsilon}^p\,|\nabla\ln\phi_\epsilon|^{p-2}\,\nabla\ln\phi_\epsilon.\left(\nabla\ln\phi_{1,\epsilon} - \nabla\ln\phi_\epsilon\right)dx \\ &\qquad - p\int_\Omega \phi_\epsilon^p\,|\nabla\ln\phi_{1,\epsilon}|^{p-2}\,\nabla\ln\phi_{1,\epsilon}.\left(\nabla\ln\phi_\epsilon - \nabla\ln\phi_{1,\epsilon}\right)dx. \end{aligned}$$

Clearly

$$\lim_{\epsilon\to 0}\int_\Omega \left(\lambda_1\frac{\phi_1^{p-1}}{\phi_{1,\epsilon}^{p-1}} - \lambda\frac{\phi^{p-1}}{\phi_\epsilon^{p-1}}\right)(\phi_{1,\epsilon}^p - \phi_\epsilon^p)\,dx = (\lambda_1 - \lambda)\int_\Omega (\phi_1^p - \phi^p)\,dx.$$

Using the convexity of $X \mapsto |X|^p$,

$$p\,|\nabla \ln \phi_{1,\epsilon}|^{p-2}\,\nabla \ln \phi_{1,\epsilon}.(\nabla \ln \phi_{1,\epsilon} - \nabla \ln \phi_\epsilon)$$
$$\geq |\nabla \ln \phi_{1,\epsilon}|^p - |\nabla \ln \phi_\epsilon|^p$$
$$\geq p\,|\nabla \ln \phi_\epsilon|^{p-2}\,\nabla \ln \phi_\epsilon.(\nabla \ln \phi_{1,\epsilon} - \nabla \ln \phi_\epsilon).$$

Put $\Omega' = \{x \in \Omega : \phi_{1,\epsilon} \geq \phi_\epsilon\}$, $\Omega'' = \Omega \setminus \Omega$, then

$$\left(\phi_{1,\epsilon}^p - \phi_\epsilon^p\right)\left(|\nabla \ln \phi_{1,\epsilon}|^p - |\nabla \ln \phi_\epsilon|^p\right)$$
$$\geq p\phi_\epsilon^p\left(|\nabla \ln \phi_{1,\epsilon}|^{p-2}\,\nabla \ln \phi_{1,\epsilon} - |\nabla \ln \phi_\epsilon|^{p-2}\,\nabla \ln \phi_\epsilon\right).\nabla\left(\ln \phi_{1,\epsilon} - \ln \phi_\epsilon\right),$$

on $\Omega'$ and

$$\left(\phi_{1,\epsilon}^p - \phi_\epsilon^p\right)\left(|\nabla \ln \phi_{1,\epsilon}|^p - |\nabla \ln \phi_\epsilon|^p\right)$$
$$\geq p\phi_{1,\epsilon}^p\left(|\nabla \ln \phi_{1,\epsilon}|^{p-2}\,\nabla \ln \phi_{1,\epsilon} - |\nabla \ln \phi_\epsilon|^{p-2}\,\nabla \ln \phi_\epsilon\right).\nabla\left(\ln \phi_{1,\epsilon} - \ln \phi_\epsilon\right),$$

on $\Omega''$. The right-hand side terms in the two above inequalities are non-negative, thus

$$(\lambda_1 - \lambda)\int_\Omega (\phi_1^p - \phi^p)\,dx \geq 0.$$

Since we can replace $\phi$ by $k\phi$ for arbitrary positive $k$, the only possibility is $\lambda_1 = \lambda$.

Assume now that $\lambda_1$ is not isolated. Then there exists a decreasing sequence $\{\lambda_n\}$ of eigenvalues with corresponding eigenfunctions $\{\phi_n\}$ normalized by $\|\phi_n\|_{L^p} = 1$ converging to $\lambda_1$. By compactness we can assume that $\phi_n \to \phi'$ in $L^p(\Omega)$ and $\nabla\phi_n \rightharpoonup \nabla\phi'$ weakly in $L^p(\Omega)$ for some $\phi' \in W_0^{1,p}(\Omega)$ verifying $\|\phi'\|_{L^p} = 1$. Furthermore the convergence of $\{\phi_n\}$ holds locally uniformly in $\Omega$. Since $\|.\|_{L^p}$ is weakly lower semicontinuous,

$$\lambda_1 = \lim_{n\to\infty} \lambda_n = \liminf_{n\to\infty} R_p[\phi_n] \geq R_p[\phi'],$$

which implies $\phi' = k\phi$ for some nonzero $k$. Since $\lambda_n > \lambda_1$, $\phi_n$ has at least two nodal domains $\Omega'_n$ and $\Omega''_n$. By the previous remark,

$$\inf\left\{|\Omega'_n|^{\frac{p}{N}}, |\Omega''_n|^{\frac{p}{N}}\right\} \geq \lambda_n^{-1} p\omega_N^{\frac{p}{N}} = \lambda^{-1} p\omega_N^{\frac{p}{N}}(1 + \circ(1)).$$

By a selection lemma, $\limsup_{n\to\infty} \inf\{|\Omega'_n|, |\Omega''_n|\} > 0$, which contradicts $\phi_n \to \phi$ in $L^p(\Omega)$ with $\|\phi_n\|_{L^p} = 1$. □

The question of the variation of the domain $\Omega$ gives rise to the same type of results as if $p = 2$ [Courant, Hilbert (1953)], with quite similar proofs.

**Proposition 2.1.4** *Assume $p > 1$ and denote $\lambda_1 = \lambda_1(p, \Omega)$. Then*
*(i) $\Omega_1 \subsetneq \Omega_2$ implies $\lambda_1(p, \Omega_2) < \lambda_1(p, \Omega_1)$.*
*(ii) If $\Omega_1 \subset \Omega_2 \subset \Omega_3 \subset ....$ is an increasing sequence of domains in $\mathbb{R}^N$, then*

$$\lim_{j\to\infty} \lambda_1(p, \Omega_j) = \lambda_1(p, \Omega) \quad \textit{where } \Omega = \bigcup_{j\geq 1} \Omega_j. \tag{2.1.12}$$

### 2.1.2 *The case $p \to \infty$*

**Proposition 2.1.5** *Let $\Omega$ be a bounded domain of $\mathbb{R}^N$. For any $1 < p < s$ there holds*

$$p\sqrt[p]{\lambda_1(p, \Omega)} \leq s\sqrt[s]{\lambda_1(s, \Omega)}. \tag{2.1.13}$$

*Proof.* Let $\psi \in C_c^1(\Omega)$, $\psi \geq 0$ and $\phi = \psi^{\frac{s}{p}}$, then $\phi \in C_c^1(\Omega)$ and

$$\begin{aligned}\sqrt[p]{\lambda_1(p, \Omega)} &\leq \frac{\left(\int_\Omega |\nabla\phi|^p dx\right)^{\frac{1}{p}}}{\left(\int_\Omega \phi^p dx\right)^{\frac{1}{p}}} = \frac{s}{p}\frac{\left(\int_\Omega \psi^{s-p}|\nabla\psi|^p dx\right)^{\frac{1}{p}}}{\left(\int_\Omega \psi^s dx\right)^{\frac{1}{p}}} \\ &\leq \frac{s}{p}\frac{\left(\int_\Omega \psi^s dx\right)^{\frac{1}{p}-\frac{1}{s}}\left(\int_\Omega |\nabla|^s dx\right)^{\frac{1}{s}}}{\left(\int_\Omega \psi^s dx\right)^{\frac{1}{p}}} = \frac{s}{p}\frac{\left(\int_\Omega |\nabla|^s dx\right)^{\frac{1}{s}}}{\left(\int_\Omega \psi^s dx\right)^{\frac{1}{s}}}.\end{aligned}$$

If we take the infimum over all the $\psi \in C_c^1(\Omega)$ we derive (2.1.13). □

This implies that the function $s \mapsto s\sqrt[s]{\lambda_1(s, \Omega)}$ is monotone nondecreasing. Therefore there exists

$$\lim_{s\uparrow p} s\sqrt[s]{\lambda_1(s, \Omega)} \leq p\sqrt[p]{\lambda_1(p, \Omega} \leq \lim_{s\downarrow p} s\sqrt[s]{\lambda_1(s, \Omega)}. \tag{2.1.14}$$

We denote by $p_-{}^{p}\!\!\sqrt{\lambda_1(p_-, \Omega)}$ and $p_+{}^{p}\!\!\sqrt[+]{\lambda_1(p_+, \Omega)}$ respectively the left-hand side and the right-side of the above limits.

**Corollary 2.1.6** *Under the assumptions of Proposition 2.1.5, there holds*

$$\lim_{s\uparrow p} \lambda_1(s, \Omega) \leq \lambda_1(p, \Omega) \leq \lim_{s\downarrow p} \lambda_1(s, \Omega). \tag{2.1.15}$$

**Theorem 2.1.7** *For any bounded domain and $p > 1$, one has*

$$\lim_{s \downarrow p} \lambda_1(s, \Omega) = \lambda_1(p, \Omega). \tag{2.1.16}$$

*Proof.* For any $\phi \in C_c^1(\Omega)$ there holds

$$\lambda_1(s, \Omega) \leq \frac{\displaystyle\int_\Omega |\nabla \phi|^s dx}{\displaystyle\int_\Omega |\phi|^s dx}.$$

Using Lebesgue convergence theorem and (2.1.15) yields

$$\lim_{s \downarrow p} \lambda_1(s, \Omega) \leq \lim_{s \downarrow p} \frac{\displaystyle\int_\Omega |\nabla \phi|^s dx}{\displaystyle\int_\Omega |\phi|^s dx} = \frac{\displaystyle\int_\Omega |\nabla \phi|^p dx}{\displaystyle\int_\Omega |\phi|^p dx},$$

and if we take the infimum over $\phi$ we obtain the claim. □

*Remark.* The continuity to the left of $s \mapsto \lambda_1(s, \Omega)$ is not always verified and some delicate assumptions on the regularity of $\partial\Omega$ are needed (see [Lindqvist (1993)]).

Let us denote

$$\Lambda_\infty = \inf \left\{ \frac{\|\nabla \phi\|_{L^\infty}}{\|\phi\|_{L^\infty}} : \phi \in C_c^1(\Omega) \right\}. \tag{2.1.17}$$

**Proposition 2.1.8** *For any bounded domain $\Omega$ of $\mathbb{R}^N$, there holds*

$$\Lambda_\infty = \frac{\|\nabla \rho\|_{L^\infty}}{\|\rho\|_{L^\infty}} = \frac{1}{\max\limits_{x \in \Omega} \rho(x)}, \tag{2.1.18}$$

*where* $\rho(x) = \operatorname{dist}(x, \partial\Omega)$.

*Proof.* By the triangle inequality the function $\rho$ is a contraction. If $\phi \in C_c^1(\Omega)$ there holds by the mean value theorem,

$$|\phi(x)| \leq \rho(x) \|\nabla \phi\|_{L^\infty}.$$

The claim follows since

$$\frac{\|\nabla \phi\|_{L^\infty}}{|\phi(x)|} \geq \frac{1}{\rho(x)} \geq \frac{1}{\|\rho\|_{L^\infty}} = \frac{\|\nabla \rho\|_{L^\infty}}{\|\rho\|_{L^\infty}}. \tag{2.1.19}$$

□

**Proposition 2.1.9** *There holds*

$$\lim_{p\to\infty} \sqrt[p]{\lambda_1(p,\Omega)} = \Lambda_\infty. \tag{2.1.20}$$

*Proof.* Taking $\rho$ as test function in the minimization problem (2.1.1) yields

$$\sqrt[p]{R_p[\rho]} = \frac{\|\nabla\rho\|_{L^p}}{\|\rho\|_{L^p}} = \frac{1}{\left(|\Omega|^{-1}\int_\Omega (\rho(x))^p dx\right)^{\frac{1}{p}}}. \tag{2.1.21}$$

Hence

$$\limsup_{p\to\infty} \sqrt[p]{R_p[\rho]} \le \limsup_{p\to\infty} \frac{1}{\left(|\Omega|^{-1}\int_\Omega (\rho(x))^p dx\right)^{\frac{1}{p}}} = \frac{1}{\|\rho\|_L^\infty} = \Lambda_\infty. \tag{2.1.22}$$

Let $\{\phi_p\}$ be the sequence of positive first eigenfunctions of $-\Delta_p$ normalized by $|\Omega|^{-\frac{1}{p}}\|\phi_p\|_{L^p} = 1$, then

$$\left(|\Omega|^{-1}\int_\Omega |\nabla\phi_p|^p dx\right)^{\frac{1}{p}} = \sqrt[p]{\lambda_1(p,\Omega)}.$$

Let $N < m < p$. By Hölder's inequality

$$\left(|\Omega|^{-1}\int_\Omega |\nabla\phi_p|^m dx\right)^{\frac{1}{m}} \le \sqrt[p]{\lambda_1(p,\Omega)}.$$

Thus the set $\{\phi_p\}_{p\ge m}$ is uniformly bounded in $W_0^{1,m}(\Omega)$. By Cantor diagonal process there exists a sequence $\{\phi_{p_k}\}$ which converges to some function $\phi_\infty$, weakly in $W_0^{1,m}(\Omega)$ for any $m > N$ and strongly in any $C^\alpha(\overline{\Omega})$ for $\alpha = 1 - \frac{N}{m}$ by Morrey inequality (1.2.35), and thus uniformly in $\Omega$. Since $|\Omega|^{-\frac{1}{p}}\|\phi_p\|_{L^p} = 1$, it follows $\sup_{x\in\Omega}\phi_\infty(x) \ge 1$ by Hölder's inequality and the uniform convergence. Using the weak lower semicontinuity of the norm in $W^{1,m}$ and Hölder's inequality, yields

$$\frac{\|\nabla\phi_\infty\|_{L^m}}{\|\phi_\infty\|_{L^m}} \le \liminf_{p_k\to\infty} \frac{\|\nabla\phi_{p_k}\|_{L^m}}{\|\phi_{p_k}\|_{L^m}} \le \liminf_{p_k\to\infty} \Lambda_{p_k}\frac{\|\phi_{p_k}\|_{L^\infty}}{\|\phi_{p_k}\|_{L^m}}.$$

This implies

$$\frac{\|\nabla\phi_\infty\|_{L^m}}{\|\phi_\infty\|_{L^m}} \le \left(\liminf_{p_k\to\infty} \Lambda_{p_k}\right)\frac{\|\phi_\infty\|_{L^\infty}}{\|\phi_\infty\|_{L^m}}.$$

Therefore, for any $m > N$,

$$\frac{\|\nabla\phi_\infty\|_{L^m}}{\|\phi_\infty\|_{L^\infty}} \leq \liminf_{p_k\to\infty} \Lambda_{p_k};$$

henceforth this inequality holds also with $m = \infty$ and the claim follows using the minimizing property (2.1.17) and the reverse inequality (2.1.22). □

*Remark.* Using Harnack inequality, it is proved in [Lindqvist, Manfredi (1997)] that the function $\phi_\infty$ is positive in $\Omega$. This fact is important for application to eigenvalue problem (see e.g. Theorem 2.1.12).

Contrary to the case $p < \infty$, the minimization problem (2.1.17) has too many solutions and no possibility to select the function $\rho$ among them. If one expand the equation $\Delta_p\phi + \lambda_1(p,\Omega)\phi^{p-1} = 0$, one find that outside the set of critical points of $\phi$, there holds

$$\frac{|\nabla\phi|^2\,\Delta\phi}{p-2} + H_\phi(\nabla\phi).\nabla\phi + \frac{\lambda_1(p,\Omega)}{p-2}\,|\nabla\phi|^{4-p}\,\phi^{p-1} = 0. \tag{2.1.23}$$

By definition the *infinite Laplacian* is expressed by

$$\Delta_\infty\phi = H_\phi(\nabla\phi).\nabla\phi = \sum_{j,k}\frac{\partial^2\phi}{\partial x_j\partial x_k}\frac{\partial\phi}{\partial x_j}\frac{\partial\phi}{\partial x_k}. \tag{2.1.24}$$

It is proved in [Juutinen, Lindqvist, Manfredi (1999)] that the limit of (2.1.23) when $p \to \infty$ exists, in *the viscosity sense* and the limit equation is

$$\max\left\{\Lambda_\infty - \frac{|\nabla\phi(x)|}{|\phi(x)|}, \Delta_\infty\phi(x)\right\} = 0. \tag{2.1.25}$$

**Definition 2.1.10** A nonnegative function $\phi \in C(\Omega)$ is a viscosity solution of (2.1.25) if

(i) For every $x_0 \in \Omega$ and $\eta \in C^2(\Omega)$ such that $\eta(x_0) = \phi(x_0)$ and $\phi(x) < \eta(x)$ if $x \neq x_0$ in a neighborhood of $x_0$, there holds

$$\Lambda_\infty - \frac{|\nabla\eta(x_0)|}{|\eta(x_0)|} \geq 0 \quad \textbf{or} \quad \Delta_\infty\eta(x_0) \geq 0. \tag{2.1.26}$$

(ii) For every $x_0 \in \Omega$ and $\eta \in C^2(\Omega)$ such that $\eta(x_0) = \phi(x_0)$ and $\phi(x) > \eta(x)$ if $x \neq x_0$ in a neighborhood of $x_0$, there holds

$$\Lambda_\infty - \frac{|\nabla\eta(x_0)|}{|\eta(x_0)|} \leq 0 \quad \textbf{and} \quad \Delta_\infty\eta(x_0) \leq 0. \tag{2.1.27}$$

It is shown in [Juutinen, Lindqvist, Manfredi (1999)] that if $\Omega = \{(x_1, x_2) \in \mathbb{R}^N : |x_1| + |x_2| < 1\}$, the function $d(x) = 1 - \frac{|x_1|+|x_2|}{\sqrt{2}}$ is not a viscosity solution of (2.1.25). Actually, the viscosity solutions have to be found as limit of the normalized eigenfunctions $\phi_p$ used in the proof of Proposition 2.1.9 and the following deep result is proved in [Juutinen, Lindqvist, Manfredi (1999), Th 1.21].

**Theorem 2.1.11** *For any sequence of positive normalized eigenfunctions $\{\phi_{p_k}\}$, converging weakly in $W_0^{1,m}(\Omega)$ for any $m > N$, the limit function $\phi_\infty = \lim_{p_k \to \infty} \phi_{p_k}$ is a viscosity solution of (2.1.25).*

It is noticeable that even if the uniqueness of the normalized viscosity solution of (2.1.25) is not known, the parameter $\Lambda_\infty$ is uniquely determined.

**Theorem 2.1.12** *Assume $\Omega$ is a bounded domain such that $\partial\Omega = \partial\overline{\Omega}$ and there exists a continuous positive solution $\phi \in C(\overline{\Omega})$ vanishing on $\partial\Omega$ which is a viscosity solution of*

$$\max\left\{\Lambda - \frac{|\nabla\phi(x)|}{|\phi(x)|}, \Delta_\infty\phi(x)\right\} = 0. \tag{2.1.28}$$

*Then $\Lambda = \Lambda_\infty$.*

### 2.1.3 *Higher order eigenvalues*

It is well-known that the spectrum of the Laplacian is completely obtained by the theory of compact (linear) operators in Hilbert spaces. It can also be constructed by induction with the following minimax formulation

$$\lambda_k = \min\{R_2[v] : v \in \mathcal{H}_{k-1}^\perp\} \quad \text{for } k = 1, ... \tag{2.1.29}$$

where $\mathcal{H}_1 = H_1 = N(\Delta + \lambda_1 I)$ and $\mathcal{H}_j = \oplus_{\ell=1}^j H_\ell$ where $H_\ell = N(\Delta + \lambda_\ell I)$, noticing that the spaces $H_\ell$ are finite dimensional and mutually orthogonal in $L^2(\Omega)$. This construction can be adapted to construct a variational spectrum of $-\Delta_p$ for $p \neq 2$ in the following way. Let $A$ be a closed subset of a Banach space $X$ such that $A = -A$. The Krasnoselskii genus (see e.g. [Struwe (1996)]) of $A$, $\gamma(A)$, is the smallest integer $k$ such that there exists a continuous and odd mapping from $A$ into $\mathbb{R}^k \setminus \{0\}$. Let $S_k$ be the collection of all closed symmetric subsets of $W_0^{1,p}(\Omega$ such that $\gamma(A) \geq k$ and $A \cap S$ is compact where $S = \{u \in W^{1,p} : \|u\|_{L^p} = 1\}$. The number

$$\lambda_k = \inf_{A \in S_k} \max_{v \in A} R_p[v] \tag{2.1.30}$$

is an eigenvalue of $-\Delta_p$. The set of all such $\lambda_k$ is the variational spectrum of $-\Delta_p$ denoted by $\sigma_{var}[-\Delta_p; W_0^{1,p}(\Omega)]$. It is a deep open problem whether $\sigma[-\Delta_p; W_0^{1,p}(\Omega)]$ and $\sigma_{var}[-\Delta_p; W_0^{1,p}(\Omega)]$ coincide.

The asymptotic distribution of the $\lambda_k$ is described by the following result [Garcia Azorero, Peral (1988)], [Friedlander (1989)].

**Proposition 2.1.13** *There exist two positive constants, $c_1$, $c_2$ such that, for all integer $k$,*

$$c_1 \left(\frac{k}{|\Omega|}\right)^{\frac{p}{N}} \leq \lambda_k \leq c_2 \left(\frac{k}{|\Omega|}\right)^{\frac{p}{N}} \qquad \forall \lambda_k \in \sigma_{var}[-\Delta_p; W_0^{1,p}(\Omega)]. \tag{2.1.31}$$

The proof is quite involved, but we have the next easy-to-prove result.

**Proposition 2.1.14** *If an eigenfunction $\phi \in W_0^{1,p}(\Omega)$ has $k$ nodal domains, i.e. connected components of $\{x \in \Omega : \phi(x) \neq 0\}$, and corresponding eigenvalue $\lambda \in \sigma[-\Delta_p; W_0^{1,p}(\Omega)]$, the following inequality holds*

$$\lambda \leq p \left(\frac{k\omega_N}{|\Omega|}\right)^{\frac{p}{N}}. \tag{2.1.32}$$

*Proof.* Let $G_j$, $j = 1, ..., k$ be the nodal domains of $\phi$. Then, by (2.1.10)

$$|G_j| \geq \left(\frac{p}{\lambda}\right)^{\frac{N}{p}} \omega_N \Longrightarrow |\Omega| \geq k \left(\frac{p}{\lambda}\right)^{\frac{N}{p}} \omega_N.$$

This implies the claim. □

*Remark.* It is a classical result due to Courant that an eigenfunction of $-\Delta$ in $W_0^{1,2}(\Omega)$ corresponding to the k-th eigenvalue $\lambda_k$ has at most $k$ nodal domains. This result admits an extension to the $p$-Laplacian [Dràbek, Robinson (2004)]

**Proposition 2.1.15** *Any eigenfunction $\phi \in W_0^{1,p}(\Omega)$ corresponding to the eigenvalue $\lambda_k \in \sigma_{var}[-\Delta_p; W_0^{1,p}(\Omega)]$ has at most $2k-2$ nodal domains. Furthermore if there exists an eigenfunction with $k+\ell$ nodal domains for some $\ell \in \mathbb{N}$, there exists an eigenfunction corresponding to the same eigenvalue $\lambda_k$ with $k-\ell$ nodal domains.*

Eigenfunctions associated to a nodal domain $G$ satisfies a universal estimate [Lindqvist (1995)].

**Proposition 2.1.16** *If $G \subset \Omega$ is a nodal domain of an eigenfunction $\phi \in W_0^{1,p}(\Omega)$ with corresponding eigenvalue $\lambda \in \sigma[-\Delta_p; W_0^{1,p}(\Omega)]$, there* *holds if* $\phi \geq 0$ *in* $G$,

$$\sup_G \operatorname{ess} \phi \leq \frac{2^{N+1}\lambda^{\frac{N}{p}}}{S_{N,p}^N} \int_G \phi dx, \tag{2.1.33}$$

*where* $S_{N,p}$ *is the Sobolev constant.*

*Proof.* Let $k > 0$ and $\phi_k = \max\{\phi - k, 0\}$. Since $v_k \in W_0^{1,p}(G)$,

$$\int_{G_k} |\nabla\phi|^p dx = \lambda \int_{G_k} \phi^{p-1}\phi_k dx,$$

where $G_k = \{x \in G : \phi > k\}$. Since $a^{p-1} \leq 2^{p-1}(a-k)_+^{p-1} + 2^{p-1}k^{p-1}$,

$$\int_{G_k} \phi^{p-1}\phi_k dx \leq 2^{p-1}\int_{G_k} \phi_k^p dx + 2^{p-1}k^{p-1}\int_{G_k} \phi_k dx.$$

By Sobolev inequality

$$S_{N,p}^p \int_{G_k} \phi_k^p dx \leq |G_k|^{\frac{p}{N}} \int_{G_k} |\nabla\phi|^p dx,$$

where $S_{N,p}$ is defined in (1.2.33). Because

$$\lambda^{-1}\int_{G_k} |\nabla\phi|^p dx = \int_{G_k} \phi^{p-1}\phi_k dx \leq 2^{p-1}\int_{G_k} \phi_k^p dx, +2^{p-1}k^{p-1}\int_{G_k} \phi_k dx,$$

we obtain

$$\lambda^{-1}2^{1-p}\, |G_k|^{-\frac{p}{N}}\, S_{N,p}^p \int_{G_k} \phi_k^p dx \leq \int_{G_k} \phi_k^p dx + k^{p-1}\int_{G_k} \phi_k dx,$$

and finally

$$\begin{aligned} k^{p-1}\int_{G_k} \phi_k dx &\geq \left(\lambda^{-1}2^{1-p}\, |G_k|^{-\frac{p}{N}}\, S_{N,p}^p - 1\right)\int_{G_k} \phi_k^p dx \\ &\geq \left(\lambda^{-1}2^{1-p}\, |G_k|^{-\frac{p}{N}}\, S_{N,p}^p - 1\right)|G_k|^{1-p}\left(\int_{G_k} \phi_k dx\right)^p. \end{aligned}$$

If $k$ is such that $|G_k| \leq 2^{-N}\lambda^{-\frac{N}{p}} S_{N,p}^N$, which holds true if $k \geq k_1$, where $k_1 = \lambda^{\frac{N}{p}} 2^N S_{N,p}^{-N} \|\phi\|_{L^1}$, there holds

$$k^{p-1}|G_k|^{\frac{N+1}{N}p-1} \geq \lambda^{-1}2^{-p}S_{N,p}^p\left(\int_{G_k} \phi_k dx\right)^{p-1}.$$

Set $f(k) = \int_{G_k} \phi_k dx = \int_k^\infty |G_s| \, ds$. Then $f'(k) = -|G_k|$ and

$$\begin{aligned}(f(k_1))^{\frac{p}{(N+1)p-N}} &\geq (f(k))^{\frac{p}{(N+1)p-N}} \\ &+ \left(\frac{S^p_{N,p}}{2^p \lambda}\right)^{\frac{N}{(N+1)p-N}} \left(k^{\frac{p}{(N+1)p-N}} - k_1^{\frac{p}{(N+1)p-N}}\right),\end{aligned} \tag{2.1.34}$$

on the set of the $k$ larger or equal to $k_1$ such that $f(k) > 0$. Therefore, if we set

$$k_2 = k_1 + \frac{2^N \lambda^{\frac{N}{p}}}{S^N_{N,p}} f(k_1),$$

inequality (2.1.34) cannot hold. This implies that $f(k) = 0$ for $k \geq k_2$ and (2.1.33) follows. □

If $N = 1$ and $\Omega = (a, b)$, the eigenvalue problem

$$(|\phi'|^{p-2} \phi')' + \lambda |\phi|^{p-2} \phi = 0, \quad \phi(a) = \phi(b) = 0, \tag{2.1.35}$$

is thouroughly studied in [Otani (1984)]. This equation is integrable since

$$|\phi'|^p + \frac{\lambda}{p-1} |\phi|^p = C.$$

The explicit value of $\lambda_1 := \lambda_1(p, b-a)$ is

$$\sqrt[p]{\lambda_1} = \frac{2\sqrt[p]{(p-1)}}{b-a} \int_0^1 \frac{ds}{\sqrt[p]{1-s^p}} = \frac{2\pi \sqrt[p]{(p-1)}}{(b-a)p \sin \frac{\pi}{p}}. \tag{2.1.36}$$

The spectrum $\sigma(-\Delta_p; W_0^{1,p}((a,b)))$ is the set of $\lambda_k = k^p \lambda_1$ for $k \in \mathbb{N}^*$. It coincides with the variational spectrum $\sigma_{var}(-\Delta_p; W_0^{1,p}((a,b)))$. The first eigenvalue depends heavily on $p$, but the remarkable and mysterious duality identity $\sqrt[p]{\lambda_1(p, b-a)} = \sqrt[p']{\lambda_1(p', b-a)}$ holds if $\frac{1}{p} + \frac{1}{p'} = 1$.

### 2.1.4 *Eigenvalues on a compact manifold*

If $(M^d, g)$ is a compact Riemannian manifold with empty boundary, the divergence operator on $X \in TM$ is denoted by $div_g$. The $p$-Laplacian is defined by

$$\begin{aligned}-\Delta_{p,g} u &= -div_g(|\nabla_g u|^{p-2} \nabla_g u) \\ &= -|\nabla_g u|^{p-2} \Delta_g u - (p-2) |\nabla_g u|^{p-2} \langle D^2 u \nabla_g u, \nabla_g u\rangle_g,\end{aligned} \tag{2.1.37}$$

whenever $D^2u$ exists, thanks to the identification of $\delta u$ with the tangential gradient of $u$ (it is one of the consequences of Nash imbedding theorems). When there is no ambiguity we denote $\Delta_{p,g} = \Delta_p$ and $\Delta_{2,g} = \Delta_g$. Note also that $2\langle D^2u\nabla u, \nabla u\rangle_g = \langle \nabla|\nabla u|^2, \nabla u\rangle_g$.

Since $\partial M = \emptyset$, the spectrum of $-\Delta_p$ is considered for $W^{1,p}(M)$ functions. It is denoted by $\sigma(\Delta_p; W^{1,p}(M)$. Clearly $\min \sigma(\Delta_p; W^{1,p}(M) = 0$ and the corresponding eigenfunction are constant. If $\lambda$ is a nonzero eigenvalue, necessarily positive, and $\phi$ an associated eigenfunction, then $\phi$ has zero average. The following characterization of the first nonzero eigenvalue $\lambda_1$ holds [Véron (1991)].

**Theorem 2.1.17** *One has*

$$\lambda_1 = \inf\left\{\int_M |\nabla_g u|^p \, dv_g : u \in \Sigma\right\}, \tag{2.1.38}$$

*where*

$$\Sigma = \left\{u \in W^{1,p}(M) : \int_M |u|^p \, dv_g = 1, \int_M |u|^{p-2} u dv_g = 0\right\}. \tag{2.1.39}$$

*Proof.* Relation (2.1.38) defines $\lambda_1$ as the minimum of the p-energy functional under two constraints. The set $\Sigma$ is a $C^0$ submanifold of $W^{1,p}(M)$ which happens to be $C^1$ only if $p \geq 2$. If $B_k$ is the closed ball of $W^{1,p}(M)$ of radius $k$, $\Sigma \cap B_k$ is compact. Thus the infimum is achieved by some $\phi \in \Sigma$. If $p \geq 2$ there exist two Lagrange multipliers $\alpha$ and $\beta$ such that

$$-\Delta_p \phi = \alpha \, |\phi|^{p-2} \, u + \beta \, |\phi|^{p-2}, \tag{2.1.40}$$

in the weak sense. Integrating (2.1.40) yields $\beta = 0$ and $\alpha = \lambda_1$.
When $1 < p < 2$, we set for $\epsilon > 0$ small enough,

$$\Sigma_\epsilon = \left\{u \in W^{1,p}(M) : \int_M (u^2 + \epsilon^2)^{\frac{p}{2}} \, dv_g = 1, \int_M (u^2 + \epsilon^2)^{\frac{p-2}{2}} u dv_g = 0\right\}, \tag{2.1.41}$$

and

$$\lambda_{1,\epsilon} = \inf\left\{\int_M |\nabla_g u|^p \, dv_g : u \in \Sigma_\epsilon\right\}. \tag{2.1.42}$$

There exists a Lagrange multiplier $\mu_\epsilon$ (the other multiplier is zero) and $\phi_\epsilon \in \Sigma_\epsilon$ such that

$$-\Delta_p \phi_\epsilon = \mu_\epsilon (\phi_\epsilon^2 + \epsilon^2)^{\frac{p-2}{2}} \phi_\epsilon, \tag{2.1.43}$$

which verifies

$$\lambda_{1,\epsilon} = \int_M |\nabla_g \phi_\epsilon|^p \, dv_g = \mu_\epsilon \int_M (\phi_\epsilon^2 + \epsilon^2)^{\frac{p-2}{2}} \phi_\epsilon^2 dv_g.$$

Hereafter we assume $1 < p < 2$. The following inequality holds

$$(\epsilon^2 + r^2)^{\frac{p}{2}} - \epsilon^p \le r^2(\epsilon^2 + r^2)^{\frac{p-2}{2}} \le (\epsilon^2 + r^2)^{\frac{p}{2}},$$

therefore

$$(1 - \epsilon^p Vol(M))\mu_\epsilon \le \lambda_{1,\epsilon} \le \mu_\epsilon.$$

Set $\tilde{\lambda}_1 = \liminf_{\epsilon \to 0} \lambda_{1,\epsilon} = \lim_{n\to\infty} \lambda_{1,\epsilon_n} = \lim_{n\to\infty} \mu_{\epsilon_n}$ for some sequence $\{\epsilon_n\}$ converging to 0. Identity (2.1.43) implies that the set of functions $\phi_\epsilon$ remains bounded in $W^{1,p}(M)$, and thus in $L^\infty(M)$ and finally in $C^{1,\alpha}(M)$ for some $\alpha \in (0,1)$ by Theorem 1.3.9. Then there exists a subsequence, extracted from $\{\epsilon_n\}$ and still denoted by the same designation and function $\tilde{\phi} \in W^{1,p}(M)$ such that $\{\phi_{\epsilon_n}\}$ converges to $\tilde{\phi}$ in $C^{1,\beta}(M)$ for any $\beta \in (0,\alpha)$. Clearly $\tilde{\phi} \in \Sigma$ and

$$-\Delta_p \tilde{\phi} = \tilde{\lambda}_1 |\tilde{\phi}|^{p-2} \tilde{\phi}, \tag{2.1.44}$$

which indicates that $\tilde{\lambda}_1 \in \sigma(-\Delta_p; W^{1,p}(M))$. Let us assume that $\lambda_1 < \tilde{\lambda}_1$ and denote by $\phi$ an element of $\Sigma$ for which the infimum in (2.1.42) is achieved, thus

$$\int_M |\nabla_g \phi|^p dv_g = \lambda_1.$$

We write $\phi = \phi^+ - \phi^-$. These two functions belong to $W^{1,p}(M)$ and,

$$\int_M (\phi^+)^{p-1} dv_g = \int_M (\phi^-)^{p-1} dv_g,$$

$$\int_M \left((\phi^+)^p + (\phi^-)^p\right) dv_g = 1.$$

Using the mean value theorem we see that for any $\epsilon > 0$ there is a unique $\sigma(\epsilon) = \sigma > 1$ such that

$$\int_M \sigma\phi^+(\epsilon^2 + (\sigma\phi^+)^2)^{\frac{p-2}{2}} dv_g = \int_M (\phi^-)^{p-1} dv_g = \int_M (\phi^+)^{p-1} dv_g. \tag{2.1.45}$$

There exists $c_1 > 0$ such that, for $r \geq 0$,

$$0 \leq r^{p-1} - r(\epsilon^2 + r^2)^{\frac{p-2}{2}} \leq c_1 \epsilon^{p-1},$$

therefore

$$-c_1\epsilon^{p-1} \leq \int_M \sigma\phi^+ \left(\epsilon^2 + (\sigma\phi^+)^2\right)^{\frac{p-2}{2}} dv_g - \int_M (\sigma\phi^+)^{p-1} dv_g \leq 0,$$

where $c_2 = c_1 Vol(M)$ and

$$0 \leq (\sigma^{p-1} - 1) \int_M (\phi^+)^{p-1} dv_g \leq c_2 \epsilon^{p-1}. \tag{2.1.46}$$

Hence there exists $\epsilon_0$ such that for $0 < \epsilon \leq \epsilon_0$ one has $0 < \sigma - 1 \leq \frac{1}{2}$, thus

$$0 \leq \sigma - 1 \leq c_3 \epsilon^{p-1}, \tag{2.1.47}$$

for some $c_3 = c_3(p, \phi, \epsilon_0) > 0$. By the mean value theorem and (2.1.47), we infer

$$\int_M (\epsilon^2 + (\phi^-)^2)^{\frac{p}{2}} dv_g - \int_M (\phi^-)^p dv_g \leq \int_M (\epsilon + \phi^-)^p dv_g - \int_M (\phi^-)^p dv_g \leq c_4 \epsilon,$$

and finally

$$\left| \int_M (\phi^+)^p dv_g - \int_M (\epsilon^2 + (\sigma\phi^+)^2)^{\frac{p}{2}} dv_g \right| \leq c_5 \epsilon.$$

Denote $M^+ = \{x \in M : \phi^+(x) > 0\}$ and $M^- = \{x \in M : \phi^-(x) > 0\}$, hence

$$\begin{aligned}\int_M \left(\epsilon^2 + (\sigma\phi^+ - \phi^-)^2\right)^{\frac{p}{2}} dv_g &= \int_{M^+} \left(\epsilon^2 + (\sigma\phi^-)^2\right)^{\frac{p}{2}} dv_g \\ &+ \int_{M^-} \left(\epsilon^2 + (\phi^-)^2\right)^{\frac{p}{2}} dv_g + \epsilon^p Vol(M \setminus (M^+ \cup M^-)).\end{aligned}$$

We deduce that there exists $c_6 = c_6(p, M, \phi) > 0$ such that

$$\left| 1 - \int_M \left(\epsilon^2 + (\sigma\phi^+ - \phi^-)^2\right)^{\frac{p}{2}} dv_g \right| \leq c_6 \epsilon^{p-1}, \tag{2.1.48}$$

and therefore one can find $\eta > 0$ satisfying

$$\int_M \left((\epsilon\eta)^2 + (\sigma\eta\phi^+ - \eta\phi^-)^2\right)^{\frac{p}{2}} dv_g = 1, \tag{2.1.49}$$

and

$$|1 - \eta| \leq c_7 \epsilon^{p-1}. \tag{2.1.50}$$

By combining (2.1.45) and (2.1.49) we derive $\sigma\eta\phi^+ - \eta\phi^- \in \Sigma_{\epsilon\eta}$. Moreover

$$\lambda_1 = \int_M |\nabla_g\phi|^p \, dv_g = \int_{M^+} |\nabla_g\phi^+|^p \, dv_g + \int_{M^-} |\nabla_g\phi^-|^p \, dv_g, \tag{2.1.51}$$

and

$$\int_M |\nabla_g(\sigma\eta\phi^+ - \eta\phi^-)|^p \, dv_g = (\sigma\eta)^p \int_{M^+} |\nabla_g\phi^+|^p \, dv_g + \eta^p \int_{M^-} |\nabla_g\phi^-|^p \, dv_g. \tag{2.1.52}$$

We finally infer from (2.1.51), (2.1.52) that

$$\left|\lambda_1 - \int_M |\nabla_g(\sigma\eta\phi^+ - \eta\phi^-)|^p \, dv_g\right| \leq c_8\epsilon^{p-1}, \tag{2.1.53}$$

which yields

$$\lambda_{1,\epsilon\eta} \leq \lambda_1 + c_8\epsilon^{p-1}, \tag{2.1.54}$$

and $\tilde{\lambda}_1 \leq \lambda_1$, a contradiction. Therefore $\tilde{\lambda}_1 = \lambda_1$ and (2.1.38) holds. □

Since an eigenfunction $\phi$ is continuous, $M^+ = \phi^{-1}((0,\infty))$ and $M^{=}\phi^{-1}((-\infty,0))$ are open subsets of $M$ and any connected component of $M \setminus \phi^{-1}(0)$ is a nodal domain of $\phi$. If $G$ is a connected component of $M \setminus \phi^{-1}(0)$, $\mu(G)$ denotes the first eigenvalue of $-\Delta_p$ in $W^{1,p}(G)$.

**Theorem 2.1.18** *An eigenfunction corresponding to the first nonzero eigenvalue $\lambda_1$ has two nodal domains. Furthermore*

$$\lambda_1 = \min_{(\omega,\tilde{\omega})\in\mathcal{S}} \max\{\mu(\omega), \mu(\tilde{\omega})\} \tag{2.1.55}$$

*where $\mathcal{S}$ is the set of couples of disjoint nonempty open subsets of $M$.*

*Proof.* Let $\omega$ and $\tilde{\omega}$ be two disjoint subdomains of $M$ and $\psi$ and $\tilde{\psi}$ the two first eigenfunctions (positive) of $-\Delta_p$ in $W_0^{1,p}(\omega)$ and $W_0^{1,p}(\tilde{\omega})$ with corresponding eigenvalues $\mu(\omega)$ and $\mu(\tilde{\omega})$ respectively, normalized by

$$\max_\omega \psi = \max_{\tilde{\omega}} \tilde{\psi} = 1.$$

These functions are extended by 0 to become elements of $W^{1,p}(M)$. We can find $\tau$ and $\tilde{\tau}$ such that $\tau\psi + \tilde{\tau}\tilde{\psi} \in \Sigma$. Since

$$\int_M \left|\nabla_g(\tau\psi + \tilde{\tau}\tilde{\psi})\right|^p dv_g = \mu(\omega)\int_\omega |\tau\nabla_g\psi|^p \, dv_g + \mu(\tilde{\omega})\int_{\tilde{\omega}} \left|\tilde{\tau}\nabla_g\tilde{\psi}\right|^p dv_g,$$

we obtain by convexity

$$\lambda_1 \leq \max\{\mu(\omega), \mu(\tilde{\omega})\} \Longrightarrow \lambda_1 \leq \min_{(\omega,\tilde{\omega})\in\mathcal{S}} \max\{\mu(\omega), \mu(\tilde{\omega})\}. \quad (2.1.56)$$

Let $\phi \in \Sigma$ be an eigenfunction of $-\Delta_p$ corresponding to $\lambda_1$. Assume $M\setminus\{0\}$ has at least three connected components $\omega_j$, $j = 1, 2, 3$, then

$$\lambda_1 \leq \min_{j\in\mathbb{Z}/3\mathbb{Z}} \max\{\mu(\omega_j), \mu(\omega_{j+1})\},$$

where the indices are numbered modulo 3. The restriction of $\psi$ to $\omega_j$ belongs to $W_0^{1,p}(\omega_j)$, thus

$$\int_{\omega_j} |\nabla_g \phi|^p \, dv_g = \lambda_1 \int_{\omega_j} |\phi|^p \, dv_g \geq \mu(\omega_j) \int_{\omega_j} |\phi|^p \, dv_g.$$

This implies that, at least for two indices $j$, say $1, 2$, $\lambda_1 = \mu(\omega_1) = \mu(\omega_2)$. By Hopf boundary lemma, at any point $x \in \partial\omega_j$ where the interior sphere to $\omega_j$ is satisfied, $\frac{\partial\psi}{\partial \mathbf{n}_x} < 0$ where $\mathbf{n}_x$ is the unit outward normal vector to any sphere contained in $\omega_j$ and tangent to $\partial\omega_j$ at $x$. Notice the set $\mathcal{R}_{\omega_j}$ of *regular* points form a dense subset of $\partial\omega_j$ since any point $z$ in $\omega_j$ admits *proximum* points in $\partial\omega_j$ which are the points where the distance from $z$ to $\partial\omega_j$ is achieved; clearly $\partial\omega_j$ satisfies the interior sphere condition at each of these proximum points . Therefore, if $\phi$ is positive in $\omega_1$ and $\omega_2$, there exist at least two points $y_1 \in \partial\omega_1 \setminus \partial\omega_2$ and $y_2 \in \partial\omega_2 \setminus \partial\omega_1$ and we can replace $\omega_1$ and $\omega_2$ by two larger disjoint domains $\tilde{\omega}_1$ and $\tilde{\omega}_2$, for which $\mu(\tilde{\omega}_1) < \mu(\omega_1)$ and $\mu(\tilde{\omega}_2) < \mu(\omega_2)$, which would lead to $\lambda_1 > \max\{\mu(\tilde{\omega}_1), \mu(\tilde{\omega}_2)\}$, a contradiction with (2.1.56). If $\phi$ is positive in $\omega_1$ and $\omega_3$ and negative in $\omega_2$ and if $\mu(\omega_3) = \mu(\omega_1)$, we proceed similarly, replacing only $\omega_2$ by $\omega_3$. Finally, if $\mu(\omega_1) = \mu(\omega_2) > \mu(\omega_3)$, we have two possibilities:

(i) either $\mathcal{R}_{\omega_1} = \mathcal{R}_{\omega_2}$, which implies $\partial\omega_1 = \partial\omega_2$, in which case $M = \overline{\omega}_1 \cup \overline{\omega}_2$, which is impossible.

(ii) or there exists $z \in \mathcal{R}_{\omega_1} \setminus \mathcal{R}_{\omega_2}$ (or $z \in \mathcal{R}_{\omega_2} \setminus \mathcal{R}_{\omega_1}$ in the same way). Then we can replace $\omega_1$ by a larger domain $\tilde{\omega}_1$ such that $\tilde{\omega}_1 \cap \omega_2 = \emptyset$ and $\mu(\omega_3 \setminus \overline{\tilde{\omega}_1}) < \mu(\tilde{\omega}_1) < \mu(\omega_1)$, and then we can replace $\omega_2$ by a larger domain $\tilde{\omega}_2$ such that $\tilde{\omega}_2 \cap \tilde{\omega}_1 = \emptyset$ and $\mu(\omega_3 \setminus (\overline{\tilde{\omega}_1 \cup \tilde{\omega}_1}) < \mu(\tilde{\omega}_2) < \mu(\omega_2)$. Therefore $\tilde{\omega}_1$, $\tilde{\omega}_2$ and $\omega_3 \setminus (\overline{\tilde{\omega}_1 \cup \tilde{\omega}_1}$ are three disjoint domains and $\max\{\mu(\tilde{\omega}_1), \mu(\tilde{\omega}_2), \mu(\omega_3 \setminus (\overline{\tilde{\omega}_1 \cup \tilde{\omega}_1}))\} < \mu(\omega_1) = \lambda_1$ which is a contradiction. □

When $p = 2$ there are many estimates of the first nonzero eigenvalue of the Laplacian due to Lichnerowitz, Obata, Cheeger, Buser (see [Berger, Gauduchon, Mazet (1971) ] and the references therein). Many of these estimates have been extended to the $p$-Laplacian by Matei [Matei (2000)] and Valtorta [Valtorta (2014)].

**Theorem 2.1.19** *Suppose $Ricc_g \geq (d-1)k > 0$ and $p \geq 2$. If $\lambda_1 = \lambda_1(p, M)$ is the first nonzero eigenvalue of $-\Delta_p$, it satisfies*

$$\lambda_1 \geq \left(\frac{(d-1)k}{p-1}\right)^{\frac{p}{2}}. \tag{2.1.57}$$

*Proof.* Let $u \in C^3(M)$. By Wiestenböck's formula (1.1.14),

$$\begin{aligned}\int_M \Delta_p u \Delta_g u dv_g &= -\int_M \langle \nabla_g u, \nabla_g \Delta_g u \rangle_g |\nabla_g u|^{p-2} dv_g \\ &= \int_M \left(\left\|D^2 u\right\|^2 + Ricc_g(\nabla_g u, \nabla_g u) - \frac{1}{2}\Delta_g |\nabla_g u|^2\right) |\nabla_g u|^{p-2} dv_g.\end{aligned}$$

Since

$$\begin{aligned}\int_M |\nabla_g u|^{p-2} \Delta_g |\nabla_g u|^2 dv_g &= -\int_M \langle \nabla_g |\nabla_g u|^{p-2}, \nabla_g |\nabla_g u|^2 \rangle_g dv_g \\ &= \frac{2-p}{2}\int_M |\nabla_g |\nabla_g u|^2 |^2 |\nabla_g u|^{p-4} dv_g\end{aligned}$$

and $4\left\|D^2 u\right\|^2 |\nabla_g u|^2 \geq \left|\nabla_g |\nabla_g u|^2\right|^2$, it follows an inequality valid for any $C^3$ function (and by approximation $C^3$ can be replaced by $C^2$),

$$\begin{aligned}\int_M \Delta_p u \Delta_g u dv_g &\geq \int_M \Big(Ricc_g(\nabla_g u, \nabla_g u) |\nabla_g u|^2 \\ &\qquad + (p-1)|\nabla_g |\nabla_g u|^2 |^2\Big) |\nabla_g u|^{p-4} dv_g \\ &\geq \int_M Ricc_g(\nabla_g u, \nabla_g u) |\nabla_g u|^{p-2} dv_g \\ &\geq (d-1)k \int_M |\nabla_g u|^p dv_g.\end{aligned} \tag{2.1.58}$$

If $u = \phi_\epsilon$ is a minimizer for problem (2.1.42), then $-\Delta_p u = \mu_\epsilon (u^2 + \epsilon^2)^{\frac{p-2}{2}} u$,

$$\begin{aligned}\int_M \Delta_p u \Delta_g u dv_g &= (p-1)\mu_\epsilon \int_M (u^2+\epsilon^2)^{\frac{p-2}{2}} |\nabla_g u|^2 dv_g \\ &\quad - \epsilon^2 (p-1)\mu_\epsilon \int_M (u^2+\epsilon^2)^{\frac{p-4}{2}} |\nabla_g u|^2 dv_g,\end{aligned}$$

thus

$$(p-1)\mu_\epsilon \int_M (u^2+\epsilon^2)^{\frac{p-2}{2}} |\nabla_g u|^2 \, dv_g \geq (d-1)k \int |\nabla_g u|^p \, dv_g. \quad (2.1.59)$$

Letting $\epsilon \to 0$ we obtain from the previous convergence, with $\mu_\epsilon \to \lambda_1$ and $u = \phi_\epsilon \to \phi$,

$$(p-1)\lambda_1 \int_M |\phi|^{p-2} |\nabla_g \phi|^2 \, dv_g \geq (d-1)k \int |\nabla_g \phi|^p \, dv_g. \quad (2.1.60)$$

We derive (2.1.55) by Hölder's inequality. □

Other estimates involve either the diameter of $M$ or the Cheeger's isometric constant $h_M$ defined by

$$h_M = \inf \left\{ \frac{Per(S)}{\inf\{Vol(M'), Vol(M'')\}} : S \in \mathfrak{S} \right\}, \quad (2.1.61)$$

where $\mathfrak{S}$ is the set of all closed hypersurfaces $S$ (or imbedded $(d-1)$-dim submanifolds) which divide $M$ into two open submanifolds $M'$ and $M''$ and $Per(S)$ is the $(d-1)$-volume for the induced volume form $v_{g\lfloor_S}$.

**Theorem 2.1.20** *Let $p > 1$, then $\lambda_1 = \lambda_1(p, M)$ verifies*

$$\lambda_1 \geq \left(\frac{h_M}{p}\right)^p. \quad (2.1.62)$$

*Proof.* Let $\phi$ be a first eigenfunction. It is easy to construct a sequence $\{\phi_n\} \subset C^2(M)$ converging to $\phi$ in $W^{1,p}(M)$ such that, for each $n \in \mathbb{N}$, $|\nabla_g \phi_n| + |det(H_{\phi_n})| > 0$ and $\nabla_g \phi_n \neq 0$ on the set $S = \phi_n^{-1}(0)$. We say that $\phi_n$ is a *Morse function* and 0 a regular value of $\phi_n$. Notice that, since $M$ is compact, the set of critical points of $\phi_n$ is finite. We put $M'_n = \phi_n^{-1}((0,\infty))$ and $M''_n = \phi_n^{-1}((-\infty, 0))$. Since $\lambda_1 = \mu(M') = \mu(M')$,

$$\lim_{n\to\infty} \frac{\displaystyle\int_{M'_n} |\nabla_g \phi_n|^p \, dv_g}{\displaystyle\int_{M'_n} |\phi_n|^p \, dv_g} = \frac{\displaystyle\int_{M'} |\nabla_g \phi|^p \, dv_g}{\displaystyle\int_{M'} |\phi,|^p \, dv_g} = \lambda_1. \quad (2.1.63)$$

The same relation holds if $M'_n$ is replaced by $M''_n$, and without loss of generality we can assume $Vol(M_n') \leq Vol(M_n'')$. By Hölder's inequality

$$\int_{M_n'} |\nabla_g \phi_n^p|\, dv_g = p \int_{M_n'} |\nabla_g \phi_n|\, \phi_n^{p-1} dv_g$$
$$\leq p \left( \int_{M_n'} \phi_n^p dv_g \right)^{1-\frac{1}{p}} \left( \int_{M_n'} |\nabla_g \phi_n|^p\, dv_g \right)^{\frac{1}{p}},$$

which yields

$$\frac{\displaystyle\int_{M_n'} |\nabla_g \phi_n|^p\, dv_g}{\displaystyle\int_{M_n'} |\phi_n|^p\, dv_g} \geq \frac{1}{p^p} \left( \frac{\displaystyle\int_{M_n'} |\nabla_g \phi_n^p|\, dv_g}{\displaystyle\int_{M_n'} |\phi_n|^p\, dv_g} \right)^p. \tag{2.1.64}$$

Set $S_{\phi_n^p}(t) = \{x \in M_n' : \phi_n^p(x) > t\}$, $s'_{\phi_n^p}(t) = \{x \in M_n' : \phi_n(x) = t^p\}$. By the co-area formula (1.2.39)

$$\int_{M_n'} |\nabla_g \phi_n^p|\, dv_g = \int_0^{\sup \phi_n^p} \left( \int_{\partial M_n'(t)} d\mathcal{H}_{N-1} \right) dt = \int_0^{\sup \phi_n^p} Per(s'_{\phi_n}(t)) dt.$$

Since $Vol(M_n') \leq Vol(M_n'')$, for every $t > 0$, $Vol(S_{\phi_n}(t)) \leq Vol(M_n''(t))$ where $M_n''(t) = \{x \in M : (\phi_n |\phi_n|^{p-1})(x) < t^p\}$, and because $\phi_n$ has only a finite number of critical points, there holds by the definition of $h_M$,

$$\int_{\partial M_n'(t)} d\mathcal{H}_{N-1} \leq h_M Vol(S_{\phi_n^p}(t)) = h_M |S_{\phi_n^p}(t)| := \lambda_{\phi_n^p}(t),$$

except at most for a finite set of values of $t$ (see Section 1.2.1). This implies

$$\int_{M_n'} |\nabla_g \phi_n^p|\, dv_g \leq h_M \int_0^{\sup \phi_n^p} \lambda_{\phi_n^p}(t) dt = h_M \int_{M_n'} \phi_n^p dv_g. \tag{2.1.65}$$

Inequality (2.1.62) follows from (2.1.63), (2.1.64) and (2.1.65). □

*Remark.* If the Ricci curvature of $M$ is nonnegative, it is proved in [Gallot (1983)] that

$$h_M \geq \frac{2}{\operatorname{diam}(M)}, \tag{2.1.66}$$

where $\operatorname{diam}(M)$ is the maximal geodesic distance between two points of $M$. Therefore (2.1.62) implies the more explicit estimate

$$\lambda_1(p, M) \geq \frac{2^p}{p^p (\operatorname{diam}(M))^p}. \tag{2.1.67}$$

When $p = 2$ Zhong and Yang [Yang, Zhong (1984)] proved the following result:

**Theorem 2.1.21** *Assume $Ricc_g \geq 0$, then*

$$\lambda_1(2, M) \geq \left(\frac{\pi}{\operatorname{diam}(M)}\right)^2. \tag{2.1.68}$$

This estimate improves a previous one due to Li and Yau [Li, Yau (1980)]. Their proof is based upon an estimate of the function $F(\phi) = \frac{|\nabla_g \phi|^2}{1-\phi^2}$, assuming that $|\phi| < 1$, and on the Weitzenböck formula. Recently [Naber, Valtorta (2014)], [Valtorta (2014)] gave a remarkable extension of this inequality to the $p$-Laplacian.

**Theorem 2.1.22** *Assume $Ricc_g \geq 0$ and $p > 1$, then*

$$\frac{\lambda_1(p, M)}{p-1} \geq \left(\frac{\pi_p}{\operatorname{diam}(M)}\right)^p, \tag{2.1.69}$$

*where*

$$\pi_p = \int_{-1}^{1} \frac{ds}{\sqrt[p]{1-s^p}} = \frac{2\pi}{p \sin(\frac{\pi}{p})}. \tag{2.1.70}$$

It is noticeable that this lower bound is much larger that the one obtained in (2.1.67). The proof is long and involves a delicate gradient comparison theorem with one-dimensional model. One of the intermediate result which is of intrinsic interest is the so-called $p$-Bochner formula. To state it, for any $u \in C^2(M)$ we define the operator $P_u^{II}$ by

$$P_u^{II}(\eta) = |\nabla_g u|^{p-2} \Delta_g \eta + (p-2)|\nabla_g u|^{p-2} \langle D^2\eta(\nabla_g u), \nabla_g u\rangle_g \quad \forall \eta \in C^2(M), \tag{2.1.71}$$

and set

$$A_u = \frac{\langle D^2 u \nabla_g u, \nabla_g u\rangle_g}{|\nabla_g u|^2}. \tag{2.1.72}$$

**Proposition 2.1.23** *Let $p > 1$ and $u \in C^3(M)$. At any point where $\nabla_g u \neq 0$ there holds*

$$\begin{aligned} \frac{1}{p} P_u^{II}(|\nabla_g u|^p) &= |\nabla_g u|^{p-2} \left(\langle \nabla_g \Delta_p u, \nabla_g u\rangle_g - (p-2)A_u\right) \\ &+ |\nabla_g u|^{2p-4} \left(\left\|D^2 u\right\|^2 + p(p-2)A_u^2 + Ricc_g(\nabla_g u, \nabla_g u)\right). \end{aligned} \tag{2.1.73}$$

## 2.2 Spherical $p$-harmonic functions

### 2.2.1 *The spherical p-harmonic eigenvalue problem*

Let $S^{N-1} := (S^{N-1}, g_0)$ be the unit sphere in $\mathbb{R}^N$ endowed with the canonical metric $g_0$ induced by the isometric imbedding into $\mathbb{R}^N$. We denote by $(r, \sigma)$ the spherical coordinates in $\mathbb{R}^N \setminus \{0\}$, $r > 0$ and $\sigma \in S^{N-1}$, by $\nabla'\omega$ the covariant gradient of a function $\omega$ defined on $S^{N-1}$, by $div' = div_{g_0}$ the divergence of vector fields on $S^{N-1}$ and by $\Delta'$ the Laplacian (or Laplace-Beltrami operator). If one looks for separable $p$-harmonic $C^2$ functions $(p > 1)$ under the form

$$u(x) = u(r, \sigma) = r^{-\beta}\omega(\sigma), \tag{2.2.1}$$

then $\omega$ satisfies

$$div'\left(\left(\beta^2\omega^2 + |\nabla'\omega|^2\right)^{\frac{p-2}{2}} \nabla'\omega\right) + \beta\Lambda_p(\beta)\left(\beta^2\omega^2 + |\nabla'\omega|^2\right)^{\frac{p-2}{2}} \omega = 0, \tag{2.2.2}$$

on $S^{N-1}$ with

$$\Lambda_p(\beta) = \beta(p-1) + p - N. \tag{2.2.3}$$

The *spherical p-harmonic eigenvalue problem* is to find the couples $(\beta, \omega)$ which solve (2.2.2)-(2.2.3). Clearly $\beta\Lambda_p(\beta)$ must be nonnegative. Constant functions $\omega$ satisfy (2.2.2) if and only if $\beta = 0$ or $\beta = \frac{N-p}{p-1}$, in which case the function $u$ in (2.2.1) is called a *fundamental solution*. A necessary condition for the existence of non-constant solutions $\omega$ is

$$\begin{array}{lll} (i) & \beta \in (-\infty, 0) \cup (\frac{N-p}{p-1}, \infty) & \text{if } 1 < p < N \\ (ii) & \beta \in (-\infty, -\frac{p-N}{p-1}) \cup (0, \infty) & \text{if } p \geq N. \end{array} \tag{2.2.4}$$

If $p = 2$, (2.2.2) becomes

$$-\Delta'\omega = \beta\Lambda_2(\beta)\omega. \tag{2.2.5}$$

The set of $\{\beta\Lambda_2(\beta)\}$ is the set of integers $\{k(N + k - 2)\}_{k\in\mathbb{N}}$ and the $\omega$ are the restriction to $S^{N-1}$ of the $k$-homogeneous harmonic polynomials.

**Theorem 2.2.1** *Let $S \subsetneq S^{N-1}$ be a $C^2$ domain and $p > 1$. Then there exist two couples $(\beta, \omega) = (\beta_S, \omega_S)$ and $(\beta, \omega) = (\beta'_S, \omega'_S)$ where $\beta_S > 0$,*

$\beta'_S < 0$, $\omega_S$, $\omega'_S \in W_0^{1,p}(S)$ *are positive solutions of*

$$div'\left(\left(\beta^2\omega^2 + |\nabla'\omega|^2\right)^{\frac{p-2}{2}} \nabla'\omega\right) \\ + \beta\Lambda_p(\beta)\left(\beta^2\omega^2 + |\nabla'\omega|^2\right)^{\frac{p-2}{2}} \omega = 0 \quad in\ S \\ \omega = 0 \quad in\ \partial S. \tag{2.2.6}$$

*Furthermore* $\omega_S$ *and* $\omega'_S$ *are* $C^\infty$ *in* $S$ *and unique up to a multiplication by a constant.*

If $\omega \in W_0^{1,p}(S)$ is a positive solution of (2.2.6) with $\beta > 0$, $v = -\frac{1}{\beta}\ln\omega$ satisfies

$$-div'\left(\left(1 + |\nabla' v|^2\right)^{\frac{p-2}{2}} \nabla' v\right) + \beta(p-1)\left(1 + |\nabla' v|^2\right)^{\frac{p-2}{2}} |\nabla' v|^2 \\ + \Lambda_p(\beta)\left(1 + |\nabla' v|^2\right)^{\frac{p-2}{2}} |\nabla' v|^2 = 0 \qquad \text{in } S \\ \lim_{\rho(\sigma)\to 0} v(\sigma) = \infty \qquad \text{in } \partial S. \tag{2.2.7}$$

where $\rho(\sigma)$ is the geodesic distance from $\sigma$ to $\partial S$. Note that this equation is never degenerate, therefore the solutions are $C^\infty$ in $S$. The proof of Theorem 2.2.1 is a consequence of a more general result

**Theorem 2.2.2** *Let* $(M^d, g)$ *be a d-dimensional Riemannian manifold with covariant gradient* $\nabla$ *and divergence operator* $div_g$ *and let* $p > 1$. *For any relatively compact subdomain* $S \subset M$ *with a nonempty boundary* $\partial S$ *and* $\beta > 0$, *there exists a unique* $\lambda(\beta) > 0$ *such that the problem*

$$-div_g\left(\left(1 + |\nabla v|^2\right)^{\frac{p-2}{2}} \nabla v\right) + \beta(p-1)\left(1 + |\nabla v|^2\right)^{\frac{p-2}{2}} |\nabla v|^2 \\ + \lambda(\beta)\left(1 + |\nabla v|^2\right)^{\frac{p-2}{2}} |\nabla v|^2 = 0 \qquad in\ S \\ \lim_{\rho(x)\to 0} v(\sigma) = \infty \qquad in\ \partial S, \tag{2.2.8}$$

*where* $\rho(x)$ *is the geodesic distance from* $x \in S$ *to* $\partial S$, *admits a positive solution* $v \in C^2(S) \cap C(\overline{S})$. *Furthermore* $v$ *is unique up to the addition of a constant.*

When $p = 2$ this statement is customary in ergodic problems and the constant $\lambda(\beta)$ is called an *ergodic constant*. By analogy this name remains in the quasilinear case even if no link with probability theory has been found so far. If we expand the equation the previous statement becomes

**Theorem 2.2.3** *Let $p > 1$ and $\Delta_g$ be the Laplacian on $(M^d, g)$. For any relatively compact subdomain $S \subset M$ with a nonempty boundary $\partial S$ and $\beta > 0$, there exists a unique $\lambda(\beta) > 0$ such that the problem*

$$\begin{aligned} -\Delta_g v - \frac{p-2}{2}\frac{\langle \nabla |\nabla v|^2, \nabla v\rangle_g}{1+|\nabla v|^2} + \beta(p-1)|\nabla v|^2 &= -\lambda(\beta) \quad \text{in } S \\ \lim_{\rho(x)\to 0} v(x) &= \infty \quad \text{in } \partial S, \end{aligned} \tag{2.2.9}$$

*admits a positive solution $v \in C^2(S) \cap C(\overline{S})$, unique up to the addition of a constant.*

*Proof of Theorem 2.2.3. Existence.* For $\epsilon > 0$ consider the problem

$$\begin{aligned} -\Delta_g v - \frac{p-2}{2}\frac{\langle \nabla |\nabla v|^2, \nabla v\rangle_g}{1+|\nabla v|^2} + \beta(p-1)|\nabla v|^2 + \epsilon v &= 0 \quad \text{in } S \\ \lim_{\rho(x)\to 0} v(x) &= \infty \quad \text{in } \partial S. \end{aligned} \tag{2.2.10}$$

*Construction of supersolutions and subsolutions.* We denote by $\dot\rho(x)$ the signed geodesic distance from $x \in M$ to $\partial S$ which is equal to $\rho(x)$ if $x \in S$ and to $-\rho(x)$ when $x \in M \setminus S$. Since $\partial S$ is compact and $C^2$, there exists $\delta > 0$ such that for any $x$ such that $|\dot\rho(x)| \geq \delta$, there exists a unique $\sigma_x \in \partial S$ such that $|x - \sigma_x| = \rho(x)$. We set $\mathcal{N}_\delta = \{x \in M : |\dot\rho(x)| < \delta\}$. We extend $\dot\rho$ to $M \setminus \mathcal{N}_\delta$ as a smooth function $\tilde\rho$ and put

$$\tilde u(x) = -\frac{1}{\beta}\ln\tilde\rho(x) - M_0\tilde\rho(x) + \frac{M_1}{\epsilon}, \tag{2.2.11}$$

where the $M_j$ will be chosen later on. Since $|\nabla\tilde\rho| = 1$ in $\overline{\mathcal{N}_\delta}$,

$$|\nabla \tilde u(x)|^2 = \frac{1 + 2\beta M_0\rho(x) + O(\rho^2(x))}{\beta^2\rho^2(x)} \quad \text{as } \rho(x) \to 0,$$

and $\Delta_g\tilde\rho(x)$ is bounded in $M$. Therefore

$$\begin{aligned} &-\Delta_g\tilde u - \frac{p-2}{2}\frac{\langle \nabla |\nabla \tilde u|^2, \nabla \tilde u\rangle_g}{1+|\nabla \tilde u|^2} + \beta(p-1)|\nabla\tilde u|^2 + \epsilon\tilde u \\ &= \frac{1}{\tilde\rho}\left(\frac{\Delta_g}{\beta} - \frac{\epsilon}{\beta}\tilde\rho\ln\tilde\rho + 2(p-1)M_0|\nabla\tilde\rho|^2\right) + \psi_\beta + M_1, \end{aligned} \tag{2.2.12}$$

where $\psi_\beta$ is a bounded function depending on $\beta$ and $M_0$. Thus we can choose $M_0$ and $M_1$ to ensure the positivity of the right-hand side term in

(2.2.12). Notice also that the $M_j$ and the bound for $\psi_\beta$ can be chosen independent of $\beta$ when it remains in a compact subset of $(0,\infty)$. Similarly the function $\underline{u}$ defined by

$$\underline{u}(x) = -\frac{1}{\beta}\ln\tilde{\rho}(x) + M_0\tilde{\rho}(x) - \frac{M_1}{\epsilon}, \tag{2.2.13}$$

can be constructed as being a subsolution for suitable $M_j$ compatible with the ones in $\tilde{u}$. For $0 < h < \delta$ we approximate $\tilde{u}$ and $\underline{u}$ in the following way

$$\begin{aligned} &(i) \qquad \tilde{u}_h(x) = -\frac{1}{\beta}\ln(\tilde{\rho}(x) - h) - M_0(\tilde{\rho}(x) - h) + \frac{M_1}{\epsilon} \\ &(ii) \qquad \underline{u}_h(x) = -\frac{1}{\beta}\ln(\tilde{\rho}(x) + h) - M_0(\tilde{\rho}(x) + h) + \frac{M_1}{\epsilon}, \end{aligned} \tag{2.2.14}$$

and they are respectively supersolution in $S \setminus \mathcal{N}_\delta$ and subsolution in $S$. The functions $g$ and $h$ are increasing functions of the nonnegative variable $z$. Consequently we can apply the comparison principle between supersolutions and subsolutions.

*Estimates.* In this paragraph, we use the equation satisfied by the function $z = |\nabla u|^2$, where $u$ is a solution of (2.2.12), combined with Theorem 1.1.2. Set $m = \inf_S\{Ricc_g(\xi,\xi) : |\xi| \leq 1\}$. Any $C^2$ function defined on $M$ satisfies

$$\frac{1}{2}\Delta_g|\nabla u|^2 \geq \frac{1}{d}(\Delta_g u)^2 + \langle\nabla\Delta_g u, \nabla u\rangle_g + m|\nabla u|^2, \tag{2.2.15}$$

by the Weitzenböck formula Theorem 1.1.2 and since $|D^2u|^2 \geq \frac{1}{d}(\Delta_g u)^2$ by Schwarz inequality. Identity (2.2.12) becomes

$$-\Delta_g z = -\frac{p-2}{2}\frac{\langle\nabla z, \nabla\tilde{u}\rangle_g}{1+|\nabla u|^2} + \beta(p-1)z + \epsilon\tilde{u} = 0. \tag{2.2.16}$$

Furthermore

$$\begin{aligned}\langle\nabla\Delta_g\tilde{u}, \nabla\tilde{u}\rangle_g = &-\frac{p-2}{4}\frac{\langle D^2z\nabla\tilde{u}, \nabla\tilde{u}\rangle_g}{1+|\nabla u|^2} - \frac{p-2}{4}\frac{|\nabla z|^2}{1+|\nabla\tilde{u}|^2} + \frac{p-2}{2}\frac{(\langle\nabla z, \nabla\tilde{u}\rangle_g)^2}{(1+|\nabla\tilde{u}|^2)^2} \\ &+\beta(p-1)\langle\nabla z, \nabla\tilde{u}\rangle_g + \epsilon z.\end{aligned}$$

From (2.2.16) there exist two positive constants $c_0, c_1$, independent of $\epsilon$ and bounded with $\beta$ such that

$$(\Delta_g u)^2 \geq c_0 z^2 - c_1\left((\epsilon\tilde{u}^-)^2 + \frac{(\langle\nabla z, \nabla\tilde{u}\rangle_g)^2}{(1+|\nabla\tilde{u}|^2)^2}\right).$$

Using (2.2.16) we obtain

$$\Delta_g z + (p-2)\frac{\langle D^z \nabla\tilde{u}, \nabla\tilde{u}\rangle_g}{1+|\nabla\tilde{u}|^2} \geq \frac{2c_0 z^2}{d}$$
$$- \frac{2c_1}{d}\left((\epsilon\tilde{u}^-)^2 + \frac{(\langle\nabla z, \nabla\tilde{u}\rangle_g)^2}{(1+|\nabla\tilde{u}|^2)^2}\right) + 2(m+\epsilon)z$$
$$- \frac{p-2}{2}\frac{|\nabla z|^2}{1+|\nabla\tilde{u}|^2} + (p-2)\frac{(\langle\nabla z, \nabla\tilde{u}\rangle_g)^2}{(1+|\nabla\tilde{u}|^2)^2} + 2\beta(p-1)\langle\nabla z, \nabla\tilde{u}\rangle_g.$$

Since $\frac{2c_0 z^2}{d} + 2(m+\epsilon)z \geq \frac{c_0 z^2}{d} - c_2$ by Young's inequality, we derived that there exist positive constants $C_j$, $j = 0, 1, 2$, such that

$$-\Delta_g z - (p-2)\frac{\langle D^z \nabla\tilde{u}, \nabla\tilde{u}\rangle_g}{1+|\nabla\tilde{u}|^2} + C_0 z^2 \leq C_1\frac{|\nabla z|^2}{1+|\nabla\tilde{u}|^2} + C_2. \qquad (2.2.17)$$

Let us define the operator $\mathcal{A}$ by

$$\mathcal{A}z = -\Delta_g z - (p-2)\frac{\langle D^z \nabla\tilde{u}, \nabla\tilde{u}\rangle_g}{1+|\nabla\tilde{u}|^2}. \qquad (2.2.18)$$

In local coordinates $\mathcal{A}z = -\sum_{i,j} a_{i,j} z_{x_i x_j} + \sum_{i,j} b_i z_{x_i}$ where the coefficients $a_{i,j}$ and $b_i$ are bounded. The ellipticity condition is satisfied since

$$\min\{1, p-1\}\langle\xi,\xi\rangle_g \leq \sum_{i,j} a_{i,j}\xi_i\xi_j \leq \max\{1, p-1\}\langle\xi,\xi\rangle_g \quad \forall\xi = (\xi_1, ..., \xi_d).$$

It follows from (2.2.17) that $z$ is a positive subsolution of an equation of the type

$$\mathcal{A}z + h(z) + g(z)|\nabla z|^2 = f, \qquad (2.2.19)$$

where $g(z) = -C_1/(1+z)$, $h(z) = C_0 z^2$ and $f = C_2$. The two functions $g$ and $h$ are increasing with respect to $z \geq 0$, therefore the comparison principle applies for a subsolution and a solution of

$$-\Delta_g z - (p-2)\frac{\langle D^z \nabla\tilde{u}, \nabla\tilde{u}\rangle_g}{1+|\nabla\tilde{u}|^2} + C_0 z^2 - C_1\frac{|\nabla z|^2}{1+z} = C_2. \qquad (2.2.20)$$

It is easy to verify that there exist two large enough positive constants $\lambda$ and $\mu$ such that the function

$$\bar{z}(x) = \frac{\lambda}{\tilde{\rho}^2(x)} + \mu, \qquad (2.2.21)$$

is a supersolution of (2.2.20) which blows-up on the boundary. Henceforth, any bounded subsolution of the same equation is bounded from above by $\bar{z}$. For $0 < h < \delta$ we replace $\bar{z}$ by $\bar{z}_h$ defined in $S \setminus \mathcal{N}_h$ by

$$\bar{z}_h(x) = \frac{\lambda}{(\tilde{\rho}(x) - h)^2} + \mu, \tag{2.2.22}$$

where, accordingly to the definition of $\mathcal{N}_\delta$, $\mathcal{N}_h = \{x \in M : \dot{\rho}(x) < h\}$. The function $\bar{z}_h$ is still a supersolution of (2.2.20) in $S \setminus \mathcal{N}_h$ which blows-up on the boundary. Therefore it dominates in this domain the restriction of any subsolution in $S$. Letting $h \to 0$ implies that $\bar{z}$ is an absolute dominant for any subsolution in $S$. As a consequence if $u \in C^2(S)$ satisfies (2.2.15) there holds

$$|\nabla \tilde{u}(x)| \leq \frac{L_0}{\tilde{\rho}(x)} + L_1, \tag{2.2.23}$$

in $S$, for some positive constants $L_0$, $L_1$ depending on $\epsilon \left\| u^- \right\|_L^\infty$. Furthermore $L_0$ and $L_1$ can be chosen independent of $\beta$ provided it remains in a bounded interval. Once uniform Lipschitz estimates hold, the regularity theory Theorem 1.3.9 and Theorem 1.3.10 presented in Chapter 1 yields local $C^{2,\alpha}$ estimates.

*Approximate solutions.* By using Schauder's fixed point theorem it is easy to prove that for any $n \in \mathbb{N}^*$ there exists a solution $v := v_{n,\epsilon}$ to

$$\begin{aligned} -\Delta_g v - \frac{p-2}{2} \frac{\langle \nabla |\nabla v|^2, \nabla v \rangle_g}{1 + |\nabla v|^2} + \beta(p-1)|\nabla v|^2 + \epsilon v = 0 \qquad &\text{in } S \\ v = n \qquad &\text{in } \partial S. \end{aligned} \tag{2.2.24}$$

By comparison this solution is bounded from above by $\tilde{u}_h$ for any $0 < h < \delta$, this leads to

$$v_{n,\epsilon}(x) \leq -\frac{1}{\beta} \ln \tilde{\rho}(x) - M_0 \tilde{\rho}(x) + \frac{M_1}{\epsilon}. \tag{2.2.25}$$

Notice also that $n \mapsto v_{n,\epsilon}$ is increasing by the maximum principle and it converges to some function $v_\epsilon$ which satisfies also (2.2.25). Since $\{v_{n,\epsilon}\}$ remains locally bounded in $C^{2,\alpha}(S)$, it converges in the local $C^2$ topology to $v_\epsilon$ which is a $C^{2,\alpha}$ solution of (2.2.10). Furthermore $v_\epsilon$ satisfies the same estimate (2.2.23) as $u$.

*The ergodic limit.* Comparing $v_\epsilon$ with $\underline{u}_h$ and then to $\underline{u}$ we obtain the lower bound. Combining it with the upper bound, we see that $v_\epsilon$ satisfies

$$-\frac{1}{\beta}\ln\tilde{\rho}(x) + M_0\tilde{\rho}(x) - \frac{M_1}{\epsilon} \leq v_\epsilon(x) \leq -\frac{1}{\beta}\ln\tilde{\rho}(x) - M_0\tilde{\rho}(x) + \frac{M_1}{\epsilon}. \tag{2.2.26}$$

This implies that $\epsilon v_\epsilon$ is locally bounded in $S$. Since $\nabla v_\epsilon$ is also locally bounded, it implies that there exist a subsequence $\{\epsilon_n\}$ and a nonnegative constant $\lambda_0$ such that $\epsilon_n v_{\epsilon_n}$ converges to $\lambda_0$ locally uniformly in $S$. We fix $X_0 \in S$, set $w_\epsilon = v_\epsilon - v_\epsilon(x_0)$. Since $w_\epsilon(x_0) = 0$, $\nabla w_\epsilon$ is locally bounded and there holds

$$-\Delta_g w_\epsilon - \frac{p-2}{2}\frac{\langle\nabla\left|\nabla w_\epsilon\right|^2, \nabla w_\epsilon\rangle_g}{1+|\nabla w_\epsilon|^2} + \beta(p-1)|\nabla w_\epsilon|^2 + \epsilon w_\epsilon = -\epsilon v_\epsilon(x_0),$$

there exist a subsequence $\{\epsilon_{n_k}\}$ and a $C^{2+\alpha}$ function $w$ such that $\{w_{\epsilon_{n_k}}\}$ converges to $w$ locally in the $C^2(S)$ topology. Thus $w$ satisfies

$$-\Delta_g w - \frac{p-2}{2}\frac{\langle\nabla\left|\nabla w\right|^2, \nabla w\rangle_g}{1+|\nabla w|^2} + \beta(p-1)|\nabla w|^2 = -\lambda_0 \qquad \text{in } S. \tag{2.2.27}$$

At end we prove that $w(x)$ tends to infinity on the boundary. We set

$$\underline{\psi}(x) = -\frac{1}{\beta}\ln\tilde{\rho}(x) + M_0\tilde{\rho}(x).$$

Then

$$\begin{aligned}-\Delta_g\underline{\psi} - \frac{p-2}{2}\frac{\langle\nabla\left|\nabla\underline{\psi}\right|^2, \nabla\underline{\psi}\rangle_g}{1+|\nabla\underline{\psi}|^2} + \beta(p-1)|\nabla\underline{\psi}|^2 + \epsilon\underline{\psi} \\ = \frac{1}{\tilde{\rho}}\left(\frac{\Delta_g\tilde{\rho}}{\beta} - \epsilon\frac{\tilde{\rho}\ln\tilde{\rho}}{\beta}\right) - 2(p-1)M_0\left|\nabla\tilde{\rho}\right|^2 + a_\beta,\end{aligned}$$

where $a_\beta$ is a bounded function. We can choose $M_0 > 0$ and $\rho_0 \in (0, \delta)$ such that $\underline{\psi}$ is a subsolution of (2.2.10) in $\Omega \cap \mathcal{N}_{\rho_0}$. Next we have that $w_\epsilon(x) \geq -c_0$ for some constant $c_0 > 0$ on the set of $x \in S$ such that $\rho(x) = \rho_0$ since $\nabla w_\epsilon$ is locally bounded. Because $\underline{\psi} - c$ is still a subsolution for $c \geq 0$ and independent of $\epsilon$, we derive

$$w_\epsilon(x) \geq -\frac{1}{\beta}\ln\tilde{\rho}(x) + M_0\tilde{\rho}(x) - c \quad \text{in } \Omega \cap \mathcal{N}_{\rho_0}. \tag{2.2.28}$$

Letting $\epsilon \to 0$ implies that $w(x) \to \infty$ as $\rho(x) \to 0$. In the same way the function

$$\overline{\psi}(x) = -\frac{1}{\beta} \ln \tilde{\rho}(x) - M_0 \tilde{\rho}(x) + c,$$

is a supersolution and $w$, as any solution of (2.2.27) which blows-up on the boundary, satisfies

$$-\frac{1}{\beta} \ln \tilde{\rho}(x) + M_0 \tilde{\rho}(x) - c \leq w(x) \leq -\frac{1}{\beta} \ln \tilde{\rho}(x) - M_0 \tilde{\rho}(x) + c \quad \text{in } \Omega \cap \mathcal{N}_{\rho_0}. \tag{2.2.29}$$

*Proof of Theorem 2.2.3. Uniqueness.* It will follow from the next statement. *There exists a unique positive constant $\lambda_0 > 0$ such that there is a function $v \in C^2(S)$ solution of*

$$\begin{aligned} -\Delta_g v - \frac{p-2}{2} \frac{\langle \nabla |\nabla v|^2, \nabla v \rangle_g}{1 + |\nabla v|^2} + \beta(p-1)|\nabla v|^2 + \lambda_0 = 0 \qquad & \text{in } S \\ \lim_{\rho(x) \to 0} v(x) = \infty. & \end{aligned} \tag{2.2.30}$$

The role of the ergodic constant $\lambda_0$ is enlighted by the following equivalence result.

**Lemma 2.2.4** *A function $v \in C^2(S)$ is a solution of (2.2.30) if and only if $\omega := e^{-\beta v} \in C^2(S) \cap C(\overline{S})$ satisfies*

$$\begin{aligned} -\mathrm{div}_g \left( \left( \beta^2 \omega^2 + |\nabla \omega|^2 \right)^{\frac{p-2}{2}} \nabla \omega \right) &= \beta \lambda_0 \left( \beta^2 \omega^2 + |\nabla \omega|^2 \right)^{\frac{p-2}{2}} \omega \qquad \textit{in } S \\ w &= 0 \qquad \textit{in } \partial S. \end{aligned} \tag{2.2.31}$$

*Moreover $\omega \in C^{1+\alpha}(\overline{S})$ and $\langle \nabla \omega, \mathbf{n} \rangle_g < 0$ on $\partial S$ where $\mathbf{n} \in TS$ is the outward normal unit vector to $\partial S$.*

*Proof.* A straightforward computation shows that the two problems (2.2.30) and (2.2.31) are exchanged by the setting $\omega := e^{-\beta v}$: if $v \in C^2(S)$ is a solution of (2.2.30), $\omega \in C(\overline{S}) \cap W^{1,\infty}(S)$ is a weak solution of (2.2.31). By Theorem 1.3.10, $\omega \in C^1(\overline{S})$ and since (2.2.29) holds,

$$\left| v(x) + \frac{1}{\beta} \ln \tilde{\rho}(x) \right| \leq c \qquad \text{in } \Omega \cap \mathcal{N}_{\rho_0}, \tag{2.2.32}$$

which implies

$$\tilde{\rho}(x)e^{-\beta c} \le \omega(x) \le \tilde{\rho}(x)e^{\beta c} \qquad \text{in } \Omega \cap \mathcal{N}_{\rho_0}. \tag{2.2.33}$$

Because $w$ satisfies (2.2.23),

$$\beta(L_0 + \tilde{\rho}(x)L_1)e^{-\beta c} \le |\nabla\omega(x)| \le \beta(L_0 + \tilde{\rho}(x)L_1)e^{\beta c} \tag{2.2.34}$$

near the boundary, and thus $\langle \nabla\omega, \mathbf{n}\rangle_g < 0$ on $\partial S$. Finally, since $\beta^2\omega^2 + |\nabla\omega|^2 > 0$, the equation is not degenerate and thus the solutions are smooth. Conversely, any solution of (2.2.31) with the above indicated regularity is transformed into a $C^2(S)$ solution of (2.2.30). □

*End of the proof of Theorem 2.2.3.*

*Uniqueness of the ergodic limit.* Assume $\lambda_j$, $j = 0, 1$, are two ergodic constants associated to solutions $v_j = e^{-\beta\omega_j}$ of (2.2.30) and (2.2.31). By energy estimates $\lambda_j > 0$. Assume $\lambda_0 > \lambda_1$. By (2.2.33), there exists $\theta > 0$ such that $\omega_1 \le \theta\omega_0$, and since the equation is homogeneous we can assume that $\theta = 1$. The tangency conditions of the two graphs of the functions $\omega_0 = e^{-\beta v_0}$ and $\omega_1 = e^{-\beta v_1}$ implies that either there exists some $x_0 \in S$ such that $\omega_1(x_0) = \omega_0(x_0)$ and $\nabla\omega_1(x_0) = \nabla\omega_1(x_0)$, or $\omega_1 < \omega_0$ in $S$ and there exists $x_0 \in \partial S$ such that $\langle \nabla\omega_0(x_0), \mathbf{n}\rangle_g = \langle \nabla\omega_1(x_0), \mathbf{n}\rangle_g < 0$. In the first case the function $z = v_1 - v_0$ is nonnegative in $S$, achieves its minimal value 0 at $x_0$ and satisfies at this point

$$-\Delta_g z(x_0) - (p-2)\frac{\langle D^2 z(x_0)\nabla v_0(x_0), \nabla v_0(x_0)\rangle_g}{1 + |\nabla v_0(x_0)|^2} = \lambda_0 - \lambda_1 > 0,$$

which is impossible because of ellipticity. In the second case the tangency condition at $x_0$ yields $\langle \nabla(\omega_0(x_0) - \omega_1(x_0)), \mathbf{n}\rangle_g = 0$, which implies $|\nabla\omega_0(x_0)|^2 = |\nabla\omega_0(x_0)|^2$, whereas the function $\zeta = \omega_1 - \omega_0$ is negative in $S$ and vanishes at $x_0$. Since $\beta^2\omega_j^2 + |\nabla\omega_j|^2 > 0$, the problem (2.2.31) is uniformly elliptic. Since $\omega_0(x_0) = \omega_1(x_0)$, this is a contradiction with Hopf maximum principle. Because the ergodic limit $\lambda_0$ is uniquely determined by $\beta$, we denote it by $\lambda(\beta)$.

*Uniqueness of the eigenfunction up to homothety.* Similarly to the previous argument, we consider two tangent solutions positive $\omega_0$ and $\omega_1$ (by homothety). We set $\zeta = \omega_1 - \omega_0 \ge 0$ and assume that $\zeta$ is not identically zero. We apply the strong maximum principle to $\zeta$ at $x_0$ and derive that $\zeta$ has to vanish everywhere. Similarly Hopf maximum principle yields $\zeta \equiv 0$, a contradiction. Note that this uniqueness implies that the whole sequence $\{v_\epsilon\}$ converges to $v$ as $\epsilon \to 0$. □

Next we study the properties of the ergodic constant.

**Proposition 2.2.5** *Under the assumptions of Theorem 2.2.3 the mapping $\beta \mapsto \lambda(\beta)$ is continuous and decreasing on $(0,\infty)$. Furthermore*

$$\lim_{\beta\to 0} \lambda(\beta) = \infty. \tag{2.2.35}$$

*Proof. Monotonicity* Assume $0 < \beta_1 < \beta_2$ and let $v_{\epsilon,1}$ and $v_{\epsilon,2}$ be the corresponding solutions of (2.2.10) with $\beta = \beta_1$ and $\beta_2$ respectively. Using the maximum principle and letting $n \to \infty$ in (2.2.24) yields $v_{\epsilon,1} > v_{\epsilon,2}$ and thus

$$\lambda(\beta_1) = \lim_{\epsilon\to 0} \epsilon v_{\epsilon,1} \geq \lim_{\epsilon\to 0} \epsilon v_{\epsilon,2} = \lambda(\beta_2).$$

If we assume that there exists $0 < \beta_1 < \beta_2$ such that $\lambda(\beta_1) = \lambda(\beta_2) = \lambda$ and denote by $\omega_1$ and $\omega_2$ the corresponding solutions of (2.2.31) with $\beta = \beta_1$ and $\beta_2$ and $\lambda_0 = \lambda(\beta_1)$ and $\lambda(\beta_2)$ respectively, then $\omega_1 \leq \omega_2$ and by (2.2.33),

$$m^{-1}\tilde{\rho}(x) \leq \omega_1(x) \leq \omega_2(x) \leq m\tilde{\rho}(x) \qquad \forall x \in \overline{S}. \tag{2.2.36}$$

Put $\tilde{\omega} = \omega^{\frac{\beta_2}{\beta_1}}$. Then

$$-div_g\left(\left(\beta_2^2\tilde{\omega}^2 + |\nabla\tilde{\omega}|^2\right)^{\frac{p-2}{2}} \nabla\tilde{\omega}\right) - \beta_2\lambda\left(\beta_2^2\tilde{\omega}^2 + |\nabla\tilde{\omega}|^2\right)^{\frac{p-2}{2}}\tilde{\omega}$$
$$= (p-1)\left(1 - \frac{\beta_2}{\beta_1}\right)\left(\frac{\beta_2}{\beta_1}\right)^{p-1} \omega_1^{(p-1)\left(\frac{\beta_2}{\beta_1}-1\right)} \left(\beta_1^2\omega_1^2 + |\nabla\omega_1|^2\right)^{\frac{p-2}{2}} \frac{|\nabla\omega_1|^2}{\omega_1} < 0. \tag{2.2.37}$$

Equivalently $\tilde{w}$ is a strict supersolution of the equation satisfied by $w_2$. By homogeneity, we can assume that $\tilde{\omega} \leq \omega_2$ and that the coincidence set $\mathcal{S}$ of $\tilde{\omega}$ and $\omega_2$ is nonempty. Thus it must be a compact subset of $S$ because $\nabla\tilde{\omega} = 0$ on $\partial S$. Put

$$z = -\frac{1}{\beta_2}\left(\ln\omega_2 - \ln\tilde{\omega}\right) = w_2 - \tilde{w}.$$

Clearly $z \leq 0$ is not identically zero and there exists $x_0 \in \mathcal{S}$ where $z(x_0) = 0$. Because of (2.2.29), $z(x) \to -\infty$ as $\tilde{\rho}(x) \to 0$. If we expand (2.2.37), we obtain at $x = x_0$,

$$-\Delta_g z - (p-2)\frac{\langle D^2 z\nabla v_2, \nabla v_2\rangle_g}{1 + |\nabla v_2|^2} + \beta_2(p-1)(|\nabla v_2|^2 - |\nabla\tilde{v}|^2)$$
$$+ (p-2)\left(\frac{\langle H_{\tilde{v}}(\nabla\tilde{v}), \nabla\tilde{v}\rangle_g}{1 + |\nabla\tilde{v}|^2} - \frac{\langle H_{\tilde{v}}(\nabla v_2), \nabla v_2\rangle_g}{1 + |\nabla v_2|^2}\right) \leq 0.$$

This contradicts the strong maximum principle. Thus $\lambda(\beta_1) > \lambda(\beta_2)$.

*Continuity.* Let $\{\beta_n\} \subset \mathbb{R}_+$ be a sequence converging to $\beta > 0$, and $v_{\beta_n}$ and $v_{\epsilon,\beta_n}$ be the corresponding solutions to (2.2.9) and (2.2.10) respectively with $\beta = \beta_n$. Since $\epsilon v_{\epsilon,\beta_n}$ remains locally bounded in $S$ and converges locally in $S$ to $\lambda(\beta_n)$ as $\epsilon \to 0$, the set $\{\lambda(\beta_n)\}$ remains bounded. Up to a subsequence, still denoted by $\{\beta_n\}$, $\{\lambda(\beta_n)\}$ converges to some $\overline{\lambda} \geq 0$ as $n \to \infty$. Estimates (2.2.29) and (2.2.12) are valid with $v_{\beta_n}$ and the constants therein uniform with respect to $\beta_n$. Therefore $v_{\beta_n}$ remains locally bounded in $W^{1,\infty}(S)$ and thus in $C^{2,\alpha}(S)$. Up to a subsequence, not relabeled, $\{v_{\beta_n}\}$ converges to $\bar{v}$ as $n \to \infty$, and $\bar{v}$ satisfies

$$-\Delta_g \bar{v} - (p-2)\frac{\langle H_{\bar{v}}(\nabla \bar{v}), \nabla \bar{v}\rangle_g}{1+|\nabla \bar{v}|^2} + \beta_0(p-1)|\nabla \bar{v}|^2 = -\bar{\lambda}. \tag{2.2.38}$$

Because of the uniqueness of the ergodic constant $\bar{\lambda} = \lambda(\beta_0)$ and the whole sequence $\{\lambda(\beta_n)\}$ converges to $\lambda(\beta_0)$.

*Limit.* By (2.2.31),

$$\int_S \left(\beta^2\omega^2 + |\nabla\omega|^2\right)^p dv_g = \beta(\lambda(\beta)+\beta)\int_S \left(\beta^2\omega^2 + |\nabla\omega|^2\right)^{\frac{p-2}{2}} \omega^2 dv_g.$$

Let $\lambda_1(p,S)$ be the first eigenvalue of $-\Delta_p$ in $W_0^{1,p}(S)$ and normalize $\omega$ by

$$\int_S |\nabla\omega|^p dv_g = 1.$$

Then there holds

$$\int_S |\omega|^p dv_g \leq \frac{1}{\lambda_1(p,S)}.$$

Clearly

$$\int_S \left(\beta^2\omega^2 + |\nabla\omega|^2\right)^p dv_g \geq \int_S |\nabla\omega|^p dv_g = 1.$$

If $p \geq 2$,

$$\begin{aligned}\int_S \left(\beta^2\omega^2 + |\nabla\omega|^2\right)^{\frac{p-2}{2}} \omega^2 dv_g &\leq 2^{\frac{p}{2}}\int_S (\beta^{p-2}|\omega|^p + \omega^2|\nabla\omega|^{p-2})dv_g \\ &\leq 2^{\frac{p}{2}}\left(\beta^{p-2} + \frac{2}{p}\right)\int_S |\omega|^p dv_g + 2^{\frac{p}{2}}\left(1-\frac{2}{p}\right) \\ &\leq \frac{2^{\frac{p}{2}}}{\lambda_1(p,S)}\left(\beta^{p-2}+\frac{2}{p}\right) + 2^{\frac{p}{2}}.\end{aligned}$$

We derive

$$\beta + \lambda(\beta) \geq \frac{2^{\frac{p}{2}}}{\beta}\left(\frac{1}{\lambda_1(p,S)}\left(\beta^{p-2} + \frac{2}{p}\right) + 1\right). \tag{2.2.39}$$

If $1 < p < 2$, then

$$\int_S \frac{\omega^2}{(\beta^2\omega^2 + |\nabla\omega|^2)^{\frac{2-p}{2}}} dv_g \leq \beta^{p-2}\int_S |\omega|^p dv_g \leq \frac{\beta^{p-2}}{\lambda_1(p,S)}.$$

Since $\beta^{p-1}(\lambda(\beta) + \beta) \geq \lambda_1(p,S)$

$$\lambda(\beta) \geq \frac{\lambda_1(p,S)}{\beta^{p-1}} - \beta. \tag{2.2.40}$$

In the two cases, (2.2.35) follows. □

*Proof of Theorem 2.2.1: The singular case.* For $\beta > 0$ we set $\omega_\beta = e^{-\beta v_\beta}$ where $v_\beta$ is the solution of (2.2.9). Then $\omega_\beta$ satisfies

$$\begin{aligned} -div_g\left(\left(\beta^2\omega^2 + |\nabla\omega|^2\right)^{\frac{p-2}{2}} \nabla\omega\right) &= \beta\lambda(\beta)\left(\beta^2\omega^2 + |\nabla\omega|^2\right)^{\frac{p-2}{2}} \omega && \text{in } S \\ \omega &= 0 && \text{on } \partial S. \end{aligned} \tag{2.2.41}$$

The conclusion will follow if we prove that there exists a unique $\beta > 0$ such that $\lambda(\beta) = \Lambda(\beta) := \beta(p-1) + p - N$. The mapping $\beta \mapsto (p-1)\beta - \lambda(\beta)$ is continuous and increasing with

$$\lim_{\beta\to 0} (p-1)\beta - \lambda(\beta) = -\infty \quad \text{and} \quad \lim_{\beta\to\infty} (p-1)\beta - \lambda(\beta) = \infty.$$

Therefore there exists a unique $\beta_s > 0$ such that $(p-1)\beta - \lambda(\beta) = N - p$, which ends the proof.

*The regular case.* We set $\beta = -\beta'$ with $\beta' > 0$. Then the spherical $p$-harmonic equation becomes

$$\begin{aligned} -div_g\left(\left(\beta'^2\omega^2 + |\nabla\omega|^2\right)^{\frac{p-2}{2}} \nabla\omega\right) &= \beta'\Lambda'(\beta')\left(\beta'^2\omega^2 + |\nabla\omega|^2\right)^{\frac{p-2}{2}} \omega && \text{in } S \\ \omega &= 0 && \text{on } \partial S \end{aligned} \tag{2.2.42}$$

where $\Lambda'(\beta') = \beta'(p-1) + N - p$. If we set $v = -\frac{1}{\beta'}\ln\omega$, then (2.2.9) becomes

$$\begin{aligned}
&-div_g\left(\left(1+|\nabla v|^2\right)^{\frac{p-2}{2}}\nabla v\right) + \beta'(p-1)\left(1+|\nabla v|^2\right)^{\frac{p-2}{2}}|\nabla v|^2 \\
&\qquad = -\Lambda'(\beta')\left(1+|\nabla v|^2\right)^{\frac{p-2}{2}} && \text{in } S \\
&\lim_{\rho(x)\to 0} v(x) = \infty && \text{in } \partial S.
\end{aligned} \tag{2.2.43}$$

Thanks to Theorem 2.2.3, the remaining of the proof in the regular case goes as in the singular case, except that we conclude in proving, by the same limit argument, that there exists a unique $\beta' := \beta'_S > 0$ such that $(p-1)\beta' - \lambda(\beta') = p - N$. This ends the proof of Theorem 2.2.1. □

*Remark.* There is no known relation between $\beta_S$ and $\beta'_S$ except in the case $p = N$ where the $N$-Laplace equation is invariant by conformal transformations (see Section 1.4). Thus if $x \mapsto |x|^{-\beta_S}\omega(\frac{x}{|x|})$ is $N$-harmonic in the cone $C_S = \{x \in \mathbb{R}^N \setminus \{0\} : \frac{x}{|x|} \in S\}$ and vanishes on $\partial C_S$, $x \mapsto |x|^{\beta_S}\omega(\frac{x}{|x|})$ endows the same properties, thus $\beta_S = -\beta_S$.

### 2.2.2 *The 2-dimensional case*

When $N = 2$, (2.2.2) becomes

$$-\left(\left(\beta^2\omega^2 + \omega'^2\right)^{\frac{p-2}{2}}\omega'\right)' = \beta(\beta(p-1) + p - 2)\left(\beta^2\omega^2 + \omega_\theta^2\right)^{\frac{p-2}{2}}\omega, \tag{2.2.44}$$

on $S^1 \approx \mathbb{R}/2\pi$. If we set $Y = \omega'/\omega$, then $Y$ satisfies

$$\left(\frac{\beta^2}{Y^2+\beta^2} - \frac{\beta+1}{Y^2+\beta(\beta-\beta_0)}\right)Y_\theta = 1, \tag{2.2.45}$$

with $\beta_0 = \frac{2-p}{p-1}$. This equation is completely integrable by quadratures and the research of $2\pi$-periodic solutions yields the following two results corresponding to singular and regular solutions.

**Theorem 2.2.6** *Assume $N = 2$ and $p > 1$. Then for any $k \in \mathbb{N}^*$ there exist $\beta_k > 0$ and $\omega_k : \mathbb{R} \mapsto \mathbb{R}$, with least period $2\pi/k$ such that*

$$u_k(x) \sim u_k(r,\theta) = r^{-\beta_k}\omega_k(\theta) \tag{2.2.46}$$

*is p-harmonic in* $\mathbb{R}^2 \setminus \{0\}$; $\beta_k$ *is the positive root of*

$$(\beta+1)^2 = \left(1+\frac{1}{k}\right)^2 \left(\beta^2 - \beta\beta_0\right). \tag{2.2.47}$$

*The couple* $(\omega_k, \beta_k)$ *is unique up to an homothety over* $\omega_k$.

**Theorem 2.2.7** *Assume* $N = 2$ *and* $p > 1$. *Then for any* $k \in \mathbb{N}^*$ *there exist* $\beta'_k \geq 1$ *and* $\omega'_k : \mathbb{R} \mapsto \mathbb{R}$, *with least period* $2\pi/k$ *such that*

$$u_k(x) \sim u_k(r,\theta) = r^{\beta'_k}\omega'_k(\theta) \tag{2.2.48}$$

*is p-harmonic in* $\mathbb{R}^2$; $\beta'_k$ *is the root larger or equal to* 1 *of*

$$(\beta-1)^2 = \left(1-\frac{1}{k}\right)^2 \left(\beta^2 + \beta\beta_0\right). \tag{2.2.49}$$

*The couple* $(\omega'_k, \beta'_k)$ *is unique up to an homothety over* $\omega'_k$.

*Remark.* We notice that there always hold $\beta'_1 = 1$. Actually whatever is the dimension, the function $x \mapsto x_j$ is $p$-harmonic in $\mathbb{R}^N$ and separable. Thus, if $S$ is an hemisphere of the sphere $S^{N-1}$, $\beta'_S = 1$.

The regular 2-dimensional equation plays a fundamental role in constructing $N$-dimensional separable $p$-harmonic functions in $\mathbb{R}^N$. We consider the following representation of spherical coordinates in $\mathbb{R}^N := \{x = (x_1, ..., x_N)\}$.

$$\begin{aligned} x_1 &= r\sin\theta_{N-1}\sin\theta_{N-2}...\sin\theta_2\sin\theta_1 \\ x_2 &= r\sin\theta_{N-1}\sin\theta_{N-2}...\sin\theta_2\cos\theta_1 \\ &\vdots \\ x_{N-1} &= r\sin\theta_{N-1}\cos\theta_{N-2} \\ x_N &= r\cos\theta_{N-1}, \end{aligned} \tag{2.2.50}$$

where $\theta_1 \in [0, 2\pi]$ and $\theta_j \in [0, \pi]$ for $j = 2, ..., N-1$. If $f$ is defined in $\mathbb{R}^N$, we write $f(x) \sim f(r, \theta_1, ..., \theta_{N-1})$. We can also denote a point $\sigma$ on $S^{N-1}$ as $\sigma = (\theta_1, ..., \theta_{N-2}, \theta_{N-1}) = (\sigma', \theta_{N-1})$. In that case $\sigma'$ stands for a general point on $S^{N-2}$.

**Definition 2.1** A spherical $p$-harmonic eigenfunction $\omega$ on $S^{N-1}$ is separable if there exist a system of coordinates $(\theta_1, ..., \theta_{N-2}, \theta_{N-1} = (\sigma', \theta_{N-2})$ with $\theta_1 \in [0, 2\pi]$ and $\theta_j \in [0, \pi]$ for $j = 2, ..., N-1$, $b > 0$ and $\phi \in C^2(S^{N-2})$ such that $\omega(\sigma) = (\sin\theta_{N-1})^b\phi(\sigma')$.

The next theorem shows how separable spherical $p$-harmonic eigenfunctions on $S^{N-1}$ can be generated from spherical $p$-harmonic eigenfunctions on $S^{N-2}$.

**Proposition 2.2.8** *Assume $N \geq 3$, $p > 1$ and $\omega$ is a spherical $p$-harmonic eigenfunction on $S^{N-1}$ with exponent $\beta' > 0$. If $\omega(\sigma) = (\sin\theta_{N-1})^{\beta'}\phi(\sigma')$ for some system of coordinates and $\phi$ is a spherical $p$-harmonic eigenfunction on $S^{N-2}$ with the same exponent $\beta' > 0$.*

*Proof.* The gradient of a function $\omega$ is expressed by

$$\nabla_{S^{N-1}}\omega = -\omega_{\theta_{N-1}}\mathbf{e} + \frac{1}{\sin\theta_{N-1}}\nabla_{S^{N-2}}\omega,$$

where $\mathbf{e}$ is derived from $\frac{x}{|x|}$ by the rotation of center 0, angle $\frac{\pi}{2}$ in the plan defined by 0, $\frac{x}{|x|}$ and $\mathbf{e}_N$. If we set

$$Q = \beta'^2\omega^2 + \omega^2_{\theta_{N-1}} + \frac{1}{\sin^2\theta_{N-1}}\left|\nabla_{S^{N-2}}\omega\right|^2,$$

equation (2.2.42) reads

$$\begin{aligned}-\frac{1}{(\sin\theta_{N-1})^{N-2}}&\left((\sin\theta_{N-1})^{N-2}Q^{\frac{p-2}{2}}\omega_{\theta_{N-1}}\right)_{\theta_{N-1}}\\ &-\frac{1}{\sin^2\theta_{N-1}}div_{S^{N-2}}\left(Q^{\frac{p-2}{2}}\nabla_{S^{N-2}}\omega\right) = \beta'\Lambda'(\beta',N)Q^{\frac{p-2}{2}}\omega,\end{aligned} \tag{2.2.51}$$

where $\Lambda'(\beta',N) = \beta'(p-1) + N - p$. If $\omega$ is a solution under the form $\omega(\sigma',\theta_{N-1}) = (\sin\theta_{N-1})^{\beta'}\phi(\sigma')$, then

$$Q^{\frac{p-2}{2}} = (\sin\theta_{N-1})^{(\beta'-1)(p-2)}\left(\beta'^2\phi^2 + |\nabla_{S^{N-2}}\phi|^2\right)^{\frac{p-2}{2}},$$

hence

$$\begin{aligned}&\frac{1}{(\sin\theta_{N-1})^{N-2}}\left((\sin\theta_{N-1})^{N-2}Q^{\frac{p-2}{2}}\omega_{\theta_{N-1}}\right)_{\theta_{N-1}}\\ &= \left(N-2+(\beta'-1)(p-1) - (N-1+(\beta'-1)(p-1))\sin^2\theta_{N-1})\right)\\ &\qquad \times \beta'(\sin\theta_{N-1})^{(\beta'-1)(p-1)-1}\left(\beta'^2\phi^2 + |\nabla_{S^{N-2}}\phi|^2\right)^{\frac{p-2}{2}}\phi,\end{aligned}$$

and

$$\frac{1}{\sin^2\theta_{N-1}} div_{S^{N-2}}\left(Q^{\frac{p-2}{2}}\nabla_{S^{N-2}}\omega\right)$$
$$= (\sin\theta_{N-1})^{(\beta'-1)(p-1)-1} div_{S^{N-2}}\left(\left(\beta'^2\phi^2+|\nabla_{S^{N-2}}\phi|^2\right)^{\frac{p-2}{2}}\nabla_{S^{N-2}}\phi\right).$$

Finally $\phi$ satisfies

$$\begin{aligned}-div_{S^{N-2}}&\left(\left(\beta'^2\phi^2+|\nabla_{S^{N-2}}\phi|^2\right)^{\frac{p-2}{2}}\nabla_{S^{N-2}}\phi\right)\\ &=\beta'\Lambda'(\beta',N-1)\left(\beta'^2\phi^2+|\nabla_{S^{N-2}}\phi|^2\right)^{\frac{p-2}{2}}\phi,\end{aligned}\tag{2.2.52}$$

on $S^{N-2}$. The proof follows. □

*Remark.* We can iterate this decomposition and construct separable spherical $p$-harmonic eigenfunctions under the form

$$w(\theta_1,...\theta_{N-1})=(\sin\theta_{N-1})^{\beta'}(\sin\theta_{N-2})^{\beta'}...(\sin\theta_2)^{\beta'}\omega(\theta_1),\tag{2.2.53}$$

where $\omega$ in a solution of (2.2.44) (with $\beta=-\beta'$). Therefore $\omega=\omega'_k$ and $\beta'=\beta'_k$ for some $k\in\mathbb{N}^*$ given in Theorem 2.2.7. This result is not surprising since

$$\sin\theta_{N-1}\sin\theta_{N-2}...\sin\theta_2=\frac{\sqrt{x_1^2+x_2^2}}{|x|}\Longrightarrow\sin\theta_1=\frac{x_1}{\sqrt{x_1^2+x_2^2}},$$

thus the corresponding separable $p$-harmonic function $u$ in $\mathbb{R}^N$ is the trivial extension to $\mathbb{R}^N$ of the 2-dimensional separable $p$-harmonic function. Furthermore, in the case $p=N$ we can construct separable $N$-harmonic functions regular or singular at 0 in a similar way by performing conformal transformation.

If $S$ is the hemisphere $S_+^{N-1}$, the exponent $\beta_S$, although unknown explicitly, satisfies a remarkable inequality [Bidaut-Véron, Garcia-Huidobro, Véron (2015)].

**Theorem 2.2.9** *Assume $N\geq 3$ and $p>1$. Then there holds*

$$\begin{aligned}&(i)\quad \textit{if } 1<p<2,\quad \textit{then } \beta_{S_+^{N-1}}>\frac{N-1}{p-1}\\ &(ii)\quad \textit{if } 2<p<N,\quad \textit{then } \max\left\{1,\frac{N-p}{p-1}\right\}<\beta_{S_+^{N-1}}<\frac{N-1}{p-1}.\end{aligned}\tag{2.2.54}$$

*Proof. The eigenvalue identity.* Let $\omega > 0$ be the solution of (2.2.42) corresponing to $\beta = \beta_{s^{N-1}}$. Because of uniqueness it depends only on the variable $\theta_{N-1}$. For simplicity put $\theta = \theta_{N-1}$, then $\omega := \omega(\theta)$ satisfies

$$\begin{aligned} -\frac{1}{\sin^{N-2}\theta}\left(\sin^{N-2}\theta\left(\beta^2\omega^2+\omega'^2\right)^{\frac{p-2}{2}}\omega'\right)' \\ = \beta\Lambda(\beta)\left(\beta^2\omega^2+\omega'^2\right)^{\frac{p-2}{2}}\omega \quad \text{in } (0,\tfrac{\pi}{2}) \end{aligned} \tag{2.2.55}$$

$$\omega(\tfrac{\pi}{2}) = 0 \text{ and } \omega'(0) = 0.$$

This equation can be written

$$-\omega'' - (N-2)\cot\theta\,\omega' - (p-2)\frac{\beta^2\omega+\omega''}{\beta^2\omega^2+\omega'^2} = \beta\Lambda(\beta)\omega. \tag{2.2.56}$$

Since

$$-\int_0^{\frac{\pi}{2}}(\omega''+(N-2)\cot\theta\,\omega')\cos\theta\sin^{N-2}\theta d\theta = (N-1)\int_0^{\frac{\pi}{2}}\omega\cos\theta\sin^{N-2}\theta d\theta$$

and

$$\beta\Lambda(\beta)+1-N = (p-1)\left(\beta-\frac{N-1}{p-1}\right)(\beta+1),$$

we obtain

$$\begin{aligned} (2-p)\int_0^{\frac{\pi}{2}}\frac{\beta^2\omega+\omega''}{\beta^2\omega^2+\omega'^2}\omega'^2\omega\cos\theta\sin^{N-2}\theta d\theta \\ = (p-1)\left(\beta-\frac{N-1}{p-1}\right)(\beta+1)\int_0^{\frac{\pi}{2}}\omega\cos\theta\sin^{N-2}\theta d\theta. \end{aligned} \tag{2.2.57}$$

*Reduction.* It is clear that $\omega$ decreases in a right neighborhood of 0 and that it cannot have a local minimum in $(0,\frac{\pi}{2})$. We introduce elliptic coordinates associated to the arc of ellipse,

$$E_r = \{(x,y)\in\mathbb{R}^2 : x>0,\ y<0,\ x^2+\beta^{-2}y^2 = r^2\},$$

by setting $\omega = r\cos\phi$ and $-\omega' = \beta r\sin\phi$ with $\phi = \phi(\theta)$ and $r = r(\theta)$. The functions $r$ and $\phi$ are $C^2$, hence $r_\theta\cos\phi - r\phi_\theta\sin\phi = -\beta r\sin\phi$ and $r_\theta = r(\phi_\theta-\beta)\tan\phi$. Replacing into (2.2.56) we derive first,

$$-\left((p-1)\frac{r_\theta}{r}+\phi_\theta\cot\phi+(N-2)\cot\theta\right)+\Lambda(\beta)\cot\phi = 0, \tag{2.2.58}$$

and next,

$$(p-1)(\phi_\theta - \beta)\tan\phi + (\phi_\theta - \Lambda(\beta))\cot\phi = (2-N)\cot\theta. \tag{2.2.59}$$

*Estimate of* $\phi_\theta$. The previous equation can be written

$$(p-1)(\phi_\theta - \beta)\tan^2\phi + \phi_\theta - \Lambda(\beta) = (2-N)\frac{\cos\theta\sin\phi}{\sin\theta\cos\phi}. \tag{2.2.60}$$

Since

$$\lim_{\theta\to 0}\frac{\sin\phi}{\sin\theta} = \lim_{\theta\to 0}\frac{\cos\phi}{\cos\theta}\phi_\theta = \phi_\theta(0),$$

we derive $\phi_\theta(0) = \frac{\Lambda(\beta)}{N-1}$ from (2.2.60). Because $p < N$, then $\Lambda(\beta) < \beta(N-1)$ and thus $\phi_\theta(0) < \beta$. Similarly, we expand $\phi$ near $\theta = \frac{\pi}{2}$ and obtain $\phi_\theta(\frac{\pi}{2}) = \beta$. *The key estimate is the following:*

$$\phi_\theta(\theta) \leq \beta \qquad \forall\theta \in [0, \tfrac{\pi}{2}], \tag{2.2.61}$$

and we distinguish two cases:
(i) If $\Lambda(\beta) \leq \beta$, then

$$\begin{aligned}(2-N)\cot\theta &= (p-1)(\phi_\theta - \beta)\tan\phi + (\phi_\theta - \Lambda(\beta))\cot\phi \\ &\geq ((p-1)\tan\phi + \cot\phi)(\phi_\theta - \beta),\end{aligned}$$

and (2.2.61) follows.
(ii) If $\Lambda(\beta) > \beta$ then $0 < (p-2)\beta - (N-p)$, and this is possible only if $p > 2$. We claim that

$$\beta \leq \frac{N-2}{p-2}. \tag{2.2.62}$$

Equivalently $\beta(\Lambda(\beta) - \beta) \leq N-2$. Since

$$\lim_{\theta\to\frac{\pi}{2}}\cot\theta\tan\phi = \lim_{\theta\to\frac{\pi}{2}}\frac{\cos\theta}{\cos\phi} = \lim_{\theta\to\frac{\pi}{2}}\frac{\sin\theta}{\phi_\theta\sin\phi} = \frac{1}{\beta},$$

there holds

$$\begin{aligned}(p-1)(\phi_\theta(\theta) - \beta)\tan^2\phi &= \Lambda(\beta) - \phi_\theta(\theta) + (2-N)\frac{\cos\theta\sin\phi}{\cos\phi\sin\theta} \\ &= \frac{1}{\beta}(\beta(\Lambda(\beta) - \beta) + 2 - N) + o(1) \quad \text{as } \theta \to \frac{\pi}{2}.\end{aligned} \tag{2.2.63}$$

At a point $\bar{\theta}$ where $\phi_\theta(\bar{\theta}) = \beta$ we have

$$\begin{aligned}
A(\bar{\theta}) &:= (p-1)\phi_{\theta\theta}(\bar{\theta})\left((p-1)\tan\phi(\bar{\theta}) + \cot\phi(\bar{\theta})\right)\\
&= \left(1 + \left(\frac{N-2}{\Lambda(\beta)-\beta}\right)^2 \cot^2\bar{\theta}\right)\beta(\beta-\Lambda(\beta)) + (N-2)(1+\cot^2\bar{\theta})\\
&= \beta(\beta-\Lambda(\beta)) + N - 2 - \left(\frac{(N-2)^2}{\Lambda(\beta)-\beta} + 2 - N\right)\cot^2\bar{\theta}\\
&= (2-p)(\beta+1)\left(\beta - \frac{N-2}{p-2}\right) - \frac{N-2}{\Lambda(\beta)-\beta}\left(\beta(N-1) - \Lambda(\beta)\right)\cot^2\bar{\theta}.
\end{aligned} \tag{2.2.64}$$

If (2.2.62) does not hold, there exists $\epsilon > 0$ such that $\phi_\theta > \beta$ on $[\frac{\pi}{2} - \epsilon, \frac{\pi}{2})$. Therefore there exists $\bar{\theta} \in (0\frac{\pi}{2})$ such that $\phi_\theta(\bar{\theta}) = \beta$ and $\phi_{\theta\theta}(\bar{\theta}) \geq 0$. Hence $0 \leq A(\bar{\theta}) < 0$, a contradiction. Thus (2.2.62) holds.

Next, if $\beta < \frac{N-2}{p-2}$, it follows from (2.2.63) that there exists $\epsilon > 0$ such that $\phi_\theta < \beta$ on $[\frac{\pi}{2} - \epsilon, \frac{\pi}{2})$. Therefore, if (2.2.61) does not hold, there exists $0 < \theta_1 < \theta_2 \leq \frac{\pi}{2} - \epsilon$ such that $\phi_\theta(\theta_1) = \phi_\theta(\theta_2) = \beta$ and $\phi_{\theta\theta}(\theta_2) \leq 0 \leq \phi_{\theta\theta}(\theta_1)$. Since

$$A(\theta_i) = (2-p)(\beta+1)\left(\beta - \frac{N-2}{p-2}\right) - \frac{N-2}{\Lambda(\beta)-\beta}\left(\beta(N-1) - \Lambda(\beta)\right)\cot^2\theta_i,$$

we have $A(\theta_2) \leq 0 \leq A(\theta_1)$ and

$$A(\theta_2) - A(\theta_1) = \frac{N-2}{\Lambda(\beta)-\beta}\left(\beta(N-1) - \Lambda(\beta)\right)(\cot^2\theta_1 - \cot^2\theta_2 > 0,$$

since the function cot is decreasing. Contradiction, thus (2.2.61) holds.

Finally, if $\beta = \frac{N-2}{p-2}$ and (2.2.61) does not hold, the maximum of $\phi_\theta$ on $[0, \frac{\pi}{2})$ is larger than $\beta$ and achieved at some $\theta_1 \in (0, \frac{\pi}{2})$. Then there exists $\bar{\theta} \in (0, \frac{\pi}{2})$ such that $\phi_\theta(\bar{\theta}) = \beta$ and $\phi_{\theta\theta}(\bar{\theta}) \geq 0$. In this case,

$$0 \leq A(\bar{\theta}) = -\frac{N-2}{\Lambda(\beta)-\beta}(\beta(N-1) - \Lambda(\beta))\cot^2\bar{\theta} < 0,$$

which is again a contradiction, and (2.2.61) holds in all the cases.
*End of the proof.* Because $r^2 = \beta^2\omega^2 + \omega_\theta^2$ and $r_\theta = r(\phi_\theta - \beta)\tan\phi$, we derive

$$rr_\theta = (\beta^2\omega + \omega_{\theta\theta})\omega_\theta = r^2(\phi_\theta - \beta)\tan\phi.$$

By (2.2.61) $\beta^2\omega + \omega_{\theta\theta} \geq 0$ and since $\omega_\theta \leq 0$,

$$\frac{\beta^2\omega + \omega_{\theta\theta}}{\beta^2\omega^2 + \omega_\theta^2}\omega_\theta^2\omega \geq 0 \Longrightarrow \int_0^{\frac{\pi}{2}} \frac{\beta^2\omega + \omega_{\theta\theta}}{\beta^2\omega^2 + \omega_\theta^2}\omega_\theta^2\omega \cos\theta \sin^{N-2}\theta d\theta > 0. \quad (2.2.65)$$

This implies the right-hand part of inequalities (2.2.54) (i)-(ii). Jointly with (2.2.4) (i) we obtain the result in (ii). □

### 2.2.3 *The case $p \to \infty$*

#### 2.2.3.1 *The spherical infinite harmonic eigenvalue problem*

The infinite Laplacian $\Delta_\infty$ admits the following form in spherical coordinates in $\mathbb{R}^N$,

$$-\nabla |\nabla u|^2 .\nabla u(x) = -\left(u_r^2 + \frac{1}{r^2}|\nabla' u|^2\right)_r - \frac{1}{r^2}\langle\nabla'\left(u_r^2 + \frac{1}{r^2}|\nabla' u|^2\right), \nabla' u\rangle. \quad (2.2.66)$$

Then separable solutions under the form $u(x) = u(r,\sigma) = r^{-\beta}\omega(\sigma)$ may exist provided $\omega$ satisfies

$$\frac{1}{2}\nabla' |\nabla'\omega|^2 .\nabla\omega + \beta(2\beta+1)|\nabla'\omega|^2\omega + \beta^3(\beta+1)\omega^3 = 0, \quad (2.2.67)$$

on $S^{N-1}$, an expression which can be deduced directly from the expression (2.2.2). If $\omega > 0$ we set $\omega = e^{-\beta v}$ and $v$ satisfies

$$-\frac{1}{2}\nabla' |\nabla' v|^2 .\nabla v + \beta|\nabla' v|^4 + (2\beta+1)|\nabla' v|^2 + \beta + 1 = 0. \quad (2.2.68)$$

The next result [Bidaut-Véron, Garcia-Huidobro, Véron (2017)] shows that there exist separable infinite harmonic functions in $\mathbb{R}^N \setminus \{0\}$.

**Theorem 2.2.10** *Assume $S \subsetneq S^{N-1}$ is a $C^2$ domain. Then for any $\gamma > 0$ there exist $\lambda(\gamma) > 0$ and a positive function $v \in C^1(S)$ viscosity solution*

$$\begin{aligned} -\frac{1}{2}\nabla' |\nabla' v|^2 .\nabla v + \gamma|\nabla' v|^4 + (2\gamma+1)|\nabla' v|^2 + \lambda(\gamma) = 0 \quad &\text{in } S \\ \lim_{\rho(\sigma)\to 0} v(\sigma) = \infty. & \end{aligned} \quad (2.2.69)$$

*Proof.* The proof is a delicate adaptation of Theorem 2.2.3 and we indicate only the two main estimates. For $\epsilon, \theta > 0$ we introduce the uniformly elliptic problem

$$-\epsilon\Delta' v - \frac{1}{2}\nabla' \left|\nabla' v\right|^2 . \nabla v + \gamma \left|\nabla' v\right|^4 + (2\gamma+1)\left|\nabla' v\right|^2 + \theta v = 0 \quad \text{in } S$$
$$\lim_{\rho(\sigma)\to 0} v(\sigma) = \infty, \tag{2.2.70}$$

and set

$$\mathcal{P}_{\epsilon,\gamma} v = -\epsilon\Delta' v - \frac{1}{2}\nabla' \left|\nabla' v\right|^2 . \nabla v + \gamma \left|\nabla' v\right|^4 + (2\gamma+1)\left|\nabla' v\right|^2 + \theta v.$$

If a solution $v$ of (2.2.70) depends only on the distance $\rho$ to $\partial S$, in a neighborhood $\mathcal{N}_{\rho_0} = \{\sigma \in S : \rho(\sigma) < \rho_0\}$ of $\partial S$, then it satisfies

$$-\epsilon v'' + (N-2)Hv' - v'^2 v'' + \gamma v'^4 + (2\gamma+1)v'^2 + \theta v = 0 \quad \text{in } \mathcal{N}_{\rho_0}$$
$$\lim_{\rho\to 0} v(\rho) = \infty, \tag{2.2.71}$$

where $H$ is the mean curvature of the submanifold $S_\rho = \{\sigma \in S : \rho(\sigma) = \rho\}$.
*Step 1: Construction of supersolutions and subsolutions.* If $a$ is a positive parameter, we introduce the singular ODE

$$\begin{aligned} \epsilon y'' + y'^2 y'' - \gamma y'^4 - a y'^2 = 0 \qquad & \text{in } (0,1) \\ \lim_{\rho\to 0} y'(\rho) = -\infty. \end{aligned} \tag{2.2.72}$$

It can be completely integrated in $y'$ (which is negative). If we set $y' = -\zeta^{-1}$, there holds

$$\frac{\epsilon\zeta}{a} + \left(\frac{1}{\gamma} - \frac{\epsilon}{a}\right)\sqrt{\frac{\gamma}{a}} \arctan\left(\zeta\sqrt{\frac{a}{\gamma}}\right) = \rho, \tag{2.2.73}$$

provided $a > \gamma\epsilon$. Since $0 \le \arctan x \le x$ for $x \ge 0$, we infer

$$\zeta(\rho) \ge \gamma\rho \Longleftrightarrow 0 \ge y'(\rho) \ge -\frac{1}{\gamma\rho}. \tag{2.2.74}$$

Using this estimate, we derive from (2.2.73)

$$\zeta(\rho) \le \sqrt{\frac{\gamma}{a}} \tan(\rho\sqrt{a\gamma}) \Longleftrightarrow y'(\rho) \le -\sqrt{\frac{a}{\gamma}} \cot(\rho\sqrt{a\gamma}). \tag{2.2.75}$$

From this estimate we obtain that for $\rho_0 \in (0,1)$,

$$\frac{1}{\gamma}\ln\left(\frac{\sin\rho_0\sqrt{a\gamma}}{\sin\rho\sqrt{a\gamma}}\right) \le y(\rho) - y(\rho_0) \le \frac{1}{\gamma}\ln\left(\frac{\rho_0}{\rho}\right) \quad \forall \rho \in (0,\rho_0]. \tag{2.2.76}$$

Assume $(N-2)|H| \leq m$ in $\mathcal{N}_{\rho_0}$. Then for all $\tau \in (0, \rho_0]$, any positive solution $v_\tau$ of (2.2.71) in $\mathcal{N}_\tau$ vanishing for $\rho = \tau$ satisfies

$$\begin{aligned}\mathcal{P}_{\epsilon,\gamma}(v_\tau) &= \epsilon(N-2)Hv'_\tau + (2\gamma+1-a)v'^2_\tau + \theta v_\tau \\ &\geq |v'_\tau|\left((2\gamma+1-a)|v'_\tau|\right) - m) \\ &\geq |v'_\tau|\left((2\gamma+1-a)\sqrt{\frac{a}{\gamma}}\cot(\tau\sqrt{a\gamma}) - m\right).\end{aligned}$$

We choose $a = a_1 \in (1, 2\gamma+1)$ and $\tau = \tau_1 \in (0, \rho_0)$ such that

$$(2\gamma+1-a_1)\sqrt{\frac{a_1}{\gamma}}\cot(\tau\sqrt{a\gamma}) \geq m.$$

Then $v_\tau = v_{\tau,a_1}$ is a supersolution in $\mathcal{N}_\tau$. We choose now $a = a_2 > 2\gamma+1$, then

$$\begin{aligned}\mathcal{P}_{\epsilon,\gamma}(v_\tau) &\leq |v'_\tau|\left((2\gamma+1-a)|v'_\tau|\right) + m) + \theta v_\tau \\ &\leq |v'_\tau|\left(m - (a-2\gamma-1)\sqrt{\frac{a}{\gamma}}\cot(\tau\sqrt{a\gamma})\right) + \theta v_\tau.\end{aligned}$$

If we fix $a_2 = 4\gamma + 2 - a_1$, then

$$m - (a-2\gamma-1)\sqrt{\frac{a}{\gamma}}\cot(\tau\sqrt{a\gamma}) \leq -c < 0 \qquad \forall \tau \in (0, \tau_1],$$

which implies that $v_\tau = v_{\tau,a_2}$ satisfies

$$\mathcal{P}_{\epsilon,\gamma}(v_\tau) \leq v_\tau\left(\theta + \frac{v'_\tau}{v_\tau}\right).$$

Therefore there exists $\theta_0$ such that for any $0 < \theta \leq \theta_0$ there holds

$$\mathcal{P}_{\epsilon,\gamma}(v_\tau) \leq -\nu v_\tau \qquad \text{in } \mathcal{N}_{\tau_1}, \tag{2.2.77}$$

for some $\nu > 0$ depending on $\theta_0$. Hence $v_{\tau,a_1}$ and $v_{\tau,a_2}$ are respectively supersolutions and subsolutions of $\mathcal{P}_{\epsilon,\gamma}(v) = 0$ in $\mathcal{N}_{\tau_1}$ and we extend them to whole $S$ by $v^*_{\tau,a_1}$ and $v^*_{\tau,a_2}$ in order $\left|\mathcal{P}_{\epsilon,\gamma}(v^*_{\tau,a_j})\right|$ remain bounded by some $M_1 > 0$ in $S \backslash \mathcal{N}_{\tau_1}$. Thus there exists $M > 0$ independent of $\epsilon$ and $\theta$ such that $\overline{v} = v^*_{\tau,a_1} + \theta^{-1}M_1$ and $\underline{v} = v^*_{\tau,a_2} - \theta^{-1}M_1$ are respectively supersolutions and subsolutions of $\mathcal{P}_{\epsilon,\gamma}(v) = 0$ in $S$. As in the proof of Theorem 2.2.3, for $h > 0$ small enough, we replace $\underline{v}$ and $\overline{v}$ by $\underline{v}_h(\sigma) = \underline{v}(\dot{\rho}(\sigma) + h)$ and $\overline{v}_h(\sigma) = \overline{v}(\rho(\sigma) - h)$ which are respectively bounded subsolution in $S$ and supersolutions in $S \setminus \mathcal{N}_h$. By the maximum principle any $C^2$ solution $v$ of problem (2.2.70) is larger than $\underline{v}_h$ in $S$ and dominated by $\overline{v}_h$ in $S \setminus \mathcal{N}_h$.

Letting $h$ to zero we infer that there exists $M > 0$ independent of $\epsilon, \delta$ such that

$$\begin{aligned}
&(i) \qquad v^*_{\tau,a_2} - \frac{M}{\theta} \leq v \leq v^*_{\tau,a_1} + \frac{M}{\theta} \qquad \text{in } S \\
&(ii) \qquad \left| v^*_{\tau,a_2} + \frac{1}{\gamma} \ln \rho \right| \leq M \ \text{ and } \left| v^*_{\tau,a_1} + \frac{1}{\gamma} \ln \rho \right| \leq M.
\end{aligned} \tag{2.2.78}$$

*Step 2: Gradient estimates.* We claim that any $C^3$ solution of $\mathcal{P}_{\epsilon,\gamma}(v) = 0$ in the geodesic ball $B_R(\sigma_0)$ satisfies

$$|\nabla' v(\sigma_0)| \leq \frac{c}{\gamma R}, \tag{2.2.79}$$

where $c = c(N) > 0$. We set $z = |\nabla v|^2$ and define formally the linearized operator of $\Delta_\infty$ at $v$ by

$$B_v(h) = \frac{d}{dt}\Delta_\infty(v + th)\lfloor_{t=0} = \frac{1}{2}\langle \nabla' v, \nabla' z\rangle_g + \langle \nabla'(\langle \nabla' v, \nabla' h\rangle_g), \nabla' v\rangle_g, \tag{2.2.80}$$

and

$$\mathcal{L}_v(h) = B_v(h) + \epsilon \Delta h. \tag{2.2.81}$$

If $h = z$ we get

$$\mathcal{L}_v(z) = \frac{1}{2}\left|\nabla' z\right|^2 + \langle \nabla'(\langle \nabla' v, \nabla' z\rangle_g), \nabla' v\rangle_g + \epsilon \Delta z.$$

Since the equation satisfied by $v$ can be written

$$\frac{1}{2}\langle \nabla' v, \nabla' z\rangle_g = \gamma z^2 + (2\gamma + 1)z + \theta v - \epsilon \Delta v,$$

we obtain, by taking the gradient and multiplying by $\nabla' v$,

$$\frac{1}{2}\langle \nabla' v, \nabla'(\langle \nabla' v, \nabla' z\rangle_g)\rangle_g = (2\gamma z + 2\gamma + 1)\langle \nabla' z, \nabla' v\rangle_g + \theta z - \epsilon\langle \nabla'(\Delta v), \nabla' v\rangle_g.$$

Since $Ricc\lfloor_{S^{N-1}} = (N-2)g_0$, Weinszenböck's formula reads

$$\frac{1}{2}\Delta z = \left|D^2 v\right|^2 + \langle \nabla'(\Delta v), \nabla' v\rangle_g + (N-2)z.$$

This yields, after some computation and using $(\Delta v)^2 \leq (N-1)\left|D^2 v\right|^2$,

$$\begin{aligned}\mathcal{L}_v(z) = \frac{1}{2}\left|\nabla' z\right|^2 + 2\left((2\gamma z + 2\gamma + 1)\langle\nabla' z, \nabla' v\rangle_g + 2\theta z\right) \\ + 2\epsilon\left|D^2 v\right|^2 + 2\epsilon(N-2)z \\ \geq \frac{1}{2}\left|\nabla' z\right|^2 + \frac{\epsilon}{N-1}(\Delta w)^2 + 4\gamma^2 z^3 - c_0,\end{aligned} \tag{2.2.82}$$

for some $c_0 = c_0(N,\gamma) > 0$. If $\xi \in C_c^2(\overline{B}_R(\sigma_0))$, $Z = \xi^2 z$ satisfies

$$\mathcal{L}_v(Z) = \xi^2\mathcal{L}_v(z) + z\mathcal{L}_v(\xi^2) + 2\langle\nabla' v, \nabla'\xi^2\rangle_g\langle\nabla' v, \nabla' z\rangle_g + 2\epsilon\langle\nabla' z, \nabla'\xi^2\rangle_g, \tag{2.2.83}$$

where

$$\begin{aligned}\mathcal{L}_v(\xi^2) &= \frac{1}{2}\langle\nabla' z, \nabla'\xi^2\rangle_g + \langle\nabla' v, \nabla'(\langle\nabla' v, \nabla'\xi^2\rangle_g)\rangle_g + \epsilon\Delta\xi^2 \\ &\geq \langle\nabla' z, \nabla'\xi^2\rangle_g - z\left|D^2\xi^2\right| + \epsilon\Delta\xi^2.\end{aligned}$$

Next we choose $\xi$ such that $0 \leq \xi \leq 1$ and $R\left|\nabla\xi\right| + R^2\left|D^2\xi\right| \leq c_1$, then

$$(z+2\epsilon)\left|\langle\nabla' z, \nabla'\xi^2\rangle_g\right| \leq c_1\frac{(z+2\epsilon)\xi}{R}\left|\nabla' z\right| \leq \frac{\xi^2}{8}\left|\nabla' z\right|^2 + c_2\frac{(z+2\epsilon)^2}{R^2},$$

$$\left|z\left(\epsilon\Delta\xi^2 - z\left|D^2\xi^2\right|\right)\right| \leq \frac{c_3(z+2\epsilon)^2}{R^2},$$

$$\left|\langle\nabla' v, \nabla' z\rangle_g\right| \leq \sqrt{z}\left|\nabla' z\right|,$$

$$\left|\langle\nabla' v, \nabla'\xi^2\rangle_g\right| \leq 2\xi\left|\langle\nabla' v, \nabla'\xi\rangle_g\right| \leq \frac{2c_1\xi\sqrt{z}}{R},$$

and

$$\left|2\langle\nabla' v, \nabla'\xi^2\rangle_g\langle\nabla' v, \nabla' z\rangle_g\right| \leq \frac{4c_1\xi z\left|\nabla' z\right|}{R} \leq \frac{\xi^2}{8}\left|\nabla' z\right|^2 + c_3\frac{z^2}{R^2}.$$

At a point $\sigma_1 \in B_R(\sigma_0)$ where $Z$ achieves its maximum, we have $\mathcal{L}_v(z)(\zeta_0) \leq 0$, which implies

$$\xi^2\left(\frac{1}{2}\left|\nabla' z\right|^2 + \frac{\epsilon}{N-1}(\Delta v)^2 + 4\gamma^2 z^3 - c_0\right) \leq \frac{\xi^2}{4}\left|\nabla' z\right|^2 + \frac{c_4(z+2\epsilon)^2}{R^2}.$$

Imposing $R \leq 1$ and $\epsilon < \frac{1}{2}$, we multiply by $\xi^4$ and derive

$$\frac{1}{4}\left|\xi^3\nabla' z\right|^2 + \frac{\epsilon\xi^6(\Delta v)^2}{N-1} + 4\gamma^2(\xi^2 z)^3 \leq \frac{c_5\left((\xi^2 z)^2 + 1\right)}{R^2} + c_0.$$

Because

$$4\gamma^2(\xi^2 z)^3 \leq \frac{c_5(\xi^2 z)^2}{R^2} + \frac{c_6}{R^2},$$

we derive, if we assume $\xi(\sigma_0) = 1$,

$$z(\sigma_0) \leq Z(\sigma_1) \leq \frac{c_7}{R^2} \quad \text{with } c_7 = \max\left\{c_6, \frac{c_5}{\gamma}\right\},$$

which is (2.2.79).

The remaining of the proof of the theorem is based upon the existence of an ergodic constant and follows of the ideas developed in [Porretta, Véron (2009)]. □

From this result it is possible to derive the existence of separable infinite harmonic functions in the cone $C_S$ generated by $S$. The next result is the extension of Theorem 2.2.1 in the case $p = \infty$.

**Theorem 2.2.11** *Let $S \subsetneq S^{N-1}$ be a $C^2$ domain. Then there exist two real numbers $\beta_s > 0$ and $\beta'_s < 0$ and at least two positive Lipschitz continuous functions defined in $\overline{S}$, $\omega_s$ and $\omega'_s$ such that $(r,\sigma) \mapsto u(r,\sigma) := r^{-\beta_s}\omega_s(\sigma)$ and $(r,\sigma) \mapsto u'(r,\sigma) : r^{-\beta'_s}\omega'_s(\sigma)$ are infinite harmonic in $C_S$ and vanish on $\partial C_S \setminus \{0\}$. The two functions $\omega_s$ and $\omega'_s$ are viscosity solutions of (2.2.67) in $S$ with zero boundary value on $\partial S$ where $\beta = \beta_s$ and $\beta = \beta'_s$ respectively. Furthermore $\beta_s$ and $\beta'_s$ are unique.*

*Remark.* 1- Uniqueness of $\omega_s$ and $\omega'_s$, up to homothety is unknown.

2- Separable infinite harmonic functions can be generated as in Proposition 2.2.8.

2.2.3.2 *The 1-dim equation*

If $N = 2$, (2.2.67) becomes

$$-\omega_{\theta\theta}\omega_\theta^2 = \beta^3(\beta+1)\omega^3 + (2\beta+1)\omega_\theta^2\omega, \tag{2.2.84}$$

in $S^1 \sim (0, 2\pi)$. This equation can be completely integrated by putting $Y = \frac{\omega_\theta}{\omega}$, see [Bhattacharya (2005)]. The next result characterize the separable infinite harmonic functions in $\mathbb{R}^2 \setminus \{0\}$.

**Theorem 2.2.12** *For any $k \in \mathbb{N}^*$ set $\beta_k = \frac{k^2}{1+2k}$ and $\tilde{\beta}_k = \frac{k^2}{2k-1}$. There exist two $\frac{\pi}{k}$ anti-periodic function $\omega_k$ and $\tilde{\omega}_k$, positive on $(0, \frac{\pi}{k})$, such that*

*$x \mapsto |x|^{-\beta_k}\omega_k(\frac{x}{|x|})$ and $x \mapsto |x|^{\tilde\beta_k}\tilde\omega_k(\frac{x}{|x|})$ are infinite-harmonic respectively in $\mathbb{R}^2 \setminus \{0\}$ and $\mathbb{R}^2$. The functions $\omega_k$ and $\tilde\omega_k$ are unique up to the multiplication by a constant.*

*Remark.* 1- It is noticeable that the exponents $\beta_k$ or $\tilde\beta_k$ are rational numbers, contrary to the case $p \in (1,\infty) \setminus \{2\}$ where they are quadratic real numbers.

2- It is proved in [Aronson (1984)] that the function

$$(x_1, x_2) \mapsto x_1^{\frac{4}{3}} - x_2^{\frac{4}{3}},$$

is infinite harmonic in $\mathbb{R}^2$. It corresponds to $k = 2$, $\tilde\beta_k = \frac{4}{3}$ and $\tilde\omega_2(\theta_1) = (\cos\theta_1)^{\frac{4}{3}} - (\sin\theta_1)^{\frac{4}{3}}$, $\theta \in [0, 2\pi]$.

Contrary to the case $p < \infty$, equation (2.2.84) is also valid for separable infinite harmonic functions in a cone with vertex 0 generated by a spherical annulus or a spherical cap for functions $\omega$ satisfying some symmetry. If we use the representation of $S^{N-1}$ (2.2.50) we define for $0 \leq \kappa \leq \alpha \leq \pi$ the spherical annulus $S_{\kappa,\alpha}$ centered at the north pole $\sigma_N$ by

$$\begin{aligned} S_{\kappa,\alpha} = \{\sigma \approx (\theta_1, ..., \theta_{N-1}) : 0 \leq \theta_1 < 2\pi,\ 0 \leq \theta_j < \pi \\ \text{if } 2 \leq j \leq N-2 \text{ and } \kappa < \theta_{N-1} < \alpha\}, \end{aligned} \tag{2.2.85}$$

and the spherical cap $S_\alpha$

$$\begin{aligned} S_{\alpha} = \{\sigma \approx (\theta_1, ..., \theta_{N-1}) : 0 \leq \theta_1 < 2\pi,\ 0 \leq \theta_j < \pi \\ \text{if } 2 \leq j \leq N-2 \text{ and } 0 \leq \theta_{N-1} < \alpha\}. \end{aligned} \tag{2.2.86}$$

The following results are proved in [Bidaut-Véron, Garcia-Huidobro, Véron (2017)].

**Proposition 2.2.13** *Assume $0 \leq \kappa \leq \alpha \leq \pi$ and set $\nu = \frac{1}{2}(\alpha - \kappa)$. Then there exist two positive separable infinite hamonic functions in the cone $C_{S_{\kappa,\alpha}}$ under the form $u(r,\sigma) = r^{-\beta}\omega(\theta_{N-1})$ vanishing on $\partial C_{S_{\kappa,\alpha}} \setminus \{0\}$ where $\omega$ is a solution of (2.2.84) in $(\kappa, \alpha)$ with $\theta = \theta_{N-1}$ satisfying*

$$\omega(0) = \omega(\alpha) = 0, \tag{2.2.87}$$

*with $\beta = \beta_{S_{\kappa,\alpha}}$ or $\beta = \beta'_{S_{\kappa,\alpha}}$, where*

$$\beta_{S_{\kappa,\alpha}} = \frac{\pi^2}{4\nu(\pi+\nu)} \quad \textit{and } \beta_{S_{\kappa,\alpha}} = -\frac{\pi^2}{4\nu(\pi-\nu)}. \tag{2.2.88}$$

*The exponents $\beta_{S_{\kappa,\alpha}}$ and $\beta'_{S_{\kappa,\alpha}}$ are unique and the solutions $\omega$ of (2.2.84), (2.2.87) unique up to an homothety.*

**Proposition 2.2.14** *Assume $0 < \alpha \leq \pi$. Then there exist two positive separable infinite hamonic functions in the cone $C_{S_\alpha}$ under the form $u(r,\sigma) = r^{-\beta}\omega(\theta_{N-1})$ vanishing on $\partial C_{S_\alpha} \setminus \{0\}$ where $\omega$ is a solution of (2.2.84) in $(0,\alpha)$ with $\theta = \theta_{N-1}$ satisfying*

$$\omega_\theta(0) = 0 \text{ and } \omega(\alpha) = 0, \tag{2.2.89}$$

*with $\beta = \beta_{S_{\kappa,\alpha}}$ or $\beta = \beta'_{S_{\kappa,\alpha}}$ if $\alpha < \pi$, where*

$$\beta_{S_{\kappa,\alpha}} = \frac{\pi^2}{4\alpha(\pi+\alpha)} \quad \text{and} \quad \beta_{S_{\kappa,\alpha}} = -\frac{\pi^2}{4\alpha(\pi-\alpha)}. \tag{2.2.90}$$

*The exponents $\beta_{S_\alpha}$ and $\beta'_{S_\alpha}$ are unique and the solutions $\omega$ of (2.2.84), (2.2.89) unique up to an homothety.*

*Remark.* It is noticeable that in the case $\alpha = \pi$ there exists a separable infinite hamonic function $u_D$ in $\mathbb{R}^N \setminus D$ where $D = \{\lambda \mathbf{e}_N : \lambda < 0\}$ under the form

$$u_D(r,\sigma) = r^{-\frac{1}{8}}\omega(\theta_{N-1}) \qquad r > 0\,,\ 0 \leq \theta_{N-1} < \pi, \tag{2.2.91}$$

where $\omega$ is computable (not easy) since $Y = \frac{8\omega_\theta}{3\omega}$ satisfies

$$\frac{8Y^3}{3Y^4 - 6Y^2 - 1} = \tan\theta.$$

## 2.3 Boundary singularities of $p$-harmonic functions

### 2.3.1 *Boundary singular solutions*

In all this section we assume that $p > 1$ and $\Omega$ is a $C^2$ domain with $0 \in \partial\Omega$ and we consider equations with singular potential and shift of the type

$$-\Delta_p u + |\nabla u|^{p-2} \langle c(x), \nabla u\rangle + d(x)\,|u|^{p-2}\,u = 0 \qquad \text{in } \Omega. \tag{2.3.1}$$

We assume that $c$, $d$ are measurable and locally bounded functions in $\overline{\Omega}\setminus\{0\}$. In many applications to boundary singularity problems we assume that $c$ and $d$ satisfy

$$|x|\,|c(x)| + |x|^p\,|d(x)| \leq C_0 \qquad \forall x \in \overline{\Omega} \setminus \{0\}, \tag{2.3.2}$$

for some $C_0 \geq 0$.

The following Carleson estimate is an extension of Proposition 1.5.12.

**Proposition 2.3.1** *Let $R_0 > 0$, $u \in C^1(\overline{\Omega} \setminus \{0\})$ be a positive solution of (2.3.2) which vanishes on $\partial\Omega \setminus \{0\} \cap B_{2R_0}$ and $c$, $d$ satisfy (2.3.2). Then there exist positive constants $c_1$ and $R_1 < R_0$ depending on $p$, $C_0$ and $\Omega$ such that*

$$\frac{1}{c_1}\frac{u(x)}{\rho(x)} \leq \frac{u(y)}{\rho(y)} \leq c_1 \frac{u(x)}{\rho(x)}, \tag{2.3.3}$$

*for all $x, y \in \Omega$ such that $\frac{1}{2}|x| \leq |y| \leq 2|x| \leq R_1$.*

As a consequence we obtain the classical boundary Harnack inequality.

**Corollary 2.3.2** *Let $R_0 > 0$. If $u, \tilde{u} \in C^1(\overline{\Omega} \setminus \{0\})$ are positive solutions of (2.3.2) vanishing on $\partial\Omega \setminus \{0\} \cap B_{2R_0}$ and $c, d$ satisfy (2.3.2). Then there exist positive constants $c_2 = c_2(\Omega, p, C_0)$ and $R_1 = R_1(\Omega, p, C_0) < R_0$ such that for any $a \in \partial\Omega \setminus \{0\} \cap B_{R_1}$ there holds*

$$\frac{1}{c_2}\frac{u(x)}{\tilde{u}(x)} \leq \frac{u(y)}{\tilde{u}(y)} \leq c_2 \frac{u(x)}{\tilde{u}(x)} \qquad \forall x, y \in \Omega \cap B_{\frac{2|a|}{3}}(a). \tag{2.3.4}$$

As a variant of (2.3.4) it is easy to derive from (2.3.3) that for any $a \in \partial\Omega \setminus \{0\}$ there holds,

$$\frac{1}{c_3}\frac{u(x)}{\rho(x)} \leq \frac{u(y)}{\rho(y)} \leq c_3 \frac{u(x)}{\rho(x)} \qquad \forall x, y \in \Omega \cap B_{\frac{2|a|}{3}}(a). \tag{2.3.5}$$

If we combine this result with Harnack inequality we infer,

**Corollary 2.3.3** *Let $u, \tilde{u} \in C^1(\overline{\Omega} \setminus \{0\})$ be positive solutions of (2.3.2) which vanishes on $\partial\Omega \setminus \{0\}$ and $c, d$ satisfy (2.3.2). Then there exist positive constants $c_4$ and $r_0$, depending on $\Omega$ and $p$ such that for all $r \in (0, r_0]$ there holds*

$$\sup\left\{\frac{u(x)}{\tilde{u}(x)} : x \in \Omega \cap \Gamma_r\right\} \leq c_4 \inf\left\{\frac{u(x)}{\tilde{u}(x)} : x \in \Omega \cap \Gamma_r\right\}, \tag{2.3.6}$$

*where $\Gamma_r = \overline{B}_r \setminus B_{\frac{r}{2}}$.*

*Proof.* Since $c, d$ satisfy (2.3.2) the classical Harnack inequality stipulates that for any $b \in \Omega$, any $\theta > 1$ and $s > 0$ such that $\overline{B}_{\theta s}(b) \subset \Omega$ and any nonnegative solution $v$ of (2.3.2), there holds

$$v(x) \leq cv(y) \qquad \forall x, y \in B_s(b),$$

where $c = c(p, N, C_0, \theta) > 0$. Thus

$$\frac{u(x)}{\tilde{u}(x)} \leq c^2 \frac{u(y)}{\tilde{u}(y)} \qquad \forall x, y \in B_s(b).$$

A simple geometric construction shows that there exist $r_0 > 0$ and $k_0 \in \mathbb{N}_*$ such that for any $r \in (0, r_0]$ one can find at most $k_0 + 1$ points $b_j \in \overline{\Omega}$ such that $|b_j| = \frac{3r}{4}$, $b_0, b_{k_0} \in \partial\Omega$, $|b_j - b_{j+1}| \leq \frac{r}{4}$ for $1 \leq j < k_0$, $|b_0 - b_1| = |b_{k_0-1} - b_{k_0}| \geq \frac{r}{2}$, $\rho(b_j) \geq \frac{5r}{12}$ and

$$\Gamma_r \subset \overline{B}_{\frac{r}{2}}(b_0) \bigcup \overline{B}_{\frac{r}{2}}(0) \bigcup_{j=1}^{b_0-1} \overline{B}_{\frac{r}{3}}(b_j).$$

We set $r_j = \frac{r}{3}$ if $j = 1, ..., b_0 - 1$, $r_0 = r_{b_0} = \frac{r}{2}$. If $x, y \in \Gamma_r$, there exist indices $j_x$, $j_y$ in $\{0, 1, ..., b_0\}$ such that $x \in B_{r_{j_x}}(b_{j_x})$ and $y \in B_{r_{j_y}}(b_{j_y})$. Therefore

$$\frac{u(x)}{\tilde{u}(x)} \leq c^{2k_0} \frac{u(y)}{\tilde{u}(y)}. \tag{2.3.7}$$

□

**Definition 2.3.4** Let $\Omega$ be bounded. We say that $d \in L^\infty_{loc}(\overline{\Omega} \setminus \{0\})$ satisfies the *comparison hypothesis* if for any $\epsilon > 0$ and any two solutions $u, \tilde{u} \in C^1(\overline{\Omega} \setminus \{0\})$ of (2.3.1) there holds

$$u \geq \tilde{u} \quad \text{in } (\partial\Omega \cap B^c_\epsilon) \cup (\Omega \cap \partial B_\epsilon) \Longrightarrow u \geq \tilde{u} \quad \text{in } \Omega \cap B^c_\epsilon. \tag{2.3.8}$$

This assumption is satisfied if $d \geq 0$. If $d$ is signed and satisfied (2.3.2) the comparison hypothesis holds provided $|x|^p \, d_-(x)$ is small enough. The main tool for estimating positive singular solutions is the following

**Theorem 2.3.5** *Assume that $\Omega$ is bounded and $d$ satisfies the comparison hypothesis. If $u, \tilde{u} \in C^1(\overline{\Omega} \setminus \{0\})$ are nonzero nonnegative functions satisfying (2.3.1) with $c = 0$ and vanishing on $\partial\Omega \setminus \{0\}$, there exists $\kappa > 0$ such that $\tilde{u} \leq \kappa u$.*

*Proof.* Since a nonnegative solution is either positive or identically zero by Harnack inequality, $u$ and $\tilde{u}$ are positive in $\Omega$. Let us assume that no such $\kappa$ exists. Then for any $n \in \mathbb{N}^*$ there exists $x_n \in \Omega$ such that $\tilde{u}(x_n) \geq nu(x_n)$. Up to a subsequence $x_n \to \xi \in \overline{\Omega}$. Clearly $\xi \in \Omega$ is impossible because $u$ and $\tilde{u}$ are continuous and positive in $\Omega$. We denote by $\xi_n$ the projection of

$x_n$ on the boundary. We first assume that $\xi \in \partial\Omega \setminus \{0\}$ thus $\xi_n \neq 0$ for $n$ large enough. Then

$$\frac{\tilde{u}(x_n) - \tilde{u}(\xi_n)}{\rho(x_n)} \geq n \frac{u(x_n) - u(\xi_n)}{\rho(x_n)}.$$

Since

$$\frac{\tilde{u}(x_n) - \tilde{u}(\xi_n)}{\rho(x_n)} \to -\frac{\partial \tilde{u}(\xi)}{\partial \mathbf{n}} > 0 \text{ and } \frac{u(x_n) - u(\xi_n)}{\rho(x_n)} \to -\frac{\partial u(\xi)}{\partial \mathbf{n}} > 0,$$

by Hopf lemma, we derive a contradiction. At the end we assume that $\xi = 0$ and we set $r_n = |x_n|$. By Corollary 2.3.3 there holds

$$n \leq \frac{\tilde{u}(x_n)}{u(x_n)} \leq c_4 \min \left\{ \frac{\tilde{u}(x)}{u(x)} : x \in \Omega\,,\ |x| = r_n \right\}.$$

By the local comparison theorem, it yields $\tilde{u} \geq nu$ in $\Omega \cap B^c_{r_n}$ which leads to a contradiction. □

**Proposition 2.3.6** *Under the assumption of Theorem 2.3.5 any nonnegative unbounded solution $u \in C^1(\overline{\Omega} \setminus \{0\})$ of (2.3.1) tends to infinity at 0 in every closed cone $C_0$ with vertex 0 such that $C_0 \cap B_{r_0} \subset \Omega \cup \{0\}$.*

*Proof.* Actually we prove a stronger result, namely

$$\lim_{x \to 0} \frac{|x| u(x)}{\rho(x)} = \infty. \tag{2.3.9}$$

Assume it is not true, then there exists a sequence $\{x_n\}$ converging to 0 and a constant $m > 0$ such that $r_n u(x_n) \leq \rho(x_n) m$. By Corollary 2.3.3

$$\sup \left\{ \frac{|x| u(x)}{\rho(x)} : |x| = r_n \right\} \leq c_4 m,$$

where $r_n = |x_n|$. Thus $u(x) \leq cm$ for all $x \in \Omega$ such that $|x| = r_n$. By the maximum principle $u(x) \leq cm$ for all $x \in \Omega$ such that $r_n \leq |x| \leq r_1$. Letting $n$ to infinity yields a contradiction. □

Such a function $u$ is called a *boundary singular solution.* In the next theorem we prove that whenever it is nonempty the set of positive boundary singular solutions is a half straight line.

**Theorem 2.3.7** *Assume that the assumptions of Theorem 2.3.5 are fulfilled and there exists a positive boundary singular solution $u$. Assume also* *either* $1 < p \leq 2$ *and* $d \leq 0$ *or* $p \geq 2$ *and*

$$\liminf_{x \to 0} \frac{\rho(x)\,|\nabla u(x)|}{u(x)} > 0. \tag{2.3.10}$$

*Then any other positive boundary singular solution is proportional to* $u$.

*Proof.* We assume that $v$ is a nonzero nonnegative singular solution. Then there exists $k > 0$ such that $v \leq ku := u_k$. For $\epsilon > 0$ we define $\kappa_\epsilon$ by

$$\kappa_\epsilon = \inf\{c > 0 : v(x) \leq cu(x) : \ \forall x \in \Omega, |x| = \epsilon\}.$$

Thus $v \leq \kappa_\epsilon u$ on $|x| = \epsilon$. Since $d$ satisfies the comparison hypothesis, $v \leq \kappa_\epsilon u$ in $\Omega \cap B^c_\epsilon$, the mapping $\epsilon \mapsto \kappa_\epsilon$ is nonincreasing with limit $\kappa$ as $\epsilon \to 0$ and $v \leq \kappa u$ in $\Omega$. We proceed by contradiction in assuming $v \neq \kappa u$ in $\Omega$.

Step 1: *We claim that* $v < \kappa u$ *in* $\Omega$. Because of (2.3.10) there exists $\delta > 0$ such that $\nabla u$ never vanishes in $\Omega \cap B_\delta$. We first prove that $v(x) < \kappa u(x) := u_\kappa(x)$ in $\Omega \cap B_\delta$. If it is not true the contact set $\Xi$ of the graphs of $v$ and $u_\kappa$ over $\Omega \cap B_\delta$ is not empty. The function $w = u_\kappa - v$ satisfies

$$\mathcal{L}w + Dw = 0, \tag{2.3.11}$$

where $\mathcal{L}$ is a linear elliptic operator and $D = \frac{u_\kappa^{p-1} - v^{p-1}}{w}$. There exists a neighborhood of any $\xi \in \Xi$ where $\mathcal{L}$ is not degenerate and $D$ is bounded. By the strong maximum principle $w$ is identically zero in this neighborhood. Therefore the set $\Xi$ is open. This implies $\Xi = \Omega \cap B_\delta$ and in particular $v = \kappa u$ in $\Omega \cap \partial B_\delta$. Since uniqueness follows from the comparison hypothesis, equality holds in whole $\Omega$, a contradiction. Up to replacing $\delta$ by a smaller value, we have in particular $v(x) < \kappa u(x)$ for all $x$ in $\Omega \cap \partial B_\delta$. Let $\xi \in \partial\Omega \cap B_\delta$. Then $\frac{\partial u_\kappa}{\partial \mathbf{n}}(\xi)$ and $\frac{\partial v}{\partial \mathbf{n}}(\xi)$ are both negative. If $p \geq 2$, $D$ is bounded. Thus Hopf boundary lemma applies to $w$ wich satisfies $\frac{\partial w}{\partial \mathbf{n}}(\xi) < 0$. If $1 < p < 2$ and $d$ is nonpositive, $w$ is $\mathcal{L}$-superharmonic, thus we still get $\frac{\partial w}{\partial \mathbf{n}}(\xi) < 0$. Because $v(x) < \kappa u(x)$ on $\Omega \cap \partial B_\delta$ and $\frac{\partial u_\kappa}{\partial \mathbf{n}} < \frac{\partial v}{\partial \mathbf{n}}$ on $\partial\Omega \cap \partial B_\delta$ there exists $\kappa' < \kappa$ such that $v(x) \leq \kappa' u(x)$ on $\Omega \cap \partial B_\delta$. By the comparison principle it implies $v(x) \leq \kappa' u(x)$ in $\Omega \cap B^c_\delta$. Since we can replace $\delta$ by a smaller $\delta' \in (0, \delta)$ we conclude that the inequalities $v < \kappa u$ and $\frac{\partial u_\kappa}{\partial \mathbf{n}} < \frac{\partial v}{\partial \mathbf{n}}$ hold in $\Omega$ and $\partial\Omega \setminus \{0\}$ respectively.

Step 2: *End of the proof.* By the definition of $\kappa$, there exists a sequence

$\{x_n\}$ converging to 0 such that

$$\frac{v(x_n}{u(x_n)} = \max\left\{\frac{v(x}{u(x)} : |x| = |x_n| := r_n\right\} \to \kappa \quad \text{as } n \to \infty.$$

Set $a_n = \max\{u(x) : |x| = r_n\}$ and $\ell_n = -r_n\mathbf{n}_0$ where $\mathbf{n}_0$ is the normal outward unit vector to $\partial\Omega$ at 0. For $n$ large enough $\rho(\ell_n) \leq r_n \leq \frac{5}{4}\rho(\ell_n)$. By Proposition 2.3.1

$$\frac{1}{c_1}\frac{u(y)}{\rho(y)} \leq \frac{u(\ell_n)}{\rho(\ell_n)} \leq \frac{a_n}{\rho(\ell_n)} \qquad \forall y \in \Omega \text{ s.t. } \frac{r_n}{2} \leq |y| \leq 2r_n,$$

which implies

$$\max\left\{u(y) : y \in \Omega \cap (\overline{B}_{2r_n} \setminus B_{\frac{r_n}{2}})\right\} \leq \frac{5ca_n}{2}. \tag{2.3.12}$$

We define $u_n(x) = a_n^{-1}u(r_nx)$ and $v_n(x) = a_n^{-1}v(r_nx)$ and set $d_n(x) = r_n^p d(r_nx)$ for $x \in \Omega^{r_n} := r_n^{-1}\Omega$. Then $u_n$ and $v_n$ are positive solutions of

$$-\Delta_p w + d_n w^p = 0 \qquad \text{in } \Omega_n, \tag{2.3.13}$$

and they vanish on $\Omega_n \setminus \{0\}$. By (2.3.12) both $u_n$ and $v_n$ are uniformly bounded in $\tilde{\Omega}_n := \Omega_n \cap (\overline{B}_2 \setminus B_{\frac{1}{2}})$. By the regularity results of Section 1-3, there exists a subsequence $\{n_j\}$ such that $u_{n_j}$ and $v_{n_j}$ converge in the $C^1_{loc}$ topology in $\Omega_n \cap (\overline{B}_2 \setminus B_{\frac{1}{2}}$ to some functions $U$ and $V$ and $d_{n_j}$ converges in the weak-star topology of $L^\infty_{loc}(\Omega_n \cap (\overline{B}_2 \setminus B_{\frac{1}{2}})$ to some $\tilde{d}$ and $U$ and $V$ are solutions of

$$-\Delta_p W + \tilde{d}W^p = 0 \qquad \text{in } H \cap B_2 \setminus \overline{B}_{\frac{1}{2}}.$$

Noticing that

$$|\nabla u_n(x)| = \frac{r_n\,|\nabla u(r_nx)|}{a_n} = \frac{\rho(r_nx)\,|\nabla u(r_nx)|}{u(r_nx)}, \frac{r_nu(r_nx)}{a_n\rho(r_nx)},$$

and since by Corollary 2.3.3

$$\frac{r_nu(r_nx)}{a_n\rho(r_nx)} \geq \frac{r_n}{a_nc_4}\frac{u(x_n)}{\rho(x_n)} = \frac{|x_n|}{c_4\rho(x_n)} \geq \frac{1}{c_4},$$

it follows from (2.3.10) that

$$|\nabla u_n(x)| \geq \frac{\gamma}{c_4} \quad \forall x \in H \cap B_2 \setminus \overline{B}_{\frac{1}{2}}. \tag{2.3.14}$$

Set $\xi_n = \frac{x_n}{r_n}$ and

$$\min\left\{\frac{v(x}{u(x)} : |x| = r_n\right\} = m_n \leq \kappa.$$

We can assume that $\xi_n \to \xi \in H \cap S^{N-1}$ and $m_n \to m \leq \kappa$. Furthermore $V(\xi) = \kappa U(\xi)$ and, if $\xi \in \partial H$,

$$\frac{\partial V}{\partial \mathbf{n}}(\xi) = \kappa \frac{\partial U}{\partial \mathbf{n}}(\xi) < 0.$$

The non-degeneracy inequality (2.3.14) implies that the strong maximum principle or Hopf boundary lemma are valid. Thus $V$ and $\kappa U$ coincide in a neighborhood of $\xi$. By connectedness the coincidence set of $V$ and $\kappa U$ is the whole set $H \cap B_2 \setminus \overline{B}_{\frac{1}{2}}$ which yields $m = \kappa$. Therefore for any $\epsilon > 0$ there exists $n_\epsilon \in \mathbb{N}$ such that $n \geq n_\epsilon$ implies

$$(\kappa - \epsilon)u(x) \leq v(x) \leq \kappa u(x) \qquad \forall x \in \Omega \text{ s.t. } |x| = r_n.$$

The comparison hypothesis implies $(\kappa-\epsilon)u(x) \leq v(x)$ in $\Omega \cap B_{r_n^c}$ and finally, letting $r_n \to 0$ and $\epsilon \to 0$, $v = \kappa u$. □

*Remark.* The existence of positive $p$-harmonic functions with a boundary singularity at 0 can be seen as a particular case of Theorem 3.3.11 in Chapter 3, with a geometric constraint on the domain $\Omega$ when $1 < p < N$, since we have to assume that $\Omega$ is contain in a half-space and that $\partial\Omega \cap B_\delta = \partial H \cap B_\delta$ for some $\delta > 0$.

We end this section in proving an *a priori* estimate valid for positive solutions of (2.3.1) with a boundary singularity.

**Theorem 2.3.8** *Assume $p > 1$, $\Omega$ is bounded and $c, d$ satisfy (2.3.2). Then there exist $\theta > 0$, $r_0 > 0$ and $c = c(\Omega, p)$ such that if $A \in \Omega$ verifies $\rho(A) \geq r_0$, any positive solution $u \in C^1(\overline{\Omega} \setminus \{0\})$ of (2.3.1) satisfies*

$$u(x) \leq cu(A)\rho(x)|x|^{-\theta-1} \qquad \forall x \in \Omega. \tag{2.3.15}$$

*Proof.* We recall that for $\beta > 0$, $\Omega_\beta = \{x \in \Omega : \rho(x) \leq \beta\}$, $\Omega'_\beta = \Omega \setminus \Omega_\beta$. There exists $r_0 > 0$ such that every $x \in \Omega_\beta$ admits a unique projection on $\partial\Omega$ denoted by $\sigma(x)$, thus $\rho(x) = |x - \sigma(x)|$. If $a \in \partial\Omega$ and $t \leq r_0$ we set $N_t(a) = a - t\mathbf{n}_a$ where $\mathbf{n}_a$ is the normal outward unit vector at $a$. Then $\rho(N_t(a)) = t$ and $\sigma(N_t(a)) = a$. We put $A = -r_0\mathbf{n}_0$, $A_1 = -2^{-1}r_0\mathbf{n}_0$,..., $A_k = -2^{-k}r_0\mathbf{n}_0$.

Step 1: *Estimate in* $B_{r_0}$. By Proposition 2.3.1 there holds

$$\frac{u(x)}{\rho(x)} \leq c_1 \frac{u(A_k)}{\rho(A_k)} \qquad \forall x \in \overline{B}_{2^{-k}r_0} \setminus B_{2^{-k-1}r_0}, \tag{2.3.16}$$

and in particular

$$\frac{u(A_{k+1})}{\rho(A_{k+1})} \leq c_1 \frac{u(A_k)}{\rho(A_k)} \leq c_1^k \frac{u(A)}{r_0} \quad \forall k = 1, \ldots. \tag{2.3.17}$$

Therefore

$$\frac{u(x)}{\rho(x)} \leq c_1 \frac{u(A_k)}{\rho(A_k)} \leq c_1^k \frac{u(A)}{r_0} \qquad \forall x \in \overline{B}_{2^{-k}r_0} \setminus B_{2^{-k-1}r_0}. \tag{2.3.18}$$

If $x \in \overline{B}_{r_0}$, there exists $k \in \mathbb{N}$ such that $2^{-k-1}r_0 < |x| \leq 2^{-k}r_0$. Thus

$$\log_2\left(\frac{r_0}{2|x|}\right) < k \leq \log_2\left(\frac{r_0}{|x|}\right).$$

Combined with (2.3.18) it yields

$$\frac{u(x)}{\rho(x)} \leq \left(\frac{r_0}{|x|}\right)^{\frac{\ln c}{\ln 2}} \frac{u(A)}{r_0} \qquad \forall x \in \overline{B}_{r_0} \cap \Omega. \tag{2.3.19}$$

Step 2: *Estimate in* $\Omega'_{r_1}$. We assume that $0 < r_1 \leq \frac{r_0}{3}$. Since $\Omega$ is bounded, there exists an integer $k_1$ such that any two points $x$ and $y$ in $\Omega'_{r_1}$ can be joined by a connected chain of at most $k_0$ closed balls $B_{\frac{r_1}{2}}(b_j)$ of radius $\frac{r_1}{2}$ and center $b_j$. Since $B_{\frac{2r_1}{3}}(b_j) \subset \Omega$, we have by Harnack inequality

$$u(x) \leq c^{k_0} u(y).$$

This implies in particular

$$u(x) \leq c^{k_1} u(A). \tag{2.3.20}$$

Step 3: *Estimate in* $\Omega_{r_1} \setminus B_{r_0}$. If $x \in \Omega_{r_1} \setminus B_{r_0}$, there exists $a \in \partial\Omega$ such that $|x - a| \leq r_1$. Since $r_1 \leq \frac{r_0}{3}$, then $|a| > \frac{2r_0}{3}$. Hence, by (2.3.5) and (2.3.20),

$$\frac{u(x)}{\rho(x)} \leq c_3 \frac{u(a - r_1\mathbf{n}_a)}{r_0} \leq \frac{c_3 c^{k_1}}{r_0} u(A). \tag{2.3.21}$$

Combining (2.3.19), (2.3.20) and (2.3.21) we obtain (2.3.15). □

### 2.3.2 *Rigidity results for singular p-harmonic functions in the half-space*

Let $1 < p \leq N$ and $u \in C^1(\overline{\mathbb{R}}^N_+ \setminus \{0\})$ be a positive solution of

$$\begin{aligned} -\Delta_p u &= 0 && \text{in } \mathbb{R}^N_+ \\ u &= 0 && \text{on } \partial\mathbb{R}^N_+ \setminus \{0\}. \end{aligned} \tag{2.3.22}$$

By Theorem 2.2.1 there exists a positive solution of (2.3.22) under the form

$$V_{\mathbb{R}^N_+}(r,\sigma) = r^{-\beta_{S^{N-1}_+}} \omega_{S^{N-1}_+}(\sigma) \qquad \forall (r,\sigma) \in \mathbb{R}_+ \times S^{N-1}_+. \tag{2.3.23}$$

The next results are extensions of Theorem 2.3.7 when the domain is a half-space. We prove that under some growth assumptions, all the positive $p$-harmonic functions vanishing $\partial\mathbb{R}^N_+ \setminus \{0\}$ are multiple of $V_{\mathbb{R}^N_+}$.

**Theorem 2.3.9** *Assume $p > 1$ and $u \in C^1(\overline{\mathbb{R}}^N_+ \setminus \{0\})$ is a nonnegative $p$-harmonic function in $\mathbb{R}^N_+$ which vanishes on $\partial\mathbb{R}^N_+ \setminus \{0\}$ and such that*

$$(i) \limsup_{x\to 0} |x|^{\beta_{S^{N-1}_+}} u(x) < \infty \text{ and } (ii) \lim_{|x|\to\infty} |x|^{-1} u(x) = 0. \tag{2.3.24}$$

*Then there exists $k \geq 0$ such that $u = kV_{\mathbb{R}^N_+}$.*

*Proof. Step 1: We claim that $u(x) \to 0$ when $|x| \to \infty$.* We recall that $\beta'_{S^{N-1}_+} = -1$ and $x \mapsto W(x) := x_N$ is the regular nonnegative separable $p$-harmonic function in $\mathbb{R}^N_+$ which vanishes on $\partial\mathbb{R}^N_+$. We put $\beta_+ := \beta_{S^{N-1}_+}$. By (2.3.4)

$$u(x) \leq c_2 \frac{u(|x|\,\mathbf{e}_N)}{W(|x|\,\mathbf{e}_N)} W(x) = c_2 \frac{u(|x|\,\mathbf{e}_N)}{|x|} x_N \quad \text{if } |x| \geq 1.$$

This implies that for any $\epsilon > 0$ there exists $R_\epsilon > 0$ such that for all $R \geq R_\epsilon$,

$$u(x) \leq \sup_{|y|=1} u(y) + \epsilon x_N \qquad \forall 1 \leq |x| \leq R.$$

Letting $R \to \infty$ and $\epsilon \to 0$ yields

$$u(x) \leq \sup_{|y|=1} u(y) = C \quad \forall |x| \geq 1.$$

The proof of Lemma 1.5.7 admits several generalizations and Lemma 3.3.2 and Lemma 3.4.15 in next section provide very useful and general ones.

However it is easy to see, using the usual scaling techniques and regularity results of Section 1-3, that there holds for some $\alpha \in (0,1)$,

$$\begin{aligned} &(i) \qquad |\nabla u(x)| \leq c(N,p)C|x|^{-1} \\ &(ii) \qquad |\nabla u(x) - \nabla u(x')| \leq c(N,p,\alpha)C|x|^{-1-\alpha}\,|x-x'|^{\alpha}, \end{aligned} \tag{2.3.25}$$

for all $x, x'$ such that $2 \leq |x| \leq |x'|$. If $\{r_n\}$ is an increasing sequence tending to infinity, we set $u_n(x) = u(r_n x)$. Then $0 \leq u_n \leq C$ in $\mathbb{R}^N_+ \cap B_{r_n^{-1}}$, and from (2.3.25),

$$\begin{aligned} &(i) \qquad |\nabla u_n(x)| \leq c(N,p)C|x|^{-1} \\ &(ii) \qquad |\nabla u_n(x) - \nabla u_n(x')| \leq c(N,p,\alpha)C|x|^{-1-\alpha}\,|x-x'|^{\alpha}, \end{aligned} \tag{2.3.26}$$

if $2r_n^{-1} \leq |x| \leq |x'| \leq 2\,|x|$. The sequence $\{u_n\}$ is relatively compact in the $C^1_{loc}$ topology of $\mathbb{R}^N_+ \setminus \{0\}$ and bounded by $C$. Up to a subsequence it converges in this topology to a bounded nonnegative $p$-harmonic function $Z$ in $\mathbb{R}^N_+$ which vanishes on $\partial\mathbb{R}^N_+ \setminus \{0\}$. Since $Z$ is bounded its possible singularity at 0 is removable (this can be seen by performing a reflexion through $\partial\mathbb{R}^N_+$). By Theorem 1.5.13, $Z(x) = ax_N$ for some $a \geq 0$. The boundedness of $Z$ implies $a = 0$. Therefore $u(x) \to 0$ when $|x| \to \infty$.

*Step 2: End of the proof.* We apply again boundary Harnack inequality to $u$ and $V_{\mathbb{R}^N_+}$ in $B_1^+ \setminus \{0\}$. Because of (2.3.24)-(i), we see that there exists $C' > 0$ such that

$$u(x) \leq C' V_{\mathbb{R}^N_+}(x), \tag{2.3.27}$$

for all $x \in B_1^+ \setminus \{0\}$. Comparing $u$ and $(C' + \epsilon)V_{\mathbb{R}^N_+} + \epsilon$ in $B_R^+$ for $\epsilon > 0$ and $R \geq R_\epsilon$ large enough, we derive from the previous estimate, Step 1 and the comparison principle that $u \leq (C' + \epsilon)V_{\mathbb{R}^N_+} + \epsilon$. Letting $\epsilon \to 0$ implies that (2.3.27) holds in $\mathbb{R}^N_+$.

Finally, we use a variant of Lemma 1.5.7 valid if $\mu_p$ is replaced by $|.|^{-\beta}$, and obtain the following estimates with $c = c(N,p,\alpha) > 0$:

$$\begin{aligned} &(i) \qquad |\nabla u(x)| \leq c|x|^{-\beta-1} \\ &(ii) \qquad |\nabla u(x) - \nabla u(x')| \leq c|x|^{-\beta-1-\alpha}\,|x-x'|^{\alpha}, \end{aligned} \tag{2.3.28}$$

for all $x, x' \in \mathbb{R}^N_+$ such that $|x| \leq |x'|$. Let $k \geq 0$ be the supremum of the $c \leq C'$ such that (2.3.27) holds with $c$ instead of $C'$. We can assume that $k > 0$ otherwise $u = 0$. If the graphs of the two functions $u$ and $kV_{\mathbb{R}^N_+}$ are tangent at some point $x_0 \in \overline{\mathbb{R}}^N_+ \setminus \{0\}$, it follows from the strong maximum principle if $x_0 \in \mathbb{R}^N_+$ or Hopf lemma if $x_0 \in \partial\mathbb{R}^N_+ \setminus \{0\}$ that $u = kV_{\mathbb{R}^N_+}$. Notice that $\nabla V_{\mathbb{R}^N_+}$ never vanishes. If the two graphs are not tangent in $\overline{\mathbb{R}}^N_+ \setminus \{0\}$, they are asymptotically tangent either at 0 or at infinity.

In the first case there exist two sequences $\{k_n\}$ decreasing to $k$ and $\{x_n\}$ converging to 0 such that

$$\frac{u(x_n)}{V_{\mathbb{R}^N_+}(x_n)} = k_n.$$

We set $r_n = |x_n|$ and $u_n(x) = r_n^\beta u(r_n x)$. Then $u_n$ is positive and $p$-harmonic in $\mathbb{R}^N_+$, vanishes on $\partial\mathbb{R}^N_+ \setminus \{0\}$ and satisfies the same estimates (2.3.28) as $u$. Up to a subsequence $\frac{x_n}{r_n} \to \xi \in \overline{S}^{N-1}_+$ and $u_n \to U$ in the $C^1_{loc}(\overline{\mathbb{R}}^N_+ \setminus \{0\})$ topology, where $U$ is a positive $p$-harmonic function in $\mathbb{R}^N_+$ which vanishes on $\partial\mathbb{R}^N_+ \setminus \{0\}$ and such that $U \leq kV_{\mathbb{R}^N_+}$. Furthermore the graphs of $U$ and $kV_{\mathbb{R}^N_+}$ are tangent at $\xi \in \overline{S}^{N-1}_+$. As above it implies $U = kV_{\mathbb{R}^N_+}$. Then, for $\epsilon > 0$, there exists $n_\epsilon \in \mathbb{N}$ such that, for $n \geq n_\epsilon$,

$$(k-\epsilon)V_{\mathbb{R}^N_+}(x) \leq u(x) \leq (k+\epsilon)V_{\mathbb{R}^N_+}(x) \qquad \forall x \text{ s.t. } |x| = r_n.$$

By the maximum principle the above inequality holds for all $x$ such that $|x| \geq r_n$; notice that here we use also the fact that both $u$ and $V_{\mathbb{R}^N_+}$ vanish at infinity. Letting $n \to \infty$ yields $(k-\epsilon)V_{\mathbb{R}^N_+} \leq u \leq (k+\epsilon)V_{\mathbb{R}^N_+}$ which implies the claim in this case.

We are left with the case where the graphs of $u$ and $kV_{\mathbb{R}^N_+}$ are tangent only at infinity. It means that if we define $k_\ell$ by

$$k_\ell = \sup_{|x|\leq \ell} \frac{u(x)}{V_{\mathbb{R}^N_+}(x)} \qquad \forall \ell > 0,$$

then $k_\ell < k$ and it is a increasing function of $\ell$. Put

$$k' = \limsup_{x\to 0} \frac{u(x)}{V_{\mathbb{R}^N_+}(x)} = \lim_{n\to\infty} \frac{u(r_n\xi_n)}{V_{\mathbb{R}^N_+}(r_n\xi_n)} < k,$$

where $|\xi_n| = 1$, and the sequence $\{r_n\}$ decreases to 0 while $\xi_n \to \xi \in \overline{S}^{N-1}_+$. We set $u_n(x) = r_n^\beta u(r_n x)$. As above there exists a positive $p$-harmonic function $U$ in $\mathbb{R}^N_+$ which vanishes on $\partial\mathbb{R}^N_+ \setminus \{0\}$ such that $u_n \to U$ and $\xi_n = \frac{x_n}{r_n} \to \xi \in \overline{S}^{N-1}_+$. As in the first case it implies that $U = k'V_{\mathbb{R}^N_+}$ and for every $\epsilon > 0$ there exists $n_\epsilon \in \mathbb{N}$ such that for $n \geq n_\epsilon$,

$$(k'-\epsilon)V_{\mathbb{R}^N_+}(x) \leq u(x) \leq (k'+\epsilon)V_{\mathbb{R}^N_+}(x) \qquad \forall x \text{ s.t. } |x| = r_n.$$

Therefore this inequality holds for $|x| \geq r_n$ and thus in $\mathbb{R}^N_+$. This implies that $u = k'V_{\mathbb{R}^N_+}$ contradicting the tangency assumption. □

If $p = N$, $\beta_{S^{N-1}_+} = 1$, $\beta_{S^{N-1}_-} = -1$ and the result holds without the upper estimate at $x = 0$.

**Theorem 2.3.10** *Assume $p = N$ and $u \in C^1(\overline{\mathbb{R}}^N_+ \setminus \{0\})$ is a nonnegative $N$-harmonic function in $\mathbb{R}^N_+$ which vanishes on $\partial\mathbb{R}^N_+ \setminus \{0\}$ and satisfies*

$$\lim_{|x|\to\infty} |x|^{-1} u(x) = 0. \tag{2.3.29}$$

*Then there exists $k \geq 0$ such that $u = kV_{\mathbb{R}^N_+}$.*

*Proof.* We recall that $V_{\mathbb{R}^N_+}(x) = |x|^{-2} x_N$. As in the proof of Theorem 2.3.9-Step 1, $u(x) \to 0$ when $|x| \to \infty$. Set $\tilde{u}(y) = u(x)$ with $y = |x|^{-2} x$. Then $\tilde{u}$ is $N$-harmonic in $\mathbb{R}^N_+$ and vanishes on $\partial\mathbb{R}^N_+ \setminus \{0\}$. Furthermore $\tilde{u}(y) \to 0$ when $y \to 0$. By Theorem 1.5.13 $\tilde{u}(y) = ky_N$ for some $k \geq 0$. This implies the claim. □

*Remark.* In the case $p = 2$ the Kelvin transform $\tilde{u}(y) = |x|^{2-N} u(x)$ with $y = |x|^{-2} x$ leaves the Laplace equation invariant. Therefore the above proof can be adapted to this case. This yields the following: *any nonnegative harmonic function $u$ in $\mathbb{R}^N_+$ which vanishes on $\partial\mathbb{R}^N_+ \setminus \{0\}$ and satisfies (2.3.29) is a multiple of the Poisson kernel $x \mapsto |x|^{-N} x_N$.*

## 2.4 Notes and open problems

*2.4.1.* The uniqueness of the first eigenfunction of the p-Laplacian is due to Lindqvist [Lindqvist (1990-1992)]. Other proofs either involving some regularity conditions on the boundary or topological condition on $\Omega$ have been obtained by Anane [Anane (1987)], Barles [Barles (1988)] or Guedda and Véron [Guedda, Véron (1988)]. The uniqueness proof presented here is an adaptation of Belloni and Kawohl's argument [Belloni, Kawhol (2002)].

*2.4.2.* There exist geometric lower bounds on the first positive eigenvalue $\lambda_1(p, M)$ of a compact complete manifold. It is proved in [Matei (2000)] that if the Ricci curvature of $M$ satisfies $Ric_g \geq (1 - N)$ there exists a constant $c = c(k, p) > 0$ such that

$$\lambda_1(p, M) \leq c_{N,p}(h_M + h_M^p) \tag{2.4.1}$$

where $h_M$ is defined in (2.1.61). If the sectional curvature of $M$ is nonnegative, then

$$\lambda_1(p,M) \leq \frac{2^{2p+2}(N+p)^{N+p}}{(N-1)^{N-1}p^p(\mathrm{diam}(M))^p}. \tag{2.4.2}$$

*2.4.3.* The existence of the regular couple $(\beta'_S, \omega'_S)$ in Theorem 2.2.3 when $S$ is a smooth subdomain of $S^{N-1}$ is due to Tolksdorf [Tolksdorff (1983)]. His proof of existence is specific to the case the equation on a domain of$S^{N-1}$. An other proof, more flexible, is obtained by Porretta and Véron [Porretta, Véron (2009)].

*2.4.4.* If $S \subset S^{N-1}$ is a domain such that $c_{1,p}^{\mathbb{R}^{N-1}}(S^c) > 0$, [Gkikas, Véron (2017)] prove the existence of a singular couple $(\overline{\beta}_S, \omega_S)$ (and a regular couple $(\bar{\beta}'_S, \omega'_S)$ as well) in approximating $S$ by an increasing sequence of smooth subdomain $\{S_k\}$. They prove that a subsequence of normalized solutions $\{\overline{\omega}_k\}$ of the corresponding spherical $p$-harmonic eigenvalue problem in $S_k$ with $\beta = \kappa_k$ converges weakly in $W_0^{1,p}(S)$ and that $\{\nabla\overline{\omega}_k\}$ converges in $L^p_{loc}(S)$ to some solution $\overline{\omega} \in W_0^{1,p}(S)$ of (2.2.6), with $\overline{\beta} = \lim_{k\to\infty} \beta_k$.

If $S$ coincides with the interior of its closure it is possible to approximate $S$ from outside by a decreasing sequence of smooth subdomains $\{\underline{S}_k\}$. A subsequence of the corresponding spherical $p$-harmonic eigenvalue problem in $\underline{S}_k$ with $\beta = \underline{\beta}_k$ converges weakly in $W^{1,p}_{loc}(S)$ to some solution $\underline{\omega} \in W^{1,p}(S)$ of (2.2.2) in $S$ with $\beta =: \underline{\beta} = \lim_{k\to\infty} \underline{\beta}_k$. The fact that $\underline{\omega} \in W_0^{1,p}(S)$ is a consequence of Netrusov's theorem [Adams, Hedberg (1999), Th 10.1.1]. The exponents $\overline{\beta}$ and $\underline{\beta}$ are respectively the maximal and the minimal possible exponents corresponding to singular separable solutions in $C_S$. If $S$ is Lipschitz then $\overline{\beta} = \underline{\beta}$ and there exists $c > 0$ such that $c\overline{\omega} \leq \underline{\omega} \leq c^{-1}\overline{\omega}$. The fact that $\overline{\omega} = \lambda\underline{\omega}$ is an open problem. Deep results in connexion with the characterization of the Martin boundary for $p$-harmonic functions in a domain can be found in [Lewis, Nyström (2010)] and [Lewis (2012)].

*2.4.5.* It is proved by Ancona [Ancona (2012)] that the quotient of two positive harmonic functions in a cone $C_S$ generated by any spherical domain $S$, and vanishing on $\partial C_S$ is constant. As a consequence, any such harmonic function $u$ is separable under the form

$$u(r,\sigma) = \lambda r^{b_S}\omega'_S(\sigma), \tag{2.4.3}$$

where $\lambda_{1,S}$ is the first eigenvalue of $-\Delta'$ in $W_0^{1,2}(S)$, $b_S > 0$ depends explicitely of $\lambda_{1,S}$ and $\lambda$ is an arbitrary positive real number. When $p \neq 2$ it is conjectured that any positive $p$-harmonic function $u$ in $C^S$ vanishing on $\partial C_S$ is separable and under the form

$$u(r,\sigma) = \lambda r^{-\beta'_S}\omega'_S(\sigma). \tag{2.4.4}$$

If $u$ is singular at 0, $u$ it is conjectured that $u(r,\sigma) = \lambda r^{-\beta_S}\omega_S)$ at least in the case where the $\omega_S$ are unique up to homothety.

*2.4.6.* The eigenvalue problem for the infinite Laplacian is solved by [Juutinen, Lindqvist, Manfredi (1999)] and then extended by [Juutinen (2007)]. Local gradient estimates of infinite harmonic functions are obtained by [Evans, Smart (2011)] where it is proved the differentiability of the viscosity first eigenfunction; uniqueness is an open problem. The existence of separable infinite harmonic function in a cone is proved by [Bidaut-Véron, Garcia-Huidobro, Véron (2017)]. They prove existence of such solutions in *any* cone, regular or not.

*2.4.7.* When $p = 1$, existence of separable 1-harmonic functions in $\mathbb{R}^N \setminus \{0\}$ under the form $u(r,\sigma) = r^{-\beta}\omega(\sigma)$ leads to the spherical 1-harmonic eigenvalue problem on $S^{N-1}$ that satisfies $\omega$

$$-div'\frac{\nabla'\omega}{\sqrt{\beta^2\omega^2 + |\nabla'\omega|^2}} = (1-N)\beta\frac{\omega}{\sqrt{\beta^2\omega^2 + |\nabla'\omega|^2}} \tag{2.4.5}$$

on $S^{N-1}$. A necessary condition for existence of a solution in $S \subset S^{N-1}$ vanishing on $\partial S$ is $\beta < 0$. The construction of solutions seems an open problem.

*2.4.8.* Given a $d$- dimensional compact Riemannian manifold without boundary $(M^d, g)$, it is an open problem to prove the existence of couples $(\beta, \omega)$ where $\beta \in \mathbb{R}$ and $\omega \in C^1(M)$ which satisfies

$$-div_g\left(\left(\beta^2\omega^2 + |\nabla\omega|^2\right)^{\frac{p-2}{2}}\nabla\omega\right) = \beta\Lambda(\beta)\left(\beta^2\omega^2 + |\nabla\omega|^2\right)^{\frac{p-2}{2}}\omega \tag{2.4.6}$$

in $M$. If $\beta = 0$ any constant satisfies (2.4.6). If $\beta \neq 0$ any solution satisfies

$$\int_M \left(\beta^2\omega^2 + |\nabla\omega|^2\right)^{\frac{p-2}{2}}\omega dv_g = 0. \tag{2.4.7}$$

If there exist an involutive isometry $\tau$ of $M$ and two relatively open smooth domains $M'$ and $M'' = \tau(M')$ such that $\partial M' = \partial M'' = \{\sigma \in M : \tau(\sigma = \sigma)\}$ and $M = M \cup M' \cup \partial M'$, it is possible to construct such a couple in considering $(\beta, \omega')$ where $\omega'$ is a positive solution of (2.4.6) in $M'$ vanishing on $\partial M'$ and then by extending $\omega'$ to whole $M$ is putting

$$\omega(\sigma) = \begin{cases} \omega'(\sigma) & \text{if } \sigma \in \overline{M}' \\ -\omega' \circ \tau(\sigma) & \text{if } \sigma \in M''. \end{cases}$$

*2.4.9.* The study of $p$-Laplace equations perturbed by a potential of the following type

$$-\Delta_p u + d(x)|u^{p-1}|u = 0 \tag{2.4.8}$$

is a natural problem. When the potential is written $d(x) = D(\frac{x}{|x|})|x|^{-p}$ it is possible to look for singular separable solutions under the form $u(r, \sigma) = r^{-\beta}\omega(\sigma)$ and $\omega$ is a solution of

$$\begin{aligned} div'\left(\left(\beta^2\omega^2 + |\nabla'\omega|^2\right)^{\frac{p-2}{2}} \nabla'\omega\right) + \beta\Lambda_p(\beta)\left(\beta^2\omega^2 + |\nabla'\omega|^2\right)^{\frac{p-2}{2}} \omega \\ - D(\sigma)|\omega^{p-1}|\omega = 0. \end{aligned} \tag{2.4.9}$$

The study of this spherical equation in a domain $\Omega$ which boundary contains 0 could be a way to analyze the boundary behaviour of solutions of (2.4.8) vanishing on $\partial\Omega \setminus \{0\}$.

*2.4.10.* A similar problem is the research of separable solutions of

$$\begin{aligned} -\Delta_p u = 0 \qquad & \text{in } \mathbb{R}^N_+ \\ |\nabla u|^{p-2}\frac{\partial u}{\partial \mathbf{n}} + d(x)\,|u|^{p-2}\,u = 0 \qquad & \text{in } \partial\mathbb{R}^N_+ \setminus \{0\}, \end{aligned} \tag{2.4.10}$$

assuming that $d(x) = D(\frac{x}{|x|})|x|^{1-p}$. If $u(r, \sigma) = r^{-\beta}\omega(\sigma)$, then

$$\begin{aligned} div'\left(\left(\beta^2\omega^2 + |\nabla'\omega|^2\right)^{\frac{p-2}{2}} \nabla'\omega\right) & \\ + \beta\Lambda_p(\beta)\left(\beta^2\omega^2 + |\nabla'\omega|^2\right)^{\frac{p-2}{2}} \omega = 0 \ \text{ in } S^{N-1}_+ & \\ \left(\beta^2\omega^2 + |\nabla'\omega|^2\right)^{\frac{p-2}{2}} \langle\nabla'\omega, \mathbf{e}_N\rangle + D(\sigma)|\omega^{p-1}|\omega = 0 \ \text{ in } \partial S^{N-1}_+. & \end{aligned} \tag{2.4.11}$$

*2.4.11.* The construction of $p$-harmonic functions in a bounded $C^2$ domain with a boundary singularity when $N > p > 1$ and $\Omega$ is not flat near this singularity is still an open problem. We conjecture existence and uniqueness of such singular solutions (up to homothety). It could be also interesting to find a way different from super and sub solutions to construct such solutions. In this direction, it is worth noting that $V_{\mathbb{R}^N_+}(r,\sigma) := r^{-\beta}\omega(\sigma)$ (where $\beta = \beta_{S^{N-1}_+}$ and $\omega = \omega_{S^{N-1}_+}$) defined in $\mathbb{R}^N_+$ satisfies

$$PV\int_H \left|\nabla V_{\mathbb{R}^N_+}\right|^{p-2} \langle \nabla V_{\mathbb{R}^N_+}, \nabla\zeta\rangle dx = \mathcal{T}(\zeta), \tag{2.4.12}$$

where $PV$ stands for principal value, for all $\zeta \in C^1_0(\overline{\mathbb{R}}^N_+)$ such that

$$\zeta(x) = \zeta(r,\sigma) = r^{\Lambda(\beta)}\psi(\sigma) \quad \text{where } \psi \in C^1_0(\overline{S}^{N-1}_+),$$

(we recall that $\Lambda(\beta) = \beta(p-1) + p - N$) and where $\mathcal{T}$ is the functional defined by

$$\begin{aligned}\mathcal{T}(\zeta) &= \lim_{\epsilon\to 0}\int_{\partial B^+_\epsilon} \left|\nabla V_{\mathbb{R}^N_+}\right|^{p-2} \langle \nabla V_{\mathbb{R}^N_+}, \mathbf{e}_N\rangle \zeta dS \\ &= \int_{S^{N-1}_+} \left(\beta^2\omega^2 + |\nabla'\omega|^2\right)^{\frac{p-2}{2}} \left(\beta\omega\sigma.\mathbf{e}_N - \langle \nabla'\omega, \mathbf{e}_N\rangle_g\right)\psi d\sigma.\end{aligned} \tag{2.4.13}$$

In [Lewis (2012), Th 3.9] it is shown that in a Reifenberg-flat domain the set of positive boundary singular solutions is a half-line. The proof is based on a sharp form of boundary Harnack inequality for $p$-harmonic functions.

*2.4.12.* In [Bhattacharya (2005)] it is proved that all the nonnegative infinity harmonic functions in a half-space vanishing on the boundary except one point $a$ are comparable to the singular one $x \mapsto |x-a|^{-\frac{1}{3}}$. His proof is based upon Harnack inequality and boundary Harnack inequality. Other results concerning blow-up estimates of infinity harmonic functions with a boundary singularity are obtained in [Bhattacharya (2004)].

# Chapter 3

# Quasilinear equations with absorption

In this section we present several aspects of the isolated singularities problem for solutions of

$$-\Delta_p u + g(u, |\nabla u|) = 0, \tag{3.0.1}$$

either in a punctured domain $\Omega' := \Omega \setminus \{\text{one point}\}$ (*internal singularity*), or lying at the boundary $\partial\Omega \setminus \{\text{one point}\}$ of a domain $\Omega$ (*boundary singularity*). The function $g$ is either $|u|^{q-1}u$ or $|\nabla u|^q$ and the interest lies in the case where the reaction term $g(u, \nabla u)$ dominates the diffusion term $\Delta_p u$ by its growth at infinity, which means $q > p-1 > 0$. In the cases we consider here the reaction is called an absorption term. The common thread underlying the description is the existence of explicit separable solutions of (3.0.1), although in many cases the proof of their existence is far from simple. The second step relies on existence of *a priori* estimates of the possible blow-up near the singularity.

## 3.1 Singular solutions with power absorption in $\mathbb{R}^N$

Let $q > p-1 > 0$. Existence of separable solutions of

$$-\Delta_p u + |u|^{q-1}u = 0, \tag{3.1.1}$$

in $\mathbb{R}^N \setminus \{0\}$ under the form (in spherical coordinates) $u(x) = u(r,\sigma) = r^{-\beta}\omega(\sigma)$ yields $\beta = \beta_{p,q} = \frac{p}{q+1-p}$ and $\omega$ satisfies

$$\begin{aligned} -div'\left(\left(\beta_{p,q}^2\omega^2 + |\nabla'\omega|^2\right)^{\frac{p-2}{2}}\nabla'\omega\right) + |\omega|^{q-1}\omega \\ - \beta_{p,q}\Lambda(\beta_{p,q})\left(\beta_{p,q}^2\omega^2 + |\nabla'\omega|^2\right)^{\frac{p-2}{2}}\omega = 0 \qquad \text{in } S^{N-1}, \end{aligned} \tag{3.1.2}$$

where, with the notations of Chapter 2,

$$\beta_{p,q}\Lambda(\beta_{p,q}) = \beta_{p,q}((p-1)\beta_{p,q} + p - N) = \frac{p}{q+1-p}\left(\frac{pq}{q+1-p} - N\right). \tag{3.1.3}$$

A necessary condition for the existence of a solution to (3.1.2) is $\Lambda(\beta_{p,q}) > 0$ which is equivalent to

$$\begin{array}{lll} (i) & 0 < p-1 < q < q_1 := \frac{N(p-1)}{N-p} & \text{if } 1 < p < N \\ (ii) & 0 < p-1 < q & \text{if } p \geq N. \end{array} \tag{3.1.4}$$

**Lemma 3.1.1** *Assume $p - 1 < q < q_1$. Then equation (3.1.2) admits a unique positive solution*

$$\omega_c = \left(\beta_{p,q}^{p-1}\Lambda(\beta_{p,q})\right)^{\frac{1}{q+1-p}}. \tag{3.1.5}$$

*Proof.* Notice that $\beta_{p,q}^2\omega^2 + |\nabla'\omega|^2 > 0$, thus any solution is $C^2$. Then at the point of maximum $x_0$ (resp. minimum $x_1$) of a positive solution there holds $\omega(x_0) \leq \omega_c$ (resp. $\omega(x_0) \geq \omega_c$), thus $\omega \equiv \omega_c$. □

The restriction $q < q_1$ is crucial for the existence of solutions with isolated singularities. The exponent $q_1$ depends also on $p$, we note $q_{1,p} = q_1$ except when it is necessary to point out the dependence with respect to $p$.

### 3.1.1 *Removable singularities in* $\mathbb{R}^N$

The following *a priori* estimate due to [Vazquez (1981)] is the key of the description of singular solutions of (3.1.1).

**Lemma 3.1.2** *Let $q > p - 1 > 0$ and $u \in L^\infty_{loc}(B_{R_0}) \cap W^{1,p}_{loc}(B_{R_0})$ which satisfies, for some constants $a > 0$, $b \geq 0$,*

$$-\Delta_p u_+ + au_+^q \leq b \qquad in\ \mathcal{D}'(B_{R_0}). \tag{3.1.6}$$

*Then*

$$u(x) \leq \left(\frac{c}{aR_0^p}\right)^{\frac{1}{q+1-p}} + \left(\frac{b}{a}\right)^{\frac{1}{q}}, \tag{3.1.7}$$

*where $c = c(N, p, q) > 0$.*

*Proof.* Let $R < R_0$ and

$$v(x) = \lambda(R^{p'} - |x|^{p'})^{-\frac{p}{q+1-p}} + \mu.$$

Since $v(x) = v(r)$ is radial with respect to 0, a straightforward calculation yields

$$-\Delta_p v + av^q \geq b \qquad \text{in } B_R,$$

provided $a\mu^q = b$ and $a\lambda^{q+1-p} = c(N,p,q)R^{p'}$. By the comparison principle $v \geq u$ in $B_R$ and in particular

$$v(0) = \left(\frac{c(N,p,q)}{aR^p}\right)^{\frac{1}{q+1-p}} + \left(\frac{b}{a}\right)^{\frac{1}{q}}.$$

The result follows by letting $R \to R_0$. □

**Theorem 3.1.3** *Assume $\Omega$ is a domain containing 0, $g : \mathbb{R} \to \mathbb{R}$ is a continuous function satisfying*

$$\liminf_{r\to\infty} r^{-q_1} g(r) > 0, \quad \limsup_{r\to-\infty} |r|^{-q_1} g(r) < 0, \tag{3.1.8}$$

*and $u \in L^\infty_{loc}(\Omega \setminus \{0\}) \cap W^{1,p}_{loc}(\Omega \setminus \{0\})$ satisfies*

$$-\Delta_p u + g(u) = 0, \tag{3.1.9}$$

*in $\Omega \setminus \{0\}$. Then $u$ can be extended as a solution of (3.1.9) in $\Omega$.*

The proof is based upon the following result.

**Lemma 3.1.4** *Let $\Omega$ be a domain containing 0, $1 < p < N$ and $q \geq q_1$. If $u \in L^\infty_{loc}(\Omega \setminus \{0\}) \cap W^{1,p}_{loc}(\Omega \setminus \{0\})$ satisfies*

$$-\Delta_p u + au^q \leq b \qquad \text{a.e. in } \{x \in \Omega : u(x) \geq 0\}, \tag{3.1.10}$$

*for some constants $a > 0$, $b \geq 0$. Then $u^+ \in L^\infty_{loc}(\Omega)$*

*Proof.* Set $R_0 = \operatorname{dist}(0, \partial\Omega)$ and we can assume $R_0 > 1$. It follows from Lemma 3.1.2 that for any $x \in B_R$ there holds

$$u(x) \leq \left(\frac{c}{a|x|^p}\right)^{\frac{1}{q+1-p}} + \left(\frac{b}{a}\right)^{\frac{1}{q}}. \tag{3.1.11}$$

Let $\zeta \in C_0^\infty(\Omega)$, $\zeta \geq 0$, $\theta \in C^1(\mathbb{R}) \cap L^\infty(\mathbb{R})$ be a nondecreasing function, vanishing on $(-\infty, 0]$ and increasing on $[0, \infty)$ such that $0 \leq \theta \leq \operatorname{sign}_+$, and $\mu \geq \left(\frac{b}{a}\right)^{\frac{1}{q}}$, then

$$\int_\Omega \left(|\nabla u|^{p-2} \nabla u.\nabla(\zeta^p \theta(u-\mu)) + (au^q - b)\zeta^p \theta(u-\mu)\right) dx \leq 0.$$

Therefore

$$\int_{\Omega} \zeta^p |\nabla(u-\mu)|^p \theta'(u-\mu)dx$$
$$+p\int_{\Omega} \theta(u-\mu)\zeta^{p-1} |\nabla(u-\mu)|^{p-2} \nabla(u-\mu)\nabla\zeta dx \leq 0.$$

Using Hölder's inequality and letting $\theta$ to converge to $\mathrm{sign}_+$ yields,

$$\int_{\Omega} \zeta^p |\nabla(u-\mu)_+|^p dx \leq p\int_{\Omega} (u-\mu)_+^p |\nabla\zeta|^p dx. \tag{3.1.12}$$

Next, let $\{\eta_n\} \subset C^\infty(\Omega)$ be a sequence of *mollifiers*, i.e. nonnegative functions such that $0 \leq \eta_n \leq 1$,

$$\eta_n = \begin{cases} 0 & \text{if } |x| \leq \frac{1}{2n} \text{ or } |x| \geq \frac{2R_0}{3} \\ 1 & \text{if } \frac{1}{n} \leq |x| \leq \frac{R_0}{2}, \end{cases}$$

and $|\nabla\eta_n| \leq cn$. Let $\theta$ be as above and $\mu \geq \max\left\{ \left(\frac{b}{a}\right)^{\frac{1}{q}}, \max_{\frac{R_0}{2}\leq|x|\leq\frac{2R_0}{3}} u(x)\right\}$. We set $\Delta_=\{x : \frac{1}{2n} \leq |x| \leq \frac{1}{n}\}$ and

$$X_n = -\int_{\Omega} \theta(u-\mu)\, |\nabla u|^{p-2} \nabla u.\nabla\eta_n dx.$$

Then $|X_n| \leq c_1 \|\theta\|_{L^\infty} n^{1-\frac{N}{p}} \|\nabla(u-\mu)_+\|_{L^p(\Delta_n)}^{p-1}$ by Hölder's inequality, and therefore

$$0 \leq \int_{\Omega} \theta(u-\mu)(au^q - b)\eta_n dx \leq X_n.$$

Using (3.1.12) with $\zeta = \eta_n$ and the *a priori* estimate (3.1.7), we obtain

$$|X_n| \leq c_1 p n^{\frac{p^2-N}{p}} \|(u-\mu)_+\|_{L^p(\Delta_n)}^{p-1} \leq c n^{\frac{p^2-N}{p}} n^{\frac{p^2-p}{q+1-p} - \frac{N(p-1)}{p}}. \tag{3.1.13}$$

But

$$\frac{p^2-N}{p} + \frac{p^2-p}{q+1-p} - \frac{N(p-1)}{p} = \frac{pq}{q+1-p} - N$$
$$= \frac{pq}{q+1-p} - \frac{pq_1}{q_1+1-p} \leq 0,$$

hence $X_n$ remains bounded independently of $n$. Letting $\theta$ converge to $\text{sign}_+$ we derive

$$0 \le \int_{B_{\frac{R_0}{2}}} \text{sign}_+(u-\mu)(au^q - b)dx \le \liminf_{n\to\infty} |X_n| < \infty. \tag{3.1.14}$$

Therefore $u_+ \in L^q(B_{\frac{R_0}{2}})$. Notice that if $q > q_1$, $X_n \to 0$ when $n \to \infty$, which implies $u \le \mu$, however the proof of the upper bound on $u_+$ differs according the position of $q$ with respect to $p$.
In the case $q \ge p$, then

$$\begin{aligned}\|(u-\mu)_+\|^{p-1}_{L^p(\Delta_n)} &\le \|(u-\mu)_+\|^{p-1}_{L^q(\Delta_n)} |\Delta_n|^{\frac{(q-p)(p-1)}{pq}} \\ &\le c\,\|(u-\mu)_+\|^{p-1}_{L^q(\Delta_n)}\, n^{\frac{N(p-q)(p-1)}{pq}}.\end{aligned}$$

This implies $X_n \to 0$ and $u \le \mu$.
In the case $q < p$, then

$$\begin{aligned}\|(u-\mu)_+\|_{L^p(\Delta_n)} &\le \|(u-\mu)_+\|^{\frac{q_1}{p}}_{L^{q_1}(\Delta_n)} \|(u-\mu)_+\|^{1-\frac{q}{p}}_{L^\infty(\Delta_n)} \\ &\le c\,\|(u-\mu)_+\|^{p-1}_{L^q(\Delta_n)}\, n^{N(p-1)(\frac{1}{q}-\frac{1}{p})} \\ &\le c\,\|(u-\mu)_+\|^{p-1}_{L^q(\Delta_n)}\, n^{N(p-1)(\frac{1}{q}-\frac{1}{q_1})},\end{aligned}$$

by (3.1.13). Then $X_n \to 0$ and again $u \le \mu$. □

*Proof of Theorem 3.1.3.* If $g$ satisfies (3.1.8), then (3.1.6) holds for some constants $a > 0$ and $b \ge 0$. Thus $u_+ \in L^\infty_{loc}(\Omega)$ by Lemma 3.1.4. Similarly $u_- \in L^\infty_{loc}(\Omega)$. It is straightforward that $u \in W^{1,p}_{loc}(\Omega)$ by using the same test functions as above and the result follows by Theorem 1.3.9. □

Estimate (3.1.7) is a particular case of a more general result proved in [Vazquez (1981)].

**Proposition 3.1.5** *Assume $g : \mathbb{R} \mapsto \mathbb{R}_+$ is a continuous nondecreasing function such that $g(0) > 0$ which satisfies $\lim_{s\to-\infty} g(s) = 0$ and*

$$\int_a^\infty \frac{ds}{\sqrt[p]{J(s)}} < \infty \qquad \text{where } J(s) = \int_0^s g(t)dt, \tag{3.1.15}$$

*for some $a \ge 0$. There there exists a nonincreasing function $h : \mathbb{R}_+ \mapsto \mathbb{R}$ such that*

$$\lim_{t\to 0} h(t) = \infty \quad \text{and} \quad \lim_{t\to\infty} h(t) = -\infty, \tag{3.1.16}$$

*with the property that any function* $u \in W^{1,p}_{loc}(\Omega) \cap C(\Omega)$ *solution of*

$$-\Delta_p u + g(u) = 0 \qquad in\ \Omega, \tag{3.1.17}$$

*satisfies*

$$u(x) \leq h(\rho(x)) \qquad \forall x \in \Omega. \tag{3.1.18}$$

As a variant, the following statement holds.

**Proposition 3.1.6** *Assume* $g : \mathbb{R} \mapsto \mathbb{R}$ *is a continuous nondecreasing function such that* $g(0) = 0$ *which satisfies (3.1.15) and*

$$\int_{-\infty}^{a} \frac{ds}{\sqrt[p]{\tilde{J}(s)}} < \infty \qquad where\ \ \tilde{J}(s) = -\int_{s}^{0} g(t)dt, \tag{3.1.19}$$

*for some* $a \leq 0$. *There there exists a nonincreasing function* $\tilde{h} : \mathbb{R}_+ \mapsto \mathbb{R}_+$ *such that*

$$\lim_{t\to 0} \tilde{h}(t) = \infty \quad and \quad \lim_{t\to 0} \tilde{h}(t) = 0, \tag{3.1.20}$$

*with the property that any function* $u \in W^{1,p}_{loc}(\Omega) \cap C(\Omega)$ *solution of (3.1.17) satisfies*

$$|u(x)| \leq \tilde{h}(\rho(x)) \qquad \forall x \in \Omega. \tag{3.1.21}$$

### 3.1.2 *Classification of isolated singularities*

We introduce the class of functions $g$ continuous in $\mathbb{R} \times \mathbb{R}_+$ satisfying *the dissipativity assumption*

$$rg(r,s) \geq 0 \qquad \forall (r,s) \in \mathbb{R} \times \mathbb{R}_+, \tag{3.1.22}$$

and *the weak singularity assumption*: for all $(c,d) \in \mathbb{R} \times \mathbb{R}_+$:

$$\begin{aligned} &(i) \qquad \lim_{t\to\infty} t^{-\frac{N(p-1)}{N-p}} g(ct, dt^{\frac{N-1}{p-1}}) = 0 \quad \text{if } p \neq N \\ &(ii) \qquad \lim_{t\to\infty} e^{-Nt} g(ct, de^{t}) = 0 \qquad\quad \text{if } p = N, \end{aligned} \tag{3.1.23}$$

uniformly when $(c,d)$ remains bounded in $\mathbb{R}\times\mathbb{R}_+$. A variant of this assumption is *the integral form* of the weak singularity assumption:

$$\begin{array}{lll}(i) & \displaystyle\int_1^\infty \left|g\left(ct, dt^{\frac{N-1}{N-p}}\right)\right| t^{-\frac{p(N-1)}{N-p}}dt<\infty & \text{if } p\neq N\\ (ii) & \displaystyle\int_1^\infty \left|g(ct,de^t)e^{-Nt}\right|dt<\infty & \text{if } p=N,\end{array} \tag{3.1.24}$$

also uniformly with respect to $(c,d)$ bounded in $\mathbb{R}\times\mathbb{R}_+$.

**Theorem 3.1.7** *Let $\Omega\subset\mathbb{R}^N$ be a domain containing 0 and $1<p\leq N$. If $g\in C^{1,\delta}(\mathbb{R}\times\mathbb{R}_+)$ for some $\delta\in(0,1]$, satisfies (3.1.22)-(3.1.23) and $u$ is a solution of (3.0.1) such that $u/\mu_p$ and $|x|^{\frac{N-1}{p-1}}\nabla u$ are locally bounded near 0, then there exists $\gamma\in\mathbb{R}$ such that*

$$\lim_{x\to 0}\frac{u(x)}{\mu_p(x)}=\gamma. \tag{3.1.25}$$

*Furthermore*

$$\lim_{\substack{x\to 0\\ \frac{x}{|x|}\to \mathbf{e}}}|x|^{\frac{N-1}{p-1}}\nabla u(x)=\begin{cases}-\frac{N-p}{p-1}\gamma_{N,p}\gamma\mathbf{e} & \text{if } 1<p<N\\ -\gamma_N\gamma\mathbf{e} & \text{if } p=N.\end{cases} \tag{3.1.26}$$

*If we assume also that (3.1.24) holds, then $g(u,|\nabla u|)\in L^1_{loc}(\Omega)$ and the following equation holds in the sense of distributions*

$$-\Delta_p u+g(u,|\nabla u|)=|\gamma|^{p-2}\gamma\delta_0 \qquad \text{in } \mathcal{D}'(\Omega). \tag{3.1.27}$$

*Proof.* We recall that $c_{N,p}$ (resp. $c_N$) is a normalization constant for the fundamental solution $\mu_p$ (resp. $\mu_N$). We can assume that $\overline{B}_1\subset\Omega$. Set

$$\gamma^+=\limsup_{x\to 0}\frac{u(x)}{\mu_p(x)},\qquad \gamma^-=\liminf_{x\to 0}\frac{u(x)}{\mu_p(x)}. \tag{3.1.28}$$

If $(\gamma^+,\gamma^-)=(0,0)$, let $\epsilon>0$, $M_+$ be the supremum of $u_+$ on $\partial R$, $R<1$ and $w_\epsilon=\epsilon\mu_p+M_+$. Let $\rho$ be a $C^1$ function vanishing on $(-\infty,0]$, increasing on $[0,\infty)$ and $0\leq\rho'(r)\leq\operatorname{sign}_+(r)$. Then

$$\int_{B_R}\langle|\nabla u|^{p-2}\nabla u-|\nabla w_\epsilon|^{p-2}\nabla w_\epsilon,\nabla(u-w_\epsilon)\rangle\rho'(u-w_\epsilon)dx$$
$$+\int_{B_R}\rho(u-w_\epsilon)g(u,|\nabla u|)dx=0.$$

By the dissipativity assumption the second term is nonnegative. If we let $\rho(r)$ converge $r_+$ we get

$$\int_{B_R} \langle |\nabla u|^{p-2}\nabla u - |\nabla w_\epsilon|^{p-2}\nabla w_\epsilon, \nabla(u-w_\epsilon)_+\rangle dx \leq 0,$$

which implies $u \leq w_\epsilon$ by the strict monotonicity of $-\Delta_p$. Similarly $u \geq -w_\epsilon - M_-$ where $M_-$ is the infimum of $u_-$ on $\partial R$. We derive that $u$ is bounded by letting $\epsilon \to 0$.

Assume now that $|\gamma^+| + |\gamma^-| > 0$, and $\gamma^+ > 0$ for example.

*Case 1: $p \neq N$.* Set

$$u_r(x) = \frac{u(rx)}{\mu_p(r)} \qquad \forall x \in B_{\frac{1}{r}} \setminus \{0\}. \tag{3.1.29}$$

Then

$$-\Delta_p u_r + r^N g(r^{-\frac{N-p}{p-1}} u_r, r^{-\frac{N-1}{p-1}} |\nabla u_r|) = 0 \qquad \text{in } B_{\frac{1}{r}} \setminus \{0\}, \tag{3.1.30}$$

and let $\xi_r$ such that $|\xi_r| = 1$ and $\gamma(r)\mu_p(r\xi_r) = u(r\xi_r)$. By assumption $|u_r(x)| \leq c\mu_p(x)$ and $|\nabla u_r(x)| \leq c|x|^{-\frac{N-1}{p-1}}$, therefore it follows from the weak singularity assumption that

$$r^N g(r^{-\frac{N-p}{p-1}} u_r, r^{-\frac{N-1}{p-1}} |\nabla u_r|) \to 0 \quad \text{as } r \to 0,$$

uniformly on any compact set $K \in \mathbb{R}^N \setminus \{0\}$, provided $r$ is smaller than $r_K := (\sup\{|z|, z \in K\})^{-1}$. Because $u_r$ and $r^N g(r^{-\frac{N-p}{p-1}} u_r, r^{-\frac{N-1}{p-1}} |\nabla u_r|)$ remain eventually locally bounded in $\mathbb{R}^N \setminus \{0\}$, Theorem 1.3.9 applies and we conclude that $\nabla u_r$ remains locally bounded in $C^\alpha(\mathbb{R}^N \setminus \{0\})$. Therefore the set of functions $\{u_r\}$ is relatively compact in the $C^{1,\alpha'}$ local topology of $\mathbb{R}^N \setminus \{0\}$ for all $0 < \alpha' < \alpha$. If $\{r_n\}$ is a sequence converging to 0 and $U \in C^{1,\alpha}$ is such that $\{u_{r_n}\}$ converges to $U$ in $C^{1,\alpha'}_{loc}(\mathbb{R}^N \setminus \{0\})$, then $U$ is $p$-harmonic in $\mathbb{R}^N \setminus \{0\}$ and it satisfies

$$|U(x)| \leq c\mu_p(x) \qquad \text{in } \mathbb{R}^N \setminus \{0\}. \tag{3.1.31}$$

It follows from Corollary 1.5.9 that $U = \lambda\mu_p$ for some $\lambda \in [-c, c]$. If we assume that the sequence $\{r_n\}$ is the one for which the lim sup is achieved, then

$$\limsup_{x \to 0} \frac{u(x)}{\mu_p(x)} = \lim_{r_n \to 0} \sup_{|\xi|=1} \frac{u(r_n\xi)}{\mu_p(r_n)} = \gamma_+ = \gamma, \tag{3.1.32}$$

and the convergence

$$\gamma_+ = \lim_{r_n \to 0} \frac{u(r_n \xi)}{\mu_p(r_n)}, \tag{3.1.33}$$

holds uniformly on $\{\xi : |\xi| = 1\}$. Since $g$ satisfies (3.1.22), it follows by the maximum principle that $u$ is positive in the annulus $\{x : r_p \leq |x| \leq r_n\}$ for any $p > n$; hence there exists $0 < R_0 < 1$ such that $u > 0$ in $B_{R_0} \setminus \{0\}$. Since $\gamma\mu_p$ is a supersolution of equation (3.0.1), for any $\epsilon > 0$, there exists $r_\epsilon > 0$ such that $r_\epsilon \to 0$ when $\epsilon \to 0$ and

$$u(x) \leq (\gamma + \epsilon)\mu_p(x) + \max_{|y|=1} u(y) \qquad \forall x \in \overline{B}_1 \setminus \{B_{r_\epsilon}\}.$$

Letting $\epsilon \to 0$ yields

$$u \leq \gamma\mu_p + \max_{|y|=1} u(y) \qquad \text{in } \overline{B}_1 \setminus \{0\}.$$

Next, there exists a sequence $\{r'_n\}$ converging to 0 such that

$$\liminf_{x \to 0} \frac{u(x)}{\mu_p(x)} = \lim_{r'_n \to 0} \inf_{|\xi|=1} \frac{u(r'_n \xi)}{\mu_p(r_n)} = \gamma_- \geq 0. \tag{3.1.34}$$

By the compactness of $\{u_r\}$ in $C^{1,\alpha}$ locally in $\mathbb{R}^N \setminus \{0\}$, we can assume that the sequence $\{r'_n\}$ is such that the convergence

$$\gamma_- = \lim_{r_n \to 0} \frac{u(r_n \xi)}{\mu_p(r_n)},$$

holds uniformly on $\{\xi : |\xi| = 1\}$. Therefore, since $u$ is $p$-subharmonic and nonnegative in $B_{R_0} \setminus \{0\}$ we derive from the maximum principle that for any $\epsilon > 0$, $u$ is bounded from above in $B_{R_0} \setminus B_{r'_n}$ by the $p$-harmonic function $v_\epsilon = (\gamma_- + \epsilon)\mu_p + \max\{u(y) : |y| = R_0\}$. Letting $r'_n \to 0$ and $\epsilon \to 0$ yields

$$u(x) \leq \gamma_- \mu_p + \max\{u(y) : |y| = R_0\} \qquad \forall x \in B_{R_0} \setminus \{0\}. \tag{3.1.35}$$

Since $\gamma_- \leq \gamma_+$ it follows from (3.1.33) and (3.1.35) that $\gamma_+ = \gamma_- = \gamma$ and (3.1.25) holds. Since the convergence holds in the $C^1_{loc}$ topology, (3.1.26) follows.

Next, assume (3.1.24)-(i) holds and let $\zeta \in C_c^\infty(\Omega)$, then

$$\int_{\Omega \setminus B_{\frac{1}{n}}} \left( \langle |\nabla u|^{p-2} \nabla u, \nabla \zeta \rangle + g(u, |\nabla u|)\zeta \right) dx = -\int_{\partial B_{\frac{1}{n}}} \zeta \, |\nabla u|^{p-2} \langle \nabla u, \mathbf{n} \rangle dS.$$

Because of (3.1.25) and (3.1.26), $g(u,|\nabla u|)$ is integrable in $B_1$ and

$$\lim_{n\to\infty}\int_{\partial B_{\frac{1}{n}}}\zeta\,|\nabla u|^{p-2}\,\langle\nabla u,\mathbf{n}\rangle dS=|\gamma|^{p-2}\gamma\zeta(0),$$

which yields (3.1.27).

*Case 1:* $p=N$. We put

$$u_r(x)=\frac{u(rx)}{\mu_N(r)}\qquad\forall x\in B_{(2r)^{-1}}\setminus\{0\}.\tag{3.1.36}$$

Then $|u_r(x)|\leq c$ if $0<r|x|\leq\frac{1}{2}$, and

$$-\Delta_N u_r+r^N g\left(\mu_N(r)u_r,\mu_N(r)|x||\nabla u_r|\right)=0\qquad\text{in } B_{(2r)^{-1}}\setminus\{0\}.\tag{3.1.37}$$

Furthermore, since we can assume $r\leq\frac{1}{2}$,

$$|u_r(\xi)|\leq c\left(1+\frac{|\mu_N(\xi)|}{|\mu_N(r)|}\right).$$

Notice the right-hand side of the above expression remains locally bounded in $\mathbb{R}^N\setminus\{0\}$ and it converges to $c$ when $r\to 0$. Next

$$|\nabla u_r(\xi)|\leq\frac{c}{\mu_N(r)|\xi|}.$$

Because of assumption (3.1.24), the term $r^N g\left(\mu_N(r)u_r,\mu_N(r)|x||\nabla u_r|\right)$ remains locally bounded in $\mathbb{R}^N\setminus\{0\}$. Therefore, proceeding as in the previous case, we conclude that the set of functions $\mathcal{U}:=\{u_r\}_{0<r<2^{-1}}$ is relatively compact in the $C^1_{loc}$ topology of $\mathbb{R}^N\setminus\{0\}$ and if $U$ is any cluster point of $\mathcal{U}$ at $r=0$, then it is a bounded $N$-harmonic function in $\mathbb{R}^N$. By Theorem 1.3.8, $U$ is equal to some constant $\gamma$. Hence there exists a sequence $\{r_n\}$ converging to 0 such that

$$\lim_{r_n\to 0}\frac{u(r_n x)}{\mu_N(r_n)}=\gamma,$$

uniformly for $|x|=1$. This isotropy relation yields (3.1.26) and (3.1.27) as in the case $p\neq N$. The assumption (3.1.24)-(ii) implies that $g(u,|\nabla u|)$ is integrable in $B_1$ and we end the proof as above. □

**Proposition 3.1.8** *Let* $1<p\leq N$ *and* $g\in C^{1,\delta}(\mathbb{R}\times\mathbb{R}_+)$ *for some* $\delta\in(0,1]$, *satisfies (3.1.22)-(3.1.24). Then for any* $R>0$ *and* $\gamma>0$ *there* *exists a radial function $u \in C^1(\overline{B}_R \setminus \{0\})$ satisfying*

$$\begin{aligned} -\Delta_p u + g(u, |\nabla u|) &= |\gamma|^{p-2}\gamma\delta_0 \qquad && \text{in } \mathcal{D}'(B_R) \\ u &= 0 && \text{on } \partial B_R. \end{aligned} \tag{3.1.38}$$

*Furthermore u is nonnegative and it satisfies*

$$\lim_{|x|\to 0} \frac{u(x)}{\mu_p(x)} = c_{N,p}\gamma. \tag{3.1.39}$$

*If we assume moreover that $(t,s) \mapsto g(s,t)$ is nondecreasing, then $u$ is the unique solution satisfying (3.1.39); we denote it by $u_\gamma$. It is a nondecreasing function of $\gamma$ and $R$.*

*Proof.* Since $u(x) = u(|x|) = u(r)$, we look for positive solutions of

$$\begin{aligned} -|u'|^{p-2}\left((p-1)u'' + \frac{N-1}{r}u'\right) + g(u, |u'|) &= 0 \qquad \text{in } (0,R) \\ \lim_{r\to 0} \frac{u(r)}{\mu_p(r)} &= c_{N,p}\gamma \\ u(R) &= 0. \end{aligned} \tag{3.1.40}$$

*Step 1: Existence when $p = N$.* Set $t = -\ln r$, $\tau := -\ln R$; then $\tau < t < \infty$, and if $u(r) = v(t)$, (3.1.40) is equivalent to

$$\begin{aligned} -(|v'|^{N-2}v')' + e^{-Nt}g(v, e^t|v'|) &= 0 \qquad \text{in } (\tau,\infty) \\ \lim_{r\to 0} \frac{v(t)}{t} &= \gamma_N\gamma \\ v(\tau) &= 0. \end{aligned} \tag{3.1.41}$$

The function $w_1(t) = \gamma_N\gamma(t-\tau)$ is a supersolution for (3.1.41). For $\theta > 0$, let $v_\theta$ be the maximal solution of

$$\begin{aligned} (|v_\theta'|^{N-2}v_\theta')' &= e^{-Nt}g(v_\theta, e^t|v_\theta'|) \qquad \text{in } (\tau, T_{max}) \\ v_\theta(\tau) &= 0 \text{ and } v_\theta'(\tau) = \theta. \end{aligned}$$

The function $v_\theta'$ is increasing and larger than $\theta(t-\tau)$. By the maximum principle the curves of $w_1$ and $v_\theta$ cannot intersect, thus $v_\theta \leq w_1$. The Cauchy-Lipschitz theorem and the fact that $\theta(t-\tau) \leq v_\theta(t) \leq w_1(t)$ imply that the maximal solution is defined on $[\tau,\infty)$ and there exists $\lim_{t\to\infty} v_\theta(t) = \ell(\theta) \leq \gamma_N\gamma$. More precisely

$$(\ell(\theta))^{p-1} = \theta^{p-1} + \int_\tau^\infty e^{-Nt}g(v_\theta(t), e^t|v_\theta(t)|)dt. \tag{3.1.42}$$

Because $\theta \mapsto v_\theta(t)$ is continuous and increasing for any $t \geq \tau$, we infer from (3.1.24)-(ii) that the integral term in (3.1.42) is a continuous function of $\theta$. Since $\ell(c_N\gamma) \geq \gamma_N\gamma$, it follows from the mean value theorem that there exists $\theta^* \leq c_N\gamma$ such that $\ell(\theta^*) = \gamma_N\gamma$.

*Step 2: Existence when* $1 < p < N$. Set $t = r^{\frac{p-N}{p-1}}$, $\tau := R^{\frac{p-N}{p-1}}$; thus $\tau < t < \infty$, and if $v(t) = u(r)$, (3.1.40) is equivalent to

$$\begin{aligned} -(|v'|^{p-2}v')' + \left(\tfrac{N-p}{p-1}\right)^p t^{-\frac{p(N-1)}{N-p}} g\left(v, \tfrac{N-p}{p-1} t^{-\frac{N-1}{p-1}}|v'|\right) = 0 \quad \text{in } (\tau,\infty) \\ \lim_{r\to 0} \frac{v(t)}{t} = \gamma_{N,p}\gamma \\ v(\tau) = 0. \end{aligned} \tag{3.1.43}$$

The remaining of the proof, based upon convexity arguments and the fact that linear functions are supersolutions, is the same as in Step 1. The function $v_\theta$ is the solution of the initial value problem associated to equation (3.1.43) with $v_\theta(\tau) = 0$ and $v'_\theta(\tau) = \theta$, and $\ell(\theta)$ is the limit of $v'_\theta(t)$ when $t \to \infty$. In particular

$$(\ell(\theta))^{p-1} = \theta^{p-1} + \left(\tfrac{N-p}{p-1}\right)^p \int_\tau^\infty \tau^{-\frac{p(N-1)}{N-p}} g\left(v_\theta, \tfrac{N-p}{p-1} t^{-\frac{N-1}{p-1}}|v'_\theta|\right) d\tau, \tag{3.1.44}$$

and we infer that problem (3.1.43) admits a solution.

*Step 3: Identity (3.1.38) holds.* In both cases $p = N$ and $1 < p < N$, relation (3.1.39) holds by construction. As in the proof of Theorem 3.1.7, the validity of (3.1.39)-(3.1.44) implies that the convergence of $\frac{u(rx)}{\mu_p(r)}$ holds locally in the $C^1$ topology. Thus (3.1.26) holds. This implies (3.1.38).

*Step 4: Uniqueness.* The two mappings $\theta \mapsto v_\theta$ and $\theta \mapsto v'_\theta$ are nondecreasing. Since $(s,t) \mapsto g(s,t)$ is nondecreasing in $s$ and $t$, the correspondence $\theta \mapsto \ell(\theta)$ defined by (3.1.42) is increasing. It is continuous, hence there exists a unique $\theta^* \in \mathbb{R}_+$ such that $\ell(\theta^*) = \gamma_{N,p}\gamma$. Notice that $\theta^* < \gamma_N\gamma$. □

*Remark.* Uniqueness of solutions of (3.1.38) in a bounded domain with prescribed boundary value (non necessarily radially symmetric) can be obtained if one can prove that (3.1.39) holds. This can be obtained either with a sign assumption on $u$ or with the assumption that $|u| \leq c\mu_p$, using Theorem 3.1.7. In that case the method is to use the maximum principle in order to prove that two solutions $u$ and $u^*$ of the same problem satisfy $u \leq (1+\epsilon)u^*$ and $u^* \leq (1+\epsilon)u$ for any $\epsilon > 0$.

**Proposition 3.1.9** *Under the assumptions of Proposition 3.1.8, we assume moreover that there exists* $\alpha > \frac{N-p}{p-1}$ *such that* $g$ *satisfies*

$$g(s,t) = \lambda^{(\alpha+1)(p-1)+1} g(\lambda^{-\alpha}s, \lambda^{-\alpha-1}t) \qquad \forall s > 0,\, t > 0\, \lambda > 0, \quad (3.1.45)$$

*that* $s \mapsto s^{1-p}g(s,\alpha s)$ *is increasing on* $\mathbb{R}_+$ *and the following limits hold:*

$$\begin{aligned} &(i) & \lim_{s\to 0} s^{1-p}g(s,\alpha s) = 0, \\ &(ii) & \lim_{s\to\infty} s^{1-p}g(s,\alpha s) = \infty. \end{aligned} \qquad (3.1.46)$$

*We denote by* $u_{\gamma,R}$ *the solution of (3.1.38). Then* $(\gamma, R) \mapsto u_{\gamma,R}$ *is increasing and* $u_{\gamma,R}$ *converges to* $u_{\infty,R}$ *as* $\gamma \to \infty$*. Furthermore* $u_{\infty,R}$ *converges to* $u_{\infty,\infty} : u_\infty$ *when* $R \to \infty$, $u_\infty(x) = \kappa |x|^{-\alpha}$ *where* $\kappa$ *is the only positive zero of the mapping*

$$c \mapsto c^{1-p}g(c,\alpha c) - \alpha^{p-1}((p-1)\alpha + p - N), \qquad (3.1.47)$$

*and there holds*

$$0 \leq u_\infty - u_{\infty,R} \leq \kappa R^{-\alpha} \qquad \forall R > 0. \qquad (3.1.48)$$

*Proof.* For $\lambda > 0$, set $T_\lambda[u](x) = \lambda^\alpha u(\lambda x)$. Then $T_\lambda[u_{\gamma,R}] = u_{\lambda^{\frac{p-N}{p-1}+\alpha}\gamma, \frac{R}{\lambda}}$. Since $(\gamma, R) \mapsto u_{\gamma,R}$ is increasing, with limit $u_{\infty,R}$ at infinity when $\gamma \to \infty$, we obtain

$$T_\lambda[u_{\infty,R}] = u_{\infty,\frac{R}{\lambda}} \qquad \forall \lambda \in \mathbb{R}_+.$$

When $R \to \infty$, $u_{\gamma,R} \to u_{\gamma,\infty} := u_\gamma$ and $u_{\infty,R} \to u_{\infty,\infty} := u_\infty$. Thus

$$T_\lambda[u_\infty] = u_\infty \qquad \forall \lambda \in \mathbb{R}_+. \qquad (3.1.49)$$

This proves that $u_\infty$ is self-similar and in particular,

$$u_\infty(r) = u_\infty(1)r^{-\alpha} := \kappa r^{-\alpha} \qquad \forall r \in \mathbb{R}_+. \qquad (3.1.50)$$

Next, $\kappa r^{-\alpha}$ is a particular radial solution of (3.0.1), therefore, using (3.1.45),

$$\begin{aligned} \kappa^{p-1} r^{-(\alpha+1)(p-1)+1} \alpha^{p-1}((p-1)\alpha + p - N) &= g(\kappa r^{-\alpha}, \alpha\kappa r^{-\alpha-1}) \\ &= r^{-(\alpha+1)(p-1)+1} g(\kappa, \alpha\kappa). \end{aligned}$$

This implies that $\kappa$ satisfies the identity

$$\alpha^{p-1}\kappa^{p-1}((p-1)\alpha + p - N) = g(\kappa, \alpha\kappa). \qquad (3.1.51)$$

Because of (3.1.46) and $\alpha > \frac{N-p}{p-1}$, $\kappa$ is uniquely defined. Notice also that the functions $u_{\gamma,R}$ are smaller than $\kappa r^{-\alpha}$. This implies that the limits of $u_{\gamma,R}$ at infinity exist and for any $\gamma > 0$ and $R' > R > 0$

$$0 < u_{\gamma,R}(r) < u_{\gamma,R'}(r) < u_{\gamma,R}(r) + u_{\gamma,R'}(R) \qquad \forall r \in (0,R].$$

This implies (3.1.48). □

The next result, typical of quasilinear equation with strong absorption, shows the dichotomy between the two types of possible behaviour for a singular solution of (3.1.1) when $g(u,|\nabla u|) = |u|^{q-1}u$. This phenomenon will be encountered in many other situations.

**Theorem 3.1.10** *Let $\Omega$ and $p$ be as in Theorem 3.1.7 and $p-1 < q < q_1$. If $u \in C^1(\Omega \setminus \{0\})$ is a positive solution of (3.1.1) in $\Omega \setminus \{0\}$, then the following alternative occurs:*

*(i) either*

$$\lim_{x\to 0} |x|^{\beta_{p,q}} u(x) = \omega_c, \tag{3.1.52}$$

*where $\beta_{p,q} = \frac{p}{q+1-p}$ and $\omega_c = (\beta_{p,q}^{p-1}\Lambda(\beta_{p,q}))^{\frac{1}{q+1-p}}$,*

*(ii) or there exists $\gamma \geq 0$ such that*

$$\lim_{x\to 0} \frac{u(x)}{\mu_p(x)} = \gamma, \tag{3.1.53}$$

*and there holds*

$$-\Delta_p u + u^q = \gamma^{p-1}\delta_0 \qquad \text{in } \mathcal{D}'(\Omega). \tag{3.1.54}$$

*Proof.* Without any loss of generality we can assume $\overline{B}_1 \subset \Omega$. Since $g(u,|\nabla u|) = |u|^{q-1}u$ and $p-1 < q < q_1$, the assumptions (3.1.23), (3.1.22), (3.1.24) and 3.1.37) with $\alpha = \frac{p}{q+1-p} = \beta_{p,q}$ hold. Furthermore $\kappa = \omega_c$ (see (3.1.2), (3.1.3)). If $u \leq c\mu_p$, assertion (ii) follows from Theorem 3.1.7, thus we assume that

$$\limsup_{x\to 0} \frac{u(x)}{\mu_p(x)} = \infty. \tag{3.1.55}$$

We write (3.0.1) under the form

$$-\Delta_p u + D(x)u^{p-1} = 0 \qquad \text{with } D(x) = (u(x))^{q+1-p}. \tag{3.1.56}$$

Since $u$ satisfies (3.1.7), there holds $D(x) \leq c|x|^{-p}$ in $B_1 \setminus \{0\}$. By Harnack inequality, (see e.g. Theorem 1.3.6) there holds

$$\max_{|x-y|\leq \frac{|x|}{2}} u(y) \leq c \min_{|x-y|\leq \frac{|x|}{2}} u(y) \qquad \forall x \in B_{\frac{1}{2}} \setminus \{0\},$$

for some $c = c(N,p) > 0$. This implies

$$u(x) \leq c^{13} u(y) \qquad \forall x, y \in B_{\frac{1}{2}} \setminus \{0\} \text{ s.t. } |x| = |y|. \tag{3.1.57}$$

Therefore (3.1.55) implies

$$\liminf_{x\to 0} \frac{u(x)}{\mu_p(x)} = \infty, \tag{3.1.58}$$

and there exist sequences $\{r_n\}$ converging to 0 and $\{\gamma_n\}$ converging to $\infty$ such that

$$\inf_{|x|=r_n} u(x) = \gamma_n \mu_p(r_n). \tag{3.1.59}$$

By the comparison principle and with the notations of Proposition 3.1.9, we obtain

$$u(x) \geq u_{\gamma_n,1}(x) \qquad \forall\, r_n \leq |x| \leq 1. \tag{3.1.60}$$

If we fix $|x|$ and let $n \to \infty$, we finally infer

$$u(x) \geq u_{\infty,1}(x) \qquad \forall x \in \overline{B}_1 \setminus \{0\}, \tag{3.1.61}$$

which implies, using (3.1.48) and (3.1.7),

$$\omega_c |x|^{-\beta_{p,q}} - \omega_c \leq u(x) \leq c|x|^{-\beta_{p,q}} \qquad \forall x \in \overline{B}_1 \setminus \{0\}. \tag{3.1.62}$$

There are several ways for obtaining the sharp upper bound for $u$, but all are based upon the uniqueness of positive separable solutions of (3.0.1) in $\mathbb{R}^N \setminus \{0\}$. For $\epsilon > 0$ we denote by $U_\epsilon$ the solution of

$$\begin{aligned} -\Delta_p u + u^q &= 0 \qquad \text{in } \overline{B}_\epsilon^c \\ \lim_{|x|\to\epsilon} u(x) &= \infty. \end{aligned} \tag{3.1.63}$$

The existence of this solution follows from (3.1.7); it is radially symmetric and constructed as the increasing limit, when $n$ and $R$ tend to $\infty$, of the sequence $\{U_{\epsilon,n,R}\}$ of the minimal solutions of

$$\begin{aligned} -\Delta_p u + u^q &= 0 \qquad && \text{in } B_R\overline{B}_\epsilon^c \\ u(x) &= n && \text{on } \partial B_\epsilon \\ u(x) &= 0 && \text{on } \partial B_R. \end{aligned} \tag{3.1.64}$$

Although it is easy to prove that the solutions of (3.1.63) is unique, this property will not be used. Because of the minimality, the function $T_r[U_\epsilon](x) := U_{\epsilon,r}(x) = r^{\frac{p}{q+1-p}} U_\epsilon(rx)$ satisfies (3.1.63) with $\epsilon$ replaced by $\frac{\epsilon}{r}$, hence it is equal to $U_{\frac{\epsilon}{r}}(x)$. Then

$$\omega_c|x|^{-\beta_{p,q}} - \omega_c \leq u(x) \leq U_\epsilon + \max_{|z|=1}\{u(z)\} \qquad \forall x \in \overline{B}_1 \setminus B_\epsilon. \tag{3.1.65}$$

Put $u_r(x) = r^{\beta_{p,q}} u(rx)$. Then $u_r$ satisfies the same equation (3.0.1) as $u$ in $\Omega_r := \frac{1}{r}\Omega$, and (3.1.65) is transformed into

$$\omega_c|x|^{-\beta_{p,q}} - \omega_c r^{\beta_{p,q}} \leq u_r(x) \leq U_{\frac{\epsilon}{r}}(x) + r^{\beta_{p,q}} \max_{|z|=1}\{u(z)\} \;\; \forall x \in \overline{B}_{\frac{1}{r}} \setminus B_{\frac{\epsilon}{r}}. \tag{3.1.66}$$

If we let $\epsilon \to 0$, $U_\epsilon$ converges to $U_0$ and there holds

$$\omega_c|x|^{-\beta_{p,q}} - \omega_c r^{\beta_{p,q}} \leq u_r(x) \leq U_0(x) + r^{\beta_{p,q}} \max_{|z|=1}\{u(z)\} \qquad \forall x \in \overline{B}_{\frac{1}{r}} \setminus \{0\}. \tag{3.1.67}$$

Moreover, since $T_r[U_\epsilon] = U_{\frac{\epsilon}{r}}$, we derive that $T_r[U_0] = U_0$, by letting $\epsilon \to 0$. This means that $U_0$ is a positive self-similar solution of (3.0.1) in $\mathbb{R}^N \setminus \{0\}$. Hence it is separable, and since it is bounded from below by $\omega_c|x|^{-\beta_{p,q}}$, it follows from Lemma 3.1.1 that it is equal to $\omega_c|x|^{-\beta_{p,q}}$. Relation (3.1.52) follows now from (3.1.67). Notice that the gradient of $u_r$ satisfies

$$\begin{aligned} &|\nabla u_r(x)| \leq c|x|^{-\beta_{p,q}-1} && \forall x \in \overline{B}_{\frac{1}{r}} \setminus \{0\} \\ &|\nabla u_r(x) - \nabla u_r(x)| \leq c|x-x'|^\alpha |x|^{-\beta_{p,q}-1-\alpha} && \forall 0 < |x| \leq |x'| \leq \tfrac{1}{r}. \end{aligned} \tag{3.1.68}$$

Combined with Ascoli's theorem, these estimates imply that the limit of $r^{\beta_{p,q}} u(r,.)$, when $r \to 0$, occurs in the $C^1(S^{N-1})$ topology. □

In the case $p = 2$, $q_1 = q_{1,2} = \frac{N}{N-2}$; the expression of Theorem 3.1.10 gives a sharper result since not only positive solutions can be classified, but also signed solutions, provided $q \geq \tilde{q}_{1,2} := \tilde{q}_1 = \frac{N+1}{N-1}$. This exponent is called the isotropy threshold. In this case

$$\beta_{p,q} = \beta_{2,q} = \frac{2}{q-1} \quad \text{and} \quad \omega_c = (\beta_{2,q}(\beta_{2,q} + 2 - N))^{\frac{1}{q-1}}.$$

The exponent $\tilde{q}_1$ corresponds to the threshold of isotropic behaviour in the problem. The separable solutions are written under the form $u(x) = u(r,\sigma) = r^{-\beta_{2,q}}\omega(\sigma)$ and $\omega$ satisfies

$$-\Delta'\omega - \beta_{2,q}(\beta_{2,q}+2-N)\omega + |\omega|^{q-1}\omega = 0 \qquad \text{in } S^{N-1}. \tag{3.1.69}$$

**Lemma 3.1.11** *Assume $\tilde{q}_1 \le q < q_1$, then the only solutions of (3.1.69) are the three constants $\omega_c$, $-\omega_c$ and 0.*

*Proof.* The importance of $\tilde{q}_1$ comes from the fact that if $q \ge \tilde{q}_1$, then $\beta_{2,q}(\beta_{2,q}+2-N) \le N-1 = \lambda_1(2, S^{N-1})$. For any function $v = v(r,\sigma)$, we set

$$\bar{v}(r) = \frac{1}{|S^{N-1}|}\int_{S^{N-1}} v(r,\sigma)dS(\sigma),$$

and $\eta = \omega - \bar{\omega}$. Then

$$-\Delta'\eta - \beta_{2,q}(\beta_{2,q}+2-N)\eta + |\omega|^{q-1}\omega - \overline{|\omega|^{q-1}\omega} = 0.$$

Since $\overline{\omega - \bar{\omega}} = \bar{\eta} = 0$,

$$-\int_{S^{N-1}} \eta\Delta'\eta dv_{g_0} \ge (N-1)\int_{S^{N-1}} \eta^2 dv_{g_0}.$$

Moreover

$$\begin{aligned}\int_{S^{N-1}} (|\omega|^{q-1}\omega - \overline{|\omega|^{q-1}\omega})\eta dv_{g_0} &= \int_{S^{N-1}} (|\omega|^{q-1}\omega - |\bar{\omega}|^{q-1}\bar{\omega})\eta dv_{g_0}\\ &\quad + \int_{S^{N-1}} (|\bar{\omega}|^{q-1}\bar{\omega} - \overline{|\omega|^{q-1}\omega})\eta dv_{g_0}\\ &= \int_{S^{N-1}} (|\omega|^{q-1}\omega - |\bar{\omega}|^{q-1}\bar{\omega})\eta dv_{g_0}\\ &\ge 2^{1-q}\int_{S^{N-1}} |\eta|^{q+1} dv_{g_0}.\end{aligned} \tag{3.1.70}$$

Therefore

$$(N-1-\beta_{2,q}(\beta_{2,q}+2-N))\int_{S^{N-1}} \eta^2 dv_{g_0} + 2^{1-q}\int_{S^{N-1}} |\eta|^{q+1} dv_{g_0} \le 0.$$

This yields $\eta = 0$, thus $\omega$ is constant and the result follows. □

**Theorem 3.1.12** *Let $\Omega$ be as in Theorem 3.1.7 and $\tilde{q}_1 \le q < q_1$. If $u \in C^1(\Omega \setminus \{0\})$ is a solution of*

$$-\Delta u + |u|^{q-1}u = 0, \tag{3.1.71}$$

*in* $\Omega \setminus \{0\}$, *then there exists* $\ell = \lim_{x\to 0} |x|^{\beta_{p,q}} u(x)$, *and* $\ell \in \{\omega_c, -\omega_c, 0\}$. *If* $\ell = 0$ *there exists* $\gamma \in \mathbb{R}$ *such that*

$$\lim_{x\to 0} \frac{u(x)}{\mu_2(x)} = \gamma, \tag{3.1.72}$$

*and there holds*

$$-\Delta u + |u|^{q-1}u = \gamma\delta_0 \qquad \text{in } \mathcal{D}'(\Omega). \tag{3.1.73}$$

*Proof.* *Step 1: We claim that* $\ell \in \{\omega_c, -\omega_c, 0\}$. The method of proof is completely different from the one of Theorem 3.1.10 and is based upon the reduction of the problem to an autonomous equation in a cylinder where the techniques associated to dynamical systems analysis can be used. We still assume $\overline{B}_1 \subset \Omega$ and we put $u(r,\sigma) = r^{-\beta_{2,q}} v(t,\sigma)$, with $\ln r = t \in (-\infty, 0]$. Then $v$ satisfies

$$v_{tt} + \left(N - \frac{2q+2}{q-1}\right) v_t + \Delta' v + \beta_{2,q}(\beta_{2,q} + 2 - N)v - |v|^{q-1} v = 0, \tag{3.1.74}$$

in $(-\infty, 0] \times S^{N-1}$. Since $|u(x)| \leq c|x|^{-\beta_{2,q}}$, $v$ is bounded in $(-\infty, 0] \times S^{N-1}$. By standard *a priori* estimates in $L^p$ and Hölder spaces, all the derivatives of $v$ up to the order 3 are bounded in $C^\delta((-\infty, 1] \times S^{N-1})$ for some $\delta \in (0,1)$ (precisely, if $1 < q < 2$, $\delta = \mathbb{E}[q-1]$, the integer part of $q-1$; if $q \geq 2$, then $\delta = 1$). We set

$$\mathcal{E}(v) = \frac{1}{2} \int_{S^{N-1}} \left( |\nabla' v|^2 - \beta_{2,q}(\beta_{2,q} + 2 - N)v^2 - v_t^2 + \frac{2}{q+1} |v|^{q+1} \right) dv_{g_0}.$$

The next identity is the energy estimate:

$$\frac{d\mathcal{E}(v)}{dt} = \left(N - \frac{2q+2}{q-1}\right) \int_{S^{N-1}} v_t^2 dv_{g_0}. \tag{3.1.75}$$

Therefore

$$\left(N - \frac{2q+2}{q-1}\right) \int_\tau^{-1} \int_{S^{N-1}} v_t^2 dv_{g_0} dt = \mathcal{E}(v)(-1) - \mathcal{E}(v)(\tau),$$

for $\tau < -1$. Thanks to the previous regularity estimates, $\mathcal{E}(v)(t)$ is uniformly bounded on $(-\infty, -1]$. Since $\left(N - \frac{2q+2}{q-1}\right) \neq 0$, it follows

$$\int_\tau^{-1} \int_{S^{N-1}} v_t^2 dv_{g_0} dt < \infty. \tag{3.1.76}$$

Differentiating (3.1.74) with respect to $t$ yields

$$v_{ttt} + \left(N - \frac{2q+2}{q-1}\right) v_{tt} + \Delta' v_t + \beta_{2,q}(\beta_{2,q} + 2 - N)v_t - q\,|v|^{q-1}\, v_t = 0.$$

Then

$$\left(N - \frac{2q+2}{q-1}\right) \int_\tau^{-1} \int_{S^{N-1}} v_{tt}^2 dv_{g_0}\, dt - q\int_\tau^{-1} \int_{S^{N-1}} v_t^2\,|v|^{q-1}\, dv_{g_0}\, dt$$
$$= \mathcal{E}(v_t)(-1) - \mathcal{E}(v_t)(\tau).$$

Using (3.1.76) and the boundedness of $v$, we derive

$$\int_\tau^{-1} \int_{S^{N-1}} v_{tt}^2 dv_{g_0}\, dt < \infty. \tag{3.1.77}$$

Finally, since $v_t$ and $v_{tt}$ are uniformly continuous we have that

$$\lim_{t\to\infty} \left(|v_t(t,.)| + |v_{tt}(t,.)|\right) = 0, \quad \text{uniformly in } S^{N-1}. \tag{3.1.78}$$

We define the negative trajectory of $v$ in $C^2(S^{N-1})$ by

$$\mathcal{T}_-[v] = \bigcup_{t\leq 0}\{v(t,.)\},$$

and its $\alpha$-limit set by

$$A[\mathcal{T}_-[v]] = \bigcap_{\tau<0} \overline{\bigcup_{t\leq\tau}\{v(t,.)\}}.$$

Since $\overline{\bigcup_{t\leq\tau}\{v(t,.)\}}$ is a connected compact subset of $C^2(S^{N-1})$ by the previous regularity estimates, $A[\mathcal{T}_-[v]]$ is a non-empty connected compact subset of $C^2(S^{N-1})$. Furthermore

$$A[\mathcal{T}_-[v]] = \Big\{\omega \in C^2(S^{N-1}) : \exists\{t_n\},\ t_n \to -\infty \text{ as } n\to\infty$$
$$\text{s.t. } \|v(t_n,.) - \omega\|_{C^2(S^{N-1})} \to 0\Big\}.$$

Then (3.1.78) implies that $A[\mathcal{T}_-[v]]$ is a subset of the set of solutions of (3.1.69), which means $A[\mathcal{T}_-[v]] \subset \{\omega_c, -\omega_c, 0\}$ if $\tilde{q}_1 \leq q < q_1$. Next, if $\omega_c \in A[\mathcal{T}_-[v]]$, there exists a sequence $\{t_n\}$ such that $v(t_n,.) \to \omega$ uniformly on $S^{N-1}$. Therefore there exists $k \in \mathbb{N}$ such that for any $n \in \mathbb{N}$, $n \leq k$ implies $\min\{v(t_n,\sigma) \geq 2^{-1}\omega_c\}$. Since the constant function $2^{-1}\omega_c$ is a subsolution of (3.1.74), it follows that $v(t,\sigma) \geq 2^{-1}\omega_c$ on $[t_n,t_k]\times S^{N-1}$. Letting $n\to\infty$ yields $v(t,\sigma) \geq 2^{-1}\omega_c$ on $(-\infty,t_k]\times S^{N-1}$. This implies that $A[\mathcal{T}_-[v]] = \{\omega_c\}$ and $v(t,.) \to \omega_c$ when $t\to-\infty$. Similarly, if $-\omega_c \in A[\mathcal{T}_-[v]]$, we infer $v(t,.) \to -\omega_c$ when $t \to -\infty$. Finally, if $0 \in A[\mathcal{T}_-[v]]$, a reasoning by contradiction using the previous arguments shows that $v(t,.) \to 0$ when $t \to -\infty$.

*Step 2: If $\ell = 0$, then there holds*

$$\|u(t,.) - \bar{u}(t)\|_{L^\infty(S^{N-1})} \leq ce^{a^*t} \qquad \forall t \leq -1, \tag{3.1.79}$$

where $a^*$ is an explicit positive constant. The proof is based upon a linearization procedure. Using (3.1.70), we derive from (3.1.74) that the $L^2(S^{N-1})$-norm $y(t)$ of $w(t,.) = v(t,.) - \bar{v}(t)$ satisfies

$$y'' + \left(N - \frac{2q+2}{q-1}\right) y' - (N - 1 - \beta_{2,q}(\beta_{2,q} + 2 - N)y \geq 0, \tag{3.1.80}$$

in $\mathcal{D}'((-\infty,-1))$.
We first assume that $q > \tilde{q}_1$. Since $y(t) \to 0$ when $t \to -\infty$, we obtain by the maximum principle $y(t) \leq y(-2)e^{a_+(t+2)}$ for $t \leq -2$, where $a_+$ is the positive root of

$$x^2 + \left(N - \frac{2q+2}{q-1}\right) x - (N - 1 - \beta_{2,q}(\beta_{2,q} + 2 - N) = 0.$$

The expression of $a_+$ is complicated, but it is easy to check that it is a monotone decreasing function of $q$, which ranges from $N$, when $q = \tilde{q}_1$ to $N-1$ when $q = q_1$. Furthermore

$$a_+ - \beta_{2,q} = a^* > 0.$$

Since $w$ satisfies the elliptic equation

$$w_{tt} + \left(N - \frac{2q+2}{q-1}\right) w_t + \Delta' v = h, \tag{3.1.81}$$

where $h$ can be written as

$$h = (N - 1 - \beta_{2,q}(\beta_{2,q} + 2 - N)w + v|v|^{q-1} - \bar{v}|\bar{v}|^{q-1} + \bar{v}|\bar{v}|^{q-1} - \overline{v|v|^{q-1}},$$

there holds

$$\|h\|_{L^2(S^{N-1})} \leq c\,\|w\|_{L^2(S^{N-1})} \leq c'e^{a_+t}.$$

By $L^2$-linear estimates for elliptic equations

$$\begin{aligned}\|w\|_{H^2((t-1,t+1)\times S^{N-1})} &\leq c\left(\|v\|_{L^2((t-2,t+2)\times S^{N-1})} + \|h\|_{L^2((t-2,t+2)\times S^{N-1})}\right)\\ &\leq c(c'+1)e^{a_+t}.\end{aligned} \tag{3.1.82}$$

If $N > 4$ we apply twice Gagliardo-Nirenberg inequality (1.2.32) and get,

$$\|w\|_{L^{2^{**}}((t-1,t+1)\times S^{N-1})} \leq c\|w\|_{H^2((t-1,t+1)\times S^{N-1})},$$

where $2^{**} = \frac{2N}{N-4}$. Then

$$\|w\|_{L^{2^{**}}((t-1,t+1)\times S^{N-1})} \leq ce^{a_+ t}.$$

It implies

$$\|h\|_{L^{2^{**}}((t-1,t+1)\times S^{N-1})} \leq c\|w\|_{L^{2^{**}}((t-1,t+1)\times S^{N-1})} \leq c'e^{a_+ t}.$$

Using the $L^p$-linear estimates for elliptic equations with $p = 2^{**}$ we obtain as above

$$\|w\|_{W^{2,2^{**}}((t-\frac{1}{2},t+\frac{1}{2})\times S^{N-1})} \leq ce^{a_+ t}.$$

At this moment we use again Gagliardo-Nirenberg or Morrey inequalities (1.2.32)-(1.2.41) and either obtain an estimate of $w$ in $L^s((t-\frac{1}{2},t+\frac{1}{2})\times S^{N-1}))$ with $s = \frac{2^{**}N}{N-2^{**}}$, or in any $L^s$-space if $N = 2^{**}$, or in $L^\infty$, if $N < 2^{**}$. Iterating this bootstrap a finite number of times we finally derive,

$$\|w(t,.)\|_{L^\infty(S^{N-1})} \leq ce^{a_+ t}. \tag{3.1.83}$$

If $N < 4$, we obtain this estimate immediately, while if $N = 4$, we first obtain an estimate in any $L^s$ space, for $s < \infty$, from which we infer (3.1.83). Finally, we assume that $q = \tilde{q}_1$. Hence, with the same notations,

$$y'' - Ny' - cy^{\tilde{q}_1} \geq 0, \qquad \text{in } \mathcal{D}'((-\infty,-1)), \tag{3.1.84}$$

where $c = c(N) > 0$. Then for any $\epsilon$, there exists $a = a(N)$ such that $t \mapsto ae^{Nt} + \epsilon$ is a supersolution for (3.1.84), larger that $y$ at $-\infty$ and at $t = -1$ (and at this point, it is independent of $\epsilon$). Therefore $y(t) \leq ae^{Nt} + \epsilon'$ and finally $y(t) \leq ae^{Nt}$. From this estimate in $L^2$, we derive as in the first case an estimate in $L^\infty(S^{N-1})$, which takes on the same aspect as in (3.1.83) except that $a_+$ is replaced by $N$. Returning to $u$ we infer (3.1.79).

*Step 3: If $\ell = 0$, we claim that (3.1.72).* If $u/\mu_2$ remains bounded, then (3.1.72) follows by Theorem 3.1.7. Henceforth we assume that this is not the case, and for example

$$\limsup_{x\to 0} \frac{u(x)}{\mu_2(x)} = \infty. \tag{3.1.85}$$

If $N \geq 3$, we set $s = \frac{r^{N-2}}{N-2}$, $u(r,\sigma) = r^{2-N} z(s,\sigma)$. Then $z$ satisfies

$$s^2 z_{ss} + \frac{1}{N-2}\Delta' z = bs^{\frac{N}{N-2}-q}|z|^{q-1}z \qquad \text{in } (0,1] \times S^{N-1}, \tag{3.1.86}$$

with $b = b(N,q) > 0$. Since estimate (3.1.79) holds, we infer

$$\|z(s,.) - \bar{z}(s)\|_{L^\infty(S^{N-1})} \leq cs^{b^*}, \tag{3.1.87}$$

where

$$b^* = \frac{a_+ + 2 - N}{N-2} > \frac{a_+ - \beta_{2,q}}{N-2} = \frac{a^*}{N-2} > 0.$$

From ((3.1.85)) and ((3.1.87)) there exists a sequence $\{s_n\}$ decreasing to 0 such that $z(s_n,\sigma) > 0$ for $n \in \mathbb{N}$ and $\sigma \in S^{N-1}$. If $r_n = \sqrt[N-2]{(N-2)s_n}$, then $u(r_n,\sigma) > 0$ for all $\sigma \in S^{N-1}$. This implies that $u > 0$ on $(r_p, r_n) \times S^{N-1}$ and finally $u > 0$ in $B_{r_0} \setminus \{0\}$. By Theorem 3.1.10, if $u > 0$ and $\ell = 0$, then $u/\mu_2$ remains bounded, hence (3.1.85) cannot hold.
If $N = 2$ we set $\tau = \ln r$, $u(r,\sigma) = z(\tau,\sigma)$. Then

$$z_{\tau\tau} + z_{\theta\theta} = e^{2\tau}|z|^{q-1}z \qquad \text{in } (-\infty, 0] \times S^1. \tag{3.1.88}$$

Since there holds

$$\|z(\tau,.) - \bar{z}(\tau)\|_{L^\infty(S^1)} \leq ce^{a^* t}, \tag{3.1.89}$$

the proof follows as in the previous case from Theorem 3.1.10. □

*Remark.* The energy method used in the Step 1 of the proof of Theorem 3.1.12 is valid for any $q \in (1, q_1)$. We denote by $\mathcal{H}$ the set of solutions of (3.1.69), then the analysis therein asserts that there exists a compact and connected component $\mathcal{H}_\ell$ of $\mathcal{H}$ such that

$$\lim_{r\to 0} \operatorname{dist}_{C^2(S^{N-1})}\left(r^{\frac{2}{q-1}}u(r,.), \mathcal{H}_\ell\right) = 0. \tag{3.1.90}$$

However if $1 < q < \tilde{q}_1$ there exist connected components of $\mathcal{H}$ which are not reduced to a singleton. If $N = 2$ a more precise result can be proved. In that case the structure of $\mathcal{H}$ is known and it is composed of the three elements $0, \omega_c, -\omega_c$ plus a finite number of subsets $\mathcal{H}_k$ where $k$ runs from 1 to $\mathcal{B}_{2,q}$ where $\mathcal{B}_{2,q}$ is the larger integer smaller that $\beta_{2,q}$ and $\mathcal{H}_k$ is the set of functions $\omega(.+\alpha)$ where $\alpha \in \mathbb{R}$ and $\omega$ is a $\frac{\pi}{k}$-anti-periodic solution of

$$-\omega_{\theta\theta} - \left(\frac{2}{q-1}\right)^2 \omega + |\omega|^{q-1}\omega = 0 \qquad \text{on } \mathbb{R}. \tag{3.1.91}$$

The following result is proved in [Chen, Matano, Véron (1989)].

**Theorem 3.1.13** *Assume $N = 2$, $\Omega \subset \mathbb{R}^2$ is a domain containing 0 and $1 < q < 3$. If $u$ is a solution of (3.1.71), there exists a solution $\omega$ of (3.1.91) such that $r^{\frac{2}{q-1}}u(r,.)$ converges to $\omega$ in $C^2(S^1)$ when $r \to 0$. If $\omega = 0$, either there exists $\gamma \in \mathbb{R}$ such that*

$$\lim_{r\to 0} \frac{u(x)}{\ln(|x|^{-1})} = \gamma, \tag{3.1.92}$$

*or there exist an integer $k$ smaller than $\mathcal{B}_{2,q}$ and two real numbers $A$ and $\theta_0$ such that*

$$\lim_{r\to 0} r^k u(r,\theta) = A\sin(k\theta + \theta_0), \tag{3.1.93}$$

*uniformly with respect to $\theta \in S^1$.*

The proof of this result is based upon an elliptic variant of the classical theory of intersections of nodal lines of solutions of parabolic equations developed by Sturm. In this proof Jordan separation theorem plays a fundamental role.

### 3.1.3 *Global solutions in $\mathbb{R}^N$*

Since estimate (3.1.7) with $b = 0$ is valid for any $R > 0$, any solution of (3.1.1) in $\mathbb{R}^N$ is identically 0. If we look at the same equation in $\mathbb{R}^N \setminus \{0\}$, when $q \geq q_1$ it follows from Theorem 3.1.3 that any solution of (3.1.1) can be extended as a solution of (3.1.1) in $\mathbb{R}^N$, thus it is identically 0.

**Theorem 3.1.14** *Assume $1 < p \leq N$, $p - 1 < q < q_1$ and $u$ is a positive solution of (3.1.1) in $\mathbb{R}^N \setminus \{0\}$. Then*

*(i) either $u(x) \equiv \omega_c|x|^{-\beta_{p,q}}$,*

*(ii) or there exists $\gamma \geq 0$ such that (3.1.53) holds and $u$ is the unique solution of (3.1.54) with $\Omega = \mathbb{R}^N$.*

*Proof.* By (3.1.7) with $b = 0$ any solution in $\mathbb{R}^N \setminus \{0\}$ decays to 0 at infinity. From Theorem 3.1.10 $u$ satisfies either (3.1.52) or (3.1.53) for some $\gamma > 0$. For any $\epsilon > 0$, $v_\epsilon(x) = (1+\epsilon)\omega_c|x|^{-\beta_{p,q}} + \epsilon$ is a supersolution of (3.1.1) which is larger than $u$ at $x = 0$ or at infinity, thus it is larger than $u$ in $\mathbb{R}^N \setminus \{0\}$. Letting $\epsilon \to 0$ yields

$$u(x) \leq \omega_c|x|^{-\beta_{p,q}} \qquad \forall x \in \mathbb{R}^N \setminus \{0\}. \tag{3.1.94}$$

Assume first that (3.1.52) holds, then $u_\epsilon(x) = (1+\epsilon)u+\epsilon$ is a supersolution which dominates $\omega_c|x|^{-\beta_{p,q}}$ both at 0 and at $\infty$. Thus it is larger than $\omega_c|x|^{-\beta_{p,q}}$. Letting $\epsilon \to 0$ yields $u(x) \geq \omega_c|x|^{-\beta_{p,q}}$. Combined with (3.1.94) it implies $u(x) \equiv \omega_c|x|^{-\beta_{p,q}}$.
Assume now that (3.1.53) holds with $\gamma \geq 0$. By Proposition 3.1.8, for any $R > 0$ there exists a positive radially symmetric solution $u := u_{\gamma,R}$ to

$$\begin{aligned} -\Delta_p u + u^q &= \gamma^{p-1}\delta_0 && \text{in } \mathcal{D}'(B_R) \\ u &= 0 && \text{on } \partial B_R, \end{aligned} \tag{3.1.95}$$

and it satisfies

$$\lim_{x\to 0} \frac{u_{\gamma,R}(x)}{\mu_p(x)} = c_{N,p}\gamma. \tag{3.1.96}$$

Furthermore $R \mapsto u_{\gamma,R}$ is increasing and it converges, as $R \to \infty$ to the unique solution $u_{\gamma,\infty} := u_\gamma$ of

$$-\Delta_p u + u^q = \gamma^{p-1}\delta_0 \qquad \text{in } \mathcal{D}'(\mathbb{R}^N), \tag{3.1.97}$$

and $u_{\gamma,\infty}$ still satisfies (3.1.96) and (3.1.94). Proceeding as in the first case, for any $\epsilon > 0$ $u_{(1+\epsilon)\gamma} + \epsilon$ is a supersolution which dominates $u$ while $(1+\epsilon)u + \epsilon$ is a supersolution which dominates $u_\gamma$ at 0 and at infinity. At end $u = u_\gamma$. □

When $p = 2$, the situation for signed solutions is understood up to the isotropy threshold $\tilde{q}_1$. The following result is easy to prove, using Theorem 3.1.12 and the construction of supersolutions as in Theorem 3.1.14.

**Theorem 3.1.15** *Assume $1 < q < \tilde{q}_1$ and $u$ is a nonzero solution of (3.1.71) in $\mathbb{R}^N \setminus \{0\}$. Then*
*(i) either $u(x) = \omega_c|x|^{-\beta_{2,q}}$,*
*(ii) either $u(x) = -\omega_c|x|^{-\beta_{2,q}}$,*
*(iii) or there exists $\gamma \neq 0$ such that (3.1.72) holds and $u$ is the unique solution of (3.1.73) with $\Omega = \mathbb{R}^N$.*

When $N = 2$ all the solutions of (3.1.71) in $\mathbb{R}^2 \setminus \{0\}$ with $q > 1$ are classified. The asymptotic behaviour of solutions in an exterior domain is obtained in [Chen, Matano, Véron (1989)] by a similar technique based upon intersection theory as the one for Theorem 3.1.13.

**Theorem 3.1.16** *Assume $N = 2$, $q > 1$ and $u$ is a nonzero solution of (3.1.71) in $\mathbb{R}^2 \setminus K$, where $K$ is compact. Then there exists a nonzero* *solution $\omega$ of (3.1.73) such that $r^{\frac{2}{q-1}}u(r,.)$ converges to $\omega$ in $C^2(S^1)$ when $r \to \infty$.*

*Remark.* If $u$ is a solution of (3.1.71) in $\mathbb{R}^2 \setminus \{0\}$, the function $= v(t,\sigma) = r^{\beta_{2,q}}u(r,\sigma)$, with $\ln r = t$ is a uniformly bounded solution of

$$v_{tt} - \frac{4}{q-1}v_t + \left(\frac{2}{q-1}\right)^2 v + v_{\theta\theta} - |v|^{q-1}v = 0 \tag{3.1.98}$$

in $\mathbb{R} \times S^1$. The trajectory of $v$

$$\mathcal{T}[v] = \bigcup_{-\infty<t<\infty} \{v(t,.)\},$$

is a bounded subset of $C^2(S^1)$. It admits an $\alpha$-limit $\mathcal{A}[\mathcal{T}[v]]$ defined in the proof of Theorem 3.1.12 and an $\omega$- limit set $\Omega[\mathcal{T}[v]]$ defined by

$$\Omega[\mathcal{T}[v]] = \bigcap_{\tau>0} \overline{\bigcap_{t>\tau}}\{v(t,.)\}.$$

By Theorem 3.1.13 and Theorem 3.1.16, both $\mathcal{A}[\mathcal{T}[v]]$ and $\Omega[\mathcal{T}[v]]$ are reduced to a single solution of (3.1.91), that we denote by $\omega_-$ and $\omega_+$. Moreover the variation of the energy functional $\mathcal{E}$ is expressed by

$$\mathcal{E}(\omega_+) - \mathcal{E}(\omega_-) = -\frac{4}{q-1}\int_{-\infty}^{\infty}\int_0^{2\pi} v_t^2 d\theta dt. \tag{3.1.99}$$

Furthermore, if $\omega$ is a solution of (3.1.91), the energy of $\omega$, which also the same as the energy of the whole connected component $\mathcal{H}_\omega$ of $\mathcal{H}$ containing $\omega$, is expressed by

$$\mathcal{E}(\omega) = \frac{1-q}{2(q+1)}\int_0^{2\pi} |\omega|^{q+1} d\theta. \tag{3.1.100}$$

Finally, it is easy to prove that the energy is quantized on $\mathcal{H}$, i.e. has different values on each connected component $\mathcal{H}_k$ of $\mathcal{H}$. The minimal energy is achieved on the two nontrivial constant solutions

$$\min\{\mathcal{E}(\omega) : \omega \in \mathcal{F}\} = \mathcal{E}(\pm\omega_c) = -\frac{\pi(q-1)}{2(q+1)}\left(\frac{2}{q-1}\right)^{\frac{2q+2}{q-1}}. \tag{3.1.101}$$

Since the other connected components $\mathcal{H}_k$ where $1 < k \leq \mathcal{B}_{2,q}$ are generated by a $\frac{\pi}{k}$-anti-periodic solution $\omega_k$, it is easy to verify that

$$0 = \mathcal{E}(0) > \mathcal{E}(\omega_{\mathcal{B}_{2,q}}) > \mathcal{E}(\omega_{\mathcal{B}_{2,q}-1}) > ... > \mathcal{E}(\omega_2) > \mathcal{E}(\omega_1) > \mathcal{E}(\pm\omega_c). \tag{3.1.102}$$

A solution $u$ of (3.1.71) in $\mathbb{R}^2 \setminus \{0\}$ can be seen as an heteroclinic connecting orbit $\mathcal{T}[v]$ between two elements $\omega$ and $\omega'$ of $\mathcal{H}$. Therefore a necessary condition in order two elements $\omega_-$ and $\omega_+$ of $\mathcal{H}$ be connected by a non constant solution is

$$\mathcal{E}(\omega_+) = \lim_{t\to\infty} \mathcal{E}(v(t,.)) < \mathcal{E}(\omega_-) = \lim_{t\to-\infty} \mathcal{E}(v(t,.)). \qquad (3.1.103)$$

This condition turns out to be also sufficient [Matano (1990)]:

**Theorem 3.1.17** *Assume $N = 2$ and $q > 1$. If $\omega$ and $\omega'$ are two elements of $\mathcal{H}$ such that $\mathcal{E}(\omega') < \mathcal{E}(\omega)$, there exists a solution $u$ of (3.1.71) in $\mathbb{R}^2 \setminus \{0\}$ such that*

$$\lim_{r\to 0} r^{\beta_{2,q}} u(r,.) = \omega \quad \text{and} \quad \lim_{r\to\infty} r^{\beta_{2,q}} u(r,.) = \omega', \qquad (3.1.104)$$

*uniformly on $S^1$.*

### 3.1.4 *Large solutions*

Assume $\Omega$ is a proper subdomain of $\mathbb{R}^N$, $1 < p < \infty$ and $g : \mathbb{R} \mapsto \mathbb{R}$ a continuous nondecreasing function, and let $\rho(x)$ denote the distance from $x \in \Omega$ to $\partial\Omega$. A large solution of (3.1.17) is a solution in $\Omega$ which satisfies

$$\lim_{\rho(x)\to 0} u(x) = \infty. \qquad (3.1.105)$$

Existence of large solutions is linked to existence of uniform *a priori* bounds for solutions. The assumptions on $g$ can be either $g(0) > 0$ and (3.1.15) holds, or $g(0) = 0$ and (3.1.19) holds. In the sequel we only consider bounded domains, therefore there is no real loss of generality to assume that $g$ satisfies $g(0) = 0$ and (3.1.19). There are several ways to construct functions $u$ candidate for being large solutions.

**Definition 3.1.18** Let $\Omega$ be a bounded domain. A sequence of smooth subdomains $\{\Omega_n\}$ is called a smooth exhaustion of $\Omega$ if

$$\Omega_n \subset \overline{\Omega_n} \subset \Omega_{n+1} \quad \text{and} \quad \bigcup_{n\in\mathbb{N}} \Omega_n = \Omega.$$

Smooth exhaustions always exist. A sequence of smooth subdomain $\{\Omega'_n\}$ is called a smooth exterior approximation of $\overline{\Omega}$ if

$$\Omega_n \subset \overline{\Omega_n} \subset \Omega_{n-1} \quad \text{and} \quad \bigcap_{n\in\mathbb{N}} \Omega_n = \overline{\Omega}.$$

A smooth exterior approximation exists if and only if $\partial\Omega = \partial\overline{\Omega^c}$ or, equivalently if $\Omega$ coincides with the interior of its closure. This is equivalent to the property that for any $a \in \partial\Omega$ there exists a sequence $\{x_n\} \subset \Omega^c$ which converges to $a$.

**Proposition 3.1.19** *Assume $p > 1$, $\Omega \subset \mathbb{R}^N$ is a bounded domain and $g : \mathbb{R} \mapsto \mathbb{R}$ a continuous nondecreasing function satisfying $g(0) = 0$ and (3.1.15). Then there exists a maximal solution $\bar{u}_\Omega$ of (3.1.17) in $\Omega$.*

*Proof.* Let $\{\Omega_n\}$ be a smooth exhaustion of $\Omega$ and for any $n, k \in \mathbb{N}^*$ let $u := u_{n,k}$ be the solution of

$$\begin{aligned} -\Delta_p u + g(u) &= 0 && \text{in } \Omega_n \\ u &= k && \text{in } \partial\Omega_n. \end{aligned} \tag{3.1.106}$$

Since $g$ is monotone, the sequence $\{u_{n,k}\}$ is increasing with $k$. If $\sigma_n = \inf\{\operatorname{dist}(x, y) : x \in \Omega_n, y \in \Omega^c\}$, there holds

$$0 \leq u_{n,k}(x) \leq \max\{h(\rho(x) - \sigma_n), k\} \qquad \forall x \in \Omega_n, \tag{3.1.107}$$

where $h$ is defined in Proposition 3.1.5. Therefore $u_{n+1,k} \leq u_{n,k}$. If $v$ is any other solution of (3.1.17) in any domain containing $\overline{\Omega_n}$, it is bounded in $\Omega_n$ thus it is eventually smaller than $u_{n,k}$ on $\partial\Omega_n$, hence $v \leq u_{n,k}$. This estimate, combined with Theorem 1.3.9, implies that $u_{n,k}$ converges in $C^1_{loc}(\Omega_n)$, when $k \to \infty$, to some function $\overline{u}_{\Omega_n}$ which is a large solution of (3.1.17) in $\Omega_n$. Furthermore $\overline{u}_{\Omega_n}$ satisfies (3.1.107). As a consequence of inequality $u_{n+1,k} \leq u_{n,k}$ in $\Omega_n$, we infer $\overline{u}_{\Omega_{n+1}} \leq \overline{u}_{\Omega_n}$. Therefore the sequence $\{\overline{u}_{\Omega_n}\}$ decreases and converges in $C^1_{loc}(\Omega)$ to a solution $\bar{u}_\Omega$ of (3.1.17) in $\Omega$ which is larger than any other solution. □

*Remark.* The main question concerning the maximal solution is whether it is a large solution or not. When it is the case, another important question is whether this large solution is unique or not. We will see below some conditions which ensure these properties.

**Proposition 3.1.20** *Let $1 < p \leq N$, $\Omega$ be a bounded domain and $g : \mathbb{R} \mapsto \mathbb{R}$ a continuous nondecreasing function satisfying $g(0) = 0$, (3.1.15) and*

$$\begin{aligned} &(i) && \int_1^\infty g(t)\, t^{-\frac{p(N-1)}{N-p}} dt < \infty && \text{if } p \neq N \\ &(ii) && \int_1^\infty g(t) e^{-Nt} dt < \infty && \text{if } p = N. \end{aligned} \tag{3.1.108}$$

*Then the maximal solution of (3.1.17) is a large solution.*

*Proof.* Since $\Omega$ is bounded, we denote by $u_{\gamma,R}$ the solution of (3.1.95) where $R > \operatorname{diam}(\Omega)$ and $\gamma > 0$. If $a \in \partial\Omega$, we have

$$\overline{u}_{\Omega_n}(x) \geq u_{\gamma,R}(x-a) \qquad \forall x \in \Omega_n.$$

Letting $n \to \infty$ and $\gamma \to \infty$, we derive

$$\overline{u}_{\Omega}(x) \geq u_{\infty,R}(x-a) = u_{\infty,R}(|x-a|). \tag{3.1.109}$$

Next, if $x \in \Omega$ we take for $a$ any projection $a = a_x$ of $x$ onto $\partial\Omega$. Then $\overline{u}_{\Omega}(x) \geq u_{\infty,R}(\rho(x))$, hence $\overline{u}_{\Omega}$ is a large solution. If we assume moreover that $g$ satisfies the scaling property (3.1.45) for some $\alpha > \frac{N-p}{p-1}$ (notice that this implies (3.1.108)) and the growth properties (3.1.46), then

$$\overline{u}_{\Omega}(x) \geq u_{\infty,R}(|x-a|) \geq \kappa\left((\rho(x))^{-\alpha} - R^{-\alpha}\right), \tag{3.1.110}$$

where $\kappa$ is defined by (3.1.47). □

**Definition 3.1.21** A bounded domain $\Omega$ is regular for the $p$-Laplacian if for any $\phi \in C(\bar{\Omega}) \cap W^{1,p}(\Omega)$ the unique $p$-harmonic function $u$ in $\Omega$ such that $u - \phi \in W_0^{1,p}(\Omega)$ satisfies, for any $a \in \partial\Omega$,

$$\lim_{\substack{x \to a \\ x \in \Omega}} u(x) = \phi(a). \tag{3.1.111}$$

The following result, proved by [Wiener (1924)] in the case $p = 2$, [Maz'ya (1970)] and [Kilpeläinen, Malý (1994)] in the case $p \neq 2$ characterizes regular domains.

**Theorem 3.1.22** *Assume $1 < p \leq N$. Then a bounded domain $\Omega$ is regular for the p-Laplacian if and only if*

$$\int_0^1 \left(\frac{c_{1,p}^{\mathbb{R}^N}(\Omega^c \cap B_r(a))}{r^{N-p}}\right)^{p'-1} \frac{dr}{r} = \infty \qquad \forall a \in \partial\Omega, \tag{3.1.112}$$

*where $c_{1,p}^{\mathbb{R}^N}$ denotes the Sobolev capacity associated to $W^{1,p}(\mathbb{R}^N)$ and $p'$ is the conjugate of $p$.*

*Remark.* Note that any Lipschitz domain satisfies (3.1.112). The sufficiency part of the proof [Maz'ya (1970)] shows that the result remains true if $u$ is a solution of

$$-\Delta_p u = H \qquad \text{in } \Omega,$$

where $H$ is a bounded function.

**Theorem 3.1.23** *Assume $1 < p \leq N$ and $\Omega$ is a bounded domain regular for the p-Laplacian. Then the maximal solution of (3.1.17) in $\Omega$ is a large solution.*

*Proof.* Let $k \in \mathbb{N}^*$. We consider the problem

$$\begin{aligned} -\Delta_p u + g(u) &= 0 \qquad \text{in } \Omega \\ u &= k \qquad \text{on } \partial\Omega. \end{aligned} \tag{3.1.113}$$

Clearly the constant $k$ is a supersolution of the problem. Furthermore, by the maximum principle, any solution of (3.1.113) is smaller than $k$. By Theorem 3.1.22 and the remark hereafter, there exists a solution $v := v_k$ to the problem

$$\begin{aligned} -\Delta_p v + g(k) &= 0 \qquad \text{in } \Omega \\ v &= k \qquad \text{on } \partial\Omega. \end{aligned} \tag{3.1.114}$$

Since $v_k \leq k$ and $g$ is nondecreasing, $v_k$ is a subsolution for problem (3.1.113). Therefore there exists a solution $u := u_k$ of (3.1.113) such that $v_k \leq u_k \leq k$. By the maximum principle $k \mapsto u_k$ is increasing, $v_k \leq u_k \leq k$ and $u_k(x) \leq h(\rho(x))$ in $\Omega$. Letting $k \to \infty$ we infer that $u_k$ converges to a large solution of (3.1.17) in $\Omega$. □

**Theorem 3.1.24** *Let $1 < p < \infty$. Assume $\Omega$ is a bounded domain star-shaped with respect to some point a and $g : \mathbb{R} \mapsto \mathbb{R}$ is a continuous nondecreasing function which vanishes at 0, is not identically zero and satisfies*

$$g(s) \leq \lambda^{(\alpha+1)(p-1)+1} g(\lambda^{-\alpha} s) \qquad \forall s > 0,\ 0 < \lambda \leq 1, \tag{3.1.115}$$

*for some $\alpha > 0$. Then (3.1.17) admits at most one large solution in $\Omega$.*

*Proof.* We can assume that $\Omega$ is star-shaped with respect to 0. If $u$ and $v$ are two large solutions and $\lambda > 0$, we set $v_\lambda(x) = \lambda^\alpha v(\lambda x)$. Then

$$\Delta_p v_\lambda = \lambda^{(p-1)(\alpha+1)+1} g(\lambda^{-\alpha} v_\lambda) \geq g(v_\lambda) \qquad \text{in } \Omega_\lambda := \tfrac{1}{\lambda}\Omega,$$

thus $v_\lambda$ is a subsolution in $\Omega_\lambda$. If we assume that $\lambda < 1$, $\bar{\Omega} \subset \Omega_\lambda$, thus $v_\lambda$ is bounded in $\Omega$ and therefore smaller than $u$. If we let $\lambda \to 1$, we infer $v \leq u$. Similarly $u \leq v$. □

If the domain $\Omega$ is the complement of a bounded star-shaped domain, a similar uniqueness result holds.

**Theorem 3.1.25** *Let $1 < p < \infty$, $G$ be a bounded domain star-shaped with respect to some point $a$ and set $\Omega = \overline{G}^c$. If $g : \mathbb{R} \mapsto \mathbb{R}$ is a continuous nondecreasing function which vanishes at 0, is not identically zero and satisfies*

$$g(s) \geq \lambda^{(\alpha+1)(p-1)+1} g(\lambda^{-\alpha} s) \qquad \forall s > 0,\, 0 < \lambda \leq 1. \tag{3.1.116}$$

*for some $\alpha > 0$. Then the problem*

$$\begin{aligned} -\Delta_p u + g(u) &= 0 \qquad \text{in } \Omega \\ \lim_{\rho(x)\to 0} u(x) &= \infty, \end{aligned} \tag{3.1.117}$$

*admits at most one solution.*

*Proof.* If $G$ is star-shaped, with respect to some point $a$ that we may assume to be 0, then for any $x \in \Omega$ and $\kappa > 1$, $\kappa x \in \Omega$. Notice that (3.1.116) implies $g(s) \geq s^{p(1+\frac{1}{\alpha}))-1} g(1)$ (we can assume $g(1) > 0$), thus Lemma 3.1.2 holds with $b = 0$, hence there exists a maximal solution in $\Omega$ and any positive solution tends to 0 at infinity. If $u$ and $v$ are two solutions of (3.1.117), $v_\lambda$ which is defined by the same transformation as in the proof of Theorem 3.1.24 is a solution of (3.1.117) where $\Omega$ is replaced by $\frac{1}{\lambda}\Omega$. If we assume that $\lambda < 1$, then for any $\epsilon > 0$, $v_\lambda + \epsilon$ is a supersolution of (3.1.117) in $\frac{1}{\lambda}\Omega$, thus $u \leq v_\lambda + \epsilon$. Letting $\epsilon \to 0$ and $\lambda \to 1$ yields $u \leq v$. □

*Remark.* If $\Omega$ is an exterior domain the results of Proposition 3.1.19 and Proposition 3.1.20 are still valid.

**Theorem 3.1.26** *Let $1 < p < \infty$. Assume $\Omega$ is a bounded domain satisfying a uniform interior and exterior sphere condition. If $g : \mathbb{R} \mapsto \mathbb{R}$ is a continuous nondecreasing function such that $g(0) = 0$, $r \mapsto r^{1-p} g(r)$ is nondecreasing on $\mathbb{R}_+$ and*

$$\lim_{r\to\infty} r^{-q} g(r) = \gamma, \tag{3.1.118}$$

*for some $q > p - 1$ and $\gamma > 0$. Then $\overline{u}_\Omega$ is the unique large solution of (3.1.17).*

*Proof.* Without loss of generality we can assume that $r_0 = \gamma = 1$. The two-sphere condition assumption on the boundary asserts that there exists some $r_0 > 0$ such that for any $a \in \partial\Omega$ one can find $z_a \in \Omega$ and $\tilde{z}_a \in \overline{\Omega}^c$ such that $|a - z_a| = |a - \tilde{z}_a| = r_0$ and the balls $B_{r_0}(z_a)$ and $B_{r_0}(\tilde{z}_a)$ are respectively included into $\Omega$ and $\overline{\Omega}^c$.

*Step 1.* Let $0 < R \leq 1$ and $y$ be the maximal solution of

$$\begin{aligned} -\Delta_p y + y^q &= 0 \qquad \text{in } B_R \\ \lim_{|x|\to R} y(x) &= \infty. \end{aligned} \tag{3.1.119}$$

Then $y$ is the unique large solution, it is radial and it satisfies

$$y(r) = \omega_{p,q}(R-r)^{-\frac{p}{q+1-p}}(1+o(1)) \quad \text{as } r \uparrow R \tag{3.1.120}$$

where

$$\omega_{p,q} = \left(\left(\frac{p}{q+1-p}\right)^{p-1}\left(\frac{pq}{q+1-p} - 1\right)\right)^{\frac{1}{q+1-p}}.$$

Since $y$ is radial, it satisfies

$$\begin{aligned} & -|y'|^{p-2}\left((p-1)y'' + \frac{N-1}{r}y'\right) + y^q = 0 \quad \text{in } (0,R) \\ & y'(0) = 0, \quad \lim_{r\to R} y(r) = \infty. \end{aligned} \tag{3.1.121}$$

By the maximum principle, $y' > 0$ on $(0,R)$ and by (3.1.7), there holds

$$y(r) \leq c(R-r)^{-\frac{p}{q+1-p}}, \tag{3.1.122}$$

$c = c(N,p,q) > 0$. Actually, since $(p-1)y'^{p-2}y'' \leq y^q$, we obtain

$$y'(r) \leq \left(\frac{p}{(p-1)(q+1)}\right)^{\frac{1}{p}} (y(r))^{\frac{q+1}{p}},$$

and

$$y(r) \geq \omega_{p,q}(R-r)^{-\frac{p}{q+1-p}} \qquad \forall r \in [0,R). \tag{3.1.123}$$

We deduce also from the above upper estimate on $y'$ that for any $\epsilon > 0$, there exists $r_\epsilon \in (0,R)$ such that $(p-1)y'^{p-2}y'' \geq (1-\epsilon)y^q$ on $[r_\epsilon, R)$. This implies

$$\limsup_{r\to 1}(R-r)^{\frac{p}{q+1-p}} y(r) \leq (1-\epsilon)^{-\frac{1}{q+1-p}}\omega_{p,q}, \tag{3.1.124}$$

and finally (3.1.120) holds.

*Step 2.* Let $0 < R < S < 1$ and $\tilde{y}$ be the maximal solution of

$$\begin{aligned} -\Delta_p \tilde{y} + \tilde{y}^q &= 0 && \text{in } B_S \setminus \overline{B}_R \\ \lim_{|x| \to R} \tilde{y}(x) &= \infty && \\ \tilde{y}(x) &= 0 && \text{on } \partial B_S. \end{aligned} \tag{3.1.125}$$

The function $\tilde{y}$ is radially symmetric and satisfies (3.1.121) in $(R, S)$. Furthermore

$$\tilde{y}(r) \leq c(r - R)^{-\frac{p}{q+1-p}}, \tag{3.1.126}$$

and $y$ is decreasing and $(p-1)(-\tilde{y}')^{p-2}\tilde{y}'' \geq y^q$. Then

$$\tilde{y}(r) \leq \omega_{p,q}(r - R)^{-\frac{p}{q+1-p}} \qquad \forall r \in (R, S]. \tag{3.1.127}$$

The use of the estimate on $y'$ implies, as in Step 1, that for any $\epsilon > 0$ one can find $r_\epsilon \in (R, S)$, such that $(p-1)\tilde{y}'^{p-2}\tilde{y}'' \leq (1 + \epsilon)\tilde{y}^q$ on $(R, r_\epsilon]$. This implies

$$\liminf_{r \to R} (r - R)^{\frac{p}{q+1-p}} \tilde{y}(r) \geq (1 + \epsilon)^{-\frac{1}{q+1-p}} \omega_{p,q}, \tag{3.1.128}$$

and finally

$$\tilde{y}(r)(r) = \omega_{p,q}(r - 1)^{-\frac{p}{q+1-p}}(1 + o(1)) \quad \text{as } r \downarrow R. \tag{3.1.129}$$

*Step 3.* Let $u_1$ and $u_2$ be two large solutions of (3.1.17). Then we claim that

$$\lim_{\rho(x) \to 0} \frac{u_1(x)}{u_2(x)} = 1. \tag{3.1.130}$$

Since $u_1$ and $u_2$ are large solutions, for any $\delta \in (0, 1)$ there exists $0 < R_\delta < 1$ such that

$$(1 - \delta)(u_i(x))^q \leq g(u_i(x)) \leq (1 + \delta)(u_i(x))^q \qquad \forall x \in \Omega_{2R_\delta}, \tag{3.1.131}$$

(we recall that $\Omega_{2R_\epsilon} := \{x \in \Omega : \rho(x) < 2R_\epsilon\}$). Thus, for $i = 1, 2$,

$$-\Delta_p u_i + (1 - \delta)u_i^q \leq -\Delta_p u_i + g(u_i) \leq -\Delta_p u_i + (1 + \delta)u_i^q \quad \text{in } \Omega_{2R_\delta}.$$

Let $a \in \Omega$ such that $\rho(a) = R_\delta$. If $y$ is the solution of (3.1.119) with $R = R_\delta$, we have (by rescaling and translation)

$$u_i(x) \leq (1 - \delta)^{-\frac{1}{q+1-p}} y(x - a) \qquad \forall x \in B_{\rho(a)}(a).$$

Since $a$ and $\delta$ are arbitrary, it follows from Step 1 that

$$\limsup_{\rho(x)\to 0}(\rho(x))^{\frac{p}{q+1-p}}u_i(x) \leq \omega_{p,q}. \tag{3.1.132}$$

For proving the estimate from below, let $\tilde{a} \in \Omega^c$ be such that $\dot{\rho}(\tilde{a}) = \text{dist}\,(\tilde{a}, \partial\Omega) = R_\delta$ and $\tilde{y}$ be the solution of (3.1.125) with $R = R_\delta$ and $S = 2R_\delta$. Then

$$u_i(x) \geq (1+\delta)^{-\frac{1}{q+1-p}}\tilde{y}(x-\tilde{a}) \qquad \forall x \in B_{2\dot{\rho}(\tilde{a})}(\tilde{a}) \cap \Omega.$$

Again, since $\tilde{a}$ and $\delta$ are arbitrary, we infer from Step 1 that

$$\liminf_{\rho(x)\to 0}(\rho(x))^{\frac{p}{q+1-p}}u_i(x) \geq \omega_{p,q}. \tag{3.1.133}$$

Combining (3.1.132) and (3.1.133) yields (3.1.130).

*Step 4.* End of the proof. Since $r \mapsto r^{1-p}g(r)$ is nondecreasing on $\mathbb{R}_+$, for $\epsilon > 0$, the function $(1+\epsilon)u_1$ is a supersolution of (3.1.17). Moreover $(u_2 - (1+\epsilon)u_1)_+$ has compact support in $\Omega$. By monotonicity it implies $u_2 \leq (1+\epsilon)u_1$ and thus $u_2 \leq u_1$. Similarly $u_1 \leq u_2$. Since the maximal solution is a large solution, it is the unique large solution. □

*Remark.* In the previous result, the assumption that $\Omega$ bounded can be easily replaced by $\partial\Omega$ bounded.

When $p = 2$, the regularity assumption on $\partial\Omega$ can be relaxed. It is proved in [Marcus, Véron (1997)], [Marcus, Véron (2004)] the next two results which, up to now, are specific to the semilinear case.

**Theorem 3.1.27** *Let $\Omega \subset \mathbb{R}^N$ be a bounded domain and $g : \mathbb{R} \mapsto \mathbb{R}$ be a nondecreasing convex function satisfying $g(0) \geq 0$ and (3.1.15) with $p = 2$. Let $u := \overline{u}_\Omega$ be the maximal solution of equation (3.1.134) in $\Omega$. If there exist two constants $K = K(\Omega, g) > 1$ and $\delta = \delta(\Omega, g) > 0$ such that any large solution $u$ of*

$$-\Delta u + g(u) = 0, \tag{3.1.134}$$

*in $\Omega$ satisfies*

$$0 \leq \overline{u}_\Omega(x) \leq Ku(x) \qquad \forall\, x \in \Omega \;\; s.t. \;\; \rho(x) \leq \delta, \tag{3.1.135}$$

*then the equation admits at most one large solution in $\Omega$.*

*Proof.* If $\overline{u}_\Omega$ is not a large solution the result holds since no large solution exists. Henceforth let us assume that $\overline{u}_\Omega$ is a large solution. If $\overline{\Omega} \subset B_R$, then any large solution is bounded from below by the solution $\phi$ of

$$\begin{aligned} -\Delta\phi + g(\phi) &= 0 && \text{in } B_R \\ \phi &= 0 && \text{on } \partial B_R, \end{aligned} \tag{3.1.136}$$

and we denote by $\ell_\phi$ the minimum of $\phi$; it is a nonpositive real number. Let $u$ be a large solution and we suppose that $u \neq \overline{u}_\Omega$. We set

$$w = u - \frac{1}{2K}(\overline{u}_\Omega - u).$$

Then

$$-\Delta w + g(w) = g\left(\left(1+\frac{1}{2K}\right)u - \frac{\overline{u}_\Omega}{2K}\right) - \left(1+\frac{1}{2K}\right)g(u) + \frac{1}{2K}g(\overline{u}_\Omega).$$

Since

$$\frac{2K}{1+2K}\left[\left(1+\frac{1}{2K}\right)u - \frac{\overline{u}_\Omega}{2K}\right] + \left(1+\frac{1}{2K}\right)\overline{u}_\Omega = u,$$

by the convexity assumption

$$\frac{2K}{1+2K}g\left(\left(1+\frac{1}{2K}\right)u - \frac{\overline{u}_\Omega}{2K}\right) + \frac{1}{1+2K}g(\overline{u}_\Omega) \geq g(u).$$

Therefore $w$ is a supersolution for (3.1.134), and it is larger than $\frac{K+1}{2K}u$. For any $\theta \in (0,1)$, the function $w_\theta = \theta u + (1-\theta)\phi$ satisfies

$$-\Delta w_\theta + g(w_\theta) = g(\theta u + (1-\theta)\phi) - (1-\theta)g(\phi) \leq 0,$$

because of the convexity assumption. If we fix $\theta < \frac{K+1}{2K}$, the function $x \mapsto (w_\theta - w)_+(x)$ has compact support in $\Omega$. Since $g$ is decreasing there holds $w_\theta \leq w$ in $\Omega$. Since $w_\theta$ is a subsolution of (3.1.135) smaller than the supersolution $w$, there exists a solution $u_1$ satisfying $w_\theta \leq u_1 \leq w$. Therefore $u_1$ is a large solution, consequently it satisfies (3.1.135). Therefore

$$\begin{aligned} u_1 \leq w = u - \frac{1}{2K}(\overline{u}_\Omega - u) \Longrightarrow \\ \overline{u}_\Omega - u_1 \geq \left(1+\frac{1}{2K}\right)(\overline{u}_\Omega - u). \end{aligned} \tag{3.1.137}$$

Replacing $u$ by $u_1$ which is a large solution, we set

$$w_1 = u_1 - \frac{1}{2K}(\overline{u}_\Omega - u_1) \quad \text{and} \quad w_{1,\theta} = \theta u_1 + (1-\theta)\phi.$$

The functions $w_1$ and $w_{1,\theta}$ are respectively a supersolution and a subsolution of (3.1.135) and $w_{1,\theta} \leq w_1$. Therefore there exists a solution $u_2$ of (3.1.135) in $\Omega$ where it satisfies $w_{1,\theta} \leq u_2 \leq w_1$. Furthermore

$$\begin{aligned} &u_2 \leq u_1 - \frac{1}{2K}(\overline{u}_\Omega - u_1) \Longrightarrow \\ &\overline{u}_\Omega - u_2 \geq \left(1 + \frac{1}{2K}\right)(\overline{u}_\Omega - u_1). \end{aligned} \tag{3.1.138}$$

Iterating this process, we construct a sequence $\{u_n\}$ of large solutions of (3.1.135) in $\Omega$ satisfying (3.1.135), $u_0 = u$ and such that

$$\begin{aligned} &u_n \leq u_{n-1} - \frac{1}{2K}(\overline{u}_\Omega - u_{n-1}) \Longrightarrow \\ &\overline{u}_\Omega - u_n \geq \left(1 + \frac{1}{2K}\right)(\overline{u}_\Omega - u_{n-1}). \end{aligned} \tag{3.1.139}$$

Therefore

$$\overline{u}_\Omega - u_n \geq \left(1 + \frac{1}{2K}\right)^n (\overline{u}_\Omega - u). \tag{3.1.140}$$

Since $u_n \geq \phi$, we infer a contradiction by letting $n \to \infty$. □

**Definition 3.1.28** The boundary $\partial\Omega$ of a domain $\Omega \subset \mathbb{R}^N$ has *the local continuous graph property* if for every boundary point $P$, there exists an open neighborhood $Q_P$ of $P$ and a set of coordinates $y = \{y_1, ..., y_N\}$, obtained from $x$ by translation and rotation, and a function $F_P \in C(\mathbb{R}^{N-1})$ such that

$$Q_P \cap \Omega = Q_P \cap G(F_P) \quad \text{where } G(F_P) = \{y : y_N < F_P(y_1, ..., y_{N-1})\}.$$

The local continuous graph property which can be viewed as a "local star-shapedness" property with respect to points at infinity.

**Theorem 3.1.29** *Let $\Omega \subset \mathbb{R}^N$ be a bounded domain the boundary of which has the local continuous graph property and $g : \mathbb{R} \mapsto \mathbb{R}$ be a nondecreasing function such that $g(0) \geq 0$, satisfying (3.1.15) with $p = 2$ and:*

*(i) either $g$ is convex and for any real number $m$ there exists some constant $L_m \geq 0$ such that*

$$g(a + b) \geq g(a) + g(b) - L_m \qquad \forall a, b \geq m, \tag{3.1.141}$$

*(ii) or for any real number $m$ there exists some constant $L_m \geq 0$ such that*

$$g(ra+sb) \geq rg(a)+sg(b)-(r+s-1)L_m \quad \forall a,b \geq 0\,,\ \forall r,s \geq 1. \tag{3.1.142}$$

*Then the equation (3.1.134) admits at most one large solution in $\Omega$.*

*Proof.* Up to choosing $R$ large enough there is no loss of generality in assuming that $\overline{Q}_P \subset B_R$ is a bounded circular cylinder with symmetry center $P \in \partial\Omega$, with axis parallel to the $y_N$ axis:

$$Q_P = \{z = (z', z_N) \in \mathbb{R}^{N-1} \times \mathbb{R} : |z'| < R_P, -T_P < z_N < T_P\},$$

where $z = y - P$. We assume also that $\partial\Omega$ is bounded away from the top and the bottom of the cylinder $Q_P$ and that $\partial\Omega \cap \overline{Q}_P = \overline{\partial\Omega \cap Q_P}$ and we denote

$$\Theta = \Omega \cap Q_P\,,\ \Gamma_1 = Q_P \cap \partial\Omega\,,\ \Gamma_2 = \partial Q_P \cap \overline{\Omega}\,,\ \overset{\circ}{\Gamma}_2 = \partial Q_P \cap \Omega. \tag{3.1.143}$$

We assume that $u$ is a large solution. Then $u$ is bounded from below by the solution $\phi$ of (3.1.136).

*Step 1.* We claim that there exists a function $v \in W^{2,s}_{loc}(\Theta)$ for any $s < \infty$, solution of in $\Theta$, such that

(i) $v(x) \to \infty$ locally uniformly as $x \to \Gamma_1$,

(ii) $v(x) \to 0$ locally uniformly as $x \to \overset{\circ}{\Gamma}_2$,

where $v(x) \to 0$ (resp. $v(x) \to \infty$) means that for every compact set $K$ contained in the relative interior of $\overset{\circ}{\Gamma}_2$ (resp. $\Gamma_1$), there holds $v(x) \to 0$ (resp. $v(x) \to \infty$) as dist $(x, K) \to 0$. Let $D_P := \{z' = (z_1, ..., z_{N-1} : |z'| \leq R_P\}$. Set $\tilde{F}_P(z') = F_P(y' + P')$ and let $\{f_n\}$ be an increasing sequence of smooth positive functions defined on $D_P$ which converges to $\tilde{F}_P$ uniformly on $D_P$. We put $\Theta_n = \{z : |z'| < R_P, -T_P < z_N < f_n(z')\}$, $\Gamma_{2,n} = \partial Q_P \cap \partial\Theta_n$ and $\Gamma_{1,n} = \partial\Theta_n \setminus \partial Q_P$. Let $v_{n,k}$ be a positive solution of (3.1.134) in $\Theta_n$ such that

$$v_{n,k}(z', z_N) = \begin{cases} k & \text{if } (z', z_N) \in \Gamma_{1,n} \\ 0 & \text{if } (z', z_N) \in \Gamma_{2,n} \\ & \text{s.t. } -T_P \leq z_N \leq f_n(z') - \frac{1}{n} \\ n(z_N - f_n(z') + \frac{1}{n})k & \text{if } |z'| = R_P \\ & \text{s.t. } f_n(z') - \frac{1}{n} \leq z_N \leq f_n(z'). \end{cases} \tag{3.1.144}$$

For fixed $n$, the sequence $\{v_{n,k}\}$ is increasing with respect to $k$ and it converges, as $n \to \infty$, to some function $v$ which is continuous in $\Theta_n$, satisfies (3.1.134) in $\Theta_n$ and has the property that (i-n) $v_n(x) \to \infty$ locally uniformly as $x \to \Gamma_{1,n}$,

(ii-n) $v(x) \to 0$ locally uniformly as $x \to \overset{\circ\,*}{\Gamma}_{2,n}$,

where $\overset{\circ\,*}{\Gamma}_{2,n} = \{z \in \overset{\circ}{\Gamma}_{2,n} : z_n \leq f_n(z') - \frac{1}{n}\}$. The sequence $\{v_n\}$ is decreasing and converges locally uniformly in $\Theta$ to some continuous function $v$ which satisfies (3.1.134) in $\Theta$, and clearly (ii) holds. Next we have to prove that (i) holds. Since $\partial Q_P$ is Lipschitz continuous, there exists a solution $v^*$ to the following problem

$$\begin{aligned} -\Delta v^* + g(v^*) &= 0 && \text{in } Q_P \\ v^* &= 0 && \text{on } \partial Q_P, \end{aligned} \tag{3.1.145}$$

$v^*$ is nonpositive since $g(0) \geq 0$, and we set $\ell_{v^*} = \min\{v^*(x) : x \in \overline{Q}_P\}$. Then $v_{n,k}$, $v_n$ and ultimately $v$ are bounded from below by $\ell_{v^*}$ in their respective domains of definition. There exists also a large solution $w$ to (3.1.134) in $Q_P$ and it is bounded from below by $\ell_\phi$ (defined to be the minimum of $\phi$ as we have seen it in the proof of Theorem 3.1.27). We put $\ell = \min\{\ell_\phi, \ell_{v^*}, 0\}$, $v_\ell = v - \ell$, $v_{n,\ell} = v_n - \ell$ and $w_\ell = w - \ell$. Then $v_{n,\ell}$ and $w_\ell$ are nonnegative and

$$\begin{aligned} -\Delta(v_{n,\ell} + w_\ell) + g(v_{n,\ell} + w_\ell) &= g(v_{n,\ell} + w_\ell) - g(v_{n,\ell} + \ell) - g(w_\ell + \ell) \\ &\geq -L_\ell. \end{aligned}$$

Let $\psi$ be the solution of

$$\begin{aligned} -\Delta\psi &= L_\ell && \text{in } B_R \\ \psi &= 0 && \text{on } \partial B_R. \end{aligned} \tag{3.1.146}$$

Then $\psi \geq 0$ and the function $W_n = v_{n,\ell} + w_\ell + \psi$ satisfies

$$-\Delta W_n + g(W_n) \geq 0 \qquad \text{in } \Theta_n,$$

and it is larger than $u$ on $\partial\Theta_n$, hence $W_n \geq u$ in $\Theta_n$. Letting $n \to \infty$ yields $v_\ell + w_\ell \geq u$, which implies (i). Furthermore, if we set $\epsilon_n = \max\{F_P(z') - f_n(z') : z' \in D_P\}$, $\tilde{v}_n(z', z_N) = v(z', z_N + \epsilon_n$ and $\Theta_n = \{(z', z_N) \in Q_P : z_N > -T_P + \epsilon_n)\}$, then $u \geq \tilde{v}_n + \ell$ in $\Theta_n$. Letting $n \to \infty$ we infer

$$v \leq u \leq v_\ell + w_\ell \qquad \text{in } \Theta. \tag{3.1.147}$$

*Step 2.* If $\overline{u}_\Omega$ is the maximal solution of (3.1.134) and $u$ any large solution, it follows from (3.1.147) and the compactness of $\partial\Omega$ that

$$\lim_{\rho(x)\to 0} \frac{u(x)}{\overline{u}_\Omega} = 1. \tag{3.1.148}$$

If assumption (i) holds, then

$$-\Delta(2u) + g(2u) \geq -L_\ell,$$

thus

$$-\Delta(2u+\psi) + g(2u+\psi) \geq 0,$$

which yields

$$\overline{u}_\Omega \leq 2u+\psi \leq 3u \qquad \text{in } \Omega_\delta := \{x \in \Omega : \rho(x) < \delta\}, \tag{3.1.149}$$

for some $\delta > 0$. Then the proof follows from Theorem 3.1.27.
If assumption (ii) holds, then for any $\epsilon > 0$ the function $u_\epsilon = (1+\epsilon)u$ satisfies

$$-\Delta u_\epsilon + g(u_\epsilon) \geq -\epsilon L_\ell,$$

which implies

$$-\Delta(u_\epsilon + \epsilon\psi) + g(u_\epsilon + \epsilon\psi) \geq 0.$$

Therefore $u_\epsilon + \epsilon\psi \geq \overline{u}_\Omega$ by (3.1.148) and the proof follows by letting $\epsilon \to 0$.
□

**Applications.** Let $\Omega \subset \mathbb{R}^N$ be a domain which satisfies $\partial\Omega = \partial\overline{\Omega}^c$. Then Theorem 3.1.27 applies in particular to

$$-\Delta u + e^{au} = 0 \quad \text{in } \Omega \subset \mathbb{R}^2, \tag{3.1.150}$$

where $a > 0$, since any large solution $u$ satisfies

$$u(x) = -\frac{2}{a}\ln\rho(x)\,(1+o(1)) \qquad \text{as } \rho(x) \to 0.$$

It applies also to

$$-\Delta u + |u|^{q-1}\,u = 0 \quad \text{in } \Omega, \tag{3.1.151}$$

provided $1 < q < \frac{N}{N-2}$, since for any large solution $u$, there holds

$$c(\rho(x))^{-\frac{2}{q-1}} \leq u(x) \leq c^{-1}(\rho(x))^{-\frac{2}{q-1}},$$

near $\partial\Omega$, for some $c = c(N,q,\Omega) > 0$. It holds also for weakly super linear equation such as

$$-\Delta u + u(\ln_+ u)^\alpha = 0 \quad \text{in } \Omega, \tag{3.1.152}$$

for $\alpha > 2$ since any large solution satisfies

$$u(x) = \gamma(\alpha, N) e^{\gamma(\alpha)|x|^{\frac{2}{2-\alpha}}} \left(1 + O\left(|x|^{\frac{2}{\alpha-2}}\right)\right) \quad \text{as } \rho(x) \to 0,$$

where

$$\gamma(\alpha) = \left(\frac{2}{\alpha - 2}\right)^{\frac{2}{\alpha-2}} \quad \text{and} \quad \gamma(\alpha, N) = e^{\frac{\alpha-(N-1)(\alpha-2)}{2\alpha}}.$$

## 3.2 Singular solutions in $\mathbb{R}^N_+$

In this section, we study isolated boundary singularities of solutions of quasilinear equations with power absorption. The model problem under consideration could be summarized as follows: assume $q > p-1 > 0$ and $\Omega$ is a smooth domain in $\mathbb{R}^N$ the boundary of which contains 0, is it possible to describe the behavior near 0 of any function $u \in C(\overline{\Omega} \setminus \{0\})$ which satisfies

$$\begin{aligned} -\Delta_p u + |u|^{q-1}u &= 0 \qquad \text{in } \Omega \\ u &= 0 \qquad \text{in } \partial\Omega \setminus \{0\}. \end{aligned} \tag{3.2.1}$$

### 3.2.1 *Separable solutions in* $\mathbb{R}^N_+$

If we look for functions $u \in C(\overline{\mathbb{R}^N_+} \setminus \{0\})$ under the form $u(x) = u(r, \sigma) = r^{-\beta}\omega(\sigma)$ satisfying

$$\begin{aligned} -\Delta_p u + |u|^{q-1}u &= 0 \qquad \text{in } \mathbb{R}^N_+ \\ u &= 0 \qquad \text{in } \partial\mathbb{R}^N_+ \setminus \{0\}, \end{aligned} \tag{3.2.2}$$

then $\beta = \beta_{p,q} = \frac{p}{q+1-p}$ and $\omega$ satisfies

$$\begin{aligned} -div'\left(\left(\beta^2_{p,q}\omega^2 + |\nabla'\omega|^2\right)^{\frac{p-2}{2}} \nabla'\omega\right) + |\omega|^{q-1}\omega & \\ - \beta_{p,q}\Lambda(\beta_{p,q})\left(\beta^2_{p,q}\omega^2 + |\nabla'\omega|^2\right)^{\frac{p-2}{2}} \omega = 0 & \qquad \text{in } S^{N-1}_+ \\ \omega = 0 & \qquad \text{on } \partial S^{N-1}_+, \end{aligned} \tag{3.2.3}$$

where $\Lambda(\beta_{p,q}) = (\beta_{p,q}(p-1) + p - N)$ as defined in see (3.1.3). The homogeneous problem without absorption which corresponds to separable $p$-harmonic functions $u = r^{-\beta}\phi(\sigma)$ in $\mathbb{R}^N_+$ which vanish on $\partial\mathbb{R}^N_+ \setminus \{0\}$ is the following

$$\begin{aligned} -div'\left(\left(\beta^2\phi^2+|\nabla'\phi|^2\right)^{\frac{p-2}{2}}\nabla'\phi\right)\qquad\qquad&\\ -\beta\Lambda(\beta)\left(\beta^2\phi^2+|\nabla'\phi|^2\right)^{\frac{p-2}{2}}\phi=0&\quad\text{in } S_+^{N-1}\\ \phi=0&\quad\text{on } \partial S_+^{N-1}. \end{aligned}\tag{3.2.4}$$

The exponent $\beta_S = \beta_{S_+^{N-1}}$ is defined in Theorem 2.2.1 and estimates are given in Theorem 2.2.9. The function $V_{\mathbb{R}_+^N}(r,\sigma) = r^{-\beta_{S_+^{N-1}}}\phi_{S_+^{N-1}}(\sigma)$, where $\phi_{S_+^{N-1}}$ is a positive solution of (3.2.4) plays the role of the Poisson kernel with pole at 0 in the study of boundary isolated singularities. The critical exponent $\tilde{q}_1$ is implicitly defined by

$$\beta_{S_+^{N-1}} = \frac{p}{\tilde{q}_1+1-p} \Longleftrightarrow \tilde{q}_1 = \frac{p}{\beta_{S_+^{N-1}}}+p-1. \tag{3.2.5}$$

This exponent corresponds to the isotropy threshold already introduced in Lemma 3.1.11. This exponent depends on $p$ and should be denoted by $\tilde{q}_{1,p}$, but we will drop this extra index except if necessary.

**Theorem 3.2.1** *Assume $p>1$. Then*

*(i) If $q \geq \tilde{q}_1$, problem (3.2.3) admits no positive solution.*

*(ii) If $p-1<q<\tilde{q}_1$, problem (3.2.3) admits one and only one positive solution denoted by $\omega_s$.*

*Proof.* We define the operator $\mathcal{T}$ by

$$\begin{aligned} \mathcal{T}(\eta) = -div'\left(\left(\beta_{p,q}^2\eta^2+|\nabla'\eta|^2\right)^{\frac{p-2}{2}}\nabla'\eta\right)+|\eta|^{q-1}\eta&\\ -\beta_{p,q}\Lambda(\beta_{p,q})\left(\beta_{p,q}^2\eta^2+|\nabla'\eta|^2\right)^{\frac{p-2}{2}}\eta.& \end{aligned}\tag{3.2.6}$$

*Non-existence.* Assume $q \geq \tilde{q}_1$ and $\omega$ is a solution of (3.2.3). Set $\theta = \beta_{p,q}/\beta_{S_+^{N-1}}$, thus $0<\theta\leq 1$, and $\eta = \delta\phi^{\theta}_{S_+^{N-1}}$ for some $\delta>0$. Then

$$\begin{aligned} \delta^{1-p}\mathcal{T}(\eta) &= \delta^{q+1-p}\phi^{\theta q}_{S_+^{N-1}}\\ &+(p-1)(1-\theta)\theta^{p-1}\phi_{S_+^{N-1}}^{(\theta-1)(p-1)-1}\left(\beta^2_{S_+^{N-1}}\phi^2_{S_+^{N-1}}+|\nabla'\phi_{S_+^{N-1}}|^2\right)^{\frac{p}{2}}. \end{aligned}\tag{3.2.7}$$

We choose the smallest possible $\delta=\delta_0$ in order $\eta := \eta_\delta \geq \omega$. Such a choice is possible since $\omega$ is $C^1$ in $\overline{S_+^{N-1}}$ and the normal derivative to $\partial S_+^{N-1}$

(relative to $S^{N-1}$) is negative by Hopf boundary lemma. We will distinguish according there exists some $\sigma_0 \in S_+^{N-1}$ such that

$$\eta(\sigma) \geq \omega(\sigma) \quad \forall \sigma \in \overline{S_+^{N-1}} \quad \text{and} \quad \eta(\sigma_0) = \omega(\sigma_0), \tag{3.2.8}$$

or not. If (3.2.8) holds true, which is always the case when $\beta_{p,q} < \beta_{S_+^{N-1}}$, the function $w = \eta - \omega$ is nonnegative in $\overline{S_+^{N-1}}$, not identically 0 and achieves its minimum at $\sigma_0 \in S_+^{N-1}$. Denote by $g_{ij}$ the components of the metric tensor $g_0$ on $S^{N-1}$, with inverse $g^{ij}$. Then (see Chapter 1), if $(\sigma_1, ..., \sigma_{N-1})$ is a system of local coordinates near $\sigma_0$, we have

$$|\nabla w|^2 = \sum_{j,k} g^{ij} \frac{\partial w}{\partial \sigma_j} \frac{\partial w}{\partial \sigma_k},$$

and for any vector field $X$ with components $(X^1, ..., X^{N-1})$

$$div' X = \frac{1}{\sqrt{|g|}} \sum_k \frac{\partial}{\partial \sigma_k} \left( \sqrt{|g|} X^k \right).$$

Hence, by the mean value theorem,

$$\left(\beta_{p,q}^2 \eta^2 + |\nabla' \eta|^2\right)^{\frac{p-2}{2}} \frac{\partial \eta}{\partial \sigma_i} - \left(\beta_{p,q}^2 \omega^2 + |\nabla' \omega|^2\right)^{\frac{p-2}{2}} \frac{\partial \omega}{\partial \sigma_i} = \textstyle\sum_j \alpha_j^i \frac{\partial w}{\partial \sigma_i} + b^i w,$$

where

$$b^i = (p-2)\left(\beta_{p,q}^2 (\omega + tw)^2 + |\nabla'(\omega + tw)|^2\right)^{\frac{p-4}{2}} (\omega + tw) \frac{\partial(\omega + tw)}{\partial \sigma_i},$$

and

$$\begin{aligned} \alpha_j^i = (p-2)\left(\beta_{p,q}^2 (\omega + tw)^2 + |\nabla'(\omega + tw)|^2\right)^{\frac{p-4}{2}} \frac{\partial(\omega + tw)}{\partial \sigma_i} \sum_k g^{jk} \frac{\partial(\omega + tw)}{\partial \sigma_k} \\ + \delta_j^i \left(\beta_{p,q}^2 (\omega + tw)^2 + |\nabla'(\omega + tw)|^2\right)^{\frac{p-2}{2}}, \end{aligned}$$

where $\delta_i^j$ is the Kronecker's symbol. Since the graphs of $\eta$ and $\omega$ are ordered and tangent at $\sigma_0$, $\eta(\sigma_0) = \omega(\sigma_0) = P_0 > 0$ and $\nabla' \eta(\sigma_0) = \nabla' \omega(\sigma_0) = Q$. Thus

$$b^i(\sigma_0) = (p-2)\left(\beta_{p,q}^2 P_0^2 + |Q|^2\right)^{\frac{p-4}{2}} P_0 Q_i$$

and

$$\alpha_j^i(\sigma_0) = \left(\beta_{p,q}^2 P_0^2 + |Q|^2\right)^{\frac{p-4}{2}} \left( \delta_i^j \left(\beta_{p,q}^2 P_0^2 + |Q|^2\right) + (p-2) Q_i \sum_k g^{jk} Q_k \right).$$

Next

$$
\begin{aligned}
&\mathcal{T}(\eta)-\mathcal{T}(\omega)\\
&=\frac{-1}{\sqrt{g}}\sum_{\ell,i}\frac{\partial}{\partial\sigma_\ell}\left(\sqrt{g}g^{\ell i}\left(\beta_{p,q}^2\eta^2+|\nabla'\eta|^2\right)^{\frac{p-2}{2}}\frac{\partial\eta}{\partial\sigma_i}-\left(\beta_{p,q}^2\omega^2+|\nabla'\omega|^2\right)^{\frac{p-2}{2}}\frac{\partial\omega}{\partial\sigma_i}\right)\\
&\quad+\beta_{p,q}\Lambda(\beta_{p,q})\left(\left(\beta_{p,q}^2\eta^2+|\nabla'\eta|^2\right)^{\frac{p-2}{2}}\eta-\left(\beta_{p,q}^2\omega^2+|\nabla'\omega|^2\right)^{\frac{p-2}{2}}\omega\right)+\eta^q-\omega^q\\
&=\frac{-1}{\sqrt{g}}\sum_{\ell,i}\frac{\partial}{\partial\sigma_\ell}\left(\sqrt{g}g^{\ell i}\sum_i\alpha_j^i\frac{\partial w}{\partial\sigma_j}+b^iw\right)+\sum_i C_i\frac{\partial w}{\partial\sigma_i}+Cw\\
&=\frac{-1}{\sqrt{g}}\sum_{\ell,i}\frac{\partial}{\partial\sigma_\ell}\left(a_j^\ell\frac{\partial w}{\partial\sigma_j}\right)+\sum_i C_i\frac{\partial w}{\partial\sigma_i}+Cw,
\end{aligned}
\tag{3.2.9}
$$

where

$$a_j^\ell=\sqrt{g}\sum_i g^{\ell i}\alpha_j^i.$$

The matrix $(\alpha_j^i(\sigma_0))$ is symmetric, definite and positive since it is the Hessian at $Q$ of the function

$$X=(X_1,...,X_{N-1})\mapsto\frac{1}{p}\left(P_0^2+|X|^2\right)^{\frac{p}{2}}=\frac{1}{p}\left(P_0^2+\sum_{j,k}g^{jk}X_jX_k\right)^{\frac{p}{2}}.$$

Finally, since $w$ is nonnegative, vanishes at $\sigma_0$ and satisfies

$$\frac{-1}{\sqrt{g}}\sum_{\ell,i}\frac{\partial}{\partial\sigma_\ell}\left(a_j^\ell\frac{\partial w}{\partial\sigma_j}\right)+\sum_i C_i\frac{\partial w}{\partial\sigma_i}+Cw\geq 0,\tag{3.2.10}$$

it follows that it is locally identically zero and thus identically zero since $S_+^{N-1}$ is connected.

If (3.2.8) does not hold, then $\theta=1$ and the graphs of $\eta$ and $\omega$ are separated in $S_+^{N-1}$ and tangent at some point $\sigma_0\in\partial S_+^{N-1}$. Since the normal derivative of $\eta$ on $\partial S_+^{N-1}$ and relative to $S_+^{N-1}$ is negative, it follows from the tangency condition that the normal derivative of $w$ vanishes at $\sigma_0$ which contradicts Hopf maximum principle which is a consequence of (3.2.10). Therefore such a solution $\omega$ cannot exist.

*Existence.* Clearly any constant $k\geq\omega_c$ is a supersolution for (3.2.3). If $\theta=\beta_{p,q}/\beta_{S_+^{N-1}}$, then $\theta>1$. We consider again $\eta=\delta\phi^\theta_{S_+^{N-1}}$ ($\delta>0$). We compute $\mathcal{T}(\eta)$ as above. Since $\theta q>(\theta-1)(p-1)-1$ and

$\left(\beta^2_{S^{N-1}_+}\phi^2_{S^{N-1}_+} + \left|\nabla'\phi_{S^{N-1}_+}\right|^2\right)^{\frac{p}{2}} \geq \alpha_0 > 0$ we see from the expression of (3.2.7) that, for $\delta$ small enough, $\mathcal{T}(\eta) \leq 0$ and $0 \leq \eta < \omega_c$. We set

$$A(r,\xi) = \left(\beta^2_{p,q}r^2 + |\xi|^2\right)^{\frac{p-2}{2}}\xi,$$

for $(r,\xi) \in \mathbb{R} \times \mathbb{R}^{N-1} \sim \mathbb{R} \times TS^{N-1}$, then

$$\langle A(r,\xi) - A(r,\xi'), \xi - \xi'\rangle_g > 0 \qquad \text{if } \xi \neq \xi',$$

and

$$B(r,\xi) = |r|^{q-1}r - \left(\beta^2_{p,q}r^2 + |\xi|^2\right)^{\frac{p-2}{2}} r.$$

*Case 1:* $p \geq 2$. There holds

$$\begin{aligned} &(i) && \langle A(r,\xi),\xi\rangle_g \geq |\xi|^p \\ &(ii) && |A(r,\xi)| \leq p|\xi|^{p-1} + p\beta^{p-2}_{p,q}|r|^{p-1} \\ &(iii) && |B(r,\xi)| \leq (\beta^{p-2}_{p,q}+1)(|r|^q+1)(1+|\xi|^{p-1}). \end{aligned} \tag{3.2.11}$$

By Theorem 1.4.7 there exists $\omega \in W^{1,p}_0(S^{N-1}_+) \cap L^\infty(S^{N-1}_+)$ solution of

$$\begin{aligned} -div'&\left(\left(\beta^2_{p,q}\omega^2 + |\nabla'\omega|^2\right)^{\frac{p-2}{2}}\nabla'\omega\right) + |\omega|^{q-1}\omega \\ &- \beta_{p,q}\Lambda(\beta_{p,q})\left(\beta^2_{p,q}\omega^2 + |\nabla'\omega|^2\right)^{\frac{p-2}{2}}\omega = 0 \quad \text{in } \mathcal{D}'(S^{N-1}_+) \end{aligned} \tag{3.2.12}$$

such that $\delta\phi_{S^{N-1}_+} \leq \omega \leq \omega_c$.

*Case 2:* $1 < p < 2$. In this case the operators $A$ and $B$ present singularities when $r = \xi = 0$. Although for any solution of (3.2.12) the quantity $\beta^2_{p,q}\omega^2 + |\nabla'\omega|^2$ remains bounded from below by some positive constant, the value of this constant is not clear. The idea is to desingularize the operators in order to apply Theorem 1.4.8, but this in turn makes the boundary value singular. If we set $v = -\frac{1}{\beta_{p,q}}\ln\omega$, then $v$ satisfies

$$\begin{aligned} \mathcal{L}(v) := -div'&\left(\left(1+|\nabla'v|^2\right)^{\frac{p-2}{2}}\nabla'v\right) + (p-1)\beta_{p,q}\left(1+|\nabla'v|^2\right)^{\frac{p-2}{2}}|\nabla'v|^2 \\ &+ \Lambda(\beta_{p,q})\left(1+|\nabla'v|^2\right)^{\frac{p-2}{2}} - \beta^{1-p}_{p,q}e^{-pv} = 0 \quad \text{in } S^{N-1}_+ \\ &\lim_{\rho(\sigma)\to 0} v(\sigma) = \infty, \end{aligned} \tag{3.2.13}$$

where $\rho(\sigma)$ is the geodesic distance from $x$ to the equator on $S_+^{N-1}$. Therefore we define $\tilde{A}$ and $\tilde{B}$ by

$$\tilde{A}(r,\xi) = \left(1+|\xi|^2\right)^{\frac{p-2}{2}}\xi,$$

for $(r,\xi)\in\mathbb{R}\times\mathbb{R}^{N-1}$, thus

$$\langle \tilde{A}(r,\xi)-\tilde{A}(r,\xi'),\xi-\xi'\rangle_g > 0 \qquad \text{if } \xi\neq\xi',$$

and

$$\tilde{B}(r,\xi) = (p-1)\beta_{p,q}\left(1+|\xi|^2\right)^{\frac{p-2}{2}}|\xi|^2+\Lambda(\beta_{p,q})\left(1+|\xi|^2\right)^{\frac{p-2}{2}}-\beta_{p,q}^{1-p}e^{-pT_k(r)},$$

where $T_k(r)=\min\{|r|,k\}\text{sgn}(r)$ for $k>\omega_c$. Then

$$\begin{array}{ll} (i) & \langle \tilde{A}(r,\xi),\xi\rangle_g \geq |\xi|^p - c(p) \\ (ii) & |A(r,\xi)| \leq |\xi|^{p-1}+c(p) \\ (iii) & |B(r,\xi)| \leq (p-1+k^{q+1-p})(\beta_{p,q}+1)(1+|\xi|^{p-1}). \end{array} \tag{3.2.14}$$

Clearly

$$\tilde{\phi} = -\frac{1}{\beta_{p,q}}\ln\left(\delta\phi^{\theta}_{S_+^{N-1}}\right) = -\frac{1}{\beta_{S_+^{N-1}}}(\ln\phi_{S_+^{N-1}}+\ln\delta),$$

is a supersolution of $\mathcal{L}$ for $\delta>0$ small enough. The construction of a subsolution of $\mathcal{L}$ which tends to infinity on $\partial S_+^{N-1}$ is a little more delicate (notice that the subsolution $v_c = -\frac{1}{\beta_{p,q}}\ln\omega_c$ does not blow-up on the equator). We fix $\tau\in(0,\frac{\pi}{4})$, denote by $S_\tau$ the spherical cap centered at the north pole with opening $\tau$, that is $S_\tau=\{\sigma\in S_+^{N-1}:\tau<\rho(\tau)\leq\frac{\pi}{4}\}$ and by $C_{S_\tau}$ the closed cone generated by $S_\tau$, i.e. $C_{S_\tau}=\{x\in\mathbb{R}_+^N:|x|^{-1}x\in S_\tau\}\cup\{0\}$. We construct the minimal solution $U:=U_{S_\tau}$ of the following problem

$$\begin{array}{rll} -\Delta_p U + U^q = 0 & & \text{in } \mathbb{R}_+^N\setminus C_{S_\tau} \\ U = \infty & & \text{on } \partial C_{S_\tau} \\ U = 0 & & \text{on } \partial\mathbb{R}_+^N. \end{array} \tag{3.2.15}$$

This solution is minimal since it is the limit, when $n\to\infty$, of the increasing sequence of solutions $\{U_n\}$ of the same equation, in the same domain, with boundary data 0 on $\partial\mathbb{R}_+^N$ and $n$ ($n\in\mathbb{N}^*$) on $\partial C_{S_\tau}$. The function $U_n$ is uniquely determined. By Proposition 3.1.6 the solution $U_n$ remains locally bounded in $\overline{\mathbb{R}}_+^N\setminus C_{S_\tau}$, thus the boundary conditions is preserved on $\partial\mathbb{R}_+^N$

when $n \to \infty$. Denoting

$$T_\lambda[u](x) = \lambda^{\beta_q} u(\lambda x) \qquad \forall \lambda > 0,\ \forall x \in \mathbb{R}^N_+ \setminus C_{S_\tau},$$

then $T_\lambda[U_n] = U_{n\lambda^{\beta_q}}$, for any $\lambda > 0$, which implies, letting $n \to \infty$,

$$T_\lambda[U] = U \qquad \forall \lambda > 0. \tag{3.2.16}$$

Therefore $U_{S_\tau}$ is self-similar and thus $U_{S_\tau}(x) = U_{S_\tau}(r, \sigma) = r^{-\beta_q}\omega_\tau(\sigma)$, where $\omega_\tau(\sigma)$ is the minimal solution of equation (3.2.12) but in $\mathcal{D}'(S^{N-1}_+ \setminus S_\tau)$ instead of $\mathcal{D}'(S^{N-1}_+)$, subject to the boundary conditions $\omega_\tau = 0$ on $\partial S^{N-1}_+$ and $\omega_\tau(\sigma) \to \infty$ when $\operatorname{dist}(\sigma, \partial S_\tau) \to 0$. Finally, the function $\tilde{\omega}_\tau = \inf\{\omega_\tau, \omega_c\}$ is a Lipschitz continuous supersolution of (3.2.12) in $\mathcal{D}'(S^{N-1}_+)$ and it belongs to $W^{1,p}_0(S^{N-1}_+)$. Next, $\tilde{\psi} = -\frac{1}{\beta_{p,q}} \ln \tilde{\omega}_\tau$ is a Lipschitz continuous subsolution smaller than $\tilde{\psi}$. By Theorem 1.4.8 there exists a solution $\tilde{\omega}$ of $\mathcal{L}(\tilde{\omega}) = 0$ in $S^{N-1}_+$ satisfying $\tilde{\psi} \le \tilde{\omega} \le \tilde{\phi}$. The proof of the existence ends with Lemma 2.2.4.

*Uniqueness.* Assume $\omega_1$ and $\omega_2$ are nonnegative nontrivial solutions. Then they are positive and $\frac{\partial \omega_i}{\partial \mathbf{n}} < 0$ on $\partial S^{N-1}_+$. Either $\omega_2 \ge \omega_1$ or the graphs of $\omega_2$ and $\omega_1$ intersect. We define

$$\tau = \inf\{s > 1 : s\omega_1 \ge \omega_2\} \tag{3.2.17}$$

and set $\omega^* = \tau\omega_1$. Either the graphs of $\omega^*$ and $\omega_2$ are tangent at $\sigma_0 \in S^{N-1}_+$ or they are separated in $S^{N-1}_+$, but tangent at some $\sigma_0 \in \partial S^{N-1}_+$, thus $\frac{\partial \omega_2}{\partial \mathbf{n}}(\sigma_0) = \frac{\partial \omega^*}{\partial \mathbf{n}}(\sigma_0)$. Then $\mathcal{T}(\omega^*) \ge 0$, thus $\mathcal{T}(\omega^*) - \mathcal{T}(\omega_2)$. Applying either the strong maximum principle or Hopf boundary lemma to $\omega^* - \omega_2$ at $\sigma_0$ we derive a contradiction.

□

## 3.3 Classification of boundary singularities

Let $\Omega \subset \mathbb{R}^N$ be a domain and $g : \mathbb{R} \mapsto \mathbb{R}$ a continuous nondecreasing function. The aim of this section is to describe the behavior near some point $x_0 \in \partial\Omega$ of the functions $u \in C(\overline{\Omega} \setminus \{x_0\})$ which satisfy

$$\begin{aligned} -\Delta_p u + g(u, |\nabla u|) &= 0 \qquad \text{in } \Omega \\ u &= 0 \qquad \text{on } \partial\Omega \setminus \{x_0\}, \end{aligned} \tag{3.3.1}$$

where $p > 1$. In the general case the results are obtained under a geometric constraint on $\partial\Omega$, except in the conformal case $p = N$ and the semi linear case $p = 2$. The next statement is an *a priori* estimate for any solution of (3.3.1) provided $g$ has some growth at infinity.

**Lemma 3.3.1** *Assume $0 < p - 1 < q$, $\Omega \subset \mathbb{R}^N$ is a domain, $x_0 \in \partial\Omega$ and $u \in C(\overline{\Omega} \setminus \{x_0\}) \cap W^{1,p}(B_R(x_0) \cap \Omega)$ is nonnegative and satisfies*

$$\begin{aligned} -\Delta_p u + au^q &\leq b \qquad \text{in } \mathcal{D}'((B_R(x_0) \cap \Omega) \\ u &= 0 \qquad \text{on } (B_R(x_0) \cap \partial\Omega) \setminus \{x_0\}, \end{aligned} \tag{3.3.2}$$

*for some constants $a > 0$ and $b \geq 0$. Then*

$$u(x) \leq \left(\frac{c}{a|x - x_0|^p}\right)^{\frac{1}{q+1-p}} + \left(\frac{b}{a}\right)^{\frac{1}{q}} + \max_{\partial B_R(x_0)\cap\Omega} u, \quad \forall x \in B_R(x_0) \cap \Omega. \tag{3.3.3}$$

*Proof.* We can assume that $x_0 = 0$ and set $m = \max_{\partial B_R(x_0)\cap\Omega} u$. For $\epsilon > 0$ the function $j_\epsilon(u) = (u - \epsilon - m)_+$ belongs to $C(\overline{\Omega} \setminus \{x_0\}) \cap W^{1,p}(B_R(x_0) \cap \Omega)$. It is a subsolution and since it vanishes in a neighborhood of $(\partial B_R(x_0) \cap \Omega) \cup (B_R(x_0) \cap \partial\Omega) \setminus \{x_0\}$, its extension by 0 $\tilde{j}_\epsilon$ in $(B_R(x_0) \cap \Omega)^c$ is a subsolution therein. By Lemma 3.1.2 it satisfies

$$\tilde{j}_\epsilon(x) \leq \left(\frac{c}{a|x - x_0|^p}\right)^{\frac{1}{q+1-p}} + \left(\frac{b}{a}\right)^{\frac{1}{q}} \quad \forall x \in \mathbb{R}^N \setminus \{0\}.$$

Letting $\epsilon \to 0$ yields (3.3.1). □

### 3.3.1 *The general gradient estimate near a singularity*

In what follows, we show that an *a priori* estimate on a singular solution implies an estimate of the gradient of this solution whenever the solutions of the regular problem are $C^1$ up to the boundary. It is important to notice that the statements are valid for any *signed solution*, a property which will be much used in the sequel.

**Definition 3.1** Let $g : \mathbb{R} \times \mathbb{R}_+ \mapsto \mathbb{R}$ be a continuous function and $a : \mathbb{R}_+^* \mapsto \mathbb{R}_+^*$ a continuous increasing function. We say that $g$ satisfies the *$(a, p)$-scaling-growth property* if there exist a continuous function $C : \mathbb{R}_+ \mapsto \mathbb{R}_+$ and $\lambda_0 > 0$ such that for any $0 < \lambda \leq \lambda_0$, there holds

$$\begin{aligned} &(i) \quad \inf_{\lambda\in(0,\lambda_0]} \lambda^p(a(\lambda))^{1-p} > 0 \\ &(ii) \quad \lambda^p(a(\lambda))^{1-p} \left|g\left(a(\lambda)r, \lambda^{-1}a(\lambda)s\right)\right| \leq C(|r|)(1+s^p) \quad \forall (r,s) \in \mathbb{R}\times\mathbb{R}_+. \end{aligned} \tag{3.3.4}$$

**Example**. If $|g(r,s)| = |r|^q s^m$, $q, m \geq 0$, (3.3.4) means

$$\lambda^{p-m}(a(\lambda))^{q+m+1-p}\,|r|^q\,s^m \leq C(|r|)(1+s^p) \quad \forall (r,s) \in \mathbb{R}\times\mathbb{R}_+. \tag{3.3.5}$$

This necessitates $m \leq p$. In the limit case $m = p$ the above inequality becomes $(a(\lambda))^{q+1}\,|r|^q\,s^p \leq C(|r|)(1+s^p)$.

**Lemma 3.3.2** *Let $\Omega \subset \mathbb{R}^N$ be a bounded $C^2$ domain with $0 \in \partial\Omega$ and $u \in C^1(\overline{\Omega}\setminus\{0\})$ be a solution of (3.0.1) which vanishes on $\partial\Omega\setminus\{0\}$ and satisfies*

$$|u(x)| \leq \phi(|x|) \qquad \forall x \in \Omega, \tag{3.3.6}$$

*where $\phi : \mathbb{R}_+^* \mapsto \mathbb{R}_+$ is continuous, nonincreasing and*

$$\phi(rs) \leq \gamma\phi(r)\phi(s) \qquad \forall r, s \leq r_0, \tag{3.3.7}$$

*for some $r_0, \gamma > 0$. If $g$ satisfies the $(\phi, p)$-scaling-growth property, then there exist $c_1 > 0$ and $\alpha \in (0,1)$ such that*

$$\begin{array}{lll} (i) & |\nabla u(x)| \leq c_1 \dfrac{\phi(|x|)}{|x|} & \forall x \in \Omega \\[2ex] (ii) & |\nabla u(x) - \nabla u(x')| \leq c_1 \dfrac{\phi(|x|)\,|x-x'|^\alpha}{|x|^{1+\alpha}} & \forall x, x' \in \Omega,\ |x| \leq |x'|. \end{array} \tag{3.3.8}$$

*Furthermore*

$$|u(x)| \leq c_1 \frac{\phi(|x|)\rho(x)}{|x|} \quad \forall x \in \Omega. \tag{3.3.9}$$

*Proof. 1- The flat case.* We recall that $\mathbb{R}_+^N = \{x = (x', x_N) : x' \in \mathbb{R}^{N-1}, x_N > 0\}$. For the sake of clarity we give first the proof when $\Omega \cap B_R = \mathbb{R}_+^N \cap B_R := B_R^+$ for some $R > 0$. We extend $u$ by reflexion through $\Gamma_R := \Omega \cap B_R$ and define $\tilde{v} \in C^1(\overline{B}_R \setminus \{0\})$ by

$$\tilde{u}(x) = \begin{cases} u(x', x_N) & \text{if } x_N > 0 \\ -u(x', -x_N) & \text{if } x_N < 0, \end{cases}$$

and

$$\tilde{g}(x, r, s) = \begin{cases} g(r,s) & \text{if } x_N > 0 \\ -g(r,s)) & \text{if } x_N < 0. \end{cases}$$

Hence $\tilde{u} \in C^1(\overline{B}_{\frac{R}{\lambda}})$ is a solution of

$$-\Delta_p \tilde{u} + \tilde{g}(x, \tilde{u}, |\nabla \tilde{u}|) = 0 \qquad \text{in } B_{\frac{R}{\lambda}} \setminus \{0\}. \tag{3.3.10}$$

Furthermore $|\tilde{u}(x)| \leq \phi(|x|)$ holds in $\overline{B}_R \setminus \{0\}$. For $\lambda > 0$, we set

$$\Phi_\lambda[\tilde{u}](x) = \frac{1}{\phi(\lambda)} \tilde{u}(\lambda x) := \tilde{u}_\lambda(x) \qquad \forall x \in B_{\frac{R}{\lambda}}. \tag{3.3.11}$$

Then

$$-\Delta_p \tilde{u}_\lambda + (\phi(\lambda))^{1-p} \lambda^p \tilde{g}\left(\phi(\lambda)\tilde{u}_\lambda, \lambda^{-1}\phi(\lambda)|\nabla \tilde{u}_\lambda|\right) = 0, \tag{3.3.12}$$

in $\mathcal{D}'(B_{\frac{R}{\lambda}})$. Put $R' = \min\{\frac{R}{2}, \frac{\lambda_0}{2}, \frac{r_0}{2}\}$. For $0 < \lambda \leq R'$ and $|z| = \lambda$, the function $\tilde{u}_\lambda$ satisfies (3.3.12) in $B_{\frac{1}{2}}(\frac{z}{\lambda})$ where there holds also

$$|\tilde{u}_\lambda(x)| \leq \frac{\phi(\lambda|x|)}{\phi(\lambda)} \leq \gamma\phi(|x|) \leq \gamma\phi\left(\frac{1}{2}\right). \tag{3.3.13}$$

By assumption

$$(\phi(\lambda))^{1-p}\lambda^p \left|\tilde{g}\left(\phi(\lambda)\tilde{u}_\lambda, \lambda^{-1}\phi(\lambda)|\nabla \tilde{u}_\lambda|\right)\right| \leq C(|\tilde{u}_\lambda|)(1 + |\nabla \tilde{u}_\lambda|^p). \tag{3.3.14}$$

Therefore, by Theorem 1.3.9 and Theorem 1.3.10, one has

$$\begin{array}{lll} (i) & |\nabla \tilde{u}_\lambda(x)| \leq c_1 & \forall x \in B_{\frac{1}{4}}(\frac{z}{\lambda}) \\ (ii) & |\nabla(\tilde{u}_\lambda(x) - \tilde{u}_\lambda(x'))| \leq c_1 |x - x'|^\alpha & \forall x, x' \in B_{\frac{1}{4}}(\frac{z}{\lambda}). \end{array} \tag{3.3.15}$$

Going back to the original function we obtain, with $y = \lambda x = |z|\, x$,

$$|\nabla \tilde{u}(y)| \leq c_1 \frac{\phi(|z|)}{|z|} \qquad \forall y \in B_{\frac{|z|}{4}}(z).$$

We derive 3.3.8-(i) by taking $x = z$. Similarly, with $\lambda x = y$, $\lambda x' = y'$,

$$\begin{aligned} |\nabla(\tilde{u}_\lambda(x) - \tilde{u}_\lambda(x'))| &= \lambda\phi(\lambda)\, |\nabla(\tilde{u}(y) - \tilde{u}(y'))| \\ &\leq c_1 \lambda^{-\alpha} |y - y'|^\alpha . \end{aligned}$$

If $|y| \leq |y|'$, $y$ and $y'$ belonging to $B_{\frac{|z|}{4}}(z)$, we infer (3.3.8)-(ii).

*2- The general case.* The idea is to flatten the boundary of the dilated $\Omega$, $\Omega^\lambda := \frac{1}{\lambda}\Omega$. Since $\lambda \in (0, \lambda_0]$ the curvature of $\partial\Omega^\lambda$ remains uniformly bounded and up to scaling we can suppose that $\lambda_0 = 1$. There exists a diffeomorphism $\psi_\lambda$ from $\overline{\Omega}^\lambda \cap \overline{B}_{\frac{1}{\lambda}}$ into $(\overline{\Omega}^\lambda)^c \cap B_{\frac{1}{\lambda}}$ which is the identity on $\partial\Omega^\lambda \cap \overline{B}_{\frac{1}{\lambda}}$ and such that $D\psi_\lambda(\xi)$ is locally the symmetry with respect to the tangent hyperplane $T_\xi\partial\Omega^\lambda$ for any $\xi \in \partial\Omega^\lambda \cap \overline{B}_{\frac{1}{\lambda}}$. This diffeomorphism can be explicitly constructed in the following way: up to a rotation we can suppose that near 0, $\partial\Omega$ is a graph over the closed ball $\overline{B}'_1$ of $\mathbb{R}^{N-1}$:

$$\partial\Omega\cap\overline{B}_1=\{x=(x',x_N):|x'|\leq 1,\,\phi(x')<x_N\leq 1\}$$

where $\phi\in C^2(\overline{B}'_1)$ satisfies, $\max_{|x'|\leq 1}|\phi(x')|\leq\frac{1}{2}$, $\phi(0)=0$, $D\phi(0)=0$ and there exist $a_j,b_{j,k}\in C(\overline{B}'_1)$ (j, k=1,...,N-1) such that

$$\phi(x')=\sum_j a_j(x')x_j^2 \text{ and } \frac{\partial\phi}{\partial x_j}(x')=\sum_k b_{j,k}(x')x_k \quad \forall x'\in\overline{B}'_1. \tag{3.3.16}$$

Then

$$\Omega^\lambda\cap B_{\frac{1}{\lambda}}=\left\{(x',x_N):|x'|<\lambda^{-1},\,\lambda^{-1}\phi(\lambda x')<x_N\leq\lambda^{-1}\right\}. \tag{3.3.17}$$

If we set

$$\psi_\lambda(x)=(x',\lambda^{-1}\phi(\lambda x')-x_N) \qquad \forall x\in\Omega^\lambda\cap B_{\frac{1}{\lambda}}, \tag{3.3.18}$$

then $\psi_\lambda$ has the above mentioned properties. We extend any function $v$ defined in $\overline{\Omega}^\lambda\cap\overline{B}_{\frac{1}{\lambda}}\setminus\{0\}$ and vanishing on $\partial\Omega^\lambda\cap\overline{B}_{\frac{1}{\lambda}}\setminus\{0\}$ into a function $\tilde{v}$ defined in $\overline{B}_{\frac{1}{\lambda}}\setminus\{0\}$ by

$$\tilde{v}(x)=\begin{cases} v(x) & \text{if } x\in\overline{\Omega}^\lambda\cap\overline{B}_{\frac{1}{\lambda}} \\ -v\circ\psi_\lambda(x) & \text{if } x\in(\overline{\Omega}^\lambda)^c\cap\overline{B}_{\frac{1}{\lambda}}. \end{cases} \tag{3.3.19}$$

If $v\in C^1(\overline{\Omega}^\lambda\cap\overline{B}_{\frac{1}{\lambda}})\setminus\{0\}$ is a solution of (3.0.1) in $\Omega^\lambda\cap B_{\frac{1}{\lambda}}$ which vanishes on $\partial\Omega^\lambda\cap\overline{B}_{\frac{1}{\lambda}}\setminus\{0\}$, then $\tilde{v}$ satisfies

$$-\sum_j\frac{\partial}{\partial x_j}\tilde{A}_j(x,\nabla\tilde{v})+B(x,\nabla\tilde{v})=0 \quad \text{in } B_{\frac{1}{\lambda}}\setminus\{0\}, \tag{3.3.20}$$

and the $A_j$ and $B$ satisfy

$$\begin{array}{ll} (i) & \tilde{A}_j(x,0)=0 \\ (ii) & \sum_{i,j}\frac{\partial}{\partial\eta_i}\tilde{A}_j(x,\eta)\xi_i\xi_j\geq C_1|\eta|^{p-2}|\xi|^2 \\ (iii) & \left|\frac{\partial}{\partial\eta_i}\tilde{A}_j(x,\eta)\right|\leq C_2|\eta|^{p-2} \\ (iv) & |B(x,\eta)|\leq C_3(1+|\eta|^p). \end{array} \tag{3.3.21}$$

Notice that, even if $A_j$ and $B$ depend on $\lambda$, the expression of $\phi$ and $\nabla\phi$ in (3.3.18) shows that the constants $C_j$, j=1,2,3, are independent of $\lambda\in(0,\lambda_0]$.

We define $u_\lambda := \Phi_\lambda[u]$ by (3.3.11) for all $x \in \overline{\Omega}^\lambda \cap \overline{B}_{\frac{1}{\lambda}}$. Then it satisfies (3.3.12) in $\mathcal{D}'(\Omega^\lambda)$ and it vanishes on $\partial\Omega^\lambda$. Since $D\psi_\lambda(0) = Id$, inequality (3.3.13) holds in $\Omega^\lambda \cap B_{\frac{1}{\lambda}}$ up to some multiplicative constant $\gamma^*$ instead of $\gamma$. We extend $u_\lambda$ by the reflexion (3.3.19) as a function $\tilde{u}_\lambda$ which satisfies (3.3.20) in $B_{\frac{1}{\lambda}} \setminus \{0\}$ and is locally bounded. If $z \in \overline{\Omega}^\lambda \cap \overline{B}_{\frac{1}{\lambda}}$ is such that $|z| \leq \frac{1}{4}$ we apply the *a priori* estimate of Theorem 1.3.9 in $B_{\frac{1}{8}}(z)$ as in Case 1 and derive that $\nabla\tilde{u}$ is uniformly bounded in $C^\alpha(\overline{B}_{\frac{1}{12}}(z))$. This yields (3.3.8)-(i)-(ii).

Next we prove (3.3.10). Let $0 < \delta_1 \leq \delta_0$ be such that at any point $z \in \partial\Omega$ there exist two closed balls with radius $\delta_1$, tangent at $\partial\Omega$ at the point $z$ and respectively included into $\overline{\Omega} \cup \{z\}$ and $\overline{\Omega}^c \cup \{z\}$. Let $x \in \Omega$ such that $\rho(x) \leq \delta_1$ and $z_x$ be the projection of $x$ onto $\partial\Omega$. We assume first that $x$ does not belong to the cone $C_{\frac{\pi}{4},\mathbf{n}_0}(0)$ with vertex 0, opening $\frac{\pi}{4}$ and axis $\mathbf{n}_0$, the normal direction to $\partial\Omega$ at 0. Set $\zeta(t) = tx + (1-t)z_x$, then

$$u(x) = \int_0^1 \frac{d}{dt} u \circ \zeta(t) dt = \int_0^1 \langle \nabla u \circ \zeta(t), x - z_x \rangle dt.$$

If $x \notin C_{\frac{\pi}{4},\mathbf{n}_0}(0)$, $\zeta(t) \notin C_{\frac{\pi}{4},\mathbf{n}_0}(0)$ and there exists $c > 0$ depending only on $\Omega$ such that $c|x| \leq |\zeta(t)| \leq c^{-1}|x|$ for all $0 < t \leq 1$. Thus $\phi(|\zeta(t)|) \leq \phi(c|x|) \leq \gamma\phi(c)\phi(|x|)$. By Cauchy-Schwarz inequality and using (3.3.8)-(i), we get

$$|u(x)| \leq \gamma c^{-1} c_1 \phi(c) \frac{\rho(x)\phi(|x|)}{|x|}.$$

Thus (3.3.10) holds. When $x \in C_{\frac{\pi}{4},\mathbf{n}_0}(0)$, $\rho(x) \leq |x| \leq c'\rho(x)$, where $c'$ depends on the curvature of $\partial\Omega$. This implies (3.3.10). □

*Remark.* A variant of this result dealing with internal singularities is stated at Lemma 3.4.15.

### 3.3.2 *Removable singularities*

We assume that $\Omega$ is a bounded $C^2$ domain and that $x_0 \in \partial\Omega$. The general weak removability result is associated to the exponent $\beta_{S_+^{N-1}}$ of the singular spherical $p$-harmonic function in $\mathbb{R}_+^N$. The critical exponent for solution of (3.2.2) with an isolated boundary singularity is $\tilde{q}_1$ defined by relations (3.2.5).

**Proposition 3.3.3** *Let $p > 1$ and $u \in C^1(\overline{\mathbb{R}}^N_+ \setminus \{0\})$ be a solution of (3.2.2) with $q \geq \tilde{q}_1$. Then $u = 0$.*

*Proof.* For $\epsilon > 0$ we denote by $u_\epsilon$ the minimal solution of

$$\begin{aligned} -\Delta_p u + |u|^{q-1}u &= 0 \qquad \text{in } \mathbb{R}^N_+ \setminus \overline{B}^+_\epsilon \\ u &= 0 \qquad \text{on } \partial\mathbb{R}^N_+ \setminus \overline{B}^+_\epsilon \\ \lim_{|x|\to\epsilon} u(x) &= \infty. \end{aligned} \tag{3.3.22}$$

The function $u_\epsilon$ is minimal among the other solutions of (3.3.22) since it is the limit of the solutions of the same problem with finite value on $\partial B_\epsilon \cap \mathbb{R}^N_+$. Furthermore $u_\epsilon \leq u_{\epsilon'}$ for $\epsilon > \epsilon'$. This implies that there exists $u_0 = \lim_{\epsilon\to 0} u_\epsilon$, and $u_0$ is a solution of (3.2.2). If $u$ is a solution of (3.2.2), its extension $\tilde{u}$ by reflexion through $x_N = 0$ is a solution of (3.1.1) in $\mathbb{R}^N_+ \setminus \{0\}$. By Lemma 3.1.2 it satisfies

$$|\tilde{u}(x)| \leq \left(\frac{c}{|x|^p}\right)^{\frac{1}{q+1-p}} \qquad \forall x \in \mathbb{R}^N \setminus \{0\}, \tag{3.3.23}$$

and $u$ satisfies the same estimate in $\overline{\mathbb{R}}^N_+ \setminus \{0\}$. Because $u_\epsilon + \delta$ is a supersolution for every $\delta > 0$, $u \leq u_\epsilon + \delta$ in $\mathbb{R}^N_+ \setminus \overline{B}^+_\epsilon$. Letting $\delta \to 0$ yields $u \leq u_\epsilon$, from which follows $u \leq u_0$. We also notice that the scaling transformation $v \mapsto T_\lambda[v]$ defined by $T_\lambda[v](x) = \lambda^{\frac{p}{q+1-p}} v(\lambda x)$, for $\lambda > 0$, leaves equation (3.3.22) invariant, except that $\epsilon$ is transform into $\lambda^{-1}\epsilon$. Thus $T_\lambda[u_\epsilon] = u_{\lambda^{-1}\epsilon}$. Letting $\epsilon \to 0$ yields $T_\lambda[u_0] = u_0$ for all $\lambda > 0$. Therefore $u_0$ is a self-similar solution of (3.2.2) which happens to be zero by Theorem 3.2.1. Therefore $u \leq 0$, and similarly $u \geq 0$. □

**Theorem 3.3.4** *Assume $g : \mathbb{R} \mapsto \mathbb{R}$ is a continuous function such that*

$$\liminf_{r\to\infty} r^{-\tilde{q}_1} g(r) > 0 \quad \limsup_{r\to-\infty} |r|^{-\tilde{q}_1} g(r) < 0. \tag{3.3.24}$$

*If $u \in C^1(\overline{\Omega} \setminus \{x_0\})$ is a solution of (3.3.1), then*

$$\lim_{x\to x_0} |x - x_0|^{1+\beta_{S^{N-1}_+}} \frac{u(x)}{\rho(x)} = 0. \tag{3.3.25}$$

*Proof.* Without loss of generality we can assume that $x_0 = 0$. Because of (3.3.24), there exist two constants $a > 0$ and $b \geq 0$ such that

$$-\Delta_p u_+ + a u_+^{\tilde{q}_1} \leq b \quad \text{in } \mathcal{D}'(\Omega). \tag{3.3.26}$$

As in Proposition 3.3.3, for $\epsilon > 0$ we denote by $u_\epsilon$ the minimal solution of

$$\begin{array}{rl} -\Delta_p u + a|u|^{q-1}u = 0 & \text{in } \Omega \cap B_\epsilon^c \\ u = 0 & \text{on } \partial\Omega \cap B_\epsilon^c \\ \lim_{|x| \to \epsilon} u(x) = \infty. & \end{array} \tag{3.3.27}$$

Then $u_\epsilon \geq 0$ and $u_\epsilon + \left(\frac{b}{a}\right)^{\frac{1}{q}}$ is a supersolution of (3.3.26), thus

$$u \leq u_\epsilon + \left(\frac{b}{a}\right)^{\frac{1}{\tilde{q}_1}} \Longrightarrow u \leq u_0 + \left(\frac{b}{a}\right)^{\frac{1}{\tilde{q}_1}}, \tag{3.3.28}$$

where $u_0 = \lim_{\epsilon \to 0} u_\epsilon$ satisfies

$$\begin{array}{rl} -\Delta_p u + a|u|^{\tilde{q}_1-1}u = 0 & \text{in } \Omega \\ u = 0 & \text{on } \partial\Omega \setminus \{0\}. \end{array} \tag{3.3.29}$$

By Lemma 3.3.2 $u_0$ satisfies the following estimates in $\Omega$:

$$0 \leq u_0(x) \leq c\rho(x)|x|^{-\frac{\tilde{q}_1+1}{\tilde{q}_1+1-p}} = c\rho(x)|x|^{-\beta_{S_+^{N-1}}-1}, \tag{3.3.30}$$

and

$$\begin{array}{ll} (i) & |\nabla u_0(x)| \leq c|x|^{-\frac{\tilde{q}_1+1}{\tilde{q}_1+1-p}} \\ (ii) & |\nabla u_0(x) - \nabla u_0(x')| \leq c|x|^{-\frac{\tilde{q}_1+1}{\tilde{q}_1+1-p}-\alpha}|x-x'|^\alpha \quad \text{if } |x| \leq |x'|. \end{array} \tag{3.3.31}$$

We consider again the scaling $T_\lambda : u \mapsto T_\lambda[u]$ and $T_\lambda[u](x) = \lambda^{\frac{p}{\tilde{q}_1+1-p}} u(\lambda x)$, then $T_\lambda[u_0] = u_{0\,\lambda}$ satisfies

$$\begin{array}{rl} -\Delta_p u + a|u|^{\tilde{q}_1-1}u = 0 & \text{in } \Omega^\lambda \\ u = 0 & \text{on } \partial\Omega^\lambda \setminus \{0\}. \end{array} \tag{3.3.32}$$

We compose $u_{0\,\lambda}$ with the diffeomorphism $\phi_\lambda$ (see (3.3.16) and define $\tilde{v} = \tilde{v}_\lambda$ accordingly by (3.3.19) in $\Omega^\lambda \cap B_{\frac{1}{\lambda}} \setminus \{0\}$. Since estimate (3.3.13) holds with $\phi(r) = r^{\frac{p}{\tilde{q}_1+1-p}}$, we derive that $\tilde{u}_{0\,\lambda}$ is eventually locally bounded in $C^{1,\alpha}(\mathbb{R}^N \setminus \{0\})$. Therefore there exists a subsequence $\{\lambda_k\}$ and a function $u^* \in C^1(\mathbb{R}^N \setminus \{0\})$ such that $\tilde{u}_{0\,\lambda_k} \to u^*$ and $\nabla\tilde{u}_{0\,\lambda_k} \to \nabla u^*$ locally uniformly in $\mathbb{R}^N \setminus \{0\}$. By (3.3.18) $\psi_\lambda(x', x_N) \to (x', -x_N)$ and a straightforward computation shows that the limit equation of (3.3.20) that satisfies $u^*$ in $\mathbb{R}^N \setminus \{0\}$ is (3.1.1) with $q = \tilde{q}_1$. Furthermore, $u^*$ is nonnegative on $\mathbb{R}_+^N$ and vanishes on $\partial\mathbb{R}_+^N \setminus \{0\}$. By Proposition 3.3.3, $u^* = 0$. Since the limit does not depend on the subsequence, we get

$$\lambda^{\frac{p}{\tilde{q}_1+1-p}} \tilde{u}_{0\,\lambda}(x) \to 0 \text{ and } \lambda^{1+\frac{p}{\tilde{q}_1+1-p}} \nabla\tilde{u}_{0\,\lambda}(x) \to 0, \tag{3.3.33}$$

locally uniformly in $\mathbb{R}^N_+ \setminus \{0\}$ when $\lambda \to 0$. This implies in particular

$$\lim_{x\to 0} |x|^{\frac{p}{\tilde{q}_1+1-p}} u(x) = 0 \text{ and } \lim_{x\to 0} |x|^{\frac{\tilde{q}_1+1}{\tilde{q}_1+1-p}} \frac{u(x)}{\rho(x)} = 0. \tag{3.3.34}$$

Therefore $u_+$ satisfies (3.3.25) since we have $\frac{p}{\tilde{q}_1+1-p} = \beta_{S^{N-1}_+}$. Similarly $u_-$ satisfies (3.3.25). □

In order to make more precise the previous result, one possibility is to construct a *p-harmonic analog* of the Poisson kernel with a point singularity on $\partial\Omega$. When $1 < p \leq N$ and $p \neq 2$ this construction is done under global geometric and local flatness assumptions.

**Proposition 3.3.5** *Let $p > 1$ and $\Omega$ be a $C^2$ domain such that $0 \in \partial\Omega$. We assume that $\Omega$ is contained into the half space $H^+$ bordered by the tangent hyperplane $T_0\partial\Omega$ to $\partial\Omega$ at 0, and that*

$$|x| = O\left((\rho(x))^{\beta_{S^{N-1}_+}+1}\right) \quad \text{on } \partial\Omega \text{ near } x = 0. \tag{3.3.35}$$

*Then there exists a unique positive p-harmonic function $V_\Omega \in C^1(\overline{\Omega} \setminus \{0\})$, vanishing on $\partial\Omega \setminus \{0\}$ such that*

$$V_\Omega(x) = V_{\mathbb{R}^N_+}(x)\,(1 + \circ(1)) \quad \text{as } x \to 0, \tag{3.3.36}$$

*where $V_{\mathbb{R}^N_+}(x) = V_{\mathbb{R}^N_+}(r,\sigma) = r^{-\beta_{S^{N-1}_+}} \phi_{S^{N-1}_+}(\sigma)$ is the singular spherical p-harmonic function in $H^+$ which is continuous in $\overline{H^+} \setminus \{0\}$, vanishes on $T_0\partial\Omega \setminus \{0\}$, achieves its maximum 1 on $|x| = 1$ constructed in Theorem 2.2.1 with, up to a rotation, $S = S^{N-1}_+$.*

*Proof.* Assume that $H^+ = \mathbb{R}^N_+ = \{x = (x', x_N) : x' \in \mathbb{R}^{N-1}, x_N > 0\}$. We write $\beta = \beta_{S^{N-1}_+}$ and $\phi = \phi_{S^{N-1}_+}$ with the notations of Theorem 2.2.1. We denote by $w$ the separable $p$-harmonic function expressed by $w(x) = w(r,\sigma) = r^{-\beta}\omega(\sigma)$. First we give the proof when $\Omega$ is bounded. Because of (3.3.35) there exists $m > 0$ such that

$$w(x) \leq m \qquad \forall x \in \partial\Omega \setminus \{0\}.$$

The function $w_m = (w - m)_+$ is $p$-subharmonic in $\Omega$ and vanishes on $\partial\Omega \setminus \{0\}$. Next, for $\epsilon > 0$ small enough, let $v_\epsilon$ be the $p$-harmonic function in $\Omega \cap \overline{B}^c_\epsilon$ with value $\epsilon^{-\beta}\omega(\sigma)$ on $|x| = \epsilon$ and 0 on $\partial\Omega \cap B^c_\epsilon$. It is obtained by minimization, is nonnegative and larger than $w_m$ in $\Omega \cap \overline{B}^c_\epsilon$ and smaller than the restriction of $w$. If $0 < \epsilon' < \epsilon$, $w_{\epsilon'} \lfloor_{\partial B^+_\epsilon} \leq w \lfloor_{\partial B^+_\epsilon} = v_\epsilon \lfloor_{\partial B^+_\epsilon}$. Therefore

$v_{\epsilon'} \leq v_\epsilon$. As a consequence of Theorem 1.3.9, the set $\{v_\epsilon\}$ remains locally bounded in $C^{1+\alpha}(\overline{\Omega} \setminus \{0\})$. If we set $v = \lim_{\epsilon \to 0} w_\epsilon$, then $v$ is $p$-harmonic in $\Omega$ and larger than $w$. It vanishes on $\partial\Omega \setminus \{0\}$ and satisfies

$$\lim_{x \to 0} \frac{v(x)}{w(x)} = 1. \tag{3.3.37}$$

If $\Omega$ is unbounded, we fix $r_0 > 0$ and denote by $m$ the maximum of $w$ on $(\partial\Omega \cap \overline{B}_{r_0}) \cup (\overline{\Omega} \cap \partial B_{r_0})$. Then $w_m = (w - m)_+$ is $p$-subharmonic in $\Omega \cap B_{r_0}$. It vanishes on $B_{r_0} \cap \partial\Omega \setminus \{0\}$ and on $\overline{\Omega} \cap \partial B_{r_0}$. Its extension by 0 in $\Omega \cap (B^+_{r_0})^c$, still denoted by $w_m$, is a subsolution in $\Omega$. The remaining of the proof is as above.

The $C^{1+\alpha}$ local estimates and (3.3.37) imply that any other solution $\tilde{v}$ satisfies

$$\lim_{x \to 0} \frac{v(x)}{\tilde{v}(x)} = 1, \tag{3.3.38}$$

which implies uniqueness. □

When $p = N$ the geometric and flatness assumptions on $\Omega$ are unnecessary since the $N$-Laplace equation is invariant under conformal transformations in $\mathbb{R}^N$.

**Proposition 3.3.6** *Let $\Omega$ be a $C^2$ domain such that $0 \in \partial\Omega$. Then there exists a unique positive $N$-harmonic function $v \in C^1(\overline{\Omega} \setminus \{0\})$, vanishing on $\partial\Omega \setminus \{0\}$ such that*

$$v(x) = |x|^{-1} \omega \left( \frac{x}{|x|} \right) (1 + \circ(1)) \quad \text{as } x \to 0, \tag{3.3.39}$$

*where $|.|^{-1} \omega$ is the singular spherical $N$-harmonic function already introduced in Proposition 3.3.6 with general $p$.*

*Proof.* Since $\partial\Omega$ is $C^2$, it satisfies the interior and exterior sphere conditions. Up to a dilation with center 0, we assume that the balls $B_{\frac{1}{2}}(a)$ and $B_{\frac{1}{2}}(a')$ are respectively the interior and exterior spheres to $\partial\Omega$ at 0 (and $a = -\mathbf{n}\frac{1}{2}$, $a' = \mathbf{n}\delta = -a$ and $\mathbf{n}$ is the normal outward unit vector to $\partial\Omega$ at 0). Let $\tilde{a} = 2a$, $\tilde{a}' = 2a'$, and denote by $\mathcal{J}_1^{\tilde{a}}$ and $\mathcal{J}_1^{\tilde{a}'}$ the inversions of center $\tilde{a}$ and $\tilde{a}'$ and power 1. The origin is invariant under these two transformations. The balls $B_{\frac{1}{2}}(a)$ and $B_{\frac{1}{2}}(a')$ are transformed by $\tilde{\mathcal{J}}_1^{\tilde{a}} = -\mathcal{J}_1^{\tilde{a}}$ and $\mathcal{J}_1^{\tilde{a}'}$ into the same half space $H^+$ going through 0, with boundary $T_0\partial\Omega$ and containing $-\mathbf{n}$. As above we can assume that $H^+ = \mathbb{R}^N_+$ and $T_0\partial\Omega = \partial\mathbb{R}^N_+$. If $w$ is the fundamental separable $N$-harmonic function defined in $H^+$ with singularity at 0 already introduced in the proof of Proposition 3.3.5, then $\tilde{w} = w \circ \tilde{\mathcal{J}}_1^{\tilde{a}}$ and $\tilde{w}' = w \circ \mathcal{J}_1^{\tilde{a}'}$ are positive $N$-harmonic functions respectively in $B_{\frac{1}{2}}(a)$ and $B_{\frac{1}{2}}(a')$, which vanish respectively on $\partial B_{\frac{1}{2}}(a) \setminus \{0\}$ and $\partial B_{\frac{1}{2}}(a') \setminus \{0\}$ and satisfy

$$\begin{aligned} &(i) \quad \lim_{\substack{x\to 0\\ |x|^{-1}x\to\sigma}} |x|\,\tilde{w}(x) = \omega(\sigma) \quad \text{locally uniformly on } S_+^{N-1} \\ &(ii) \quad \lim_{\substack{x\to 0\\ |x|^{-1}x\to\sigma}} |x|\,\tilde{w}'(x) = \omega(\sigma) \quad \text{uniformly on } S_+^{N-1}. \end{aligned} \tag{3.3.40}$$

If we extend $\tilde{w}$ by 0 in $\Omega$, the new function, still denoted by $\tilde{w}$, is $p$-subharmonic, nonnegative and vanishes on $\partial\Omega \setminus \{0\}$. For $\epsilon > 0$ we construct by minimization the function $v_\epsilon$ which is $p$-harmonic in $\Omega \cap \overline{B}_\epsilon$, vanishes on $\partial\Omega \cap \overline{B}_\epsilon$ and takes the value $\tilde{w}$ on $\Omega \cap \partial B_\epsilon$. Then $v_\epsilon \geq \tilde{w}$ in $\Omega \cap \overline{B}_\epsilon$. Thus $v_\epsilon \leq v_{\epsilon'}$ if $0 < \epsilon' < \epsilon$. On the other hand, $\tilde{w} \leq c\tilde{w}'$ on $\partial B_\epsilon \cap B_{\frac{1}{2}}(a)$ for some $c > 1$ independent of $\epsilon$ which can be made as close as we want to 1. Therefore $v_\epsilon \leq \tilde{w}'$ in $\Omega \cap \overline{B}_\epsilon$. We derive from Lemma 3.3.2 that the set of functions $\{v_\epsilon\}$ is eventually relatively compact in the $C^1_{loc}$ topology of $\overline{\Omega} \setminus \{0\}$ and thus there exists a subsequence $\{v_{\epsilon_n}\}$ which converges in this topology to a nonnegative $N$-harmonic function $v$. Moreover $v$ vanishes on $\partial\Omega \setminus \{0\}$ because of (3.3.9) and it satisfies

$$\tilde{w} \leq v \quad \text{in } B_{\frac{1}{2}}(a) \text{ and } v \leq \tilde{w}' \quad \text{in } \Omega, \tag{3.3.41}$$

hence (3.3.39) holds. Actually, the $C^{1,\alpha}$ estimate of Lemma 3.3.2 implies that $v$ satisfies

$$\lim_{\substack{x\to 0\\ |x|^{-1}x\to\sigma}} r^2 \nabla v(r, |x|^{-1}x) = -\beta\omega(\sigma)\vec{\sigma} + \nabla'\omega(\sigma), \tag{3.3.42}$$

locally uniformly on $S_+^{N-1}$, where $\nabla'$ denotes the tangential component of the gradient and $\vec{\sigma}$ is the unit vector corresponding to $\sigma \in S^{N-1}$. Consequently if $v'$ is any other $N$-harmonic function vanishing on $\partial\Omega \setminus \{0\}$ and satisfying (3.3.39), there holds

$$\lim_{x\to 0} \frac{v(x)}{v'(x)} = 1. \tag{3.3.43}$$

This implies uniqueness. □

Thanks to Proposition 3.3.5 and Proposition 3.3.6 we can make more precise the result of Theorem 3.3.4.

**Corollary 3.3.7** *Let $1 < p \leq N$, $\Omega$ be a bounded $C^2$ domain with $0 \in \partial\Omega$, $g : \mathbb{R} \mapsto \mathbb{R}$ a continuous function which satisfies (3.3.24) and $u \in C^1(\bar{\Omega} \setminus \{0\})$ a solution of (3.3.1) with $x_0 = 0$. Then $u$ can be extended as a $C^1(\bar{\Omega})$ function if one of the following statement holds:*

*(i) either $1 < p < N$ and the assumptions on $\Omega$ of Proposition 3.3.5 hold,*

*(ii) or $p = N$.*

*Proof.* We denote by $v$ the singular $p$-harmonic functions constructed in Proposition 3.3.5 and Proposition 3.3.6. With the notations of Theorem 3.3.4, $u$ satisfies (3.3.28) where $u_0$ is a nonnegative solution of (3.3.29). Thus $u_0$ satisfies (3.3.25) with $x_0 = 0$ which implies

$$u_0(x) = o(w(x)) \qquad \text{as } x \to 0. \tag{3.3.44}$$

Since $\epsilon v$ is a supersolution for (3.3.25), this yields $u_0 \leq \epsilon v$ for any $\epsilon > 0$ and finally $u_0 = 0$. Therefore

$$u(x) \leq \left(\frac{b}{a}\right)^{\frac{1}{\tilde{q}_1}} \qquad \forall x \in \overline{\Omega} \setminus \{0\}.$$

Similarly $u$ is uniformly bounded from below in $\overline{\Omega} \setminus \{0\}$. As in the proof of Lemma 3.3.2 we extend $u$ by reflexion through $\partial\Omega$ near 0 into a new function $\tilde{u}$ which remains bounded near 0 and satisfies

$$-\sum_j \frac{\partial}{\partial x_j} \tilde{A}_j(x, \nabla\tilde{u}) + g(\tilde{u}) = 0 \quad \text{in } B_{\epsilon_0} \setminus \{0\}, \tag{3.3.45}$$

where the $A_j$ verify the conditions (3.3.21). Since $\tilde{u}$ is bounded, it can be extended as a weak solution of (3.3.45) in $B_{\epsilon_0}$. By Theorem 1.3.10, $\tilde{u} \in C^1(B_{\epsilon_0})$, which ends the proof. □

### 3.3.3 *Classification of boundary singularities*

We take $\Omega$ and $x_0$ as in Section 3.3.1 and we assume that $x_0 = 0$. We consider the following problem

$$\begin{aligned} -\Delta_p u + |u|^{q-1} u &= 0 && \text{in } \Omega \\ u &= 0 && \text{on } \partial\Omega \setminus \{0\}, \end{aligned} \tag{3.3.46}$$

where $u \in C^1(\overline{\Omega} \setminus \{0\})$. Since $q \geq \tilde{q}_1$ implies that any solution of (3.3.46) is identically 0 by Corollary 3.3.7, henceforth we assume

$$0 < p - 1 < q < \tilde{q}_1. \tag{3.3.47}$$

We construct specific solutions of problem (3.3.46) either with a strong singularity or with prescribed weak singularities at 0 and the description of their boundary behavior need the following definition.

**Definition 3.3.8** Let $\Omega \subset \mathbb{R}^N$ be a $C^1$ open set with $0 \in \partial\Omega$ and such that $\mathbf{e}_N$ is the normal inward unit vector at 0. We say that a function $x \mapsto f(x) = f(r,\sigma)$ defined in a neighborhood of 0 *converges eventually uniformly* on $S_+^{N-1}$ to 0 when $r \to 0$ if for any $\epsilon > 0$ there exists $r_\epsilon > 0$ such that

$$\sup\left\{|f(r,\sigma)| : 0 < r \leq r_\epsilon, \sigma \in S_{+r}^{N-1}\right\} \leq \epsilon, \qquad (3.3.48)$$

where $S_{+r}^{N-1} = \{\sigma \in S_+^{N-1} : x \sim (r,\sigma) \in \Omega\}$. A function $f$, $C^1$ in neighborhood of 0 converges eventually in $C^1(S_+^{N-1})$ to 0 if $f$ and $\nabla' f$ converges eventually uniformly to 0.

In this definition, we notice that $S_{+r}^{N-1} = S_+^{N-1}$ for $0 < r \leq \delta_0$ if $\Omega \cap B_{\delta_0} = B_{\delta_0}^+$ for some $\delta_0 > 0$, thus this convergence is the uniform convergence. In general $S_{+r}^{N-1} \to S_+^{N-1}$ in the sense of Hausdorff distance, this convergence implies the uniform convergence on compact subsets of $S_+^{N-1}$, but it is actually much stronger. If we flatten the boundary near 0 by a diffeomorphism $\Phi : B_{\delta_0} \mapsto B_{\delta_0}$ such that $\Phi(0) = 0$, $\Phi(\Omega \cap B_{\delta_0} = B_{\delta_0}^+$ and $D\Phi(0) = I$ and set $\tilde{f} = f \circ \Phi$, then $\tilde{f}(r,\sigma) \to 0$ uniformly on $S_+^{N-1}$.

**Definition 3.3.9** Under the assumptions of Definition 3.3.8 we say that the set of functions $f(r,.) : S_{+r}^{N-1} \mapsto \mathbb{R}$ defined in $\{(r,\sigma) : 0 < r \leq r_0, \sigma \in S_{+r}^{N-1}\}$ is *eventually uniformly continuous* if for all $\epsilon > 0$ there exists $\delta > 0$ such that for all $r \in (0, r_0]$, for any couple $(\sigma, \sigma') \in S_{+r}^{N-1} \times S_{+r}^{N-1}$,

$$\ell(\sigma,\sigma') \leq \delta \Longrightarrow |f(r,\sigma) - f(r,\sigma')| \leq \epsilon. \qquad (3.3.49)$$

The following scholie ($\sigma\chi\acute{o}\lambda\iota o\nu$) will be very useful in applications.

**Proposition 3.3.10** *Let $\Omega \subset \mathbb{R}^N$ be a $C^1$ domain with $0 \in \partial\Omega$ and $f$ a function defined in a neighborhood of 0 and such that the set of functions $\{f(r,.)\}_{0<r<r_0}$ is eventually uniformly continuous. Then there exists a sequence $\{r_n\} \subset (0, r_0]$ and a continuous function $\omega : S_+^{N-1} \mapsto \mathbb{R}$ such that $f(r_n,.)$ converges eventually uniformly to $\omega$.*

**Theorem 3.3.11** *Assume that (3.3.47) holds and*

*(i) either $1 < p < N$, $q \geq 1$ if $p < 2$ and $\Omega$ is a $C^2$ domain contained into the half space $H^+$ bordered by the tangent hyperplane $T_0\partial\Omega$ to $\partial\Omega$ at 0 and furthermore $\partial\Omega \cap B_{\delta_0} = \partial H^+ \cap B_{\delta_0}$ for some $\delta_0 > 0$,*

*(ii) or* $p = N$.

*Then for any* $\gamma > 0$ *there exists a unique positive function* $u_\gamma \in C^1(\overline{\Omega} \setminus \{0\})$ *which satisfies (3.3.46) and*

$$\lim_{r\to 0} r^\beta \left( \left| (u_\gamma - \gamma V_{\mathbb{R}^N_+}) \right| + \left| \nabla'(u_\gamma - \gamma V_{\mathbb{R}^N_+}) \right| \right)(r,\sigma) = 0, \qquad (3.3.50)$$

*eventually uniformly on* $S^{N-1}_+$, *where* $V_{\mathbb{R}^N_+}$ *is the fundamental singular p-harmonic function constructed in Proposition 3.3.5 and Proposition 3.3.6.*

*Proof.* We assume that $H_+ = \mathbb{R}^N_+ = \{(x', x_N) : x' \in \mathbb{R}^{N-1}, x_N > 0\}$. Uniqueness is a consequence of the comparison principle and (3.3.50). We set $\beta = \beta_{S^{N-1}_+}$, $\phi = \phi_{S^{N-1}_+}$ so that

$$(r,\sigma) \mapsto V_{\mathbb{R}^N_+}(r,\sigma) = r^{-\beta_{S^{N-1}_+}} \phi_{S^{N-1}_+}(\sigma) = r^{-\beta}\phi(\sigma),$$

is the fundamental positive spherical $p$-harmonic function in $\mathbb{R}^N_+$ with a pole at 0 vanishing on $\partial\mathbb{R}^N_+ \setminus \{0\}$.

*Case* $1 < p < N$. In this proof, we use again the singular $p$-harmonic functions $v$ constructed in Proposition 3.3.5 and Proposition 3.3.6. Clearly the function $\gamma V_{\mathbb{R}^N_+}$ is a supersolution for problem (3.3.46) which vanishes on $\partial\Omega \setminus \{0\}$. The technical difficulty is to construct a subsolution. We define

$$\begin{aligned} \mathcal{H}u &= -\Delta u - \frac{p-2}{2} \frac{\langle \nabla |\nabla u|^2, \nabla u\rangle}{|\nabla u|^2} + |\nabla u|^{2-p} |u|^{q-1} u \\ &= \mathcal{P}u + |\nabla u|^{2-p} |u|^{q-1} u. \end{aligned} \qquad (3.3.51)$$

For $\alpha > 0$, set $w = (1-r^\alpha)\phi$ with $r = |x|$, then $\nabla w = (1-r^\alpha)\nabla\phi - \alpha r^{\alpha-2}\phi x$,

$$|\nabla w|^2 = (1-r^\alpha)^2 |\nabla\phi|^2 + \alpha^2 r^{2(\alpha-1)}\phi^2 - 2\alpha r^{\alpha-2}(1-r^\alpha)\phi\langle\nabla\phi, x\rangle,$$

$$\Delta w = (1-r^\alpha)\Delta\phi - 2\alpha r^{\alpha-2}\langle\nabla\phi, x\rangle - \alpha(N+\alpha-2)r^{\alpha-2}\phi,$$

$$\begin{aligned} \langle \nabla |\nabla w|^2, \nabla w\rangle &= (1-r^\alpha)^3 \langle \nabla |\nabla\phi|^2, \nabla\phi\rangle - \alpha r^{\alpha-2}(1-r^\alpha)^2\phi\langle\nabla|\nabla\phi|^2, x\rangle \\ &\quad - 2\alpha r^{\alpha-2}(1-r^\alpha)^2 |\nabla\phi|^2 \langle\nabla\phi, x\rangle + 2\alpha^2 r^{2(\alpha-1)}(1-r^\alpha)\phi|\nabla\phi|^2 \\ &\quad + 2\alpha^2(\alpha-1)r^{2(\alpha-2)}(1-r^\alpha)\phi^2\langle\nabla\phi, x\rangle - 2\alpha^3(\alpha-1)r^{3\alpha-4}\phi^3 \\ &\quad + 2\alpha^2 r^{2(\alpha-1)}(1-r^\alpha)\phi|\nabla\phi|^2 - 2\alpha^3 r^{3\alpha-4}\phi^2\langle\nabla\phi, x\rangle \end{aligned}$$

$$- 2\alpha r^{\alpha-2}(1-r^{\alpha})^2\left(\phi\left(\langle D^2\phi(x),\nabla\phi\rangle+|\nabla\phi|^2\right)+\langle\nabla\phi,x\rangle\,|\nabla\phi|^2\right)$$

$$+ 2\alpha^2 r^{2(\alpha-2)}(1-r^{\alpha})\left(\phi^2\left(\langle D^2\phi(x),x\rangle+\langle\nabla\phi,x\rangle\right)+\phi\langle\nabla\phi,x\rangle^2\right)$$

$$- 2\alpha\left((\alpha-2)r^{\alpha-4}-2(\alpha-1)r^{2(\alpha-2)}\right)(1-r^{\alpha})\phi\langle\nabla\phi,x\rangle^2$$

$$+ 2\alpha^2\left((\alpha-2)r^{2(\alpha-2)}-2(\alpha-1)r^{3\alpha-4}\right)\phi^2\langle\nabla\phi,x\rangle.$$

Because $\nabla\phi = r^{-\beta-1}\left(-\beta\phi\,\mathbf{e}+\nabla'\phi\right)$, where $\mathbf{e}=\frac{x}{|x|}$, there holds

(*i*) $$|\nabla\phi|^2 = r^{-2(\beta+1)}\left(\beta^2\phi^2+|\nabla'\phi|^2\right)$$

(*ii*) $$\langle\nabla\phi,x\rangle = -\beta r^{-\beta}\phi$$

(*iii*) $$\langle\nabla|\nabla\phi|^2,x\rangle = 2\langle D^2\phi(x),\nabla\phi\rangle = -2(\beta+1)\,|\nabla\phi|^2,$$

and hence

$$\begin{aligned}|\nabla w|^2 &= (1-r^{\alpha})^2\,|\nabla\phi|^2\,(1+o(1))\\ &= (1-r^{\alpha})^2 r^{-2\beta-2}\left(\beta^2\phi^2+|\nabla'\phi|^2\right)(1+o(1)).\end{aligned}$$

Since $\mathcal{P}\phi = 0$, we obtain after a lengthy computation,

$$\begin{aligned}\mathcal{P}w = \alpha\phi\Bigg(&-2\beta(p-1)+N+\alpha-p\\ &+(p-2)(\alpha-2)\frac{\beta^2\phi^2}{\beta^2\phi^2+|\nabla'\phi|^2}\Bigg)r^{\alpha-\beta-2}(1+o(1)).\end{aligned}\tag{3.3.52}$$

Because there always holds $\beta \geq \frac{N-p}{p-1}$ for any $1 < p \leq N$ and $\beta \geq \frac{N-1}{p-1}$ if $1 < p \leq 2$, we infer that

$$\mathcal{P}w \leq -(\kappa-\alpha)\alpha\phi r^{\alpha-\beta-2}(1+o(1)),\tag{3.3.53}$$

for some $\kappa > 0$. Next

$$|\nabla w|^{2-p}\,w^q = r^{(p-2)(\beta+1)-q\beta}\phi^q\left(\beta^2\phi^2+|\nabla'\phi|^2\right)^{\frac{2-p}{2}}(1+o(1)).\tag{3.3.54}$$

By (3.3.50) there holds $\beta_{p,q} > \beta$, then $-\beta-2 < (p-2)(\beta+1)-q\beta$, hence for $\alpha > 0$ small enough, $r^{(p-2)(\beta+1)-q\beta} = o(r^{\alpha-\beta-2})$ near $r = 0$. Since $q \geq 1$, there exists $r_0 > 0$ such that $\mathcal{H}w \leq 0$ in $B^+_{r_0} \subset \Omega$. Since $w(x) \leq r_0^{-\beta}$ for $|x| = r_0$, the function $\tilde{w} = (w - r_0^{-\beta})_+$ is a subsolution which vanishes on $\partial B^+_{r_0} \setminus \{0\}$. Thus it is smaller than $V_{\mathbb{R}^N_+}$. If we extend it by 0 in $\Omega \setminus B^+_{r_0}$, we obtain a subsolution, still denoted by $\tilde{w}$. By Theorem 1.4.6 there exists a function $u_1 \in C^1(\overline{\Omega} \setminus \{0\})$ which satisfies (3.3.46) and

$$\lim_{r \to 0} r^\beta u_1(r, \sigma) = \phi(\sigma), \tag{3.3.55}$$

locally uniformly on $S_+^{N-1}$. The $C^{1,\alpha}$ regularity which follows from the *a priori* estimate implies that $r^\beta u_1(r,\sigma)$ converges in $C^1(\overline{S}_+^{N-1})$. Thus (3.3.50) holds with $\gamma = 1$. When $\gamma \neq 1$ the function $w_\gamma = \gamma(1 - r^\alpha)\phi$ satisfies

$$\mathcal{P}w_\gamma \leq -\gamma(\kappa - \alpha)\alpha\phi r^{\alpha-\beta-2}(1 + o(1)), \tag{3.3.56}$$

while

$$|\nabla w_\gamma|^{2-p} w_\gamma^q = \gamma^{2-p+q} r^{(p-2)(\beta+1)-q\beta} \phi^q \left(\beta^2\phi^2 + |\nabla'\phi|^2\right)^{\frac{2-p}{2}} (1 + o(1)). \tag{3.3.57}$$

The result follows up to changing $r_0$. Notice that the construction is valid if $\Omega$ is the half space $H_+$.

*Case* $p = N$. The condition on $q$ is

$$N - 1 < q < 2N - 1. \tag{3.3.58}$$

The previous construction is still valid and in particular, for any $\gamma > 0$ there exists a unique function $u_\gamma \in C^1(\overline{\mathbb{R}_+^N} \setminus \{0\})$ which satisfies

$$\begin{aligned} -\Delta_N u + u^q &= 0 && \text{in } \mathbb{R}_+^N \\ u &= 0 && \text{in } \partial\mathbb{R}_+^N \setminus \{0\} \\ \lim_{r \to 0} ru(r, .) &= \gamma\phi(.) && \text{in } C^1(\overline{S_+^{N-1}}). \end{aligned} \tag{3.3.59}$$

We first construct a solution in a ball. As in the proof of Proposition 3.3.6 we set $a = -\frac{1}{2}\mathbf{e}_N$, $\tilde{a} = -\mathbf{e}_N$ and denote by $\mathcal{J}_1^{\tilde{a}}$ the inversion of center $a$ and power 1. Then $\mathcal{J}_1^{\tilde{a}}(\mathbb{R}_+^N) = B_{\frac{1}{2}}(a)$, $\mathcal{J}_1^{\tilde{a}}(0) = 0$ and the function $\tilde{w}_\gamma = w_\gamma \circ \mathcal{J}_1^{\tilde{a}}$ satisfies

$$\begin{aligned} -\Delta_N \tilde{w} + |x - \tilde{a}|^{-2N} \tilde{w}^q &= 0 && \text{in } B_{\frac{1}{2}}(a) \\ \tilde{w} &= 0 && \text{in } \partial B_{\frac{1}{2}}(a) \setminus \{0\} \\ \lim_{r \to 0} r\tilde{w}(r, .) &= \gamma\omega(.) && \text{in } C^1_{loc}(\overline{S_+^{N-1}}). \end{aligned} \tag{3.3.60}$$

Since $|x - \tilde{a}| \leq 1$ in $B_{\frac{1}{2}}(a)$, $\tilde{w}_\gamma$ is a subsolution for problem (3.3.46) and it satisfies (3.3.50. Since the fundamental $N$-harmonic function $\gamma v$ defined in $B_{\frac{1}{2}}(a)$ with a pole at 0 constructed in Proposition 3.3.6 is a supersolution which vanishes on $\partial\Omega\setminus\{0\}$, the existence of $v_\gamma$ follows from Theorem 1.4.6.

Next, up to a scaling transformation, we assume that the ball $B_{\frac{1}{2}}(a)$ is included in $\Omega$ and tangent to $\partial\Omega$ at 0. The extension of the function $v_\gamma$ solution of problem (3.3.46) by 0 in $\Omega\setminus B_{\frac{1}{2}}(a)$ is a subsolution. Since $\gamma v$ is a supersolution there exists a solution $u_\gamma$ problem (3.3.46) in $\Omega$ which satisfies $v_\gamma\leq u_\gamma\leq\gamma v$. As in the previous case $u_\gamma$ satisfies (3.3.50). □

**Theorem 3.3.12** *Assume that (3.3.47) holds and*

*(i) either* $1<p<N$ *and* $\Omega$ *is a* $C^2$ *domain contained into the half space* $H^+$ *bordered by the tangent hyperplane* $T_0\partial\Omega$ *to* $\partial\Omega$ *at* 0 *and furthermore* $\partial\Omega\cap B_{\delta_0}=\partial H^+\cap B_{\delta_0}$ *for some* $\delta_0>0$,

*(ii) or* $p=N$.

*Then there exists a unique positive function* $u_\infty\in C^1(\overline{\Omega}\setminus\{0\})$ *which satisfies (3.3.46) and*

$$\lim_{r\to 0}\left(\left|r^{\beta_{p,q}}u_\infty-\omega_s\right|-\left|\nabla'\left(r^{\beta_{p,q}}u_\infty-\omega_s\right)\right|\right)(r,\sigma)=0 \qquad (3.3.61)$$

*eventually uniformly on* $S_+^{N-1}$, *where* $\omega_s$ *is the unique positive solution of (3.2.3) obtained in Theorem 3.2.1.*

*Proof.* We consider first the case $1<p<N$ although the construction is still valid if $p=N$. The restriction to $\Omega$ of the function $(r,\sigma)\mapsto u_s(r,\sigma)=r^{-\beta_{p,q}}\omega_s(\sigma)$ is a supersolution for

$$\begin{aligned}
-\Delta_p u+|u|^{q-1}u&=0 && \text{in } \Omega\\
u&=0 && \text{in } \partial\Omega\setminus\{0\}\\
\lim_{r\to 0} r^{\beta_{p,q}}u(r,\sigma)&=\omega_s(\sigma) && \text{uniformly on } S_+^{N-1}.
\end{aligned} \qquad (3.3.62)$$

If $m$ is the maximum of $u_s$ on $\partial\Omega\cap H^+$, the function $(u_s-m)_+$ is a subsolution of (3.3.62). For $0<\epsilon<\delta_0$ we denote by $u_\epsilon$ the solution of

$$\begin{aligned}
-\Delta_p u+|u|^{q-1}u&=0 && \text{in } \Omega\setminus\overline{B}_\epsilon\\
u&=0 && \text{in } \partial\Omega\cap B_\epsilon^c\\
u(\epsilon,.)&=\epsilon^{-\beta_{p,q}}\omega_s(.) && \text{on } \partial B_\epsilon\cap\Omega.
\end{aligned} \qquad (3.3.63)$$

The function $u_\epsilon$ is unique and obtained by minimization. Furthermore $(u_s-m)_+\leq u_\epsilon\leq u_s$ in $\Omega\setminus\overline{B}_\epsilon$. For $0<\epsilon'<\epsilon$, $u_{\epsilon'}\lfloor_{\partial B_\epsilon\cap\Omega}\leq u_\epsilon$, thus $u_{\epsilon'}\leq u_\epsilon$ in $\Omega\setminus\overline{B}_\epsilon$. Hence $u_\epsilon$ converges to some function that we denote $u$

which satisfies $(u_s - m)_+ \leq u \leq u_s$ in $\Omega$ and which vanishes on $\partial\Omega \setminus \{0\}$. For reasons which will be made clearer later on we will denote $u = u_\infty$. Since $u_\epsilon(x) \leq C|x|^{-\beta_{p,q}}$ in $\Omega \setminus \overline{B}_\epsilon$, it follows from Lemma 3.3.2 that the set of functions $\{r^{\beta_{p,q}} u_\epsilon(r,.)\}$ is relatively compact in $C^1_{loc}(\overline{S}^{N-1}_+)$. Therefore $u_\infty$ satisfies $-\Delta_p u + |u|^{q-1}u = 0$ in $\Omega$. Furthermore, the set $\{u_\infty(r,.)\}$ is also relatively compact in $C^1_{loc}(\overline{S}^{N-1}_+)$, which implies that $r^{\beta_{p,q}} u_\infty(r,.) \to \omega$ in $C^1(\overline{S}^{N-1}_+)$.

In the case $p = N$, the function $u_s(r,\sigma) = r^{-\beta_{p,q}}\omega_s(\sigma)$ is a solution of

$$\begin{aligned} -\Delta_N u + u^q &= 0 && \text{in } \mathbb{R}^N_+ \\ u &= 0 && \text{in } \partial\mathbb{R}^N_+ \setminus \{0\} \\ \lim_{r\to 0} r^{\beta_{p,q}} u(r,.) &= \omega_s(.) && \text{in } C^1(\overline{S^{N-1}_+}). \end{aligned} \tag{3.3.64}$$

First we construct a solution in the ball $B_{\frac{1}{2}}(a)$. With the notations of the proof of Theorem 3.3.11 $\tilde{v}_\infty = u_s \circ \mathcal{J}_1^{\tilde{a}}$ satisfies

$$\begin{aligned} -\Delta_N \tilde{v} + \tilde{v}^q &\leq 0 && \text{in } B_{\frac{1}{2}}(a) \\ \tilde{v} &= 0 && \text{in } \partial B_{\frac{1}{2}}(a) \setminus \{0\} \\ \lim_{r\to 0} r^{\beta_{p,q}} \tilde{v}(r,.) &= \omega_s(.) && \text{in } C^1_{loc}(\overline{S^{N-1}_+}). \end{aligned} \tag{3.3.65}$$

It satisfies $\tilde{v}_\infty \leq u_s$. For $0 < \epsilon < \frac{1}{2}$ we denote by $v_\epsilon$ the solution, obtained by minimization, of

$$\begin{aligned} -\Delta_N v + v^q &= 0 && \text{in } B_{\frac{1}{2}}(a) \setminus B_\epsilon \\ v &= 0 && \text{in } \partial B_{\frac{1}{2}}(a) \cap B^c_\epsilon \\ v(r,.) &= u_s(\epsilon,\sigma)(.) && \text{in } B_{\frac{1}{2}}(a) \cap \partial B_\epsilon. \end{aligned} \tag{3.3.66}$$

By the maximum principle, $\tilde{v}_\infty \leq v_\epsilon \leq u_s$ in $B_{\frac{1}{2}}(a) \setminus B_\epsilon$. Moreover, since $v_{\epsilon'} \lfloor_{B_{\frac{1}{2}}(a)\cap\partial B_\epsilon} \leq v_\epsilon \lfloor_{B_{\frac{1}{2}}(a)\cap\partial B_\epsilon} = u_s(\epsilon,\sigma)$ for $0 < \epsilon' < \epsilon$, it follows that $v_{\epsilon'} \leq v_\epsilon$. Therefore there exists $v_0 = \lim_{\epsilon\to 0} v_\epsilon$; $v_0$ vanishes on $\partial B_{\frac{1}{2}}(a) \setminus \{0\}$ and satisfies $\tilde{v}_\infty \leq v_0 \leq u_s$. Furthermore the convergence holds in $C^1_{loc}(\overline{B}_{\frac{1}{2}}(a) \setminus \{0\})$ and $r^{\beta_{p,q}} v_0(r,.) \to \omega_s(.)$ in $C^1_{loc}(\overline{S^{N-1}_+})$. Hence $v_0$ is the solution of (3.3.46) in $B_{\frac{1}{2}}(a)$ with $p = N$ which satisfies (3.3.61).

In order to construct a supersolution in the complement of a given unit ball

$B_1(b)$ such that $0 \in \partial B_1(b)$, we denote by $v_0'$ the solution of

$$\begin{aligned} -\Delta_N v + v^q &= 0 && \text{in } B_1(b) \\ v &= 0 && \text{in } \partial B_1(b) \setminus \{0\} \\ \lim_{r \to 0} r^{\beta_{p,q}} v(r,.) &= \omega_s(.) && \text{in } C^1_{loc}(\overline{S_+^{N-1}}). \end{aligned} \tag{3.3.67}$$

If $\mathcal{J}_1^b$ denotes the inversion of center $b$ and power 1, then $\tilde{v}_0' = v_0' \circ \mathcal{J}_1^b$ satisfies

$$\begin{aligned} -\Delta_N v + |x-b|^{-2N} v^q &= 0 && \text{in } B_1^c(b) \\ v &= 0 && \text{in } \partial B_1(b) \setminus \{0\} \\ \lim_{r \to 0} r^{\beta_{p,q}} v(r,.) &= \omega_s(.) && \text{in } C^1(\overline{S_+^{N-1}}). \end{aligned} \tag{3.3.68}$$

Next we set $a' = -a$ and we can assume that the two balls $B_{\frac{1}{2}}(a)$ and $B_{\frac{1}{2}}(a')$ are respectively included in $\Omega$ and $\Omega^c$ and tangent at $\partial\Omega$ at 0. Up to a change of scale, we can assume that there exists a $C^{1,\alpha}$ function $\tilde{v}_0'$ defined in $B^c_{\frac{1}{2}}(a')$ and such that

$$\begin{aligned} -\Delta_N v' + v'^q &\geq 0 && \text{in } B^c_{\frac{1}{2}}(a') \\ v' &= 0 && \text{in } B^c_{\frac{1}{2}}(a') \setminus \{0\} \\ \lim_{r \to 0} r^{\beta_{p,q}} v'(r,.) &= \omega_s(.) && \text{in } C^1(\overline{S_+^{N-1}}). \end{aligned} \tag{3.3.69}$$

For $\epsilon > 0$ small enough, let $u_\epsilon$ be the solution of

$$\begin{aligned} -\Delta_N u + u^q &= 0 && \text{in } \Omega \cap B^c_\epsilon \\ u &= 0 && \text{in } \partial\Omega \cap B^c_\epsilon \\ u(\epsilon,.) &= \tilde{v}_0'(\epsilon,\sigma)(.) && \text{in } \Omega \cap \partial B_\epsilon. \end{aligned} \tag{3.3.70}$$

By the maximum principle $v_0 \leq u_\epsilon \leq \tilde{v}_0'$ in the respective domains where two of the functions are simultaneously defined. Furthermore $u_\epsilon \leq u_{\epsilon'}$ for $0 < \epsilon' < \epsilon$. Therefore $u_\infty := \lim_{\epsilon \to 0} u_\epsilon$ exists and, by the previous compacity results already used, it is the unique solution of (3.3.46) which satisfies (3.3.61). □

The connexion between $u_\gamma$ and $u_\infty$ is the following

**Theorem 3.3.13** *Let the assumptions of Theorem 3.3.12 be satisfied. Then the mapping $\gamma \mapsto u_\gamma$ is increasing and*

$$\lim_{\gamma \to \infty} u_\gamma = u_\infty. \tag{3.3.71}$$

*Proof.* Let $0 < \gamma < \gamma'$. It follows from (3.3.50) that $(u_\gamma - u_{\gamma'})_+$ vanishes in a neighborhood of 0; thus $u_\gamma \leq u_{\gamma'}$. Similarly $u_\gamma \leq u_\infty$ because of the monotonicity of $\gamma \mapsto u_\gamma$, as the fact that $u_\gamma \leq u_\infty$, is a consequence of the maximum principle and (3.3.61). We recall that $T_\lambda[u](x) = \lambda^{\beta_{p,q}} u(\lambda x)$, but since the domain is modified upon dilations, we denote by $u_\gamma^\Omega$ and $u_\infty^\Omega$ the functions constructed in Theorem 3.3.11 and Theorem 3.3.12 relative to $\Omega$. Then $T_\lambda[u_\gamma^\Omega] = u^{\Omega_\lambda}_{\gamma\lambda^{\beta_{p,q}-\beta}}$, therefore

$$\lim_{\gamma\to\infty} T_\lambda[u_\gamma^\Omega] = T_\lambda[\lim_{\gamma\to\infty} u_\gamma^\Omega] = T_\lambda[u^\Omega_{\text{"}\infty\text{"}}] = u^{\Omega_\lambda}_{\text{"}\infty\text{"}}.$$

We recall that the transformation $T_\lambda$ leaves equation (3.1.1) invariant. Furthermore

$$\left|T_\lambda[u_\gamma^\Omega](x)\right| \leq c|x|^{-\beta_{p,q}} \qquad \text{in } \Omega_\lambda,$$

and, by the regularity results of Section 1-3 (valid also if $\gamma = \infty$),

$$\begin{aligned} &(i) \quad \left|\nabla T_\lambda[u_\gamma^\Omega](x)\right| \leq c|x|^{-\beta_{p,q}-1} \\ &(ii) \quad \left|\nabla\left(T_\lambda[u_\gamma^\Omega](x) - T_\lambda[u_\gamma^\Omega](x')\right)\right| \leq c|x|^{-\beta_{p,q}-1-\alpha}|x-x'|^\alpha \text{ if } |x| \leq |x'|. \end{aligned} \tag{3.3.72}$$

Then there exists a subsequence $\{\lambda_n\}$ converging to 0 and a nonnegative function $U \in C^1(\overline{\mathbb{R}}^N_+ \setminus \{0\})$, solution of (3.1.1) and vanishing on $\partial\mathbb{R}^N_+ \setminus \{0\}$, such that $T_{\lambda_n\mu}[u^\Omega_{\text{"}\infty\text{"}}] \to U$ in the $C^1_{loc}(\overline{\mathbb{R}}^N_+ \setminus \{0\})$ topology for any $\mu \in \mathbb{Q}_+$. Since $T_{\lambda_n\mu}[u^\Omega_{\text{"}\infty\text{"}}] = T_\mu[T_{\lambda_n}[u^\Omega_{\text{"}\infty\text{"}}]]$ and $T_\mu$ is continuous, we obtain that $T_\mu[U] = U$ for all $\mu \in \mathbb{Q}_+$. Using again the continuity we derive that this identity holds for any $\mu \in \mathbb{R}_+$. Therefore $U$ is a nonnegative self-similar solution of (3.1.1). Therefore it is separable, that is $U(r,\sigma) = r^{-\beta_{p,q}}\omega$ and either $\omega = 0$, or $\omega = \omega_s$ is the unique positive solution of (3.2.4) (see Theorem 3.2.1).

In order to see that it is not zero, we use the fact that since $u_\gamma^\Omega$ is both increasing with respect to $\gamma$ and $\Omega$, we consider a ball $B \subset \Omega$ such that $0 \in \partial B \cap \partial\Omega$. Therefore $B^\lambda \subset \Omega^\lambda$ and $0 \in \partial B^\lambda \cap \partial\Omega^\lambda$. Then $u_\gamma^B \leq u_{\gamma'}^{\Omega^\lambda}$ for all $\lambda > 0$ and $\gamma' > 0$. This implies $u_\gamma^B \leq u^{\Omega^\lambda}_{\text{"}\infty\text{"}} = T_\lambda[u^\Omega_{\text{"}\infty\text{"}}]$ and finally $u_\gamma^B \leq U$. We denote by $U_s$ the function defined in $\mathbb{R}^N_+$ by

$$U_s(x) = U_s(r,\sigma) = r^{-\beta_{p,q}}\omega_s(\sigma) \qquad \forall (r,\sigma) \in \mathbb{R}_+ \times S^{N-1}_+. \tag{3.3.73}$$

We infer from the previous steps that $T_\lambda[u^\Omega_{\text{"}\infty\text{"}}] \to U_s$ in $C^1_{loc}(\overline{\mathbb{R}}^N_+ \setminus \{0\})$ topology. By taking $|x| = 1$ and putting $r = \lambda$ this implies that $u^\Omega_{\text{"}\infty\text{"}}$

satisfies the same convergence property (3.3.61) than $u_\infty$. By uniqueness the proof follows. □

Next we prove a general classification result for positive singular solutions.

**Theorem 3.3.14** *Let $\Omega$ be a bounded $C^2$ domain contained into the half space $H^+$ bordered by the tangent hyperplane $T_0\partial\Omega$ at 0 and furthermore $\partial\Omega \cap B_{\delta_0} = \partial H^+ \cap B_{\delta_0}$ for some $\delta_0 > 0$. Assume $1 < p < N$ and $p - 1 < q < \tilde{q}_1$. If $u \in C^1(\overline{\Omega} \setminus \{0\})$ is a positive solution of (3.3.46) the following alternative holds*

*(i) either there exists $\gamma \geq 0$ and $u = u_\gamma$, hence it is the unique positive solution which satisfies (3.3.50).*

*(ii) or $u = u_\infty$ and it is the unique positive solution which satisfies (3.3.61).*

*Proof.* Without loss of generality we assume that $\mathbf{e}_N$ is the normal inward unit vector to $\partial\Omega$ at 0. We first assume that $|x|^\beta u(x)$ is bounded by some constant $c_0$, where $\beta = \beta_{S_+^{N-1}}$. By Lemma 3.3.2 there holds

$$\begin{array}{lll} (i) & |\nabla u(x)| \leq c_1 |x|^{-\beta-1} & \forall x \in \Omega \\ (ii) & |\nabla u(x) - \nabla u(x')| \leq c_1 |x|^{-\beta-1-\alpha} |x - x'|^\alpha & \forall x, x' \in \Omega,\ |x| \leq |x'|. \end{array} \tag{3.3.74}$$

This implies that the set of functions $\{u_\lambda\}$ defined by $u_\lambda(x) = \lambda^\beta u(\lambda x)$ is relatively compact in the $C^1_{loc}(\Omega^\lambda \setminus \{0\})$ topology (remember that $\Omega^\lambda = \frac{1}{\lambda}\Omega$). This local compactness is usually not enough to prove fine property of the limit, but since $\partial\Omega \cap B_{\delta_0} = \partial H^+ \cap B_{\delta_0}$, there always holds $\Omega^\lambda \cap B_{\frac{\delta_0}{\lambda}} = B^+_{\frac{\delta_0}{\lambda}} := H^+ \cap B_{\frac{\delta_0}{\lambda}}$. Therefore, for any compact subset $K \subset \overline{H}^+$ the convergence holds eventually in the $C^1(K)$ topology. Furthermore

$$0 \leq u_\lambda(x) \leq c_0 |x|^{-\beta} \quad \forall x \in \overline{\Omega}^\lambda \setminus \{0\}. \tag{3.3.75}$$

Notice that $u_\lambda$ is a positive solution of the problem

$$\begin{aligned} -\Delta_p u + \lambda^{(q+1-p)(\beta_{p,q}-\beta)} |u|^{q-1} u &= 0 && \text{in } \Omega^\lambda \\ u &= 0 && \text{in } \partial\Omega^\lambda \setminus \{0\}. \end{aligned} \tag{3.3.76}$$

Since $\beta_{p,q} - \beta > 0$ any cluster-point function of the set $\{u_\lambda\}$ when $\lambda \to 0$, is a nonnegative $p$-harmonic function in $\mathbb{R}^N_+$ which vanishes on $\partial\mathbb{R}^N_+ \setminus \{0\}$ and satisfies (3.3.75) with $\overline{\mathbb{R}}^N_+ \setminus \{0\}$ instead of $\overline{\Omega}^\lambda \setminus \{0\}$. Let $U = \lim_{\lambda_n \to 0} u_{\lambda_n}$

be any of these cluster-points, then

$$U(x) = \gamma V_{\mathbb{R}_+^N}(x) = \gamma |x|^{-\beta} \phi_{S_+^{N-1}}(\tfrac{x}{|x|}), \tag{3.3.77}$$

by Theorem 2.3.9 for some $\gamma \in [0, c_0]$. Because the gradient of $\{u_\lambda\}$ is also converging,

$$\lim_{r_n \to 0} r_n^\beta \frac{u(r_n, \sigma)}{\phi(\sigma)} = \gamma \quad \text{uniformly on } S_+^{N-1}, \tag{3.3.78}$$

where $\phi = \phi_{S_+^{N-1}}$. If $\gamma > 0$, it follows from Theorem 3.3.11-(i) and (3.3.50) that

$$\lim_{r_n \to 0} \frac{u(r_n, \sigma)}{u_\gamma(r_n, \sigma)} = 1 \quad \text{uniformly on } S_+^{N-1}. \tag{3.3.79}$$

If $\gamma > 0$, then for $\epsilon \in (0,1)$ and $n \geq n_\epsilon \in \mathbb{N}$, we derive from the maximum principle $(1-\epsilon)u_\gamma \leq u \leq (1+\epsilon)u_\gamma$ in $\Omega \cap B_{r_n}^c$. Letting $n \to \infty$ and $\epsilon \to 0$ yields $u = u_\gamma$. If $\gamma = 0$, then

$$\lim_{r \to 0} r^\beta \frac{u(r, \sigma)}{\omega(\sigma)} = 0,$$

from the step above. Thus $u \leq u_\epsilon$ for any $\epsilon > 0$ by the maximum principle, which yields $u = 0$.

Next we assume that $|x|^\beta u(x)$ is unbounded and there exists a sequence $\{r_n\}$ converging to 0 such that

$$\limsup_{x \to 0} |x|^\beta u(x) = \lim_{n \to \infty} r_n^\beta u(r_n, \sigma_n) = \lim_{r_n \to 0} \gamma_n = \infty, \tag{3.3.80}$$

where $\sigma_n \in S_+^{N-1}$. The function $u$ satisfies

$$-\Delta_p u + d(x)|u|^{p-2}u = 0,$$

with $d = |u|^{q+1-p}$. By (3.3.3) we have $d(x) \leq C_0 |x|^{-p}$, therefore, using boundary Harnack inequality under the form (2.3.5), and since, if $x \sim (r, \sigma)$, there holds $\rho(x) = r\mathbf{e}_N.\sigma$, then

$$\frac{u(r_n, \sigma)}{r_n \mathbf{e}_N.\sigma} \geq \frac{1}{c} \frac{u(r_n, \sigma_n)}{r_n \mathbf{e}_N.\sigma_n} \quad \forall \sigma \in S_+^{N-1}.$$

Therefore

$$u(r_n, \sigma) \geq \frac{1}{c} \frac{\mathbf{e}_N.\sigma}{\mathbf{e}_N.\sigma_n} u(r_n, \sigma_n) \geq \frac{\mathbf{e}_N.\sigma}{c} u(r_n, \sigma_n) = \frac{\gamma_n r_n^{-\beta}}{c} \mathbf{e}_N.\sigma. \tag{3.3.81}$$

Furthermore, because $\omega_{S_+^{N-1}}$ is $C^1$ in $\overline{S}_+^{N-1}$ and Hopf boundary lemma applies on $\partial S_+^{N-1}$, there exists $c_1 > 0$ such that

$$\frac{1}{c_1}\mathbf{e}_N.\sigma \leq \omega_{S_+^{N-1}}(\sigma) \leq c_1\mathbf{e}_N.\sigma.$$

This inequality combined with (3.3.81), yields

$$u(r_n,\sigma) \geq \frac{\gamma_n}{cc_1} r_n^{-\beta}\omega_{S_+^{N-1}}(\sigma) \geq u_{\tilde{\gamma}_n}(r_n,\sigma) \qquad \forall \sigma \in S_+^{N-1}, \tag{3.3.82}$$

where we have set $\tilde{\gamma}_n = \frac{\gamma_n}{cc_1}$ and used the fact that $u_\gamma \leq V_{\mathbb{R}_+^N}$, see Theorem 3.3.11. By the maximum principle it implies that $u \geq u_{\tilde{\gamma}_n}$ in $\Omega \cap B_{r_n}^c$, for any $n \in \mathbb{N}$. Letting $n \to \infty$ and using Theorem 3.3.13, we obtain

$$u(x) \geq \lim_{\gamma\to\infty} u_\gamma(x) = u_\infty(x) \qquad \forall x \in \Omega. \tag{3.3.83}$$

In order to prove the estimate from above we write $u_\infty = u_\infty^\Omega$ and for $\epsilon > 0$ small enough we define $u_\infty^{\Omega,B_\epsilon}$ as being the minimal solution of

$$\begin{aligned} -\Delta_p u + |u|^{q-1}u = 0 \qquad & \text{in } \Omega \setminus B_\epsilon \\ u = 0 \qquad & \text{on } \partial\Omega \setminus B_\epsilon \\ u = \infty \qquad & \text{on } \Omega \cap \partial B_\epsilon. \end{aligned} \tag{3.3.84}$$

Uniqueness could be proved but it is of no use here. By the maximum principle $u(x) \leq u_\infty^{\Omega,B_\epsilon}$ in $\Omega \setminus B_\epsilon$. Moreover $\epsilon \mapsto u_\infty^{\Omega,B_\epsilon}$ is decreasing, thus

$$u(x) \leq u_\infty^{\Omega,B_0} := \lim_{\epsilon\to 0} u_\infty^{\Omega,B_\epsilon} \qquad \text{in } \Omega \setminus B_\epsilon, \tag{3.3.85}$$

and $u_\infty^{\Omega,B_0}$ is a positive solution in $\Omega$ which vanishes on $\partial\Omega \setminus \{0\}$. Since $T_\lambda[u_\infty^{\Omega,B_\epsilon}] = u_\infty^{\Omega^\lambda,B_{\frac{\epsilon}{\lambda}}}$, it follows that $T_\lambda[u_\infty^{\Omega,B_0}] = u_\infty^{\Omega^\lambda,B_0}$. The function $u_\infty^{\Omega^\lambda,B_0}$ satisfy the estimates (3.3.72) and it is a positive solution of (3.1.1) in $\Omega^\lambda$, continuous in $\overline{\Omega}^\lambda \setminus \{0\}$ and vanishing on $\partial\Omega^\lambda \setminus \{0\}$. Similarly as in the proof of Theorem 3.3.13, $T_\lambda[u_\infty^{\Omega,B_0}]$ converges locally uniformly in $\overline{\mathbb{R}}_+^N \setminus \{0\}$, as well as its gradient, to a positive solution $\hat{U}$ of (3.1.1) continuous on $\overline{\mathbb{R}}_+^N \setminus \{0\}$ which vanishes on $\partial\mathbb{R}_+^N \setminus \{0\}$ and satisfies $T_\mu[\hat{U}] = \hat{U}$. Since $\hat{U}$ is nonzero, it coincides with the self-similar solution $U_s$. This implies that $u_\infty^{\Omega,B_0} = u_{"\infty"}^\Omega = u_\infty$, which ends the proof. □

The proof of the following result is similar.

**Theorem 3.3.15** *Let $\Omega$ be a bounded $C^2$ domain such that $0 \in \partial\Omega$ and $N-1 < q < 2N-1$. If $u \in C^1(\overline{\Omega} \setminus \{0\})$ is a positive solution*

$$\begin{aligned} -\Delta_N u + u^q &= 0 && \text{in } \Omega \\ u &= 0 && \text{on } \partial\Omega \setminus \{0\}, \end{aligned} \tag{3.3.86}$$

*then the following dichotomy occurs:*

*(i) either there exists $\gamma \geq 0$ and $u = u_\gamma$, thus it is the unique positive solution which satisfies (3.3.50),*

*(ii) or $u = u_\infty$ and it is the unique positive solution which satisfies (3.3.61).*

Using the methods developed above it is easy to prove the following rigidity theorem.

**Theorem 3.3.16** *Assume $0 < p-1 < \tilde{q}_1$. If $u \in C^1(\overline{\mathbb{R}}^N_+ \setminus \{0\})$ is a positive solution of (3.1.1) in $\mathbb{R}^N_+$ which vanishes on $\partial\mathbb{R}^N_+ \setminus \{0\}$, then either $u = u_\gamma$ for some $\gamma > 0$, or $u = U_s$.*

## 3.4 Singularities of quasilinear Hamilton-Jacobi type equations

Let $\Omega \subset \mathbb{R}^N$ be a domain, $p > 1$ and $q > p-1$ two real numbers. This section deals with the properties of solutions of the following equation

$$-\Delta_p u + |\nabla u|^q = 0 \tag{3.4.1}$$

in $\Omega$ or in $\Omega \setminus \{a\}$ for some $a \in \Omega$. The equation is not invariant under the transformation $u \mapsto -u$ but many of the results below are valid for signed solutions. Therefore we will speak of problem with absorption if $u > 0$ and problem with source if $u < 0$. If $u$ is radial with respect 0, $r = |x|$ and $u(x) = u(r)$, then it satisfies the following ODE,

$$\left(|u'|^{p-2}u'\right)' + \frac{N-1}{r}|u'|^{p-2}u' - |u'|^q = 0. \tag{3.4.2}$$

### 3.4.1 *Comparison principles*

In this section it is crucial to quote some comparison principles between regular solutions of (3.4.1).

**Proposition 3.4.1** *Let $q > p-1 > 0$. If $f$ is a continuous radial function in $\overline{B}_R$ and $a \in \mathbb{R}$ there exists a unique radial $C^1$ function $u$ defined in $\overline{B}_R$*

*which satisfies*

$$\begin{aligned} -\Delta_p u + |\nabla u|^q &= f \qquad \text{in } B_R \\ u &= a \qquad \text{on } \partial B_R. \end{aligned} \tag{3.4.3}$$

*Assume* $u_1$ *and* $v_2$ *are radial* $C^1$ *functions in* $\overline{B}_R$ *which satisfy respectively*

$$-\Delta_p u_1 + |\nabla u_1|^q = f_1, \tag{3.4.4}$$

*and*

$$-\Delta_p u_2 + |\nabla u_2|^q = f_2. \tag{3.4.5}$$

*If* $f_1 \geq f_2$ *in* $B_R$*,* $u_1 \geq u_2$ *on* $\partial B_K$*, then* $u_1 \geq u_2$ *in* $\overline{B}_R$*.*

*Proof. Step 1: Existence and uniqueness.* For a given $f \in C([0,R])$ we define the mapping $T_f$ on the ball $\mathcal{B}_M$ on $C([0,R])$ endowed with the uniform convergence norm by

$$T_f[v](r) = r^{1-N} \int_0^r s^{N-1} \left( |v(s)|^{\frac{q}{p-1}} - f(s) \right) ds. \tag{3.4.6}$$

Then

$$\sup_{0<s\leq r} |T_f[v](r) - T_f[w](r)| \leq \frac{qM^{\frac{q+1-p}{p-1}} r}{N(p-1)} \sup_{0<s\leq r} |v(s) - w(s)|.$$

An easy induction shows that the $k$-iterate $T_f^k$ of $T_f$ satisfies

$$\sup_{0<s\leq r} \left| T_f^k[v](s) - T_f^k[w](s) \right| \leq \left( \frac{qM^{\frac{q+1-p}{p-1}} r}{N(p-1)} \right)^k \frac{1}{k!} \sup_{0<s\leq r} |v(s) - w(s)|.$$

Therefore, for any $M > 0$, there exists $k = k(M) \in \mathbb{N}^*$ such that $T^k$ satisfies

$$\sup_{0<s\leq R} \left| T_f^k[v](s) - T_f^k[w](s) \right| \leq \frac{1}{2} \sup_{0<s\leq R} |v(s) - w(s)| \qquad \forall v, w \in \mathcal{B}_M. \tag{3.4.7}$$

Hence $T_f$ admits a unique fixed point in $\mathcal{B}_M$. This means that for a given $f \in C([0,R])$, there is a unique $v \in C([0,R])$ such that

$$v(r) = r^{1-N} \int_0^r s^{N-1} \left( |v(s)|^{\frac{q}{p-1}} - f_j(s) \right) ds \qquad \forall r \in [0,R], \tag{3.4.8}$$

and $v \in C^1([0,R])$ with $v'(0) = -\frac{1}{N} f(0)$. Next we set $v = |u'|^{p-2} u'$. Equivalently, $u' = F_p(v)$. If $u(R) = a$ is given, then $u$ is uniquely determined on

$[0, R]$ by

$$u(r) = a - \int_r^R F_p\left(s^{1-N}\int_0^s t^{N-1}\left(|u'(s)|^q - f(s)\right)dt\right)ds. \tag{3.4.9}$$

*Step 2: Comparison.* Since the radial solution $u$ of (3.4.3) is unique, it can be obtained as the limit, when $\epsilon \to 0$ of the solutions $u_\epsilon$ of the regularized problem

$$\begin{aligned} -div\left((\epsilon^2 + |\nabla u|^2)^{\frac{p-2}{2}}\nabla u\right) + (\epsilon^2 + |\nabla u|^2)^{\frac{q}{2}} &= f \qquad \text{in } B_R \\ u &= a \qquad \text{on } \partial B_R, \end{aligned} \tag{3.4.10}$$

which are also solutions of

$$-\left(r^{N-1}(\epsilon^2 + u'^2)^{\frac{p-2}{2}}u'\right)' + r^{N-1}(\epsilon^2 + u'^2)^{\frac{q}{2}} = r^{N-1}f, \tag{3.4.11}$$

and it can be expressed by

$$u_\epsilon(r) = a - \int_r^R F_{p,\epsilon}\left(s^{1-N}\int_0^s t^{N-1}\left((\epsilon^2 + u'^2)^{\frac{q}{2}} - f(s)\right)dt\right)ds, \tag{3.4.12}$$

where $F_{p,\epsilon}(\xi) = \eta$ means $\xi = (\epsilon^2 + \eta^2)^{\frac{p-2}{2}}\eta$. If $u_1$ and $u_2$ are the solutions of (3.4.4) and (3.4.5) respectively, we define $u_{1,\epsilon}$ and $u_{2,\epsilon}$ by

$$u_{j,\epsilon}(r) = u_j(R) - \int_r^R F_{p,\epsilon}\left(s^{1-N}\int_0^s t^{N-1}\left((\epsilon^2 + u_{j,\epsilon}'^2)^{\frac{q}{2}} - f_j(s)\right)dt,\right)ds, \tag{3.4.13}$$

for $j = 1, 2$. Then we claim that $u_{1,\epsilon} + \epsilon \geq u_{2,\epsilon}$. If it is not true, the set of $x \in B_R$ such that $u_{1,\epsilon}(|x|) + \epsilon < u_{2,\epsilon}(|x|)$ is not empty. Let $G$ be one of its connected component. It is either a spherical shell $B_{\nu_1} \setminus \overline{B}_{\nu_1}$ where $0 \leq \nu_2 < \nu_1 < R$, or a ball $B_\nu$. Then the function $r \mapsto w = u_{2,\epsilon}(r) - u_{1,\epsilon}(r) - \epsilon$ achieves its maximum at some interior point $r_0$. If we set $\Gamma_{r_0} = \{x \in B_R : |x| = r_0\}$, then for any $x_0 \in \Gamma_{r_0}$, $\nabla u_{2,\epsilon}(x_0) = \nabla u_{1,\epsilon}(x_0)$. Hence

$$\begin{aligned} -div\left((\epsilon^2 + |\nabla u_{2,\epsilon}|^2)^{\frac{p-2}{2}}\nabla u_{2,\epsilon}\right) &+ div\left((\epsilon^2 + |\nabla u_{1,\epsilon}|^2)^{\frac{p-2}{2}}\nabla u_{1,\epsilon}\right) \\ &+ (\epsilon^2 + |\nabla u_{2,\epsilon}|^2)^{\frac{q}{2}} - (\epsilon^2 + |\nabla u_{1,\epsilon}|^2)^{\frac{q}{2}} \\ &= -\mathcal{L}w + bw' \\ &\leq 0, \end{aligned}$$

where

$$-\mathcal{L}_\epsilon(r)w = -\left(\left(\epsilon^2 + U'^2(r)\right)^{\frac{p-4}{2}}\left(\epsilon^2 + (p-1)U'^2(r)\right)w'\right)',$$

and $U'(r) = (1-\theta)u'_{2,\epsilon}(r) + \theta u'_{1,\epsilon}(r)$ for some $\theta \in (0,1)$, and

$$b(r) = q\left(\epsilon^2 + U'^2(r)\right)^{\frac{q-2}{2}} U'(r).$$

Since $\left(\epsilon^2 + u'^2(r_0)\right)^{\frac{p-4}{2}}\left(\epsilon^2 + (p-1)u'^2(r_0)\right) > 0$, $\mathcal{L}_\epsilon$ is a uniformly elliptic operator. Furthermore $r \mapsto b(r) = q\left(\epsilon^2 + U'^2(r)\right)^{\frac{q-2}{2}} U'(r)$ is continuous at $r = r_0$. By the strong maximum principle $u_{2,\epsilon} - u_{1,\epsilon} - \epsilon$ is constant, which is a contradiction. Thus $u_{2,\epsilon} \le u_{1,\epsilon} + \epsilon$ letting $\epsilon \to 0$ yields $u_2 \le u_1$. □

In the non-radial case we have a similar result under an additional assumption. If $u \in C^1(\Omega)$, we set

$$Crit_\Omega(u) = \{x \in \Omega : \nabla u(x) = 0\}. \tag{3.4.14}$$

**Theorem 3.4.2** *Assume $p > 1$ and $q > 0$ and $\Omega \subset \mathbb{R}^N$ is a bounded $C^2$ domain. Let $f_j \in C(\overline{\Omega})$ and $u_j \in C^1(\overline{\Omega})$ for $j = 1, 2$ such that*

$$\begin{aligned} -\Delta_p u_j + |\nabla u_j|^q &= f_j && \text{in } \Omega \\ u_j &= \psi_j && \text{on } \partial\Omega. \end{aligned} \tag{3.4.15}$$

*Suppose $f_1 \ge f_2$, $\psi_1 \ge \psi_2$ and $Crit_\Omega(u_1) \cap Crit_\Omega(u_2) = \emptyset$. Then $u_1 \ge u_2$ in $\Omega$.*

*Proof.* Notice that the assumption implies that for any compact $K \subset \Omega$ there exists $\alpha_K > 0$ such that

$$\max\{|\nabla u_1(x)|, |\nabla u_2(x)|\} \ge \alpha_K \qquad \forall x \in K. \tag{3.4.16}$$

Suppose that the assertion $u_1 \ge u_2$ is not true. Then for $\epsilon > 0$ small enough the set $\mathcal{O} := \mathcal{O}_\epsilon = \{x \in \Omega : u_2(x) > u_1(x) + \epsilon\}$ is not empty. Let $G$ be a connected component of $\mathcal{O}$ and $x_0 \in G$ where the function $u_2 - u_1 - \epsilon$ achieves its maximum. Then $\nabla u_1(x_0) = \nabla u_2(x_0) \neq 0$. The function $w = u_2 - u_1 - \epsilon$ satisfies

$$-(\Delta_p u_2 - \Delta_p(u_1 + \epsilon)) + |\nabla u_2|^q - |\nabla(u_1 + \epsilon)|^q \le 0,$$

and

$$\begin{aligned}-(\Delta_p u_2 - \Delta_p(u_1+\epsilon)) + |\nabla u_2|^q - |\nabla(u_1+\epsilon)|^q &= -\mathcal{L}w + \langle B, \nabla w\rangle \\ &= \sum_{i,j} \frac{\partial}{\partial x_i}\left(a_{ij}\frac{\partial w}{\partial x_j}\right) + \sum_i b_i \frac{\partial w}{\partial x_i},\end{aligned}$$

where $a_{ij}$ and $b_i$ are expressed by the following expressions since $\nabla u_1(x_0) = \nabla u_2(x_0)$,

$$a_{ij}(x_0) = |\nabla u_1(x_0)|^{p-4}\left(\delta_{ij}\,|\nabla u_1(x_0)|^2 + (p-2)\frac{\partial u_1(x_0)}{\partial x_i}\frac{\partial u_1(x_0)}{\partial x_j}\right), \tag{3.4.17}$$

and

$$b_i(x_0) = q\,|\nabla u_1(x_0)|^{q-2}\,\frac{\partial u_1(x_0)}{\partial x_i}. \tag{3.4.18}$$

The vector field $B$ is continuous at $x_0$, as for the $a_{ij}$ they satisfy

$$|\nabla u_1(x_0)|^{p-2}\min\{1,p-1\}\,|\xi|^2 \le \sum_{i,j} a_{ij}\xi_i\xi_j \le |\nabla u_1(x_0)|^{p-2}\max\{1,p-1\}\,|\xi|^2.$$

Therefore $\mathcal{L}$ is uniformly elliptic in a neighborhood of $x_0$. By the strong maximum principle $w$ is constant in a neighborhood of $x_0$. Hence the subset of $G$ where $u_2 - u_1 - \epsilon = w(x_0)$ is also open. By connectedness it yields $w = w(x_0) > 0$ in $G$, which contradicts $w = 0$ on $\partial G$. Thus $\mathcal{O} = \emptyset$ and $u_2 \le u_1 + \epsilon$ in $\Omega$ which yields the result. □

**Corollary 3.4.3** *Assume $p > 1$ and $q > 0$ and $\Omega \subset \mathbb{R}^N$ is a bounded $C^2$ domain. Let $f_j \in C(\overline{\Omega})$ and $u_j \in C^1(\overline{\Omega})$ for $j = 1, 2$ satisfy (3.4.15) for $j = 1, 2$ and $f_1 \ge f_2$, $\psi_1 \ge \psi_2$. If $Crit_\Omega(u_1)$ is discrete and $Crit_\Omega(u_2)$ has empty interior or vice versa, then $u_1 \ge u_2$.*

*Proof.* For $h \in \mathbb{R}^N$, we denote by $\tau_h u$ the function $x \mapsto u(x+h)$. Then $\tau_h u_1$ is defined on $\tau_{-h}(\Omega) := \Omega - h$ where it satisfies (3.4.15) (with $j = 1$) and

$$\sup_{\Omega\cap\tau_{-h}(\Omega)} (|\tau_h u_1 - u_1| + |\nabla \tau_h u_1 - \nabla u_1|) \to 0 \quad \text{as } h \to 0.$$

Let $\epsilon > 0$, there exists $h_0 > 0$ such that for $|h| \le h_0$, $|\tau_h u_1(x) - u_1(x)| \le \epsilon$ for all $x \in \Omega \cap \tau_{-h}(\Omega)$. Since $Crit_\Omega(u_1)$ is discrete and $Crit_\Omega(u_2)$ has empty interior, there exists a sequence $\{h_k\}$ converging to 0 such that

$Crit_\Omega(\tau_{h_k}u_1)\cap Crit_\Omega(u_2)=\emptyset$. Since $u_1$ and $u_2$ are continuous and $u_2\leq u_1$ in $\partial\Omega$ there exists $k_0\in\mathbb{N}$ such that

$$u_2(x)\leq\tau_{h_k}u_1(x)+\epsilon\qquad\forall x\in\partial(\tau_{-h}(\Omega)\cap\Omega),\ \forall k\geq k_0.$$

By Theorem 3.4.2 $u_2(x)\leq\tau_{h_k}u_1(x)+\epsilon$ in $\tau_{-h_k}(\Omega)\cap\Omega$. Letting successively $h_k\to 0$ and $\epsilon\to 0$ yields the result. □

*Remark.* It is noticeable that no sign assumption on the functions $u_j$, $f_j$ is needed. Actually this comparison result is valid if $|\nabla u|^q$ is replaced by $g(\nabla u)$ where $g:\mathbb{R}^N\mapsto\mathbb{R}$ is any $C^1$ function, and equation (3.4.15) by

$$-\Delta_p u_j+g(\nabla u_j)=f_j.\tag{3.4.19}$$

### 3.4.2 *Radial solutions*

If $u$ is a radial solution of (3.4.1) in $B_1^*=B_1\setminus\{0\}$, then (3.4.2) holds in $(0,1)$. We put

$$\gamma_{p,q}=\frac{p-q}{q+1-p}.$$

There exists an explicit solution under the form

$$U_s(r)=\mu_{N,p,q}r^{-\gamma_{p,q}},\tag{3.4.20}$$

with

$$\mu_{N,p,q}=\gamma_{p,q}^{-1}\left((p-1)\gamma_{p,q}+p-N\right)^{\frac{1}{p+q-1}},\tag{3.4.21}$$

provided $p-1<q<q_c$ where

$$q_c=\frac{N(p-1)}{N-1}.\tag{3.4.22}$$

If we set $w=r^{N-1}|u'|^{p-2}u'$ then

$$w'(r)=r^{-\frac{(q+1-p)(N-1)}{p-1}}|w|^{\frac{q}{p-1}}.\tag{3.4.23}$$

It is an easy-to-integrate first order separable equation (linear if $q=p-1$). When we deal with nonnegative solutions the following statements, valid also in the non-radial case, hold.

**Proposition 3.4.4** *Assume $1<p\leq N$, $p-1<q<p$ and $u\in C^1((0,1))$ is a nonnegative solution of (3.4.2).*

*I- If $p-1<q<q_c$ the following alternative holds:*

*(i) either there exists $\gamma \geq 0$ such that*

$$\lim_{r\to 0}\frac{u(r)}{\mu_p(r)}=\gamma, \tag{3.4.24}$$

*where $\mu_p$ is the fundamental solution of the p-Laplacian in $\mathbb{R}^N$, see (1.5.14),*
*(ii) or*

$$\lim_{r\to 0} r^{\gamma_{p,q}}u(r)=\mu_{N,p,q}. \tag{3.4.25}$$

*II- If $p>q\geq q_c$ $u(r)$ admits a finite limit and $u'(r)\to 0$ when $r\to 0$.*

A simple consequence is the following

**Corollary 3.4.5** *Assume $1<p\leq N$, $p-1<q<q_c$. Then for any $\gamma>0$ and $c\in\mathbb{R}$) there exists a unique solution $u_{\gamma,c}$ of (3.4.2) on $(0,\infty)$ which satisfies*

$$\begin{array}{ll}(i) & \displaystyle\lim_{r\to 0}\frac{u(r)}{\mu_p(r)}=\gamma\\ (ii) & \displaystyle\lim_{r\to\infty}u(r)=c.\end{array} \tag{3.4.26}$$

*Furthermore*

$$\lim_{\gamma\to\infty}u_\gamma(r)=\mu_{N,p,q}r^{-\gamma_{p,q}}+c. \tag{3.4.27}$$

*Proof.* If $w(r)=r^{N-1}|u'|^{p-2}u'$ satisfies (3.4.23), then we can look for $w_\gamma$ such that

$$w_\gamma(0)=-\tilde{\gamma}:=\begin{cases}-\left(\frac{N-p}{p-1}\right)^{p-1}\gamma^{p-1} & \text{if } 1<p<N\\ \gamma^{N-1} & \text{if } p=N.\end{cases} \tag{3.4.28}$$

Then

$$w_\gamma(r)=-\left(\frac{q+1-p}{(N-1)(q_c-q)}r^{\frac{(N-1)(q_c-q)}{p-1}}+|\tilde{\gamma}|^{-\frac{q+1-p}{p-1}}\right)^{-\frac{p-1}{q+1-p}}, \tag{3.4.29}$$

hence

$$u_{\gamma,c}(r)=c+\int_r^\infty\left(\frac{q+1-p}{(N-1)(q_c-q)}s^{\frac{(N-1)(q_c-q)}{p-1}}+|\tilde{\gamma}|^{-\frac{q+1-p}{p-1}}\right)^{-\frac{1}{q+1-p}}s^{-\frac{N-1}{p-1}}ds,$$

and

$$\lim_{\gamma\to\infty}u_{\gamma,c}(r)=u_{\infty,c}(r)=c+\mu_{N,p,q}r^{-\gamma_{p,q}}=c+U_s(r).$$

□

When we consider nonpositive solutions, one gets easily from (3.4.23).

**Proposition 3.4.6** *Assume* $1 < p \leq N$, $p-1 < q < p$ *and* $u \in C^1((0,1))$ *is a negative solution of (3.4.2).*

*I- If* $p-1 < q < q_c$ *there exists* $c \leq 0$ *such that (3.4.24) holds.*
*II- If* $q = q_c$

$$\lim_{r\to 0} \frac{(-\ln r)^{\frac{N-1}{p-1}}\, u(r)}{\mu_p(r)} = \ell_{N,p} < 0, \tag{3.4.30}$$

*II- If* $q < q_c$,
*(i) either*

$$\lim_{r\to 0} r^{\gamma_{p,q}} u(r) = -\tilde{\mu}_{N,p,q}, \tag{3.4.31}$$

*where* $\tilde{\mu}_{N,p,q} = \gamma_{p,q}^{-1}\,(N-p-(p-1)\gamma_{p,q})^{\frac{1}{p+q-1}}$,
*(ii) or* $u(r)$ *admits a finite limit and* $u'(r) \to 0$ *when* $r \to 0$.

In the non-radial case only case I is already known (see Chapter 6).

### 3.4.3 *Gradient estimates and applications*

The following estimate, dealing with signed solutions, is fundamental in the study of the properties of solutions of (3.4.1).

**Theorem 3.4.7** *Let* $q > p-1 > 0$. *Assume* $u \in C^1(\Omega)$ *is a solution of (3.4.1), then*

$$|\nabla u(x)| \leq c_{N,p,q}\,(\rho(x))^{-\frac{1}{q+1-p}} \quad \forall x \in \Omega. \tag{3.4.32}$$

*Proof.* Let $G \subset \Omega$ be an open subset of $\Omega$ where $\nabla u \neq 0$. Then $u \in C^2(G)$ and

$$-\Delta u - (p-2)\frac{\langle D^2 u \nabla u, \nabla u\rangle}{|\nabla u|^2} + |\nabla u|^{q+2-p} = 0. \tag{3.4.33}$$

Since

$$\frac{1}{2}\Delta\left|\nabla u\right|^2 = \left|D^2 u\right|^2 + \langle \nabla\Delta u, \nabla u\rangle \geq \frac{1}{N}(\Delta u)^2 + \langle \nabla \Delta u, \nabla u\rangle, \tag{3.4.34}$$

by Schwartz inequality, we derive from (3.4.33) with $z = \left|\nabla u\right|^2$

$$\Delta u = -\frac{p-2}{2}\frac{\langle \nabla z, \nabla u\rangle}{z} + z^{\frac{q+2-p}{2}},$$

$$\langle\nabla\Delta u,\nabla u\rangle=-\frac{p-2}{2}\frac{\langle D^2z\nabla u,\nabla u\rangle}{z}-\frac{p-2}{4}\frac{|\nabla z|^2}{z}+\frac{p-2}{2}\frac{\langle\nabla z,\nabla u\rangle^2}{z^2}$$
$$+\frac{q+2-p}{2}\langle\nabla z,\nabla u\rangle.$$

Furthermore, from the previous inequalities,

$$(\Delta u)^2\geq\frac{1}{2}z^{q+2-p}-\frac{(p-2)^2}{4}\frac{\langle\nabla z,\nabla u\rangle^2}{z^2},$$

and

$$\Delta z+(p-2)\frac{\langle D^2z\nabla u,\nabla u\rangle}{z}\geq\frac{1}{N}z^{q+2-p}-\frac{(p-2)^2}{2N}\frac{\langle\nabla z,\nabla u\rangle^2}{z^2}-\frac{p-2}{2}\frac{|\nabla z|^2}{z}$$
$$+(p-2)\frac{\langle\nabla z,\nabla u\rangle^2}{z^2}+(q+2-p)z^{q+2-p}\langle\nabla z,\nabla u\rangle.\tag{3.4.35}$$

Since

$$\frac{\langle\nabla z,\nabla u\rangle^2}{z^2}\leq\frac{|\nabla z|^2}{z},$$

and, for $\epsilon>0$,

$$z^{q+2-p}\left|\langle\nabla z,\nabla u\rangle\right|\leq z^{\frac{q+2-p}{2}}\frac{|\nabla z|}{\sqrt{z}}\leq\epsilon z^{q+2-p}+\frac{1}{4\epsilon}\frac{|\nabla z|^2}{z},$$

we obtain that the right-hand side of (3.4.35) is bounded from below by the quantity $\frac{1}{2N}z^{q+2-p}-D\frac{|\nabla z|^2}{z}$, where $D=D(N,p,q)>0$. We define the linear operator $\mathcal{A}$ by

$$\mathcal{A}(v)=-\Delta v-(p-2)\frac{\langle D^2v\nabla u,\nabla u\rangle}{|\nabla u|^2}=-\sum_{i,j=1}^{N}a_{ij}v_{x_i,x_j},\tag{3.4.36}$$

where the $a_{ij}$ depend on $\nabla u$. Because $z=|\nabla u|^2$, there holds for all $\xi=(\xi_1,...,\xi_N)\in\mathbb{R}^N$

$$\min\{1,p-1\}|\xi|^2:=\theta|\xi|^2\leq\sum_{i,j=1}^{N}a_{ij}\xi_i\xi_j\leq\max\{1,p-1\}|\xi|^2:=\Theta|\xi|^2.\tag{3.4.37}$$

Consequently $\mathcal{A}$ is uniformly elliptic in $G$ and $z$ satisfies

$$\mathcal{L}(z):=\mathcal{A}(z)+\frac{1}{2N}z^{q+2-p}-D\frac{|\nabla z|^2}{z}\leq 0\qquad\text{in }\Omega.\tag{3.4.38}$$

Let $a \in G$ and $R > 0$ be such that $\overline{B}_R(a) \subset \Omega$. We set $w(x) = \lambda\left(R^2 - |x-a|^2\right)^{-\frac{2}{q+1-p}}$ and $|x-a| = r$, then

$$w_{x_i} = \frac{4\lambda}{q+1-p}\left(R^2 - |x-a|^2\right)^{-\frac{2}{q+1-p}-1} x_i,$$

$$\begin{aligned} w_{x_ix_j} &= \frac{4\lambda}{q+1-p}\left(R^2 - |x-a|^2\right)^{-\frac{2}{q+1-p}-1}\delta_{ij} \\ &\quad + \frac{8(3+q-p)\lambda}{(q+1-p)^2}\left(R^2 - |x-a|^2\right)^{-\frac{2}{q+1-p}-2} x_ix_j. \end{aligned}$$

Thus

$$\frac{|\nabla w|^2}{w} = \frac{16\lambda}{(q+1-p)^2}\left(R^2 - |x-a|^2\right)^{-\frac{2}{q+1-p}-2},$$

$$w^{q+2-p} = \lambda^{q+2-p}(R^2-r^2)^{-\frac{q+2-p}{q+1-p}} = \lambda^{q+2-p}(R^2-r^2)^{-\frac{2}{q+1-p}-2},$$

and finally

$$\mathcal{A}(w) \geq -\frac{\Theta\lambda}{q+1-p}(R^2-r^2)^{-\frac{2}{q+1-p}-2}\left(NR^2 + \left(\frac{q+3-p}{q+1-p} - N\right)r^2\right).$$

Since $r < R$, we obtain, for some $c = c(N,p,q) > 0$

$$\mathcal{L}(w) \geq \frac{\lambda}{2N}(R^2-r^2)^{-\frac{2}{q+1-p}-2}(\lambda^{q+1-p} - cR^2).$$

We choose $\lambda = (cR^2)^{\frac{1}{q+1-p}}$, therefore $\mathcal{L}(w) \geq 0$. Next we take for $G$ a connected component of the set $\{x \in B_R(a) : z(x) > w(x)\}$. Then $z(x) > 0$ in $G$ and $\overline{G} \subset \overline{B}_R(a)$. Let $x_0 \in G$ such that $z(x_0) - w(x_0) = \max\{z(x) - w(x) : x \in G\}$. Then $\nabla z(x_0) = \nabla w(x_0)$, $z(x_0) > w(x_0) > 0$ and

$$\begin{aligned} \mathcal{A}\left(z(x_0) - w(x_0)\right) &+ \frac{1}{2N}\left(z(x_0)^{q+2-p} - w(x_0)^{q+2-p}\right) \\ &- D\left|\nabla z(x_0)\right|^2\left(\frac{1}{z(x_0)} - \frac{1}{w(x_0)}\right) \leq 0. \end{aligned}$$

This inequality contradicts the fact that all the involve terms are non-negative with the exception of $z(x_0)^{q+2-p} - w(x_0)^{q+2-p}$ which is positive. Therefore $G = \emptyset$ and $z \leq w$ in $B_R(a)$. In particular

$$z(a) \leq w(a) = c'_{N,p,q}R^{-\frac{2}{q+1-p}}. \tag{3.4.39}$$

We obtain (3.4.32) by letting $R \to \rho(a)$. □

As a consequence we prove an upper estimate on a signed solution in $\Omega$ with a possible isolated singularity.

**Corollary 3.4.8** *Assume $q > p-1 > 0$, $\Omega$ is a domain containing 0 and $R^* > 0$ is such that $\rho(0) \geq 2R^*$. Then for any $x \in B_{R^*} \setminus \{0\}$ and $0 < R \leq R^*$, any $u \in C^1(\Omega \setminus \{0\})$ solution of (3.4.1) in $\Omega \setminus \{0\}$ satisfies*

$$|u(x)| \leq c_{N,p,q} \left||x|^{-\gamma_{p,q}} - R^{-\gamma_{p,q}}\right| + \max\{|u(z)| : |z| = R\}, \tag{3.4.40}$$

*if $p \neq q$, and*

$$|u(x)| \leq c_{N,p,q} \left|\ln R - \ln |x|\right| + \max\{|u(z)| : |z| = R\}, \tag{3.4.41}$$

*if $p = q$.*

*Proof.* Let $X = \frac{R}{|x|}x$. For $0 < t < 1$,

$$\text{dist}\,(tx + (1-t)X, \partial B_R(X)) = t|x| + (1-t)R.$$

Then, using Theorem 3.4.7,

$$\begin{aligned} |u(x)| &= \left|u(X) + \int_0^1 \frac{d}{dt} u(tx + (1-t)X dt\right| \\ &\leq |u(X)| + c_{N,p,q} |x - X| \int_0^1 (t|x| + (1-t)R)^{-\frac{1}{q+1-p}}. \end{aligned}$$

Integrating this relation we obtain (3.4.40) or (3.4.41). Notice that if $p > q$ and $|x| \leq \frac{R}{2}$, we derive

$$|u(x)| \leq c_{N,p,q}|x|^{-\gamma_{p,q}} + \max\{|u(z)| : |z| = R\}.$$

□

The gradient estimate allows to obtain upper estimates for solutions in a domain with respect to the distance to the boundary.

**Corollary 3.4.9** *Assume $q > p-1 > 0$ and $\Omega$ is a domain with a $C^2$ compact boundary. Then there exists $\delta^* > 0$ such that any solution $u \in C^1(\Omega)$ of (3.4.1) in $\Omega$ satisfies for all $x \in \Omega_{\delta^*} := \{x \in \Omega : \rho(x) < \delta^*\}$,*

$$|u(x)| \leq c_{N,p,q} \left((\rho(x))^{-\gamma_{p,q}} - \delta^{*\,-\gamma_{p,q}}\right), + \max\{|u(z)| : \rho(z) = \delta^*\}, \tag{3.4.42}$$

*if $p \neq q$, and*

$$|u(x)| \leq c_{N,p} (\ln \delta - \ln \rho(x)) + \max\{|u(z)| : \rho(z) = \delta^*\}, \tag{3.4.43}$$

*if* $p = q$.

*Proof.* Let $\delta^*$ be the largest $\delta > 0$ such that any boundary point $a$ belongs to the boundary of a ball $B_r(a_i)$ such that $B_r(a_i) \subset \Omega$ and to the boundary a ball $B_r(a_s)$ such that $B_r(a_i) \subset \overline{\Omega}^c$. If $x \in \Omega_{\delta^*}$ let $\sigma(x)$ be its projection on $\partial\Omega$, $\mathbf{n}_{\sigma(x)}$ the normal outward unit vector to $\partial\Omega$ at $\sigma(x)$ and $z^*(x) = x - 2\delta^*\mathbf{n}_{\sigma(x)}$. Then

$$\begin{aligned} u(x) &= u(z^*(x)) + \int_0^1 \frac{d}{dt} u(tx + (1-t)z^*(x))dt \\ &= u(z^*(x)) + \int_0^1 \langle \nabla u(tx + (1-t)z^*(x)), x - z^*(x) \rangle dt. \end{aligned}$$

Combined with (3.4.32) it implies

$$|u(x)| \leq |u(z^*(x))| + c_{N,p,q}(\delta^* - \rho(x)) \int_0^1 (t\rho(x) + (1-t)\rho(z^*(x))^{-\frac{1}{q+1-p}} dt,$$

which gives (3.4.42) by integration. □

The following Liouville-type result follows from Theorem 3.4.7. Notice that statement II has been proved by [Reshetniak (1966)] Theorem 1.3.8 using Harnack inequality.

**Corollary 3.4.10** *I- Assume* $q > p - 1 > 0$. *Then any* $C^1$ *solution of (3.4.1) in* $\mathbb{R}^N$ *is a constant.*

*II- Let* $p > 1$, *then any positive p-harmonic function in* $\mathbb{R}^N$ *is a constant.*

*Proof.* For I we apply Theorem 3.4.7 in $B_R$ and let $R \to \infty$. For II, if $v$ is a positive $p$-harmonic function in $\mathbb{R}^N$, it is $C^{1,\alpha}$ for some $\alpha \in (0,1)$ and the function $u = \frac{1}{p-1} \ln v$, has the same regularity and solves

$$-\Delta_p u + |\nabla u|^p = 0, \tag{3.4.44}$$

in $\mathbb{R}^N$. □

### 3.4.4 *Isolated singularities*

#### 3.4.4.1 *Removable isolated singularities*

The next theorem is a particular case of a more general one which will be proved in Chapter 4.

**Theorem 3.4.11** *Let $\Omega \subset \mathbb{R}^N$ be an open subset containing $0$, $1 < p \leq N$ and $g : \mathbb{R}_+ \mapsto \mathbb{R}_+$ a continuous function such that*

$$\inf_{s>0} \frac{g(s)}{s^{q_c}} > 0. \tag{3.4.45}$$

*If $u \in C^1(\Omega \setminus \{0\})$ is nonnegative and satisfies*

$$-\Delta_p u + g(|\nabla u|) = 0, \tag{3.4.46}$$

*in $\Omega \setminus \{0\}$, then $g(|\nabla u|) \in L^1_{loc}(\Omega)$, $u$ belongs to $W^{1,p}_{loc}$ and it can be extended to $\Omega$ as a weak solution of (3.4.46) in $\Omega$.*

*Proof.* Assume $\overline{B}_R \subset \Omega$. Because of (3.4.45) $u$ satisfies

$$-\Delta_p u + \gamma |\nabla u|^{q_c} \leq 0 \quad \text{in } \Omega \setminus \{0\}, \tag{3.4.47}$$

for some $c, \gamma > 0$ and we can assume $\gamma = 1 + \delta$ for some $\delta > 0$. We notice that $\frac{(q_c+1-p)(N-1)}{p-1} = 1$ and, for $\epsilon > 0$, we denote by $w_\epsilon$ the solution of the problem

$$\begin{aligned} rw' - |w|^{\frac{q_c}{p-1}} &= 0 \qquad \text{in } (\epsilon, 1] \\ w(\epsilon) &= -\infty. \end{aligned} \tag{3.4.48}$$

Then

$$w_\epsilon(r) = -\left(\frac{1}{(N-1)\ln\left(\frac{r}{\epsilon}\right)}\right)^{\frac{1}{N-1}}.$$

We set

$$u_\epsilon(r) = (\tau + \max_{|x|=R} u(x)) + (N-1)^{\frac{1-N}{p-1}} \epsilon^{\frac{p-N}{p-1}} \int_{\frac{r}{\epsilon}}^{\frac{R}{\epsilon}} (\ln s)^{-\frac{N-1}{p-1}} s^{-\frac{N-1}{p-1}} ds,$$

for some $\tau > 0$. By integration by parts

$$\begin{aligned} &\epsilon^{\frac{p-N}{p-1}} \int_{\frac{r}{\epsilon}}^{\frac{R}{\epsilon}} (\ln s)^{-\frac{N-1}{p-1}} s^{-\frac{N-1}{p-1}} ds \\ &\quad \leq \frac{p-1}{N-p}\left(\left(\ln\left(\frac{r}{\epsilon}\right)\right)^{-\frac{N-1}{p-1}} r^{-\frac{N-p}{p-1}} - \left(\ln\left(\frac{R}{\epsilon}\right)\right)^{-\frac{N-1}{p-1}} R^{-\frac{N-p}{p-1}}\right). \end{aligned}$$

Therefore, for any $r > 0$,

$$\lim_{\epsilon \to 0} u_\epsilon(r) = \max_{|x|=R} u(x).$$

The function $w = u - u_\epsilon$ is negative on $|x| = \epsilon$ and $|x| = R$. Assume it achieves a maximum at some $x_0 \in B_R \setminus B_\epsilon$. Then $\nabla w(x_0) = \nabla u(x_0) - \nabla u_\epsilon(x_0) = 0$, and thus $\nabla u(x_0) \neq 0$ because $\nabla u_\epsilon$ never vanishes. We have

$$-(\Delta_p u - \Delta_p u_\epsilon) + |\nabla u| - |\nabla u_\epsilon| \leq -\delta\,|\nabla u|^q. \tag{3.4.49}$$

At $x_0$ the right-hand side of (3.4.49) is negative while the left-hand side is nonnegative since

$$-(\Delta_p u - \Delta_p u_\epsilon) = -\mathcal{L}w$$

where $\mathcal{L}$ is an elliptic operator, uniformly elliptic in a neighborhood of $x_0$. There we get a contradiction. Thus $w$ achieves its maximum on $\partial B_R$, which implies $u \leq u_\epsilon$ in $B_R \setminus B_\epsilon$. If we let $\epsilon \to 0$ and $\tau \to 0$ we derive $u \leq \max\limits_{|x|=R} u(x)$ in $B_R \setminus \{0\}$.

Let $\eta_n \in C_c^\infty(\mathbb{R}^N)$, $0 \leq \eta_n \leq 1$, $\eta_n(x) \equiv 0$ if $|x| \leq \frac{1}{n}$ or $|x| \geq R$, $\eta_n(x) \equiv 1$ if $\frac{2}{n} \leq |x| \leq \frac{R}{2}$ and $|\nabla \eta_n| \leq 2n\chi_{B_{\frac{2}{n}}\setminus B_{\frac{1}{n}}} + R\chi_{B_R\setminus B_{\frac{R}{2}}}$. Then

$$\int_\Omega \left(p\eta_n^{p-1}\,|\nabla u|^{p-2}\,\langle \nabla u, \nabla \eta_n\rangle + |\nabla u|^{q_c}\,\eta_n^p\right)dx \leq 0.$$

Put $\Theta_n = B_R \setminus B_{\frac{R}{2}} \bigcup B_{\frac{2}{n}} \setminus B_{\frac{1}{n}}$

$$\begin{aligned}\left|\int_\Omega \eta_n^{p-1}\,|\nabla u|^{p-2}\,\langle \nabla u, \nabla \eta_n\rangle dx\right| &\leq \left(\int_\Omega |\nabla u|^{q_c}\,\eta_n^p dx\right)^{\frac{p-1}{q_c}} \left(\int_{\Theta_n} |\nabla \eta_n|^N\,dx\right)^{\frac{1}{N}}\\ &\leq c_N \left(\int_\Omega |\nabla u|^{q_c}\,\eta_n^p dx\right)^{\frac{p-1}{q_c}}.\end{aligned}$$

Which yields, by Beppo-Levi theorem,

$$\int_\Omega |\nabla u|^{q_c}\,\eta_n^p dx \leq c_{N,p} \Longrightarrow \int_{B_{\frac{R}{2}}} |\nabla u|^{q_c}\,dx \leq c_{N,p}. \tag{3.4.50}$$

Furthermore

$$\int_\Omega \left(p\eta_n^{p-1} u\,|\nabla u|^{p-2}\,\langle \nabla u, \nabla \eta_n\rangle + |\nabla u|^p\,\eta_n^p\right)dx + \int_\Omega |\nabla u|^{q_c}\,\eta_n^p dx \leq 0.$$

Since

$$\int_\Omega \left(p\eta_n^{p-1}u\,|\nabla u|^{p-2}\langle\nabla u,\nabla\eta_n\rangle + |\nabla u|^p\,\eta_n^p\right)dx$$
$$\geq \int_\Omega |\nabla u|^p\,\eta_n^p dx - p\left(\int_\Omega |\nabla u|^p\,\eta_n^p dx\right)^{\frac{p-1}{p}}\left(\int_{B_R} u^p dx\right)^{\frac{1}{p}},$$

we derive

$$\int_\Omega \left(|\nabla u|^{q_c} + |\nabla u|^p\right)\eta_n^p dx \leq c_{N,p}R^{\frac{N}{p}}\left(\int_\Omega |\nabla u|^p\,\eta_n^p dx\right)^{\frac{p-1}{p}}\max_{|x|=R} u(x),$$

hence $u \in W^{1,p}(B_R)$. We also have

$$\int_\Omega \left(p\eta_n^{p-1}\,|\nabla u|^{p-2}\langle\nabla u,\nabla\eta_n\rangle + g(|\nabla u|)\eta_n^p\right)dx = 0.$$

By the previous estimate the term $\displaystyle\int_\Omega p\eta_n^{p-1}\,|\nabla u|^{p-2}\langle\nabla u,\nabla\eta_n\rangle dx$ is uniformly bounded. Thus $\displaystyle\int_\Omega g(|\nabla u|)\eta_n^p dx$ is so; this implies that $g(|\nabla u|)$ is locally integrable in $\Omega$.

We replace $\eta_n$ by $\tilde\eta_n \in C^\infty(\mathbb{R}^N)$ such that $0 \leq \tilde\eta_n \leq 1$, $\tilde\eta_n(x) \equiv 0$ if $|x| \leq \frac{1}{n}$, $\tilde\eta_n(x) \equiv 1$ if $|x| \geq \frac{2}{n}$ and $|\nabla\tilde\eta_n| \leq 2n\chi_{B_{\frac{2}{n}}\setminus B_{\frac{1}{n}}}$. Let $\zeta \in C_c^1(\Omega)$, then

$$\int_\Omega \left(|\nabla u|^{p-2}\langle\nabla u,\nabla\tilde\eta_n\rangle\zeta + |\nabla u|^{p-2}\langle\nabla u,\nabla\zeta\rangle\tilde\eta_n\right)dx$$
$$+\int_\Omega g(|\nabla u|)\eta_n\zeta dx = 0. \tag{3.4.51}$$

Since

$$\left|\int_\Omega |\nabla u|^{p-2}\langle\nabla u,\nabla\tilde\eta_n\rangle\zeta dx\right| \leq 2c_N n^{1-\frac{N}{p}}\,\|\zeta\|_{L^\infty}\left(\int_{B_{\frac{2}{n}}} |\nabla u|^p\,dx\right)^{1-\frac{1}{p}},$$

the right-hand side term of which tends to 0 when $n \to \infty$, and the two other terms in (3.4.51) have limits, we finally infer

$$\int_\Omega \left(|\nabla u|^{p-2}\langle\nabla u,\nabla\zeta\rangle + g(|\nabla u|)\zeta\right)dx = 0. \tag{3.4.52}$$

□

### 3.4.4.2 *Isolated singularities of positive solutions*

The next result provides a complete classification of isolated singularities for positive solutions of (3.4.1) when $q$ is subcritical.

**Theorem 3.4.12** *Let $\Omega \subset \mathbb{R}^N$ be a domain containing $0$, $1 < p \leq N$ and $p - 1 < q < q_c$. If $u \in C^1(\Omega \setminus \{0\})$ is a nonnegative solution of (3.4.1) in $\Omega \setminus \{0\}$, the following alternative holds:*

*(i) either there exists $\gamma \geq 0$ such that*

$$\lim_{x \to 0} \frac{u(x)}{\mu_p(x)} = \gamma, \tag{3.4.53}$$

*where $\mu_p$ is defined at (1.5.14), and the following equation*

$$-\Delta_p u + |\nabla u|^q = c_{N,p}\delta_0, \tag{3.4.54}$$

*holds in $\mathcal{D}'(\Omega)$,*

*(ii) or*

$$\lim_{x \to 0} |x|^{\gamma_{p,q}} u(x) = \mu_{N,p,q}. \tag{3.4.55}$$

The proof relies on several intermediate results which have their own interest.

**Proposition 3.4.13** *Assume $\Omega \subset \mathbb{R}^N$ is as in Theorem 3.4.12, $1 < p \leq N$, and $q > p-1$. Then there exists a constant $c = c(N, p, q) > 0$ such that any nonnegative $u \in C^1(\Omega)$ solution of (3.4.1) in $\Omega$ satisfies*

$$\max\{u(x) : x \in B_{\frac{R}{2}}(a)\} \leq c \min\{u(x) : x \in B_{\frac{R}{2}}(a)\}. \tag{3.4.56}$$

*Proof.* We assume that $a = 0$ and write (3.4.1) as

$$-\Delta_p u + C(x)|\nabla u(x)|^{p-1} = 0, \tag{3.4.57}$$

with $C(x) = |\nabla u(x)|^{q+1-p}$. By (3.4.41), $|C(x)| \leq cR^{-1}$ with $c = c(N, p, q) > 0$. Next we set $u_R(y) = u(Ry)$. Then

$$-\Delta u_R + RC(Ry)|\nabla u_R(x)|^{p-1} = 0 \qquad \text{in } B_1.$$

Since $RC(Ry)$ is bounded in $B_1$ independently of $R$ and $y$, we derive from Theorem 1.3.6

$$\max\{u(x) : x \in B_{\frac{1}{2}}\} \leq c \min\{u(x) : x \in B_{\frac{1}{2}}\}.$$

Then (3.4.56) follows. □

As a consequence of Proposition 3.4.13 we have

**Corollary 3.4.14** *Assume $\Omega$ is as in Theorem 3.4.12, $q > p-1$. Then there exist a constant $c = c(N,p,q) > 0$ such that for any $R > 0$ such that $R < \rho(0)$ and for any nonnegative solution $u \in C^1(\Omega \setminus \{0\})$ solution of (3.4.1) in $\Omega \setminus \{0\}$, there holds*

$$\max\{u(z) : \tfrac{R}{8} \leq |z| \leq \tfrac{R}{8}\} \leq c \min\{u(z) : \tfrac{R}{8} \leq |z| \leq \tfrac{R}{8}. \tag{3.4.58}$$

*Remark.* Since the estimate of the gradient does not depend on the sign of the solution of (3.4.1), the Harnack inequality holds also for a negative solution $u$ in $\Omega \setminus \{0\}$. It reads

$$\max\{-u(z) : \tfrac{R}{8} \leq |z| \leq \tfrac{R}{8}\} \leq c \min\{-u(z) : \tfrac{R}{8} \leq |z| \leq \tfrac{R}{8}. \tag{3.4.59}$$

The next result is a version of Lemma 3.3.2 adapted to the study of isolated singularities is a domain. Its proof is much easier since we need not perform a reflexion and we can use directly scaling and the regularity results of Theorem 1.3.9. Again, the result holds for any signed solution.

**Lemma 3.4.15** *Assume $R > 0$, $1 < p \leq N$ and $p - 1 < q \leq p$. Let $u \in C^1(\overline{B}_R \setminus \{0\})$ be a solution of (3.4.1) in $B_R \setminus \{0\}$ which satisfies*

$$|u(x)| \leq \phi(|x|) \qquad \forall x \in B_R \setminus \{0\}, \tag{3.4.60}$$

*where $\phi : \mathbb{R}_+^* \mapsto \mathbb{R}_+$ is continuous, nonincreasing and verifies*

$$\phi(rs) \leq \gamma\phi(r)\phi(s) \qquad \forall r, s \in (0, r_0], \tag{3.4.61}$$

*for some $\gamma, r_0 > 0$. Then there exist constants $c_1 > 0$ and $\alpha \in (0,1)$ such that*

$$\begin{aligned}
&(i) \qquad |\nabla u(x)| \leq c_1 \frac{\phi(|x|)}{|x|} \qquad \forall x \in B_{\frac{R}{2}} \\
&(ii) \qquad |\nabla u(x) - \nabla u(x')| \leq c_1 \frac{|x - x'|^{\alpha}\phi(|x|)}{|x|^{1+\alpha}} \qquad \forall x, x \in B_{\frac{R}{2}}, |x| \leq |x'|.
\end{aligned} \tag{3.4.62}$$

*Proof of Theorem 3.4.12.* By scaling we can assume that $\overline{B}_1 \subset \Omega$.
*Step 1: We claim that if $u$ satisfies*

$$\liminf_{|x|\to 0} \frac{u(x)}{\mu_p(x)} < \infty, \tag{3.4.63}$$

*then statements (i) of Theorem 3.4.12 hold.* Because of (3.4.63) there exist $k \geq 0$ and a sequence $\{r_n\}$ converging to 0 such that

$$\lim_{n\to\infty} \inf_{|x|=r_n} \frac{u(x)}{\mu_p(r_n)} = k, \tag{3.4.64}$$

(remember that $\mu_p$ is a radial function). Using (3.4.58), we derive

$$\lim_{n\to\infty} \sup_{|x|=r_n} \frac{u(x)}{\mu_p(r_n)} = \tilde{k} \leq ck. \tag{3.4.65}$$

For $\epsilon > 0$ and $M = \max\{u(x) : |x| = 1\}$, the $p$-harmonic function $v = (\tilde{k}+\epsilon)\mu_p + M$ dominates the sub-$p$-harmonic function $u$ on $|x| = r_n$ for $n$ large enough and on $|x| = 1$. By the maximum principle $u \leq v$ in $\overline{B}_1 \setminus B_{r_n}$. Letting $n \to \infty$ and $\epsilon \to 0$ implies, up to changing the multiplicative constant $ck$,

$$0 \leq u(x) \leq m\mu_p \qquad \text{in } B_1 \setminus \{0\}. \tag{3.4.66}$$

By Lemma 3.4.15,

$$\begin{aligned}
&(i) \quad |\nabla u(x)| \leq c_1 \frac{m\mu_p(|x|)}{|x|} \qquad \forall x \in B_{\frac{1}{2}} \\
&(ii) \quad |\nabla u(x) - \nabla u(x')| \leq c_1 \frac{m|x-x'|^\alpha \mu_p(|x|)}{|x|^{1+\alpha}} \quad \forall x, x \in B_{\frac{1}{2}}, |x| \leq |x'|.
\end{aligned} \tag{3.4.67}$$

Next we define $y \mapsto u_r(y)$ by

$$u_r(y) = \frac{u(ry)}{\mu_p(r)} \qquad \forall y \in B_{\frac{1}{r}} \setminus \{0\}.$$

It satisfies in the above domain,

$$-\Delta_p u_r + (\mu_p(r))^{q+1-p} r^{p-q} |\nabla u_r|^q = 0. \tag{3.4.68}$$

Furthermore the following estimates hold from (3.4.66)-(3.4.67),

$$\begin{aligned}
&(i) \qquad 0 \leq u_r(y) \leq \frac{m\mu_p(r|y|)}{\mu_p(r)} \qquad \forall y \in B_{\frac{1}{r}} \setminus \{0\} \\
&(ii) \qquad |\nabla u_r(y)| \leq c_1 \frac{m\mu_p(r|y|)}{|y|\mu_p(r)} \qquad \forall y \in B_{\frac{1}{2r}} \setminus \{0\} \\
&(iii) \qquad |\nabla u_r(y) - \nabla u_r(y')| \leq c_1 \frac{m\mu_p(r|y|)|y-y'|^\alpha}{|y|^{1+\alpha}\mu_p(r)} \\
&\qquad\qquad \forall y, y \in B_{\frac{1}{2r}}, 0 < |y| \leq |y'|.
\end{aligned} \tag{3.4.69}$$

When $y$ remains in a compact subset $K$ of $\mathbb{R}^N_*$, there exists $r_K > 0$ such that $\frac{\mu_p(r|y|)}{\mu_p(r)}$ remains bounded for $0 < r \leq r_K$. By estimates (3.4.69) (ii)-(iii) the set of functions $\{u_r\}_{0<r\leq r_K}$ is relatively compact in the $C^1(K)$ topology. Furthermore $(\mu_p(r))^{q+1-p}r^{p-q} \to 0$ as $q < q_c$ when $r \to 0$. Therefore, by Cantor diagonal process, there exist a subsequence $\{r_{n_j}\}$ of $\{r_n\}$ converging to 0 and a nonnegative $p$-harmonic function in $\mathbb{R}^N \setminus \{0\}$ $v$ such that $u_{r_{n_j}}$ converges to $v$ in the $C^1_{loc}(\mathbb{R}^N \setminus \{0\})$-topology. By (3.4.69) (i) the function $v$ is bounded if $p = N$ and it is bounded from above by $m|y|^{-\frac{N-p}{p-1}}$ if $1 < p < N$. In the first case $v$ is equal to some nonnegative constant $\theta$, while if $1 < p < N$ $v = \theta\mu_p + \tau$ for some $\theta, \tau \geq 0$ by Corollary 1.5.9. However the estimate $0 \leq v(y) \leq m|y|^{-\frac{N-p}{p-1}}$ for all $y \neq 0$ imposes $\tau = 0$. When $1 < p < N$, we have in particular

$$\lim_{r_{n_j}\to 0} \frac{u(r_{n_j}\xi)}{\mu_p(r_{n_j})} = \theta\mu(1) \qquad \text{uniformly for } |\xi| = 1.$$

In view of (3.4.64), it yields $k = \theta\mu(1)$; furthermore, for any $\epsilon > 0$ there exists $n_{j_\epsilon}$ such that if $r_{n_j} \leq r_{n_{j_\epsilon}}$ we have $u(x) \leq (k+\epsilon)\mu_p(x)$ for $|x| = r_{n_j}$. By comparison,

$$u(x) \leq (k+\epsilon)\mu_p(x) + M \qquad \forall x \in B_1 \setminus B_{r_{n_{j_\epsilon}}}.$$

Letting $r_{n_j} \to 0$ we obtain that the above inequality holds in $B_1 \setminus \{0\}$. Letting $\epsilon \to 0$ and using (3.4.65) yields $k = \tilde{k}$ and again $u \leq \tilde{\mu}_p + M$ in $B_1 \setminus \{0\}$. Therefore

$$\lim_{x\to 0} \frac{u(x)}{\mu_p(x)} = k. \tag{3.4.70}$$

The $C^1$ convergence implies

$$\lim_{\substack{x \to 0 \\ \frac{x}{|x|} \to \mathbf{e}}} |x|^{\frac{N-1}{p-1}} \nabla u(x) = \begin{cases} k\frac{p-N}{p-1}\mathbf{e} & \text{if } 1 < p < N \\ -k\mathbf{e} & \text{if } p = N, \end{cases} \tag{3.4.71}$$

and as a consequence $|\nabla u|^q \in L^1_{loc}(\Omega)$.
When $p = N$, there holds

$$\lim_{r_{n_j}\to 0} \frac{u(r_{n_j}\xi)}{\mu_p(r_{n_j})} = \theta \qquad \text{uniformly for } |\xi| = 1.$$

Hence $k = \theta$ and we conclude as in the previous case. Finally, for any $\delta > 0$ and $\zeta \in C_c^\infty(\Omega)$, we have

$$\int_{\Omega\setminus B_\delta} \left(|\nabla u|^{p-2}\langle\nabla u, \nabla\zeta\rangle + |\nabla u|^q\right) dx = \int_{|x|=\delta} |\nabla u|^{p-2}\frac{\partial u}{\partial \mathbf{n}}\zeta dS.$$

Using (3.4.71), we infer that (3.4.54) holds.

*Step 2: We claim that if $u$ satisfies*

$$\liminf_{|x|\to 0} \frac{u(x)}{\mu_p(x)} = \infty, \tag{3.4.72}$$

*then statement (ii) of Theorem 3.4.12 holds.* Put $m = \min_{|x|=1} u(x)$. Since for any $\gamma > 0$ the function $u_{\gamma,m}$ (see definition in Corollary 3.4.5) is a solution of (3.4.1) in $B_1^*$ which has no critical points, is smaller than $u$ on $\partial B_1$ and at $x = 0$ because is satisfies (3.4.24)-(i), it is smaller than $u$. Letting $\gamma \to \infty$ yields

$$u(x) \geq U_s(|x|) + m = \mu_{N,p,q}|x|^{-\gamma_{p,q}} + m \qquad \forall x \in B_1^*. \tag{3.4.73}$$

The upper bound is obtained by a scaling technique. We adapt the argument used in the proof of Theorem 3.4.11, and for $\epsilon > 0$ we denote by $\tilde{w}_\epsilon$ the solution of

$$\begin{aligned} w' - r^{\frac{(q+1-p)(p-1)}{N-1}} |w|^{\frac{q}{p-1}} &= 0 \qquad \text{in } (\epsilon, \infty) \\ w(\epsilon) &= -\infty. \end{aligned} \tag{3.4.74}$$

Then

$$\tilde{w}_\epsilon = -\left(\frac{q+1-p}{(N-1)(q_c-q)}\left(r^{\frac{(N-1)(q_c-q)}{p-1}} - \epsilon^{\frac{(N-1)(q_c-q)}{p-1}}\right)\right)^{-\frac{p-1}{q+1-p}}. \tag{3.4.75}$$

Therefore, the function $\tilde{u}_\epsilon$ defined by

$$\tilde{u}_\epsilon(r) = M + \int_r^\infty \left(\frac{q+1-p}{(N-1)(q_c-q)} r^{\frac{(N-1)(q_c-q)}{p-1}} - \epsilon^{\frac{(N-1)(q_c-q)}{p-1}}\right)^{-\frac{1}{q+1-p}} s^{-\frac{N-1}{p-1}} ds,$$

has no critical point and dominates $u$ for $|x| = 1$ and $|x| = \epsilon$; hence it dominates $u$ in $B_1 \setminus B_\epsilon$. Letting $\epsilon \to 0$ leads to

$$\begin{aligned} u(x) \leq \tilde{u}_0(|x|) &= M + \int_{|x|}^\infty \left(\frac{q+1-p}{(N-1)(q_c-q)} r^{\frac{(N-1)(q_c-q)}{p-1}}\right)^{-\frac{1}{q+1-p}} s^{-\frac{N-1}{p-1}} ds \\ &\leq U_s(x) + M. \end{aligned}$$

This yields (3.4.55). □

### 3.4.4.3 *Global singular solutions*

The following rigidity result holds:

**Theorem 3.4.16** *Assume $0 < p-1 < q < p \leq N$ and $u \in C^1(\mathbb{R}^N \setminus \{0\})$ is a nonnegative solution of (3.4.1) in $\mathbb{R}^N \setminus \{0\}$.*

*I- If $q \geq q_c$, then $u$ is a constant.*

*II- If $p-1 < q < q_c$, then*
*(i) either $u$ is a constant,*
*(ii) or there exist $\gamma > 0$ and $c \geq 0$ such that*

$$\begin{aligned} u(x) &= u_{\gamma,c}(|x|) \\ &= c + \int_{|x|}^{\infty} \left( \frac{q+1-p}{(N-1)(q_c-q)} s^{\frac{(N-1)(q_c-q)}{p-1}} + |\tilde{\gamma}|^{-\frac{q+1-p}{p-1}} \right)^{-\frac{1}{q+1-p}} s^{-\frac{N-1}{p-1}} ds. \end{aligned} \tag{3.4.76}$$

*(ii) or there exists $c \geq 0$ such that*

$$u(x) = U_s(x) + c = \mu_{N,p,q}|x|^{-\gamma_{p,q}} + c. \tag{3.4.77}$$

*Proof.* If $q \geq q_c$ the result follows from Theorem 3.4.11 and Corollary 3.4.10. If $p-1 < q < q_c$, we have the following estimate by Theorem 3.4.7

$$|\nabla u(x)| \leq C_{N,p,q}|x|^{-\frac{1}{q+1-p}} \qquad \forall x \in \mathbb{R}^N \setminus \{0\}.$$

Since for any $a \in S^{N-1}$,

$$\left| \frac{d}{dt} u(ta) \right| \leq C_{N,p,q} t^{-\frac{1}{q+1-p}} \Longrightarrow \frac{d}{dt} u(ta) \in L^1(1,\infty),$$

there exists $c_a = \lim_{t\to\infty} u(ta)$. Next, if $|x| = t \neq 0$,

$$|u(x) - u(ta)| = \left| u(t\tfrac{x}{|x|}) - u(ta) \right| \leq \pi C_{N,p,q} t^{-\frac{1}{q+1-p}+1} \to 0 \text{ as } t \to \infty.$$

Then $c_{\frac{x}{|x|}} = c_a$, which implies that there exists $c = \lim_{|x|\to\infty} u(x)$. By Theorem 3.4.12 either (3.4.53) or (3.4.55) holds. In the first case $\gamma$ must be positive other while $u$ would be regular and therefore constant by Corollary 3.4.10. Then, for any $\epsilon \in (0,\gamma)$, the function $u_{\gamma+\epsilon,c+\epsilon}$ (see (3.4.76)) dominates $u$ at $x = 0$ and $|x| \to \infty$. It has no critical values so it dominates $u$ in $\mathbb{R}^N \setminus \{0\}$. Then $u \leq u_{\gamma,c}$. In the same way $u \geq u_{\gamma,c}$. In the second case we notice that $(1+\epsilon)U_s$ is a supersolution while $(1-\epsilon)U_s$ is a subsolution. Hence

$$(1-\epsilon)U_s + c - \epsilon \le u \le (1+\epsilon)U_s + c + \epsilon.$$

This yields $u = U_s + c$. □

3.4.4.4 *Isolated singularities of negative solutions*

If $u$ is a negative solution of (3.4.1) we set $\tilde{u} = -u$ and deal with positive solutions of

$$-\Delta_p \tilde{u} = |\nabla \tilde{u}|^q, \tag{3.4.78}$$

in $B_1 \setminus \{0\}$.

**Theorem 3.4.17** *Assume $\Omega \subset \mathbb{R}^N$ is a domain containing $0$, $1 < p \le N$ and $p-1 < q < q_c$. If $\tilde{u} \in C^1(\Omega \setminus \{0\})$ is a nonnegative solution of (3.4.78) in $\Omega \setminus \{0\}$, there exists $\gamma \ge 0$ such that*

$$\lim_{x \to 0} \frac{\tilde{u}(x)}{\mu_p(x)} = \gamma, \tag{3.4.79}$$

*and the following equation*

$$-\Delta_p \tilde{u} - |\nabla \tilde{u}|^q = c_{N,p} \gamma \delta_0, \tag{3.4.80}$$

*holds in $\mathcal{D}'(\Omega)$ with $c_{N,p} = N\omega_N \left(\frac{N-p}{p-1}\right)^{p-1}$.*

*Proof.* We can assume that $\overline{B}_1 \subset \Omega$.
*Step 1: We claim that (3.4.80) holds.* If $1 < p < N$, it follows from Theorem 1.5.14 that $\Delta_p \tilde{u} \in L^1_{doc}(\Omega)$; hence $|\nabla \tilde{u}|^q \in L^1_{loc}(\Omega)$ and there exists $\gamma \ge 0$ such that (3.4.80) holds. If $p = N$, then $u \in L^r(B_1)$ for all $r < \infty$ and $\nabla u \in L^s(B_1)$ for all $s < N$. Since $q_c = N$, it implies that $-\Delta_p \tilde{u} = g \in L^m(B_1)$ for all $m \in [1, \frac{N}{q})$. We assume first that

$$\liminf_{x \to 0} \tilde{u}(x) = m < \infty. \tag{3.4.81}$$

Then there exists a sequence $\{r_n\}$ converging to 0 such that

$$\lim_{n \to \infty} \inf_{|x| = r_n} \tilde{u}(x) = m.$$

By (3.4.59) it follows

$$\lim_{n \to \infty} \sup_{|x| = r_n} \tilde{u}(x) = M \le cm < \infty,$$

where $c = c(N, q) > 0$. Let $\epsilon > 0$, there exists $n_\epsilon \in \mathbb{N}$ such that for $n \geq n_\epsilon$ there holds $\tilde{u}(x) \leq M_\epsilon := M + \epsilon$ if $|x| = r_n$. Then

$$\begin{aligned}\int_{B_1\setminus B_{r_n}} |\nabla(\tilde{u} - M_\epsilon)_+|^N dx &\leq \int_{B_1\setminus B_{r_n}} |\nabla \tilde{u}|^q (\tilde{u} - M_\epsilon)_+ dx \\ &\quad + \int_{\partial B_1} |\nabla \tilde{u}|^{p-1} (\tilde{u} - M_\epsilon)_+ dS \\ &\leq \left(\int_{B_1\setminus B_{r_n}} |\nabla(\tilde{u} - M_\epsilon)_+|^N dx\right)^{\frac{q}{N}} \left(\int_{B_1\setminus B_{r_n}\}} (\tilde{u} - M_\epsilon)_+^{\frac{N}{N-q}} dx\right)^{1-\frac{q}{N}} \\ &\quad + \int_{\partial B_1} |\nabla \tilde{u}|^{p-1} (\tilde{u} - M_\epsilon)_+ dS.\end{aligned}$$

Letting $n \to \infty$ and $\epsilon \to 0$ yields

$$\begin{aligned}\int_{B_1} |\nabla(\tilde{u} - M)_+|^N dx &\leq \int_{B_1} (\tilde{u} - M)_+^{\frac{N}{N-q}} dx \\ &\quad + \int_{\partial B_1} |\nabla \tilde{u}|^{p-1} (\tilde{u} - M)_+ dS \qquad (3.4.82) \\ &< \infty.\end{aligned}$$

If we assume now that

$$\limsup_{x\to 0} \tilde{u}(x) \geq cM, \qquad (3.4.83)$$

there exists a sequence $\{r'_n\}$ converging to 0 such that

$$\lim_{n\to\infty} \inf_{|x|=r_n} \tilde{u}(x) \geq M.$$

Set $P_M(\tilde{u}) = M - \inf\{\tilde{u}, M\}$, then $P'_M(\tilde{u}) = -\chi_{\tilde{u}\leq M}$ and

$$\begin{aligned}\int_{(B_1\setminus B_{r'_n})\cap\{\tilde{u}\leq M\}} |\nabla \tilde{u}|^N dx + \int_{(B_1\setminus B_{r'_n})\cap\{\tilde{u}\leq M\}} |\nabla \tilde{u}|^q P_M(\tilde{u}) dx \\ = -\int_{\partial B_1} |\nabla \tilde{u}|^{N-2} \frac{\partial \tilde{u}}{\partial \mathbf{n}} P_M(\tilde{u}) dS.\end{aligned}$$

Hence

$$\int_{B_1\cap\{\tilde{u}\leq M\}} |\nabla \tilde{u}|^N dx \leq M \int_{\partial B_1} |\nabla \tilde{u}|^{N-1} dS. \qquad (3.4.84)$$

Therefore, if (3.4.81) and (3.4.83) hold, $\nabla \tilde{u} \in L^p(B_1)$. If (3.4.83) does not hold, it implies in particular that $\tilde{u}$ remains uniformly bounded in $B_1 \setminus \{0\}$.

Let $\eta_\epsilon \in C^\infty 1(\mathbb{R}^N)$ such that $0 \le \eta_\epsilon \le 1$, $\eta_\epsilon = 0$ in $B_\epsilon$, $\eta_\epsilon = 1$ in $B^c_{2\epsilon}$ and $|\nabla \eta_\epsilon| \le 3\chi_{B_{2\epsilon}\setminus B_\epsilon}$, then

$$\int_{B_1} |\eta_\epsilon \nabla \tilde{u}|^N dx - N\int_{B_1} \tilde{u}\, |\eta_\epsilon \nabla \tilde{u}|^{N-1} |\nabla \eta_\epsilon|\, dx \\ \le \int_{B_1} |\eta_\epsilon|^N |\nabla \tilde{u}|^q \tilde{u} dx + \int_{\partial B_1} |\nabla \tilde{u}|^{N-2} \frac{\partial \tilde{u}}{\partial \mathbf{n}} \tilde{u} dS.$$

Hence

$$\int_{B_1} |\eta_\epsilon \nabla \tilde{u}|^N dx \le c_N \|\tilde{u}\|_{L^\infty} \left(\int_{B_{2\epsilon}\setminus B_\epsilon} |\eta_\epsilon \nabla \tilde{u}|^N dx\right)^{1-\frac{1}{N}} + \int_{B_1} |\eta_\epsilon|^N |\nabla \tilde{u}|^q \tilde{u} dx \\ + \int_{\partial B_1} |\nabla \tilde{u}|^{N-2} \frac{\partial \tilde{u}}{\partial \mathbf{n}} \tilde{u} dS.$$

This implies that $\nabla \tilde{u} \in L^N(B_1)$ and

$$\int_{B_1} |\eta_\epsilon \nabla \tilde{u}|^N dx \le \int_{B_1} |\nabla \tilde{u}|^q \tilde{u} dx + \int_{\partial B_1} |\nabla \tilde{u}|^{N-2} \frac{\partial \tilde{u}}{\partial \mathbf{n}} \tilde{u} dS. \tag{3.4.85}$$

Next we denote by $T$ the distribution associated to the locally integrable function $-\Delta_p \tilde{u} - |\nabla \tilde{u}|^q$. Its support is the point 0. Let $\zeta \in C_0^\infty(\Omega)$ and $\eta_\epsilon$ as above. Taking $\eta_\epsilon \zeta$ as test function, we get

$$T(\eta_\epsilon \zeta) = \int_\Omega |\nabla \tilde{u}|^{N-2} \langle \nabla \tilde{u}, \nabla \zeta \rangle \eta_\epsilon dx + \int_\Omega |\nabla \tilde{u}|^{N-2} \langle \nabla \tilde{u}, \nabla \eta_\epsilon \rangle \zeta dx \\ - \int_\Omega \eta_\epsilon \zeta\, |\nabla \tilde{u}|^q dx.$$

Since

$$\left|\int_\Omega |\nabla \tilde{u}|^{N-2} \langle \nabla \tilde{u}, \nabla \eta_\epsilon \rangle \zeta dx\right| \le c(N) \|\zeta\|_{L^\infty} \left(\int_{B_{2\epsilon}\setminus B_\epsilon} |\nabla \tilde{u}|^N dx\right)^{1-\frac{1}{N}},$$

which tends to 0 when $\epsilon \to 0$, then

$$|T(\zeta)| \le |G|^{\frac{1}{N}} \left(\int_G |\nabla \tilde{u}|^N dx\right)^{1-\frac{1}{N}} \|\nabla \zeta\|_{L^\infty} + \left(\int_G |\nabla \tilde{u}|^q dx\right) \|\zeta\|_{L^\infty},$$

where $G$ denotes the support of $\zeta$. Therefore $T$ is a distribution of order 0 or 1 and it can be written as

$$T = a\delta_0 - \sum_{j=1}^N a_j \frac{\partial \delta_0}{\partial x_j}.$$

Take $\zeta \in C_0^\infty(B_1)$ such that $\zeta(0) = a$ and $\dfrac{\partial \zeta}{\partial x_j}(0) = -a_j$, and for $\epsilon > 0$, set $\zeta_\epsilon(x) = \zeta(\frac{x}{\epsilon})$. Then

$$T(\zeta_\epsilon) = \int_\Omega |\nabla \tilde{u}|^{N-2} \langle \nabla \tilde{u}, \nabla \zeta_\epsilon \rangle dx - \int_\Omega \zeta_\epsilon |\nabla \tilde{u}|^q \, dx = a^2 + \sum_{j=1}^N \frac{a_j^2}{\epsilon}.$$

But

$$\left| \int_\Omega |\nabla \tilde{u}|^{N-2} \langle \nabla \tilde{u}, \nabla \zeta_\epsilon \rangle dx \right| \leq c(N) \left( \int_{B_\epsilon} |\nabla \tilde{u}|^N \, dx \right)^{\frac{1}{N}} \|\nabla \zeta\|_{L^\infty} \to 0 \text{ as } \epsilon \to 0.$$

Similarly

$$\int_\Omega \zeta_\epsilon |\nabla \tilde{u}|^q \, dx \to 0 \text{ as } \epsilon \to 0.$$

Therefore $a = a_j = 0$ for all $j = 1, ..., N$. This implies that (3.4.80) holds with $\gamma = 0$.

Finally we are left with

$$\liminf_{x \to 0} \tilde{u}(x) = \infty, \tag{3.4.86}$$

in which case (3.4.80) holds by Theorem 1.5.14.

*Step 2: We claim that (3.4.79) holds.* With $\tilde{C}(x) = -|\nabla \tilde{u}(x)|^{q+1-p}$, equation (3.4.78) becomes

$$-\Delta_p \tilde{u} + \tilde{C} |\nabla \tilde{u}(x)|^{p-1} = 0 \qquad \text{in } \Omega \setminus \{0\}. \tag{3.4.87}$$

When $1 < p < N$, $\tilde{C} \in L_{loc}^{\frac{N(p-1)}{q+1-p}, \infty}(\Omega) \subset L_{loc}^{\frac{N(p-1)}{q+1-p} - \epsilon}(\Omega)$ for any $\epsilon > 0$. Since $q < q_c$, $\frac{N(p-1)}{q+1-p} > N$, thus $\tilde{C} \in L_{loc}^{N+\epsilon}(\Omega)$ for $\epsilon > 0$ small enough. As a consequence the positive singularities are isotropic by Theorem 1.3.7 which means that either $\tilde{u}$ is regular near 0, in which case the coefficient $\gamma$ of the Dirac mass in (3.4.78) is zero, or there exists $c > 0$ such that

$$c \leq |x|^{\frac{N-p}{p-1}} \tilde{u}(x) \leq \frac{1}{c} \qquad \forall x \in \overline{B}_1 \setminus \{0\}. \tag{3.4.88}$$

If $p = N$, $\tilde{C} \in L_{loc}^{\frac{N-\epsilon}{q+1-N}, \infty}(\Omega) \subset L_{loc}^{N+\epsilon'}(\Omega)$ for $\epsilon, \epsilon'$ small enough since $N - 1 < q < N$. Then

$$c \ln \left( \frac{1}{|x|} \right) \leq \tilde{u}(x) \leq \frac{1}{c} \ln \left( \frac{1}{|x|} \right) \qquad \forall x \in \overline{B}_{\frac{1}{2}} \setminus \{0\}. \tag{3.4.89}$$

The proof of (3.4.77) presents some similarity with the one of Theorem 3.4.12, with a major modification at the end since the equation has no comparison principle. Let $\underline{\gamma} \in [c, c^{-1}]$ such that there exists a sequence $\{x_n\}$ tending to 0 verifying

$$\underline{\gamma} = \lim_{n\to\infty} \frac{\tilde{u}(x_n)}{\mu_p(x_n)}. \tag{3.4.90}$$

We denote $r_n = |x_n|$. If $1 < p < N$ we set

$$\tilde{u}_r(x) = \frac{\tilde{u}(rx)}{\mu_p(r)}.$$

Then $\tilde{u}_r$ satisfies

$$-\Delta_p \tilde{u}_r = (\mu_p(r))^{q+1-p} r^{p-q} |\nabla \tilde{u}_r|^q \qquad \text{in } B_{\frac{1}{r}} \setminus \{0\},$$

and the estimates (3.4.69) are valid provided $u_r$ is replaced by $\tilde{u}_r$. Up to a subsequence $\{r_{n_k}\}$ of $\{r_n\}$ tending to 0, $\{\tilde{u}_{r_{n_k}}\}$ converges in the $C^1_{loc}$ topology of $\mathbb{R}^N \setminus \{0\}$ to some nonnegative $p$-harmonic function $v$ which satisfies the same upper and lower bounds as $\tilde{u}$, but in whole $\mathbb{R}^N \setminus \{0\}$. Then it is a positive multiple of $\mu_p$. Since (3.4.90) holds, $v = \underline{\gamma}\mu_p$ and $\tilde{u}_{r_{n_k}} \to \underline{\gamma}\mu_p$. Because the convergence is locally in $C^1$, we have

$$\lim_{r_n\to 0} r_{n_k}^{\frac{N-1}{p-1}} \nabla u(r_{n_k}\xi) = -\frac{N-p}{p-1}\underline{\gamma}\xi \quad \text{uniformly for } |\xi| = 1. \tag{3.4.91}$$

Using (3.4.78)

$$\int_{\Omega\setminus B_{r_{n_k}}} \left(|\nabla u|^{p-2} \langle \nabla u, \nabla \zeta\rangle - |\nabla u|^q\right) dx = \int_{|x|=r_{n_k}} |\nabla u|^{p-2} \frac{\partial u}{\partial \mathbf{n}} \zeta dS,$$

where $\zeta \in C_0^\infty(\Omega)$. By (3.4.91) the right-hand side of the above identity converges to

$$N\omega_N \left(\frac{N-p}{p-1}\right)^{p-1} \underline{\gamma}\zeta(0) = c_{N,p}\underline{\gamma}\zeta(0).$$

By (3.4.80), this limit is $c_{N,p}\gamma\zeta(0)$. We obtain $\gamma = \underline{\gamma}$; finally (3.4.79) holds since the limit is uniquely defined.
The proof when $p = N$ is based upon similar ideas with the help of the scaling used in the proof of Theorem 3.4.12. □

### 3.4.5 *Geometric estimates*

#### 3.4.5.1 *Gradient estimates on a Riemannian manifold*

The technique used in proving the gradient estimates extends to a geometric setting. We assume that $(M, g)$ is $N$-dimensional Riemannian manifold, $TM$ its tangent bundle, $\nabla_g$ is the covariant gradient, $\langle ., . \rangle_g$ the scalar product expressed in the metric $g = (g_{ij})$, $Ricc_g$ the Ricci tensor and $Sec_g$ the sectional curvature, and $\Delta_g$ the Laplacian. We consider the following equation on $M$

$$-\Delta_p u + |\nabla_g u|^q = 0, \tag{3.4.92}$$

which expands in

$$-\Delta_g u - \frac{\langle D^2 u \nabla_g u, \nabla_g u \rangle}{|\nabla_g u|^2} + |\nabla_g u|^{q+2-p} = 0. \tag{3.4.93}$$

In order to obtained bound from above on $z = |\nabla_g u|^2$ we recall Weitzenböck formula (combined with Schwarz inequality),

$$\begin{aligned} \frac{1}{2}\Delta_g |\nabla u|^2 &= \left\|D^2 u\right\|^2 + \langle \nabla u, \nabla_g \Delta_g u \rangle_g + Ricc_g(\nabla u, \nabla u) \\ &\geq \frac{1}{N}\left(\Delta_g u\right)^2 + \langle \nabla u, \nabla_g \Delta_g u \rangle_g + Ricc_g(\nabla_g u, \nabla_g u). \end{aligned} \tag{3.4.94}$$

We write (3.4.93) under the form

$$\Delta_g u = -\frac{1}{2}\frac{\langle \nabla_g z, \nabla_g u \rangle}{z} + z^{\frac{q+2-p}{2}} = 0. \tag{3.4.95}$$

With the help of Hölder's inequality (3.4.35) is replaced by

$$\begin{aligned} \Delta_g z + (p-2)\frac{\langle D^2 z \nabla_g u, \nabla_g u \rangle_g}{z} &\geq \frac{2a^2}{N} z^{q+2-p} - \frac{1}{Na^2}\frac{\langle \nabla_g z, \nabla_g u \rangle_g^2}{z^2} \\ &\quad - \frac{p-2}{2}\frac{|\nabla_g z|^2}{z} + (p-2)\frac{\langle \nabla_g z, \nabla_g u \rangle_g^2}{z^2} \\ &\quad + (q+2-p) z^{q+2-p} \langle \nabla_g z, \nabla_g u \rangle_g + Ricc_g(\nabla_g u, \nabla_g u), \end{aligned} \tag{3.4.96}$$

where $a > 0$. The operator $\mathcal{A}$ is defined by (3.4.36) (in the geometric setting) and ellipticity condition (3.4.37) holds. A natural assumption on the curvature of the manifold is to assume

$$Ricc_g(\xi, \xi) \geq (1-N)B^2|\xi|^2 \qquad \forall \xi \in TM, \tag{3.4.97}$$

for some $B \geq 0$. Then we define $\mathcal{L}$ by

$$\mathcal{L}(z) = \mathcal{A}(z) + c_1 z^{q+2-p} - c_2 \frac{|\nabla_g z|^2}{z} - (N-1)B^2 z, \tag{3.4.98}$$

where $c_1, c_2 > 0$ are chosen such that $\mathcal{L}(z) \leq 0$, which is possible by (3.4.96), (3.4.97).

**Lemma 3.4.18** *Assume $q > p - 1 > 0$ and let $a \in M$, $R > 0$ and $B \geq 0$ such that (3.4.97) holds in $B_R(a)$. Assume also $Sec_g \geq -\tilde{B}^2$ for some $\tilde{B} \geq 0$ if $p > 2$, or $r_M(a) \geq R$ if $1 < p < 2$ where $r_M(a)$ is the convexity radius of $a$. We set $r(x) = d_g(x, a)$. Then there exists $c = c(N, p, q) > 0$ such that the function*

$$w(x) = \lambda \left(R^2 - r^2(x)\right)^{-\frac{2}{q+1-p}} + \mu, \tag{3.4.99}$$

*satisfies*

$$\mathcal{L}(w) \geq 0 \qquad \text{in } B_R(a), \tag{3.4.100}$$

*provided*

$$\lambda \geq cR^{\frac{2}{q+1-p}} \left(B^2 R^2 + B_p R + 1\right)^{\frac{1}{q+1-p}}, \tag{3.4.101}$$

*where $B_p = B + (p-2)_+ \tilde{B}$, and*

$$\mu \geq \left((N-1)\, B^2\right)^{\frac{1}{q+1-p}}. \tag{3.4.102}$$

*Proof.* The proof is a combination of some geometric inequalities and standard computations. The following classical one can be found in [Ratto, Rigoli, Véron (1994), Lemma 1]

$$\Delta_g r \leq (N-1)\coth(Br) \leq \frac{N-1}{r}(1 + Br). \tag{3.4.103}$$

Then

$$\begin{aligned} \Delta_g w \leq \left(\frac{8N}{q+1-p}\right) & \left(R^2 - r^2\right)^{-\frac{2(q+2-p)}{q+1-p}} \\ & \times \left(\frac{q+3-p}{q+1-p} r^2 + \left(R^2 - r^2\right)(1 + Br)\right). \end{aligned} \tag{3.4.104}$$

If $r(x) \leq r_M(a)$ the ball $B_{r(x)}$ is convex and $D^2 r \geq 0$ (see e.g. [Sakai (1997), IV-5]). Furthermore, if $Sec_g \geq -\tilde{B}^2$, there holds [Sakai (1997), Lemma 2.9]

$$D^2 r \leq \tilde{B}\coth(\tilde{B}r) \leq \frac{\tilde{B}}{r}(1 + \tilde{B}r). \tag{3.4.105}$$

Therefore

$$0 \leq \frac{\langle D^2 w \nabla_g w, \nabla_g w \rangle}{|\nabla_g w|^2}$$
$$\leq \frac{8\left(R^2 - r^2\right)^{-\frac{2(q+2-p)}{q+1-p}}}{q+1-p}\left(\frac{r^2(q+3-p)}{q+1-p} + \left(R^2 - r^2\right)\left(1 + \tilde{B} r\right)\right). \tag{3.4.106}$$

It follows

$$\begin{aligned}\mathcal{A}(w) &= -\Delta_g w - (p-2)\frac{\langle D^2 w \nabla_g w, \nabla_g w \rangle}{|\nabla_g w|^2} \\ &\geq -k\lambda\left(R^2 - r^2\right)^{-\frac{2(q+2-p)}{q+1-p}}\left(R^2 + (R^2 - r^2)B_p r\right),\end{aligned} \tag{3.4.107}$$

where $k = k(n, p, q) > 0$. Since

$$w^{q+2-p} \geq \lambda^{q+2-p}(R^2 - r^2)^{-\frac{2(q+2-p)}{q+1-p}} + \mu^{q+2-p},$$

there holds

$$\begin{aligned}\mathcal{L}(w) \geq{}& \lambda\left(R^2 - r^2\right)^{-\frac{2(q+2-p)}{q+1-p}}\left(-k(R^2 + (R^2 - r^2)B_p r + c_1\lambda^{q+1-p}\right) \\ &- c_3\lambda r^2 (R^2 - r^2)^{-2\frac{q+2-p}{q+3-p}} - (N-1)B^2\lambda\left(R^2 - r^2\right)^{-\frac{2}{q+1-p}} \\ &+ c_2\mu^{q+2-p} - (N-1)B^2\mu,\end{aligned} \tag{3.4.108}$$

where $c_3 = c_3(N, p, q) > 0$. We fix

$$c_2\mu^{q+1-p} \geq (N-1)B^2, \tag{3.4.109}$$

and then $\lambda$ in order to have

$$\frac{c_1\lambda^{q+1-p}}{2} \geq k(R^2 + (R^2 - r^2)B_p r + c_3 r^2 \quad \forall r \in [0, R).$$

It suffices to choose

$$\lambda^{q+1-p} \geq c_4 R^2(1 + RB_p), \tag{3.4.110}$$

for some $c_4 = c_4(N, p, q) > 0$. At end we require also

$$\frac{c_1\lambda^{q+1-p}}{2}\left(R^2 - r^2\right)^{-\frac{2(q+2-p)}{q+1-p}} \geq (N-1)B^2\left(R^2 - r^2\right)^{-\frac{2}{q+1-p}},$$

and we impose, for some $c_5 = c_5(N, p, q) > 0$,

$$\lambda^{q+1-p} \geq c_5 B^2 R^4. \tag{3.4.111}$$

Combining (3.4.110) and (3.4.111) we fix

$$\lambda^{q+1-p} = c_6 R^2(1 + B^2 + RB_p) \tag{3.4.112}$$

and we get (3.4.100) with (3.4.100) and (3.4.100). □

This result allows to have an upper bound on the norm of the gradient of a solution of (3.4.92).

**Proposition 3.4.19** *Let $q > p-1 > 0$ and $\Omega$ be an open subset of $(M,g)$ in which there holds $Ricc_g \geq (1-N)B^2$ for some $B \geq 0$. Assume also $Sec_g \geq -\tilde{B}^2$ in $\Omega$ if $p > 2$, or $r_M(x) \geq \operatorname{dist}_g(x, \partial\Omega)$ for any $x \in \Omega$ if $1 < p < 2$. Then any $C^1$ solution $u$ of (3.4.92) in $\Omega$ satisfies*

$$|\nabla_g u(x)| \leq c\left(\frac{1}{\rho^2_{\partial\Omega}(x)} + \frac{B_p}{\rho_{\partial\Omega}(x)} + B^2\right)^{\frac{1}{2(q+1-p)}}, \tag{3.4.113}$$

*where $\rho_{\partial\Omega}(x) = \operatorname{dist}_g(x, \partial\Omega)$ and $c = c(N,p,q) > 0$.*

*Proof.* Let $a \in \Omega$ and $R < \operatorname{dist}_g(a, \partial\Omega)$. If $w$ is as in Lemma 3.4.18, then

$$\begin{aligned}\mathcal{A}(z-w) &+ c_1(z^{q+2-p} - w^{q+2-p}) - (N-1)B^2(z-w)\\ &- c_2\left(\frac{|\nabla_g z|^2}{z} - \frac{|\nabla_g w|^2}{w}\right) \leq 0,\end{aligned} \tag{3.4.114}$$

in any connected component $G$ of the set of $x \in \Omega$ such that $z(x) > w(x)$. If $c_1(q+2-p)(w(a))^{q+2-p} > (N-1)B^2$, it follows by the mean value theorem and the fact that $w$ is minimal at $a$, that

$$c_1(z^{q+2-p} - w^{q+2-p}) - (N-1)B^2(z-w) > 0 \qquad \text{in } G. \tag{3.4.115}$$

Since $w(a) \geq \mu \geq ((N-1)B^2)^{\frac{1}{q+1-p}}$ and $q+2-p > 1$, this condition is fulfilled by choosing the right $\mu$ as in (3.4.109). We conclude as in the proof of Theorem 3.4.71 that $G = \emptyset$. Therefore $z \leq w$ in $B_R(a)$. In particular,

$$z(a) \leq w(a) = \lambda R^{-\frac{4}{q+1-p}} + \mu, \tag{3.4.116}$$

with $\mu$ and $\lambda$ given respectively by (3.4.101) and (3.4.102) with equality, which yields (3.4.113). □

*Remark.* Since $Ricc_g(x)(\xi,\xi) = (N-1)\sum_V Sec_g(x)(V)$ where $V$ denotes the set of two planes in $T_xM$ which contain the unit vector $\xi$, there holds

$$Sec_g \geq \tilde{B}^2 \Longrightarrow Ricc_g(1-N)\tilde{B}^2.$$

However, in the previous estimate, the long range estimate on $\nabla_g u$ depends only on the Ricci curvature.

3.4.5.2 *Growth of solutions and Liouville type results*

**Corollary 3.4.20** *Assume $(M,g)$ is a complete noncompact N-dimensional Riemannian manifold such that $Ricc_g \geq (1-N)B^2$ and let $q > p-1 > 0$. Assume also $r_M(x) = \infty$ if $1 < p < 2$, or that the sectional curvature $Sec_g$ satisfies for some $a \in M$*

$$\lim_{\operatorname{dist}_g(x,a)\to\infty} \frac{|Sec_g(x)|}{\operatorname{dist}_g(x,a)} = 0, \tag{3.4.117}$$

*if $p > 2$. Then any solution $u$ of (3.4.92) satisfies*

$$|\nabla_g u(x)| \leq c_{N,p,q} B^{\frac{1}{q+1-p}} \qquad \forall x \in M. \tag{3.4.118}$$

*In particular, $u$ is constant if $Ricc_g \geq 0$, while in the general case $u$ has at most a linear growth with respect to $\operatorname{dist}_g(x,a)$.*

*Proof.* The proof follows from (3.4.113). □

**Definition 3.4.21** A N-dimensional Riemannian manifold $(M,g)$ is asymptotically flat if there exist a compact set $K \subset M$ and two positive constants $\tau$ and $A$ such that $M \setminus K$ is diffeomorphic to $\mathbb{R}^N \setminus B$ for some ball $B$, and in this chart,

$$\begin{array}{ll} (i) & g_{ij}(x) \to \delta_{ij} \quad \text{as } \rho_B(x) \to \infty \\ (ii) & \rho_B^{\tau}(x)\,|\nabla_g g_{ij}(x)| + \rho_B^{\tau+1}(x)\,|\nabla_{k\ell} g_{ij}(x)| \leq A, \end{array} \tag{3.4.119}$$

where $\rho_B(x) = \operatorname{dist}_g(x,B)$, for all $x \in M \setminus K$ and all $i,j,k,\ell = 1,...,N$. This implies (see Chapter 1) that the Riemann tensor satisfies

$$|Riem_g(x)| \leq c(N) A \rho_B^{-\tau-1}(x) \qquad \forall x \in M \setminus K, \tag{3.4.120}$$

and all the curvatures have the same decay at infinity. In many applications $\tau > 1$, therefore (3.4.119) (i) is a consequence of (3.4.119) (ii).

A classical example of asymptotically flat metric on $\mathbb{R}^N \setminus \{a\}$ for $a \in \mathbb{R}^N$, $N \geq 3$, is the Schwarzchild metric defined by

$$g_{ij}(x) = \left(1 + \frac{m}{(N-2)|x-a|}\right)^{\frac{4}{N-2}} \delta_{ij}, \tag{3.4.121}$$

where $m > 0$ is an important geometric invariant called the ADM mass (see e.g. [Arnowitt, Deser, Misner (1961)]).

**Corollary 3.4.22** *Assume $(M,g)$ is an asymptotically flat N-dimensional Riemannian manifold with exponent $\tau$ and $0 < p-1 < q < \min\{p+\frac{\tau-1}{2}, p\}$. Then any solution $u \in C^1(M)$ of (3.4.92) in $M$ is asymptotically constant.*

*Proof.* We fix a point $a \in K$ and denote $S_R(a) = \{x \in M : \mathrm{dist}_g(x,a) = R\}$. Since $M$ is asymptotically Euclidean, there exists $R_0$ such that $K \subset \overline{B}_{R_0}(a) := \mathbf{B}$ and any point $x \in M \setminus \mathbf{B}$ is connected to some $a_x \in S_{R_0}(a) := \mathbf{S}$ by a unique geodesic curve issued of $a_x$. This geodesic curve depends on $a_x$ and its initial speed $X$ at $t = 0$. In Chapter 1 it is denoted by $t \mapsto \exp_{a_x}(tX)$ with $x = \exp_{a_x}(\mathrm{dist}_g(x,a_x)X) = \exp_{a_x}(\rho_{\mathbf{B}}(x)X)$ since $\rho_{\mathbf{B}}(x) = \mathrm{dist}_g(x,a_x)$. Furthermore, for any $x \in M \setminus \mathbf{B}$, $r_M(x) \geq \rho_{\mathbf{B}}(x)$. Let $x$ such that $\rho_{\mathbf{B}}(x) > 2R_0$ and $R \leq \frac{1}{2}\rho_{\mathbf{B}}(x)$, we have from (3.4.120)

$$\begin{array}{lll} (i) & Ricc_g(y) \geq (1-N)b^2 R^{-1-\tau} & \forall y \in B_R(x) \\ (ii) & Sec_g(y) \geq -\tilde{b}^2 R^{-1-\tau} & \forall y \in B_R(x), \end{array} \tag{3.4.122}$$

for some $b, \tilde{b} > 0$. Therefore we can apply estimates (3.4.113) in $B_R(x)$ and obtain

$$|\nabla_g u(x)| \leq c\left(\rho_{\mathbf{B}}^{-2}(x) + \rho_{\mathbf{B}}^{-\frac{3+\tau}{2}}(x) + \rho_{\mathbf{B}}^{-(1+\tau)}(x)\right)^{\frac{1}{2(q+1-p)}}. \tag{3.4.123}$$

Therefore

$$|\nabla_g u(x)| \leq c \max\left\{\rho_{\mathbf{B}}^{-\frac{1}{q+1-p}}(x), \rho_{\mathbf{B}}^{-\frac{1+\tau}{2(q+1-p)}}(x)\right\}. \tag{3.4.124}$$

Let $x$ such that $\rho_{\mathbf{B}}(x) > 2R_0$. If $y$ is a point on the curve $\{\exp_{a_x}(tX)\}_{t>0}$ such that $\rho_{\mathbf{B}}(x) \geq \rho_{\mathbf{B}}(y) > 2R_0$, there holds

$$\begin{aligned} u(x) &= u(y) + \int_{\rho_{\mathbf{B}}(y)}^{\rho_{\mathbf{B}}(x)} \frac{d}{dt} u \circ \exp_{a_x}(tX) dt \\ &= u(y) + \int_{\rho_{\mathbf{B}}(y)}^{\rho_{\mathbf{B}}(x)} \langle \nabla_g u(\exp_{a_x}(tX)), \frac{d \exp_{a_x}(tX)}{dt} \rangle_g dt. \end{aligned}$$

Using (3.4.123) and integrating yields

$$\mathrm{dist}_g(x,y) \leq c \max\left\{\rho_{\mathbf{B}}^{1-\frac{1}{q+1-p}}(y), \rho_{\mathbf{B}}^{1-\frac{1+\tau}{2(q+1-p)}}(y)\right\}.$$

If $q < \min\{p+\tau, p+1\}$, $\text{dist}_g(x,y) \to 0$ when $\rho_{\mathbf{B}}(y) \to \infty$ on $\tilde{\gamma}_x$, and there exists

$$m(x) = \lim_{t\to\infty} u \circ \exp_{a_x}(tX). \tag{3.4.125}$$

For $R > 2R_0$, the sphere $S_R(a)$ is endowed with the induced metric $g_R$. If $(x,y) \in S_R(a)$, there exists a geodesic $\gamma_{x,y,R}$ on $S_R(a)$ which connect them. We denote by $d_{S_R(a)}(x,y)$ the geodesic distance between $x$ and $y$ induced on $S_R(a)$. Then $\gamma_{x,y,R}(0) = x$ and $\gamma_{x,y,R}(d_{S_R(a)}(x,y)) = y$ and clearly

$$d_{S_R(a)}(x,y) \leq cR, \tag{3.4.126}$$

for some $c$ depending on $g$. Next

$$\begin{aligned} u(x) - u(y) &= \int_0^{d_{S_R(a)}(x,y)} \frac{d}{dt} u \circ \gamma_{x,y,R}(t)dt \\ &= \int_0^{d_{S_R(a)}(x,y)} \langle \nabla_g u(\exp_{a_x}(tX), \frac{d\gamma_{x,y,R}(t)}{dt} \rangle_g dt. \end{aligned}$$

Then

$$d_{S_R(a)}(x,y) \leq \max\left\{R^{1-\frac{1}{q+1-p}}, R^{1-\frac{1+\tau}{2(q+1-p)}}\right\},$$

which tends to 0 when $R \to \infty$, hence $m(x) = m(y)$. □

The previous results applies to positive $p$-harmonic functions in considering the transformation

$$v = \exp(-\tfrac{u}{p-1}), \tag{3.4.127}$$

which is a one to one correspondence between $C^1$ positive $p$-harmonic functions $v$ on $M$ and $C^1$ solutions of

$$-\Delta_p u + |\nabla_g u|^p = 0 \quad \text{on } M. \tag{3.4.128}$$

**Proposition 3.4.23** *Let the assumptions of Corollary 3.4.20 be satisfied. Then any positive p-harmonic function v satisfies*

$$v(y)e^{-c_{N,p}B\text{dist}_g(x,y)} \leq v(x) \leq v(y)e^{c_{N,p}B\text{dist}_g(x,y)} \qquad \forall x,y \in M. \tag{3.4.129}$$

*In particular, if $Ricc_g \geq 0$, $v$ is constant.*

*Proof.* The second assertion follows from (3.4.118) with $B = 0$. For the first assertion, we consider a geodesic curve $\gamma$ from $x$ to $y$ and define $u = (1-p)\ln v$. Then

$$u(x) - u(y) = \int_0^{\operatorname{dist}_g(x,y)} \frac{d}{dt} u \circ \gamma(t) dt = \int_0^{\operatorname{dist}_g(x,y)} \langle \nabla_g u(\gamma(t)), \frac{d\gamma(t)}{dt} \rangle_g dt.$$

Since

$$\left| \langle \nabla_g u(\gamma(t)), \frac{d\gamma(t)}{dt} \rangle_g \right| \leq |\nabla_g u(\gamma(t))| \leq c_{N,p} B,$$

we derive (3.4.129). □

*Remark.* In the particular case of $p$-harmonic functions $(1 < p \leq N)$ on an asymptotically flat manifold Corollary 3.4.22 does not apply. The next result shows that two types of behaviors might occur.

**Proposition 3.4.24** *Assume $(M, g)$ is asymptotically flat with $\tau \geq 1$. If $v$ is a positive p-harmonic function on $M$ with $1 < p \leq N$, then $v(x)$ admits a limit, finite or infinite, when $x$ tends to infinity on $M$.*

*Proof.* We apply (3.4.123) in $B_R(x)$ for $|x| \geq 2R \geq 2R_0$ and obtain that $u = (1-p)\ln v$ satisfies

$$|\nabla_g u(x)| \leq c(\operatorname{dist}_g(x, a))^{-1}, \tag{3.4.130}$$

where $a \in B_{R_0}$ is fixed. It implies that $v$ satisfies Harnack inequality at infinity. Therefore there exists $c > 0$ such that

$$v(x) \leq cv(y) \quad \forall x, y \text{ s.t. } R_0 < \frac{\operatorname{dist}_g(x,a)}{2} \leq \operatorname{dist}_g(y,a) \leq 2\operatorname{dist}_g(x,a). \tag{3.4.131}$$

Assume first

$$\limsup_{\operatorname{dist}_g(x,a)\to\infty} v(x) = \infty. \tag{3.4.132}$$

Then for any $k > 0$ there exists a sequence $\{x_{k,n}\} \subset M$ such that $\operatorname{dist}_g(x_{k,n}, a) \uparrow \infty$ and $v(x_{k,n}) \geq k$. For $|x_{k,n}| \geq 2R_0$, there holds

$$v(x) \geq c^{-1}k \quad \forall x \text{ s.t. } \operatorname{dist}_g(x_{k,n}, a) = \operatorname{dist}_g(x, a). \tag{3.4.133}$$

In particular, if $n > 1$, (3.4.132) is valid if $\operatorname{dist}_g(x,a) = \operatorname{dist}_g(x_{k,n}, a)$ and if $\operatorname{dist}_g(x,a) = \operatorname{dist}_g(x_{k,n}, a)$. Since $c^{-1}k$ is p-harmonic, we derive by the maximum principle

$$v(x) \geq c^{-1}k \text{ if } \operatorname{dist}_g(x_{k,1}, a) \leq \operatorname{dist}_g(x, a) \leq \operatorname{dist}_g(x_{k,n}, a).$$

Letting $n \to \infty$ and then $k \to \infty$ we obtain

$$\lim_{\operatorname{dist}_g(x,a)\to\infty} v(x) = \infty.$$

Next we assume

$$\limsup_{\operatorname{dist}_g(x,a)\to\infty} v(x) = m < \infty. \tag{3.4.134}$$

Then for any $\epsilon > 0$ exists $r_\epsilon$ such that

$$m - \epsilon \leq \sup\{v(x) : \operatorname{dist}_g(x, a) \geq r\} < m + \epsilon \qquad \forall r \geq r_\epsilon, \tag{3.4.135}$$

and we can impose $r_\epsilon \geq 2R_0$. Then there exists a sequence $\{x_{n,\epsilon}\}$ such that $\operatorname{dist}_g(x_{n,\epsilon}, a) = r_{n,\epsilon} \geq 2^n r_\epsilon$,

$$\begin{aligned} &(i) \qquad v(x_{n,\epsilon}) = \sup_{\operatorname{dist}_g(x,a)=r_{n,\epsilon}} v(x) \\ &(ii) \qquad m - \epsilon \leq v(x_{n,\epsilon}) < m + \epsilon \qquad \forall n \in \mathbb{N}^*. \end{aligned} \tag{3.4.136}$$

The function $x \mapsto m + 2\epsilon - v$ is $p$-harmonic and positive in $\{x \in M : \operatorname{dist}_g(x, a) \geq r_{1,\epsilon}\}$. By Harnack inequality (3.4.131) there holds

$$m + \epsilon - v(y) \leq c\,(m + \epsilon - v(x_{n,\epsilon})) \leq 2c\epsilon \tag{3.4.137}$$

for all $y$ such that $\operatorname{dist}_g(y, a) = r_{n,\epsilon}$ with $n \geq 2$. This implies

$$v(y) \geq m - (2c+1)\epsilon \quad \forall y \text{ s.t. } \operatorname{dist}_g(y, a) = r_{n,\epsilon}. \tag{3.4.138}$$

From the maximum principle in $\Theta_{\epsilon,n} := \{x \in M : r_{1,\epsilon} \leq \operatorname{dist}_g(y, a) \leq r_{n,\epsilon}\}$ we derive that $v(y) \geq m - (2c+1)\epsilon$ in $\Theta_{\epsilon,n}$. Letting $n \to \infty$ yields

$$v(y) \geq m - (2c+1)\epsilon \quad \forall y \text{ s.t. } \operatorname{dist}_g(y, a) \geq r_{2,\epsilon}. \tag{3.4.139}$$

Thus

$$\liminf_{\operatorname{dist}_g(x,a)\to\infty} v(x) \geq m - (2c+1)\epsilon. \tag{3.4.140}$$

Since $\epsilon$ is arbitrary, combining (3.4.134) and (3.4.140) we infer the claim. □

## 3.5 Boundary singularities of quasilinear Hamilton-Jacobi type equations

Let $\Omega \subset \mathbb{R}^N$ be a domain with a $C^2$ boundary containing 0. This section is concerned with the properties of the functions $u \in C(\overline{\Omega} \setminus \{0\}) \cap C^1(\Omega)$ which satisfy

$$\begin{aligned} -\Delta_p u + |\nabla u|^q &= 0 && \text{in } \Omega \\ u &= 0 && \text{on } \partial\Omega \setminus \{0\}, \end{aligned} \tag{3.5.1}$$

in the range of exponents $0 < p-1 < q < p$.

### 3.5.1 *Separable solutions*

The Ariane's shread for this study is the understanding of the same problem when $\Omega = \mathbb{R}^N_+$. Separable solutions have the form $u(x) = u(r,\sigma) = r^{-\gamma_{p,q}}\omega(\sigma)$ with $r > 0$, $\sigma \in S^{N-1}_+$. Then $\omega$ is a solution of

$$\begin{aligned} -div'\left(\left(\gamma_{p,q}^2\omega^2 + |\nabla'\omega|^2\right)^{\frac{p-2}{2}}\nabla'\omega\right) + \left(\gamma_{p,q}^2\omega^2 + |\nabla'\omega|^2\right)^{\frac{q}{2}} & \\ - \gamma_{p,q}\Lambda(\gamma_{p,q})\left(\gamma_{p,q}^2\omega^2 + |\nabla'\omega|^2\right)^{\frac{p-2}{2}}\omega = 0 \quad & \text{in } S^{N-1}_+ \\ \omega = 0 \quad & \text{on } \partial S^{N-1}_+. \end{aligned} \tag{3.5.2}$$

The critical exponent $\tilde{q}_c$ is the value of $q$ which verifies the identity

$$\beta_{S^{N-1}_+} = \gamma_{p,q} = \frac{p-q}{q+1-p} \Longleftrightarrow q := \tilde{q}_c = p - \frac{\beta_{S^{N-1}_+}}{\beta_{S^{N-1}_+} + 1}. \tag{3.5.3}$$

**Theorem 3.5.1** *Assume $p > 1$, then the following holds.*

*(i) If $p > q \geq \tilde{q}_c$ there exists no nontrivial nonnegative solution of (3.5.2).*

*(ii) If $p - 1 < q < \tilde{q}_c$ there exists only one positive solution $\omega := \omega_h$ of (3.5.2).*

*Proof.* We recall some ideas and statements introduced in the proof of Theorem 3.2.1.

$$\begin{aligned} \mathcal{T}(\eta) = -div'\left(\left(\gamma_{p,q}^2\eta^2 + |\nabla'\eta|^2\right)^{\frac{p-2}{2}}\nabla'\eta\right) + \left(\gamma_{p,q}^2\eta^2 + |\nabla'\eta|^2\right)^{\frac{q}{2}} & \\ - \gamma_{p,q}\Lambda(\gamma_{p,q})\left(\gamma_{p,q}^2\eta^2 + |\nabla'\eta|^2\right)^{\frac{p-2}{2}}\eta. & \end{aligned} \tag{3.5.4}$$

Any solution is $C^1$ up to the boundary. By the strong maximum principle $\omega > 0$ in $S_+^{N-1}$ and $\frac{\partial \omega}{\partial \mathbf{n}} < 0$ on $\partial S_+^{N-1}$. Thus $\gamma_{p,q}^2\omega^2 + |\nabla'\omega|^2 > 0$ in $\overline{S}_+^{N-1}$.

*Non-existence.* We set $\theta = \gamma_{p,q}/\beta_{S_+^{N-1}} \in (0,1]$ and assume that there exists a nontrivial positive solution $\omega$. Put $\beta = \beta_{S_+^{N-1}}$ and $\eta = \phi^\theta_{S_+^{N-1}} = \phi^\theta$, then

$$\begin{aligned}\mathcal{T}(\eta) = &-\theta^{p-1}\phi^{(\theta-1)(p-1)}div'\left(\left(\beta^2\phi^2 + |\nabla'\phi|^2\right)^{\frac{p-2}{2}}\nabla'\phi\right)\\ &- \theta^p(p-1)(\theta-1)\phi^{(\theta-1)(p-1)-1}\left(\beta^2\phi^2 + |\nabla'\phi|^2\right)^{\frac{p-2}{2}}|\nabla'\phi|^2\\ &- \gamma_{p,q}\Lambda(\gamma_{p,q})\theta^{p-2}\phi^{(\theta-1)(p-1)}\left(\beta^2\phi^2 + |\nabla'\phi|^2\right)^{\frac{p-2}{2}}\phi\\ &+ \theta^q\phi^{(\theta-1)q}\left(\beta^2\phi^2 + |\nabla'\phi|^2\right)^{\frac{q}{2}}.\end{aligned} \tag{3.5.5}$$

Using the equation satisfied by $\phi_{S_+^{N-1}}$ we have

$$\begin{aligned}\mathcal{T}(\eta) = &\,\theta^{p-1}\phi^{(\theta-1)(p-1)}\beta\left(\Lambda(\beta) - \Lambda(\gamma_{p,q})\right)\left(\beta^2\phi^2 + |\nabla'\phi|^2\right)^{\frac{p-2}{2}}\phi\\ &- \theta^p(p-1)(\theta-1)\phi^{(\theta-1)(p-1)-1}\left(\beta^2\phi^2 + |\nabla'\phi|^2\right)^{\frac{p-2}{2}}|\nabla'\phi|^2\\ &+ \theta^q\phi^{(\theta-1)q}\left(\beta^2\phi^2 + |\nabla'\phi|^2\right)^{\frac{q}{2}},\end{aligned} \tag{3.5.6}$$

since $\gamma_{p,q}\Lambda(\gamma_{p,q}) = \beta\Lambda(\gamma_{p,q})\theta$. One has $\Lambda(\gamma_{p,q}) \leq \Lambda(\beta)\theta^{p-1}$ because $\gamma_{p,q} \leq \beta$, and therefore $\mathcal{T}(\eta) > 0$. The function $\phi$ being defined up to an homothety, we can assume that it is the smallest one larger than $\omega$. Therefore, the graphs of $\phi$ and $\omega$ are tangent, either in $S_+^{N-1}$ or only on $\partial S_+^{N-1}$. Using the strict maximum principle or Hopf boundary lemma in the same way as in the proof of Theorem 3.2.1, we reach a contradiction.

*Uniqueness.* The proof based either on the strong maximum principle or Hopf boundary lemma is similar to the one in Theorem 3.2.1.

*Existence.* We write (3.5.2) under the form

$$\begin{aligned}\mathbf{A}[\omega] &= \mathbf{B}(\omega, \nabla\omega) && \text{in } S_+^{N-1}\\ \omega &= 0 && \text{on } \partial S_+^{N-1},\end{aligned} \tag{3.5.7}$$

where $\mathbf{A}[\omega] = -div'\mathbf{a}(\omega, \nabla\omega)$,

$$\mathbf{a}(r,\xi) = (\gamma_{p,q}r^2 + |\xi|^2)^{\frac{p-2}{2}}\xi,$$

and

$$\mathbf{B}(r,\xi) = \gamma_{p,q}\Lambda(\gamma_{p,q})(r^2 + |\xi|^2)^{\frac{p-2}{2}}r - (\gamma_{p,q})(r^2 + |\xi|^2)^{\frac{q}{2}}.$$

The operator $\mathbf{A}$ is a Leray-Lions operator as in Theorem 3.2.1 and $\mathbf{B}$ is a Caratheodory function in $S_+^{N-1} \times TS^{N-1}$ which satisfies

$$\begin{aligned}|\mathbf{B}(r,\xi)| &\leq (\gamma_{p,q}+1)\Lambda(\gamma_{p,q})(\gamma_{p,q}r^2+|\xi|^2)^{\frac{p}{2}} + (\gamma_{p,q}r^2+|\xi|^2)^{\frac{q}{2}} \\ &\leq C(r)(1+|\xi|^p)\end{aligned} \tag{3.5.8}$$

for some increasing function $C$. Clearly the constant function $\mu_{N,p,q} = \gamma_{p,q}^{-1}\Lambda(\gamma_{p,q})^{\frac{1}{q+1-p}}$ is a positive solution of (3.5.2). The subsolution is looked for under the form $\eta = \phi^\theta_{S_+^{N-1}} := \phi^\theta$ where $\theta = \gamma_{p,q}/\beta_{S_+^{N-1}} := \gamma_{p,q}/\beta > 1$ and $\phi_{S_+^{N-1}}$ is a positive solution of (3.2.4). Using the expression of $\mathcal{T}(\eta)$ in (3.5.6) we see that it is negative provided $\phi$ is replaced by $\epsilon\phi$, for $\epsilon > 0$ small enough, chosen also such that $\epsilon\phi < \mu_{N,p,q}$. By Theorem 1.4.7 there exist a solution $\omega$ of (3.5.7) such that $\epsilon\phi \leq \omega \leq \mu_{N,p,q}$. □

### 3.5.2 *Boundary isolated singularities*

#### 3.5.2.1 *Removable singularities*

**Theorem 3.5.2** *Assume $\tilde{q}_c \leq q < p \leq N$, $g : \mathbb{R}_+ \mapsto \mathbb{R}_+$ is a continuous function which satisfies*

$$\inf_{s>0} \frac{g(s)}{s^{\tilde{q}_c}} > 0, \tag{3.5.9}$$

*and $\Omega \subset \mathbb{R}^N$ is a bounded $C^2$ domain such that $0 \in \partial\Omega$ and $\Omega \cap B_\delta \subset B_\delta^+$ for some $\delta > 0$. If $u \in C(\overline{\Omega} \setminus \{0\}) \cap C^1(\Omega)$ is a nonnegative solution of*

$$\begin{aligned} -\Delta_p u + g(|\nabla u|) &= 0 && \text{in } \Omega \\ u &= 0 && \text{on } \partial\Omega \setminus \{0\}, \end{aligned} \tag{3.5.10}$$

*it is identically zero.*

*Proof.* We recall that we have denoted $H^+ = \mathbb{R}_+^N$. We proceed by contradiction and assume that such a function $u$ is nontrivial. First we can assume that $u$ satisfies

$$\begin{aligned} -\Delta_p u + |\nabla u|^{\tilde{q}_c} &\leq 0 && \text{in } \Omega \\ u &= 0 && \text{on } \partial\Omega \setminus \{0\}, \end{aligned} \tag{3.5.11}$$

and that $\Omega \subset \mathbb{R}_+^N$. We denote by $M$ the supremum of $u(x)$ for $|x| = \delta$. By the maximum principle $u \leq M$ in $\Omega \setminus B_\delta^+$. The function $\tilde{u} = (u - M)_+$ is defined by extension to $\overline{\mathbb{R}}_+^N \setminus \{0\}$ by 0 outside $B_\delta^+$ is a subsolution of problem (3.5.10) in $\overline{\mathbb{R}}_+^N \setminus \{0\}$. For $\epsilon > 0$ and $n \in \mathbb{N}$, large enough so that

$\Omega \subset B_n$, there exists $k_\epsilon > 0$ such that $\tilde{u}(x) \leq kV_n(x)$ for $x \in \partial B_\epsilon \cap H$ where $V_n$ is the $p$-harmonic function defined in $B_n^+ := H^+ \cap B_n$ which vanishes on $\partial B_n^+ \setminus \{0\}$ and satisfies

$$\lim_{r \to 0} |x|^\beta v(r,\sigma) = \phi_{S_+^{N-1}}(\sigma); \tag{3.5.12}$$

it has been constructed in Proposition 3.3.5. We put $H_\epsilon^+ = H^+ \cap B_\epsilon^c$ and consider the following problem

$$\begin{aligned} -\Delta_p v + |\nabla v|^{\tilde{q}_c} &= 0 && \text{in } H_\epsilon^+ \cap B_n \\ v &= kV_n(x) && \text{on } \partial(H_\epsilon^+ \cap B_n). \end{aligned} \tag{3.5.13}$$

The function $kV_n$ is a supersolution of this problem while $\tilde{u}$ is a subsolution and one has $\tilde{u} \leq kV_n$. By Theorem 1.4.8 there exists a solution $v := v_{\epsilon,n,k}$ of (3.5.13) and there holds $\tilde{u} \leq v_{\epsilon,n,k} \leq kV_n \leq kV_{\mathbb{R}_+^N}$. By the construction of the solution, the mapping $k, n, \epsilon) \mapsto v_{k,n,\epsilon}$ is nondecreasing in $k$ and $n$ and nonincreasing in $\epsilon$. By Corollary 3.4.9 the following estimate holds

$$|v_{k,n,\epsilon}(x)| \leq c_{N,p,q} \left(|x| - \epsilon\right)^{-\gamma_{p,q}} \qquad \forall x \in H_\epsilon \cap B_n. \tag{3.5.14}$$

The scaling transformation $v \mapsto J_\lambda[v]$ defined by $J_\lambda[v](x) = \lambda^{\gamma_{p,\tilde{q}_c}} v(\lambda x)$, for $\lambda > 0$, transform the problem (3.5.13) with parameters $(k, n, \epsilon)$ into the similar problem with parameters $(\lambda^{\gamma_{p,\tilde{q}_c}} k, \frac{n}{\lambda}, \frac{\epsilon}{\lambda})$, thus

$$J_\lambda[v_{k,n,\epsilon}] = v_{\lambda^{\gamma_{p,\tilde{q}_c}} k, \frac{n}{\lambda}, \frac{\epsilon}{\lambda}}. \tag{3.5.15}$$

Since $v_{k,n,\epsilon}$ converges eventually in $C^1_{loc}(\overline{\mathbb{R}}_+^N \setminus \{0\})$ to some function $v_{\infty,0}$ by the previous *a priori* estimate and Lemma 3.3.2, there holds $\tilde{u} \leq v_{\infty,0}$ and

$$J_\lambda[v_{\infty,0}] = v_{\infty,0}. \tag{3.5.16}$$

This implies that $v_{\infty,0}$ is self similar under the form $v_{\infty,0}(x) = v_{\infty,0}(r,\sigma) = r^{-\gamma_{p,q}}\omega(\sigma)$ where $\omega$ is a nonnegative solution of (3.5.2) with $q = \tilde{q}_c$. Then $\omega = 0$ and $u \leq M$ in $\Omega$. Since $u$ vanishes on $\partial\Omega \setminus \{0\}$ it can be extended by continuity to the zero function. □

When $p = N$ the removability holds without the geometric constraint on $\Omega$. In that case $\gamma_{N,q} = \frac{N-q}{q+1-N}$ and $\tilde{q}_c = N - \frac{1}{2}$.

**Theorem 3.5.3** *Let $\Omega \subset \mathbb{R}^N$ be a bounded $C^2$ domain such that $0 \in \partial\Omega$ and $g : \mathbb{R}_+ \mapsto \mathbb{R}_+$ is a continuous function which satisfies*

$$\inf_{s>0} \frac{g(s)}{s^{N-\frac{1}{2}}} > 0. \tag{3.5.17}$$

*If* $u \in C(\overline{\Omega} \setminus \{0\}) \cap C^1(\Omega)$ *is a nonnegative solution of*

$$\begin{aligned} -\Delta_N u + g(|\nabla u|) &= 0 && \text{in } \Omega \\ u &= 0 && \text{on } \partial\Omega \setminus \{0\}, \end{aligned} \tag{3.5.18}$$

*it is identically zero.*

*Proof.* Without loss of generality it can be supposed that $u$ satisfies

$$-\Delta_N u + |\nabla u|^{N-\frac{1}{2}} \leq 0 \qquad \text{in } \Omega. \tag{3.5.19}$$

As in the proof of Proposition 3.3.6, we assume that $B_{\frac{1}{2}}(a')$ is exterior to $\Omega$ and that $\partial B_{\frac{1}{2}}(a')$ is tangent to $\partial\Omega$ at 0. We put $a' = \mathbf{n}\delta$ where $\mathbf{n}$ is the normal outward unit vector to $\partial\Omega$ at 0, $\tilde{a}' = 2a'$, and denote by $\mathcal{J}_1^{\tilde{a}'}$ the inversion of center $\tilde{a}'$ and power 1. The ball $B_{\frac{1}{2}}(a')$ is transformed by $\mathcal{J}_1^{\tilde{a}'}$ into the half space $H^+$ passing through 0, with boundary $T_0\partial\Omega$ and containing $-\mathbf{n}$. Then $U = u \circ \mathcal{J}_1^{\tilde{a}'}$ satisfies

$$\begin{aligned} -\Delta_N U + |x - \tilde{a}'|^{-1} |\nabla U|^{N-\frac{1}{2}} &\leq 0 && \text{in } \Omega' \\ U &= 0 && \text{on } \partial\Omega' \setminus \{0\}, \end{aligned} \tag{3.5.20}$$

where $\Omega' = \mathcal{J}_1^{\tilde{a}'}(\Omega)$. Moreover $|x - \tilde{a}'|^{-1} \leq \nu$ for all $x \in \Omega'$ for some $\nu > 0$ thus the inequality in (3.5.17) can be replaced by $-\Delta_N U + \nu |\nabla U|^{N-\frac{1}{2}} \leq 0$. Because $\Omega' \subset H^- = -H^+$, it follows from Theorem 3.5.2 that $U = 0$ and thus $u = 0$. □

3.5.2.2 *Construction of singular solutions*

In the subcritical case $p - 1 < q < \tilde{q}_c$ we can classify all the nonnegative solutions of (3.4.1) with an isolated singularity on the boundary. The geometric assumptions on the domain are different according $1 < p < N$ or $p = N$. A first tool is the construction of solutions with weak singularities on the boundary.

**Theorem 3.5.4** *Let* $1 < p \leq N$, $p-1 < q < \tilde{q}_c$ *and* $\Omega \subset \mathbb{R}^N$ *be a bounded* $C^2$ *domain such that* $0 \in \partial\Omega$. *We suppose also that* $\Omega$ *is included into the half-space* $H^+$ *and that there exists* $\delta > 0$ *such that* $\Omega \cap B_\delta = \Omega \cap B_\delta$. *Then for any* $k \geq 0$ *there exists a unique positive function* $w := w_k \in C(\overline{\Omega} \setminus \{0\}) \cap C^1(\Omega)$ *satisfying (3.5.1) such that*

$$\lim_{r \to 0} r^\beta w(r, \sigma) = k\phi_{S_+^{N-1}}(\sigma) \qquad \forall \sigma \in S_+^{N-1}, \tag{3.5.21}$$

*in the* $C^1(S_+^{N-1})$ *topology.*

The proof of this result necessitates some intermediate constructions.

**Lemma 3.5.5** *Let the assumptions on $p, q$ and $\Omega$ of Theorem 3.5.4 be satisfied. If for some $k > 0$ there exists $w_k$ satisfying Theorem 3.5.4, then for any $k > 0$ such a function $w_k$ exists.*

*Proof.* We notice that for $c > 1$ (resp. $0 < c < 1$) $cw_k$ is a supersolution (resp. a subsolution) of (3.4.1). Assume that $w_k$ exists. Because $ckV_\Omega$ is a supersolution (independently of the value of $c > 0$), when $0 < c < 1$ it follows by Theorem 1.4.6 that there exists a solution $w_{ck}$ of (3.5.18)-(3.5.19) between $cw_k$ and $ckV_\Omega$. When $c > 1$ we set $\tau = (\gamma_{p,q} - \beta)^{-1}$ and $w^* = J_{c^\tau}[w_k]$ where

$$J_{c^\tau}[w_k](x) = c^{\tau\gamma_{p,q}} w_k(c^\tau x) \qquad \forall x \in \Omega^{c^\tau} := \tfrac{1}{c^\tau}\Omega.$$

Then $w^*$ is a solution of problem (3.5.18) in $\Omega^{\frac{1}{c^\tau}}$ instead of $\Omega$. In particular it satisfies the equation in $B^+_{\frac{\delta}{c^\tau}} \subset B^+_\delta$ and there holds

$$\lim_{x \to 0} \frac{w^*(x)}{V_{\mathbb{R}^N_+}} = k.$$

If $m = \sup\{w^*(x) : |x| = c^{-\tau}\delta\}$, the function $(w^* - m)_+$ is a subsolution of problem (3.5.18) in $\Omega$. Consequently, there exists a solution $w_{ck}$ of problem (3.5.1) in $\Omega$ which satisfies $(w^* - m)_+ \le w_{ck} \le ckV_\Omega$ and $w_{ck} \in C^1(\overline{\Omega} \setminus \{0\})$ by regularity. Hence

$$|x|^{1+\beta} |\nabla w_{ck}(x)| + |x|^{1+\beta+\alpha} \frac{|\nabla w_{ck}(x) - \nabla w_{ck}(y)|}{|x - y|^\alpha} \le M,$$

for some $\alpha \in (0, 1)$, $M > 0$ and all $x \neq y$, $|x| \le |y|$. This implies that the set of functions $\{r^\beta w_{ck}\}_{0<r\le\delta}$ is relatively compact in $C^1(\overline{S}^{N-1}_+)$. Since the convergence of $r^\beta w_{ck}$ to $ck\phi_{S^{N-1}_+}$ holds uniformly on compact subsets of $S^{N-1}_+$ when $r \to 0$, it implies that it also holds in $C^1(\overline{S}^{N-1}_+)$. □

The key step is the construction of a subsolution to problem (3.5.1)-(3.5.21).

**Lemma 3.5.6** *Let the assumptions on $p, q$ and $\Omega$ of Theorem 3.5.4 be satisfied. Then there exist $\delta_0 \in (0, \delta]$ and a nonnegative function $W$, locally Lipschitz continuous in $\overline{B}^+_{\delta_0} \setminus \{0\}$, vanishing on $B_{\delta_0} \cap \partial\mathbb{R}^N_+ \setminus \{0\}$, smaller than $V_{\mathbb{R}^N_+}$ and satisfying*

$$\lim_{r \to 0} r^\beta W(r, \sigma) = \phi_{S^{N-1}_+}(\sigma) \quad \textit{uniformly in } \overline{S}^{N-1}_+. \tag{3.5.22}$$

*Proof.* We have already noticed in Theorem 3.3.11 that the construction of subsolution for equation with absorption and boundary weak singularity is long. In the case of equation (3.4.1) it is even longer. The idea is to look for a subsolution $W$ under the form

$$W(r,\sigma) = r^{-\beta_{S_+^{N-1}}} \phi_{S_+^{N-1}}(\sigma) - r^{-\beta_{S_+^{N-1}}+\gamma} g(\phi_{S_+^{N-1}}(\sigma)), \tag{3.5.23}$$

where $\gamma > 0$ is small enough and the function $g$ is Lipschitz continuous. The computation, which uses many asymptotic expansions, is very delicate and technical. It is performed with all details in [Bidaut-Véron, Garcia-Huidobro, Véron (2015)] and we skip it here. □

*Proof of Theorem 3.5.4.* Set $M = \sup_{|x|=\delta_0} V_{\mathbb{R}_+^N}(x) = \delta_0^{-\beta}$. The function $\underline{W} = (W - M)_+$ initially defined in $B_{\delta_0}^+$ and extended by 0 in $\Omega \setminus B_{\delta_0}^+$ is a subsolution of problem (3.5.1). The function $V_\Omega$ (see Proposition 3.3.5) is a supersolution and it is larger than $\underline{W}$. By Theorem 1.4.6 there exists a solution $w$ of the same equation satisfying $\underline{W} \leq w \leq V_\Omega$. Thus $w$ satisfies (3.5.21) with $k = 1$. Existence for all $k > 0$ follows from Lemma 3.5.5. Uniqueness follows from (3.5.21) and the comparison principle. □

In the case $p = N$ no geometric assumption is needed for the corresponding existence result.

**Theorem 3.5.7** *Let $N - 1 < q < N - \frac{1}{2}$ and $\Omega \subset \mathbb{R}^N$ be a bounded $C^2$ domain such that $0 \in \partial\Omega$. Then for any $k \geq 0$ there exists a unique positive function $w := w_k \in C(\overline{\Omega} \setminus \{0\}) \cap C^1(\Omega)$ satisfying*

$$\begin{aligned} -\Delta_N w + |\nabla w|^q &= 0 && \text{in } \Omega \\ w &= 0 && \text{on } \partial\Omega \setminus \{0\}, \end{aligned} \tag{3.5.24}$$

*such that*

$$\lim_{r \to 0} r w(r,\sigma) = k\sigma \qquad \forall \sigma \in S_+^{N-1}, \tag{3.5.25}$$

*in the $C^1(S_+^{N-1})$ topology.*

*Proof.* Lemma 3.5.5 is immediately extendable to the case $p = N$. The main step is the construction of a subsolution $W$ which satisfies

$$\lim_{r \to 0} \frac{W(r,\sigma)}{V_\Omega(r,\sigma)} = 1 \qquad \forall \sigma \in S_+^{N-1} \tag{3.5.26}$$

locally uniformly on $S_+^{N-1}$. We use the notations of Proposition 3.3.6. If we denote by $W$ a subsolution of the equation in $\overline{B_{\delta_0}^-} \setminus \{0\}$, vanishing on

$B_{\delta_0} \cap \partial\mathbb{R}^N_- \setminus \{0\}$ and satisfying

$$\lim_{r\to 0} r^\beta W(r,\sigma) = \phi_{S^{N-1}_-}(\sigma) \quad \text{uniformly in } \overline{S}^{N-1}_-, \tag{3.5.27}$$

and we assume $\delta_0 < 1$. The inversion $\mathcal{I}^{\tilde a}_1$ of center $\tilde a = -\mathbf{n} = 2a$ and power 1 transforms $\mathbb{R}^N_-$ into the ball $B_{\frac{1}{2}}(a)$ and $\overline{B}^-_{\delta_0} \setminus \{0\}$ into a subdomain, that is called a lunula and is the intersection of $B_{\frac{1}{2}}(a)$ and a ball $B_{r_b}(b)$ for $b = \frac{\delta_0^2}{1-\delta_0^2}\mathbf{n}$ and $r_b = \frac{\delta_0}{1-\delta_0^2}$. Then $\tilde W = W \circ \mathcal{I}^{\tilde a}_1$ satisfies

$$-\Delta_N \tilde W + |x-\tilde a|^{2(q-N)} \left|\nabla \tilde W\right|^q = 0,$$

in $\Gamma$. Since $|x - \tilde a| \leq 1$ in $\Gamma$, $\tilde W$ satisfies

$$-\Delta_N \tilde W + \left|\nabla \tilde W\right|^q \leq 0. \tag{3.5.28}$$

If $M = \sup\{\tilde W(x) : x \in \partial\Gamma \cap \mathbb{R}^N_+\}$, the function $(\tilde W - M)_+$ is a subsolution of problem (3.5.24); it is smaller than the supersolution $V_\Omega$ and (3.5.27) holds. We conclude as in Theorem 3.5.4. □

The type of singularities exhibited by the solutions $u_k$ obtained in Theorem 3.5.4 and Theorem 3.5.7 are not the only ones since there exists also solution with strong singularities which are the limit of the previous ones.

**Theorem 3.5.8** *Assume $p$, $q$ and $\Omega$ satisfy the assumption of Theorem 3.5.4. Then there exists $w_\infty = \lim_{k\to\infty} w_k$, and $w_\infty$ is the unique positive solution of (3.5.1) which satisfies*

$$\lim_{r\to 0} r^{\gamma_{p,q}} w(r,\sigma) = \omega_h(\sigma) \quad \textit{uniformly on } S^{N-1}_+, \tag{3.5.29}$$

*where $\omega_h$ is the unique positive solution of (3.5.2) obtained in Theorem 3.5.1.*

*Proof.* Uniqueness follows from (3.5.29) and the comparison theorem. Furthermore the monotonicity of $k \mapsto w_k$ follows from the same reference and (3.5.21). Let $W_{\mathbb{R}^N_+}$ be the function $(r,\sigma) \mapsto W_{\mathbb{R}^N_+}(r,\sigma) = r^{-\gamma_{p,q}}\omega_h(\sigma)$. This function dominates $w_k$ since $\Omega \subset \mathbb{R}^N_+$. Therefore there exists $w_\infty = \lim_{k\to\infty} w_k$ and $w_\infty \leq W_{\mathbb{R}^N_+}$. We recall that $J_\lambda[w]$ is defined by $J_\lambda[w](x) = \lambda^{\gamma_{p,q}} w(\lambda x)$ for $\lambda > 0$. Then $W_{\mathbb{R}^N_+}$ is self-similar under $J_\lambda$. In order to specify the domain, we denote by $w_{k,B^+_\delta}$ the solution of (3.5.2) in $B^+_\delta$ instead of $\Omega$ which satisfies (3.5.21), thus $\delta \mapsto w_{k,B_\delta^+}$ is increasing and $w_{k,B_\delta^+} \leq w_k$ which implies $w_{\infty,\delta} \leq w_\infty$. Moreover, for any $\lambda$, $\delta > 0$,

$$J_\lambda[w_{k,B_\delta^+}] = w_{\lambda^{\gamma_{p,q}}k,B^+_{\lambda^{-1}\delta}}. \tag{3.5.30}$$

By letting successively $k \to \infty$ and $\delta \to \infty$, this implies

$$J_\lambda[w_{\infty,B_\delta^+}] = w_{\infty,B^+_{\lambda^{-1}\delta}} \Longrightarrow J_\lambda[w_{\infty,\mathbb{R}^N_+}] = w_{\infty,\mathbb{R}^N_+}. \tag{3.5.31}$$

Thus $w_{\infty,\mathbb{R}^N_+}$ is a nontrivial nonnegative separable solution of (3.5.1) in $\mathbb{R}^N_+$ vanishing on $\partial\mathbb{R}^N_+ \setminus \{0\}$, therefore $w_{\infty,\mathbb{R}^N_+}(r,\sigma) = r^{-\gamma_{p,q}}\omega_h(\sigma) = W_{\mathbb{R}^N_+}(r,\sigma)$. In (3.5.30) we let $\lambda \to 0$, take $|x| = 1$ and write $x = \sigma$, then

$$\lim_{\lambda\to 0} \lambda^{\gamma_{p,q}} w_{\infty,B_\delta^+}(\lambda,\sigma) = \lim_{\lambda\to 0} w_{\infty,B^+_{\lambda^{-1}\delta}}(1,\sigma) = w_{\infty,\mathbb{R}^N_+}(1,\sigma) = \omega_h(\sigma). \tag{3.5.32}$$

Furthermore this convergence holds in $C^1(S^{N-1}_+)$ as in the proof of Theorem 3.3.13. □

In the case $p = N$ we have a similar result with no geometric assumption on $\Omega$.

**Theorem 3.5.9** *Let $p = N$, $N - 1 < q < N - \frac{1}{2}$ and $\Omega \subset \mathbb{R}^N$ be a bounded $C^2$ domain such that $0 \in \partial\Omega$. Then there exists $w_\infty = \lim_{k\to\infty} w_k$ and $w_\infty$ is the unique positive solution of (3.5.28) which satisfies*

$$\lim_{r\to 0} r^{\gamma_{N,q}} w(r,\sigma) = \omega_h(\sigma) \quad \textit{eventually uniformly on } S^{N-1}_+, \tag{3.5.33}$$

*where $\omega_h$ is the unique positive solution of (3.5.2) obtained in Theorem 3.5.1.*

*Proof.* Let $B := B_{\frac{1}{2}}(a)$ and $B' := B_{\frac{1}{2}}(a')$ the inner and outer tangent balls to $\partial\Omega$ at 0 (see Proposition 3.3.6). For distinction we denote by $w_{k,\Omega}$ the solution of (3.5.1) which satisfies (3.5.29). Then

$$w_{k,B} \leq w_{k,\Omega} \leq w_{k,B'^c},$$

in the domains corresponding to each couple of solutions. Since $J_\lambda[w_{k,B}] = w_{\lambda^{\gamma_{N,q}}k,B^\lambda}$, $J_\lambda[w_{k,\Omega}] = w_{\lambda^{\gamma_{N,q}}k,\Omega^\lambda}$ and $J_\lambda[w_{k,B'^c}] = w_{\lambda^{\gamma_{N,q}}k,B'^{c\,\lambda}}$ we derive

$$J_\lambda[w_{k,B}] = w_{\lambda^{\gamma_{N,q}}k,B^\lambda} \leq J_\lambda[w_{k,\Omega}] \leq J_\lambda[w_{k,B'^c}] = w_{\lambda^{\gamma_{N,q}}k,B'^{c\,\lambda}}.$$

Furthermore $J_\lambda[w_{k,B}] \leq J_{\lambda'}[w_{k',B}]$ if $\lambda' \leq \lambda$ and $\lambda^{\gamma_{N,q}}k \leq \lambda'^{\gamma_{N,q}}k'$ and similarly $J_\lambda[w_{k,B'^c}] \leq J_{\lambda'}[w_{k',B'^c}]$ if $\lambda \leq \lambda'$ and $\lambda^{\gamma_{N,q}}k \leq \lambda'^{\gamma_{N,q}}k'$. If we denote by $w_{\infty,\Omega}$, $w_{\infty,B}$ and $w_{\infty,B'^c}$ the respective limits of $w_{k,\Omega}$, $w_{k,B}$ and $w_{k,B'^c}$ when $k \to \infty$ (which exist my monotonicity), there holds

$$w_{\infty,B} \leq w_{\infty,\Omega} \leq w_{\infty,B'^c}. \tag{3.5.34}$$

Then

$$J_\lambda[w_{\infty,B'^c}] \leq J_\lambda[w_{\infty,\Omega}] \leq J_\lambda[w_{\infty,B}], \tag{3.5.35}$$

and

$$\begin{array}{lll} (i) & J_\lambda[w_{\infty,B}] \leq J_{\lambda'}[w_{\infty,B}] & \text{if } \lambda' \leq \lambda \\ (ii) & J_\lambda[w_{k,B'^c}] \leq J_{\lambda'}[w_{k',B'^c}] & \text{if } \lambda \leq \lambda'. \end{array} \tag{3.5.36}$$

By replacing $\lambda$ by $\lambda\nu$ we can rewrite (3.5.35) as

$$J_\nu\left[J_\lambda[w_{\infty,B'^c}]\right] \leq J_\nu\left[J_\lambda[w_{\infty,\Omega}]\right] \leq J_\nu\left[J_\lambda[w_{\infty,B}]\right]. \tag{3.5.37}$$

Because of the monotonicity with respect to $\lambda$, the following two limits exist when $\lambda \to 0$,

$$\tilde{W}_B = \lim_{\lambda\to 0} J_\lambda[w_{\infty,B}] \quad \text{and} \quad \tilde{W}_{B'^c} = \lim_{\lambda\to 0} J_\lambda[w_{\infty,B'^c}]. \tag{3.5.38}$$

Applying Lemma 3.3.2 to $w_{\infty,B}$ and $w_{\infty,B'^c}$ with $\phi(|x|) = c|x|^{-\gamma_{N,q}}$ we infer the following series of estimates

$$\begin{array}{lll} (i) & |\nabla J_\lambda[w_{\infty,B}](x)| \leq c_2|x|^{-\gamma_{N,q}-1} & \forall x \in B^\lambda \\ (ii) & |\nabla J_\lambda[w_{\infty,B}](x) - \nabla J_\lambda[u^B_\infty](y)| \leq c_2\dfrac{|x-y|^\alpha}{|x|^{\gamma_{N,q}+1+\alpha}} & \forall x,y \in B^\lambda \\ & & |x| \leq |y| \\ (iii) & J_\lambda[w_{\infty,B}](x) \leq c_2|x|^{-\gamma_{N,q}-1}(\mathrm{dist}\,(x,\partial B^\lambda))^\alpha & \forall x \in B^\ell, \end{array} \tag{3.5.39}$$

and

$$\begin{array}{lll} (i) & |\nabla J_\lambda[u_{\infty,B'^c}](x)| \leq c_2|x|^{-\gamma_{N,q}-1} & \forall x \in B'^{c\,\lambda} \\ (ii) & |\nabla J_\lambda[u_{\infty,B'^c}](x) - \nabla J_\lambda[u^{B'^c}_\infty](y)| \leq c_2\dfrac{|x-y|^\alpha}{|x|^{\gamma_{N,q}+1+\alpha}} & \forall x,y \in B'^{c\,\lambda} \\ & & |x| \leq |y| \\ (iii) & J_\lambda[u_{\infty,B'^c}](x) \leq c_2|x|^{-\gamma_{N,q}-1}(\mathrm{dist}\,(x,\partial B'^{c\,\ell}))^\alpha & \forall x \in B'^{c\,\ell}. \end{array} \tag{3.5.40}$$

Therefore the sets of functions $\{J_\lambda[u_{\infty,B}]\}_{\lambda>0}$ and $\{J_\lambda[u_{\infty,B'^c}]\}_{\lambda>0}$ are equicontinuous in the $C^1_{loc}$ topology. By uniqueness the limits in (3.5.38) hold in this topology. Hence $\tilde W_B$ and $\tilde W_{B'^c}$ are positive solutions of

$$\begin{aligned} -\Delta_N w + |\nabla w|^q &= 0 \qquad \text{in } \mathbb{R}^N_+ \\ w &= 0 \qquad \text{on } \mathbb{R}^N_+ \setminus \{0\}. \end{aligned} \tag{3.5.41}$$

Since $J_\nu\left[J_\lambda[w_{\infty,B'^c}]\right] = J_{\lambda\nu}[w_{\infty,B'^c}]$ and $J_\nu\left[J_\lambda[w_{\infty,B}]\right] = J_{\lambda\nu}[w_{\infty,B}]$, it follows by letting $\lambda \to 0$ that $J_\nu\left[\tilde W_{B'^c}\right] = \tilde W_{B'^c}$ and $J_\nu\left[\tilde W_B\right] = \tilde W_B$ for all $\nu > 0$. Hence $\tilde W_{B'^c}$ and $\tilde W_B$ are nonnegative nontrivial separable solutions of (3.5.41), hence they are equal to $W_{\mathbb{R}^N_+}$.

Applying the same estimates to $J_\lambda[u_{\infty,\Omega}]$ we infer

$$\begin{aligned} &(i)\ \ |\nabla J_\lambda[u_{\infty,\Omega}](x)| \leq c_2|x|^{-\gamma_{N,q}-1} && \forall x \in B^\lambda \\ &(ii)\ \ |\nabla J_\lambda[u_{\infty,\Omega}](x) - \nabla J_\lambda[u_{\infty,\Omega}](y)| \leq c_2\frac{|x-y|^\alpha}{|x|^{\gamma_{N,q}+1+\alpha}} && \forall x,y \in B^\lambda \\ & && |x| \leq |y| \\ &(iii)\ J_\lambda[u_{\infty,\Omega}](x) \leq c_2|x|^{-\gamma_{N,q}-1}(\mathrm{dist}\,(x,\partial B^\lambda))^\alpha && \forall x \in B^\ell. \end{aligned} \tag{3.5.42}$$

By equicontinuity there exist a subsequence $\{\lambda_k\}$ and a function $W$ such that $J_{\lambda_k}[u_{\infty,\Omega}]$ converges eventually to $U$ in the $C^1_{loc}$ topology of $\mathbb{R}^N_+$. The function $U$ is a positive solution of (3.5.41) which vanishes on $\partial\mathbb{R}^N_+ \setminus \{0\}$. Using (3.5.35) we obtain that $W = W_{\mathbb{R}^N_+}$, hence (3.5.33) holds. □

### 3.5.2.3 *Global solutions*

This connexion between solutions with weak and strong boundary singularities combined with boundary Harnack inequality allows us to give a complete classification of solutions of (3.4.1) with a boundary isolated singularity.

**Theorem 3.5.10** *Assume $1 < p < N$, $p - 1 < q < \tilde q_c$ and $\Omega \subset \mathbb{R}^N$ is a $C^2$ domain such that $0 \in \partial\Omega$ and $\Omega \cap B_\delta = B^+_\delta$ for some $\delta > 0$. If $w \in C(\overline{\Omega} \setminus \{0\}) \cap C^1(\Omega)$ is a nonnegative solution of (3.5.1). Then the following alternative holds,*

*(i) either there exists $k \geq 0$ such that*

$$\lim_{r\to 0} r^{\beta_{S^{N-1}_+}} u(r,.) = k\phi_{S^{N-1}_+}, \tag{3.5.43}$$

*in $C^1(S^{N-1}_+)$, furthermore $u = 0$ if $k = 0$ and $\Omega \subset \mathbb{R}^N_+$,*

*(ii) or*

$$\lim_{r\to 0} r^{\gamma_{p,q}} u(r,.) = \omega_h, \tag{3.5.44}$$

*in* $C^1(S_+^{N-1})$.

*Proof.* We recall that $w_{k,B_\delta^+}$ denotes the solution of (3.5.1) with $\Omega = B_\delta^+$ which satisfies (3.5.29).

*Step 1.* Firstly we assume that

$$\liminf_{x\to 0} \frac{w(x)}{V_{\mathbb{R}_+^N}(x)} = k < \infty. \tag{3.5.45}$$

By Theorem 3.5.4 (3.5.43) is equivalent to

$$\liminf_{x\to 0} \frac{w(x)}{w_{1,B_\delta^+}(x)} = k. \tag{3.5.46}$$

This liminf is achieved following a sequence $\{x_n\}$ converging to 0. We write the equation under the form

$$-\Delta_p w + |\nabla w|^{p-2} \langle c(x), \nabla w\rangle = 0 \tag{3.5.47}$$

with

$$c(x) = |\nabla w|^{q-p} \nabla w.$$

By (3.4.32), $|c(x)| \leq |x|^{-1}$. Then we can apply boundary Harnack inequality separately to $w$ and $w_{1,B_\delta^+}(x)$ under the form given by Proposition 2.3.1 and derive

$$\frac{w(x_n)}{w_{1,B_\delta^+}(x_n)} = \frac{w(x_n)}{\rho(x_n)} \frac{\rho(x_n)}{w_{1,B_\delta^+}(x_n)} \geq c \frac{w(x)}{w_{1,B_\delta^+}(x)} \quad \forall x \in \Omega \text{ s.t. } |x| = |x_n|. \tag{3.5.48}$$

Therefore there exists $k' \geq k$ such that

$$w(x) \leq w_{k',B_\delta^+}(x) + m \leq k' V_{\mathbb{R}_+^N} + m \quad \forall x \in \Omega \text{ s.t. } |x| = |x_n|,$$

where $m$ is the supremum of $w$ on $|x| = \delta$. By the comparison principle this inequality holds in $B_\delta^+ \cap B_{|x_n|}^c$ for any $n$. therefore it holds in whole $B_\delta^+$. Applying again Theorem 3.5.4 and the comparison principle we obtain

$$w \leq w_{k',B_\delta^+} + m \leq k' V_{\mathbb{R}_+^N} + m \leq M |x|^{-\beta} \qquad \text{in } B_\delta^+,$$

where we set $\beta := \beta_{S_+^{N-1}}$. We also write $\phi := \phi_{S_+^{N-1}}$ and define

$$w_r(y) = r^\beta w(ry) \qquad \forall y \in B^+_{\frac{\delta}{r}}.$$

Then $w_r$ satisfies

$$-\Delta_p w_r + r^{p-q-\beta(q+1-p)} |\nabla w_r|^q \quad \text{in } B^+_{\frac{\delta}{r}}. \tag{3.5.49}$$

Applying Lemma 3.3.2 with $|w(x)| \leq M|x|^{-\beta}$, we derive the estimates

$$\begin{aligned} &(i)\ |\nabla w_r(x)| \leq c_2|x|^{-\beta-1} && \forall x \in B^+_{\frac{\delta}{2r}} \\ &(ii)\ |\nabla w_r(x) - \nabla w_r(y)| \leq c_2\frac{|x-y|^\alpha}{|x|^{\beta+1+\alpha}} && \forall x, y \in B^+_{\frac{\delta}{2r}},\ |x| \leq |y|. \end{aligned} \tag{3.5.50}$$

Consequently $\{w_r\}$ is relatively compact in the $C^1_{loc}$ topology of $\overline{B}^+_{\frac{\delta}{3r}} \setminus \{0\}$, and in particular there exist a function $W$ continuous in $\overline{\mathbb{R}}^N_+ \setminus \{0\}$, $C^1$ in $\mathbb{R}^N_+$ and vanishing on $\partial\mathbb{R}^N_+ \setminus \{0\}$ and a subsequence $\{r_{n_k}\}$ of $\{r_n\} := \{|x_n|\}$ such that $w_{r_{n_k}}$ converges eventually to $W$ in the $C^1_{loc}(\overline{\mathbb{R}}^N_+ \setminus \{0\})$ topology. Since $p - q - \beta(q+1-p) > 0$, $W$ is a positive $p$-harmonic function in $\mathbb{R}^N_+$ which vanishes on $\partial\mathbb{R}^N_+ \setminus \{0\}$ and is dominated by $M|x|^{-\beta}$. By Theorem 2.3.9, $W$ is a multiple of $V_{\mathbb{R}^N_+}$, which is $kV_{\mathbb{R}^N_+}$ by (3.5.45) and the definition of $\{r_n\}$. Uniqueness implies that $w_{r_n} \to kV_{\mathbb{R}^N_+}$ and thus

$$\lim_{r_n \to 0} \sup_{|\xi| = r_n} w(r_n, \xi) = k. \tag{3.5.51}$$

By the comparison principle for any $\epsilon > 0$ there exists $n_\epsilon \in \mathbb{N}$ such that $w(x) \leq (k+\epsilon)V_{\mathbb{R}^N_+}(x) + m$ for all $n \geq n_\epsilon$ and all $x \in B^+_\delta \setminus B_{r_n}$. Letting $n \to \infty$ and $\epsilon \to 0$, we derive

$$w(x) \leq kV_{\mathbb{R}^N_+}(x) + m \quad \forall x \in B^+_\delta. \tag{3.5.52}$$

Jointly with (3.5.45) we infer (3.5.43).

*Step 2.* Next we assume

$$\liminf_{x \to 0} \frac{w(x)}{V_{\mathbb{R}^N_+}(x)} = \infty = \lim_{x \to 0} \frac{w(x)}{V_{\mathbb{R}^N_+}(x)}. \tag{3.5.53}$$

By the comparison principle $w \geq w_{k, B^+_\delta}$ in $B^+_\delta$, for any $k > 0$. Therefore

$$w(x) \geq w_{\infty, B^+_\delta}(x) \qquad \forall x \in B^+_\delta. \tag{3.5.54}$$

Next we put $w_r(x) = J_r[w](x) = r^{\gamma_{p,q}}w(rx)$. Then $w_r$ satisfies (3.4.1) in $B^+_{\frac{\delta+}{r}}$ and it vanishes on $\partial\mathbb{R}^N_+ \cap B_{\frac{\delta+}{r}} \setminus \{0\}$. It satisfies also the regularity estimates

$$
\begin{array}{lll}
(i) \ |\nabla w_r(x)| \leq c_2|x|^{-\gamma_{p,q}-1} & \forall x \in B^+_{\frac{\delta}{2r}} & \\
(ii) \ |\nabla w_r(x) - \nabla w_r(y)| \leq c_2\dfrac{|x-y|^{\alpha}}{|x|^{\gamma_{p,q}+1+\alpha}} & \forall x, y \in B^+_{\frac{\delta}{2r}},\ |x| \leq |y|. & (3.5.55)
\end{array}
$$

Therefore there exists a subsequence $\{r_n\}$ converging to 0 such that $\{w_{r_n}\}$ converges in the $C^1_{loc}$ topology of $\overline{\mathbb{R}}^N_+ \setminus \{0\}$ to a function $W$ which is a nonnegative solution of (3.4.1) in $\mathbb{R}^N_+$ and vanishes on $\partial\mathbb{R}^N_+ \setminus \{0\}$.
In order to obtain an estimate from above of $w$ besides the *a priori* estimate (3.4.40) which is not precise enough, we see that for any $\epsilon \in (0, \delta)$, $w$ is dominated in $B^+_\delta \setminus B^+_\epsilon$ by the solution $U_{\epsilon, B^+_\delta} + m$, where $U_{\epsilon, B^+_\delta}$ is the limit when $\ell \to \infty$ of the solutions $u := U_{\epsilon, B^+_\delta, \ell}$ of

$$
\begin{array}{rll}
-\Delta_p u + |\nabla u|^q = 0 & \text{in } B^+_\delta \setminus B^+_\epsilon & \\
u = 0 & \text{on } (\partial B_\delta \cap \mathbb{R}^N_+) \cup (\partial\mathbb{R}^N_+ \cap (B_\delta \setminus B_\epsilon)) & (3.5.56) \\
u = \ell & \text{on } \partial B_\epsilon \cap \mathbb{R}^N_+. &
\end{array}
$$

Existence of $U_{\epsilon, B^+_\delta, \ell}$ follows from the fact that the constant function $\ell$ is a solution and the monotonicity with respect to $\ell$ from the comparison principle. Using (3.4.42) we have

$$U_{\epsilon', B^+_\delta, \ell}(x) \leq c_{N,p,q}(|x| - \epsilon)^{-\gamma_{p,q}},$$

therefore $U_{\epsilon, B^+_\delta, \ell} \uparrow U_{\epsilon, B^+_\delta}$ when $\ell \to \infty$ and $U_{\epsilon, B^+_\delta}$ is the minimal solution of

$$
\begin{array}{rll}
-\Delta_p u + |\nabla u|^q = 0 & \text{in } B^+_\delta \setminus B^+_\epsilon & \\
u = 0 & \text{in } (\partial B_\delta \cap \mathbb{R}^N_+) \cup (\partial\mathbb{R}^N_+ \cap (B_\delta \setminus B_\epsilon)) & (3.5.57) \\
u = \infty & \text{on } \partial B_\epsilon \cap \mathbb{R}^N_+. &
\end{array}
$$

When $\epsilon \to 0$, $U_{\epsilon, B^+_\delta} \downarrow U_{B^+_\delta}$ which is the minimal solution of

$$
\begin{array}{rll}
-\Delta_p u + |\nabla u|^q = 0 & \text{in } B^+_\delta & \\
u = 0 & \text{in } (\partial B_\delta \cap \mathbb{R}^N_+) \cup (\partial\mathbb{R}^N_+ \cap (B_\delta)) & (3.5.58) \\
\lim\limits_{x \to 0} u(x) = \infty. & &
\end{array}
$$

The minimality implies that for any $r > 0$, $J_r[U_{\epsilon, B^+_\delta}] = U_{\frac{\epsilon}{r}, B^+_{\frac{\delta}{r}}}$, then $J_r[U_{B^+_\delta}] = U_{B^+_{\frac{\delta}{r}}}$ when $\epsilon \to 0$. Furthermore, if $\delta' < \delta$, we have $U_{B^+_{\delta'}} \leq U_{B^+_\delta}$. This implies that $U_{B^+_{\delta'}}$ increases and converges to $U_{\mathbb{R}^N_+}$ which is a positive solution of (3.4.1) in $\mathbb{R}^N_+$, vanishing on $\partial\mathbb{R}^N_+ \setminus \{0\}$ and tending to infinity nontangentially at 0. Moreover

$$J_r[U_{B^+_\delta}] = U_{B^+_{\frac{\delta}{r}}} \Longrightarrow J_r[U_{\mathbb{R}^N_+}] = U_{\mathbb{R}^N_+} \qquad \forall r > 0.$$

Then $U_{\mathbb{R}^N_+}$ is a nontrivial nonnegative self-similar solution of (3.4.1) in $\mathbb{R}^N_+$, vanishing on $\partial\mathbb{R}^N_+ \setminus \{0\}$. Thus

$$U_{\mathbb{R}^N_+}(r,\sigma) = r^{-\gamma_{p,q}}\omega_h(\sigma). \tag{3.5.59}$$

Since $w \leq U_{B^+_\delta} + m$ we derive

$$w_r \leq J_r[U_{B^+_\delta} + m] = U_{B^+_{\frac{\delta}{r}}} + r^{\gamma}_{p,q} m \quad \text{in } B^+_{\frac{\delta}{r}}. \tag{3.5.60}$$

In particular, if we take $|x| = r$ and $x = r\xi$ and use the fact that $w_{\infty,B^+_\delta}$ is smaller than $w$ and satisfies (3.5.29), we infer

$$\omega_h(\xi) \leq \lim_{r\to 0} r^{\gamma_{p,q}} w(r,\xi) \leq \lim_{r\to 0} U_{B^+_{\frac{\delta}{r}}} = \omega_h(\xi), \tag{3.5.61}$$

which is (3.5.44). □

In the case $p = N$ a similar result holds and its proof follows the same ideas combined with the tools of Theorem 3.5.9.

**Theorem 3.5.11** *Let $p = N$, $N - 1 < q < N - \frac{1}{2}$ and $\Omega \subset \mathbb{R}^N$ be a bounded $C^2$ domain such that $0 \in \partial\Omega$. If $w \in C(\overline{\Omega} \setminus \{0\}) \cap C^1(\Omega)$ is a nonnegative solution of (3.5.1). Then the following alternative holds,*

*(i) either there exists $k \geq 0$ such that (3.5.43) holds in $C^1(S^{N-1}_+)$ with $\beta_{S^{N-1}_+} = 1$; furthermore $u = 0$ if $k = 0$,*
*(ii) or (3.5.44) holds with $\gamma_{N,q}$.*

### 3.5.2.4 *Entire singular solutions*

As a consequence of Theorem 3.4.11, Theorem 3.5.10 and Theorem 3.5.11 we have the following rigidity results concerning entire solutions in $\mathbb{R}^N_+$. Our first result concerns signed solutions.

**Proposition 3.5.12** *Assume $0 < p - 1 < q < p$ and $u \in C^1(\mathbb{R}^N_+) \cap C(\overline{\mathbb{R}}^N_+ \setminus \{0\})$ is a solution of (3.4.1) in $\mathbb{R}^N_+$ which vanishes on $\partial\mathbb{R}^N_+ \setminus \{0\}$.*
*(i) If $u$ satisfies*

$$\lim_{x\to 0} |x|^{\beta_{S^{N-1}_+}} u(x) = 0, \tag{3.5.62}$$

*then $u$ is bounded from above in $\mathbb{R}^N_+$.*

*(ii) If the singularity at 0 is removable, then $u \equiv 0$.*

*Proof.* Set $\mathbb{R}^N_+ = \{x = (x', x_N) : x_N > 0\} = \{(r, \sigma) : r > 0, \sigma \in S^{N-1}_+\}$. If the singularity at 0 is removable then the function $\tilde{u}$ which extends $u$ by reflexion through $\partial\mathbb{R}^N_+$ is a solution of (3.4.1). By the gradient estimate (3.4.32) it satisfies

$$|\nabla\tilde{u}(x)| \leq c_{N,p,q} \, |x - y|^{-\frac{1}{q+1-p}} \quad \forall x, y \in \mathbb{R}^N. \tag{3.5.63}$$

Letting $|y| \to \infty$ yields $\nabla\tilde{u}(x) = 0$, thus $\tilde{u}$ is constant and equal to 0.

Next we assume that the singularity at 0 is not removable; hence

$$|\nabla u(r, \sigma)| \leq c_{N,p,q} r^{-\frac{1}{q+1-p}} \quad \forall r > 0. \tag{3.5.64}$$

Therefore, for all $0 < s < r$,

$$\begin{array}{lll} (i) & |u(r,\sigma) - u(s,\sigma)| \leq c'_{N,p,q} \left(s^{-\gamma_{p,q}} - r^{-\gamma_{p,q}}\right) & \forall \sigma \in S^{N-1}_+ \\ (ii) & |u(r,\sigma) - u(r,\sigma')| \leq \pi c'_{N,p,q} r^{-\gamma_{p,q}} & \forall \sigma, \sigma' \in S^{N-1}_+. \end{array} \tag{3.5.65}$$

Consequently there exists $m = \lim_{|x|\to\infty} u(x)$ and we have

$$|u(x)| \leq m + c'_{N,p,q} |x|^{-\gamma_{p,q}} \qquad \forall x \in \mathbb{R}^N_+. \tag{3.5.66}$$

If we assume that (3.5.62) holds, then for any $\epsilon > 0$ there exists $\delta_\epsilon > 0$ such that for all $\delta \in (0, \delta_\epsilon]$,

$$|u(x)| \leq \epsilon \, |x|^{-\beta_{S^{N-1}_+}} \quad \forall x \in B^+_{\delta_\epsilon}.$$

Using Lemma 3.3.2 and the fact that $u$ vanishes on $\partial\mathbb{R}^N_+ \setminus \{0\}$ we derive

$$|u(x)| \leq c\epsilon \frac{x_N}{|x|^{\beta_{S^{N-1}_+}+1}} \leq c'\epsilon V_{R^N_+}(x) \qquad \forall x \in B^+_{\frac{\delta_\epsilon}{2}}. \tag{3.5.67}$$

Since $c'\epsilon V_{R^N_+} + m + \epsilon$ is a supersolution of (3.4.1) in $\mathbb{R}^N_+$ larger than $u$ for $|x| = \epsilon$ and $|x| = R$ large enough, we derive by the comparison principle that $u \leq c'\epsilon V_{R^N_+} + m + \epsilon$. Letting $\epsilon \to 0$, $\delta_\epsilon \to 0$ and $R \to \infty$ yields $u \leq m$.
□

**Corollary 3.5.13** *Assume $0 < p - 1$ and $u \in C^1(\mathbb{R}^N_+) \cap C(\overline{\mathbb{R}}^N_+ \setminus \{0\})$ is a nonnegative solution of (3.4.1) in $\mathbb{R}^N_+$ which vanishes on $\partial\mathbb{R}^N_+ \setminus \{0\}$.*

*I- If $p > q \geq \tilde{q}_c$, then $u \equiv 0$.*

*II- If* $p - 1 < q < \tilde{q}_c$ *and* $u(x) \to 0$ *when* $x \to \infty$, *then*
*(i) either* $u \equiv 0$
*(ii) either there exists* $k > 0$ *such that* $u = w_{k,\mathbb{R}^N_+}$,
*(iii) or* $u = W_{\mathbb{R}^N_+}$.

*Proof.* The first case follows from Theorem 3.5.2 and Proposition 3.5.12 (i). In the second case either (3.5.43) or (3.5.44) holds. If (3.5.44) holds then for any $\epsilon > 0$ $u$ is dominated near $x = 0$, at infinity and on the boundary by the supersolution $(1+\epsilon)W_{\mathbb{R}^N_+} + \epsilon$ (we recall that $W_{\mathbb{R}^N_+}(r,\sigma) = r^{-\gamma_{p,q}}\omega_h(\sigma)$). Then

$$u \le (1+\epsilon)W_{\mathbb{R}^N_+} + \epsilon \Longrightarrow u \le W_{\mathbb{R}^N_+}.$$

Similarly $u_\epsilon = (1+\epsilon)u + \epsilon$ is a supersolution which dominates $W_{\mathbb{R}^N_+}$ both at zero and at infinity. Therefore $u_\epsilon(r,\sigma) \ge W_{\mathbb{R}^N_+}$ in $\mathbb{R}^N_+$ and finally $u = W_{\mathbb{R}^N_+}$. In the case where (3.5.43) holds, then for any $\delta > 0$ we construct the function $w_{k,B^+_\delta}$ (see the definition in the proof of Theorem 3.5.8). Then $u \ge w_{k,B^+_\delta}$ which implies $u \ge w_{k,\mathbb{R}^N_+}$ by letting $\delta \to \infty$. On the other hand, since $u(x) \to 0$ when $|x| \to \infty$, there holds

$$u(x) \le w_{k,B^+_\delta}(x) + \max\{u(x) : |x| = \delta\}.$$

This implies $u \le w_{k,\mathbb{R}^N_+}$, which ends the proof. □

## 3.6 Notes and open problems

*3.6.1.* It is not known if Theorem 3.1.13 is valid in any dimension $N$, *i.e.* if the connected component $\mathcal{H}_\ell$ of $\mathcal{H}$ in (3.1.90) is a singleton $\omega$. Since (3.1.99) is valid in any dimension provided $-\frac{4}{q-1}$ be replaced by $N - \frac{2q+2}{q-1}$, it is not known if one can connect two elements of $\mathcal{H}$ with different energy $\mathcal{E}$, except in some trivial cases. It is not known if the energy $\mathcal{E}$ is quantized on $\mathcal{H}$ (one value on each connected component of $\mathcal{H}$, all different).

*3.6.2.* A necessary and sufficient conditions for the existence of large solutions to

$$-\Delta u + u^2 = 0 \qquad \text{in } \Omega \tag{3.6.1}$$

is obtained by [Dhersin, Le Gall (1997)] using tools from probability theory. Later on [Labutin (2003)] extends to the equation

$$-\Delta u + u^q = 0 \qquad \text{in } \Omega \tag{3.6.2}$$

with any $q > 1$. A local version of Labutin's conditions is given in [Marcus, Véron (2009)]. Set $F_m(x) = \{y \in \Omega^c : 2^{-(m+1)} \leq |x-y| \leq 2^{-m}\}$ for $x \in \mathbb{R}^N$ and $m \in \mathbb{Z}$, and define $W_{\Omega^c}$ by

$$W_{\Omega^c} = \sum_{-\infty}^{\infty} 2^{\frac{2n}{q-1}} C_{2,q'}\left(2^n F_n(x)\right), \tag{3.6.3}$$

where $C_{2,q'}$ is the Sobolev capacity. There is equivalence between

(i) $\overline{u}_\Omega(x) \to \infty$ as $x \to y \in \partial\Omega$,

(ii) $W_{\Omega^c}(y) \to \infty$ as $x \to y \in \partial\Omega$.

Furthermore any large solution $u$ satisfies

$$c^{-1} W_{\Omega^c}(x) \leq u(x) \leq c W_{\Omega^c}(x). \tag{3.6.4}$$

for some constant $c = c(\Omega, q) > 1$. This inequality implies uniqueness.

*3.6.3.* The local continuous graph property (see Definition 3.1.28) can be viewed as a "local star-shapedness" property with respect to points at infinity. Since uniqueness of large solutions of (3.1.17) holds when a domain is star-shaped, provided $g$ satisfies (3.1.115), we conjecture that this uniqueness holds if $\partial\Omega$ has this local graph property.

*3.6.4.* A natural problem is to consider boundary diffusion operators associated to the $p$-Laplacian. Assuming that $\Omega$ is a smooth domain and $0 \in \partial\Omega$ an interesting problem the study of properties of solutions of

$$\begin{aligned} -\Delta_p u &= 0 && \text{in } \Omega \\ |\nabla u|^{p-2} \frac{\partial u}{\partial \mathbf{n}} + |u|^{q-1} u &= 0 && \text{in } \partial\Omega \setminus \{0\}. \end{aligned} \tag{3.6.5}$$

The existence of separable solutions when $\Omega = \mathbb{R}^N_+$ and $q > p-1$ leads to interesting new quasilinear problems on $S^{N-1}_+$ with nonlinear diffusion on $\partial S^{N-1}_+$, see Note 2.6.10 in Section 2.

*3.6.5.* The study of signed solutions of (3.1.2) on $S^{N-1}$, or (3.2.3) on $S^{N-1}_+$ appears to be completely open. It is not clear whether there exists sign changing solutions whenever the exponent $q$ is below the corresponding critical exponent, even if we are unable to construct such solutions.

*3.6.6.* When the domain $\Omega$ is piecewise smooth, it is natural to study boundary singularities of solutions concentrated at the irregular parts of the boundary. It is likely that in the case of conical boundary points a large part of the previous constructions of solutions and characterization of behaviour can be performed. See [Borsuk, Kondratiev (2007)].

# Chapter 4

# Quasilinear equations with measure data

## 4.1 Equations with measure data: the framework

Let $\Omega \subset \mathbb{R}^N$ be a domain and $A \in C(\Omega \times \mathbb{R}^N) \mapsto \mathbb{R}^N$ a Caratheodory vector field which satisfies for some $p > 1$ and any $x \in \Omega$ and $\xi, \xi' \in \mathbb{R}^N$,

$$
\begin{array}{lll}
(i) & \langle A(x,\xi).\xi\rangle \geq |\xi|^p & \\
(ii) & |A(x,\xi)| \leq \lambda |\xi|^{p-1} & \qquad (4.1.1)\\
(iii) & \langle A(x,\xi) - A(x,\xi').\xi - \xi'\rangle > 0 \quad \text{if } \xi \neq \xi'. &
\end{array}
$$

We denote by $\mathfrak{M}(\Omega)$ (resp. $\mathfrak{M}_b(\Omega)$ the space of Radon measures on $\Omega$ (resp bounded Radon measure) and by $\mathfrak{M}^+(\Omega)$ (resp. $\mathfrak{M}_b^+(\Omega)$) its positive cone. Any nonnegative Radon measure is extended as a Borel measure with the same notation. If $\mu \in \mathfrak{M}(\Omega)$, we consider the following problem

$$-div\, A(x, \nabla u) = \mu \quad \text{in } \Omega. \qquad (4.1.2)$$

If one wants to define a "local" solution of (4.1.2), the most natural way is to use the weak formulation, that is to look for functions satisfying

$$\int_\Omega \langle A(x, \nabla u), \nabla\zeta\rangle dx = \int_\Omega \zeta d\mu \quad \forall \zeta \in C_0^1(\Omega). \qquad (4.1.3)$$

The validity of this expression needs $|\nabla u|^{p-1} \in L^1_{loc}(\Omega)$, but the difficulty comes from the fact that is $1 < p < 2$, $\nabla u$, in the sense of distributions, cannot be defined as a function of $L^1_{loc}(\Omega)$. For this reason two different approaches have been developed to solve (4.1.2) and the problems associated, the method of $p$-superharmonic functions and the method of renormalized solutions. The first paragraphs are devoted to present some important aspects of the two theories and to prove some of them for the sake of completeness.

In the sequel we use the term of $p$-harmonic, $p$-superharmonic and $p$-subharmonic functions when dealing with functions satisfying, in some sense,

$$-div\, A(x,\nabla u)=0\,, \quad -div\, A(x,\nabla u)\geq 0\,, \quad -div\, A(x,\nabla u)\leq 0 \quad (4.1.4)$$

with $A$ as above. We will return to the mere $p$-Laplacian when necessary.

## 4.2 $p$-superharmonic functions

### 4.2.1 *The notion of p-superharmonicity*

The following definition of $p$-superharmonic functions associated to classical potential theory, complements the notion of weakly $p$-superharmonic functions introduced in Chapter 1.

**Definition 4.1** A lower semicontinuous function $u$ defined in $\Omega$ is $p$-superharmonic with value in $\mathbb{R}\cup\{\infty\}$ and not identically equal to $\infty$, *i.e.* proper, if for any bounded subdomain $D$ of $\Omega$ such that $\overline{D}\subset\Omega$ and any $h\in C(\overline{D})\cap W^{1,p}(D)$, $u\geq h$ on $\partial D$ implies $u\geq h$ in $D$. Similarly, An upper semicontinuous function $u:\Omega\mapsto\mathbb{R}\cup\{-\infty\}$ is $p$-subharmonic if $-u$ is $p$-superharmonic.

*Remark.* I- If $u$ is $p$-superharmonic (resp. $p$-subharmonic) and finite a.e. in $\Omega$, then it is locally bounded from below (resp. from above) in $\Omega$. As a consequence, for any $k\in\mathbb{R}$, $\min\{k,u\}\in L^1_{loc}(\Omega)$ (resp. $\max\{k,u\}\in L^1_{loc}(\Omega)$). This property, easy consequence of semicontinuity, is shared by weakly $p$-superharmonic functions (resp. weakly $p$-subharmonic functions) as a consequence of weak Harnack inequality (see Theorem 1.3.1).

II- A measurable function $u$ admits a lower semicontinuous representative, still denoted by $u$, if

$$u(x)=\operatorname{ess}\liminf_{\mathrm{y}\to\mathrm{x}}\mathrm{u(y)}, \quad (4.2.1)$$

for almost all $x\in\Omega$. An important property of $p$-superharmonic functions is that (4.2.1) holds for all $x\in\Omega$.

The $c^{\Omega}_{1,p}$-capacity (Sobolev capacity), has a key role in the study of equations with measure data. It has been introduced in Section 1-4.

*Remark.* We have seen in Section 1-4 that if $\mu\in W^{-1,p'}(\Omega)=(W^{1,p}_0(\Omega))'$

there exists a unique $u_\mu \in W_0^{1,p}(\Omega)$ such that

$$\int_\Omega \langle A(x\nabla u), \nabla\zeta\rangle dx = \mu(\zeta) \qquad \forall \zeta \in W_0^{1,p}(\Omega). \tag{4.2.2}$$

If $\mu \in W^{-1,p'}(\Omega) \cap \mathfrak{M}^+(\Omega)$, then the mapping

$$\zeta \mapsto \int_\Omega \zeta d\mu \qquad \forall \zeta \in W_0^{1,p}(\Omega),$$

is an element of $W^{-1,p'}(\Omega)$. Furthermore $u_\mu$ is nonnegative and it is the weak solution of the Dirichlet problem

$$\begin{aligned} -div\, A(x, \nabla u) &= \mu && \text{in } \Omega \\ u &= 0 && \text{on } \Omega. \end{aligned} \tag{4.2.3}$$

This allows to define a dual $\tilde{c}_{1,p}^{\Omega}$-capacity by

$$\tilde{c}_{\mathcal{A}}^{\Omega}(K) = \sup\left\{\mu(K) : \mu \in W^{-1,p'}(\Omega) \cap \mathfrak{M}^+(\Omega), \mu(K^c) = 0, u_\mu \leq 1\right\}. \tag{4.2.4}$$

This set function is extended to general set by the rules (1.2.58) and (1.2.59), and although we need not go into the abstract definition of capacities (see [Choquet (1953-54)]), it is a capacity. Furthermore it is equivalent to the $c_{1,p}^{\Omega}$-capacity.

**Proposition 4.2.1** *For any Borel set $E \subset \Omega$, there holds*

$$c_{1,p}^{\Omega}(E) \leq \tilde{c}_{\mathcal{A}}^{\Omega}(E) \leq q\lambda^p c_{1,p}^{\Omega}(E). \tag{4.2.5}$$

*Proof.* If $\mathcal{A} = -\Delta_p$, we denote $\tilde{c}_{-\Delta_p}^{\Omega} = \tilde{c}_{1,p}^{\Omega}$. The set function defined for a compact set $K \subset \Omega$ by

$$c_{\mathcal{A}}^{\Omega}(K) = \inf\left\{\int_\Omega \langle A(x, \nabla\zeta), \nabla\zeta\rangle dx : \zeta \in C_0^\infty(\Omega) : \zeta \geq 1 \text{ in a neigh. of } K\right\},$$

is a capacity equivalent to the $c_{1,p}^{\Omega}$-capacity in the sense that

$$c_{1,p}^{\Omega}(K) \leq c_{\mathcal{A}}^{\Omega}(K) \leq \lambda^p c_{1,p}^{\Omega}(K).$$

The equality between $c_{1,p}^{\Omega}(K)$ and $\tilde{c}_{-\Delta_p}^{\Omega}(K)$ is a consequence of minmax theorem (see[Adams, Hedberg (1999), Th 2.5.1]). □

The following result is fundamental insofar it connects the notions of $p$-superharmonicity and weak $p$-superharmonicity.

**Proposition 4.2.2** *Let $\Omega \subset \mathbb{R}^N$ be a domain and $u$ a finite a.e. measurable function defined in $\Omega$. Then $u$ admits a $p$-superharmonic representative if and only if for any $k > 0$, $u_k = \min\{u, k\}$ is weakly $p$-superharmonic, that is $u_k \in W^{1,p}_{loc}(\Omega)$ and*

$$\int_\Omega \langle A(x, \nabla u_k), \nabla \zeta \rangle dx \geq 0 \qquad \forall \zeta \in C^1_0(\Omega),\ \zeta \geq 0. \tag{4.2.6}$$

*Proof.* We just give a sketch of the proof the details of which can be found in [Heinonen, Kilpelainen and Martio (2006)].

If $u_k$ is a supersolution, it is locally bounded from below by Theorem 1.3.1. Let $D \subset \overline{D} \subset \Omega$ be a bounded subdomain of $\Omega$ and $h \in C(\overline{D}) \cap W^{1,p}(D)$ such that $u_k \geq h$ on $\partial D$. Then for any $\epsilon > 0$, $(h - u_k - \epsilon)_+$ belongs to $W^{1,p}_0(D)$ and by the maximum principle it is identically 0. Letting $\epsilon \to 0$ yields $h \leq u_k \leq u$. Furthermore, since $u_k \in W^{1,p}_{loc}(\Omega)$, it admits a quasicontinuous representative, which means that $u_k$ coincides with a continuous function, up to a set $E_k$ of zero $c^{\Omega}_{1,p}$-capacity. The key point is to prove that $u_k$ coincides a.e. with a function, still denoted by $u_k$, which satisfies

$$u_k(x) = \operatorname{ess} \liminf_{\mathrm{y} \to \mathrm{x}} \mathrm{u_k(y)} \qquad \forall \mathrm{x} \in \Omega. \tag{4.2.7}$$

This property of supersolutions of (4.1.4) is a consequence of the weak Harnack inequality (see [Heinonen, Kilpelainen and Martio (2006), Th 3.63]). Since $u_k \uparrow u$, $u$ admits a semicontinuous representative, up to a set with zero $c^{\Omega}_{1,p}$-capacity and it is therefore $p$-superharmonic.

Assume now that $u$ is $p$-superharmonic. It is easy to see that for any $k$, $u_k$ is $p$-superharmonic and locally bounded. For $k > 0$ fixed and a bounded smooth domain $D \subset \overline{D} \subset \Omega$, there exists an increasing sequence $\{\phi_{j,k}\} \subset C^\infty_0(\mathbb{R}^N)$ converging to $u_k$ in $\overline{D}$. Define the nonempty and closed convex set of constraints

$$\mathcal{K}_{\phi_{j,k}} := \{w \in W^{1,p}(D) : w \geq \phi_{j,k} \text{ in } D,\ w - \phi_{j,k} \in W^{1,p}_0(D)\}, \tag{4.2.8}$$

and consider the following obstacle problem

$$\text{Find } v \in \mathcal{K}_{\phi_{j,k}} \text{ s.t. } \int_D \langle A(x, \nabla v), \nabla w - \nabla v) \rangle dx \geq 0 \qquad \forall w \in \mathcal{K}_{\phi_{j,k}}. \tag{4.2.9}$$

The solution $u_{j,k}$ exists and it is unique thanks to (4.1.1) (iii). Moreover it is a continuous function in $\overline{D}$ as the obstacle is; this property, far from obvious, is proved in [Heinonen, Kilpelainen and Martio (2006), Th 3.67]).

Furthermore since $u_{j,k} + \zeta \in \mathcal{K}_{\phi_{j,k}}$ for any $\zeta \in C_0^\infty(D)$, $\zeta \geq 0$, $u_{j,k}$ is weakly $p$-superharmonic in $D$ and thus $p$-superharmonic. Next let $\mathcal{U}_j$ be a connected component of the open set $\mathcal{O} := \{x \in D : u_{j,k}(x) > \phi_{j,k}(x)\}$. Then $u_{j,k}$ is $p$-harmonic in $\mathcal{U}_j$ and coincides with $\phi_{j,k}$ on $\partial\mathcal{U}_j$. Hence, if $y \in \partial\mathcal{U}_j$, we have

$$\liminf_{x\to y} u_k(x) \geq u_k(y) \geq \phi_{j,k}(y) = \lim_{x\to y} u_{j,k}(x). \tag{4.2.10}$$

By the property of $p$-superharmonic functions, it implies $u_k \geq u_{j,k}$ in $\mathcal{U}_j$ and thus in $D$. Therefore

$$u_k = \lim_{j\to\infty} \phi_{j,k} \leq \lim_{j\to\infty} u_{j,k} \leq u_k. \tag{4.2.11}$$

Since the $u_{j,k}$ are continuous, $u_k$ is lower semicontinuous. Since $u_{j,k}$ is weakly $p$-superharmonic, we take $\zeta = \eta^p(k - u_{j,k})$ as a test function where $\eta \in C_0^1(D)$ is such that $0 \leq \eta \leq 1$. Because $\zeta \geq 0$ and $u_{j,k} \leq u_k \leq k$, then

$$p\int_D \langle A(\nabla u_{j,k}), \nabla\eta\rangle\eta^{p-1}(k - u_{j,k})dx \geq \int_D \langle A(\nabla u_{j,k}), \nabla u_{j,k}\rangle\eta^p dx.$$

The function $u$ is locally bounded from below since it is lower semicontinuous, therefore $u_k$ is also locally bounded from below and thus the functions $\phi_{j,k}$, which are smooth and form an increasing sequence, are bounded from below on $\overline{D}$, uniformly with respect to $j$. Thus $u_{j,k}\lfloor_D \geq c^*$. Using (4.1.1) (i)-(ii) we obtain

$$\begin{aligned}\int_D |\nabla u_{j,k}|^p\, \eta^p dx &\leq \lambda p\int_D |\nabla u_{j,k}|^{p-1}\, \eta^{p-1}\, |\nabla\eta|\, (k - u_{j,k}) \\ &\leq \lambda p(k - c^*) \left(\int_D |\nabla u_{j,k}|^p\, \eta^p dx\right)^{1-\frac{1}{p}} \left(\int_D |\nabla\eta|^p\, dx\right)^{\frac{1}{p}}.\end{aligned}$$

This implies that $\eta\nabla u_{j,k}$ is bounded in $L^p(D)$ independently of $j$. Therefore $\eta\nabla u_k$ is also bounded in $L^p(D)$. As a consequence $u_k \in W^{1,p}_{loc}(\Omega)$. □

**Definition 4.2** The *very weak gradient* of a $p$-superharmonic function $u$ in $\Omega$ is defined by

$$\nabla u = \lim_{k\to\infty} \nabla\left(\min\{u, k\}\right). \tag{4.2.12}$$

The fact that for almost all $x \in \Omega$ there holds $\nabla\left(\min\{u, k\}\right) = \nabla\left(\min\{u, j\}\right)$ a.e. on the set $\{z \in \Omega : u(z) \leq j\}$ for any $k \geq j$, implies that for almost all $x \in \Omega$ the sequence $\{\nabla\left(\min\{u(x), k\}\right)\}_k$ is eventually constant. However this very weak gradient presents some important integrability properties proved in [Heinonen, Kilpelainen and Martio (2006), Th 7.46] and [Kilpeläinen, Li (2000)].

**Proposition 4.2.3** *Let $u$ be a $p$-superharmonic function in a domain $\Omega \subset \mathbb{R}^N$. Then*

*(i) If $1 < p < N$, $u \in L_{loc}^{\frac{N(p-1)}{N-p},\infty}(\Omega)$ and $|\nabla u|^{p-1} \in L_{loc}^{\frac{N}{N-1},\infty}(\Omega)$.*

*(ii) If $p > 2 - \frac{1}{N}$, $\nabla u$ is the distributional gradient.*

*(iii) If $p = N$, $u \in L_{loc}^{s}(\Omega)$ for $s > 0$ and $|\nabla u|^{N-1} \in L_{loc}^{\frac{N}{N-1},\infty}(\Omega)$.*

*(iv) If $p > N$, $u \in W_{loc}^{1,p}(\Omega)$.*

This result implies in particular that $A(., \nabla u) \in L_{loc}^{1}(\Omega)$. Hence $-div\, \mathcal{A}(\nabla u)$ is defined as a distribution in $\Omega$ by the identity

$$-div\, \mathcal{A}(\nabla u)(\zeta) = \int_D \langle A(x, \nabla u), \nabla \zeta \rangle dx \qquad \forall \zeta \in C_0^\infty(\Omega). \tag{4.2.13}$$

### 4.2.2 *The p-harmonic measure*

The next statement is at the core of the theory of $p$-superharmonic function.

**Proposition 4.2.4** *Let $u$ be a $p$-superharmonic function in $\Omega$. Then there exists a unique nonnegative Radon measure $\mu := \mu[u]$ such that*

$$-div\, \mathcal{A}(\nabla u) = \mu \qquad in\ \mathcal{D}'(\Omega). \tag{4.2.14}$$

*Proof.* By Proposition 4.2.2 $u_k \in W_{loc}^{1,p}(\Omega)$ is weakly $p$-superharmonic. Let $\zeta \in C_0^\infty(\Omega)$, $\zeta \geq 0$, then

$$\int_D \langle A(x, \nabla u_k), \nabla \zeta \rangle dx \geq 0.$$

Since $\nabla u_k(x) \to \nabla u(x)$ a.e. in $\Omega$ when $k \to \infty$ and $|\nabla u_k| \leq |\nabla u|$, it follows from the estimates of $\nabla u$ stated in Proposition 4.2.3 and Lebesgue dominated convergence theorem that

$$\int_D \langle A(x, \nabla u), \nabla \zeta \rangle dx = \lim_{k \to \infty} \int_D \langle A(x, \nabla u_k), \nabla \zeta \rangle dx \geq 0.$$

Thus $-div\, \mathcal{A}(\nabla u)$ is a positive distribution which can be extended by a nonnegative Radon measure by Riesz representation theorem. This measure, denoted by $\mu[u]$, is unique and called *the p-harmonic measure* of $u$. □

The following kernel plays a key role comparable to the Green kernel in the study of Poisson equation.

**Definition 4.3** Let $\mu$ be a nonnegative Borel measure on the domain $\Omega \subset \mathbb{R}^N$ and $1 < p < N$. Then the *Wolff potential* of the measure $\mu$ denoted by $\mathbf{W}_{1,p}[\mu]$ is the function defined in $\{(x,r) \in \Omega \times (0,\infty) : \overline{B}_r(x) \subset \Omega\}$ by

$$\mathbf{W}_{1,p}[\mu](x,r) = \int_0^r \left(\frac{\mu(B_s(x))}{s^{N-p}}\right)^{\frac{1}{p-1}} \frac{ds}{s}. \tag{4.2.15}$$

The fundamental estimate, which is the *deus ex machina* of the study of degenerate quasilinear equations with measure data, is the following estimate [Heinonen, Kilpelainen and Martio (2006)], [Kilpeläinen, Malý (1994)].

**Theorem 4.2.5** *Assume $1 < p \leq N$ and $u$ is a $p$-superharmonic function in some domain $\Omega$, with the $p$-harmonic measure $\mu$, then there exists $c = c(N,p,\lambda)$ such that for any $x \in \Omega$ and any $r > 0$ such that $\overline{B}_r(x) \subset \Omega$,*

$$\frac{1}{c}\mathbf{W}_{1,p}[\mu](x,r) \leq u(x) \leq c\left(\mathbf{W}_{\mu}^{1,p}(x,r) + \operatorname{ess}\inf_{\mathrm{z}\in \mathrm{B_r(x)}} \mathrm{u(z)}\right). \tag{4.2.16}$$

The following statement is a particular case of very general decomposition theorems of Radon measures with respect to a capacity (see [Fukushima, Sato, Taniguchi (1991)], [Boccardo, Gallouët, Orsina (1996)]).

**Proposition 4.2.6** *1- Let $\mu$ be a Radon measure on $\mathbb{R}^N$. Then*

$$\mu = \mu_0 + \mu_s^+ - \mu_s^-, \tag{4.2.17}$$

*where $\mu_0$, $\mu_s^+$ and $\mu_s^-$ are Radon measures, $\mu_s^+, \mu_s^- \geq 0$, $\mu_0$ satisfying $\mu_0(E) = 0$ for every Borel set $E \subset \mathbb{R}^N$ with $c_{1,p}^{\mathbb{R}^N}(E) = 0$ and $\mu_s^\pm$ with support on a Borel set $F$ such that $c_{1,p}^{\mathbb{R}^N}(F) = 0$. The measure $\mu_0$ is the absolutely continuous part of $\mu$ with respect to $c_{1,p}^{\mathbb{R}^N}$ and we write $\mu_0 \ll c_{1,p}^{\Omega}$. The measures $\mu_s^\pm$ are the singular parts of $\mu$ with respect to $c_{1,p}^{\Omega}$ and we write $\mu_s^\pm \perp c_{1,p}^{\mathbb{R}^N}$. This decomposition is unique.*

*2- Let $\mu_0$ be a Radon measure on $\mathbb{R}^N$ absolutely continuous with respect to $c_{1,p}^{\mathbb{R}^N}$. Then there exist $f \in L^1_{loc}(\mathbb{R}^N)$ and $g \in (L^{p'}_{loc}(\mathbb{R}^N))^N$ such that $\mu_0 = f - \operatorname{div} g$ in the sense that*

$$\int_{\mathbb{R}^N} \zeta d\mu_0 = \int_{\mathbb{R}^N} \zeta f dx - \int_{\mathbb{R}^N} \langle g, \nabla\zeta\rangle dx \qquad \forall \zeta \in C_0^1(\mathbb{R}^N). \tag{4.2.18}$$

*This decomposition is not unique since $L^1(\mathbb{R}^N) \cap W^{-1,p'}(\mathbb{R}^N) \neq \{0\}$. Furthermore $f$ and $g$ can be chosen such that*

$$\|f\|_{L^1} + \|\mathrm{div}\, g\|_{L^{p'}} + \|\mathrm{div}\, g\|_{\mathfrak{M}} \leq 5 \|\mu_0\|_{\mathfrak{M}}. \tag{4.2.19}$$

The following result shows that the singular part of a $p$-harmonic measure is concentrated on the set where $u$ is infinite.

**Proposition 4.2.7** *Let $u$ be a $p$-superharmonic function in some domain $\Omega$ with $p$-harmonic measure $\mu = \mu_0 + \mu_s$. Then*

$$\mu_s(\{x \in \Omega : u(x) < \infty\}) = 0. \tag{4.2.20}$$

The proof of this result is a consequence of a remarkable relation proved in [Hedberg, Wolff (1983)] which connects integrability property of the Wolff potential of a measure $\mu$ with the belonging of this measure to the dual of $W_0^{1,p}(\Omega)$.

$$\mu \in \mathfrak{M}_b^+(\Omega) \cap W^{-1,p'}(\Omega) \Longleftrightarrow \int_\Omega \mathbf{W}_{1,p}[\mu](x) d\mu(x) < \infty. \tag{4.2.21}$$

This is obtained by a series of inequalities linking Wolff potential and Bessel capacities. These arguments will be developed in next sections.

### 4.2.3 *The Peron's method*

One of the main achievements of the framework of $p$-superharmonic functions is the extension of Perron's method from the linear case to the nonlinear case to solve *completely* the Dirichlet problem for $\mathcal{A}$

$$\begin{aligned} -div\, A(x, \nabla u) &= 0 \qquad \text{in } \Omega \\ u &= f \qquad \text{on } \partial\Omega, \end{aligned} \tag{4.2.22}$$

when $\Omega \subset \mathbb{R}^N$ is a domain and $f : \partial\Omega \mapsto \mathbb{R}$ function. The classical Dirichlet problem is easily solvable by the monotonicity method.

**Proposition 4.2.8** *Assume $\Omega$ is bounded and $f$ is the restriction to $\partial\Omega$ of a function belonging to $W^{1,p}(\Omega)$ still denoted by $f$. Then there exists a unique $u$ such that $u - f \in W_0^{1,p}(\Omega)$ satisfying (4.2.22).*

*Proof.* We define the vector field

$$A_f(x, \xi) = A(x, \xi + \nabla f(x)) \qquad \text{a.e. in } \Omega \tag{4.2.23}$$

and the operator $div\,\mathcal{A}_f$ defined on $W_0^{1,p}(\Omega)$ by

$$div\,\mathcal{A}_f(v) = div\,A(x, \nabla u + \nabla f). \tag{4.2.24}$$

Then, by (4.1.1) (i)-(ii),

$$\begin{aligned}(div\,\mathcal{A}_f(v), v) &= \int_\Omega \langle A(x, \nabla v + \nabla f), \nabla v \rangle dx \\ &\geq \int_\Omega \nabla (v+f)^p dx - \lambda \int_\Omega |\nabla (v+f)|^{p-1} |\nabla f|\, dx \\ &\geq \frac{1}{2}\int_\Omega \nabla v^p dx - \frac{c}{2}\int_\Omega \nabla f^p dx,\end{aligned}$$

for some constant $c = c(p, \lambda) > 0$. This implies that $div\,\mathcal{A}_f$ is coercive in $W_0^{1,p}(\Omega)$. By (4.1.1) (iii), it is strictly monotone. It follows from Theorem 1.4.1 that there exists a unique $\tilde{v} \in W_0^{1,p}(\Omega)$ such that $div\,\mathcal{A}_f(u) = 0$. Then $u = \tilde{v} + f$ is the unique solution of (4.2.22) in $W_0^{1,p}(\Omega)$. □

The statements below are proved in [Heinonen, Kilpelainen and Martio (2006), Chap 9]. For the sake of completeness we present here the main ideas.

**Definition 4.4** Let $f : \partial\Omega \mapsto \mathbb{R} \cup \{-\infty\} \cup \{\infty\}$. The upper class $\mathcal{U}_f$ (resp. lower class $\mathcal{L}_f$) is the set of all functions $u$ defined in $\Omega$ such that

(i) $u$ is $p$-superharmonic in $\Omega$ (resp. $u$ is $p$-subharmonic in $\Omega$),

(ii) $u$ is bounded from below ($u$ is bounded from above),

(iii) there holds $\liminf_{x \to y} u(x) \geq f(y)$ (resp. $\limsup_{x \to y} u(x) \leq f(y)$) $\forall y \in \partial\Omega$.

The Perron solutions of (4.2.22) can be defined as follows.

**Definition 4.5** The function

$$\overline{H}_f = \inf\{u \in \mathcal{U}_f\} \quad (\text{resp. } \underline{H}_f = \sup\{u \in \mathcal{L}_f\}), \tag{4.2.25}$$

is the upper Perron solution (resp. the lower Perron solution) of (4.2.22).

The following result is fundamental in the construction of a solution to the Dirichlet problem (4.2.22).

**Theorem 4.2.9** *Let $f = \partial\Omega \mapsto \mathbb{R} \cup \{-\infty\} \cup \{\infty\}$ and $p > 1$. Then the following dichotomy occurs:*

*(i) either $\overline{H}_f$ is p-harmonic in $\Omega$,*

*(ii) either $\overline{H}_f \equiv \infty$ in $\Omega$,*

*(iii) either $\overline{H}_f \equiv -\infty$ in $\Omega$.*

The main question is whether $\overline{H}_f$ (or $\underline{H}_f$) achieves the value $f$ on $\partial\Omega$. Here comes the notion of *regular boundary points.*

**Definition 4.6** Let $\Omega$ be a bounded domain and $1 < p \leq N$. If $\theta \in C(\overline{\Omega}) \cap W^{1,p}(\Omega)$ we denote by $u_\theta$ the unique $p$-harmonic function in $\Omega$ such that $u_\theta - \theta \in W_0^{1,p}(\Omega)$, which exists by Proposition 4.2.8. A boundary point $y$ is called regular if

$$\lim_{x \to y} u_\theta(x) = \theta(y). \tag{4.2.26}$$

The regularity is a local property since there holds,

**Lemma 4.2.10** *A boundary point $y \in \partial\Omega$ is regular if and only if*

$$\lim_{x \to y} \overline{H}_f(x) = f(y), \tag{4.2.27}$$

*for each $f : \Omega \mapsto \mathbb{R}$, bounded and continuous at $y$.*

*Proof.* If (4.2.27) holds then $u_\theta = \overline{H}_\theta(x)$, for any $\theta \in W^{1,p}(\Omega) \cap C(\overline{\Omega})$, thus (4.2.25) holds and $y$ is regular. Conversely, if $y$ is regular and $f : \Omega \mapsto \mathbb{R}$, is bounded and continuous at $y$, then for given $\epsilon > 0$ there exists $\delta > 0$ such that $|f(x) - f(y)| \leq \epsilon$ for all $x \in \partial\Omega$ verifying $|x - y| \leq \delta$. Let $g \in C(\overline{\Omega}) \cap W^{1,p}(\Omega)$ with values in $[f(y) + \epsilon, \sup|f| + \epsilon]$ and such that $g(y) = f(y) + \epsilon$ and $g(x) \equiv \sup|f| + \epsilon$ for all $x \in \Omega \cap B_\delta^c(y)$. Since $f \leq g\lfloor_{\partial\Omega}$, then $\overline{H}_f \leq \overline{H}_{g\lfloor_{\partial\Omega}}$ and in particular

$$\limsup_{x \to y} \overline{H}_f(x) \leq \limsup_{x \to y} \overline{H}_{g\lfloor_{\partial\Omega}}(x) = \lim_{x \to y} u_{g\lfloor_{\partial\Omega}}(x) = g(y) = f(y) + \epsilon.$$

In the same way

$$\liminf_{x \to y} \overline{H}_f(x) \geq \limsup_{x \to y} \overline{H}_{\tilde{g}\lfloor_{\partial\Omega}}(x) = \lim_{x \to y} u_{\tilde{g}\lfloor_{\partial\Omega}}(x) = \tilde{g}(y) = f(y) - \epsilon,$$

where $\tilde{g} \in C(\overline{\Omega}) \cap W^{1,p}(\Omega)$ has values in $[\inf|f| - \epsilon, f(y) - \epsilon]$ and is constant with value $\inf|f| - \epsilon$ on $\Omega \cap B_\delta^c(y)$. This implies (4.2.27). □

A natural way to prove that (4.2.27) holds at a boundary point $y$ is to use a special form of $p$-superharmonic function.

**Definition 4.7** Let $\Omega$ be a domain and $1 < p \leq N$. A function $u$ defined in $\Omega$ is called a *barrier relative to* $\Omega$ at the boundary point $y$ if it satisfies:

(i) $u$ is $p$-superharmonic in $\Omega$,

(ii) $\liminf_{x \to z} u(x) > 0$ for all $z \in \partial\Omega \setminus \{y\}$,

(iii) $\lim_{x \to y} u(x) = 0$.

The role of barrier is brought to light by the following result [Heinonen, Kilpelainen and Martio (2006), Theorem 9.8].

**Theorem 4.2.11** *Let $\Omega$ be a domain and $1 < p \leq N$. A boundary point $y$ is regular if and only if there exists a barrier relative to $\Omega$ at $y$.*

*Proof.* We just present the sufficiency of the condition since the necessity needs to use the notion of *balayage* that we have not introduced and which plays an important role in abstract potential theory [Brelot (1967)]. For simplicity we assume that $\Omega$ is bounded; actually this is not a restriction since the notion of regularity of a boundary point is a local notion. Let $u$ be a barrier; since $u$ is $p$-superharmonic in $\Omega$, it cannot achieve its minimum in $\Omega$, therefore it is positive. Let $f$ be a bounded function defined on $\partial\Omega$ and continuous at $y$. We can assume that $f(y) = 0$. For $\epsilon > 0$ there exists $\delta > 0$ such that $|f(x)| \leq \epsilon$ for any $x \in \partial\Omega \cap B_\delta(y)$. Let $\Lambda > \sup\{|f(z)| : z \in \partial\Omega\}$ and $v$ be defined by

$$v = \begin{cases} \Lambda & \text{in } \Omega \cap B_\delta^c(y) \\ \frac{\Lambda}{m} \min\{u, m\} & \text{in } \Omega \cap B_\delta(y), \end{cases}$$

where $m = \inf\{u(z) : z \in \Omega \cap \partial B_\delta(y)\}$. First, observe that $v$ in $p$-superharmonic in $\Omega$. Indeed, if $D$ is a bounded domain such that $\overline{D} \subset \Omega$ and $h \in C(\overline{D}) \cap W^{1,p}(D)$ is $p$-harmonic and smaller than $v$ on $\partial D$, then, for all $x \in \partial B_\delta(y) \cap D$,

$$\lim_{\substack{z \to x \\ z \in B_\delta(y) \cap D}} h(z) \leq \Lambda = v(x).$$

Next, if $x \in B_\delta(y) \cap \partial D$,

$$\lim_{\substack{z \to x \\ z \in B_\delta(y) \cap D}} h(z) = h(x) \leq v(x) \leq \liminf_{\substack{z \to x \\ z \in B_\delta(y) \cap D}} v(z).$$

Hence, for all $x \in \partial(B_\delta(y) \cap D)$,

$$\lim_{\substack{z \to x \\ z \in B_\delta(y) \cap D}} h(z) = h(x) \leq v(x) \leq \liminf_{\substack{z \to x \\ z \in B_\delta(y) \cap D}} v(z).$$

Since $v$ is $p$-superharmonic in $D$, $h \leq v$ in $B_\delta(y) \cap D$ and because $h \leq \Lambda$ on $\partial D$, the maximum principle implies that the inequality holds in $D$. Therefore $h \leq v$ in $D$. Furthermore, by the choice of $\Lambda$ and $\delta$,

$$\lim_{\substack{z \to x \\ z \in \Omega}} (v(z) + \epsilon) \geq f(z) \quad \forall z \in \partial\Omega.$$

Thus $v+\epsilon \in \mathcal{U}_f$. Since $v(z) \to 0$ when $z \to y$, it follows that

$$\limsup_{z \to y} \overline{H}_f(z) \leq \lim_{z \to y} (v(z)+\epsilon) = \epsilon.$$

Similarly $-v-\epsilon \in \mathcal{L}_f$, hence

$$\liminf_{z \to y} \overline{H}_f(z) \geq \liminf_{z \to y} \underline{H}_f(z) \geq \lim_{z \to y} (-v(z)-\epsilon) = -\epsilon.$$

Combining the last two inequalities yields

$$\lim_{z \to y} \overline{H}_f(z) = 0 = f(y).$$

□

The characterization of regular boundary points is expressed by a Wiener type test involving the $c_{1,p}$-capacity.

**Definition 4.8** Assume $1 < p \leq N$. A set $E \subset \mathbb{R}^N$ is thick (resp. thin) at $y$ if

$$\int_0^1 \left( \frac{c_{1,p}^{B_{2t}(y)}(E \cap B_t(y))}{t^{N-p}} \right)^{\frac{1}{p-1}} \frac{dt}{t} = \infty \qquad (\text{resp. } < \infty). \tag{4.2.28}$$

The next result characterizes regular points in terms of the geometry of $\partial\Omega$ (see [Maz'ya (1970)], [Kilpeläinen, Malý (1994)]).

**Theorem 4.2.12** *Assume $1 < p \leq N$ and $\Omega \subset \mathbb{R}^N$ is a domain. A boundary point $y$ is regular if and only if $\Omega^c$ is thick at $y$.*

*Remark.* A classical result of the theory of fine topology associated to a capacity [Adams, Hedberg (1999), Chap. 6] asserts that the set $e_{1,p}(E)$ of thin points of a set $E$ has zero capacity. Since the set of irregular boundary points points of $\Omega$ is included in $e_{1,p}(\Omega^c)$, it has zero $c_{1,p}^{\mathbb{R}^N}$-capacity.

## 4.3 Renormalized solutions

### 4.3.1 *Locally renormalized solutions*

The concept of renormalized solutions is the major step for studying general degenerate elliptic equations with measure data. The initial definition is given in [Dal Maso, Murat, Orsina, Prignet (1999)] and extended in a local and very useful form in [Bidaut-Véron (2003)]. If $k > 0$ we set $T_k(u) = \min\{k, |u|\}\mathrm{sgn}(u)$.

**Definition 4.9** Let $\Omega \subset \mathbb{R}^N$ be a domain and $\mu$ a Radon measure in $\Omega$ and $1 < p \leq N$. A measurable function $u : \Omega \mapsto \mathbb{R} \cup \{\infty, -\infty\}$ and a.e. finite is a *locally renormalized solution* of (4.1.2) if

$$\begin{array}{lll} (i) & T_k(u) \in W^{1,p}_{loc}(\Omega) & \forall k > 0 \\ (ii) & u \in L^s_{loc}(\Omega) & \forall 1 \leq s < \frac{N(p-1)}{N-p} \\ (iii) & |\nabla u|^{p-1} \in L^r_{loc}(\Omega) & \forall 1 \leq r < \frac{N}{N-1}, \end{array} \tag{4.3.1}$$

with the convention $\frac{N(p-1)}{N-p} = \infty$ if $p = N$, and

**D1-loc**: for any $h \in W^{1,\infty}(\mathbb{R})$ such that $h'$ has compact support, and any $\phi \in W^{1,m}(\Omega)$ for some $m > N$ with compact support and such that $h(u)\phi \in W^{1,p}(\Omega)$, there holds

$$\begin{aligned} \int_\Omega \langle A(x, \nabla u), \nabla(h(u)\phi)\rangle dx = \int_\Omega h(u)\phi d\mu_0 \\ + h(\infty)\int_\Omega \phi d\mu_s^+ - h(-\infty)\int_\Omega \phi d\mu_s^-. \end{aligned} \tag{4.3.2}$$

There are several equivalent definitions of locally renormalized solutions which turn out to be useful in different occasions.

**Theorem 4.3.1** *Let $\Omega$, $\mu$ and $p$ be as in Definition 4.4 and $u$ a function satisfying (4.3.1). Then* **D1-loc** *is equivalent to any of the following conditions:*

**D2-loc**: *if $\omega \in W^{1,p}(\Omega) \cap L^\infty(\Omega)$ has compact support in $\Omega$ and if there exists $k > 0$ and $\omega^+, \omega^- \in W^{1,r}(\Omega) \cap L^\infty(\Omega)$ with $r > N$ such that $\omega = \omega^+$ on $\{x \in \Omega : u(x) > k\}$ and $\omega = \omega^-$ on $\{x \in \Omega : u(x) < -k\}$, then*

$$\int_\Omega \langle A(x, \nabla u), \nabla \omega\rangle dx = \int_D \omega d\mu_0 + \int_\Omega \omega^+ d\mu_s^+ - \int_D \omega^- d\mu_s^-. \tag{4.3.3}$$

**D3-loc**: *for any $k > 0$ there exist two nonnegative measures $\alpha_k$ and $\beta_k$ in $\Omega$ absolutely continuous with respect to $c^\Omega_{1,p}$, with support in the sets $\{x \in \Omega : u(x) = k\}$ and $\{x \in \Omega : u(x) = -k\}$ respectively, converging in the weak topology of measures respectively to $\mu_s^+$ and $\mu_s^-$, and such that*

$$\int_{\{|u|<k\}} \langle A(x, \nabla u), \nabla \psi\rangle dx = \int_{\{|u|<k\}} \psi d\mu_0 + \int_{\{u=k\}} \psi d\alpha_k - \int_{\{u=-k\}} \psi d\beta_k, \tag{4.3.4}$$

*for all $\psi \in W^{1,p}(\Omega) \cap L^\infty(\Omega)$ with compact support in $\Omega$.*

**D4-loc**: *for any $h \in W^{1,\infty}(\mathbb{R})$ with compact support and any $\phi \in W^{1,p}(\Omega) \cap L^\infty(\Omega)$ with compact support in $\Omega$ such that $h(u)\phi \in W^{1,p}(\Omega)$,*

$$\int_\Omega \langle A(x, \nabla u), \nabla(h(u)\phi)\rangle dx = \int_\Omega h(u)\phi d\mu_0, \tag{4.3.5}$$

*and*

$$\begin{aligned} &(i) \qquad \lim_{n\to\infty} \frac{1}{n} \int_{\{n\leq u\leq 2n\}} |\nabla u|^p \phi dx = \int_\Omega \phi d\mu_s^+ \\ &(ii) \qquad \lim_{n\to\infty} \frac{1}{n} \int_{\{-2n\leq u\leq -n\}} |\nabla u|^p \phi dx = \int_\Omega \phi d\mu_s^-. \end{aligned} \tag{4.3.6}$$

The two concepts of $p$-superharmonic functions and renormalized solutions turn out to be equivalent when dealing with positive measures (see [Kilpeläinen, Kuusi, Tuhola-Kujanpää (2011)]).

**Theorem 4.3.2** *Let $1 < p \leq N$, $\Omega \subset$ be a domain and $A : \Omega \times \mathbb{R}^N \mapsto \mathbb{R}^N$ is a Caratheodory vector field satisfying (4.1.1). If $u$ is a locally renormalized solution of (4.1.2) with $\mu \geq 0$, then there exists a $p$-superharmonic function $\tilde{u}$ such that $\tilde{u} = u$ a.e. and $\tilde{u}$ satisfies*

$$\int_\Omega \langle A(x, \nabla \tilde{u}), \nabla(h(\tilde{u})\phi)\rangle dx = \int_\Omega h(u)\phi d\mu, \tag{4.3.7}$$

*for all $h \in W^{1,\infty}(\mathbb{R})$, $h'$ with compact support and all $\phi \in C_0^1(\Omega)$.*

**Lemma 4.3.3** *Under the assumptions of Theorem 4.3.2 $u$ is locally bounded from below.*

*Proof.* Since $\mu_- = 0$, $\mu = \mu_0 + \mu_s^+$. For $\epsilon > 0$ set $h_\epsilon(u) = \frac{1}{\epsilon}\min\{\epsilon, u^+\} - 1$. Let $h \in W^{1,\infty}(\mathbb{R})$, $h \geq 0$ and $h'$ with compact support, and $\phi \in C_0^1(\Omega)$, $\phi \geq 0$. Since $h_\epsilon h \in W^{1,\infty}(\mathbb{R})$ is nonpositive and vanishes on $[\epsilon, \infty)$, we have

$$\int_\Omega \langle A(x, \nabla u), \nabla\left(h(u)h_\epsilon(u)\phi\right)\rangle dx = \int_\Omega h(u)h_\epsilon(u)\phi d\mu_0 + h(\infty)h_\epsilon(\infty)\int_\Omega \phi d\mu_s.$$

Since $h(\infty)h_\epsilon(\infty) = 0$, $\nabla\left(h(u)h_\epsilon(u)\phi\right) = h_\epsilon(u)\nabla(h(u)\phi) + h(u)\phi\nabla h_\epsilon(u)$,

$$\int_\Omega \langle A(x, \nabla u), \nabla h_\epsilon(u)\rangle h(u)\phi dx \geq 0,$$

we derive

$$\int_\Omega \langle A(x, \nabla u), \nabla\left(h(u)\phi\right)\rangle h_\epsilon(u) dx \leq \int_\Omega h(u)h_\epsilon(u)\phi d\mu_0 \leq 0.$$

There holds by the dominated convergence theorem,

$$\lim_{\epsilon\to 0}\int_\Omega \langle A(x,\nabla u),\nabla\left(h(u)\phi\right)\rangle h_\epsilon(u)dx = -\int_\Omega \langle A(x,-\nabla u_-),\nabla\left(h(u_-)\phi\right)\rangle dx \leq 0.$$

We define $\tilde{A}(x,\xi)=-A(x,-\xi)$ and set $v=u_-$. Then $\tilde{A}$ satisfies the same assumptions as $A$, $T_k(v)\in W^{1,p}_{loc}$ and

$$\int_\Omega \langle \tilde{A}(x,\nabla v),\nabla\left(h(v)\phi\right)\rangle dx \leq 0. \tag{4.3.8}$$

*We claim that $v$ is locally bounded.* Assume $\overline{B}_{R'}\subset\Omega$ for some $R'>R>0$ and let $\{\phi_n\}_{n\in\mathbb{N}_*}\subset C^1_0(\Omega)$, $\phi_n\geq 0$ such that

$$\phi_n(x)=\begin{cases}1 & \text{if } |x|\leq R_{n+1}:=(1-2^{-n-1})R+2^{-n-1}R'\\ 0 & \text{if } |x|\geq R_n \quad :=(1-2^{-n})R+2^{-n}R',\end{cases}$$

and $|\nabla\phi_n|\leq 2^{n+1}(R'-R)^{-1}\chi_{B_{R_n}\setminus B_{R_{n+1}}}$. We replace $\phi$ by $\phi_n^p$ in (4.3.8) for homogeneity reason, set $v_k=\min\{v,k\}$ and take $h(v)=v_k^q$ for some $q>0$, then

$$\begin{aligned}q\int_\Omega |\nabla v_k|^p v_k^{q-1}\phi_n^p dx &\leq -p\int_\Omega \langle\tilde{A}(x,\nabla v),\nabla\phi_n\rangle\phi_n^{p-1}v_k^q dx\\ &\leq \lambda p\left(\int_\Omega |\nabla v_k|^p v_k^{q-1}\phi_n^p\right)^{\frac{1}{p'}}\left(\int_\Omega |\nabla\phi_n|^p v_k^{q+p-1}dx\right)^{\frac{1}{p}},\end{aligned}$$

which yields

$$\int_\Omega |\nabla v_k|^p v_k^{q-1}\phi_n^p dx \leq \left(\frac{p\lambda}{q}\right)^p\int_\Omega |\nabla\phi_n|^p v_k^{q+p-1}dx.$$

Next

$$\begin{aligned}\int_\Omega |\nabla v_k|^p v_k^{q-1}\phi_n^p dx &= \left(\frac{p}{q+p-1}\right)^p\int_\Omega \left|\phi_n\nabla v_k^{\frac{p+q-1}{p}}\right|^p dx\\ &\geq \left(\frac{p}{q+p-1}\right)^p\int_{B_{R_{n+1}}}\left|\nabla v_k^{\frac{p+q-1}{p}}\right|^p dx,\end{aligned}$$

and finally

$$\left\|v_k^{\frac{p+q-1}{p}}\right\|^p_{W^{1,p}(B_{R_{n+1}})}\leq 2^{np}\lambda^p c_1^p\left\|v_k^{\frac{p+q-1}{p}}\right\|^p_{L^p(B_{R_n})}, \tag{4.3.9}$$

where $c_1$ is bounded provided $q \geq q_0 > 0$ and $R' - R \geq \epsilon_0 > 0$. Assume $p < N$, then using Gagliardo-Nirenberg inequality, we deduce from (4.3.9),

$$\|v_k\|^p_{L^{\frac{N(p+q-1)}{N-p}}(B_{R_{n+1}})} \leq 2^n \lambda c_1 c_2 \left\| v_k^{\frac{p+q-1}{p}} \right\|^p_{L^p(B_{R_n})}, \tag{4.3.10}$$

and $c_2$ is independent of $n$ since $R \leq R_{n+1} \leq 2R$. Therefore

$$\|v_k\|_{L^{\frac{N(q+p-1)}{N-p}}(B_{R_{n+1}})} \leq (2^n \lambda c_1 c_2)^{\frac{p}{p+q-1}} \|v_k\|_{L^{q+p-1}(B_{R_n})}. \tag{4.3.11}$$

We choose $q < \frac{p(p-1)}{N-p}$ so that $s = p + q - 1 < \frac{N(p-1)}{N-p}$ and put $\tau = \frac{N}{N-p}$. Then the n-th iterate of the previous estimate yields

$$\|v_k\|_{L^{\tau^n s}(B_{R_{n+1}})} \leq 2^{\frac{p}{s}\sum_{k=1}^n \frac{k+1}{\tau^k}} (\lambda c_1 c_2)^{\frac{p}{s}\sum_{k=1}^n \frac{1}{\tau^k}} \|v_k\|_{L^s(B_{R_1})}. \tag{4.3.12}$$

All the series in exponent are convergent, $R_n \downarrow R$ and $R_1 < R'$, hence we derive from (4.3.1) (ii),

$$\|v_k\|_{L^\infty(B_R)} \leq C \|v_k\|_{L^s(B_{2R})} \leq C \|v\|_{L^s(B_{R'})} = m. \tag{4.3.13}$$

This implies that $\|v_k\|_{L^\infty(B_R)} \leq m$. In the case $p = N$ we use Hölder's inequality on the right-hand side of (4.3.9) and the proof follows by the same method. □

*Proof of Theorem 4.3.2.* For $k, \epsilon > 0$ we set

$$h_{k,\epsilon}(t) = \frac{1}{\epsilon}\min\{(k + \epsilon - t)_+, \epsilon\}.$$

Let $\phi \in C_0^1(\Omega)$ be nonnegative. Since $h'_{k,\epsilon}(t) \leq 0$ and $h'_{k,\epsilon}$ vanishes on $(-\infty, k) \cup (k + \epsilon, \infty)$ , we obtain

$$\begin{aligned} 0 &\leq \int_\Omega h_{k,\epsilon}(u)\phi d\mu_0 + h(\infty)\int_\Omega \phi d\mu_s \\ &= \int_\Omega \langle A(x, \nabla u), \nabla u\rangle h'_{k,\epsilon}(u)\phi dx + \int_\Omega \langle A(x, \nabla u), \nabla\phi\rangle h_{k,\epsilon}(u)dx \\ &\leq \int_\Omega \langle A(x, \nabla u), \nabla\phi\rangle h_{k,\epsilon}(u)dx. \end{aligned}$$

If we let $\epsilon \to 0$, then $h_{k,\epsilon}(u)(x) \to \chi_{(-\infty,k)}(u(x))$ and we derive

$$A(x, \nabla u), \nabla\phi\rangle h_{k,\epsilon}(u)\phi \to A(x, \nabla u_k), \nabla\phi\rangle,$$

where $u_k = \min\{u, k\}$, hence

$$\int_\Omega \langle A(x, \nabla u_k), \nabla \phi \rangle dx \geq 0. \tag{4.3.14}$$

Since $u_k$ is bounded from below it belongs to $W^{1,p}_{loc}(\Omega)$. This means that $u_k$ is weakly $p$-superharmonic. By Proposition 4.2.2 it is $p$-superharmonic. As $u_k \uparrow \tilde{u}$, $\tilde{u}$ is $p$-superharmonic and it coincides with $u$ a.e. in $\Omega$. □

The converse of Theorem 4.3.2 holds true.

**Theorem 4.3.4** *Let $p$, $\Omega$ be as in Theorem 4.3.2 and $A : \Omega \times \mathbb{R}^N \mapsto \mathbb{R}^N$ a Caratheodory vector field satisfying (4.1.1). If $u$ is a $p$-superharmonic function with Riesz measure $\mu$, then it is a locally renormalized solution of the corresponding equation (4.1.2).*

The proof which requires in particular the existence of renormalized solutions of (4.1.1) with measure data, is long and delicate and contains several intermediate results which are interesting by themselves (see [Kilpeläinen, Kuusi, Tuhola-Kujanpää (2011)]).

If $\mu \in \mathfrak{M}^+(\Omega)$, $L > 0$ and $\Omega'$ is a bounded domain such that $\overline{\Omega}' \subset \Omega$ we define the class $S_{\mu,r,L}(\Omega')$ of functions $u : \Omega' \mapsto \mathbb{R}$ by

$$S_{\mu,r,L}(\Omega') = \left\{ u : \frac{1}{c}\mathbf{W}_{1,p}[\mu](x,r) \leq u(x) \leq c\mathbf{W}_{1,p}[\mu](x,r) + L \;\; \forall x \in \Omega' \right\}. \tag{4.3.15}$$

By Theorem 4.2.5 this class is naturally associated with $p$-superharmonic functions $u$ which have $\mu$ for Riesz measure and with $L = c \operatorname{ess} \inf_{z \in B_r(x)} u(z)$. The main step is the following identity

**Proposition 4.3.5** *Let $1 < p \leq N$, $\Omega$ and $A : \Omega \times \mathbb{R}^N \mapsto \mathbb{R}^N$ be as in Theorem 4.3.4. Let $u$ and $v$ be $p$-superharmonic functions and assume that for all bounded domains $\Omega'$ verifying $\overline{\Omega}' \subset \Omega$ and all $r > 0$ small enough, there exists $L < \infty$ such that $u, v \in S_{\mu,r,L}$ where $\mu = \mu_0 + \mu_s$ is the Riesz measure of $u$. Then for any $h \in W^{1,\infty}(\mathbb{R}^2)$ such that $\nabla h$ has compact support, there holds*

$$\begin{aligned}\int_\Omega \langle A(x, \nabla u), \nabla(h(u,v)\phi)\rangle dx &= \int_\Omega h(u,v)\phi d\mu_0 \\ &\quad + h(\infty,\infty)\int_\Omega \phi d\mu_s \quad \forall \phi \in C^1_0(\Omega).\end{aligned} \tag{4.3.16}$$

The proof of Theorem 4.3.4 follows by taking $v = u$ in Proposition 4.3.5.

*Remark.* Because of Proposition 4.2.7 the support of $\mu_s$ is a subset of $\{x \in \Omega : u(x) = \infty\}$ and, since $u$ and $v$ belong to $S_{\mu,r,L}$, there holds

$$\{x \in \Omega' : u(x) = \infty\} = \{x \in \Omega' : v(x) = \infty\} = \{x \in \Omega' : \mathbf{W}_{1,p}[\mu](x,r) = \infty\}.$$

This justifies the expression of the second integral on the right-hand side of (4.3.16). Furthermore, since $h(u,v)$ is lower semicontinuous, there holds

$$\int_\Omega h(u,v)\phi d\mu_0 + h(\infty,\infty)\int_\Omega \phi d\mu_s = \int_\Omega h(u,v)\phi d\mu,$$

where the right-hand side is the integral of $h(u,v)\phi$ with respect to the measure $\mu$ (see [Bourbaki (1965)]). Similarly, if $u$ is a lower semicontinuous locally renormalized solution of (4.3.7), identity (4.3.7) can be rewritten as

$$\int_\Omega \langle A(x,\nabla u), \nabla(h(u)\phi)\rangle dx = \int_\Omega h(u)\phi d\mu^+ - \int_\Omega h(u)\phi d\mu^-, \tag{4.3.17}$$

where the integrals are taken in the Bourbaki sense (and due to A. Weil).

An important application of this equivalence is the proof of the removability of polar sets.

**Theorem 4.3.6** *Let $\Omega \subset \mathbb{R}^N$ be a domain, $1 < p \leq N$ and $u$ a nonnegative $p$-superharmonic function in $\Omega \setminus K$ where $K \subset \Omega$ is a polar set, i.e. a set such that $c_{1,p}^\Omega(K) = 0$. Then the function $\tilde{u}$ defined in whole $\Omega$ by*

$$\tilde{u}(x) = \text{ess}\lim_{y \to x} u(y), \tag{4.3.18}$$

*is $p$-superharmonic in $\Omega$.*

*Proof.* By replacing $\Omega$ by a smaller domain, we can assume that it is bounded and smooth. For $k > 0$, $u_k = \max\{u,k\}$ is $p$-superharmonic and bounded. Since $c_{1,p}^\Omega(K) = 0$ there exists a sequence $\{\eta_n\} \subset C_0^\infty(\Omega)$ of nonnegative functions such that $\eta_n \geq 1$ in a neighborhood of $K$, $\|\nabla \eta_n\|_{L^p} \to 0$ and $\eta_n \to 0$ a.e. as $n \to \infty$. Replacing $\eta_n$ by $\max\{1, \min\{\eta_1, \eta_2, ..., \eta_n\}\}$ we can suppose that the sequence is decreasing, with value 1 in a neighborhood of $K$, and we set $\tilde{\eta}_n = 1 - \eta_n$. Let $\phi \in C_0^\infty(\Omega)$, $\phi \geq 0$ and $\psi_n = \tilde{\eta}_n \phi$, then for $\alpha > 0$,

$$p\int_\Omega \langle A(x,\nabla u_k),\nabla\psi_n\rangle(1+u_k)^{-\alpha}\psi_n^{p-1}dx$$
$$\geq \alpha\int_\Omega \langle A(x,\nabla u_k),\nabla u_k\rangle(1+u_k)^{-\alpha-1}\psi_n^p dx.$$

Using (4.1.1) and Hölder's inequality

$$\int_\Omega |\nabla u_k|^p (1+u_k)^{-\alpha}\psi_n^p dx \leq \left(\frac{\lambda p}{\alpha}\right)^p \int_\Omega |\nabla\psi_n|^p (1+u_k)^{p-1-\alpha}dx.$$

Letting $n \to \infty$ yields

$$\int_\Omega |\nabla u_k|^p (1+u_k)^{-\alpha}\phi^p dx \leq \left(\frac{\lambda p}{\alpha}\right)^p \int_\Omega |\nabla\phi|^p (1+u_k)^{p-1-\alpha}dx. \quad (4.3.19)$$

If we fix $\alpha$ such that $p-1 > \alpha > 0$, we obtain

$$\int_\Omega |\nabla u_k|^p \phi^p dx \leq \left(\frac{\lambda p}{\alpha}\right)^p (1+k)^{p-1}\int_\Omega |\nabla\phi|^p dx. \quad (4.3.20)$$

Hence $u_k \in W^{1,p}_{loc}(\Omega)$. Taking $\psi_n$ as a test function, we have

$$\int_\Omega \langle A(x,\nabla u_k),\nabla\phi\rangle\tilde{\eta}_n dx \geq \int_\Omega \langle A(x,\nabla u_k),\nabla\eta_n\rangle\phi dx.$$

By Hölder's inequality and (4.3.20) the right-hand side of the above inequality converges to 0 as $n \to \infty$. Therefore

$$\int_\Omega\int_\Omega \langle A(x,\nabla u_k),\nabla\phi\rangle dx \geq 0, \quad (4.3.21)$$

which proves that $u_k$ is weakly $p$-superharmonic in $\Omega$. By Proposition 4.2.2 $u_k$ admits a lower semicontinuous $p$-superharmonic representative in $\Omega$ which is given by (4.2.7). Finally, since $u_k \uparrow u$ when $k \to \infty$ and since an increasing limit of $p$-superharmonic functions is $p$-superharmonic, it follows that $u$, given by (4.3.18), is $p$-superharmonic in $\Omega$. □

### 4.3.2 *The stability theorem*

Renormalized solutions were initially introduced for solving the Poisson equation with measure (4.2.3). The next statement is by now a classical result in the theory of renormalized solutions [Dal Maso, Murat, Orsina, Prignet (1999)].

**Theorem 4.3.7** *Let $\Omega$ be a bounded domain and $\mu \in \mathfrak{M}_b(\Omega)$ with decomposition (4.2.17). A measureable function $u$ defined in $\Omega$, finite a.e. and satisfying*

$$\begin{array}{lll} (i) & T_k(u) \in W_0^{1,p}(\Omega) & \forall k > 0 \\ (ii) & |\nabla u|^{p-1} \in L^r(\Omega) & \forall 1 \leq r < \frac{N}{N-1}, \end{array} \tag{4.3.22}$$

*is called a renormalized solution of (4.2.3) if it satisfies one of the four equivalent conditions:*

**D1**: *for any $h \in W^{1,\infty}(\mathbb{R})$ such that $h'$ has compact support, and for any $\phi \in W^{1,m}(\Omega)$ for some $m > N$ with the property that $h(u)\phi \in W_0^{1,p}(\Omega)$, the identity (4.3.2) is valid.*

**D2**: *if $\omega \in W_0^{1,p}(\Omega) \cap L^\infty(\Omega)$ and if there exist $k > 0$ and $\omega^+, \omega^- \in W^{1,r}(\Omega) \cap L^\infty(\Omega)$ with $r > N$ such that $\omega = \omega^+$ on $\{x \in \Omega : u(x) > k\}$ and $\omega = \omega^-$ on $\{x \in \Omega : u(x) < -k\}$, then (4.3.3) holds.*

**D3**: *for any $k > 0$ there exist two nonnegative bounded measures $\alpha_k$ and $\beta_k$ in $\Omega$, absolutely continuous with respect to $c_{1,p}^{\Omega}$, with support in the sets $\{x \in \Omega : u(x) = k\}$ and $\{x \in \Omega : u(x) = -k\}$ respectively and converging in the narrow topology (i.e. with test functions in $C(\Omega) \cap L^\infty(\Omega)$) to $\mu_s^+$ and $\mu_s^-$ respectively and such that (4.3.4) holds for all $\psi \in W_0^{1,p}(\Omega) \cap L^\infty(\Omega)$.*

**D4**: *for any $h \in W^{1,\infty}(\mathbb{R})$ with compact support and any $\phi \in W^{1,p}(\Omega) \cap L^\infty(\Omega)$ with compact support in $\Omega$ such that $h(u)\phi \in W_0^{1,p}(\Omega)$, identities (4.3.5) and (4.3.6) hold.*

*Remark.* It is a useful observation in the previous results to point out that the truncation of a renormalized solution satisfies an equation. Following Proposition 4.2.6 we can write $\mu = f - div\, g + \mu_s^+ - \mu_s^-$ where $f \in L^1(\Omega)$, $g \in (L^{p'}(\Omega)$ and $\mu_s^\pm$ singular with respect to the $c_{1,p}^{\mathbb{R}^N}$-capacity, then (4.3.4) becomes

$$\begin{aligned} \int_\Omega \langle A(x, T_k(u)) - g, \nabla\phi\rangle dx &= \int_{\{|u|<k\}} f\phi dx \\ &+ \int_{\{u=k\}} \phi d\alpha_k - \int_{\{u=-k\}} \phi d\beta_k \qquad \forall \phi \in C_0^1(\Omega). \end{aligned} \tag{4.3.23}$$

One of the main reason of introducing renormalized solutions is the existence result [Dal Maso, Murat, Orsina, Prignet (1999)].

**Definition 4.10** Let $\{\mu_n\}$ be a sequence of bounded measures. We say that a decomposition $\mu_n = f_n - div\, g_n + \mu_{s,n}^+ - \mu_{s,n}^-$ where $f_n \in L^1(\Omega)$,

$g_n \in (L^{p'}(\Omega))^N$ and $\mu^{\pm}_{s,n}$ are singular with respect to the $c^{\mathbb{R}^N}_{1,p}$-capacity is a *converging decomposition* having for limit $\mu = f - div\, g + \mu^+_s - \mu^-_s$, if

(i) $\{f_n\}$ converges to $f$ weakly in $L^1(\Omega)$.

(ii) $\{g_n\}$ converges to $g$ in $\left(L^{p'}(\Omega)\right)^N$ and $\{div\, g_n\}$ is bounded in $\mathfrak{M}_b(\Omega)$.

(iii) $\{\mu^{\pm}_{s,n}\}$ converges to $\mu^{\pm}_s$ in the narrow topology of $\mathfrak{M}_b(\Omega)$.

**Theorem 4.3.8** *Let $\Omega \subset \mathbb{R}^N$ be a bounded domain, $1 < p \leq N$ and $\mu$ be a bounded measure in $\Omega$. Then there exists a renormalized solution of 4.2.3. Furthermore, if $\mu_\epsilon = f_\epsilon - div\, g_\epsilon + \mu^+_{s,\epsilon} - \mu^-_{s,\epsilon}$ is a converging decomposition of the bounded measure $\mu_\epsilon$ having for limit the measure $\mu = f - div\, g + \mu^+_s - \mu^-_s$ and $u_\epsilon$ is a sequence of renormalized solutions of*

$$\begin{aligned} -div\, A(x, \nabla u_\epsilon) &= \mu_\epsilon \qquad && in\ \Omega \\ u_\epsilon &= 0 && on\ \Omega, \end{aligned} \tag{4.3.24}$$

*then there exists a subsequence of $\{u_{\epsilon_\ell}\}$ which converges a.e. to a renormalized solution $u$ of (4.2.3), and for any $k > 0$ $T_k(u_{\epsilon_\ell})$ converges to $T_k(u)$ in $W^{1,p}_0(\Omega)$.*

The proof is long and necessitates several intermediate results which present their own interest.

**Proposition 4.3.9** *Assume $\Omega$ and $p$ are as in Theorem 4.3.8. Let $\mu \in \mathfrak{M}_b(\Omega)$ such that there exists a renormalized solution $u$ to 4.2.3. Then we have*

$$\frac{1}{k}\int_{\{n \leq |u| \leq n+k\}} \langle A(x,u), \nabla u\rangle dx \leq c\, \|\mu\|_{\mathfrak{M}_b} \qquad \forall n, k \geq 0. \tag{4.3.25}$$

*Furthermore*

*I- If $1 < p < N$, then for any $k > 0$, we have for some $c = c(N,p) > 0$,*

$$meas(\{|u| > k\}) \leq c \frac{\|\mu\|_{\mathfrak{M}_b}^{\frac{N}{N-p}}}{k^{\frac{N(p-1)}{N-p}}} \qquad \forall k \geq 0, \tag{4.3.26}$$

*and*

$$meas(\{|\nabla u| > k\}) \leq c \frac{\|\mu\|_{\mathfrak{M}_b}^{\frac{N}{N-1}}}{k^{\frac{N(p-1)}{N-1}}} \qquad \forall k \geq 0. \tag{4.3.27}$$

*I- If $p = N$, then for any $r > 0$ there exists $c = c(N,r) > 0$ such that,*

$$meas(\{|u| > k\}) \leq c \frac{\|\mu\|^r_{\mathfrak{M}_b}}{k^{r(N-1)}} \qquad \forall k \geq 0. \tag{4.3.28}$$

*Furthermore, for any* $0 < s < N$ *there exists* $c = c(N,s) > 0$ *such that*

$$meas(\{|\nabla u| > k\}) \leq c\frac{\|\mu\|_{\mathfrak{M}_b}^{\frac{N}{N-1}}}{k^s} \qquad \forall k \geq 0. \tag{4.3.29}$$

*Proof.* We introduce the following truncation type functions

$$H_{n,k}(s) = \begin{cases} 0 & \text{if } s < -n-k \\ \frac{s+n+k}{k} & \text{if } -n-k \leq s < -n \\ 1 & \text{if } -n \leq s \leq n \\ \frac{n+k-s}{k} & \text{if } n \leq s \leq n+k \\ 0 & \text{if } s > n+k, \end{cases}$$

and $B_{n,k}(s) = 1 - H_{n,k}(s)$. Then using **D2** with $\omega = B_{n,k}(u^+)$, we have

$$\begin{aligned} \frac{1}{k}\int_{\{n<u<n+k\}} \langle A(x,\nabla u), \nabla u\rangle dx &= \int_\Omega B_{n,k}(u^+)d\mu_0 + \int_\Omega B_{n,k}(u^+)d\mu_s^+ \\ &\leq \left\|\mu_0\chi_{\{u>0\}}\right\|_{\mathfrak{M}_b} + \mu_s^+(\Omega). \end{aligned} \tag{4.3.30}$$

Similarly

$$\frac{1}{k}\int_{\{-n-k<u\leq -n\}} \langle A(x,\nabla u), \nabla u\rangle dx \leq \left\|\mu_0\chi_{\{u<0\}}\right\|_{\mathfrak{M}_b} + \mu_s^-(\Omega). \tag{4.3.31}$$

Adding (4.3.30) and (4.3.31) leads to (4.3.25).
*Case* $1 < p < N$. We use Gagliardo-Niremberg inequality and derive from 4.3.25) with $n = 0$,

$$\|T_k(u)\|_{L^{p*}}^p \leq c_1\int_{\{|u|<k\}} \langle A(x,u), \nabla u\rangle dx \leq cc_1 k\,\|\mu\|_{\mathfrak{M}_b}.$$

For $0 < \epsilon \leq k$, $\{x\in\Omega : |u(x)| > \epsilon\} = \{x\in\Omega : |T_k(u)(x)| > \epsilon\}$, thus

$$\begin{aligned} \text{meas}(\{|u| > \epsilon\}) &= \text{meas}(\{|T_k(u)| > \epsilon\}) \\ &\leq \left(\frac{\|T_k(u)\|_{L^{p^*}}}{\epsilon}\right)^{p^*} \\ &\leq c_2 k^{p^*}\,\|\mu\|_{\mathfrak{M}_b}^{p^*}, \end{aligned}$$

which yields (4.3.25) by taking $\epsilon = k$.
Next, for $k,\ell > 0$, we set $\Phi(\ell,\kappa) = \text{meas}(\{|u| > \ell\}\cup\{|Du|^p > \kappa\})$. Since $\Phi$

is nonincreasing in the second variable, we have

$$\Phi(0,\kappa) \leq \frac{1}{\kappa}\int_0^\kappa \Phi(0,s)ds \leq \Phi(\ell,0) + \frac{1}{\kappa}\int_0^\kappa \left(\Phi(0,s) - \Phi(\ell,s)\right)ds.$$

Notice that

$$\begin{aligned}\Phi(0,s)-\Phi(\ell,s) &= \mathrm{meas}(\{|u|>0\}\cup\{|Du|^p>s\}) - \mathrm{meas}(\{|u|>\ell\}\cup\{|Du|^p>s\})\\ &= \mathrm{meas}(\{0<|u|\leq\ell\}\cup\{|Du|^p>s\}).\end{aligned}$$

Since by (4.3.25),

$$\begin{aligned}\int_{\{0<|u|\leq\ell\}} |\nabla u|^p\,dx &= \int_0^\infty \mathrm{meas}(\{0<|u|\leq\ell\}\cup\{|Du|^p>s\})ds\\ &\leq c'\,\|\mu\|_{\mathfrak{M}_b}\,\ell,\end{aligned}$$

we derive

$$\Phi(0,\kappa) \leq c\frac{\|\mu\|_{\mathfrak{M}_b}^{\frac{N}{N-p}}}{\ell^{\frac{N(p-1)}{N-p}}} + cc'\frac{\|\mu\|_{\mathfrak{M}_b}\,\ell}{\kappa}.$$

Choosing $\ell$ such that $c'\frac{\ell}{\kappa}\|\mu\|_{\mathfrak{M}_b} = \frac{\|\mu\|_{\mathfrak{M}_b}^{\frac{N}{N-p}}}{\ell^{\frac{N(p-1)}{N-p}}}$ and putting $\kappa = k^p$, we obtain (4.3.27).

*Case* $p = N$. We combine Hölder and Gagliardo-Nirenberg inequalities and derive, with $1 < q < N$ and $q^* = \frac{Nq}{N-q}$,

$$\begin{aligned}\|T_k(u)\|_{L^{q^*}}^N &\leq c_{N,q}\,\|\nabla T_k(u)\|_{L^q}^N\\ &\leq c_{N,q}(\mathrm{meas}(\{u<k\}))^{\frac{N-q}{q}}\int_{\{|u|<k\}}\langle A(x,u),\nabla u\rangle dx\\ &\leq c_{N,q}c_1 k(\mathrm{meas}(\{u<k\}))^{\frac{N-q}{q}}\,\|\mu\|_{\mathfrak{M}_b}.\end{aligned}$$

Since, for $0 < \epsilon \leq k$ there holds

$$\begin{aligned}\mathrm{meas}(\{|u|>\epsilon\}) &= \mathrm{meas}(\{|T_k(u)|>\epsilon\})\\ &\leq \left(\frac{\|T_k(u)\|_{L^{q^*}}}{\epsilon}\right)^{q^*}\\ &\leq c_2 k^{q^*}\,\|\mu\|_{\mathfrak{M}_b}^{q^*},\end{aligned}$$

we infer, with $\epsilon = k$,

$$\mathrm{meas}(\{|u|>k\}) \leq c'_{N,q}\mathrm{meas}(\{|u|<k\})k^{-\frac{(N-1)q}{N-q}}\,\|\mu\|_{\mathfrak{M}_b}^{\frac{q}{N-q}},$$

which implies (4.3.28) by setting $r = \frac{q}{N-q}$. The proof of (4.3.29) is the same as the one of (4.3.27). □

*Remark.* Inequality (4.3.29) is not optimal in the case $p = 2$ since in that case, it is possible to obtain by using estimates on the Poisson kernel the more general one,

$$\text{meas}(\{|\nabla u| > k\}) \leq c\frac{\|\mu\|_{\mathfrak{M}_b}^{\frac{N}{N-1}}}{k^N} \qquad \forall k \geq 0. \tag{4.3.32}$$

Actually an optimal regularity result, useful in many applications, has been proved in [Dolzmann, Hungerbühler, Müller (2000)]. Its expression needs to introduce the space $BMO$ functions of bounded mean oscillation.

**Definition 4.11** A function $u \in L^N(\mathbb{R}^N)$ belongs to $BMO(\mathbb{R}^N)$ if

$$[u]_{BMO(\mathbb{R}^N)} = \left(\sup_{a\in\mathbb{R}^N}\sup_{r>0} r^{-N}\int_{Q_r(a)} |u - (u)_{a,r}|^N dx\right)^{\frac{1}{N}} < \infty, \tag{4.3.33}$$

where $Q_r(a)$ is the cube $\{x = (x_1, ..., x_N : |x_j - a_j| \leq \frac{r}{2}, \forall j = 1, ..., N\}$. This space is equipped with the norm

$$\|u\|_{BMO(\mathbb{R}^N)} = \|u\|_{L^N(\mathbb{R}^N)} + [u]_{BMO(\mathbb{R}^N)}. \tag{4.3.34}$$

All the functions defined on a domain $\Omega$ are extended by 0 in $\Omega^c$. If $u = 0$ on $\Omega^c$ and $a \in \Omega^c$ then the mean value of $u$ on $Q_r(a)$ is estimated in the following way

$$|(u)_{a,r}|\text{meas}(Q_r(a)\cap\Omega^c) \leq \int_{Q_r(a)} |u - (u)_{a,r}|\, dx \leq \text{meas}(\mathrm{Q_r(a)})[u]_{BMO(\mathbb{R}^N)}.$$

If $\Omega^c$ is of type **A**, that is there exists $K > 0$ such that

$$\text{meas}(B_r(x) \cap \Omega^c) \geq Kr^N \qquad \forall x \in \Omega^c, \tag{4.3.35}$$

then

$$\left(r^{-N}\int_{Q_r(a)} |u|^N dx\right)^{\frac{1}{N}} \leq c_{N,K}\left(r^{-N}\int_{Q_r(a)} |u - (u)_{a,r}|^N dx\right)^{\frac{1}{N}}, \tag{4.3.36}$$

and

$$|(u)_{a,r}| \leq c_K[u]_{BMO(\mathbb{R}^N)}. \tag{4.3.37}$$

One of the important property of BMO functions is the John and Nirenberg inequality [John, Nirenberg (1961)]

**Proposition 4.3.10** *Assume $\Omega \subset \mathbb{R}^N$ is a bounded domain. Then there exist constants $c_1, c_2 > 0$ depending on $N$ such that for any cube $Q_r(a) \subset \Omega$ and any $u \in BMO(\Omega)$,*

$$\text{meas}\left(\{x \in Q : |u(x) - (u)_{a,r}| > \sigma\}\right) \leq c_1 r^N e^{-\frac{c_2 \sigma}{[u]_{BMO(Q_r(a))}}}. \tag{4.3.38}$$

*If $\Omega^c$ is of type* **A** *and $u$ vanishes on $\Omega^c$ there exists a constant $c_3$ depending on $\Omega$ such that*

$$\text{meas}\left(\{x \in \Omega : |u(x) - u_\Omega| > \sigma\}\right) \leq c_3 e^{-\frac{c_2 \sigma}{[u]_{BMO(\Omega)}}}, \tag{4.3.39}$$

*where $u_\Omega$ is the average of $u$ in $\Omega$.*

**Proposition 4.3.11** *Assume $\Omega \subset \mathbb{R}^N$ is a bounded domain and $\Omega^c$ is of type* **A**. *If $\mu \in \mathfrak{M}_b(\Omega)$ is such that there exists a renormalized solution $u$ to (4.2.3) with $p = N$, then*

$$\|u\|_{BMO(\Omega)} \leq c_1 \|\mu\|_{\mathfrak{M}_b}^{\frac{1}{N-1}}, \tag{4.3.40}$$

*and*

$$meas(\{|\nabla u| > k\}) \leq c \frac{\|\mu\|_{\mathfrak{M}_b}^{\frac{N}{N-1}}}{k^N} \qquad \forall k \geq 0. \tag{4.3.41}$$

*Remark.* Estimate (4.3.41) states that $\nabla u$ belongs to the Lorentz space $L^{N,\infty}(\Omega)$ and there holds

$$\|\nabla u\|_{L^{N,\infty}(\Omega)} \leq c \|\mu\|_{\mathfrak{M}_b}^{\frac{N}{N-1}}. \tag{4.3.42}$$

Actually the two estimates are not independent since any $u$ such that $\nabla u \in L^{N,\infty}(\Omega)$ belongs to $BMO(\Omega)$ and there holds

$$\|\phi\|_{BMO(\Omega)} \leq c \|\nabla \phi\|_{L^{N,\infty}(\Omega)} \qquad \forall \phi \in C_0^1(\Omega), \tag{4.3.43}$$

where $c = c(N) > 0$. The properties of functions which gradient belong to a Lorentz space $L^{p,q}(\mathbb{R}^N)$ for $1 < p < \infty$ and $1 \leq q \leq \infty$ are clearly presented in [Tartar (1998)].

*Proof of Theorem 4.3.8.* For $\epsilon > 0$, we consider an approximation $\mu_\epsilon = f_\epsilon - \operatorname{div} g_\epsilon + \mu^+_{s,\epsilon} - \mu^-_{s,\epsilon}$ of $\mu = f - \operatorname{div} g + \mu^+_s - \mu^-_s$ by convolution of $f$, $\mu^+_s$ and $\mu^-_s$. Then $f_\epsilon$ $\mu^+_{s,\epsilon}$ and $\mu^-_{s,\epsilon}$ are smooth functions and $f_\epsilon \to f$ in $L^1(\Omega)$, $\mu^+_{s,\epsilon} \to \mu^+_s$ and $\mu^-_{s,\epsilon} \to \mu^-_s$ in the narrow topology. Since $\mu_\epsilon \in W^{-1,p'}(\Omega)$, there exists

a renormalized solution $u_\epsilon$ of $(4.2.3)_\epsilon$ with $\mu = \mu_\epsilon$. Therefore the existence part of the theorem is a particular case of the stability statement.
From now we study the stability and assume that the measure $\mu_\epsilon$ admits the decomposition $\mu_\epsilon = f_\epsilon - \operatorname{div} g_\epsilon + \lambda_\epsilon^\oplus - \lambda_\epsilon^\ominus$ where $f_\epsilon \in L^1(\Omega)$ converges weakly to $f$, $g_\epsilon \in (L^{p'}(\Omega))^N$ converges to $g$ in $(L^{p'}(\Omega))^N$ and $\operatorname{div} g_\epsilon$ remains bounded in $\mathfrak{M}_b(\Omega)$, and $\lambda_\epsilon^\oplus$ and $\lambda_\epsilon^\ominus$, belong to $\mathfrak{M}_b^+(\Omega)$ and converge to $\mu_s^+$ and $\mu_s^-$ respectively in the narrow topology. Notice that it is not assumed that $\lambda_\epsilon^\oplus$ or $\lambda_\epsilon^\ominus$ are singular with respect to $c_{1,p}^{\mathbb{R}^N}$. Actually we can write

$$\lambda_\epsilon^\oplus = \lambda_{\epsilon,0}^\oplus + \lambda_{\epsilon,s}^\oplus \quad \text{and} \quad \lambda_\epsilon^\ominus = \lambda_{\epsilon,0}^\ominus + \lambda_{\epsilon,s}^\ominus,$$

where $\lambda_{\epsilon,0}^\oplus, \lambda_{\epsilon,0}^\ominus$ (resp. $\lambda_{\epsilon,s}^\oplus, \lambda_{\epsilon,s}^\ominus$) are absolutely continuous (resp. singular) with respect to the $c_{1,p}^{\mathbb{R}^N}$-capacity, all the above measures being nonnegative. Therefore

$$\mu_\epsilon = \left(f_\epsilon - \operatorname{div} g_\epsilon + \lambda_{\epsilon,0}^\oplus - \lambda_{\epsilon,0}^\ominus\right) + \left(\lambda_{\epsilon,s}^\oplus - \lambda_{\epsilon,s}^\ominus\right) = \mu_{\epsilon,0} + \mu_{\epsilon,s}. \tag{4.3.44}$$

where $\mu_{\epsilon,0} \ll c_{1,p}^{\mathbb{R}^N}$ and $\mu_{\epsilon,s} \perp c_{1,p}^{\mathbb{R}^N}$. Note that $\mu_{\epsilon,s}^+ \leq \lambda^\oplus$ and $\mu_{\epsilon,s}^- \leq \lambda^\ominus$.

*Step 1: A priori estimates.* Since $\mu_\epsilon \to \mu$, $\|\mu_\epsilon\|_{\mathfrak{M}_b}$ is uniformly bounded by some constant $c^* > 0$; it follows by Proposition 4.3.9 that $u_\epsilon$ satisfies the same estimates (4.3.25), (4.3.26), (4.3.27) in the case $1 < p < N$, and (4.3.28) and (4.3.41) in the case $p = N$. In particular (4.3.25) holds with $u$ replaced by $T_{n+k}(u_\epsilon)$,

$$\frac{1}{k}\int_{\{n \leq |u_\epsilon| \leq n+k\}} \langle A(x, T_{n+k}(u_\epsilon)), \nabla T_{n+k}(u_\epsilon)\rangle dx \leq c^* \quad \forall n, k \geq 0, \tag{4.3.45}$$

which yields, by (4.1.2) (i) and the definition of $B_{n,k}$,

$$\frac{1}{k}\int_{\{n \leq |u_\epsilon| \leq n+k\}} |\nabla B_{n,k}(u_\epsilon))|^p \, dx \leq c^* \quad \forall n, k \geq 0. \tag{4.3.46}$$

With $n = 0$ we obtain the bound

$$\int_\Omega |\nabla T_k(u_\epsilon))|^p \, dx \leq c^* k \quad \forall k \geq 0. \tag{4.3.47}$$

Let $q < \frac{N}{N-1}$, then by (4.3.27), even if $p = N$,

$$\int_\Omega |\nabla u_\epsilon|^{q(p-1)}\,dx = \int_{\{|\nabla u_\epsilon|\le 1\}} |\nabla u_\epsilon|^{q(p-1)}\,dx + \int_{\{|\nabla u_\epsilon|>1\}} |\nabla u_\epsilon|^{q(p-1)}\,dx$$
$$\le \operatorname{meas}(\Omega) + q(p-1)\int_1^\infty t^{q(p-1)-1}\operatorname{meas}\left(\{|\nabla u_\epsilon| > t\}\right)dt$$
$$\le \operatorname{meas}(\Omega) + \frac{c^* q}{\frac{N}{N-1} - q},$$

since $\operatorname{meas}\left(\{|\nabla u_\epsilon| > t\}\right) \le c^* t^{-\frac{N(p-1)}{N-1}}$. This implies that $\nabla u_\epsilon$ is bounded in $L^{q(p-1)}$ for every $q < \frac{N}{N-1}$.

*Step 2: There exists a subsequence of $u_\epsilon$ which is a Cauchy sequence in measure.* Let $\sigma, \eta > 0$ be fixed. For $k > 0$ and $\epsilon, \delta > 0$, we have

$$\{|u_\epsilon - u_\delta| > \sigma\} \subset \{|u_\epsilon| > k\} \cup \{|u_\delta| > k\} \cup \{|T_k(u_\epsilon) - T_k(u_\delta)| > \sigma\}. \quad (4.3.48)$$

From (4.3.26) there exists $k_0$ such that for every $k \ge k_0$ and every $\epsilon, \delta > 0$,

$$\operatorname{meas}(\{|u_\epsilon| > k\}) + \operatorname{meas}(\{|u_\delta| > k\}) \le \frac{\eta}{2}.$$

If we choose $k = k_0$, then $\{T_{k_0}(u_\epsilon)\}_\epsilon$ is bounded in $W_0^{1,p}(\Omega)$ in (4.3.48). By Rellich-Kondrachov theorem there exists a subsequence, still denoted by $\{u_\epsilon\}$, such that $\{T_{k_0}(u_\epsilon)\}$ is convergent in $L^p(\Omega)$. Furthermore $\{\nabla T_{k_0}(u_\epsilon)\}$ converges weakly in $L^p(\Omega)$. This implies that $\{T_{k_0}(u_\epsilon)\}$ is a Cauchy sequence in measure. Thus there exists $\epsilon_0 > 0$ such that

$$0 < \epsilon, \delta \le \epsilon_0 \Longrightarrow \operatorname{meas}(\{|T_k(u_\epsilon) - T_k(u_\delta)| > \sigma\}) \le \frac{\eta}{2}.$$

Then there exists a measurable function $u$ such that $u_\epsilon$ converges to $u$ in measure. Up to an extracted subsequence $u_\epsilon$ converges also to $u$ a.e. in $\Omega$, but for the sake of simplicity this new subsequence is still denoted by $u_\epsilon$.

*Step 3: Weak convergence of the truncates.* Since $u_\epsilon \to u$ a.e. and $T_k$ is continuous, $T_k(u_\epsilon) \to T_k(u)$ a.e. in $\Omega$. In Step 2 we have used the fact that $\{T_k(u_\epsilon)\}$ converges to some function $v_k$ in $L^p(\Omega)$ and $\nabla T_k(u_\epsilon) \rightharpoonup \nabla v_k$ in $L^p(\Omega)$-weak. Then $v_k = T_k(u)$ and $\nabla T_k(u_\epsilon)$ converges weakly in $L^p(\Omega)$ to $\nabla T_k(u)$. Consequently $T_k(u) \in W_0^{1,p}(\Omega)$ and $\nabla u$ is well defined a.e. in $\Omega$.

*Step 4: Equation satisfied by $T_k(u_\epsilon)$.* Since $u_\epsilon$ is a renormalized solution of $(4.2.3)_\epsilon$, for every $k$ there exist two nonnegative measures $\lambda^+_{\epsilon,k}$ and $\lambda^-_{\epsilon,k}$, absolutely continuous with respect to the $c_{1,p}^{\mathbb{R}^N}$-capacity, concentrated respectively on the sets $\{u_\epsilon = k\}$ and $\{u_\epsilon = -k\}$, such that

$$\int_{\{|u_\epsilon|<k\}} \langle A(x,\nabla T_k(u_\epsilon)),\nabla\phi\rangle dx = \int_{\{|u_\epsilon|<k\}} \phi d\mu_{\epsilon,0} + \int_{\{u_\epsilon=k\}} \phi d\lambda^+_{\epsilon,k} - \int_{\{u_\epsilon=-k\}} \phi d\lambda^-_{\epsilon,k}, \tag{4.3.49}$$

for all $\phi \in L^\infty(\Omega)\cap W_0^{1,p}(\Omega)$; in this expression $\mu_{\epsilon,0}$ is the absolutely continuous part of $\mu_\epsilon$ with respect to the $c_{1,p}^{\mathbb{R}^N}$-capacity. We take $\phi = T_k(u_\epsilon)$, then

$$\int_\Omega \langle A(x,\nabla T_k(u_\epsilon)),\nabla T_k(u_\epsilon)\rangle dx = \int_{\{|u_\epsilon|<k\}} T_k(u_\epsilon)d\mu_{\epsilon,0} + k\left(\lambda^+_{\epsilon,k}(\Omega)+\lambda^-_{\epsilon,k}(\Omega)\right).$$

Using (4.1.1) (i) and (4.3.25), we derive

$$\int_{\{|u_\epsilon|<k\}} |\nabla T_k(u_\epsilon)|^p\, dx \leq kc\,\|\mu_{\epsilon,0}\|_{\mathfrak{M}_b}, \tag{4.3.50}$$

already seen in Proposition 4.3.9, and

$$k\left(\lambda^+_{\epsilon,k}(\Omega)+\lambda^-_{\epsilon,k}(\Omega)\right) \leq k(1+c)\,\|\mu_{\epsilon,0}\|_{\mathfrak{M}_b}, \tag{4.3.51}$$

from which it follows that $\lambda^+_{\epsilon,k}$ and $\lambda^-_{\epsilon,k}$ are uniformly bounded in $\mathfrak{M}_b(\Omega)$. The same boundedness holds with $\mu_{\epsilon,k} = \mu_{\epsilon,0}\lfloor_{\{|u_\epsilon|<k\}} + \lambda^+_{\epsilon,k} - \lambda^-_{\epsilon,k}$ (remember that $\mu_{\epsilon,0}$ defined in (4.3.44) is bounded since convergent). Identity (4.3.49) expresses the fact that $T_k(u_\epsilon)$ is a renormalized solution of

$$-\operatorname{div} A(x,\nabla T_k(u_\epsilon)) = \mu_{\epsilon,k} \qquad \text{in } \Omega. \tag{4.3.52}$$

*Step 5: The sequence $\{\nabla T_k(u_\epsilon)\}$ is a Cauchy sequence in measure.* Let $\sigma > 0$ and $\eta > 0$ be fixed for defining the convergence in measure. For $\gamma > 0$ and $\kappa \geq k$ we set

$$E_1 = \{\inf\{|\nabla T_k(u_\epsilon)|, |\nabla T_k(u_\delta)|\} > \kappa\},$$

$$E_2 = \{|T_k(u_\epsilon) - T_k(u_\delta)| > \gamma\},$$

$$E_3 = \{\sup\{|\nabla T_k(u_\epsilon)|, |\nabla T_k(u_\delta)|\} \leq \kappa\}\bigcap\{|T_k(u_\epsilon)-T_k(u_\delta)| \leq \gamma\} \bigcap\{|\nabla T_k(u_\epsilon) - \nabla T_k(u_\delta)| > \eta\}.$$

Then

$$\{x \in \Omega : |\nabla T_k(u_\epsilon)(x) - \nabla T_k(u_\delta)(x)| > \eta\} \subset E_1 \bigcup E_2 \bigcup E_3. \tag{4.3.53}$$

Since $T_k(u_\epsilon)$ and $T_k(u_\delta)$ are bounded in $W_0^{1,p}(\Omega)$ we fix $\kappa > 0$ such that $\mathrm{meas}(E_1) \le \eta/3$. Since $A$ is a Caratheodory function, there exists a null set $C \subset \Omega$ such that $\xi \mapsto A(x,\xi)$ is continuous for all $x \in \Omega \setminus C$. We define

$$K = \left\{(\xi,\xi') \in \mathbb{R}^N \times \mathbb{R}^N : |\xi| \le \kappa, |\xi'| \le \kappa, |\xi - \xi'| \ge \eta\right\}.$$

and, for $x \in \Omega \setminus C$,

$$\Gamma(x) = \inf\{\langle A(x,\xi) - A(x,\xi'), \xi - \xi'\rangle : (\xi,\xi') \in K\}. \tag{4.3.54}$$

By the assumption (4.1.1) (iii) the function $\Gamma$ is positive, hence for any $\theta > 0$ there exists $\tau > 0$ such that

$$\int_A \Gamma(x)dx \le \tau \Longrightarrow \mathrm{meas}\,(\mathrm{A}) \le \theta. \tag{4.3.55}$$

Next for $E_2$, we take $T_\gamma(T_k(u_\epsilon) - T_k(u_\delta))$ for test function in (4.3.52) and derive

$$\begin{aligned}\int_{\{T_k(u_\epsilon)-T_k(u_\delta)\le\gamma\}} &\langle A(x,\nabla T_k(u_\epsilon)) - A(x,\nabla T_k(u_\delta)), \nabla T_k(u_\epsilon) - \nabla T_k(u_\delta)\rangle dx \\ &\le \gamma\left(\lambda^+_{\epsilon,k}(\Omega) + \lambda^-_{\epsilon,k}(\Omega) + |\mu_{\epsilon,0}|(\Omega)\right) \\ &\le \gamma M,\end{aligned}$$

since $\lambda^+_{\epsilon,k}$, $\lambda^-_{\epsilon,k}$ and $|\mu_{\epsilon,0}|$ are uniformly bounded for $k$ fixed. Hence

$$\begin{aligned}\gamma M &\ge \int_{E_3} \langle A(x,\nabla T_k(u_\epsilon)) - A(x,\nabla T_k(u_\delta)), \nabla T_k(u_\epsilon) - \nabla T_k(u_\delta)\rangle dx, \\ &\ge \int_{E_3} \Gamma(x)dx.\end{aligned} \tag{4.3.56}$$

We fix $\theta = \eta/3$. Then there exists $\tau > 0$ such that (4.3.55) holds. Next we choose $\gamma = \tau/M$ in (4.3.56) and get $\mathrm{meas}\,(\mathrm{E}_3) \le \eta/3$. At end, we use Step 2 and derive that there exists $\epsilon_0$ such that, if $0 < \epsilon, \delta < \epsilon_0$, there holds $\mathrm{meas}\,(\mathrm{E}_2) \le \eta/3$, which implies the claim.

*Step 6: The sequence $\{\nabla u_\epsilon\}$ is a Cauchy sequence in measure.* Let $\sigma > 0$, then

$$\{|\nabla u_\epsilon - \nabla u_\delta| > \sigma\} \subset \{|\nabla T_k(u_\epsilon) - \nabla T_k(u_\epsilon)| > \sigma\} \bigcup \{|u_\epsilon| > k\} \bigcup \{|u_\delta| > k\}.$$

Using (4.3.26) or (4.3.28), we can fix $k$ so that meas$(\{T_k(u_\epsilon) > k\}) \leq \eta/3$ uniformly with respect to $\epsilon$. By Step 5 there exists $\epsilon_0$ such that, for $0 < \epsilon, \delta < \epsilon_0$, there holds meas$(\{|\nabla \mathrm{T_k(u_\epsilon)} - \nabla \mathrm{T_k(u_\epsilon)}| > \sigma\}) \leq \eta/3$, which implies the claim. This implies in particular that, up to extracting a new subsequence, $\nabla u_\epsilon(x) \to \nabla u(x)$ a.e. in $\Omega$. By Step 1 $|\nabla u_\epsilon|^{p-1}$ remains bounded in $L^q(\Omega)$ for all $1 \leq q < \frac{N}{N-1}$. By Vitali's convergence theorem we infer that $|\nabla u|^{p-1} \in L^q(\Omega)$ and $|\nabla u_\epsilon|^{p-1} \to |\nabla u|^{p-1}$ in $L^q(\Omega)$ in the same range of values of $q$. Actually $\nabla u$ satisfies estimates (4.3.29) and (4.3.41). Furthermore, it follows from (4.1.1) (ii) and the continuity of $A(x,.)$ that $A(x, \nabla u_\epsilon) \to A(x, \nabla u)$ in $(L^q(\Omega))^N$. This implies that $u$ is a solution of (4.2.3) in the sense of distributions in $\Omega$.

*Step 7: Equation satisfied by $T_k(u)$.* We analyze the different terms in (4.3.49) where the test function $\phi$ belongs to $\phi \in L^\infty(\Omega) \cap W_0^{1,p}(\Omega)$. Writing $\mu_{\epsilon,0} = f_\epsilon - \operatorname{div} g_\epsilon + \lambda^\oplus_{\epsilon,0} - \lambda^\ominus_{\epsilon,0}$ we develop (4.3.49),

$$\int_{\{|u_\epsilon|<k\}} \langle A(x, \nabla T_k(u_\epsilon)) - g_\epsilon, \nabla\phi\rangle dx = \int_{\{|u_\epsilon|<k\}} f_\epsilon \phi dx + \int_{\{|u_\epsilon|<k\}} \phi d\lambda^\oplus_{\epsilon,0}$$
$$- \int_{\{|u_\epsilon|<k\}} \phi d\lambda^\ominus_{\epsilon,0} + \int_{\{u_\epsilon=k\}} \phi d\lambda^+_{\epsilon,k} - \int_{\{u_\epsilon=-k\}} \phi d\lambda^-_{\epsilon,k}.$$

Since $A(x, \nabla T_k(u_\epsilon))$ converges to $A(x, \nabla T_k(u))$ in $L^1(\Omega))^N$ and $\chi_{\{|u_\epsilon|<k\}} \to \chi_{\{|u|<k\}}$ a.e. in $\Omega$, the convergence of the different terms is the following:

$$\int_{\{|u_\epsilon|<k\}} \langle A(x, \nabla T_k(u_\epsilon)) - g_\epsilon, \nabla\phi\rangle dx \to \int_{\{|u|<k\}} \langle A(x, \nabla T_k(u)) - g, \nabla\phi\rangle dx,$$

$$\int_{\{|u_\epsilon|<k\}} f_\epsilon \phi dx \to \int_{\{|u|<k\}} f \phi dx.$$

By (4.3.51), $\chi_{\{|u_\epsilon|<k\}} \lambda^\oplus_\epsilon + \lambda^+_{\epsilon,k}$ (resp. $\chi_{\{|u_\epsilon|<k\}} \lambda^\ominus_\epsilon + \lambda^-_{\epsilon,k}$) is bounded independently of $\epsilon$, therefore it converges weakly (up to a subsequence) to $\alpha_k$ (resp. $\beta_k$). Letting $\epsilon \to 0$ in (4.3.49) and taking $\phi \in C_0^1(\Omega)$ we infer the following identity

$$\int_{\{|u|<k\}} \langle A(x, \nabla T_k(u)) - g, \nabla\phi\rangle dx = \int_{\{|u|<k\}} f\phi dx + \int_\Omega \phi d\alpha_k - \int_\Omega \phi d\beta_k. \tag{4.3.57}$$

Because $\alpha_k - \beta_k$ coincides with $A(., \nabla T_k(u)) - g - \chi_{\{|u_\epsilon|<k\}} f$, it belongs to $W^{-1,p'}(\Omega) + L^1(\Omega)$. Therefore (4.3.57) holds with $\phi \in W_0^{1,p}(\Omega) \cap L^\infty(\Omega)$. Furthermore, since $T_k(u) \in W_0^{1,p}(\Omega)$, the sets $\{x \in \Omega : |u|(x) < k\}$ and

$\{x \in \Omega : |u|(x) > k\}$ are quasiopen. It is therefore classical [Dal Maso (1983)] that there exists an increasing sequence $\{w_n\} \subset W_0^{1,p}(\Omega)$, $w_n \geq 0$, such that $w_n(x) \uparrow \chi_{\{|u_\epsilon|>k\}}(x)$ for all $x \in \Omega$, except for a set of zero $c_{1,p}^{\mathbb{R}^N}$-capacity. If $\zeta \in C_0^1(\Omega)$, we take $\phi = w_n\zeta$ as test function and derive

$$\int_\Omega w_n \zeta d(\alpha_k - d\beta_k) = 0.$$

This implies

$$\int_{\{|u|>k\}} \zeta d(\alpha_k - d\beta_k) = 0 \qquad \forall \phi \in C_0^1(\Omega),$$

by the dominated convergence theorem, hence $\alpha_k - \beta_k$ vanishes on $\{x \in \Omega : |u(x)| > k\}$. We consider now an increasing sequence $\{\tilde{w}_n\} \subset W_0^{1,p}(\Omega)$, $\tilde{w}_n \geq 0$, such that $\tilde{w}_n \uparrow \chi_{\{|u|<k\}}$. For $\zeta \in C_0^1(\Omega)$, the function $\phi = \tilde{w}_n\zeta$ vanishes in $\{x \in \Omega : |u(x)| \geq k\}$, but for a set with zero $c_{1,p}^{\mathbb{R}^N}$-capacity. Because $\nabla u = 0$ a.e. on $\{x \in \Omega : |u(x)| = k\}$, we get,

$$\int_{\{|u|<h\}} \langle A(x, \nabla T_h(u)) - g, \nabla(\tilde{w}_n\zeta)\rangle dx = \int_{\{|u|<k\}} \langle A(x, \nabla T_h(u)) - g, \nabla(\tilde{w}_n\zeta)\rangle dx,$$

for $h > k > 0$, and

$$\int_{|u|<h} f\tilde{w}_n\zeta dx = \int_{|u|<k} f\tilde{w}_n\zeta dx.$$

Hence

$$\int_\Omega \zeta\tilde{w}_n d(\alpha_k - \beta_k) = \int_\Omega \zeta\tilde{w}_n d(\alpha_h - \beta_h).$$

Therefore, if we let $n \to \infty$,

$$\int_{\{|u|<k\}} \zeta d(\alpha_k - \beta_k) = \int_{\{|u|<k\}} \zeta d(\alpha_h - \beta_h) \qquad \forall \zeta \in C_0^1(\Omega). \tag{4.3.58}$$

This means that $\alpha_h - \beta_h$ is independent of $h > k$ and equal to some measure $\nu_0$ on the set $\{|u| < k\}$ and clearly $\nu_0 \ll c_{1,p}^{\mathbb{R}^N}$. Next we set $\nu_k^\oplus = (\alpha_k - \beta_k)\chi_{\{u=k\}}$ and $\nu_k^\ominus = -(\alpha_k - \beta_k)\chi_{\{u=-k\}}$. Then (4.3.57) reads

$$\begin{aligned}\int_{\{|u|<k\}} \langle A(x, \nabla T_k(u)) - g, \nabla\phi\rangle dx &= \int_{\{|u|<k\}} f\phi dx \\ &+ \int_{\{|u|<k\}} \phi d\nu_0 + \int_{\{u=k\}} \phi d\nu_k^\oplus - \int_{\{u=-k\}} \phi d\nu_k^\ominus,\end{aligned} \tag{4.3.59}$$

with $\phi \in W_0^{1,p}(\Omega) \cap L^\infty(\Omega)$. Notice that it follows from (4.3.50) that

$$\int_{\{|u|<k\}} |\nabla T_k(u)|^p \, dx \le kc \, \|\mu\|_{\mathfrak{M}_b} \, . \tag{4.3.60}$$

*Step 8: The measures $\nu_k^\oplus$ and $\nu_k^\ominus$ are weakly convergent to some nonnegative measures $\alpha$ and $\beta$ respectively.* For $0 < \delta < k$ we set

$$\ell_{\delta,k} = \begin{cases} 0 & \text{if } \; t < k - \delta \\ \dfrac{t-k+\delta}{\delta} & \text{if } \; k - \delta \le t \le k \\ 1 & \text{if } \; t \ge k. \end{cases} \tag{4.3.61}$$

Taking $\ell_{\delta,k}(u)\phi$ as test function in (4.3.59) with $\phi \in C_0^1(\Omega)$, we have

$$\begin{aligned} &\frac{1}{\delta}\int_{\{k-\delta<u<k\}} \langle A(x, \nabla T_k(u)) - g, \nabla T_k(u)\rangle \phi dx \\ &\quad + \int_{\{k-\delta<u<k\}} \langle A(x, \nabla T_k(u)) - g, \nabla \phi\rangle \ell_{\delta,k}(u) dx \\ &\quad = \int_{\{k-\delta\le|u|<k\}} f\ell_{\delta,k}(u)\phi dx + \int_{\{k-\delta\le|u|<k\}} \ell_{\delta,k}(u)\phi d\nu_0 \\ &\quad + \int_{\{u=k\}} \phi d\nu_k^\oplus . \end{aligned} \tag{4.3.62}$$

There exists a null set $\mathcal{E}_1 \subset \mathbb{R}_+$ such that, if $k \notin \mathcal{E}_1$, the first term on the left-hand side of (4.3.62) admits a limit when $\delta \to 0$. Moreover

$$(i) \qquad \lim_{\delta\to 0} \int_{\{k-\delta<u<k\}} \langle A(x, \nabla T_k(u)) - g, \nabla \phi\rangle \ell_{\delta,k}(u) dx = 0$$

$$(ii) \qquad \lim_{\delta\to 0} \int_{\{k-\delta\le|u|<k\}} f\ell_{\delta,k}(u)\phi dx + \int_{\{|u|<k\}} \ell_{\delta,k}(u)\phi d\nu_0 = 0.$$

This implies

$$\lim_{\delta\to 0} \frac{1}{\delta}\int_{\{k-\delta<u<k\}} \langle A(x, \nabla T_k(u)) - g, \nabla T_k(u)\rangle \phi dx = \int_{\{u=k\}} \phi d\nu_k^\oplus, \tag{4.3.63}$$

from which it follows

$$\begin{aligned} \left|\nu_k^\oplus\right|(\Omega) &= \sup\left\{\int_{\{u=k\}} \phi d\nu_k^\oplus : \phi \in C_0^1(\Omega), |\phi| \le 1\right\} \\ &\le \liminf_{\delta\to 0} \frac{1}{\delta}\int_{\{k-\delta<u<k\}} |A(x, \nabla T_k(u)) - g| \, |\nabla T_k(u)| \, dx. \end{aligned} \tag{4.3.64}$$

By (4.1.1) (ii)

$$\int_{\{k-\delta<u<k\}} |A(x, \nabla T_k(u)) - g|\, |\nabla T_k(u)|\, dx \leq \frac{1}{p'} \int_{\{k-\delta<u<k\}} |g|^{p'}\, dx + \left(\lambda + \frac{1}{p}\right) \int_{\{k-\delta<u<k\}} |\nabla T_k(u)|^p\, dx.$$

If we set

$$I(k) = \int_{\{|u|<k\}} |\nabla T_k(u)|^p\, dx \quad \text{and} \quad J(k) = \int_{\{|u|<k\}} |g|^{p'}\, dx,$$

we derive from 4.3.64 that, up to a null set $\mathcal{E}_2$ of $\mathbb{R}_+$ containing $\mathcal{E}_1$, there holds

$$\left|\nu_k^{\oplus}\right|(\Omega) \leq \left(\lambda + \frac{1}{p}\right) I'(k) + \frac{1}{p'} J'(k). \tag{4.3.65}$$

Since $g \in L^{p'}(\Omega)$ and $\nabla u$ satisfies (4.3.60), there exists a sequence $\{k_n\}$ tending to $\infty$ and $M \geq 0$ such that

$$\lim_{k_n \to \infty} I'(k_n) = \liminf_{k \to \infty} I'(k) = M \quad \text{and} \quad \lim_{k_n \to \infty} J'(k_n) = 0. \tag{4.3.66}$$

Therefore, if we choose $k = k_n$ in (4.3.65) we derive from (4.3.66) that the sequence $\{\nu_{k_n}^{\oplus}\}$ is bounded and converges, up to a subsequence denoted by $\{\nu_{k_{n_j}}\}$, to some bounded measure $\alpha$ which turns out to be nonnegative by identity (4.3.63). Furthermore, we could have considered the double sequence $\{(\nu_{k_n}^{\oplus}, \nu_{k_n}^{\ominus})\}$ and would have obtained that this sequence, which is bounded by (4.3.66) converges weakly to $(\alpha, \beta)$ where $\alpha, \beta \geq 0$.

Next we use (4.3.59) with $k = k_n$ and $\phi = h(u)\zeta$, where $h \in W^{1,\infty}(\mathbb{R})$ such that $h'$ has compact support and $\zeta \in C_0^1(\Omega)$. Since $\mu_0 = f - \operatorname{div} g$, it leads to

$$\int_\Omega \langle A(x, \nabla u), \nabla(h(u)\zeta)\rangle dx = \int_\Omega (\mu_0 + \nu_0) h(u)\zeta dx + h(\infty) \int_\Omega \zeta d\alpha - h(-\infty) \int_\Omega \zeta d\beta. \tag{4.3.67}$$

*Step 9: We claim that* $\nu_0 = 0$, $\alpha = \mu_s^+$ *and* $\beta = \mu_s^-$. If we take $h = 1$ in (4.3.67) we see that $u$ satisfies

$$-div\, A(x, \nabla u) = \mu_0 + \nu_0 + \alpha + \beta,$$

in the sense of distributions in $\Omega$. By Step 6 $u$ is also a solution of (4.2.3) in the sense of distributions, with $\mu = \mu_0 + \mu_s$. Therefore $\mu_s = \nu_0 + \alpha - \beta$. Next we consider the function $h_m$ with $m > 0$ defined by

$$h_m(t) = \begin{cases} 0 & \text{if } t < m \\ \dfrac{t-m}{m} & \text{if } m \leq t < 2m, \\ 1 & \text{if } t \geq 2m. \end{cases} \tag{4.3.68}$$

With such a choice of $h = h_m$, (4.3.67) becomes

$$\begin{aligned} \frac{1}{m}\int_{\{m<u<2m\}} \langle A(x,\nabla u),\nabla u\rangle \zeta dx = &-\int_\Omega \langle A(x,\nabla u),\nabla\zeta\rangle h_m(u)dx \\ &+\int_\Omega h_m(u)\zeta d(\mu_0+\nu_0)+\int_\Omega \zeta d\alpha. \end{aligned} \tag{4.3.69}$$

When $m \to \infty$, the first two terms on the right-hand side of (4.3.69) goes to 0 since $\langle A(x,\nabla u),\nabla u\rangle\zeta$ is integrable, $\zeta \in C_0^1(\Omega)$ and $u$ is finite a.e. in $\Omega$. Therefore

$$\lim_{m\to\infty}\frac{1}{m}\int_{\{m<u<2m\}} \langle A(x,\nabla u),\nabla u\rangle \zeta dx = \int_\Omega \zeta d\alpha. \tag{4.3.70}$$

Similarly

$$\lim_{m\to-\infty}\frac{1}{m}\int_{\{-2m<u<-m\}} \langle A(x,\nabla u),\nabla u\rangle \zeta dx = \int_\Omega \zeta d\beta. \tag{4.3.71}$$

Because $u_\epsilon$ is a renormalized solution of (4.3.24) with $\mu_\epsilon = \mu_{\epsilon,0} + \mu^+_{\epsilon,s} - \mu^-_{\epsilon,s}$, there holds also with the test function $h_m(u_\epsilon)\zeta$,

$$\begin{aligned} \frac{1}{m}\int_{\{m<u_\epsilon<2m\}} \langle A(x,\nabla u_\epsilon),\nabla u_\epsilon\rangle \zeta dx = &-\int_\Omega \langle A(x,\nabla u_\epsilon),\nabla\zeta\rangle h_m(u_\epsilon)dx \\ &+\int_\Omega h_m(u_\epsilon)\zeta d\mu_{\epsilon,0}+\int_\Omega \zeta d\mu^+_{\epsilon,s}. \end{aligned} \tag{4.3.72}$$

Letting $\epsilon \to 0$ (following the good subsequence yielding to $u$), we have by the dominated convergence theorem,

$$\lim_{\epsilon\to 0}\int_\Omega \langle A(x,\nabla u_\epsilon),\nabla\zeta\rangle h_m(u_\epsilon)dx = \int_\Omega \langle A(x,\nabla u),\nabla\zeta\rangle h_m(u)dx.$$

Next we use the expression $\mu_{\epsilon,0} = f_\epsilon - div\, g_\epsilon + \lambda^{\oplus}_{\epsilon,0} - \lambda^{\ominus}_{\epsilon,0}$, then, if $\zeta \geq 0$,

$$\int_\Omega h_m(u_\epsilon)\zeta d\mu_{\epsilon,0} \leq \int_\Omega h_m(u_\epsilon)\zeta f_\epsilon dx + \int_\Omega \langle g_\epsilon, \nabla\zeta\rangle h_m(u_\epsilon)dx$$
$$+ \frac{1}{m}\int_{\{m<u_\epsilon<2m\}} \langle g_\epsilon, \nabla T_{2m}(u_\epsilon)\rangle\zeta dx + \int_\Omega \zeta d\lambda^{\oplus}_{\epsilon,0}.$$

Since $\lambda^{\oplus}_\epsilon = \lambda^{\oplus}_{\epsilon,0} + \lambda^{\oplus}_{\epsilon,s}$ and $\mu^+_{\epsilon,s} \leq \lambda^{\oplus}_{\epsilon,s}$, we derive from the above inequality

$$\int_\Omega h_m(u_\epsilon)\zeta d\mu_{\epsilon,0} + \int_\Omega \zeta d\mu^+_{\epsilon,s} \leq \int_\Omega h_m(u_\epsilon)\zeta f_\epsilon dx + \int_\Omega \zeta d\lambda^{\oplus}_\epsilon$$
$$+ \frac{1}{m}\int_{\{m<u_\epsilon<2m\}} \langle g_\epsilon, \nabla T_{2m}(u_\epsilon)\rangle\zeta dx + \int_\Omega \langle g_\epsilon, \nabla\zeta\rangle h_m(u_\epsilon)dx. \tag{4.3.73}$$

By Dunford-Pettis theorem, $h_m(u_\epsilon)f_\epsilon \rightharpoonup h_m(u)f$ weakly in $L^1(\Omega)$, and

$$\int_{\{m<u_\epsilon<2m\}} \langle g_\epsilon, \nabla T_{2m}(u_\epsilon)\rangle\zeta dx \to \int_{\{m<u<2m\}} \langle g, \nabla T_{2m}(u)\rangle\zeta dx.$$

Therefore, we derive from (4.3.72),

$$\limsup_{\epsilon\to 0} \frac{1}{m}\int_{\{m<u_\epsilon<2m\}} \langle A(x,\nabla u_\epsilon), \nabla u_\epsilon\rangle\zeta dx \leq \int_\Omega \langle g, \nabla\zeta\rangle h_m(u)dx$$
$$+ \int_\Omega h_m(u)\zeta f dx - \int_\Omega \langle A(x,\nabla u), \nabla\zeta\rangle h_m(u)dx \tag{4.3.74}$$
$$+ \frac{1}{m}\int_{\{m<u_\epsilon<2m\}} \langle g_\epsilon, \nabla T_{2m}(u)\rangle\zeta dx + \int_\Omega \zeta d\mu^+_s.$$

By Fatou's lemma and Step 6, it follows

$$\frac{1}{m}\int_{\{m<u<2m\}} \langle A(x,\nabla u), \nabla u\rangle\zeta dx \leq \int_\Omega \langle g, \nabla\zeta\rangle h_m(u)dx$$
$$+ \int_\Omega h_m(u)\zeta f dx - \int_\Omega \langle A(x,\nabla u), \nabla\zeta\rangle h_m(u)dx \tag{4.3.75}$$
$$+ \frac{1}{m}\int_{\{m<u<2m\}} \langle g, \nabla T_{2m}(u)\rangle\zeta dx + \int_\Omega \zeta d\mu^+_s.$$

If we let $m$ go to $\infty$ and use (4.3.70) we obtain

$$\int_\Omega \zeta d\alpha \leq \int_\Omega \zeta d\mu^+_s. \tag{4.3.76}$$

Thus $\alpha \leq \mu_s^+$ and $\alpha$ is singular with respect to $c_{1,p}^{\mathbb{R}^N}$. Similarly $\beta \leq \mu_s^-$ and $\beta$ is also singular with respect to $c_{1,p}^{\mathbb{R}^N}$. Since $\mu_s^+ = \nu_0^+ + \alpha$ it implies that $\nu_0^+ = 0$ and $\alpha = \mu_s^+$. Similarly $\nu_0^- = 0$ and $\beta = \mu_s^-$.

*From Step 1-Step 9 we have that u is a renormalized solution of (4.2.3), which proves both the existence and the stability parts in Theorem 4.3.8.*

*Step 10: For any $k > 0$, $T_k(u_\epsilon)$ converges to $T_k(u)$ in $W_0^{1,p}(\Omega)$ as $\epsilon \to 0$.* We first prove that

$$\lim_{\epsilon\to 0}\int_\Omega \langle A(x,\nabla u_\epsilon),\nabla u_\epsilon\rangle dx = \int_\Omega \langle A(x,\nabla u),\nabla u\rangle dx. \tag{4.3.77}$$

Taking $\phi = T_k(u_\epsilon)$ (resp. $\phi = T_k(u)$) as test function in condition **D2** with $u_\epsilon$ (resp. with $u$), and thus $\omega^\pm = \pm k$, we obtain

$$\begin{aligned}
&\int_\Omega \Big(\langle A(x,\nabla T_k(u_\epsilon)),\nabla T_k(u_\epsilon)\,\rangle - \langle A(x,\nabla T_k(u)),\nabla T_k(u)\rangle\Big)\,dx\\
&=\int_\Omega T_k(u_\epsilon)d\mu_{\epsilon,0} - \int_\Omega T_k(u)d\mu_0 + k\left(\mu_{\epsilon,s}^+(\Omega)+\mu_{\epsilon,s}^-(\Omega)\right) - k(\mu_s^+(\Omega)+\mu_s^-(\Omega))\\
&=\int_\Omega T_k(u_\epsilon)d(f_\epsilon - div\, g_\epsilon) + \int_\Omega T_k(u_\epsilon)d\left(\lambda_{\epsilon,0}^\oplus - \lambda_{\epsilon,0}^\ominus\right) - \int_\Omega T_k(u)d(f-div\, g)\\
&+k\left(\mu_{\epsilon,s}^+(\Omega)+\mu_{\epsilon,s}^-(\Omega)\right) - k\left(\mu_s^+(\Omega)+\mu_s^-(\Omega)\right),
\end{aligned}$$

as there holds $\mu_{\epsilon,0} = f_\epsilon - div\, g_\epsilon + \lambda_{\epsilon,0}^\oplus - \lambda_{\epsilon,0}^\ominus$ and $\mu_0 = f - div\, g$. Since $f_\epsilon \to f$ in $L^1(\Omega)$-weak and $T_k(u_\epsilon)$ remains bounded (by $\pm k$) and converges a.e. to $T_k(u)$, then $f_\epsilon T_k(u_\epsilon) \rightharpoonup fT_k(u)$ weakly in $L^1(\Omega)$ by Dunford-Pettis convergence theorem. In the same way $\langle g_\epsilon, T_k(u_\epsilon\rangle \rightharpoonup \langle g, T_k(u)\rangle$ in $L^1(\Omega)$-weak, thus

$$\lim_{\epsilon\to 0}\int_\Omega T_k(u_\epsilon)d(f_\epsilon - div\, g_\epsilon) = \int_\Omega T_k(u)d(f-div\, g) = \int_\Omega T_k(u)d\mu_0.$$

Next

$$\int_\Omega T_k(u_\epsilon)d\left(\lambda_{\epsilon,0}^\oplus - \lambda_{\epsilon,0}^\ominus\right) \leq \lambda_{\epsilon,0}^\oplus(\Omega) + \lambda_{\epsilon,0}^\ominus(\Omega).$$

Therefore

$$\limsup_{\epsilon\to 0}\int_\Omega T_k(u_\epsilon)d\left(\lambda_{\epsilon,0}^\oplus - \lambda_{\epsilon,0}^\ominus\right) + k\left(|\mu_{\epsilon,s}|(\Omega) - k|\mu_s|(\Omega)\right) \leq 0.$$

Since $\lambda^{\oplus}_{\epsilon,0} - \lambda^{\ominus}_{\epsilon,0} \to |\mu_s|$ in the narrow topology and $|T_k(u_\epsilon)| \leq k$, we have

$$\lim_{\epsilon \to 0} \int_\Omega T_k(u_\epsilon) d(\lambda^{\oplus}_{\epsilon,0} - \lambda^{\ominus}_{\epsilon,0}) \leq k\, |\mu_s| \,.$$

This yields

$$\limsup_{\epsilon \to 0} \int_\Omega \langle A(x, T_k(u_\epsilon)), \nabla T_k(u_\epsilon) \rangle dx \leq \int_\Omega \langle A(x, \nabla T_k(u)), \nabla T_k(u) \rangle dx. \tag{4.3.78}$$

By (4.1.1) (iii) $\mathcal{A}$ is monotone in $W^{1,p}_0(\Omega)$, therefore

$$\begin{gathered}\int_\Omega \langle A(x, T_k(u_\epsilon)) - A(x, \nabla T_k(u)), \nabla T_k(u_\epsilon) - \nabla T_k(u) \rangle dx \geq 0 \\ \Longleftrightarrow \\ \int_\Omega \langle A(x, T_k(u_\epsilon)), \nabla T_k(u_\epsilon) \rangle dx + \int_\Omega \langle A(x, T_k(u)), \nabla T_k(u) \rangle dx \\ \geq \int_\Omega \langle A(x, T_k(u_\epsilon)), \nabla T_k(u) \rangle dx + \int_\Omega \langle A(x, T_k(u)), \nabla T_k(u_\epsilon) \rangle dx.\end{gathered} \tag{4.3.79}$$

By Step 6 all the three last integral terms in above inequality are convergent when $\epsilon \to 0$ with limit $\int_\Omega \langle A(x, T_k(u)), \nabla T_k(u) \rangle dx$, therefore

$$\liminf_{\epsilon \to 0} \int_\Omega \langle A(x, T_k(u_\epsilon)), \nabla T_k(u_\epsilon) \rangle dx \geq \int_\Omega \langle A(x, T_k(u)), \nabla T_k(u) \rangle dx, \tag{4.3.80}$$

which, combined with (4.3.78), implies (4.3.77). Since the sequence $\{\langle A(x, T_k(u_\epsilon)), \nabla T_k(u_\epsilon) \rangle\}_{\epsilon > 0}$ converges in $L^1(\Omega)$ it is uniformly integrable, but by (4.1.1) (i)-(ii) we have

$$|\nabla T_k(u_\epsilon)|^p \leq \langle A(x, T_k(u_\epsilon)), \nabla T_k(u_\epsilon) \rangle \leq \lambda\, |\nabla T_k(u_\epsilon)|^p \,,$$

this implies that the sequence $\{|\nabla T_k(u_\epsilon)|^p\}_{\epsilon > 0}$ is also uniformly integrable. Therefore

$$\liminf_{\epsilon \to 0} \int_\Omega |\nabla T_k(u_\epsilon)|^p \, dx = \int_\Omega |\nabla T_k(u)|^p \, dx. \tag{4.3.81}$$

Since the sequence $\{\nabla T_k(u_\epsilon)\}_{\epsilon > 0}$ is weakly convergent in $L^p(\Omega)$ it is strongly convergent, which is the claim. □

*Remark.* An easy adaption shows that the previous result holds if the growth of $A$ is perturbed by a zero order term and (4.1.1) (ii) replaced by

$$|A(x,\xi)| \leq \lambda |\xi|^{p-1} + b(x) \quad \forall (\xi, x) \in \mathbb{R}^N \times \Omega, \tag{4.3.82}$$

where $b \in L^{p'}(\Omega)$.

*Remark.* The problem of uniqueness of renormalized solutions remains unsolved up to now. It is proved in [Bénilan, Boccardo, Gallouët, Gariepy, Pierre, Vàzquez (1995)] if $\mu$ is an integrable function. It is a consequence of the theory of monotone operators [Lions (1969)] if $\mu \in W^{-1,p'}(\Omega)$. If $A(x,\xi) = |\xi|^{p-2}\xi$ and $\mu = \sum_{i\in J} \alpha_j \delta_{a_j}$ where $\#(J) < \infty$ and $\alpha_j > 0$ it is a consequence of the precise expression of the asymptotics near isolated singularities of nonnegative $p$-harmonic functions [Kichenassamy and Véron (1988)]. Finally, all these conditions have been gathered in a unique result [Trudinger, Wang (2009)].

**Theorem 4.3.12** *Assume $\Omega$ is a bounded domain, $1 < p \leq N$ and $A$ satisfies the conditions of Theorem 4.3.8. If $\mu = \mu_1 + \sum_{i\in J} \alpha_j \delta_{a_j}$ where $\mu_1 \ll c_{1,p}^{\mathbb{R}^N}$ is a bounded measure with $a_j \in \Omega$ and $\#(J) < \infty$, then there is unique renormalized solution to problem (4.2.3).*

### 4.3.3 *Further stability results*

The two following variants of Theorem 4.3.8 will be useful in applications.

**Corollary 4.3.13** *Assume $\Omega$ is a bounded domain, $1 < p \leq N$ and $A$ satisfies the conditions of Theorem 4.3.8. Let $\mu \in \mathfrak{M}$ with compact support, $\omega \in \mathfrak{M}_b(\Omega)$ and $\{f_n\} \subset L^1(\Omega)$ which converges to $f$ in $L^1(\Omega)$. Let $\{\rho_n\}$ be a sequence of mollifiers, $\mu_n = \mu * \rho_n$ and $\{u_n\}$ a sequence of renormalized solutions of*

$$\begin{aligned} -\operatorname{div} A(x, \nabla u_n) &= f_n + \mu_n + \omega = 0 && \text{in } \Omega \\ u_n &= 0 && \text{on } \partial\Omega. \end{aligned} \tag{4.3.83}$$

*Then there exists a subsequence $\{u_{n_j}\}$ which converges a.e. to a renormalized solution of*

$$\begin{aligned} -\operatorname{div} A(x, \nabla u) &= f + \mu + \omega = 0 && \text{in } \Omega \\ u &= 0 && \text{on } \partial\Omega. \end{aligned} \tag{4.3.84}$$

*Proof.* We write $\omega = \tilde{h} - \operatorname{div} \tilde{g} + \omega_s^+ - \omega_s^-$ and $\mu = h - \operatorname{div} g + \mu_s^+ - \mu_s^-$ with $\tilde{g}$ and $g$ in $(L^{p'}(\Omega))^N$, $\tilde{h}$ and $h$ in $L^1(\Omega)$. Because of the assumptions we can suppose that $\tilde{g}$, $g$, $\tilde{h}$, $h$, $\mu_s^+$ and $\mu_s^-$ have their support in a fixed compact subset $K$ of $\Omega$, and for $n$ large enough $\rho_n * h$, $\rho_n * g$, $\rho_n * \mu_s^+$ and $\rho_n * \mu_s^-$ have the same property. Moreover $\rho_n * h \to h$ and $\rho_n * g \to g$ respectively in $L^1(\Omega)$ and $(L^{p'}(\Omega))^N$ as $n \to \infty$, and $\operatorname{div}(\rho_n * g) \to \operatorname{div} g$ in $W^{-1,p'}(\Omega)$. Therefore

$$f_n + \mu_n + \omega = f_n + \tilde{h} + \rho_n * h - \operatorname{div}(\tilde{g} + \rho_n * g) + \omega_s^+ + \rho_n * \mu_s^+ - \omega_s^- + \rho_n * \mu_s^-,$$

is a converging decomposition with limit $f + \mu + \omega$, thus the statement follows by Theorem 4.3.8. □

**Corollary 4.3.14** *Assume $\Omega$ is a bounded domain, $1 < p \leq N$ and $A$ satisfies the conditions of Theorem 4.3.8. Let $\{\mu_{i,n}\}$ be two nondecreasing sequences of nonnegative bounded measures converging to $\mu_i \in \mathfrak{M}_b^+(\Omega)$, for i=1,2. Let $\{f_n\} \subset L^1(\Omega)$ which converges to $f$ weakly in $L^1(\Omega)$ and $\{\theta_n\} \subset \mathfrak{M}_b(\Omega)$ which converges in the narrow topology to some bounded measure $\theta$, singular with respect to the $c_{1,p}^{\mathbb{R}^N}$-capacity. For each $n \in \mathbb{N}$, let $u_n$ be a renormalized solution of*

$$\begin{aligned} -\operatorname{div} A(x, \nabla u_n) &= f_n + \mu_{1,n} - \mu_{2,n} + \theta_n = 0 && \text{in } \Omega \\ u_n &= 0 && \text{on } \partial\Omega. \end{aligned} \tag{4.3.85}$$

*Then, up to a subsequence, $\{u_n\}$ converges a.e. in $\Omega$ to a renormalized solution of*

$$\begin{aligned} -\operatorname{div} A(x, \nabla u) &= f + \mu_1 - \mu_2 + \theta = 0 && \text{in } \Omega \\ u &= 0 && \text{on } \partial\Omega. \end{aligned} \tag{4.3.86}$$

We prove first the following precision of Proposition 4.2.6 (iii).

**Lemma 4.3.15** *Let $\mu \in \mathfrak{M}_b^+(\Omega)$ and $\{\mu_n\} \subset \mathfrak{M}_b^+(\Omega)$ be a nondecreasing sequence which converges to $\mu$ in $\mathfrak{M}_b(\Omega)$. Then for $n \in \mathbb{N}$ there exist $F$ and $F_n$ in $L^1(\Omega)$, $G$ and $G_n$ in $(L^{p'}(\Omega))^N$ and $\mu_s$ and $\mu_{n,s}$ in $\mathfrak{M}_b^+(\Omega)$ which are singular with respect to the $c_{1,p}^{\mathbb{R}^N}$-capacity such that*

$$\begin{aligned} &(i) && \mu_n = \mu_{n,0} + \mu_{n,s} = F_n + \operatorname{div} G_n + \mu_{n,s} \\ &(ii) && \mu = \mu_0 + \mu_s = F + \operatorname{div} G + \mu_s, \end{aligned} \tag{4.3.87}$$

*with the property that $F_n \to F$ in $L^1(\Omega)$, $G_n \to G$ in $(L^{p'}(\Omega))^N$, $\operatorname{div} G_n \to \operatorname{div} G$ in $\mathfrak{M}_b(\Omega)$ and*

$$\|F_n\|_{L^1} + \|G_n\|_{(L^{p'})^N} + \|\operatorname{div} G_n\|_{\mathfrak{M}_b} + \|\mu_{n,s}\|_{\mathfrak{M}_b} \leq 6\mu(\Omega). \tag{4.3.88}$$

*Proof.* Since $\{\mu_n\}$ is nondecreasing it converges to $\mu$ in $\mathfrak{M}_b(\Omega)$. Moreover $\{\mu_{n,0}\}$ and $\{\mu_{n,s}\}$ share this monotony property and since

$$\|\mu-\mu_n\|_{\mathfrak{M}_b} = \|\mu_0-\mu_{n,0}\|_{\mathfrak{M}_b} + \|\mu_s-\mu_{n,s}\|_{\mathfrak{M}_b},$$

the two sequences $\{\mu_{n,0}\}$ and $\{\mu_{n,s}\}$ converge to $\{\mu_0\}$ and $\{\mu_s\}$ respectively in $\mathfrak{M}_b(\Omega)$. Furthermore $\|\mu_{n,s}\|_{\mathfrak{M}_b} \leq \|\mu_s\|_{\mathfrak{M}_b} \leq \|\mu\|_{\mathfrak{M}_b}$. We define the sequence $\{\tilde{\mu}_{n,0}\}$ by $\tilde{\mu}_{0,0}=0$ and $\tilde{\mu}_{n,0}=\mu_{n,0}-\mu_{n-1,0}$. By Proposition 4.2.6, for any $n\in\mathbb{N}$, on can find $\tilde{f}_n\in L^1(\Omega)$, $\tilde{g}_n\in(L^{p'})^N$ with $\operatorname{div}\tilde{g}_n\in\mathfrak{M}_b(\Omega)$ such that $\tilde{\mu}_{n,0}=\tilde{f}_n+\operatorname{div}\tilde{g}_n$ and

$$\left\|\tilde{f}_n\right\|_{L^1} + \|\tilde{g}_n\|_{(L^{p'})^N} + \|\operatorname{div}\tilde{g}_n\|_{\mathfrak{M}_b} \leq 5\tilde{\mu}_{n,0}(\Omega).$$

If we define

$$F_n=\sum_{j=0}^{n}\tilde{f}_n \quad\text{and}\quad G_n=\sum_{j=0}^{n}\tilde{g}_n,$$

then $\mu_{n,0}=F_n+\operatorname{div}G_n$ and

$$\|F_n\|_{L^1} + \|G_n\|_{(L^{p'})^N} + \|\operatorname{div}G_n\|_{\mathfrak{M}_b} \leq 5\mu_{n,0}(\Omega),$$

which is (4.3.88). Clearly (4.3.87) holds from the construction. □

*Proof of Corollary 4.3.14.* We write $\nu_n = f_n+\mu_{1,n}-\mu_{2,n}+\theta_n$ and $\nu = f+\mu_1-\mu_2+\theta$. By Lemma 4.3.15 we can decompose $\mu_{i,n}$ in order to have

$$\nu_n = f_n + F_{1,n} + \operatorname{div}G_{1,n} - F_{2,n} - \operatorname{div}G_{2,n} + \mu_{1,n,s} - \mu_{2,n,s} + \theta_n.$$

Accordingly

$$\nu = f + F_1 + \operatorname{div}G_1 - F_2 - \operatorname{div}G_2 + \mu_{1,s} - \mu_{2,s} + \theta.$$

We conclude by Theorem 4.3.8. □

## 4.4 Notes and open problems

*4.4.1.* The Perron's method was introduced in [Perron (1923)] for solving the Dirichlet problem associated with the Laplacian. The Wiener condition for characterizing the regular boundary points for the Laplacian has been proved in [Wiener (1924)]. Many authors contributed to its extension for more general class of second order linear operators.

*4.4.2.* The notion of entropy solutions for quasilinear degenerate elliptic equations has been developed in [Bénilan, Boccardo, Gallouët, Gariepy, Pierre, Vàzquez (1995)] to provide a framework for proving existence and uniqueness of solutions of the problem

$$\begin{aligned} -div\, A(x, \nabla u) &= f && \text{in } \Omega \\ u &= 0 && \text{on } \partial\Omega, \end{aligned} \tag{4.4.1}$$

when $f$ is an integrable function and $A$ satisfies (4.1.1). Other fruitful attempts were developed at the same moment in [Rakotoson (1994)]. These ideas gave rise to the notion of renormalized solutions which proved to a very powerfull tool for studying large classes of degenerate elliptic equations with measure data.

*4.4.3.* The first important results concerning quasilinear degenerate equations with measure data are due to [Boccardo, Gallouët (1989)]. Assuming $2-\frac{1}{N} < p \leq N$ the authors proved existence of weak solutions to (4.2.3) for any bounded measure. They made here a systematic use of the truncation method and obtain that $u \in W_0^{1,q}(\Omega)$ for $1 < q < \frac{N(p-1)}{N-1}$. In [Boccardo, Gallouët (1992)] the result is extended to a more general class of operators. In [Boccardo, Gallouët, Orsina (1996)] existence and uniqueness is obtained in the framework of entropy solutions with a measure $\mu$ absolutely continuous with respect to the $c_{1,p}^{\mathbb{R}^N}$-capacity and assuming only $p > 1$.

*4.4.4.* If $\mu$ is a Radon measure on $\partial\Omega$ it is interesting to consider the following nonlinear measure boundary data problem

$$\begin{aligned} -\, div\, A(x, \nabla u) &= 0 && \text{in } \Omega \\ \langle A(x, \nabla u), \mathbf{n} \rangle &= \mu && \text{on } \partial\Omega, \end{aligned} \tag{4.4.2}$$

where $\mathbf{n}$ is the normal outward unit vector to $\partial\Omega$. The weak formulation is

$$\int_\Omega \langle A(x, \nabla u), \nabla \zeta \rangle dx = \int_{\partial\Omega} \zeta d\mu, \tag{4.4.3}$$

for $\zeta \in C^1(\overline{\Omega})$. A constuction of renormalized solutions would likely involve the capacity $c_{1-\frac{1}{p},p}^{\mathbb{R}^{N-1}}$ instead of $c_{1,p}^{\mathbb{R}^N}$. It is noticeable that if $\Omega = \mathbb{R}_+^N$, $A(x, \xi)$ is a Caratheodory vector field in $\mathbb{R}^N \times \mathbb{R}^N$ satisfying (4.1.1) and $A(x, \xi) = A(\pi \circ x, \xi)$ where $\pi$ is the symmetry with respect to $\partial\mathbb{R}_+^N$, the function $\hat{u}$ defined in $\mathbb{R}^N$ by $\hat{u} = u$ in $\mathbb{R}_+^N$ and $\hat{u} = u \circ \pi$ in $\mathbb{R}_-^N$ satisfies

$$-\, div\, A(x, \nabla \hat{u}) = 2\tilde{\mu} \qquad \text{in } \mathbb{R}^N, \tag{4.4.4}$$

where $\tilde{\mu}(E) = \mu(E \cap \partial\Omega)$ for any Borel set $E$. For problem (4.4.2) in $\mathbb{R}^N_+$ the fundamental solution is $\mu_p \lfloor_{\mathbb{R}^N_+}$. Another question is to link the problem in $\mathbb{R}^N_+$ with the fractional $p$-Laplacian $(-\Delta_p)^s$ $(0 < s < 1)$ defined in $\Omega$ by

$$\langle(-\Delta_p)^s u, v\rangle = \int_{\mathbb{R}^N}\int_{\mathbb{R}^N} \frac{|u(x) - u(y)|^{p-2}(u(x) - u(y))(v(x) - v(y))}{|x - y|^{N+sp}} dxdy, \tag{4.4.5}$$

for all $(u, v) \in W^{s,p}(\mathbb{R}^N)$ vanishing a.e. on $\Omega^c$.

*4.4.5.* It would be interesting to investigate the perturbed nonlinear boundary value problem associated to (4.4.2),

$$\begin{aligned} -\,div\, A(x, \nabla u) &= 0 && \text{in } \Omega \\ \langle A(x, \nabla u), \mathbf{n}\rangle + g(u) &= \mu && \text{on } \partial\Omega, \end{aligned} \tag{4.4.6}$$

in the subcritical and supercritical cases, where $g$ is either an absorption or a source (see next sections).

# Chapter 5

# Quasilinear equations with absorption and measure data

In this chapter we consider quasilinear equations in a bounded domain $\Omega \subset \mathbb{R}^N$, with a lower-order pertubation involving $u$ and $\nabla u$ having a growth dominating the one of the diffusion operator and right-hand side distributions or measures,

$$\begin{aligned} -div\, A(x, \nabla u) + g(x, u, \nabla u) &= \mu \qquad \text{in } \Omega \\ u &= 0 \qquad \text{in } \partial\Omega. \end{aligned} \tag{5.0.1}$$

The functions $A : \Omega \times \mathbb{R}^N$ satisfies the assumptions (4.1.1) of Section 4.1 with $1 < p \leq N$, and we assume that $g : \Omega \times \mathbb{R} \times \mathbb{R}^N$ is a Caratheodory function which satisfies

$$\begin{aligned} &(i) \quad g(x, s, \xi)s \geq 0 \quad \forall (x, s, \xi) \in \Omega \times \mathbb{R} \times \mathbb{R}^N \\ &(ii) \quad |g(x, s, \xi)| \leq b(|s|)(|\xi|^p + c(x)) \quad \forall (x, s, \xi) \in \Omega \times \mathbb{R} \times \mathbb{R}^N, \end{aligned} \tag{5.0.2}$$

where $c$ is nonnegative measurable function and $b : \mathbb{R}_+ \mapsto \mathbb{R}_+$ a nonnegative nondecreasing continuous function.

## 5.1 The subcritical case

### 5.1.1 *General nonlinearity with data in* $W^{-1,p'}(\Omega)$

When $\mu$ belongs to $W^{-1,p'}(\Omega)$, existence of a solution in $W_0^{1,p}(\Omega)$ can be obtained by the method of super and sub solutions as it is developed in Section 1.4.2. However it is not always easy to find a super and a sub solutions in $W_0^{1,p}(\Omega)$. The following result due to [Bensoussan, Boccardo, Murat (1988)] shows how to bypass this difficulty.

**Theorem 5.1.1** *Assume $\Omega$ is a bounded domain, $1 < p \leq N$, $A$ and $g$*

*satisfy the above assumptions. Then for any* $\mu \in W^{-1,p'}(\Omega)$ *there exists* $u \in W_0^{1,p}(\Omega)$ *with* $(1+|u|)g(x,u,\nabla u) \in L^1(\Omega)$ *which satisfies*

$$\int_\Omega (\langle A(x,\nabla u), \nabla\zeta\rangle + g(u,\nabla u)\zeta)\, dx = \mu(\zeta) \quad \forall \zeta \in W_0^{1,p}(\Omega) \cap L^\infty(\Omega). \tag{5.1.1}$$

*Proof.* For $\epsilon > 0$ set

$$g_\epsilon(x,s,\xi) = \frac{g(x,s,\xi)}{1+\epsilon g(x,s,\xi)} \tag{5.1.2}$$

and consider problem $(5.0.1)_\epsilon$ where $g$ is replaced by $g_\epsilon$. This equation admits a solution $u_\epsilon \in W_0^{1,p}(\Omega)$ by Theorem 1.4.1 and the following *energy* estimates hold,

$$\begin{aligned} \max&\left\{\int_\Omega |\nabla u_\epsilon|^p dx, \int_\Omega g(x,u_\epsilon,\nabla u_\epsilon))u_\epsilon dx\right\} \\ &\leq \int_\Omega \langle A(x,\nabla u_\epsilon), \nabla u_\epsilon\rangle dx + \int_\Omega g_\epsilon(x,u_\epsilon,\nabla u_\epsilon)u_\epsilon dx \\ &\leq \mu(u_\epsilon) \leq \|\mu\|_{W^{-1,p'}} \|u_\epsilon\|_{W_0^{1,p}}, \end{aligned} \tag{5.1.3}$$

since $g_\epsilon$ satisfies (5.0.2). Therefore by compactness there exist a subsequence still denoted by $\{u_\epsilon\}$ and $u \in W_0^{1,p}(\Omega)$ such that $u_\epsilon$ converges to $u$, weakly in $W_0^{1,p}(\Omega)$, strongly in $L^q(\Omega)$ for all $q < p^* = \frac{Np}{N-p}$ and a.e. in $\Omega$. For $k > 0$ we put $u_k = T_k(u)$, $u_k^+ = T_k(u^+)$ and $z_\epsilon = u_\epsilon^+ - u_k^+$. Then $z_\epsilon$ and $z_\epsilon^+$ belong to $W_0^{1,p}(\Omega)$.

*Step 1: Estimate of* $z_\epsilon^+$. From (5.1.1) we have

$$\int_\Omega \langle A(x,\nabla u_\epsilon), \nabla z_\epsilon^+\rangle dx + \int_\Omega g(x,u_\epsilon,\nabla u_\epsilon) z_\epsilon^+ dx = \mu(z_\epsilon^+). \tag{5.1.4}$$

Notice that $g(x,u_\epsilon,\nabla u_\epsilon)z_\epsilon^+ \geq 0$ by (5.0.2) (i), the first term of the left-hand side of the identity above is smaller than the right-hand side, which yields, since $u_\epsilon = u_\epsilon^+$ on $\{x \in \Omega : z_\epsilon^+(x) > 0\}$,

$$\begin{aligned} &\int_\Omega \langle A(x,\nabla u_\epsilon^+) - A(x,\nabla u_k^+), \nabla(u_\epsilon^+ - u_k^+)^+\rangle dx \\ &\qquad \leq -\int_\Omega \langle A(x,\nabla u_k^+), \nabla(u_\epsilon^+ - u_k^+)^+\rangle dx + \mu(z_\epsilon^+). \end{aligned} \tag{5.1.5}$$

Because $z_\epsilon^+ \to (u - u_k^+)^+$ a.e. when $\epsilon \to 0$ and $z_\epsilon^+$ is bounded in $W_0^{1,p}(\Omega)$, $z_\epsilon^+$ converges weakly to $(u - u_k^+)^+$ in $W_0^{1,p}(\Omega)$. We infer from (5.1.5), that for fixed $k$,

$$\limsup_{\epsilon\to 0}\int_\Omega \langle A(x,\nabla u_\epsilon^+) - A(x,\nabla u_k^+), \nabla(u_\epsilon^+ - u_k^+)^+\rangle dx \le R_k, \tag{5.1.6}$$

where

$$R_k = -\int_\Omega \langle A(x,\nabla u_k^+), \nabla(u^+ - u_k^+)^+\rangle dx + \mu((u^+ - u_k^+)^+).$$

Notice that $u_k^+ \to u^+$ in $W_0^{1,p}(\Omega)$ when $k\to\infty$, thus $\lim_{k\to\infty} R_k = 0$.

*Step 2: Estimate of $z_\epsilon^-$.* As in the proof of Theorem 1.4.6 we introduce the special test function $v_\epsilon = \eta_\nu(z_\epsilon^-)$ fitted to obtain energy estimates where $\eta_\nu(s) = se^{\nu s^2}$ with $\nu > 0$ will be chosen later on. Since $0 \le z_\epsilon^- \le k$ and $z_\epsilon \in W_0^{1,p}(\Omega)$, $v_\epsilon \in W_0^{1,p}(\Omega)$, there holds

$$\int_\Omega \langle A(x,\nabla u_\epsilon), \nabla u_\epsilon^-\rangle \eta_\nu'(z_\epsilon^-)dx + \int_\Omega g_\epsilon(x,\nabla u_\epsilon)\eta_\nu(z_\epsilon^-)dx = \mu(\eta_\nu(z_\epsilon^-)). \tag{5.1.7}$$

We define

$$E_\epsilon = \{x\in\Omega : u_\epsilon^+(x) \le u_k^+(x)\} \quad\text{and}\quad F_\epsilon = \{x\in\Omega : 0\le u_\epsilon(x) \le u_k^+(x)\}.$$

Then $F_\epsilon \subset E_\epsilon$ and

$$\int_\Omega g(x,u_\epsilon,\nabla u_\epsilon)\eta_\nu(z_\epsilon^-)dx = \int_{E_\epsilon} g(x,u_\epsilon,\nabla u_\epsilon)\eta_\nu(z_\epsilon^-)dx.$$

Since $g(x,u_\epsilon,\nabla u_\epsilon)\le 0$ whenever $u_\epsilon \le 0$, we have

$$\begin{aligned}\int_{E_\epsilon} g(x,u_\epsilon,\nabla u_\epsilon)\eta_\nu(z_\epsilon^-)dx &\le \int_{F_\epsilon} g(x,u_\epsilon,\nabla u_\epsilon)\eta_\nu(z_\epsilon^-)dx\\ &\le b(k)\int_{F_\epsilon}\left(|\nabla u_\epsilon|^p + c(x)\right)\eta_\nu(z_\epsilon^-)dx,\end{aligned} \tag{5.1.8}$$

because of (5.0.2) (ii) and the fact that $0\le u_\epsilon \le u_k \le k$ on $F_\epsilon$. Furthermore, the last term on the right-hand side of (5.1.8) is smaller than

$$\frac{b(k)}{\lambda}\int_{F_\epsilon}\langle A(x,\nabla u_\epsilon),\nabla u_\epsilon\rangle\eta_\nu(z_\epsilon^-)dx + b(k)\int_{F_\epsilon} c(x)\eta_\nu(z_\epsilon^-)dx.$$

Notice that $z_\epsilon^- = u_k^+$ on $\{x : u_\epsilon(x)\le 0\}$, therefore

$$\begin{aligned}-\int_\Omega\langle A(x,\nabla u_\epsilon^+),\nabla z_\epsilon^-\rangle\eta_\nu'(z_\epsilon^-)dx &= -\int_\Omega\langle A(x,\nabla u_\epsilon),\nabla z_\epsilon^-\rangle\eta_\nu'(z_\epsilon^-)dx\\ &\quad + \int_\Omega\langle A(x,\nabla u_\epsilon) - A(x,\nabla u_\epsilon^+),\nabla u_k^+\rangle\eta_\nu'(z_\epsilon^-)dx.\end{aligned}$$

Using the previous inequalities we derive

$$
\begin{aligned}
&-\int_\Omega \langle A(x,\nabla u_\epsilon^+) - A(x,\nabla u_k^+), \nabla(u_\epsilon^+ - u_k^+)\rangle \eta_\nu'(z_\epsilon^-)dx \\
&\quad = \int_\Omega \langle A(x,\nabla u_\epsilon) - A(x,\nabla u_\epsilon^+), \nabla u_k^+\rangle \eta_\nu'(z_\epsilon^-)dx - \mu(\eta_\nu(z_\epsilon^-) \\
&\quad + \int_\Omega g_\epsilon(u_\epsilon,\nabla u_\epsilon)\eta_\nu(z_\epsilon^-)dx + \int_\Omega \langle A(x,\nabla u_k^+), \nabla z_\epsilon^-\rangle \eta_\nu'(z_\epsilon^-)dx \\
&\quad \leq \int_\Omega \langle A(x,\nabla u_\epsilon) - A(x,\nabla u_\epsilon^+), \nabla u_k^+\rangle \eta_\nu'(z_\epsilon^-)dx - \mu(\eta_\nu(z_\epsilon^-) \\
&\quad + \int_\Omega \langle A(x,\nabla u_k^+), \nabla u_k^+\rangle \eta_\nu'(z_\epsilon^-)dx \\
&\quad + \frac{b(k)}{\lambda}\int_\Omega \langle A(x,\nabla u_\epsilon), \nabla u_\epsilon\rangle \eta_\nu(z_\epsilon^-)dx + b(k)\int_\Omega c(x)\eta_\nu(z_\epsilon^-)dx \qquad (5.1.9) \\
&\quad \leq \int_\Omega \langle A(x,\nabla u_\epsilon) - A(x,\nabla u_\epsilon^+), \nabla u_k^+\rangle \eta_\nu'(z_\epsilon^-)dx - \mu(\eta_\nu(z_\epsilon^-) \\
&\quad + \int_\Omega \langle A(x,\nabla u_k^+), \nabla u_k^+\rangle \eta_\nu'(z_\epsilon^-)dx \\
&\quad + \frac{b(k)}{\lambda}\int_\Omega \langle A(x,\nabla u_\epsilon^+) - A(x,\nabla u_k^+, \nabla u_\epsilon^+ - \nabla u_k^+\rangle \eta_\nu(z_\epsilon^-)dx \\
&\quad + \frac{b(k)}{\lambda}\int_\Omega \langle A(x,\nabla u_\epsilon^+), \nabla u_k^+\rangle \eta_\nu(z_\epsilon^-)dx + b(k)\int_\Omega c(x)\eta_\nu(z_\epsilon^-)dx \\
&\quad + \frac{b(k)}{\lambda}\int_\Omega \langle A(x,\nabla u_k^+ e), \nabla u_\epsilon^+ - \nabla u_k^+\rangle \eta_\nu(z_\epsilon^-)dx.
\end{aligned}
$$

If we choose $\nu = \frac{b^2(k)}{4\lambda^2}$ we have that $\eta_\nu' - \frac{b(k)}{\lambda}\eta_\nu \geq \dfrac{1}{2}$. Therefore

$$
\begin{aligned}
&-\frac{1}{2}\int_\Omega \langle A(x,\nabla u_\epsilon^+) - A(x,\nabla u_k^+), \nabla u_\epsilon^+ - \nabla u_k^+\rangle dx \\
&\leq \int_\Omega \langle A(x,\nabla u_\epsilon) - A(x,\nabla u_\epsilon^+), \nabla u_k^+\rangle \eta_\nu'(u_k^+)dx - \mu(\eta_\nu(z_\epsilon^-)) \\
&+ \int_\Omega \langle A(x,\nabla u_k^+), \nabla z_\epsilon^-\rangle \eta_\nu'(z_\epsilon^-)dx + \frac{b(k)}{\lambda}\int_\Omega \langle A(x,\nabla u_\epsilon^+), \nabla u_k^+ \langle \eta_\nu(z_\epsilon^-)dx \\
&+ \frac{b(k)}{\lambda}\int_\Omega \langle A(x,\nabla u_k^+), \nabla u_\epsilon^+ - \nabla u_k^+\rangle \eta_\nu(z_\epsilon^-)dx + b(k)\int_\Omega c(x)\eta_\nu(z_\epsilon^-)dx.
\end{aligned}
\tag{5.1.10}
$$

By (5.1.4), $A(.,\nabla u_\epsilon)$ and $A(.,\nabla u_\epsilon^+)$ are bounded in $L^{p'}(\Omega)$, therefore there exist subsequences stiil denoted by $\{u_\epsilon\}$ and $\gamma, \sigma \in L^{p'}(\Omega)$ such that $A(., \nabla u_\epsilon) \rightharpoonup \gamma$ and $A(., \nabla u_\epsilon^+) \rightharpoonup \sigma$ weakly in $L^{p'}(\Omega)$. Using the fact that $z_\epsilon \to (u-k)^+$, $\eta_\nu(0) = 0$ and $\eta'_\nu(0) = 1$, this turns (5.1.10) into

$$\begin{aligned} \limsup_{\epsilon\to 0} -\int_\Omega \langle A(x, \nabla u_\epsilon^+) - A(x, \nabla u_k^+), \nabla u_\epsilon^+ - \nabla u_k^+\rangle dx \\ \leq 2\int_\Omega \langle \gamma(x) - \sigma(x), \nabla u_k^+\rangle \eta'_\nu(u_k^+) dx. \end{aligned} \tag{5.1.11}$$

We notice that $\max\{u_\epsilon^+, k\}\left(A(x, \nabla u_\epsilon^+) - A(x, \nabla u_k^+)\right) = 0$ a.e., therefore $\max\{u^+, k\}\left(\gamma(x) - \sigma(x)\right) = 0$, which yields finally

$$\limsup_{\epsilon\to 0} -\int_\Omega \langle A(x, \nabla u_\epsilon^+) - A(x, \nabla u_k^+), \nabla u_\epsilon^+ - \nabla u_k^+\rangle dx \leq 0. \tag{5.1.12}$$

*Step 3: We claim that $u_\epsilon^+ \to u^+$ in $W_0^{1,p}(\Omega)$.* By (5.1.12), (5.1.6) we have

$$\begin{aligned} &\limsup_{\epsilon\to 0} \int_\Omega \langle A(x, \nabla u_\epsilon^+) - A(x, \nabla u^+), \nabla u_\epsilon^+ - \nabla u^+\rangle dx \\ &\leq \limsup_{\epsilon\to 0} \int_\Omega \langle A(x, \nabla u_\epsilon^+) - A(x, \nabla u_k^+), \nabla (u_\epsilon^+ - u_k^+)^+\rangle dx \\ &+ \limsup\nolimits_{\epsilon\to 0} -\int_\Omega \langle A(x, \nabla u_\epsilon^+) - A(x, \nabla u_k^+), \nabla (u_\epsilon^+ - u_k^+)^-\rangle dx \\ &+ \limsup_{\epsilon\to 0} \int_\Omega \langle A(x, \nabla u_\epsilon^+) - A(x, \nabla u_k^+), \nabla u_k^+ - \nabla u^+\rangle dx \\ &+ \limsup_{\epsilon\to 0} \int_\Omega \langle A(x, \nabla u_k^+) - A(x, \nabla u^+), \nabla u_k^+ - \nabla u^+\rangle dx \\ &\leq R_k + \int_\Omega \langle \sigma(x) - A(x, \nabla u_k^+), \nabla u_k^+ - \nabla u^+\rangle dx. \end{aligned} \tag{5.1.13}$$

By Step 1, $\lim_{k\to\infty} R_k = 0$ and clearly the other term on the right-hand side of (5.1.13) tends also to 0. Hence, letting $k \to \infty$ we derive

$$\limsup_{\epsilon\to 0} \int_\Omega \langle A(x, \nabla u_\epsilon^+) - A(x, \nabla u^+), \nabla u_\epsilon^+ - \nabla u^+\rangle dx \leq 0. \tag{5.1.14}$$

By the argument of Leray and Lions used in the proof of Theorem 1.4.4-Step 6 and based upon the assumption (4.1.1) (iii), we have that $\nabla u_\epsilon^+ \to \nabla u^+$ a.e. in $\Omega$ as $\epsilon \to 0$. Combined with (5.1.3) it implies that $\nabla u_\epsilon^+ \to \nabla u^+$ in $L^r(\Omega)$ for any $r < p$. Furthermore (5.1.14) implies also

$$\limsup_{\epsilon\to 0} \int_\Omega \langle A(x, \nabla u_\epsilon^+), \nabla u_\epsilon^+\rangle dx \leq \int_\Omega \langle A(x, \nabla u^+), \nabla u^+\rangle dx.$$

We use now the monotonicity argument of the proof of Theorem 4.3.8-Step 10 and infer that

$$\lim_{\epsilon\to 0}\int_\Omega \langle A(x,\nabla u_\epsilon^+),\nabla u_\epsilon^+\rangle dx=\int_\Omega \langle A(x,\nabla u^+),\nabla u^+\rangle dx.$$

By (4.1.1) (i) this implies that $|\nabla u_\epsilon^+|^p$ in uniformly integrable in $\Omega$ and thus $\nabla u_\epsilon^+\to\nabla u^+$ in $L^p(\Omega)$.

*Step 4: We claim that $u_\epsilon^-\to u^-$ in $W_0^{1,p}(\Omega)$.* For $k>0$ we set $u_k^-=\min\{u^-,k\}=T_k(u^-)$ and $y_\epsilon=u_\epsilon^--u_k^-$. Then

$$\int_\Omega \langle A(x,\nabla u_\epsilon),\nabla y_\epsilon^-\rangle dx+\int_\Omega g(x,u_\epsilon,\nabla u_\epsilon)y_\epsilon^-\,dx=\mu(y_\epsilon^-).$$

By (5.0.2) (i) $g(x,u_\epsilon,\nabla u_\epsilon)y_\epsilon^-\le 0$ a.e. in $\Omega$. We set

$$Q_k=-\int_\Omega \langle A(x,-\nabla u_k^-),\nabla(u^--u_k^-)^+\rangle dx+\mu((u^--u_k^-)^+).$$

Then as in Step 1, we obtain

$$\limsup_{\epsilon\to 0}-\int_\Omega \langle A(x,-\nabla u_\epsilon^-)-A(x,-\nabla u_k^-),\nabla(u_\epsilon^--u_k^-)^-\rangle dx\le Q_k \tag{5.1.15}$$

and $Q_k\to 0$ as $k\to\infty$. Taking $v_\epsilon=\eta_\nu(y_\epsilon^-)$ for test function, we derive as in (5.1.14)

$$\limsup_{\epsilon\to 0}\int_\Omega \langle A(x,-\nabla u_\epsilon^-)-A(x,-\nabla u_k^+),\nabla(u_\epsilon^--u_k^-)^-\rangle dx\le 0. \tag{5.1.16}$$

As in Step 3 it yields

$$\limsup_{k\to\infty}\int_\Omega \langle A(x,-\nabla u^-)-A(x,-\nabla u_k^+),\nabla(u^--u_k^-)^-\rangle dx\le 0,$$

which in turn implies the claim by the technique used in Step 3.

*Step 5: End of the proof.* It follows from Step 3-Step 4 that, up to a subsequence, $u_\epsilon\to u$ in $W_0^{1,p}(\Omega)$. Furthermore we can choose this subsequence such that $u_\epsilon\to u$ and $\nabla u_\epsilon\to\nabla u$ a.e. in $\Omega$. Hence $g(x,u_\epsilon,\nabla u_\epsilon)\to g(x,u,\nabla u)$ and $u_\epsilon g(x,u_\epsilon,\nabla u_\epsilon)\to ug(x,u,\nabla u)$ a.e. as $\epsilon\to 0$. Since

$$0\le\int_\Omega u_\epsilon g(x,u_\epsilon,\nabla u_\epsilon)dx\le\|\mu\|_{W^{-1,p'}}\|u_\epsilon\|_{W^{1,p}}, \tag{5.1.17}$$

we derive by Fatou's lemma that the same estimate holds if $u_\epsilon$ is replaced by $u$. It remains to prove that $u_\epsilon g(., u_\epsilon, \nabla u_\epsilon) \to g(., u_\epsilon, \nabla u_\epsilon)$ in $L^1(\Omega)$. Let $E \subset \Omega$ be a Borel set. Then for $\tau > 0$, there holds

$$\begin{aligned}\int_E |g(., u_\epsilon, \nabla u_\epsilon)|\, dx &= \int_{E\cap\{|u_\epsilon|\leq\tau\}} |g(., u_\epsilon, \nabla u_\epsilon)|\, dx + \int_{E\cap\{|u_\epsilon|>\tau\}} |g(., u_\epsilon, \nabla u_\epsilon)|\, dx \\ &\leq b(\tau)\int_E (|\nabla u_\epsilon|^p + c(x))dx + \frac{1}{\tau}\int_\Omega u_\epsilon g(x, u_\epsilon, \nabla u_\epsilon)dx \\ &\leq b(\tau)\int_E (|\nabla u_\epsilon|^p + c(x))dx + \frac{1}{\tau}\|\mu\|_{W^{-1,p'}}\|u_\epsilon\|_{W^{1,p}}.\end{aligned}$$

Since $\|u_\epsilon\|_{W^{1,p}} \to \|u\|_{W^{1,p}}$, for any $\eta > 0$ there exists $\tau > 0$ such that for any $\epsilon > 0$,

$$\frac{1}{\tau}\|\mu\|_{W^{-1,p'}}\|u_\epsilon\|_{W^{1,p}} \leq \frac{\eta}{2}.$$

Then there exists $\delta > 0$ such that

$$\text{meas}\,(E) \leq \delta \Longrightarrow b(\tau)\int_E (|\nabla u_\epsilon|^p + c(x))dx \leq \frac{\eta}{2}.$$

By Vitali's convergence theorem $u_\epsilon g(., u_\epsilon, \nabla u_\epsilon) \to g(., u_\epsilon, \nabla u_\epsilon)$ in $L^1(\Omega)$. Since for any $\zeta \in W_0^{1,p}(\Omega) \cap L^\infty(\Omega)$ there holds,

$$\int_\Omega (\langle A(x, \nabla u_\epsilon), \nabla\zeta\rangle + g(u_\epsilon, \nabla u_\epsilon)\zeta)\, dx = \mu(\zeta),$$

we derive (5.1.1). □

### 5.1.2 *Subcritical nonlinearity with measure data*

We assume that $A$ and $g$ satisfy (4.1.1) and (5.0.2) and that there exists a continuous function $\tilde{g} : \mathbb{R}_+ \times \mathbb{R}_+ \mapsto \mathbb{R}_+$, nondecreasing in its two variables, such that

$$|g(x, s, \xi)| \leq \tilde{g}(|s|, |\xi|) \qquad \forall (x, s, \xi) \in \Omega \times \mathbb{R} \times \mathbb{R}^N, \tag{5.1.18}$$

and satisfying

$$\int_1^\infty \tilde{g}\left(s, s^{\frac{N-1}{N-p}}\right) s^{-\frac{p(N-1)}{N-p}}\, ds < \infty, \tag{5.1.19}$$

if $1 < p < N$, and

$$\int_1^\infty \tilde{g}\,(s, e^{\gamma s})\, e^{-\gamma N s} ds < \infty, \tag{5.1.20}$$

for some $\gamma > 0$ if $p = N$. Note that if (5.1.20) holds for some $\gamma > 0$, it holds if $\gamma$ is replaced by $\gamma' > \gamma$ since

$$\int_1^\infty \tilde{g}\left(s, e^{\gamma s}\right) e^{-\gamma N s} ds = \frac{1}{\gamma}\int_{e^\gamma}^\infty \tilde{g}\left(\frac{1}{\gamma}\ln t, t\right)\frac{dt}{t^{N+1}}, \tag{5.1.21}$$

as $\tilde{g}$ is nondecreasing in both variable.

**Definition 5.1** Let $\mu = \mu_0 + \mu_s$ be a bounded measure in $\Omega$ where $\mu_0 \ll c_{1,p}^{\mathbb{R}^N}$ and $\mu_s \perp c_{1,p}^{\mathbb{R}^N}$ and assume $A$ and $g$ satisfy (4.1.1) and (5.0.2). A measurable and finite a.e. function $u$ in $\Omega$ is a renormalized solution of (5.0.2) if it satisfies (4.3.22), $g(., u, \nabla u) \in L^1(\Omega)$ and

$$\begin{aligned}\int_\Omega \Big(\langle A(x, \nabla u), \nabla (h(u)\phi)\rangle + g(x, u, \nabla u)h(u)\phi\Big)\, dx \\ = \int_\Omega h(u)\phi d\mu_0 + h(\infty)\int_\Omega \phi d\mu_s^+ - h(-\infty)\int_\Omega \phi d\mu_s^-,\end{aligned} \tag{5.1.22}$$

for all $h \in W^{1,\infty}(\mathbb{R})$ with $h'$ having compact support, and for all $\phi \in C_0^1(\Omega)$.

*Remark.* Since $g(., u, \nabla u) \in L^1(\Omega)$, a renormalized solution $u$ of (5.0.2) satisfies also the equivalent definitions of renormalized **D2**, **D3** and **D4**.

**Theorem 5.1.2** *Assume $A$ satisfies (4.1.1) and $g$ satisfy (5.0.2) and (5.1.18).*

*(i) If $1 < p < N$ and $g$ satisfies (5.1.19), then for any bounded measure $\mu$ in $\Omega$ there exists a renormalized solution $u$ to problem (5.0.2).*

*(ii) If $p = N$ there exists a constant $c_\Omega > 0$ such that for any bounded measure $\mu$ in $\Omega$ satisfying $\|\mu\|_{\mathfrak{M}} < c_\Omega \gamma^{1-N}$ there exists a renormalized solution $u$ to problem (5.0.2).*

*Proof.* We write $\mu = f - div\, g + \mu_s^+ - \mu_s^-$ and consider a convergent smooth decomposition $\mu_\epsilon = f_\epsilon - div\, g + \mu_{s,\epsilon}^+ - \mu_{s,\epsilon}^-$ as in the proof of Theorem 4.3.8. Since $\mu_\epsilon \in W^{-1,p'}(\Omega)$ there exists $u_\epsilon \in W_0^{1,p}(\Omega)$ such that $(1 + |u_\epsilon|)g(., u_\epsilon, \nabla u_\epsilon) \in L^1(\Omega)$ which satisfies

$$\int_\Omega \Big(\langle A(x, \nabla u_\epsilon), \nabla \phi\rangle + g(x, u_\epsilon, \nabla u_\epsilon)\phi\Big)\, dx = \int_\Omega \phi d\mu_\epsilon, \tag{5.1.23}$$

for all $\phi \in W_0^{1,p}(\Omega) \cap L^\infty(\Omega)$. For $m \in \mathbb{N}_*$, $\phi = \tanh(mu_\epsilon)$, is an admisssible test function, therefore

$$m\int_\Omega \langle A(x,\nabla u_\epsilon),\nabla u_\epsilon\rangle(1-\tanh^2(mu_\epsilon))dx + \int_\Omega g(x,u_\epsilon,\nabla u_\epsilon)\tanh(mu_\epsilon)dx = \int_\Omega \tanh(mu_\epsilon)d\mu_\epsilon,$$

which implies, by (4.1.1) (i), (5.0.2) (i) and Fatou's lemma, when $m\to\infty$,

$$\int_\Omega |g(x,u_\epsilon,\nabla u_\epsilon)|\,dx \le \|\mu_\epsilon\|_{\mathfrak{M}}\,, \tag{5.1.24}$$

and the right-hand side of (5.1.24) is bounded by $c\,\|\mu\|_{\mathfrak{M}}$ for some $c>0$.

*Step 1: Convergence of* $\{u_\epsilon\}$ *and* $\{\nabla u_\epsilon\}$. The function $u_\epsilon$ is a renormalized solution of

$$\begin{aligned} -div\,A(x,\nabla u_\epsilon) &= \tilde{\mu}_\epsilon = \mu_\epsilon - g(.,u_\epsilon,\nabla u_\epsilon) && \text{in } \Omega\\ u_\epsilon &= 0 && \text{in } \partial\Omega, \end{aligned} \tag{5.1.25}$$

and there holds

$$\|\tilde{\mu}_\epsilon\|_{\mathfrak{M}} \le \|\mu_\epsilon\|_{\mathfrak{M}} + \|g(.,\nabla u_\epsilon\|_{L^1} \le 2c\,\|\mu\|_{\mathfrak{M}}\,.$$

We apply Proposition 4.3.9, Proposition 4.3.11 and Theorem 4.3.8-Step 1 and obtain estimates on $u_\epsilon$ which differ according $1<p<N$ or $p=N$:

$$\begin{array}{llll} (i) & \text{meas}\,(\{|u_\epsilon|>k\}) \le c_1 \dfrac{\|\mu\|_{\mathfrak{M}}^{\frac{N}{N-p}}}{k^{\frac{N(p-1)}{N-p}}} & \forall k>0 & \text{if } 1<p<N,\\ (ii) & \|u_\epsilon\|_{BMO} \le c_1\,\|\mu\|_{\mathfrak{M}}^{\frac{1}{N-1}} \quad \forall k>0 & & \text{if } p=N,\\ (iii) & \text{meas}\,(\{|\nabla u_\epsilon|>k\}) \le c_1 \dfrac{\|\mu\|_{\mathfrak{M}}^{\frac{N}{N-1}}}{k^{\frac{N(p-1)}{N-1}}} & \forall k>0 & \text{if } 1<p\le N, \end{array} \tag{5.1.26}$$

and, for all $1<p\le N$,

$$\int_\Omega |\nabla u_\epsilon|^{q(p-1)} \le \text{meas}\,(\Omega) + c_1\,\|\mu\|_{\mathfrak{M}} \quad \forall q<\frac{N}{N-1}. \tag{5.1.27}$$

Let $k>0$ and $u_{\epsilon,k}=T_k(u_\epsilon)$, we take $T_k(u_\epsilon)$ for test function and get

$$\int_{\{|u_\epsilon|\le k\}} \langle A(x,\nabla u_\epsilon),\nabla u_\epsilon\rangle dx + \int_\Omega g(x,u_\epsilon,\nabla u_\epsilon)T_k(u_\epsilon)dx = \int_\Omega T_k(u_\epsilon)d\mu_\epsilon. \tag{5.1.28}$$

Thus, by (4.1.1) (i),

$$\|T_k(u)\|^p_{W_0^{1,p}} \leq (k+c)\, \|\mu_\epsilon\|_{\mathfrak{M}} \quad \forall k > 0. \tag{5.1.29}$$

As in the proof of Theorem 4.3.8-Step 2 there exists a subsequence of $\{u_\epsilon\}$, still denoted by the same notation, which is a Cauchy sequence and converges a.e. to a measurable function $u$ when $\epsilon \to 0$. The fact that it is a Cauchy sequence in measure is independent of the decomposition of the right-hand side $\tilde{\mu}_\epsilon$ but depends only on its boundedness in $\mathfrak{M}_b(\Omega)$. As in Theorem 4.3.8-Step 3, this implies that $T_k(u_\epsilon) \to T_k(u)$ in $W_0^{1,p}(\Omega)$, hence $T_k(u) \in W_0^{1,p}(\Omega)$ and $\nabla u$ is well defined a.e. in $\Omega$. By Step 5 of the proof Theorem 4.3.8, for any $k > 0$, $\{\nabla T_k(u_\epsilon)\}$ is a Cauchy sequence in measure and thus, as in Step 6, $\{\nabla u_\epsilon\}$ is also a Cauchy sequence in measure and thus, up to an other extraction of subsequence ignored in our notations, $\nabla u_\epsilon \to \nabla u$ a.e. in $\Omega$.

*Step 2: Convergence of* $\{g(.,u_\epsilon,\nabla u_\epsilon)\}$. Since $g$ is continuous, $g(x,u_\epsilon,\nabla u_\epsilon) \to g(x,u,\nabla u)$ for almost all $x \in \Omega$. We prove below that $\{\tilde{g}(u_\epsilon,|\nabla u_\epsilon|)\}$ (see (5.1.18)) is uniformly integrable in $\Omega$.

*The case* $1 < p < N$. Let $E \subset \Omega$ be a Borel set. For $\sigma > 0$, we set

$$A_\sigma = \{x \in \Omega : |u_\epsilon(x)| > \sigma\} \quad \text{and} \quad a(\sigma) = \text{meas}\,(A_\sigma), \tag{5.1.30}$$

$$B_\sigma = \{x \in \Omega : |\nabla u_\epsilon(x)| > \sigma^{\frac{N-1}{N-p}}\} \quad \text{and} \quad b(\sigma) = \text{meas}\,(B_\sigma), \tag{5.1.31}$$

and

$$C_\sigma = A_\sigma \cap B_\sigma \quad \text{and} \quad c(\sigma) = \text{meas}\,(C_\sigma). \tag{5.1.32}$$

$$\begin{aligned}\int_E \tilde{g}(|u_\epsilon|,|\nabla u_\epsilon|)dx &= \int_{E\cap C_\sigma} \tilde{g}(|u_\epsilon|,|\nabla u_\epsilon|)dx + \int_{E\cap A_\sigma^c \cap B_\sigma} \tilde{g}(|u_\epsilon|,|\nabla u_\epsilon|)dx \\ &\quad + \int_{E\cap A_\sigma \cap B_\sigma^c} \tilde{g}(|u_\epsilon|,|\nabla u_\epsilon|)dx + \int_{E\cap A_\sigma^c \cap B_\sigma^c} \tilde{g}(|u_\epsilon|,|\nabla u_\epsilon|)dx.\end{aligned}$$

Since $\tilde{g}$ is nondecreasing in both variable, we have

$$\int_{E\cap A_\sigma^c \cap B_\sigma^c} \tilde{g}(|u_\epsilon|,|\nabla u_\epsilon|)dx \leq \tilde{g}(\sigma,\sigma^{\frac{N-1}{N-p}})\text{meas}\,(E),$$

and

$$\int_{E\cap A_\sigma\cap B_\sigma^c}\tilde{g}(|u_\epsilon|,|\nabla u_\epsilon|)dx \leq \int_{A_\sigma}\tilde{g}(u_\epsilon,\sigma^{\frac{N-1}{N-p}})dx$$
$$\leq -\int_\sigma^\infty \tilde{g}(s,\sigma^{\frac{N-1}{N-p}})da(s)$$
$$\leq -\int_\sigma^\infty \tilde{g}(s,s^{\frac{N-1}{N-p}})da(s).$$

Since $a(s) \leq cs^{-\frac{N(p-1)}{N-p}}$ by (5.1.26) (i),

$$\int_{E\cap A_\sigma\cap B_\sigma^c}\tilde{g}(|u_\epsilon|,|\nabla u_\epsilon|)dx \leq c\frac{N(p-1)}{N-p}\int_\sigma^\infty \tilde{g}(s,s^{\frac{N-1}{N-p}})s^{-\frac{p(N-1)}{N-p}}ds.$$

Next

$$\int_{E\cap A_\sigma^c\cap B_\sigma}\tilde{g}(|u_\epsilon|,|\nabla u_\epsilon|)dx \leq \int_{B_\sigma}\tilde{g}(\sigma,|\nabla u_\epsilon|)dx$$
$$\leq -\int_\sigma^\infty \tilde{g}(\sigma,s^{\frac{N-1}{N-p}})db(s)$$
$$\leq -\int_\sigma^\infty \tilde{g}(s,s^{\frac{N-1}{N-p}})db(s),$$

and $b(s) \leq cs^{-\frac{N(p-1)}{N-p}}$ by (5.1.30), therefore

$$\int_{E\cap A_\sigma^c\cap B_\sigma}\tilde{g}(|u_\epsilon|,|\nabla u_\epsilon|)dx \leq c\frac{N(p-1)}{N-p}\int_\sigma^\infty \tilde{g}(s,s^{\frac{N-1}{N-p}})s^{-\frac{p(N-1)}{N-p}}ds.$$

Finally

$$\int_{E\cap C_\sigma}\tilde{g}(|u_\epsilon|,|\nabla u_\epsilon|)dx \leq \int_{C_\sigma}\tilde{g}(|u_\epsilon|,|\nabla u_\epsilon|)dx$$
$$\leq -\int_\sigma^\infty \tilde{g}(s,s^{\frac{N-1}{N-p}})dc(s).$$

Since $\sup\{a(s),b(s)\} \leq cs^{-\frac{N(p-1)}{N-p}}$, it follows that $c(s) \leq cs^{-\frac{N(p-1)}{N-p}}$ and we derive again

$$\int_{E\cap C_\sigma}\tilde{g}(|u_\epsilon|,|\nabla u_\epsilon|)dx \leq c\frac{N(p-1)}{N-p}\int_\sigma^\infty \tilde{g}(s,s^{\frac{N-1}{N-p}})s^{-\frac{p(N-1)}{N-p}}ds.$$

Combining the previous estimates we infer that for any $\sigma > 0$,

$$\begin{aligned}\int_E \tilde{g}(|u_\epsilon|, |\nabla u_\epsilon|)dx &\leq \tilde{g}(\sigma, \sigma^{\frac{N-1}{N-p}})\text{meas}\,(E) \\ &\quad + 3c\frac{N(p-1)}{N-p}\int_\sigma^\infty \tilde{g}(s, s^{\frac{N-1}{N-p}})s^{-\frac{p(N-1)}{N-p}}ds.\end{aligned} \tag{5.1.33}$$

We use here (5.1.19): given $\eta > 0$ there exists $\sigma > 0$ such that

$$3c\frac{N(p-1)}{N-p}\int_\sigma^\infty \tilde{g}(s, s^{\frac{N-1}{N-p}})s^{-\frac{p(N-1)}{N-p}}ds \leq \frac{\eta}{2}.$$

Then there exists $\delta > 0$ such that

$$\text{meas}\,(E) \leq \delta \Longrightarrow \tilde{g}(\sigma, \sigma^{\frac{N-1}{N-p}})\text{meas}\,(E) \leq \frac{\eta}{2}.$$

Therefore $\{\tilde{g}(|u_\epsilon|, |\nabla u_\epsilon|)\}$ is uniformly integrable. Since $|g(., u_\epsilon, \nabla u_\epsilon)| \leq \tilde{g}(|u_\epsilon|, |\nabla u_\epsilon|)$ the sequence $\{g(., u_\epsilon, \nabla u_\epsilon)\}$ is uniformly integrable in $\Omega$ and it follows by Vitali's convergence theorem that this sequence converges to $g(., u_\epsilon, \nabla u_\epsilon)$ in $L^1(\Omega)$. Since $u_\epsilon$ is a renormalized solution of (5.1.25) with right-hand side $\tilde{\mu}_\epsilon = f_\epsilon - g(., u_\epsilon, \nabla u_\epsilon) - div\, g + \mu^+_{s,\epsilon} - \mu^-_{s,\epsilon}$ and because $g(., u_\epsilon, \nabla u_\epsilon) \to g(., u, \nabla u)$ in $L^1(\Omega)$, the above decomposition is converging to the decomposition $\mu = f - g(., u, \nabla u) - div\, g + \mu^+_s - \mu^-_s$ in the sense of Definition 4.10. Therefore $u$ is a renormalized solution of

$$\begin{aligned} -div\, A(x, \nabla u) &= \mu - g(., u, \nabla u) && \text{in } \Omega \\ u &= 0 && \text{in } \partial\Omega, \end{aligned} \tag{5.1.34}$$

which in turn implies that it satisfies (5.1.22). Notice that it follows from the proof of Theorem 4.3.8-Step 10 that $T_k(u_\epsilon) \to T_k(u)$ in $W^{1,p}_0(\Omega)$ for any $k > 0$.

*The case* $p = N$. We denote by $u_{\epsilon,\Omega}$ the average of $u_\epsilon$ in $\Omega$. By (4.3.37) and (4.3.41) there holds

$$|u_{\epsilon,\Omega}| \leq c_K[u]_{BMO(\Omega)} \leq c_1 c_K \left\|\mu_\epsilon\right\|_{\mathfrak{M}}^{\frac{1}{N-1}}, \tag{5.1.35}$$

with $K$ denoting the constant of the geometric type **A**-property in (4.3.35). If we denote

$$A_\sigma = \{x \in \Omega : |u_\epsilon(x)| > \sigma\} \text{ and } a(\sigma) = \text{meas}\,(A_\sigma),$$

then, for $\sigma > |u_{\epsilon,\Omega}|$, we have

$$A_\sigma \subset \{x \in \Omega : |u_\epsilon(x) - u_{\epsilon,\Omega}| > \sigma - |u_{\epsilon,\Omega}|\}.$$

By (4.3.39) in Proposition 4.3.10 and (4.3.40) in Proposition 4.3.11, there holds

$$a(\sigma) \leq \text{meas}\left(\{|u_\epsilon - u_{\epsilon,\Omega}| > \sigma - |u_{\epsilon,\Omega}|\}\right) \leq c_3 \exp\left(-\frac{c_4(\sigma - |u_{\epsilon,\Omega}|)}{\|\mu\|_{\mathfrak{M}}^{\frac{1}{N-1}}}\right)$$
$$\leq c_5 \exp\left(-\frac{c_4\sigma}{\|\mu\|_{\mathfrak{M}}^{\frac{1}{N-1}}}\right), \tag{5.1.36}$$

with

$$c_5 = c_3 \exp\left(\frac{c_4|u_{\epsilon,\Omega}|}{\|\mu\|_{\mathfrak{M}}^{\frac{1}{N-1}}}\right) = c_5 = c_5(\Omega),$$

and finally

$$a(\sigma) \leq c_5 e^{-c_6\sigma} \quad \text{where } c_6 = \frac{c_4}{\|\mu\|_{\mathfrak{M}}^{\frac{1}{N-1}}}. \tag{5.1.37}$$

We define

$$\tilde{B}_\sigma = \{x \in \Omega : |\nabla u_\epsilon(x)| > e^{\frac{c_6\sigma}{N}}\} \quad \text{and} \quad \tilde{b}(\sigma) = \text{meas}(\tilde{B}_\sigma).$$

Then

$$\tilde{b}(\sigma) \leq c\,\|\mu\|_{\mathfrak{M}}^{\frac{N}{N-1}}\, e^{-c_6\sigma},$$

by (4.3.41). As in the case $1 < p < N$ we put $C_\sigma = A_\sigma \cap \tilde{B}_\sigma$ and $c(\sigma) = \text{meas}(C_\sigma)$. If $E \subset \Omega$ is a Borel set we write

$$E = (E \bigcap \tilde{C}_\sigma) \bigcup (E \cap \tilde{A}_\sigma^c \cap \tilde{B}_\sigma) \bigcup (E \cap \tilde{A}_\sigma \cap B_\sigma^c) \bigcup (E \cap A_\sigma^c \cap \tilde{B}_\sigma^c),$$

and

$$\int_E \tilde{g}(|u_\epsilon|, |\nabla u_\epsilon|)dx = I(\sigma) + II(\sigma) + III(\sigma) + IV(\sigma),$$

and we estimate below the different terms. Firstly

$$I(\sigma) = \int_{E \cap A_\sigma^c \cap \tilde{B}_\sigma^c} \tilde{g}(|u_\epsilon|, |\nabla u_\epsilon|)dx \leq \tilde{g}(\sigma, e^{\frac{c_5\sigma}{N}})\text{meas}(E).$$

Then

$$II(\sigma) = \int_{E\cap A_\sigma\cap\tilde{B}_\sigma^c} \tilde{g}(|u_\epsilon|,|\nabla u_\epsilon|)dx \leq \int_{A_\sigma} \tilde{g}(|u_\epsilon|,e^{\frac{c_6\sigma}{N}})dx$$
$$\leq -\int_\sigma^\infty \tilde{g}(t,e^{\frac{c_6\sigma}{N}})da(t)$$
$$\leq c_5c_6\int_\sigma^\infty \tilde{g}(t,e^{\frac{c_6\sigma}{N}})e^{-c_6t}dt.$$

Next

$$III(\sigma) = \int_{E\cap A_\sigma^c\cap\tilde{B}_\sigma} \tilde{g}(|u_\epsilon|,|\nabla u_\epsilon|)dx \leq \int_{\tilde{B}_\sigma} \tilde{g}(\sigma,|\nabla u_\epsilon|)$$
$$\leq -\int_\sigma^\infty \tilde{g}(\sigma,e^{\frac{c_6t}{N}})d\tilde{b}(t)$$
$$\leq c_5c_6\int_\sigma^\infty \tilde{g}(t,e^{\frac{c_6t}{N}})e^{-c_6t}dt.$$

Finally, since $c(s) \leq \max\{a(\sigma),\tilde{b}(\sigma)\} \leq c_5e^{-c_6\sigma}$

$$VI(\sigma) = \int_{E\cap C_\sigma} \tilde{g}(|u_\epsilon|,|\nabla u_\epsilon|)dx \leq -\int_\sigma^\infty \tilde{g}(s,e^{\frac{c_6s}{N}})dc(s)$$
$$\leq c_5c_6\int_\sigma^\infty \tilde{g}(t,e^{\frac{c_6t}{N}})e^{-c_6t}dt.$$

We conclude as in the first case that the sequence $\{g(u_\epsilon,\nabla u_\epsilon)\}$ is uniformly integrable if $\tilde{g}$ satisfies

$$\int_1^\infty \tilde{g}(t,e^{\frac{c_6t}{N}})e^{-c_6t}dt < \infty \tag{5.1.38}$$

with $c_6$ given by (5.1.37). This holds provided $\dfrac{c_6}{N} \geq \gamma$, and in particular if

$$\|\mu\|_{\mathfrak{M}} \leq \left(\frac{c_4}{N\gamma}\right)^{N-1}. \tag{5.1.39}$$

The end of the proof is as in the previous case. □

### 5.1.3 *Applications*

If $\tilde{g}(s,t) = \tilde{g}(s)$ conditions (5.1.19) becomes

$$\int_1^\infty \tilde{g}(s)\,s^{-\frac{p(N-1)}{N-p}}ds < \infty, \tag{5.1.40}$$

if $1 < p < N$. In the case $p = 2$ this condition is proved to be a sufficient condition since [Bénilan, Brezis (1975)], [Bénilan, Brezis (2003)]. If $p = N$ (5.1.20) reads

$$\int_1^{\infty} \tilde{g}(s)\, e^{-\gamma N s} ds < \infty. \tag{5.1.41}$$

In the case $p = 2 = N$, $\gamma$ is the exponential order of growth introduced in [Vazquez (1983)].

If $\tilde{g}(s,t) = \tilde{g}(t)$ and $1 < p < N$, condition (5.1.19) can be written

$$\int_1^{\infty} \tilde{g}(t)\, t^{1-Np} dt < \infty, \tag{5.1.42}$$

and condition (5.1.20) is

$$\int_1^{\infty} \tilde{g}(t)\, t^{-1-N} dt < \infty. \tag{5.1.43}$$

When $\tilde{g}(s,t) = s^q t^r$ with $q, r \geq 0$, assumptions (5.1.19) and (5.1.20) become

$$\begin{array}{lll} (i) & (N-p)q + (N-1)r < N(p-1) & \text{if} \quad 1 < p < N \\ (ii) & q \geq 0,\ 0 \leq r < N & \text{if} \quad p = N. \end{array} \tag{5.1.44}$$

## 5.2 Removable singularities

Assume $\Omega \subset \mathbb{R}^N$ is a domain and $A : \Omega \times \mathbb{R}^N \mapsto \mathbb{R}^N$ and $g : \Omega \times \mathbb{R} \times \mathbb{R}^N \mapsto \mathbb{R}$ two Caratheodory functions satisfying assumptions (4.1.1) and (5.0.1) respectively. Let $K$ be a compact subset of $\Omega$ and $\mu$ a Radon measure in $\Omega$. The removability problem is whether a locally renormalized solution $u$ of

$$-\operatorname{div} A(x, \nabla u) + g(x, u, \nabla) = \mu \tag{5.2.1}$$

in $\Omega \setminus K$, can be extended as a locally renormalized solution of the same equation in whole $\Omega$. The answer depends on $K$ and on the term $g(x, u, \nabla)$.

### 5.2.1 *Local estimates for locally renormalized solutions*

The next estimates developed in [Bidaut-Véron (2003)] are at the core of the study of the removability problem.

**Proposition 5.2.1** *Let $u$ be a measurable function finite finite a.e. such that $T_k(u) \in W^{1,2}_{loc}(\Omega)$ for all $k > 0$ and satisfying* **D1-loc**. *If $u$ satisfies also*

$$|u|^q \in L^1_{loc}(\Omega) \qquad \text{for some } q > p-1 > 0, \tag{5.2.2}$$

*and*

$$|\nabla u|^{p-1} \in L^1_{loc}(\Omega), \tag{5.2.3}$$

*then*

$$(|u|+1)^{\alpha-1} |\nabla u|^p \in L^1_{loc}(\Omega) \qquad \text{for any } \alpha < 0, \tag{5.2.4}$$

*and $u$ satisfies (4.3.16) (ii)-(iii). Hence it is a locally renormalized solution of (4.1.2). Furthermore*

$$|\nabla u|^{p-1} \in L^\sigma_{loc}(\Omega) \qquad \text{for any } 1 \leq \sigma < \frac{p'q}{q+1}. \tag{5.2.5}$$

*Proof. Step 1: proof of (5.2.4).* For $\alpha < 0$ and $k > 0$ we put $u_k = T_k(u)$ and $h_k(u) = (1-(|u_k|+1)^\alpha)\text{sign}(u)$. We write (4.3.2) with $\phi \in C^1_0\Omega$, $\phi \geq 0$ and get from (4.1.1) (ii),

$$\begin{aligned} |\alpha| \int_\Omega \langle A(x,\nabla u), \nabla u_k\rangle (|u_k|+1)^{\alpha-1}\phi dx &= -\int_\Omega \langle A(x,\nabla u), \nabla\phi\rangle h_k(u) dx \\ &+ \int_\Omega h_k(u)\phi d\mu_0 + (1-(1+k)^\alpha)\int_\Omega \phi d|\mu_s| + \int_\Omega h_k(u)\phi d\mu_0 \\ &\leq \lambda \int_\Omega |\nabla u_k|^{p-1} |\nabla \phi|\, dx + \lambda \int_{\{|u|>k\}} |\nabla u|^{p-1} |\nabla\phi|\, dx + \int_\Omega \phi d|\mu|. \end{aligned} \tag{5.2.6}$$

By Hölder's inequality, we have

$$\begin{aligned} &\int_\Omega |\nabla u_k|^{p-1} |\nabla \phi|\, dx \\ &\leq \left(\int_\Omega (|u_k|+1)^{\alpha-1} |\nabla u_k|^p\, dx\right)^{\frac{1}{p'}} \left(\int_\Omega (|u_k|+1)^{(1-\alpha)(p-1)}\phi^{1-p}|\nabla\phi|^p dx\right)^{\frac{1}{p}}. \end{aligned}$$

Since $q > p-1$, we choose $\alpha < 0$ such that $\tau := \frac{q}{(p-1)(1-\alpha)} < 1$, therefore

$$\begin{aligned}\int_\Omega |\nabla u_k|^{p-1} |\nabla\phi|\, dx &\leq \left(\int_\Omega (|u_k|+1)^{\alpha-1} |\nabla u_k|^p\, dx\right)^{\frac{1}{p'}}\left(\int_\Omega (|u_k|+1)^q \phi dx\right)^{\frac{1}{\tau p}} \\ &\quad\times \left(\int_\Omega \phi^{1-\tau' p}|\nabla\phi|^{\tau' p} dx\right)^{\frac{1}{\tau' p}} \\ &\leq \frac{|\alpha|}{2\lambda}\int_\Omega (|u_k|+1)^{\alpha-1} |\nabla u_k|^p\, dx \\ &\quad + c(\alpha,\lambda)\left(\int_\Omega (|u_k|+1)^q \phi dx\right)^{\frac{1}{\tau}}\left(\int_\Omega \phi^{1-\tau' p}|\nabla\phi|^{\tau'} dx\right)^{\frac{1}{\tau'}}. \end{aligned} \tag{5.2.7}$$

Using (5.2.6) and (4.1.1) (i), it leads to,

$$\begin{aligned}\frac{|\alpha|}{2}\int_\Omega (|u_k|+1)^{\alpha-1} |\nabla u_k|^p\, dx &\leq \lambda \int_{\{|u|>k\}} |\nabla u|^{p-1} |\nabla\phi|\, dx + \int_\Omega \phi d|\mu| \\ &\quad + \lambda c(\alpha,\lambda)\left(\int_\Omega (|u_k|+1)^q \phi dx\right)^{\frac{1}{\tau}}\left(\int_\Omega \phi^{1-\tau' p}|\nabla\phi|^{\tau'} dx\right)^{\frac{1}{\tau'}}. \end{aligned} \tag{5.2.8}$$

Letting $k \to \infty$, we derive from (5.2.8),

$$\begin{aligned}\frac{|\alpha|}{2}\int_\Omega (|u|+1)^{\alpha-1} |\nabla u|^p\, \phi dx &\leq \lambda c(\alpha,\lambda)\left(\int_\Omega (|u|+1)^q \phi dx\right)^{\frac{1}{\tau}} \\ &\quad\times \left(\int_\Omega \phi^{1-\tau' p}|\nabla\phi|^{\tau' p} dx\right)^{\frac{1}{\tau'}} + \int_\Omega \phi d|\mu|, \end{aligned} \tag{5.2.9}$$

and from (5.2.7),

$$\begin{aligned}\int_\Omega |\nabla u|^{p-1} |\nabla\phi|\, dx &\leq \left(\int_\Omega (|u|+1)^{\alpha-1} |\nabla u|^p\, \phi dx\right)^{\frac{1}{p'}}\left(\int_\Omega (|u|+1)^q \phi dx\right)^{\frac{1}{\tau p}} \\ &\quad\times \left(\int_\Omega \phi^{1-\tau' p}|\nabla\phi|^{\tau' p} dx\right)^{\frac{1}{\tau' p}}. \end{aligned} \tag{5.2.10}$$

We take $\phi = \zeta^{p\tau'}$, with $\zeta \in C_0^1(\Omega)$, $\zeta \geq 0$; hence $\phi^{1-\tau' p}|\nabla\phi|^{\tau' p} \in L^1_{loc}(\Omega)$, and since $u \in L^q_{loc}(\Omega)$ we obtain (5.2.4), first for $\alpha$ small enough and then for all $\alpha < 0$.

*Step 2: The function u satisfies (4.3.1) (ii) and (iii) and ((5.2.5)).* Let $\mathcal{U}$ be a smooth bounded domain such that $\overline{\mathcal{U}} \subset \Omega$. If $v \in L^1_{loc}(\Omega)$ we denote by

$v_{\mathcal{U}}$ the average of $v$ on $\mathcal{U}$. By the Gagliardo-Nirenberg-Poincaré-Wirtinger inequality (1.2.37), for any $w \in W^{1,p}(\mathcal{U}) \cap L^{\ell}(\mathcal{U})$ and $0 \leq \theta \leq 1$,

$$\|w - w_{\mathcal{U}}\|_{L^{\gamma}(\mathcal{U})} \leq c \|\nabla w\|^{\theta}_{L^p(\mathcal{U})} \|w - w_{\mathcal{U}}\|^{1-\theta}_{L^{\ell}(\mathcal{U})}, \tag{5.2.11}$$

with $\frac{1}{\gamma} = \frac{\theta}{p^*} + \frac{1-\theta}{\ell}$ with $p^* = \infty$ if $N = p$. Let $\alpha \in (1-p, 0)$, $v = (1+|u|)\beta$, $\beta = \frac{\alpha+p-1}{p}$ and $\ell = \frac{p-1}{\beta}$. Then $v \in L^p(\mathcal{U})$ since $\beta p < p-1 < q$ and $u \in L^q(\mathcal{U})$. Since $T_k(u) \in W^{1,p}(\mathcal{U})$, then $v_{k+1} := T_{k+1}(v) = (1+|u_k|)^{\beta} \in W^{1,p}(\mathcal{U})$ and

$$\nabla v_{k+1} = \beta(1+|u_k|)^{\beta-1}\text{sign}\,(u_k)\nabla u_k.$$

By definition of the gradient we have $\nabla v_{k+1} = \chi_{\{|u|\leq k\}}\nabla v$, hence

$$\int_{\{|u|\leq k\}} |\nabla v|^p\,dx = \int_{\mathcal{U}} |\nabla v_{k+1}|^p\,dx = \beta^p \int_{\mathcal{U}} (1+|u_k|)^{\beta-1} |\nabla u_k|^p\,dx.$$

Letting $k \to \infty$ and using Step 1, we derive

$$\int_{\mathcal{U}} |\nabla v|^p\,dx = \beta^p \int_{\mathcal{U}} (1+|u|)^{\beta-1} |\nabla u|^p\,dx.$$

Therefore $v \in W^{1,p}(\mathcal{U})$ and the gradient a.e. of $v$ coincides with the distributional one. Furthermore, by (5.2.11) with $w = (1+|u|)^{\beta} = v$,

$$\begin{aligned}&\int_{\mathcal{U}} \left|(1+|u|)^{\beta} - v_{\mathcal{U}}\right|^{\gamma} dx\\ &\leq c\left(\int_{\mathcal{U}} (1+|u|)^{\alpha-1}|\nabla u|^p dx\right)^{\frac{\theta\gamma}{p}} \left(\int_{\mathcal{U}} \left|(1+|u|)^{\beta} - v_{\mathcal{U}}\right|^{\frac{p-1}{\beta}} dx\right)^{\frac{(1-\theta)\gamma\beta}{p-1}}.\end{aligned} \tag{5.2.12}$$

This implies that (4.3.1) (ii) holds if we take $\alpha$ small enough and $\theta$ close to 1. Finally, by Hölder's inequality we have

$$\int_{\mathcal{U}} |\nabla u|^{\eta} dx \leq \left(\int_{\mathcal{U}} (1+|u|)^{\alpha-1}|\nabla u|^p dx\right)^{\frac{\eta}{p}} \left(\int_{\mathcal{U}} (1+|u|)^{\frac{(1-\alpha)\eta}{p-\eta}} dx\right)^{\frac{p-\eta}{p}}. \tag{5.2.13}$$

Since $\frac{(1-\alpha)\eta}{p-\eta} \to \frac{\eta}{p-\eta}$ when $\alpha \to 0$, if we choose $\eta$ such that $\frac{\eta}{p-\eta} < \frac{(p-1)N}{N-p}$, equivalently $0 < \eta < \frac{(p-1)N}{N-1}$, then for $\alpha$ close enough to 0, we get (4.3.1) (iii). Notice also that by Hölder's inequality,

$$\int_{\mathcal{U}} |\nabla u|^{(p-1)\sigma} \phi dx \leq \left( \int_{\mathcal{U}} (1+|u|)^{\alpha-1} |\nabla u|^p \phi dx \right)^{\frac{\sigma}{p'}} \times \left( \int_{\mathcal{U}} (1+|u|)^{\frac{(1-\alpha)\sigma}{p'-\sigma}} \phi dx \right)^{1-\frac{\sigma}{p'}} . \tag{5.2.14}$$

If we assume $\frac{\sigma}{p'-\sigma} < q$, equivalently $\sigma < \frac{p'q}{q+1}$, we can take $\alpha$ small enough so that $\frac{(1-\alpha)\sigma}{p'-\sigma} \leq q$ in order the left-hand side of (5.2.14) be bounded. □

This result leads to a local stability result.

**Proposition 5.2.2** *Let $\{\mu_n\}$ be a locally uniformly bounded sequence of measures in $\Omega$ and $\{u_n\}$ a sequence of locally renormalized solutions of*

$$-\text{div}\, A(x, \nabla u_n) = \mu_n \qquad \text{in } \Omega, \tag{5.2.15}$$

*such that $\{|u_n|^q\}$ is uniformly bounded in $L^1_{loc}(\Omega)$ for some $q > p-1$. Then*

$$\{T_k(u_n)\} \text{ is uniformly bounded in } W^{1,p}_{loc}(\Omega) \quad \text{for any } k > 0, \tag{5.2.16}$$

$$\{|u_n|^{p-1}\} \text{ is uniformly bounded in } L^s_{loc}(\Omega) \text{ for } 1 \leq s < \tfrac{N}{N-p}, \tag{5.2.17}$$

$$\{|\nabla u_n|^{p-1}\} \text{ is uniformly bounded in } L^r_{loc}(\Omega) \quad \text{for } 1 \leq r < \max\left\{ \tfrac{N}{N-1}, \tfrac{p'q}{q+1} \right\}. \tag{5.2.18}$$

*Furthermore, up to a subsequence, $\mu_n$ converges to some measure $\mu$ in the weak sense of measures in $\Omega$, $\{u_n\}$ converges locally in measure and a.e. in $\Omega$ to some function $u$ which satisfies (4.3.1) (i)-(ii)-(iii), (5.2.2) and (5.2.4). At last $\{\nabla u_n\}$ converges locally in measure and a.e. in $\Omega$ to $\nabla u$.*

*Proof.* Estimate (5.2.9) is valid and implies that for any $\zeta \in C_0^\infty(\Omega)$, $\zeta \geq 0$, there holds

$$\int_\Omega (|u_n + 1|)^{\alpha-1} |\nabla u_n|^p \zeta^{p\tau'} dx \leq c_1(\zeta) \left( \int_\Omega (|u_n| + 1)^q \zeta^{p\tau'} dx \right)^{\frac{1}{\tau}} + \int_\Omega \zeta^{p\tau'} d|\mu_n|. \tag{5.2.19}$$

This implies

$$\int_\Omega |\nabla\zeta^{p\tau'}|\,|\nabla u_n|^{p-1}\,dx \le c'(\zeta)\left(\int_\Omega (|u_n|+1)^q\zeta^{p\tau'}dx\right)^{\frac{1}{\tau}} + c_2\int_\Omega \zeta^{p\tau'}d|\mu_n|. \tag{5.2.20}$$

Moreover, if $\phi \in C_0^1(\Omega)$, $\phi \ge 0$,

$$\begin{aligned}\int_\Omega |\nabla u_n|^{p-1}\,\phi dx \le &\left(\int_\Omega (|u_n+1|)^{\alpha-1}\,|\nabla u_n|^p\,\phi dx\right)^{\frac{1}{p'}}\\ &\times\left(\int_\Omega (|u_n+1|)^{(1-\alpha)(p-1)}\phi\right)^{\frac{1}{p}}\end{aligned} \tag{5.2.21}$$

and here we choose $\alpha < 0$ such that $(1-\alpha)(p-1) = q$. Since $\{|u_n|^q\}$ and $\{|\nabla u_n|^{p-1}\}$ are uniformly bounded in $L^1_{loc}(\Omega)$ we infer (5.2.16)-(5.2.18) from Proposition 5.2.2 and (5.2.13).

The *a priori* estimates and the convergence a.e. to some function $u$ are direct consequences of the Rellich-Kondrachov theorem and Fatou's lemma. The convergence of the gradient is based upon the formulation **D3-loc** of the equation satisfied by the truncate $u_{n,k} = T_k(u_n)$. We write $\mu_n = \mu_{n,0} + \mu^+_{n,s} - \mu^-_{n,s}$, where $\mu_{n,0} \ll c^{\mathbb{R}^N}_{1,p}$ and $\mu^\pm_{n,s} \perp c^{\mathbb{R}^N}_{1,p}$ and are nonnegative. For $k > 0$ there exists $\mu^\pm_{n,k} \ll c^{\mathbb{R}^N}_{1,p}$ with support on $\{x \in \Omega : u_n = \pm k\}$, converging to $\mu^\pm_{n,s}$ in the weak sense of measures, such that, for any $\phi \in W^{1,p}(\Omega)\cap L^\infty(\Omega)$ with compact support in $\Omega$, there holds

$$\begin{aligned}\int_{\{|u_n|<k\}} \langle A(x,\nabla u_n),\psi\rangle dx = &\int_{\{|u_n|<k\}} \psi d\mu_{n,0}\\ &+\int_{\{u_n=k\}}\psi d\mu^+_{n,k} - \int_{\{u_n=-k\}}\psi d\mu^-_{n,k}.\end{aligned} \tag{5.2.22}$$

If we take $\phi = \zeta u_{n,k}$ where $\zeta \ge 0$ has value 1 on $\mathcal{U} \subset \overline{\mathcal{U}} \subset \Omega$, then

$$d\mu^-_{n,k}(\mathcal{U}) + d\mu^-_{n,k}(\mathcal{U}) \le c(1+k),$$

where $c$ is independent of $n$ and $k$. If $\mu_{n,k} = \mu_{n,0} + \mu^+_{n,k} - \mu^-_{n,k}$, then $\{u_{n,k}\} \subset W^{1,p}_{loc}(\Omega)$ is sequence of weak solutions of

$$-\text{div}\,A(x,\nabla u_{n,k}) = \mu_{n,k} \qquad \text{in } \Omega, \tag{5.2.23}$$

uniformly bounded in $L^\infty(\Omega)$ (for $k$ fixed). As in the proof of Theorem 4.3.8-Step 5, it implies that the sequence $\{\nabla u_{n,k}\}_{n\in\mathbb{N}}$ is a Cauchy sequence in measure on any compact subset of $\Omega$. Therefore, up to extracting a diagonal subsequence still denoted by $u_{n,k}$ we conclude that $\nabla u_{n,k}$

converges to $\nabla u_k$ a.e. in $\Omega$. Furthermore the proof of Theorem 4.3.8-Step 6 applies easily and implies that $\{\nabla u_n\}$ is a Cauchy sequence in measure on any compact subset of $\Omega$. Hence, up to a subsequence, $\nabla u_n$ converges a.e. to $\nabla u$. By Vitali's convergence theorem, $|\nabla u_n|^{p-2}\nabla u_n\phi$ converges to $|\nabla u|^{p-2}\nabla u\phi$ in $L^r(\Omega)$ for any $\phi \in C_0^1(\Omega)$ and any $r$ satisfying (5.2.18). In particular $u$ is a solution of (4.1.2 in the sense of disributions in $\Omega$. □

*Remark.* The convergence of $\mu_n$ to $\mu$ is not strong enough to assert that $u$ is a locally renormalized solution of (4.1.2). The next result gives a partial answer to the stability question under a local $L^q$-bound.

**Proposition 5.2.3** *Let $\lambda$ be a Radon measure in $\Omega$ and $\{f_n\}$ a sequence of locally integrable functions which converges weakly to some $f \in L^1_{loc}(\Omega)$. Let $\{u_n\}$ be a sequence of locally renormalized solutions of*

$$-\operatorname{div} A(x, \nabla u_n) = \lambda + f_n \qquad \textit{in } \Omega, \tag{5.2.24}$$

*such that $\{|u_n|^q\}$ is uniformly bounded in $L^1_{loc}(\Omega)$ for some $q > p-1$. Then, up to a subsequence, $u_n$ converges a.e. to a locally renormalized solution of*

$$-\operatorname{div} A(x, \nabla u) = \lambda + f \qquad \textit{in } \Omega. \tag{5.2.25}$$

*Proof.* We only indicate the sketch of the proof. By Proposition 5.2.3 there exists a subsequence of $\{u_n\}$ which converges a.e. to some $u$ and $|\nabla u_n|^{p-2}\nabla u_n \to |\nabla u|^{p-2}\nabla u$ in $L^r_{loc}(\Omega)$ for all $r$ satisfying (5.2.18) (we keep the same notation for the subsequence). The analysis of Steps 7-10 which is of a local nature in the proof of Theorem 4.3.8 can be reproduced in the present framework and yields the result. □

### 5.2.2 *The power case*

If this section we assume that $A$ satisfies (4.1.1), that $g$ satisfies (5.0.2) and

$$g(x, s, \xi)s \geq a|s|^{q+1} \qquad \forall (x, s, \xi) \in \Omega \times \mathbb{R} \times \mathbb{R}^N, \tag{5.2.26}$$

for some $q > p - 1 > 0$ and $a > 0$ and we set

$$q_p = \frac{q}{q+1-p}. \tag{5.2.27}$$

**Theorem 5.2.4** *Assume $K \subset \Omega$ is relatively closed and $\mu = \mu_0 + \mu_s^+ - \mu_s^- \in \mathfrak{M}(\Omega)$. If $c^{\mathbb{R}^N}_{1,R}(K) = 0$ for some $R > pq_p$ and $\mu(K) = 0$, then any locally renormalized solution $u$ of (5.2.1) in $\Omega \setminus K$ is a locally renormalized solution of the same equation in $\Omega$.*

The proof is based upon some local estimates.

**Lemma 5.2.5** *Let $u$ be a locally renormalized solution of (5.2.1) in a domain $\Omega$. Then for any $R > pq_p$ there exists $c = c(N, p, q, \Omega, R) > 0$ such that for any $\zeta \in C_0^1(\Omega)$, $\zeta \geq 0$, there holds*

$$\int_\Omega (|u|+1)^q \zeta^R dx \leq c\left(\int_\Omega \left(\zeta^R + |\nabla\zeta|^R\right) dx + \int_\Omega \zeta^R d|\mu|\right). \tag{5.2.28}$$

*Moreover, for any $\alpha < 0$ there exists $c' = c'(\alpha, N, p, q, \Omega, R) > 0$ such that for any $\zeta$ as above,*

$$\int_\Omega (|u|+1)^{\alpha-1} |\nabla u|^p \zeta^R dx \leq c'\left(\int_\Omega \left(\zeta^R + |\nabla\zeta|^R\right) dx + \int_\Omega \zeta^R d|\mu|\right). \tag{5.2.29}$$

*Proof.* Since (5.2.26) holds and $g(., u, \nabla u)$ is locally integrable in $\Omega$, $|u|^q$ is also locally integrable in $\Omega$, and we can apply Proposition 5.2.1. Let $u_k = T_k(u)$ and $h_k(u) = (1-(|u_k|+1)^\alpha)\text{sign}\,(u)$. Then for any $\zeta \in C_0^\infty(\Omega)$, $\zeta \geq 0$, we obtain from (5.2.1), using (4.1.1) and (5.2.26),

$$\begin{aligned}
&|\alpha| \int_\Omega (|u_k|+1)^{\alpha-1} |\nabla u_k|^p \zeta^{p\tau'} dx + \int_\Omega |u|^{q-1} u h_k(u) \zeta^{p\tau'} dx \\
&\leq -\int_\Omega \langle A(x, \nabla u), \nabla\zeta^{p\tau'}\rangle h_k(u) dx + \int_\Omega h_k(u) \zeta^{p\tau'} d\mu_0 + (1-(k+1)^\alpha) \int_\Omega \zeta^{p\tau'} d|\mu_s| \\
&\leq \lambda \int_\Omega |\nabla u_k|^{p-1} |\nabla\zeta^{p\tau'}| dx + \lambda \int_{\{|u|>\lambda\}} |\nabla u|^{p-1} |\nabla\zeta^{p\tau'}| dx + \int_\Omega \zeta^{p\tau'} d|\mu|.
\end{aligned}$$

Since $|u|^{q-1} u h_k(u) \geq 0$, we let $k \to \infty$ and derive by Fatou's lemma and (5.2.20)

$$\begin{aligned}
|\alpha| &\int_\Omega (|u|+1)^{\alpha-1} |\nabla u|^p \zeta^{p\tau'} dx + \int_\Omega |u|^q (1-(|u|+1)^\alpha) \zeta^{p\tau'} dx \\
&\leq \lambda \int_\Omega |\nabla u|^{p-1} |\nabla\zeta^{p\tau'}| dx + \int_\Omega \zeta^{p\tau'} d|\mu| \\
&\leq \int_\Omega \phi d|\mu| + c_1 \left(\int_\Omega \zeta^{p\tau'(1-\tau' p)} \left|\nabla\zeta^{p\tau'}\right|^{\tau' p} dx\right)^{\frac{1}{\tau'}} \left(\int_\Omega (1+|u|)^q \zeta^{p\tau'} dx\right)^{\frac{1}{\tau}},
\end{aligned}$$

where $\tau = \frac{q}{(p-1)(1-\alpha)}$ is larger than 1 if $\alpha$ is small enough and such that $\tau' p = R$. We define

$$X = \int_\Omega (|u|+1)^q \zeta^R dx, \qquad Y = \int_\Omega (|u|+1)^{\alpha-1} |\nabla u|^p \zeta^R dx.$$

Since $|u(x)| \geq 1$ implies $1 - (|u|+1)^\alpha \geq 1 - 2^\alpha > 0$, we obtain

$$X + Y \leq c\left(\int_\Omega \zeta^R dx + \int_\Omega \zeta^R d|\mu| + C(\zeta) X^{\frac{1}{\tau}}\right), \tag{5.2.30}$$

where

$$C(\zeta) = \left(\int_\Omega |\nabla \zeta|^R \, dx\right)^{\frac{p}{R}}.$$

By Young's inequality we derive

$$X + Y \leq c'\left(\int_\Omega \zeta^R dx + \int_\Omega \zeta^R d|\mu| + \int_\Omega |\nabla \zeta|^R \, dx\right), \tag{5.2.31}$$

and $c' = c'(\alpha, p, q) > 0$. Then we obtain (5.2.28) for $\alpha < 0$ small enough and (5.2.29) for any $\alpha < 0$. □

*Proof of Theorem 5.2.4.* We recall that any renormalized solution is defined a.e. and any set with $c^{\mathbb{R}^N}_{1,R}$-capacity zero has zero Lebesgue measure. The function $u$ is such that $T_k(u) \in W^{1,p}_{loc}(\Omega \setminus K)$. Furthermore $|\nabla u|^{p-1}$ and $g(x, u, \nabla u)$ are locally integrable in $\Omega \setminus K$, hence $|u|^q \in L^1_{loc}(\Omega \setminus K)$.

*Step 1: We claim that* $|u|^q \in L^1_{loc}(\Omega)$. Let $h \in W^{1,\infty}(\mathbb{R})$, $h'$ with compact support and $\phi \in C^1_0(\Omega \setminus K)$, $\phi \geq 0$, then

$$\begin{aligned}\int_\Omega \langle A(x, \nabla u), \nabla(h(u)\phi) dx + \int_\Omega g(x, u, \nabla u) h(u) \phi dx \\ = \int_\Omega h(u) \phi d\mu_0 + h(\infty) \int_\Omega \phi d\mu_s^+ - h(-\infty) \int_\Omega \phi d\mu_s^-.\end{aligned} \tag{5.2.32}$$

From Lemma 5.2.5 with $\Omega$ replaced by $\Omega \setminus K$, we deduce

$$\int_\Omega (|u|+1)^q \phi^R dx \leq c\left(\int_\Omega \phi^R dx + \int_\Omega \phi^R d|\mu| + \int_\Omega |\nabla \phi|^R \, dx\right), \tag{5.2.33}$$

where $c = c(N, p, q, R, \Omega, K)$ and, for any $\alpha > 0$,

$$\int_\Omega (|u|+1)^{\alpha-1} |\nabla u|^p \phi^R dx \leq c'\left(\int_\Omega \phi^R dx + \int_\Omega \phi^R d|\mu| + \int_\Omega |\nabla \phi|^R \, dx\right), \tag{5.2.34}$$

with $c' = c'(N, p, q, R, \Omega, K, \alpha)$. Without loss of generality we can assume that $K$ is compact and $c^{\mathbb{R}^N}_{1,p}(K) = 0$. Then there exist a sequence $\{\xi_n\} \subset C_0^\infty(\mathbb{R}^N)$ of nonnegative functions such that $0 \leq \xi_n \leq 1$, with value 1 in a neighborhood of $K$, which converges to 0 in $W^{1,R}(\Omega)$ and everywhere in $\Omega \setminus K$ as $n \to \infty$.

Let $\phi \in C_0^\infty(\mathbb{R}^N)$ and $\phi_n = (1-\xi_n)\phi$. Then from previous lemma,

$$\int_\Omega (|u|+1)^q \phi_n^R dx \le c\left(\int_\Omega \phi_n^R dx + \int_\Omega \phi_n^R d|\mu| + \int_\Omega |\nabla \phi_n|^R dx\right)$$
$$\le c\left(\int_\Omega \phi^R dx + \int_\Omega \phi^R d|\mu| + \int_\Omega |\nabla \phi_n|^R dx\right).$$

Since the last term on the right-hand side is bounded and $\phi_n^R \to \phi^R$ as $n \to \infty$, we derive the claim by Fatou's lemma.

*Step 2: We claim that* $|\nabla u|^{p-1} \in L^1_{loc}(\Omega)$. In the same way as in Step 1,

$$\int_\Omega (|u|+1)^{\alpha-1} |\nabla u|^p \phi_n^R dx \quad \text{is bounded uniformly w.r. to } n. \tag{5.2.35}$$

By Hölder's inequality

$$\int_\Omega |\nabla u|^{p-1} \phi_n^R dx$$
$$\le \left(\int_\Omega (|u|+1)^{\alpha-1} |\nabla u|^p \phi_n^R dx\right)^{\frac{1}{p'}} \left(\int_\Omega (|u|+1)^{(1-\alpha)(p-1)} \phi_n^R dx\right)^{\frac{1}{p}}.$$

Since we can choose $\alpha < 0$ such that $(1-\alpha)(p-1) \le q$, the second term on the right-hand side is uniformly bounded, which implies that the left-hand side is uniformly bounded too, which is the claim.

*Step 3: We claim that* $u_k := T_k(u) \in W^{1,p}_{loc}(\Omega)$ *and* $\nabla u$ *is also the distributional gradient in* $\Omega$. Let $k > 0$, then

$$(k+1)^{\alpha-1} \int |\nabla u_k|^p \phi_n^R dx \le \int_{\{|u|\le k\}} (|u|+1)^{\alpha-1} |\nabla u|^p \phi_n^R dx,$$

and the right-hand side is uniformly bounded by Step 2. Put $w_n = u_k \phi_n^{\frac{R}{p}}$, it belongs to $W^{1,p}(\Omega)$ and converges a.e. to $u_k \phi^{\frac{R}{p}}$ when $n \to \infty$. Next

$$\nabla w_n = \phi_n^{\frac{R}{p}} \nabla u_k + \frac{R}{p} u_k \phi_n^{\frac{R}{p}-1} \nabla \phi_n.$$

Then $\nabla w_n$ is bounded in $L^p(\Omega)$. Thus $u_k \phi^{\frac{R}{p}} \in W^{1,p}(\Omega)$ and therefore $u_k \in W^{1,p}_{loc}(\Omega)$. Hence the two gradients coincides.

*Step 4: We claim that* $|\nabla u|^{p-1} \in L^{R'}_{loc}(\Omega)$. For $0 \le \sigma < p'$, we have

$$\int_\Omega |\nabla u|^{(p-1)\sigma} \phi_n^R dx$$
$$\le \left(\int_\Omega (|u|+1)^{\alpha-1} |\nabla u|^p \phi_n^R dx\right)^{\frac{\sigma}{p'}} \left(\int_\Omega (|u|+1)^{\frac{(1-\alpha)(\sigma-1)}{p'}} dx\right)^{1-\frac{\sigma}{p'}}.$$

The claim follows from previous steps.

*Step 5: The function* $u$ *is a locally renormalized solution in* $\Omega$. Let $h \in W^{1,\infty}(\mathbb{R})$, $h'$ with compact support, $\phi \in C_0^1(\Omega)$, $\phi \ge 0$ and $\phi_n = (1-\xi_n)\phi$. Then from (5.2.32),

$$\begin{aligned}&\int_\Omega \langle A(x,\nabla u), \nabla (h(u)\phi_n) dx + \int_\Omega g(x,u,\nabla u) h(u)\phi_n dx \\ &\quad = \int_\Omega h(u)\phi_n d\mu_0 + h(\infty)\int_\Omega \phi_n d\mu_s^+ - h(-\infty)\int_\Omega \phi_n d\mu_s^-.\end{aligned} \tag{5.2.36}$$

Since $|\mu|(K) = 0$,

$$\lim_{n\to\infty} \int_\Omega \phi_n d|\mu_0| = \int_\Omega \phi d|\mu_0|, \quad \lim_{n\to\infty} \int_\Omega \phi_n d\mu_s^\pm = \int_\Omega \phi d\mu_s^\pm.$$

Therefore, since $\psi \ge \psi_n$,

$$\left|\int_\Omega h(u)(\phi_n - \psi) d\mu_0\right| \le \|h\|_{L^\infty} \int_\Omega (\phi_n - \psi) d|\mu_0| \to 0 \text{ as } n \to \infty.$$

From Step 4, and assuming $\text{supp}(\phi) \subset \mathcal{U}$ where $\overline{\mathcal{U}} \subset \Omega$ is a compact set,

$$\left|\int_\Omega h(u)\langle A(x,\nabla u), \nabla(\phi - \phi_n)\rangle dx\right| \le c \left\| |\nabla u|^{p-1}\right\|_{L^{R'}(\mathcal{U})} \|\nabla(\phi_n - \phi)\|_{L^R(\mathcal{U})}.$$

Now

$$\nabla(\phi - \phi_n) = (1 - (1-\xi_n)^R)\nabla\phi_R + \phi(1-\xi_n)^{R-1}\nabla\xi_n,$$

thus

$$|\nabla(\phi - \phi_n)|^R \le c(|\nabla\phi|^R + |\nabla\xi_n|^R),$$

for some $c > 0$ independent of $n$. Since $\nabla(\phi_n - \phi) \to 0$ a.e., it follows from dominated convergence that $\nabla(\phi_n - \phi) \to 0$ in $L^R(\mathcal{U})$ and therefore

$$\lim_{n\to\infty} \int_\Omega h(u)\langle A(x,\nabla u), \nabla\phi_n\rangle dx = \int_\Omega h(u)\langle A(x,\nabla u), \nabla\phi\rangle dx.$$

We first assume that $h' \geq 0$ and $h(0) = 0$, then $g(x, u, \nabla u)h(u)\phi_n$ is non-negative and converges to $g(x, u, \nabla u)h(u)\phi$ a.e. in $\Omega$, and by Fatou's lemma and the fact that $0 \leq \phi_n \leq \phi$,

$$\int_\Omega g(x, u, \nabla u)h(u)\phi dx \leq \liminf_{n\to\infty} \int_\Omega g(x, u, \nabla u)h(u)\phi_n dx$$
$$\leq \int_\Omega g(x, u, \nabla u)h(u)\phi dx.$$

In particular $g(x, u, \nabla u)h(u) \in L^1_{loc}(\Omega)$. If $h(u) = T_k(u)$, then

$$\begin{aligned} 0 \leq g(x, u, \nabla u)T_k(u) &= g(x, u, \nabla u)u\chi_{\{|u|<k\}} + k\,|g(x, u, \nabla u)|\,\chi_{\{|u|\geq k\}} \\ &= g(x, u_k, \nabla u_k)u_k\chi_{\{|u|<k\}} + k\,|g(x, u, \nabla u)|\,\chi_{\{|u|\geq k\}}. \end{aligned}$$

On the set of $x \in \Omega$ such that $\{|u(x)| < k\}$ we have $u = u_k$ and $\nabla u = \nabla u_k$ a.e., and by (5.0.1) (ii),

$$|g(x, u_k, \nabla u_k)| \leq b(k)\left(|\nabla u_k|^p + c(x)\right).$$

Since $u_k \in W^{1,p}_{loc}(\Omega)$ by Step 3, we obtain that $g(., u, \nabla u)\chi_{\{|u|<k\}}$ is locally integrable. Since $g(., u, \nabla u)\chi_{\{|u|\geq k\}}$ is also locally integrable, we conclude that $g(., u, \nabla u) \in L^1_{loc}(\Omega)$. Finally, if $h$ is any function of $W^{1,\infty}(\mathbb{R})$ with $h'$ with compact support, $\phi \in C^1_0(\Omega)$ has not prescribed sign and $\phi_n = (1 - \xi_n)\phi$, we infer from dominated convergence

$$\lim_{n\to\infty} \int_\Omega g(x, u, \nabla u)h(u)\phi_n dx = \int_\Omega g(x, u, \nabla u)h(u)\phi dx.$$

Letting $n \to \infty$ in the relation

$$\begin{aligned} &\int_\Omega \langle A(x, \nabla u), \nabla(h(u)\phi_n) dx + \int_\Omega g(x, u, \nabla u)h(u)\phi_n dx \\ &\quad = \int_\Omega h(u)\phi_n d\mu_0 + h(\infty)\int_\Omega \phi_n d\mu_s^+ - h(-\infty)\int_\Omega \phi_n d\mu_s^-, \end{aligned} \tag{5.2.37}$$

we obtain (5.2.32), which ends the proof. □

*Remark.* If we assume that $p = 2$, the result asserts that any solution $u$ of

$$-\Delta u + |u|^{q-1}u = 0 \qquad \text{in } \Omega \setminus K, \tag{5.2.38}$$

can be extended to a solution in whole $\Omega$ if $c^{\mathbb{R}^N}_{1,R}(K) = 0$ for some $R > \frac{2q}{q-1}$. This result is not optimal since the necessary and sufficient condition for the removability of a relatively closed set $K$ is $c^{\mathbb{R}^N}_{2,q'}(K) = 0$ (see [Baras,

Pierre (1984)]. However, if $q \geq \frac{N}{N-2}$ there is no imbedding of $W^{2,q'}(\Omega)$ into $W^{1,R}(\Omega)$ for any $R > \frac{2q}{q-1}$.

**Theorem 5.2.6** *Assume $g$ satisfies the assumptions (5.2.26) and $\mu \in \mathfrak{M}^+(\Omega)$. If there exists a locally renormalized solution $u$ to (5.2.32) in $\Omega$, then $\mu \ll c_{1,R}^{\mathbb{R}^N}$ for any $R > pq_p$.*

*Proof.* Let $R > pq_p$ and $K \subset \Omega$ be a compact set with $c_{1,R}^{\mathbb{R}^N}$-capacity zero. There exists a sequence $\{\xi_n\} \subset C_0^\infty(\Omega)$ of functions with value between 0 and 1 and value 1 in a neighborhood $\mathcal{U} \subset \overline{\mathcal{U}} \subset \Omega$. such that $\nabla\xi_n \to 0$ in $L^R(\Omega)$ and a.e. in $\Omega$. Since $u$ is a renormalized solution, $g(.,u,\nabla u)$ and then $u^q$ are locally integrable in $\Omega$; furthermore $|\nabla u|^{p-1} \in L^s_{loc}(\Omega)$ for $s < \frac{p'q}{q+1}$. A renormalized solution being a weak solution, we have

$$\int_\Omega \langle A(x,\nabla u), \nabla\xi_n^R\rangle dx + \int_\Omega g(x,\nabla u)\xi_n^R dx = \int_\Omega \xi_n^R d\mu,$$

thus

$$\lambda R\left(\int_\Omega |\nabla u|^{(p-1)R'}\,\xi_n^R dx\right)^{\frac{1}{R'}}\left(\int_\Omega |\nabla\xi_n|^R\,dx\right)^{\frac{1}{R}} + \int_\Omega |g(x,u,\nabla u)|\,\xi_n^R dx \geq \mu(K). \tag{5.2.39}$$

If $R > pq_p$, then $R' < \frac{p'q}{q+1}$ and thus $|\nabla u|^{(p-1)R'} \in L^1_{loc}(\Omega)$. Hence the left-hand side of (5.2.39) tends to 0 as $n \to \infty$ and $\mu(K) = 0$. □

### 5.2.3 *The Hamilton-Jacobi equation*

In this section we assume

$$g(x,r,\xi) \geq a\,|\nabla\xi|^q \qquad \forall (x,r,\xi) \in \Omega\times\mathbb{R}\times\mathbb{R}^N, \tag{5.2.40}$$

for some $q > p-1$ and $a > 0$. We recall that $q_p = \frac{q}{q+1-p}$.

**Theorem 5.2.7** *Assume $K \subset \Omega$ is relatively closed and $\mu \in \mathfrak{M}(\Omega)$ such that $\mu(K) = 0$. If $c_{1,q_p}^{\mathbb{R}^N}(K) = 0$, any locally renormalized solution of (5.2.1) in $\Omega \setminus K$ is a locally renormalized solution of the same equation in $\Omega$.*

*Proof. Step 1: We claim that $g(.,u,\nabla u) \in L^1_{loc}(\Omega)$.* Let $\{\xi_n\} \subset C_0^\infty(\mathbb{R}^N)$ be a sequence of functions defining the capacity of $K$ as in the proof of Theorem 5.2.4-Step 1, and $\phi \in C_0^1(\Omega)$, $\phi \geq 0$ with support $\overline{\mathcal{U}} \subset \Omega$ and

$\phi_n = (1-\xi_n)\phi$. Since $u$ is a locally renormalized solution in $\Omega \setminus K$ it is also a weak solution, therefore

$$\int_\Omega \langle A(x,\nabla u), \nabla \phi_n^{q_p} \rangle dx + \int_\Omega g(x,u,\nabla u)\phi_n^{q_p}\, dx = \int_\Omega \phi_n^{q_p}\, d\mu. \tag{5.2.41}$$

Since

$$\begin{aligned}\int_\Omega \langle A(x,\nabla u), \nabla \phi_n^{q_p} \rangle dx &= q_p \int_\Omega \langle A(x,\nabla u, \nabla \phi_n \rangle \phi_n^{q_p - 1} dx \\ &\leq q_p \lambda \left( \int_\Omega |\nabla u|^q \, \phi_n^{q_p}\, dx \right)^{\frac{p-1}{q}} \left( \int_\Omega |\nabla \phi_n|^{q_p}\, dx \right)^{\frac{1}{q_p}},\end{aligned} \tag{5.2.42}$$

and $g$ satisfies (5.2.40), we obtain

$$a \int_\Omega |\nabla u|^q \, \phi_n^{q_p}\, dx \leq q_p \lambda \left( \int_\Omega |\nabla u|^q \, \phi_n^{q_p}\, dx \right)^{\frac{p-1}{q}} \left( \int_\Omega |\nabla \phi_n|^{q_p}\, dx \right)^{\frac{1}{q_p}} + \int_\Omega \phi^{q_p}\, d\,|\mu|,$$

from which inequality we deduce

$$\int_\Omega |\nabla u|^q \, \phi_n^{q_p}\, dx \leq c_1 \int_\Omega |\nabla \phi_n|^{q_p}\, dx + c_2 \int_\Omega \phi^{q_p}\, d\,|\mu|,$$

for some $c_1, c_2 > 0$ depending on $p$, $q$, $a$ and $\lambda$. Letting $n \to \infty$ and using the fact that $\phi_n \to \phi$ a.e. and $|\nabla \phi_n|^{q_p} \to |\nabla \phi|^{q_p}$ in $L^1(\Omega)$, we derive

$$\int_\Omega |\nabla u|^q \, \phi^{q_p}\, dx \leq c_1 \int_\Omega |\nabla \phi|^{q_p}\, dx + c_2 \int_\Omega \phi^{q_p}\, d\,|\mu|. \tag{5.2.43}$$

From (5.2.41) and the inequalities above, we have also

$$0 \leq \int_\Omega g(x,u,\nabla u)\phi_n^{q_p}\, dx \leq c_3 \int_\Omega |\nabla \phi_n|^{q_p}\, dx + c_4 \int_\Omega \phi^{q_p}\, d\,|\mu|, \tag{5.2.44}$$

for $c_3, c_4$ depending on $p$, $q$, $a$ and $\lambda$. By Fatou's lemma we derive

$$0 \leq \int_\Omega g(x,u,\nabla u)\phi^{q_p}\, dx \leq c_3 \int_\Omega |\nabla \phi|^{q_p}\, dx + c_4 \int_\Omega \phi^{q_p}\, d\,|\mu|. \tag{5.2.45}$$

*Step 2: For any $k > 0$, $u_k := T_k(u) \in W^{1,p}_{loc}(\Omega)$.* Since $q > p-1$, $|\nabla u|^{p-1} \in L^1_{loc}(\Omega)$. As usually we write $\mu = \mu_0 + \mu_s^+ - \mu_s^-$. Let $\phi, \xi_n$ and $\phi$ be as in Step 1, then

$$\begin{aligned}&\int_\Omega \langle A(x,\nabla u), \nabla u_k \rangle \phi_n^{q_p}\, dx + q_p \int_\Omega \langle A(x,\nabla u), \nabla \phi_n \rangle u_k \phi_n^{q_p - 1} dx \\ &+ \int_\Omega g(x,u,\nabla u) T_k(u) \phi_n^{q_p}\, dx = \int_\Omega T_k(u)\phi_n^{q_p}\, d\mu_0 + k \int_\Omega \phi_n^{q_p}\, d\,|\mu_s|.\end{aligned} \tag{5.2.46}$$

Using (4.1.1) (i)-(ii) we get

$$\int_\Omega |\nabla u_k|^p \, \phi_n^{q_p} dx \leq kq_p\lambda \int_\Omega |\nabla u|^{p-1} |\nabla \phi_n| \, \phi_n^{q_p-1} dx + k\int_\Omega g(x,u,\nabla u)\phi_n^{q_p} dx$$
$$+ k\int_\Omega \phi^{q_p} d\,|\mu|\,.$$

The claim follows by using the estimates on $g(.,u,\nabla u)$ and $|\nabla u|^{p-1}$ obtained in Step 1.

*Step 3: The function u is a locally renormalized solution in* $\Omega$. As usually we write $\mu = \mu_0 + \mu_s^+ - \mu_s^-$. Let $h \in W^{1,\infty}(\mathbb{R})$ with $h'$ compactly supported, $\phi \in C_0^1(\Omega)$ (not necessarily nonnegative) and $\xi_n$ and $\phi_n$ as in Step 1. Then (5.2.44) reads with $\phi = \phi_n$

$$\begin{aligned}\int_\Omega \langle A(x,\nabla u), \nabla(h(u)\phi_n)dx + \int_\Omega g(x,u,\nabla u)h(u)\phi_n dx \\ = \int_\Omega h(u)\phi_n d\mu_0 + h(\infty)\int_\Omega \phi_n d\mu_s^+ - h(-\infty)\int_\Omega \phi_n d\mu_s^-.\end{aligned} \tag{5.2.47}$$

Since

$$\langle A(x,\nabla u), \nabla(h(u)\phi_n)\rangle = \langle A(x,\nabla u), \nabla\phi_n\rangle h(u) + \langle A(x,\nabla u), \nabla u\rangle h'(u)\phi_n,$$

the integral of the two terms on the right-hand side converges by dominated convergence and steps 1-2, thus

$$\lim_{n\to\infty}\int_\Omega \langle A(x,\nabla u), \nabla(h(u)\phi_n)dx = \int_\Omega \langle A(x,\nabla u), \nabla(h(u)\phi)dx.$$

Similarly, using Step 1 and dominated convergence

$$\lim_{n\to\infty}\int_\Omega g(x,u,\nabla u)h(u)\phi_n dx = \int_\Omega g(x,u,\nabla u)h(u)\phi dxdx.$$

Since $\mu(K) = 0$, the right-hand of (5.2.47) converges to $\int_\Omega h(u)\phi d\mu_0 + h(\infty)\int_\Omega \phi d\mu_s^+ - h(-\infty)\int_\Omega \phi d\mu_s^-$. This implies that (5.2.44) holds with arbitrary test function $\phi$, which is the claim. □

A counter part of this result is a necessary condition for solving (5.2.1).

**Theorem 5.2.8** *Assume g satisfies the assumptions (5.2.40) and* $\mu \in \mathfrak{M}^+(\Omega)$. *If there exists a locally renormalized solution u to (5.2.1) in* $\Omega$, *then* $\mu \ll c_{1,q_p}^{\mathbb{R}^N}$.

*Proof.* Let $K \subset \Omega$ be a compact set and $\{\xi_n\} \subset C_0^\infty(\Omega)$ a sequence of functions such $0 \le \xi_n \le 1$, $\xi_n = 1$ in a neighborhood of $K$. Then $u$ is also a weak solution and

$$\int_\Omega \langle A(x,\nabla u), \nabla \xi_n^{q_p}\rangle dx + \int_\Omega g(x,u,\nabla u)\xi_n^{q_p} dx = \int \xi_n^{q_p} d\mu \ge \mu(K).$$

Using (5.2.42) and (5.2.40) we derive

$$q_p \lambda a^{\frac{p-1}{q}} \left(\int_\Omega g(x,u,\nabla u)\xi_n^{q_p} dx\right)^{\frac{p-1}{q}} \left(\int_\Omega |\nabla \xi_n|^{q_p} dx\right)^{\frac{1}{q_p}} + \int_\Omega g(x,u,\nabla u)\xi_n^{q_p} dx \ge \mu(K).$$

If $c_{1,q_p}^{\mathbb{R}^N}(K) = 0$, we can choose $\xi_n$ such that $\lim_{n\to\infty} \|\xi_n\|_{W^{1,q_p}} = 0$ and $\xi_n \to 0$ a.e. in $\Omega$, therefore $\mu(K) = 0$. □

## 5.3 Supercritical equations

In this section we study the problem (5.0.1) in a bounded domain $\Omega \subset \mathbb{R}^N$ with $\mu \in \mathfrak{M}_b(\Omega)$, assuming that $A$ satisfies (4.1.1) and $g$ satisfies (5.0.2) and (5.2.7) with $q > p-1$. This problem is called *supercritical* if not every bounded measure is eligible, in particular no Dirac measure is eligible. When $1 < p < N$, the critical value for $q$ is $q_1 = \frac{N(p-1)}{N-p}$ already defined in (3.1.4).

### 5.3.1 *Estimates on potentials*

We have already seen the role of Wolff potential $\mathbf{W}_{1,p}[\mu]$ for the representation of positive $p$-harmonic functions in $\mathbb{R}^N$. We introduce below a more general class of Wolff potentials. Let $\alpha > 0$, $1 < s < \alpha^{-1}N$, $R > 0$ and $\mu \in \mathfrak{M}^+(\Omega)$. The $R$-truncated $(\alpha, s)$-Wolff potential of $\mu$ is

$$\mathbf{W}_{\alpha,s}^R[\mu](x) = \int_0^R \left(\frac{\mu(B_t(x))}{t^{N-\alpha s}}\right)^{\frac{1}{s-1}} \frac{dt}{t}. \tag{5.3.1}$$

If $R = \infty$ we note $\mathbf{W}_{\alpha,s}^\infty[\mu] = \mathbf{W}_{\alpha,s}[\mu]$. We also define the $R$-truncated fractional maximal operator

$$\mathbf{M}_\alpha^R[\mu](x) = \sup_{0<t\le R} \frac{\mu(B_t(x))}{t^{N-\alpha}}, \tag{5.3.2}$$

and the fractional maximal operator $\mathbf{M}_\alpha^\infty[\mu] = \mathbf{M}_\alpha[\mu]$. For $\eta \geq 0$ we set

$$h_\eta(t) = \begin{cases} (-\ln t)^{-\eta} & \text{if } 0 < t < 2^{-1} \\ (\ln 2)^{-\eta} & \text{if } t \geq 2^{-1}. \end{cases} \tag{5.3.3}$$

The $R$-truncated fractional $\eta$-maximal operator (resp. fractional $\eta$-maximal operator) is a refinement of the $R$-truncated fractional maximal operator defined by

$$\mathbf{M}_\alpha^{\eta,R}[\mu](x) = \sup_{0<t\leq R} \frac{\mu(B_t(x))}{t^{N-\alpha}h_\eta(t)}, \tag{5.3.4}$$

(resp. $\mathbf{M}_\alpha^{\eta,\infty}[\mu] = \mathbf{M}_\alpha^\eta[\mu]$). We also recall that the Bessel potential of a measure $\mu$ in $\mathbb{R}^N$ is defined by

$$\mathbf{G}_\alpha[\mu](x) = \int_{\mathbb{R}^N} G_\alpha(x-y)d\mu(y) \qquad \text{if } x \in \mathbb{R}^N, \tag{5.3.5}$$

where $G_\alpha$ is the Bessel kernel in $\mathbb{R}^N$ (see Section 1-2-4). For $0 < R \leq \infty$, $r > 0$, $p > 1$ and $0 < \alpha p < N$, we set

$$\ell(r,R) = \begin{cases} \dfrac{N-\alpha p}{p-1}\left((\min\{r,R\})^{-\frac{N-\alpha p}{p-1}} - R^{-\frac{N-\alpha p}{p-1}}\right) & \text{if } R < \infty \\ \dfrac{N-\alpha p}{p-1} r^{-\frac{N-\alpha p}{p-1}} & \text{if } R = \infty. \end{cases} \tag{5.3.6}$$

The next technical result is called a good $\lambda$-inequality.

**Proposition 5.3.1** *Let $0 \leq \eta < p-1$, $0 < \alpha p < N$ and $r > 0$. Then exist $c_0 > 0$ depending on $N$, $\alpha$, $p$, $\eta$ and $\epsilon_0 > 0$ depending on $N$, $\alpha$, $p$, $\eta$ and $r$ such that for all $\mu \in \mathfrak{M}^+(\mathbb{R}^N)$ with $\operatorname{diam}(\operatorname{supp}\mu) \leq r$, $R \in (0,\infty]$, $\epsilon \in (0,\epsilon_0]$ and $\lambda > (\mu(\mathbb{R}^N))^{\frac{1}{p-1}}\ell(r,R)$, there holds,*

$$\begin{aligned} &\left|\left\{\mathbf{W}_{\alpha,p}^R[\mu] > 3\lambda\right\} \cap \left\{\mathbf{M}_{\alpha p}^{\eta,R}[\mu] \leq (\epsilon\lambda)^{p-1}\right\}\right| \\ &\quad \leq c_0 \exp\left(\left(\tfrac{p-1-\eta}{4(p-1)}\right)^{\frac{p-1}{p-1-\eta}} \alpha p \epsilon^{-\frac{p-1}{p-1-\eta}} \ln 2\right) \left|\left\{\mathbf{W}_{\alpha,p}^R[\mu] > \lambda\right\}\right|. \end{aligned} \tag{5.3.7}$$

*Proof. Case 1: $R = \infty$.* Since $\mathbf{W}_{\alpha,p}[\mu]$ is lower semi continuous, for any $\lambda > 0$ the set

$$D_\lambda = \left\{x \in \mathbb{R}^N : \mathbf{W}_{\alpha,p}^R[\mu](x) > \lambda\right\} := \left\{\mathbf{W}_{\alpha,p}^R[\mu] > \lambda\right\},$$

is open. By Whitney decomposition [Grafakos (2010)] there exists a countable set $\mathcal{Q}$ of closed cubes $Q_j$ such that $D_\lambda = \cup_j Q_j$, $\overset{o}{Q}_j \cap \overset{o}{Q}_k = \emptyset$ if $j \neq k$ and

$$\operatorname{diam} Q_j \leq \operatorname{dist}(Q_j, D_\lambda^c) \leq 4\operatorname{diam} Q_j.$$

For $\epsilon > 0$ we set $F_{\epsilon,\lambda} = \{\mathbf{W}_{\alpha,p}[\mu] > 3\lambda\} \cap \{\mathbf{M}_{\alpha p}^{\eta}[\mu] \leq (\epsilon\lambda)^{p-1}\}$.

The key of the proof is to show that there exist $c_0 = c_0(N,\alpha,p,\eta) > 0$ and $\epsilon_0 = \epsilon_0(N,\alpha,p,\eta,r) > 0$ such that for any $Q \in \mathcal{Q}$, $\epsilon \in (0,\epsilon_0]$ and $\lambda > (\mu(\mathbb{R}^N))^{\frac{1}{p-1}}\ell(r,\infty)$, there holds

$$|F_{\epsilon,\lambda} \cap Q| \leq \exp\left(\left(\frac{p-1-\eta}{4(p-1)}\right)^{\frac{p-1}{p-1-\eta}} \alpha p \epsilon^{-\frac{p-1}{p-1-\eta}} \ln 2\right) |Q|. \tag{5.3.8}$$

We denote

$$E_{\epsilon,\lambda} = \left\{\mathbf{W}_{\alpha,p}^{5\,diam(Q)}[\mu] > \lambda\right\} \cap \{\mathbf{M}_{\alpha p}^{\eta}[\mu] \leq (\epsilon\lambda)^{p-1}\} \cap Q.$$

*Step 1.* We claim that there exists $c_1 = c_1(N,\alpha,p,\eta) > 0$ such that for any $Q \in \mathcal{Q}$, there holds

$$F_{\epsilon,\lambda} \cap Q \subset E_{\epsilon,\lambda} \qquad \forall \epsilon \in (0, c_1]\,,\ \forall \lambda > 0. \tag{5.3.9}$$

Let $Q \in \mathcal{Q}$ verifying $Q \cap F_{\epsilon,\lambda} \neq \emptyset$ and $x_Q \in D_\lambda^c$ be such that $\operatorname{dist}(x_Q, Q) \leq 4\operatorname{diam}(Q)$ and $\mathbf{W}_{\alpha,p}^{R}[\mu](x_Q) \leq \lambda$ (by definition of $D_\lambda$). If $k \in \mathbb{N}$, $r_0 = 5\operatorname{diam}(Q)$ and $x \in F_{\epsilon,\lambda} \cap Q$, we have

$$\int_{2^k r_0}^{2^{k+1} r_0} \left(\frac{\mu(B_t(x))}{t^{N-\alpha p}}\right)^{\frac{1}{p-1}} \frac{dt}{t} = A + B,$$

where

$$A = \int_{2^k r_0}^{2^k \frac{1+2^{k+1}}{1+2^k} r_0} \left(\frac{\mu(B_t(x))}{t^{N-\alpha pq}}\right)^{\frac{1}{p-1}} \frac{dt}{t}, \ B = \int_{2^k \frac{1+2^{k+1}}{1+2^k} r_0}^{2^{k+1} r_0} \left(\frac{\mu(B_t(x))}{t^{N-\alpha p}}\right)^{\frac{1}{p-1}} \frac{dt}{t}.$$

By the definition of $\mathbf{M}_{\alpha p}^{\eta}[\mu]$,

$$\mu(B_t(x)) \leq t^{N-\alpha p} h_\eta(t) \mathbf{M}_{\alpha p}^{\eta}[\mu](x) \leq t^{N-\alpha p} h_\eta(t) (\epsilon\lambda)^{p-1}. \tag{5.3.10}$$

Plugging this inequality into the expression $B$ and using the value of $h_\eta$ yields

$$B \leq \epsilon\lambda \int_{2^k \frac{1+2^{k+1}}{1+2^k} r_0}^{2^{k+1} r_0} (h_\eta(t))^{\frac{1}{p-1}} \frac{dt}{t} \leq c_2 \epsilon\lambda 2^{-k},$$

where $c_2 = c_2(p,\eta) > 0$. Set $\delta = \left(\frac{2^k}{2^k+1}\right)^{\frac{N-\alpha p}{p-1}}$, then $1-\delta \leq c_3 2^{-k}$ where $c_3 > 0$ depends only on $\frac{N-\alpha p}{p-1}$. Therefore

$$\begin{aligned}(1-\delta)A &\leq c_3 2^{-k}\int_{2^k r_0}^{2^{k+1}r_0}\left(\frac{\mu(B_t(x))}{t^{N-\alpha p}}\right)^{\frac{1}{p-1}}\frac{dt}{t}\\ &\leq c_3 2^{-k}\int_{2^k r_0}^{2^{k+1}r_0}(h_\eta(t))^{\frac{1}{p-1}}\frac{dt}{t}\\ &\leq c_4\epsilon\lambda 2^{-k},\end{aligned}$$

where $c_4 = c_4(N,\alpha,p,\eta) > 0$. Since for any $x \in F_{\epsilon,\lambda}\cap Q$ and any $t \in [r_0(1+2^k), r_0(1+2^{k+1})]$ we have $B_{\frac{2^k t}{1+2^k}}(x) \subset B_t(x_Q)$, we get, by a suitable change of variable and using the explicit value of $\delta$,

$$\delta A = \int_{(1+2^k)r_0}^{(1+2^{k+1})r_0}\left(\frac{\mu(B_{\frac{2^k t}{1+2^k}}(x))}{t^{N-\alpha p}}\right)^{\frac{1}{p-1}}\frac{dt}{t} \leq \int_{(1+2^k)r_0}^{(1+2^{k+1})r_0}\left(\frac{\mu(B_t(x_Q))}{t^{N-\alpha p}}\right)^{\frac{1}{p-1}}\frac{dt}{t}.$$

Summing the two last inequalities combined with the estimate on $B$, we obtain

$$\int_{2^k r_0}^{2^{k+1}r_0}\left(\frac{\mu(B_t(x))}{t^{N-\alpha p}}\right)^{\frac{1}{p-1}}\frac{dt}{t} \leq c_5\epsilon\lambda 2^{-k} + \int_{(1+2^k)r_0}^{(1+2^{k+1})r_0}\left(\frac{\mu(B_t(x_Q))}{t^{N-\alpha p}}\right)^{\frac{1}{p-1}}\frac{dt}{t},$$

where $c_5 = c_5(N,\alpha,p,\eta) > 0$. If we sum over $k$ we infer

$$\begin{aligned}\int_{r_0}^{\infty}\left(\frac{\mu(B_t(x))}{t^{N-\alpha p}}\right)^{\frac{1}{p-1}}\frac{dt}{t} &\leq 2c_5\epsilon\lambda + \int_{2r_0}^{\infty}\left(\frac{\mu(B_t(x_Q))}{t^{N-\alpha p}}\right)^{\frac{1}{p-1}}\frac{dt}{t}\\ &\leq (1+2c_5\epsilon)\lambda,\end{aligned} \tag{5.3.11}$$

since $\mathbf{W}_{\alpha,p}[\mu](x_Q) \leq \lambda$. Set $c_1 = (2c_5)^{-1}$, then if $\epsilon \in (0,c_1]$ there holds

$$\int_{r_0}^{\infty}\left(\frac{\mu(B_t(x))}{t^{N-\alpha p}}\right)^{\frac{1}{p-1}}\frac{dt}{t} \leq 2\lambda. \tag{5.3.12}$$

Since $r_0 = 5\,\mathrm{diam}(Q)$ and $\mathbf{W}_{\alpha,p}[\mu] > 3\lambda$ because $x \in F_{\epsilon,\lambda}$, we derive $\mathbf{W}_{\alpha,p}^{r_0}[\mu] > 3\lambda$ which yields (5.3.9).

*Step 2.* Proof of (5.3.8). Let $\lambda > (\mu(\mathbb{R}^N))^{\frac{1}{p-1}}\ell(r,\infty)$. There exists a ball $B_r(a)$ containing the support of $\mu$. If $x \notin B_{2r}(a)$,

$$\mathbf{W}_{\alpha,p}[\mu](x) = \int_r^\infty \left(\frac{\mu(B_t(x))}{t^{N-\alpha p}}\right)^{\frac{1}{p-1}} \frac{dt}{t} \leq (\mu(\mathbb{R}^N))^{\frac{1}{p-1}}\ell(r,\infty) < \lambda,$$

therefore $x \in D_\lambda^c$. Equivalently $D_\lambda \subset B_{2r}(a)$.

Put $m_0 = \frac{\max\{1,\ln(40r)\}}{\ln 2}$ so that $2^{-m}r_0 \leq 2^{-1}$ for $m \geq m_0$ (remember that $r_0 = 5\text{diam}\,(Q) \leq 20r$). Then, for any $x \in E_{\epsilon,\lambda}$,

$$\begin{aligned}
\int_{2^{-m}r_0}^{r_0} \left(\frac{\mu(B_t(x))}{t^{N-\alpha p}}\right)^{\frac{1}{p-1}} \frac{dt}{t} &\leq \epsilon\lambda \int_{2^{-m}r_0}^{r_0} (h_\eta(t))^{\frac{1}{p-1}} \frac{dt}{t} \\
&\leq \epsilon\lambda \int_{2^{-m}r_0}^{2^{-m_0}r_0} (-\ln(t))^{-\frac{\eta}{p-1}} \frac{dt}{t} + \epsilon\lambda \int_{2^{-m_0}r_0}^{r_0} (\ln 2)^{-\frac{\eta}{p-1}} \frac{dt}{t} \\
&\leq \epsilon\lambda m_0 (\ln 2)^{-\frac{\eta}{p-1}} + \epsilon\lambda \frac{(p-1)\left((m-m_0)\ln 2\right)^{1-\frac{\eta}{p-1}}}{p-1-\eta}.
\end{aligned}$$

Notice that, in the last line, we have used inequality

$$a^{1-\frac{\eta}{p-1}} - b^{1-\frac{\eta}{p-1}} \leq (a-b)^{1-\frac{\eta}{p-1}} \quad \forall a \geq b \geq 0.$$

Therefore

$$\int_{2^{-m}r_0}^{r_0} \left(\frac{\mu(B_t(x))}{t^{N-\alpha p}}\right)^{\frac{1}{p-1}} \frac{dt}{t} \leq \frac{2(p-1)}{p-1-\eta} m^{1-\frac{\eta}{p-1}} \epsilon\lambda, \quad \forall m \geq (\ln 2)^{-\frac{\eta}{p-1}} m_0^{\frac{p-1}{p-1-\eta}}. \tag{5.3.13}$$

Set

$$g_i(x) = \int_{2^{-i+1}r_0}^{2^{-i}r_0} \left(\frac{\mu(B_t(x))}{t^{N-\alpha p}}\right)^{\frac{1}{p-1}} \frac{dt}{t}.$$

Then, for $m \geq (\ln 2)^{-\frac{\eta}{p-1}} m_0^{\frac{p-1}{p-1-\eta}}$,

$$\begin{aligned}
\mathbf{W}_{\alpha,p}^{r_0}[\mu](x) &\leq \frac{2(p-1)}{p-1-\eta} m^{1-\frac{\eta}{p-1}} \epsilon\lambda + \int_{2^{-m}r_0}^{r_0} \left(\frac{\mu(B_t(x))}{t^{N-\alpha p}}\right)^{\frac{1}{p-1}} \frac{dt}{t} \\
&\leq \frac{2(p-1)}{p-1-\eta} m^{1-\frac{\eta}{p-1}} \epsilon\lambda + \sum_{i=m+1}^{\infty} g_i(x).
\end{aligned}$$

Consequently, for any $\beta > 0$,

$$\begin{aligned}
|E_{\epsilon,\lambda}| &\le \left| Q \bigcap \left\{ \sum_{i=m+1}^{\infty} g_i > \left(1 - \frac{2(p-1)}{p-1-\eta} m^{1-\frac{\eta}{p-1}} \epsilon\right) \lambda \right\} \right| \\
&\le \left| Q \bigcap \left\{ \sum_{i=m+1}^{\infty} g_i > \sum_{i=m+1}^{\infty} (2^{-\beta(i-m-1)}(1-2^{-\beta}) \left(1 - \frac{2(p-1)}{p-1-\eta} m^{1-\frac{\eta}{p-1}} \epsilon\right) \lambda \right\} \right| \\
&\le \sum_{i=m+1}^{\infty} \left| Q \bigcap \left\{ g_i > (2^{-\beta(i-m-1)}(1-2^{-\beta}) \left(1 - \frac{2(p-1)}{p-1-\eta} m^{1-\frac{\eta}{p-1}} \epsilon\right) \lambda \right\} \right|.
\end{aligned} \tag{5.3.14}$$

We claim that

$$|\, Q \cap \{g_i > s\}| \le \frac{c_6(N,\eta) 2^{i\alpha p}}{s^{p-1}} \,|\, Q|\, (\epsilon\lambda)^{p-1}. \tag{5.3.15}$$

To see that, we pick $y \in E_{\epsilon,\lambda}$ and get, using Chebyshev's inequality,

$$\begin{aligned}
|Q \cap \{g_i > s\}| &\le \frac{1}{s^{p-1}} \int_Q (g_i(x))^{p-1} dx \\
&= \frac{1}{s^{p-1}} \left( \int_Q \int_{2^{-i+1} r_0}^{2^{-i} r_0} \left( \frac{\mu(B_t(x))}{t^{N-\alpha p}} \right)^{\frac{1}{p-1}} \frac{dt}{t} \right)^{p-1} dx \\
&\le \frac{1}{s^{p-1}} \int_Q \frac{\mu(B_{2^{-i+1} r_0}(x))}{(2^{-i} r_0)^{N-\alpha p}} dx := A.
\end{aligned}$$

By Fubini's theorem,

$$\begin{aligned}
A &= \frac{1}{s^{p-1}} \frac{1}{(2^{-i} r_0)^{N-\alpha p}} \int_Q \int_{\mathbb{R}^N} \chi_{B_{2^{-i+1} r_0}}(z) d\mu(z) dx \\
&= \frac{1}{s^{p-1}} \frac{1}{(2^{-i} r_0)^{N-\alpha p}} \int_{Q + B_{2^{-i+1} r_0}} \int_Q \chi_{B_{2^{-i+1} r_0}(z)}(x) dx d\mu(z) \\
&\le \frac{1}{s^{p-1}} \frac{1}{(2^{-i} r_0)^{N-\alpha p}} \int_{Q + B_{2^{-i+1} r_0}} \int_Q |B_{2^{-i+1} r_0}(z)|\, d\mu(z) \\
&\le \frac{c_7(N)}{s^{p-1}} 2^{-i\alpha p} r_0^{\alpha p} \mu(Q + B_{2^{-i+1} r_0}) \\
&\le \frac{c_7(N)}{s^{p-1}} 2^{-i\alpha p} r_0^{\alpha p} \mu(Q + B_{(1+2^{-i+1}) r_0}(y)),
\end{aligned}$$

since $Q + B_{2^{-i+1} r_0} \subset Q + B_{1+2^{-i+1} r_0}(y)$. Since $r_0 = 5\text{diam}\,(\text{Q})$ and

$\mu(B_t(y)) \leq (\ln 2)^{-\eta} t^{N-\alpha p}(\epsilon\lambda)^{p-1}$, we obtain

$$\begin{aligned} A &\leq c_8(N,\eta)\frac{c_8(N,\eta)}{s^{p-1}} 2^{-i\alpha p} r_0^{\alpha p} \left((1+2^{-i+1})r_0\right)^{N-\alpha p} (\epsilon\lambda)^{p-1} \\ &\leq \frac{c_6(N,\eta)}{s^{p-1}} 2^{-i\alpha p} |Q| (\epsilon\lambda)^{p-1}, \end{aligned}$$

which is (5.3.15). Therefore, (5.3.14) can be rewritten as

$$\begin{aligned} |E_{\epsilon,\lambda}| &\leq \sum_{i=m+1}^{\infty} \frac{c_6 2^{-i\alpha p}(\epsilon\lambda)^{p-1}|Q|}{\left(2^{-\beta(i-m-1)}(1-2^{-\beta})\left(1-\frac{2(p-1)}{p-1-\eta} m^{1-\frac{\eta}{p-1}}\epsilon\right)\lambda\right)^{p-1}} \\ &\leq c_6 2^{-(m+1)\alpha p}\left(\frac{\epsilon}{1-\frac{2(p-1)}{p-1-\eta} m^{1-\frac{\eta}{p-1}}\epsilon}\right)^{p-1} \\ &\qquad \times |Q|(1-2^{-\beta})^{1-p} \sum_{i=m+1}^{\infty} 2^{(\beta(p-1)-\alpha p)(i-m-1)}. \end{aligned} \tag{5.3.16}$$

If we choose $\beta < \alpha p'$ we obtain

$$|E_{\epsilon,\lambda}| \leq c_9 \left(\frac{2^{-m\alpha}\epsilon}{1-\frac{2(p-1)}{p-1-\eta} m^{1-\frac{\eta}{p-1}}\epsilon}\right)^{p-1} |Q|, \quad \forall m > (\ln 2)^{-\frac{\eta}{p-1-\eta}} m_0^{\frac{p-1}{p-1-\eta}}, \tag{5.3.17}$$

where $c_9 = c_9(N\alpha, p\eta) > 0$. We put $\epsilon_0 = \min\left\{c_1, \frac{1}{1+\frac{4(p-1)}{p-1-\eta}m_0}\right\}$, and for any $\epsilon \in (0, \epsilon_0]$ we choose $m \in \mathbb{N}$ such that

$$\left(\frac{p-1-\eta}{2(p-1)}\right)^{\frac{p-1}{p-1-\eta}} \left(\frac{1}{\epsilon}-1\right)^{\frac{p-1}{p-1-\eta}} - 1 < m \leq \left(\frac{p-1-\eta}{2(p-1)}\right)^{\frac{p-1}{p-1-\eta}} \left(\frac{1}{\epsilon}-1\right)^{\frac{p-1}{p-1-\eta}}.$$

Therefore

$$\left(\frac{\epsilon}{1-\frac{2(p-1)}{p-1-\eta} m^{1-\frac{\eta}{p-1}}\epsilon}\right)^{p-1} \leq 1,$$

and

$$\begin{aligned} 2^{-m\alpha p} &\leq 2^{\alpha p - \alpha p\left(\frac{p-1-\eta}{2(p-1)}\right)^{\frac{p-1}{p-1-\eta}}\left(\frac{1}{\epsilon}-1\right)^{\frac{p-1}{p-1-\eta}}} \\ &\leq 2^{\alpha p} \exp\left(-\alpha p \ln 2 \left(\frac{p-1-\eta}{2(p-1)}\right)^{\frac{p-1}{p-1-\eta}} \epsilon^{-\frac{p-1}{p-1-\eta}}\right). \end{aligned}$$

Combining this estimate with (5.3.17) and (5.3.9) yields (5.3.8).

Although the previous proof is valid if $\eta = 0$ it is much easier to notice that for any $\epsilon, \lambda >$ and $m \in \mathbb{N}$ we have if $x \in E_{\epsilon,\lambda}$,

$$\mathbf{W}^{r_0}_{\alpha,p}[\mu](x) \leq m\epsilon\lambda + \sum_{i=m+1}^{\infty} g_i(x).$$

Inequality (5.3.17) reads

$$|E_{\epsilon,\lambda}| \leq c_9 2^{-mp\alpha} \left(\frac{\epsilon}{1 - m\epsilon}\right)^{p-1} |Q|, \quad \forall m > m_0.$$

If we set $\epsilon_0 = \min\left\{c_1, \frac{1}{2}\right\}$, then for any $\epsilon \in (0, \epsilon_0]$ and $m \in \mathbb{N}$ such that $\epsilon^{-1} - 2 < m \leq \epsilon^{-1} - 1$ we infer from (5.3.10),

$$|Q \cap E_{\epsilon,\lambda}| \leq |E_{\epsilon,\lambda}| \leq c_{10} 2^{\alpha p} \exp(-\alpha p \ln 2\epsilon^{-1})|Q|, \tag{5.3.18}$$

which is the claim.

*Case 2:* $R < \infty$. Since for $\lambda > 0$, $D_\lambda$ is open, we use again Whitney covering $\mathcal{Q} = \{Q_k\}_{k \in \mathbb{N}}$. If $Q \in \mathcal{Q}$ with $\operatorname{diam}(Q) > \frac{R}{8}$, there exists a finite number $n_Q$ of closed dyadic cubes $\{P_{j,Q}\}_{j=1}^{n_Q}$ such that

$$Q = \bigcup_{j=1}^{n_Q} P_{j,Q}\,,\ \overset{o}{P}_{j,Q} \cap \overset{o}{P}_{i,Q} = \emptyset \text{ if } i \neq j \text{ and } \frac{R}{16} \leq \operatorname{diam}(P_{j,Q}) \leq \frac{R}{8}.$$

We set $\mathcal{F} = \mathcal{Q}' \cup \mathcal{Q}''$ where

$$\mathcal{Q}' = \left\{Q \in \mathcal{Q} : \operatorname{diam}(Q) \leq \frac{R}{8}\right\},$$

$$\mathcal{Q}'' = \left\{P_{j,Q}, 1 \leq j \leq n_Q, Q \in \mathcal{Q}, \operatorname{diam}(Q) > \frac{R}{8}\right\},$$

For $\epsilon > 0$ we denote again $F_{\epsilon,\lambda} = \{\mathbf{W}^R_{\alpha,p}[\mu] > 3\lambda\} \cap \{\mathbf{M}^{\eta,R}_{\alpha p}[\mu] \leq (\epsilon\lambda)^{p-1}\}$. Let $Q \in \mathcal{F}$ such that $F_{\epsilon,\lambda} \cap Q \neq \emptyset$ and $r_0 = 5\operatorname{diam}(Q)$.

If $\operatorname{dist}(D^c_\lambda, Q) \leq 4\operatorname{diam}(Q)$, there exists $x_Q \in D^c_\lambda$ such that $\operatorname{dist}(x_Q, Q \leq 4\operatorname{diam}(Q)$ and $\mathbf{W}^R_{\alpha,p}[\mu](x_Q) \leq \lambda$, we find, by the same argument that in the case $R = \infty$ and (5.3.11), that for any $x \in F_{\epsilon,\lambda} \cap Q$, there holds

$$\int_{r_0}^{R} \left(\frac{\mu(B_t(x))}{t^{N-\alpha p}}\right)^{\frac{1}{p-1}} \frac{dt}{t} \leq (1 + c_{11}\epsilon)\lambda, \tag{5.3.19}$$

with $c_{11} = c_{11}(N, \alpha, p, \eta) > 0$.

If dist $(D_\lambda^c, Q) > 4\text{diam}\,(Q)$, we have $\frac{R}{16} < \text{diam}\,(Q) \le \frac{R}{8}$ since $Q \in \mathcal{Q}''$. Then, for all $x \in F_{\epsilon,\lambda} \cap Q$, there holds

$$\begin{aligned}\int_{r_0}^{R} \left(\frac{\mu(B_t(x))}{t^{N-\alpha p}}\right)^{\frac{1}{p-1}} \frac{dt}{t} &\le \int_{\frac{5r}{16}}^{R} \left(\frac{t^{N-\alpha p}\ln 2^{-\eta}(\epsilon\lambda)^{p-1}}{t^{N-\alpha p}}\right)^{\frac{1}{p-1}} \frac{dt}{t} \\ &\le (\ln 2)^{-\frac{\eta}{p-1}}\epsilon\lambda \ln\left(\tfrac{16}{5}\right) \\ &\le 2\epsilon\lambda.\end{aligned} \tag{5.3.20}$$

If we take $\epsilon \in (0, c_{12}]$ where $c_{12} = \min\{1, c_{11}^{-1}\}$, we derive

$$F_{\epsilon,\lambda} \cap Q \subset E_{\epsilon,\lambda}, \tag{5.3.21}$$

where, as in case 1,

$$E_{\epsilon,\lambda} = \left\{\mathbf{W}_{\alpha,p}^R[\mu] > \lambda\right\} \bigcap \left\{\mathbf{M}_{\alpha p,R}^{\eta}[\mu] \le (\epsilon\lambda)^{p-1}\right\}.$$

Furthermore, if $x \notin B_2$,

$$\mathbf{W}_{\alpha,p}^R[\mu](x) = \int_{\min\{r,R\}}^{R} \left(\frac{\mu(B_t(x))}{t^{N-\alpha p}}\right)^{\frac{1}{p-1}} \frac{dt}{t} \le \left(\mu(\mathbb{R}^N)\right)^{\frac{1}{p-1}} \ell(r,R).$$

Hence, if $\lambda > \left(\mu(\mathbb{R}^N)\right)^{\frac{1}{p-1}} \ell(r,R)$, we infer $D_\lambda \subset B_2$, which implies again $r_0 = 5\text{diam}\,(Q) \le 20r$. The remaining of the proof is as in Case 1. □

The following estimates link the Wolff potential of a measure with its Bessel potential and the fractional maximal operator.

**Theorem 5.3.2** *Assume $\alpha > 0$, $0 < p-1 < q < \infty$, $0 < \alpha p < N$ and $0 < s \le \infty$. Then there exists a constant $c_{13} = c_{13}(N, \alpha, p, q, s) > 0$ such that for any $R \in (0, \infty]$ and $\mu \in \mathfrak{M}^+(\mathbb{R}^N)$ there holds*

$$c_{13}^{-1} \left\|\mathbf{W}_{\alpha,p}^R[\mu]\right\|_{L^{q,s}(\mathbb{R}^N)} \le \left\|\mathbf{M}_{\alpha p,R}[\mu]\right\|_{L^{\frac{q}{p-1},\frac{s}{p-1}}(\mathbb{R}^N)}^{\frac{1}{p-1}} \le c_{13} \left\|\mathbf{W}_{\alpha,p}^R[\mu]\right\|_{L^{q,s}(\mathbb{R}^N)}. \tag{5.3.22}$$

*For any $R > 0$ there exists $c_{14} = c_{14}(N, \alpha, p, q, s, R) > 0$ such that for $\mu \in \mathfrak{M}^+(\mathbb{R}^N)$ there holds*

$$c_{14}^{-1} \left\|\mathbf{W}_{\alpha,p}^R[\mu]\right\|_{L^{q,s}(\mathbb{R}^N)} \le \left\|\mathbf{G}_{\alpha p}[\mu]\right\|_{L^{\frac{q}{p-1},\frac{s}{p-1}}(\mathbb{R}^N)}^{\frac{1}{p-1}} \le c_{14} \left\|\mathbf{W}_{\alpha,p}^R[\mu]\right\|_{L^{q,s}(\mathbb{R}^N)}. \tag{5.3.23}$$

*Proof.* We denote by $\mu_n = \chi_{B_n}\mu$.

*Step 1.* We claim that

$$\left\|\mathbf{W}^R_{\alpha,p}[\mu]\right\|_{L^{q,s}(\mathbb{R}^N)} \leq c'_{13} \left\|\mathbf{M}_{\alpha p,R}[\mu]\right\|^{\frac{1}{p-1}}_{L^{\frac{q}{p-1},\frac{s}{p-1}}(\mathbb{R}^N)}. \tag{5.3.24}$$

From Proposition 5.3.1 there exist three positive constants $c_0 = c_0(N,\alpha,p)$, $a = a(N,\alpha)$ and $\epsilon_0 = \epsilon_0(N,\alpha,p)$ such that for all $n \in \mathbb{N}_*$, $t > 0$, $0 < R \leq \infty$ and $\epsilon \in (0,\epsilon_0]$, there holds

$$\left|\left\{\mathbf{W}^R_{\alpha,p}[\mu_n] > 3t\right\} \bigcap \left\{\mathbf{M}^{R,\eta}_{\alpha p}[\mu_n] \leq (\epsilon t)^{p-1}\right\}\right| \leq c_0 e^{-\frac{a}{\epsilon}} \left|\left\{\mathbf{W}^R_{\alpha,p}[\mu] > t\right\}\right|. \tag{5.3.25}$$

When $0 < s < \infty$ and $0 < q < \infty$ we have

$$\begin{aligned}\left|\left\{\mathbf{W}^R_{\alpha,p}[\mu_n] > 3t\right\}\right|^{\frac{s}{q}} &\leq c_{15} e^{-\frac{sa}{q\epsilon}} \left|\left\{\mathbf{W}^R_{\alpha,p}[\mu_n] > t\right\}\right|^{\frac{s}{q}} \\ &\quad + c_{15}\left|\left\{\mathbf{M}^{R,\eta}_{\alpha p}[\mu_n] \leq (\epsilon t)^{p-1}\right\}\right|^{\frac{s}{q}},\end{aligned}$$

with $c_{15} = c_{15}(N,\alpha,p,q,s) > 0$. Multiplying by $t^{s-1}$ and integrating over $(0,\infty)$ yields

$$\begin{aligned}\int_0^\infty t^s \left|\left\{\mathbf{W}^R_{\alpha,p}[\mu_n] > 3t\right\}\right|^{\frac{s}{q}} \frac{dt}{t} &\leq e^{-\frac{sa}{q\epsilon}} \int_0^\infty t^s \left|\left\{\mathbf{W}^R_{\alpha,p}[\mu_n] > t\right\}\right|^{\frac{s}{q}} \frac{dt}{t} \\ &\quad + c_{15} \int_0^\infty t^s \left|\left\{\mathbf{M}^{R,\eta}_{\alpha p}[\mu_n] \leq (\epsilon t)^{p-1}\right\}\right|^{\frac{s}{q}} \frac{dt}{t}.\end{aligned}$$

By a change of variable, we derive

$$\begin{aligned}\left(3^{-s} - c_{15}e^{-\frac{sa}{q\epsilon}}\right) &\int_0^\infty t^s \left|\left\{\mathbf{W}^R_{\alpha,p}[\mu_n] > t\right\}\right|^{\frac{s}{q}} \frac{dt}{t} \\ &\leq c_{15} \int_0^\infty t^s \left|\left\{\mathbf{M}^{R,\eta}_{\alpha p}[\mu_n] \leq (\epsilon t)^{p-1}\right\}\right|^{\frac{s}{q}} \frac{dt}{t}.\end{aligned}$$

We choose $\epsilon > 0$ small enough so that $3^{-s} - c_{15}e^{-\frac{sa}{q\epsilon}} > 0$. Using (1.2.8) and identity

$$\left\|t^{\frac{1}{s_1}} f^*\right\|_{L^{s_2}(\mathbb{R}^+,\frac{dt}{t})} = s_1^{\frac{1}{s_2}} \left\|t\lambda_f^{\frac{1}{s_1}}\right\|_{L^{s_2}(\mathbb{R}^+,\frac{dt}{t})} \qquad \forall f \in L^{s_1,s_2}(\mathbb{R}^N),$$

valid for $0 < s_1 < \infty$, $0 < s_2 \leq \infty$ (we recall that from Chap 1-2, $S_f(t) = \{|f| > t\}$ and $\lambda_f(t) = |S_f(t)|$), we obtain

$$\left\|\mathbf{W}^R_{\alpha,p}[\mu_n]\right\|_{L^{q,s}(\mathbb{R}^N)} \leq c'_{13} \left\|\mathbf{M}_{\alpha p,R}[\mu_n]\right\|^{\frac{1}{p-1}}_{L^{\frac{q}{p-1},\frac{s}{p-1}}(\mathbb{R}^N)}. \tag{5.3.26}$$

Letting $n \to \infty$, we derive (5.3.24) by Fatou's lemma.

*Step 2.* We claim that

$$\left\|\mathbf{W}^R_{\alpha,p}[\mu]\right\|_{L^{q,s}(\mathbb{R}^N)} \geq c''_{13} \left\|\mathbf{M}_{\alpha p,R}[\mu]\right\|^{\frac{1}{p-1}}_{L^{\frac{q}{p-1},\frac{s}{p-1}}(\mathbb{R}^N)}. \tag{5.3.27}$$

For $R > 0$ we have

$$\begin{aligned} \mathbf{W}^{2R}_{\alpha,p}[\mu_n](x) &= \mathbf{W}^R_{\alpha,p}[\mu_n](x) + \int_R^{2R} \left(\frac{\mu_n(B_t(x))}{t^{N-\alpha p}}\right)^{\frac{1}{p-1}} \frac{dt}{t} \\ &\leq \mathbf{W}^R_{\alpha,p}[\mu_n](x) + \left(\frac{\mu_n(B_{2R}(x))}{R^{N-\alpha p}}\right)^{\frac{1}{p-1}}. \end{aligned} \tag{5.3.28}$$

Thus

$$\left|\{\mathbf{W}^{2R}_{\alpha,p}[\mu_n] > 2t\}\right| \leq \left|\{\mathbf{W}^R_{\alpha,p}[\mu_n] > 2t\}\right| + \left|x : \left\{\frac{\mu_n(B_{2R}(x))}{R^{N-\alpha p}} > t^{p-1}\right\}\right|.$$

Consider a set of points $\{z_j\}_{j=1}^m \subset B_2$ such that $B_2 \subset \cup_j B_{\frac{1}{2}}(z_j)$, thus $B_{2R}(x) \subset \cup_j B_{\frac{R}{2}}(x + Rz_j)$ for all $x \in \mathbb{R}^N$ and all $R > 0$. Then

$$\begin{aligned} \left|\left\{x : \frac{\mu_n(B_{2R}(x))}{R^{N-\alpha p}} > t^{p-1}\right\}\right| &\leq \left|\left\{x : \sum_j \frac{\mu_n(B_{\frac{R}{2}}(x + Rz_j))}{R^{N-\alpha p}} > t^{p-1}\right\}\right| \\ &\leq \sum_j \left|\left\{x : \frac{\mu_n(B_{\frac{R}{2}}(x + Rz_j))}{R^{N-\alpha p}} > \frac{t^{p-1}}{m}\right\}\right| \\ &\leq \sum_j \left|\left\{x - Rz_j : \frac{\mu_n(B_{\frac{R}{2}}(x))}{R^{N-\alpha p}} > \frac{t^{p-1}}{m}\right\}\right| \\ &= m \left|\left\{x : \frac{\mu_n(B_{\frac{R}{2}}(x))}{R^{N-\alpha p}} > \frac{t^{p-1}}{m}\right\}\right|. \end{aligned}$$

Furthermore, from (5.3.28),

$$\left(\frac{\mu_n(B_{\frac{R}{2}}(x))}{R^{N-\alpha p}}\right)^{\frac{1}{p-1}} \leq 2\mathbf{W}^R_{\alpha,p}[\mu_n](x),$$

hence

$$\left|\left\{x : \frac{\mu_n(B_{2R}(x))}{R^{N-\alpha p}} > t^{p-1}\right\}\right| \leq m \left|\left\{\mathbf{W}^R_{\alpha,p}[\mu_n] > \frac{t}{2m^{p-1}}\right\}\right|.$$

We deduce

$$\left|\{\mathbf{W}^{2R}_{\alpha,p}[\mu_n] > 2t\}\right| \leq (m+1) \left|\{\mathbf{W}^R_{\alpha,p}[\mu_n] > 2t\}\right|.$$

This implies

$$\left\|\mathbf{W}^{2R}_{\alpha,p}[\mu_n]\right\|_{L^{q,s}(\mathbb{R}^N)} \leq c_{16}\left\|\mathbf{W}^{R}_{\alpha,p}[\mu_n]\right\|_{L^{q,s}(\mathbb{R}^N)},$$

and by monotone convergence,

$$\left\|\mathbf{W}^{2R}_{\alpha,p}[\mu]\right\|_{L^{q,s}(\mathbb{R}^N)} \leq c_{16}\left\|\mathbf{W}^{R}_{\alpha,p}[\mu]\right\|_{L^{q,s}(\mathbb{R}^N)}, \tag{5.3.29}$$

with $c_{16} = c_{16}(N,\alpha,p,q,s) > 0$. On the other hand, by (5.3.28), we have for any $0 < \rho < R$,

$$\mathbf{W}^{2R}_{\alpha,p}[\mu](x) \geq \mathbf{W}^{2\rho}_{\alpha,p}[\mu](x) \geq c_{17} \sup_{0<\rho\leq R}\left(\frac{\mu(B_\rho(x))}{\rho^{N-\alpha p}}\right)^{\frac{1}{p-1}},$$

with $c_{17} = c_{17}(N,\alpha,p) > 0$. This yields

$$\mathbf{W}^{2R}_{\alpha,p}[\mu](x) \geq c_{17}\left(\mathbf{M}_{\alpha p,R}[\mu](x)\right)^{\frac{1}{p-1}}. \tag{5.3.30}$$

Combining (5.3.30) and (5.3.29), we obtain (5.3.27) and finally (5.3.22).

*Step 3.* We claim that (5.3.23) holds. By (5.3.22) we have also

$$c_{18}^{-1}\left\|\mathbf{W}^{R}_{\frac{\alpha p}{2},2}[\mu]\right\|_{L^{q,s}(\mathbb{R}^N)} \leq \|\mathbf{M}_{\alpha p,R}[\mu]\|^{\frac{1}{p-1}}_{L^{\frac{q}{p-1},\frac{s}{p-1}}(\mathbb{R}^N)} \leq c_{18}\left\|\mathbf{W}^{R}_{\frac{\alpha p}{2},2}[\mu]\right\|_{L^{q,s}(\mathbb{R}^N)}, \tag{5.3.31}$$

where $c_{18} = c_{18}(N,\alpha,p,q,s) > 0$. Estimate (1.2.49) on Bessel kernel (with $\alpha$, $R$ replaced by $\alpha p$, $\frac{R}{2}$) takes the form

$$c_{19}^{-1}\frac{\chi_{B_R}(x)}{|x|^{N-\alpha p}} \leq G_{\alpha p}(x) \leq c_{19}\frac{\chi_{B_{\frac{R}{2}}}(x)}{|x|^{N-\alpha p}} + c_{19}e^{-\frac{|x|}{2}} \qquad \forall x \in \mathbb{R}^N \setminus \{0\}.$$

Therefore

$$c_{19}^{-1}\frac{\chi_{B_R}}{|.|^{N-\alpha p}} * \mu \leq \mathbf{G}_{\alpha p}[\mu] \leq c_{19}\frac{\chi_{B_{\frac{R}{2}}}}{|.|^{N-\alpha p}} * \mu + c_{19}e^{-\frac{|.|}{2}} * \mu.$$

By Fubini's theorem

$$\begin{aligned}\frac{\chi_{B_R}}{|.|^{N-\alpha p}} * \mu(x) &= (N-\alpha p)\mathbf{W}^{R}_{\frac{\alpha p}{2},2}[\mu](x) + \frac{\mu(B_R(x))}{R^{N-\alpha p}} \\ &\geq (N-\alpha p)\mathbf{W}^{R}_{\frac{\alpha p}{2},2}[\mu](x),\end{aligned}$$

which implies

$$c_{20}\left\|\mathbf{W}^{R}_{\frac{\alpha p}{2},2}[\mu]\right\|_{L^{\frac{q}{p-1},\frac{s}{p-1}}(\mathbb{R}^N)} \leq \|\mathbf{G}_{\alpha p}[\mu]\|_{L^{\frac{q}{p-1},\frac{s}{p-1}}(\mathbb{R}^N)}, \tag{5.3.32}$$

where $c_{20} = c_{20}(N, \alpha, p, q, s) > 0$. Since $e^{-\frac{|x|}{2}} \le c_{21} e^{-\frac{|.|}{2}} * \chi_{B_{\frac{R}{2}}}$ with $c_{21} = c_{21}(N, R) > 0$, we get

$$e^{-\frac{|.|}{2}} * \mu \le c_{21} \left( e^{-\frac{|.|}{2}} * \chi_{B_{\frac{R}{2}}} \right) * \mu = c_{21} e^{-\frac{|.|}{2}} * \left( \chi_{B_{\frac{R}{2}}} * \mu \right).$$

Because

$$\chi_{B_{\frac{R}{2}}} * \mu(x) = \mu(B_{\frac{R}{2}}(x)) \le c_{22} \mathbf{W}^{R}_{\frac{\alpha p}{2},2}[\mu](x),$$

where $c_{22} = c_{22}(N, \alpha, p, R) > 0$, we derive, with $c_{23} = c_{21}c_{22}$,

$$e^{-\frac{|.|}{2}} * \mu \le c_{23} e^{-\frac{|.|}{2}} * \mathbf{W}^{R}_{\frac{\alpha p}{2},2}[\mu].$$

Using Young's inequality, we obtain

$$\begin{aligned} \left\| e^{-\frac{|.|}{2}} * \mu \right\|_{L^{\frac{q}{p-1},\frac{s}{p-1}}(\mathbb{R}^N)} &\le c_{23} \left\| e^{-\frac{|.|}{2}} * \mathbf{W}^{R}_{\frac{\alpha p}{2},2}[\mu] \right\|_{L^{\frac{q}{p-1},\frac{s}{p-1}}(\mathbb{R}^N)} \\ &\le c_{24} \left\| \mathbf{W}^{R}_{\frac{\alpha p}{2},2}[\mu] \right\|_{L^{\frac{q}{p-1},\frac{s}{p-1}}(\mathbb{R}^N)} \left\| e^{-\frac{|.|}{2}} \right\|_{L^{1,\infty}(\mathbb{R}^N)} \\ &\le c_{25} \left\| \mathbf{W}^{R}_{\frac{\alpha p}{2},2}[\mu] \right\|_{L^{\frac{q}{p-1},\frac{s}{p-1}}(\mathbb{R}^N)}, \end{aligned} \tag{5.3.33}$$

where $c_{25} = c_{25}(N, \alpha, p, R) > 0$.

Since by Fubini's theorem there holds as above,

$$\begin{aligned} \left( \frac{\chi_{B_{\frac{R}{2}}}}{|.|^{N-\alpha p}} \right) * \mu(x) &= (N - \alpha p) \mathbf{W}^{\frac{R}{2}}_{\frac{\alpha p}{2},2}[\mu](x) + 2^{N-\alpha p} \frac{\chi_{B_{\frac{R}{2}}}(x)}{R^{N-\alpha p}} \\ &\le c_{26} \mathbf{W}^{R}_{\frac{\alpha p}{2},2}[\mu](x), \end{aligned}$$

we infer

$$\left\| \left( \frac{\chi_{B_{\frac{R}{2}}}}{|.|^{N-\alpha p}} \right) * \mu \right\|_{L^{\frac{q}{p-1},\frac{s}{p-1}}(\mathbb{R}^N)} \le c_{27} \left\| \mathbf{W}^{R}_{\frac{\alpha p}{2},2}[\mu] \right\|_{L^{\frac{q}{p-1},\frac{s}{p-1}}(\mathbb{R}^N)}, \tag{5.3.34}$$

where $c_{27} = c_{27}(N, \alpha, p, R, s) > 0$. This implies,

$$\|\mathbf{G}_{\alpha p}[\mu]\|_{L^{\frac{q}{p-1},\frac{s}{p-1}}(\mathbb{R}^N)} \le c_{28} \left\| \mathbf{W}^{R}_{\frac{\alpha p}{2},2}[\mu] \right\|_{L^{\frac{q}{p-1},\frac{s}{p-1}}(\mathbb{R}^N)}, \tag{5.3.35}$$

by combining (5.3.31), (5.3.33) and (5.3.34). We deduce that (5.3.23) holds from (5.3.32) and (5.3.35), and using (5.3.30) and (5.3.22). □

The next result provides an exponential estimate of the Wolff potential of a measure under a constraint over its fractional potential.

**Theorem 5.3.3** *Assume $\alpha > 0$, $0 \le \eta < p-1$, $0 < \alpha p < N$ and $r > 0$ and set $\delta_0 = \left(\frac{p-1-\eta}{12(p-1)}\right)^{\frac{p-1}{p-1-\eta}}$. Then there exists $c_{29} > 0$, depending on $N$, $\alpha$, $p$, $\eta$ and $r$ such that for any $R \in (0,\infty]$, $\delta \in (0,\delta_0]$, $\mu \in \mathfrak{M}^+$, any couple of balls $B = B_\kappa(a)$ with $\kappa \le r$ and $B' = B_{2\kappa}(a)$, there holds*

$$\frac{1}{|B'|}\int_{B'} \exp\left(\delta \frac{\left(\mathbf{W}^R_{\alpha,p}[\mu_B](x)\right)^{\frac{p-1}{p-1-\eta}}}{\left\|\mathbf{M}^\eta_{\alpha p,R}[\mu_B]\right\|^{\frac{1}{p-1-\eta}}_{L^\infty(B)}}\right) dx \le \frac{c_{29}}{\delta_0 - \delta}, \tag{5.3.36}$$

*where $\mu_B = \chi_B \mu$. Furthermore, if $\eta = 0$, $c_{29}$ is independent of $r$.*

*Proof.* Assume $M := \left\|\mathbf{M}^\eta_{\alpha p,R}[\mu_B]\right\|_{L^\infty(B)} < \infty$. By applying Proposition 5.3.1 with $\mu = \mu_B$, there exists $c_0$ depending on $N$, $\alpha$, $p$, $\eta$ and $\epsilon_0$ depending on $N$, $\alpha$, $p$, $\eta$ and $r$ such that, for all $R \in (0,\infty]$, $\epsilon \in (0,\epsilon_0]$ and $t > (\mu_B(B))^{\frac{1}{p-1}}\ell(\kappa,R)$, there holds

$$\begin{aligned}&\left|\left\{\mathbf{W}^R_{\alpha,p}[\mu_B] > 3t\right\} \cap \left\{\mathbf{M}^{\eta,R}_{\alpha p}[\mu_B] \le (\epsilon t)^{p-1}\right\}\right|\\ &\quad\le c_0 \exp\left(\left(\tfrac{p-1-\eta}{4(p-1)}\right)^{\frac{p-1}{p-1-\eta}} \alpha p \epsilon^{-\frac{p-1}{p-1-\eta}} \ln 2\right) \left|\left\{\mathbf{W}^R_{\alpha,p}[\mu_B] > t\right\}\right|.\end{aligned} \tag{5.3.37}$$

Because $(\mu_B(B))^{\frac{1}{p-1}}\ell(\kappa,R) \le \frac{N-\alpha p}{p-1}(\ln 2)^{-\frac{\eta}{p-1}} M^{\frac{1}{p-1}}$, in (5.3.7) we choose

$$\epsilon = t^{-1}\left\|\mathbf{M}^\eta_{\alpha p,R}[\mu_B]\right\|^{\frac{1}{p-1}}_{L^\infty(B)} = M^{\frac{1}{p-1}} \qquad \forall t > t_0,$$

where $t_0 = \max\left\{\epsilon_0^{-1}, \frac{N-\alpha p}{p-a}(\ln 2^{-\frac{\eta}{p-1}}\right\} M^{\frac{1}{p-1}}$. As in the proof of Proposition 5.3.1, we have that $\left\{\mathbf{W}^R_{\alpha,p}[\mu_B] > t\right\} \subset B'$. Then

$$\begin{aligned}&\left|\left\{\mathbf{W}^R_{\alpha,p}[\mu_B] > 3t\right\} \cap B'\right|\\ &\quad\le c_0 \exp\left(\left(\frac{p-1-\eta}{4(p-1)}\right)^{\frac{p-1}{p-1-\eta}} \alpha p \ln 2 M^{-\frac{1}{p-1-\eta}} t^{\frac{p-1}{p-1-\eta}}\right) |B'|.\end{aligned} \tag{5.3.38}$$

This can be re-written under the form

$$|\{F > t\} \cap B'| \le |B'|\, \chi_{(0,t_0]}(t) + e^{-\delta_0 t}\, |B'|\, \chi_{(t_0,\infty)}(t), \tag{5.3.39}$$

where

$$F = M^{-\frac{1}{p-1-\eta}} (\mathbf{W}^R_{\alpha,p}[\mu_B])^{\frac{p-1}{p-1-\eta}},$$

and

$$t_0 = \left(3\max\{\epsilon_0^{-1}, \frac{N-\alpha p}{p-a}(\ln 2^{-\frac{\eta}{p-1}}\}\right)^{\frac{p-1}{p-1-\eta}}.$$

Take $\delta \in (0, \delta_0]$, then, by Fubini's theorem

$$\begin{aligned}\int_{B'} \exp(\delta F(x))dx &= \delta \int_0^\infty e^{-\delta t} |\{F > t\} \cap B'| \, dt \\ &\delta |B'| \int_0^{t_0} e^{-\delta t} dt + c_0 \delta |B'| \int_{t_0}^\infty e^{-(\delta_0 - \delta)t} dt \\ &\leq \left(e^{\delta t_0} - 1 + \frac{c_0 \delta}{\delta_0 - \delta}\right) |B'|,\end{aligned}$$

which is (5.3.36). □

### 5.3.2 *Approximation of measures*

The following result is an improvement of Proposition 1.2.7 (iii).

**Theorem 5.3.4** *Let $\Omega \subset \mathbb{R}^N$ be a domain. Assume $p-1 < s_1 < \infty$, $p-1 < s_2 \leq \infty$, $0 < \alpha p < N$, $R > 0$ and $\mu \in \mathfrak{M}^+(\Omega)$. If $\mu$ is absolutely continuous with respect to the capacity $c^{\mathbb{R}^N}_{\alpha p, \frac{s_1}{s_1+1-p}, \frac{s_2}{s_2+1-p}}$, there exists a nondecreasing sequence $\{\mu_n\} \subset \mathfrak{M}^+(\Omega)$ with compact support in $\Omega$ which converges to $\mu$ in the weak sense of measures and such that $\mathbf{W}^R_{\alpha,p}[\mu_n] \in L^{s_1,s_2}(\mathbb{R}^N)$ for all $n$. Furthermore, if $\mu \in \mathfrak{M}^+_b(\Omega)$, the convergence of $\{\mu_n\}$ towards $\mu$ holds in the narrow topology.*

*Proof.* By Proposition 1.2.7 (iii) there exists a nondecreasing sequence of nonnegative measures $\{\mu_n\} \subset (L^{\alpha p, \frac{s_1}{s_1+1-p}, \frac{s_2}{s_2+1-p}}(\mathbb{R}^N))'$ with compact support in $\Omega$, which converges weakly to $\mu$. If $\mu \in \mathfrak{M}^+_b(\Omega)$, the convergence holds in the narrow topology. Furthermore, if $\mu$ is a nonnegative measure in $\mathbb{R}^N$,

$$\mathbf{G}_{\alpha p}[\mu] \in L^{\frac{s_1}{p-1}, \frac{s_2}{p-1}}(\mathbb{R}^N) \iff \mu \in (L^{\alpha p, \frac{s_1}{s_1+1-p}, \frac{s_2}{s_2+1-p}}(\mathbb{R}^N))'$$

and $(L^{\alpha p, \frac{s_1}{s_1+1-p}, \frac{s_2}{s_2+1-p}}(\mathbb{R}^N))' = L^{-\alpha p, \frac{s_1}{p-1}, \frac{s_2}{p-1}}(\mathbb{R}^N))$ if $s_2 < \infty$ (see (1.2.54) and references therein). Then $\mathbf{G}_{\alpha p}[\mu_n] \in L^{\frac{s_1}{p-1}, \frac{s_2}{p-1}}(\mathbb{R}^N)$ and by Theorem 5.3.2, $\mathbf{W}^R_{\alpha,p}[\mu_n] \in L^{s_1,s_2}(\mathbb{R}^N)$. □

## 5.4 Solvability with measure data revisited

Throughout this section $\Omega \subset \mathbb{R}^N$ is a bounded domain, $A : \Omega \times \mathbb{R}^N \mapsto \mathbb{R}^N$ satisfies the assumptions (4.1.1). We consider a Caratheodory function $g : \Omega \times \mathbb{R} \mapsto \mathbb{R}$ subject to part of the following set of assumptions,

$$\begin{array}{llll} (i) & sg(x,s) \geq 0 & \text{a.e. } x \in \Omega,\ \forall s \in \mathbb{R}, & \\ (ii) & s \mapsto g(x,s) & \text{is nondecreasing for a.e. } x \in \Omega & \\ (iii) & |g(x,s)| \leq \tilde{g}(|s|) & \text{a.e. } x \in \Omega,\ \forall s, \in \mathbb{R} & \\ (iv) & s \mapsto \tilde{g}(s) & \text{is nondecreasing on } \mathbb{R}_+, \tilde{g}(0) = 0. & \end{array} \tag{5.4.1}$$

We intend to prove existence of renormalized solutions to

$$\begin{aligned} -\operatorname{div} A(x, \nabla u) + g(x,u) = \mu \qquad & \text{in } \Omega \\ u = 0 \qquad & \text{on } \partial\Omega, \end{aligned} \tag{5.4.2}$$

when $\mu$ is a bounded measure and the function $g$ does not satisfies the subcriticality assumptions (5.1.19) and (5.1.20).

### 5.4.1 *The general case*

The next two-side estimate for renormalized solutions is a consequence of Theorem 4.3.8 and Corollary 4.3.14.

**Theorem 5.4.1** *There exists a constant $c = c(N,p) > 0$ such that for any bounded measure and any renormalized solution $u$ of problem (4.2.3), there holds*

$$-c\mathbf{W}_{1,p}^{2\,diam(\Omega)}[\mu^-] \leq u \leq c\mathbf{W}_{1,p}^{2\,diam(\Omega)}[\mu^+] \quad a.e. \text{ in } \Omega. \tag{5.4.3}$$

*Proof.* Let $\mu = \mu^+ - \mu^- = \mu_0^+ + \mu_s^+ - (\mu_0^- + \mu_s^-)$ the Jordan and Lebesgue decompositions of measures. By Theorem 4.3.7 the truncate $u_k = T_k(u)$ is a renormalized solution of

$$\begin{aligned} -\operatorname{div} A(x, \nabla u_k) = \chi_{\{|u_k|<k\}}\mu_0 + \alpha_k^+ - \beta_k^- \qquad & \text{in } \Omega \\ u_k = 0 \qquad & \text{on } \partial\Omega, \end{aligned} \tag{5.4.4}$$

where $\alpha_k^+$ and $\beta_k^-$ are measures absolutely continuous with respect to the $c_{1,p}^{\mathbb{R}^N}$-capacity supported by the sets $\{u = k\}$ and $\{u = -k\}$ respectively.

We also denote by $v_k$ the renormalized solution of

$$\begin{aligned} -\text{div}\, A(x,\nabla v_k) &= \chi_{\{|u_k|<k\}}\mu_0^+ + \alpha_k^+ && \text{in } \Omega \\ v_k &= 0 && \text{on } \partial\Omega. \end{aligned} \tag{5.4.5}$$

Since the measures $\chi_{\{|u_k|<k\}}\mu_0 + \alpha_k^+ - \beta_k^-$ and $\chi_{\{|u_k|<k\}}\mu_0^+ + \alpha_k^+$ are absolutely continuous with respect to $c_{1,p}^{\mathbb{R}^N}$, the renormalized solutions $u_k$ and $v_k$ are uniquely determined (see Theorem 4.3.12) and thus $u_k \leq v_k$ a.e. in $\Omega$. When $k \to \infty$, $v_k$ converges a.e. in $\Omega$ to a renormalized solution of

$$\begin{aligned} -\text{div}\, A(x,\nabla v) &= \mu_0^+ + \mu_s^+ && \text{in } \Omega \\ v &= 0 && \text{on } \partial\Omega, \end{aligned} \tag{5.4.6}$$

and $u \leq v$. By Theorem 4.3.2 and Theorem 4.2.5 $v$ is $p$-superharmonic and satisfies

$$v(x) \leq c\mathbf{W}_{1,p}^{2\,diam(\Omega)}[\mu^+] \quad \text{a.e. in } \Omega. \tag{5.4.7}$$

This yields the right-hand side of (5.4.3). We obtain the lower estimate similarly. □

**Theorem 5.4.2** *Assume $g$ satisfies (5.4.1) (i)-(ii) and let $\mu_1, \mu_2$ be two nonnegative bounded Radon measures in $\Omega$ and $\{\mu_{i,n}\}_{n\in\mathbb{N}}$, $i$=1, 2, two nondecreasing sequences of nonnegative bounded Radon measures converging weakly to $\mu_i$ and such that $\tilde{g}(c\mathbf{W}_{1,p}^{2\,diam(\Omega)}[\mu_i]) \in L^1(\Omega)$ where $c$ is the constant involved in (5.4.3). Then there exists a renormalized solution $u$ to (5.4.2) with $\mu = \mu_1 - \mu_2$. Furthermore*

$$-c\mathbf{W}_{1,p}^{2\,diam(\Omega)}[\mu_2] \leq u \leq c\mathbf{W}_{1,p}^{2\,diam(\Omega)}[\mu_1] \quad a.e. \ in \ \Omega. \tag{5.4.8}$$

The proof of this theorem is based upon the following intermediate results.

**Lemma 5.4.3** *Assume that $g$ satisfies (5.4.1) (i) and is bounded in $\Omega \times \mathbb{R}$. Let $\lambda_1, \lambda_2$ be two nonnegative bounded Radon measures with compact support in $\Omega$. Then there exist renormalized solutions $u$, $u_i$ and $v_i$, $i$=1, 2 to*

$$\begin{aligned} -\text{div}\, A(x,\nabla u) + g(x,u) &= \lambda_1 - \lambda_2 && in \ \Omega \\ u &= 0 && on \ \partial\Omega, \end{aligned} \tag{5.4.9}$$

$$\begin{aligned} -\operatorname{div} A(x,\nabla u_i) + g(x,u_i) &= (-1)^{i+1}\lambda_i && \text{in } \Omega \\ u_i &= 0 && \text{on } \partial\Omega, \end{aligned} \tag{5.4.10}$$

*and*

$$\begin{aligned} -\operatorname{div} A(x,\nabla v_i) &= (-1)^{i+1}\lambda_i && \text{in } \Omega \\ v_i &= 0 && \text{on } \partial\Omega, \end{aligned} \tag{5.4.11}$$

*for* $i$=1, 2, *and there holds*

$$\begin{aligned} -c\mathbf{W}_{1,p}^{2\,diam(\Omega)}[\lambda_2] \leq v_2 \leq \min\{u_2, -u^-\} \leq 0 \leq \max\{u^+, u_1\} \\ \leq v_1 \leq c\mathbf{W}_{1,p}^{2\,diam(\Omega)}[\lambda_1] \quad \text{a.e. in } \Omega. \end{aligned} \tag{5.4.12}$$

*Proof.* Let $\{\rho_n\}$ be a sequence of mollifiers and $\lambda_{i,n} = \lambda_i * \rho_n$, $i = 1, 2$, and $\lambda_n = \lambda_{1,n} - \lambda_{2,n}$. Let $u_n$, $u_{i,n}$ and $v_{i,n}$ be solutions of

$$\begin{aligned} -\operatorname{div} A(x,\nabla u_n) + g(x,u_n) &= \lambda_n && \text{in } \Omega \\ u_n &= 0 && \text{on } \partial\Omega, \end{aligned} \tag{5.4.13}$$

$$\begin{aligned} -\operatorname{div} A(x,\nabla u_{i,n}) + g(x,u_{i,n}) &= (-1)^{i+1}\lambda_{i,n} && \text{in } \Omega \\ u_{i,n} &= 0 && \text{on } \partial\Omega, \end{aligned} \tag{5.4.14}$$

and

$$\begin{aligned} -\operatorname{div} A(x,\nabla v_{i,n}) &= (-1)^{i+1}\lambda_{i,n} && \text{in } \Omega \\ v_{i,n} &= 0 && \text{on } \partial\Omega. \end{aligned} \tag{5.4.15}$$

Since $g(x,0) = 0$ and $sg(x,s) \geq 0$, there holds

$$v_{2,n} \leq \min\{u_{2,n}, -u_n^-\} \leq 0 \leq \max\{u_n^+, u_{1,n}\} \leq v_{1,n}. \tag{5.4.16}$$

The three sequences of measures $\{\lambda_{1,n} - \lambda_{2,n} - g(.,u_n)\}$, $\{\lambda_{i,n} - g(.,u_{i,n})\}$ and $\{\lambda_{i,n}\}$ are uniformly bounded in $\mathfrak{M}_b(\Omega)$. By Theorem 1.3.9 there exist subsequences of $u_n$, $u_{i,n}$ and $v_{i,n}$, still denoted by the same index, which converge a.e. to functions $u$, $u_i$ and $v_i$. Furthermore $g(.,u_{i,n}) \to g(.,u_i)$ and $g(.,u_n) \to g(.,u)$ in $L^1(\Omega)$. By Corollary 4.3.13 we can assume that $u$, $u_i$ and $v_i$ are renormalized solutions of (5.4.9), (5.4.10) and (5.4.11). This, combined with Theorem 4.2.5, implies (5.4.12). □

**Proposition 5.4.4** *Assume* $g$ *satisfies 5.4.1 (i)-(iii)-(iv). If* $\lambda_i$ *(for* $i$=1, 2*) are nonnegative measures with compact support such that*

*$\tilde{g}(c\mathbf{W}_{1,p}^{2\,diam(\Omega)}[\lambda_i]) \in L^1(\Omega)$ where $c$ is as in (5.4.3), there exist renormalized solutions $u$ and $u_i$ of problems (5.4.9) and (5.4.10). Furthermore, if the $v_i$ are renormalized solutions of (5.4.11), inequalities (5.4.12) hold.*

*Proof.* Set $g_n(x,r) = T_n(g(x,r)) = (\min\{n, |g(x,r)|\})\,\mathrm{sgn}\,(g(x,r))$ for all $n \in \mathbb{N}$, and denote by $u_n$ and $u_{i,n}$ renormalized solutions of

$$\begin{aligned} -\mathrm{div}\, A(x, \nabla u_n) + g_n(x, u_n) &= \lambda_1 - \lambda_2 && \text{in } \Omega \\ u_n &= 0 && \text{on } \partial\Omega, \end{aligned} \tag{5.4.17}$$

$$\begin{aligned} -\mathrm{div}\, A(x, \nabla u_{i,n}) + g_n(x, u_{i,n}) &= (-1)^{i+1}\lambda_i && \text{in } \Omega \\ u_{i,n} &= 0 && \text{on } \partial\Omega. \end{aligned} \tag{5.4.18}$$

By the previous lemma,

$$\begin{aligned} -c\mathbf{W}_{1,p}^{2\,diam(\Omega)}[\lambda_2] &\leq v_2 \leq \min\{u_{2,n}, -u_n^-\} \leq 0 \\ &\leq \max\{u_n^+, u_{1,n}\} \leq v_1 \leq c\mathbf{W}_{1,p}^{2\,diam(\Omega)}[\lambda_1] \quad \text{a.e. in } \Omega. \end{aligned} \tag{5.4.19}$$

By (5.1.24),

$$\int_\Omega |g_n(.,u_n)|\,dx \leq \lambda_1(\Omega) + \lambda_2(\Omega),$$

and

$$\int_\Omega |g_n(.,u_{i,n})|\,dx \leq \lambda_i(\Omega) \quad i = 1, 2.$$

We can choose a subsequence, still indexed by $\{.\,_n\}$, such that $u_n$ and $u_{i,n}$ converge a.e. to $u$ and $u_i$ which satisfy 5.4.12. It follows from (5.4.19) (5.4.1) (iii)-(iv) that

$$|g_n(.,u_{i,n})| \leq |g(.,u_{i,n})| \leq \tilde{g}(|u_{i,n}|) \leq \tilde{g}(c\mathbf{W}_{1,p}^{2\,diam(\Omega)}[\lambda_i]), \tag{5.4.20}$$

and

$$\begin{aligned} |g_n(.,u_n)| \leq |g(.,u_n)| &= |g(.,u_n^+)| + |g(.,u_n^-)| \leq \tilde{g}(|u_{1,n}|) + \tilde{g}(|u_{2,n}|) \\ &\leq \tilde{g}(c\mathbf{W}_{1,p}^{2\,diam(\Omega)}[\lambda_1]) + \tilde{g}(c\mathbf{W}_{1,p}^{2\,diam(\Omega)}[\lambda_2]), \end{aligned} \tag{5.4.21}$$

and $\tilde{g}(c\mathbf{W}_{1,p}^{2\,diam(\Omega)}[\lambda_i]) \in L^1(\Omega)$, we derive from the assumptions on $\lambda_i$ and the dominated convergence theorem that $g_n(.,u_{i,n}) \to g(.,u_i)$ in $L^1(\Omega)$ and $g_n(.,u_n) \to g(.,u)$ in $L^1(\Omega)$ as $n \to \infty$. By Corollary 4.3.13 we infer that

$u$ and $u_i$ are renormalized solutions of (5.4.9) and (5.4.10) respectively and (5.4.12) is valid. $\square$

*Remark.* A variant of Proposition 5.4.4 is to assume that $g$ verifies 5.4.1 (i)-(ii) and the $\lambda_i$ satisfy $g(.,(-1)^{i+1}c\mathbf{W}_{1,p}^{2\,diam(\Omega)}[\lambda_i]) \in L^1(\Omega)$. Then (5.4.19) becomes

$$\begin{aligned} -c\mathbf{W}_{1,p}^{2\,diam(\Omega)}[\lambda_2] \leq v_2 \leq u_{2,n} \leq -u_n^- \leq 0 \leq u_n^+ \leq u_{1,n} \\ \leq v_1 \leq c\mathbf{W}_{1,p}^{2\,diam(\Omega)}[\lambda_1] \quad \text{a.e. in } \Omega, \end{aligned} \tag{5.4.22}$$

and (5.4.20) and (5.4.21) are replaced by

$$\begin{aligned} (i) \qquad & |g_n(.,u_{i,n})| \leq |g(.,u_{i,n})| \\ & \qquad \leq (-1)^{i+1} g(.,(-1)^{i+1}c\mathbf{W}_{1,p}^{2\,diam(\Omega)}[\lambda_i]) \\ (ii) \qquad & |g_n(.,u_n)| \leq |g(.,u_n)| \leq g(.,u_{1,n}) - g(.,u_{2,n}) \\ & \qquad \leq g(.,c\mathbf{W}_{1,p}^{2\,diam(\Omega)}[\lambda_1]) - g(.,-c\mathbf{W}_{1,p}^{2\,diam(\Omega)}[\lambda_2]). \end{aligned} \tag{5.4.23}$$

*Proof of Theorem 5.4.2.* From Proposition 5.4.4 and the remark above there exist renormalized solutions to

$$\begin{aligned} -\operatorname{div} A(x,\nabla u_n) + g(x,u_n) &= \mu_{1,n} - \mu_{1,n} && \text{in } \Omega \\ u_n &= 0 && \text{on } \partial\Omega, \end{aligned} \tag{5.4.24}$$

and

$$\begin{aligned} -\operatorname{div} A(x,\nabla u_{i,n}) + g(x,u_{i,n}) &= (-1)^{i+1}\mu_{i,n} && \text{in } \Omega \\ u_{i,n} &= 0 && \text{on } \partial\Omega, \end{aligned} \tag{5.4.25}$$

and they satisfy

$$\begin{aligned} -c\mathbf{W}_{1,p}^{2\,diam(\Omega)}[\mu_2] \leq u_{2,n} \leq -u_n^- \leq 0 \\ \leq u_n^+ \leq u_{1,n}\} \leq c\mathbf{W}_{1,p}^{2\,diam(\Omega)}[\mu_1] \quad \text{a.e. in } \Omega. \end{aligned} \tag{5.4.26}$$

Up to a subsequence, $\{u_n\}$, $\{u_{1,n}\}$ and $\{u_{2,n}\}$ converges a.e. Furthermore

$$\int_\Omega |g(x,u_{i,n})|\,dx \leq \mu_{i,n}(\Omega) \leq \mu_i(\Omega).$$

Since $|g(x,u_n)| \leq g(x,u_{1,n}) + |g(x,u_{2,n})|$ it follows by monotonicity and the monotone convergence theorem that $(g(.,u_n), g(.,u_{1,n}), g(.,u_{2,n})) \to (g(.,u), g(.,u_1), g(.,u_2))$ in $(L^1(\Omega))^3$ as $n \to \infty$. Applying Corollary 4.3.14 with $f_n = g(.,u_n)$, we conclude that $u$ is a renormalized solution of (5.4.2). Finally (5.4.8) holds from (5.4.22). $\square$

### 5.4.2 *Power growth nonlinearities*

In Theorem 5.4.1 we have given a sufficient condition linking a specific measure $\mu$ and a specific nonlinearity $g$ in order (5.4.2) admit a renormalized solution. In the next results we characterize a class of nonlinearities and a subspace of Radon measures for solving (5.4.2). In all these results approximate solutions $u_n$ are constructed. Besides the convergence a.e. of $\{u_n\}$ to some $u$, which implies the convergence a.e. of $\{g(.,u_n)\}$ to $g(.,u)$, the main difficulty in order to apply Corollary 4.3.14 is to prove the convergence of $\{g(.,u_n)\}$ to $g(.,u)$ in $L^1(\Omega)$. As in Theorem 5.4.2 this convergence is proved using the monotone convergence theorem.

**Theorem 5.4.5** *Let $1 < p < N$ and $q > p-1$. Assume $g$ satisfies (5.4.1) (i)-(ii)-(iii) and*

$$\int_1^\infty s^{-1-q}\tilde{g}(s)ds < \infty. \tag{5.4.27}$$

*Then, for any bounded measure $\mu$ absolutely continuous with respect to the $c^{\mathbb{R}^N}_{p,\frac{q}{q+1-p},1}$-capacity, problem (5.4.2) admits a renormalized solution $u$. Furthermore (5.4.8) holds.*

*Proof.* Since $\mu \ll c^{\mathbb{R}^N}_{p,\frac{q}{q+1-p},1}$, there holds $\mu^\pm \ll c^{\mathbb{R}^N}_{p,\frac{q}{q+1-p},1}$. By Proposition 1.2.7 (iii) there exist two nondecreasing sequences $\{\mu_{i,n}\}$ of nonnegative bounded measures with compact support in $\Omega$, elements of $W^{-p,\frac{q}{p-1},\infty}(\Omega)$, converging to $\mu_+$ and $\mu_-$ respectively in the narrow topology. By Theorem 1.2.5 the operator $\mathbf{G}_p$ is an isomorphism between $W^{-p,\frac{q}{p-1},\infty}(\Omega)$ and $L^{\frac{q}{p-1},\infty}(\Omega)$, thus $\mathbf{G}_p[\mu_{i,n}] \in L^{\frac{q}{p-1},\infty}(\Omega)$. By Theorem 5.3.2, this is equivalent to $\mathbf{W}_{1,p}^{2\,\mathrm{diam}\,(\Omega)}[\mu_{i,n}] \in L^{q,\infty}(\Omega)$ and estimate (5.3.23) holds with $R = 2\,\mathrm{diam}\,(\Omega)$, $\alpha = 1$ and $s = \infty$. Therefore, with the notations of Section 1.2.1,

$$\lambda_{\mathbf{W}_{1,p}^{2\,\mathrm{diam}\,(\Omega)}[\mu_{i,n}]}(t) := \left|\left\{\mathbf{W}_{1,p}^{2\,\mathrm{diam}\,(\Omega)}[\mu_{i,n}] > t\right\}\right| \leq \min\{|\Omega|, \kappa t^{-q}\} \quad \forall t > 0,$$

where $\kappa = \kappa(n) > 0$. Let $c$ be the constant involved in (5.4.3), then

$$\begin{aligned}\int_\Omega \tilde{g}\left(c\mathbf{W}_{1,p}^{2\,\mathrm{diam}\,(\Omega)}[\mu_{i,n}]\right)dx &= -\int_0^\infty \tilde{g}(t)d\lambda_{c\mathbf{W}_{1,p}^{2\,\mathrm{diam}\,(\Omega)}[\mu_{i,n}]}(t)\\ &\leq qc^q\kappa\int_{c\kappa^{\frac{1}{q}}|\Omega|^{-\frac{1}{q}}}^\infty t^{-1-q}\tilde{g}(t)dt < \infty.\end{aligned}$$

Thus the assumptions of Proposition 5.4.4 are satisfied and for every $n \in \mathbb{N}$ there exist renormalized solutions $u_n$ and $u_{i,n}$ ($i$=1, 2) of (5.4.24) and (5.4.25). The conclusion, as well as inequality (5.4.8), follows from Theorem 5.4.2. □

If $g$ is dominated by a power function we have a more precise condition on the measure. We also recall that the Bessel-Lorentz capacity $c^{\mathbb{R}^N}_{p,\frac{q}{q+1-p},\frac{q}{q+1-p}}$ coincides with the standard Bessel one $c^{\mathbb{R}^N}_{p,\frac{q}{q+1-p}}$.

**Theorem 5.4.6** *Let $1 < p < N$ and $q > p - 1$. Assume $g : \Omega \times \mathbb{R}$ is a Caratheodory function which is nondecreasing in its second variable and satisfies*

$$0 \le rg(x,r) \le m|r|^{q+1} \qquad a.e. \ in \ \Omega, \ \forall r \in \mathbb{R}, \tag{5.4.28}$$

*for some $m > 0$. Then, for any bounded measure $\mu$ absolutely continuous with respect to the $c^{\mathbb{R}^N}_{p,\frac{q}{q+1-p},\frac{q}{q+1-p}}$-capacity, problem (5.4.2) admits a renormalized solution $u$. Furthermore (5.4.8) holds.*

*Proof.* Similarly as in Theorem 5.4.5, $\mu^{\pm} \ll c^{\mathbb{R}^N}_{p,\frac{q}{q+1-p},\frac{q}{q+1-p}}$. Then there exist nondecreasing sequences $\{\mu_{1,n}\}$, $\{\mu_{2,n}\}$ of nonnegative bounded measures with compact support in $\Omega$ such that $\mathbf{W}_{1,p}^{2\,\mathrm{diam}\,(\Omega)}[\mu_{i,n}] \in L^{q,q}(\Omega) = L^q(\Omega)$. Then the result follows from Theorem 5.4.2. □

In the previous results the assumptions on $g$ can be modified in order to include a strong singular dependence on $x$. Consider the problem

$$\begin{aligned} -\mathrm{div}\, A(x,\nabla u) + h(x)g(x,u) &= \mu \qquad \text{in } \Omega \\ u &= 0 \qquad \text{on } \partial\Omega, \end{aligned} \tag{5.4.29}$$

where $g$ is a Caratheodory function and $h : \Omega \mapsto \mathbb{R}$ is measurable and satisfies

$$0 \le h(x) \le \kappa\, |x|^{-\beta} \qquad \text{a.e. in } \Omega, \tag{5.4.30}$$

for some $\kappa, \beta \ge 0$. The next theorem holds with the same technique of proof.

**Theorem 5.4.7** *Let $1 < p < N$, $q > p-1$, $0 \le \beta < N$ and $\mu$ is a bounded Radon measure in $\Omega$. Then problem (5.4.29) admits a renormalized solution $u$ and inequalities (5.4.8) hold if one of the next two conditions is satisfied.*

*1- The function $g$ satisfies (5.4.1) (i)-(ii)-(iii) and (5.4.27), and $\mu$ is absolutely continuous with respect to the $c^{\mathbb{R}^N}_{p,\frac{Nq}{Nq-(p-1)(N-\beta)},1}$-capacity.*

*2- The function* $g$ *satisfies (5.4.28) and* $\mu$ *is absolutely continuous with respect to the* $c^{\mathbb{R}^N}_{p,\frac{Nq}{Nq-(p-1)(N-\beta)},\frac{q}{q+1-p}}$*-capacity.*

### 5.4.3 Exponential type nonlinearities

The potential method can be used to prove existence of solutions of problem (5.4.2) when $g$ satisfies

$$|g(x,r)| \le \tilde{g}(|r|) \le e^{\tau|r|^\lambda} - 1 \qquad \text{a.e. in } \Omega,\ \forall r \in \mathbb{R}. \tag{5.4.31}$$

**Theorem 5.4.8** *Assume* $1 < p < N$, $\tau > 0$, $\lambda \ge 1$ *and* $g$ *satisfies (5.4.1) (i)-(ii)-(iii) and (5.4.31). There exists* $M > 0$ *depending on* $N$, $p$, $\tau$ *and* $\lambda$ *such that for any bounded Radon measure in* $\Omega$ *which can be decomposed as* $\mu = f_1 - f_2 + \nu_1 - \nu_1$, *where* $f_i \in L^1_+(\Omega)$, $\nu_i \in \mathfrak{M}^+_b(\Omega)$ *and*

$$\left\| \mathbf{M}^{\frac{(p-1)(\lambda-1)}{\lambda}}_{p,2\,\mathrm{diam}\,(\Omega)}[\nu_i] \right\|_{L^\infty(\Omega)} \le M, \tag{5.4.32}$$

*for* $i$=1,2, *there exists a renormalized solution to problem (5.4.2) and (5.4.8) holds.*

*Proof.* Let $\{\Omega_n\}$ be an increasing sequence of subdomains of $\Omega$ such that $\overline{\Omega}_n \subset \Omega$ and $\cup_n \Omega_n = \Omega$. We define $\mu_{i,n} = T_n(\chi_{\Omega_n} f_i) + \chi_{\Omega_n}\nu_i$. Then $\{\mu_{1,n}\}$ and $\{\mu_{2,n}\}$ are nondecreasing sequence of nonnegative measures with compact support which converge to $\mu_1 = f_1 + \nu_1$ and $\mu_2 = f_2 + \nu_2$ respectively. Clearly for any $\epsilon > 0$ there exists $c_\epsilon > 0$ such that

$$\left(\mathbf{W}^{2\,\mathrm{diam}\,(\Omega)}_{1,p}[\mu_{i,n}]\right)^\lambda \le c_\epsilon n^{\frac{\lambda}{p-1}} + (1+\epsilon)\left(\mathbf{W}^{2\,\mathrm{diam}\,(\Omega)}_{1,p}[\nu_i]\right)^\lambda, \tag{5.4.33}$$

which implies

$$\exp\left(\tau\left(c\mathbf{W}^{2\,\mathrm{diam}\,(\Omega)}_{1,p}[\mu_{i,n}]\right)^\lambda\right) \le c_{\epsilon,n,c}\exp\left(\tau(1+\epsilon)\left(c\mathbf{W}^{2\,\mathrm{diam}\,(\Omega)}_{1,p}[\nu_i]\right)^\lambda\right). \tag{5.4.34}$$

If there holds

$$\left\| \mathbf{M}^{\frac{(p-1)(\lambda-1)}{\lambda}}_{p,2\,\mathrm{diam}\,(\Omega)}[\nu_i] \right\|_{L^\infty(\Omega)} \le \left(\frac{p\ln 2}{\tau(12c\lambda)^\lambda}\right)^{\frac{p-1}{\lambda}}, \tag{5.4.35}$$

we can choose $\epsilon$ small enough such that

$$\tau(1+\epsilon)c^{\lambda} \leq \frac{p\ln 2}{(12\lambda)^{\lambda}\left\|\mathbf{M}^{\frac{(p-1)(\lambda-1)}{\lambda}}_{p,2\,\mathrm{diam}\,(\Omega)}[\nu_i]\right\|^{\frac{\lambda}{p-1}}_{L^{\infty}(\Omega)}}. \tag{5.4.36}$$

By Theorem 5.3.3 with $\eta = \frac{(p-1)(\lambda-1)}{\lambda}$ we derive

$$\exp\left(\tau(1+\epsilon)\left(c\mathbf{W}^{2\,\mathrm{diam}\,(\Omega)}_{1,p}[\nu_i]\right)^{\lambda}\right) \in L^1(\Omega).$$

This implies

$$\exp\left(\tau\left(c\mathbf{W}^{2\,\mathrm{diam}\,(\Omega)}_{1,p}[\mu_{i,n}]\right)^{\lambda}\right) \in L^1(\Omega),$$

and we conclude by Theorem 5.4.2. □

*Remark.* If $\lambda = 1$, the condition on $\nu_i$ is

$$\left\|\mathbf{M}_{p,2\,\mathrm{diam}\,(\Omega)}[\nu_i]\right\|_{L^{\infty}(\Omega)} \leq M \Longleftrightarrow \sup_{x\in\Omega}\operatorname{ess}\sup_{t>0}\frac{\nu_i(B_t(x))}{t^{N-p}} \leq M. \tag{5.4.37}$$

## 5.5 Large solutions revisited

Throughout this section $\Omega \subset \mathbb{R}^N$ is a bounded domain and $A : \Omega \times \mathbb{R}^N \mapsto \mathbb{R}^N$ satisfies the assumptions (4.1.1) with $1 < p \leq N$. We assume that $g : \Omega\times\mathbb{R} \mapsto \mathbb{R}$ is a Caratheodory function on which precise growth estimates will be made later on. A solution $u$ of

$$-\mathrm{div}\,A(x,\nabla u) + g(x,u) = 0 \qquad \text{in } \Omega \tag{5.5.1}$$

is called a *large solution* $u$ if it satisfies

$$\lim_{\rho(x)\to 0} u(x) = \infty \tag{5.5.2}$$

where $\rho(x) = \mathrm{dist}\,(x,\Omega^c)$. The model problems for large solutions are

$$\begin{aligned} -\Delta_p u + |u|^{q-1}u &= 0 \qquad \text{in } \Omega \\ \lim_{\rho(x)\to 0} u(x) &= \infty, \end{aligned} \tag{5.5.3}$$

with $q > p-1$ and

$$\begin{aligned} -\Delta_p u + e^u - 1 &= 0 \qquad \text{in } \Omega \\ \lim_{\rho(x)\to 0} u(x) &= \infty. \end{aligned} \tag{5.5.4}$$

### 5.5.1 *The maximal solution*

In *any* domain there exists a *maximal solution* to (5.5.3) and (5.5.4). When $\Omega$ is smooth this maximal solution is a large solution. When $\Omega$ is no longer smooth, the maximal solution $\overline{u}_\Omega$ of (5.5.3) may not blow-up at the boundary. However, if $1 < q < q_1 := \frac{N(p-1)}{N-p}$, it is easy to prove that this maximal solution satisfies

$$\overline{u}_\Omega(x) \geq \omega_c \left(\rho(x)\right)^{-\frac{p}{q+1-p}}, \tag{5.5.5}$$

where $\omega_c = \omega_c(N,p,q)$ is some explicit positive constant which vanishes when $q = q_1$. As for (5.5.4) a large solution always exists when $p = N$. The existence of a large solutions is based upon the construction of local radial super solutions and the proof that this maximal solution is a large solution is based upon the existence of explicit radial singular subsolutions. Concerning equation (5.5.2), this approach is no longer possible. The maximal solution is constructed in proving a local universal energy estimate and the fact that this maximal solution is a large solution by potential estimates.

**Proposition 5.5.1** *Assume $N \geq 2$, $1 < p \leq N$, $A$ and the function $g$ satisfy (4.1.1) and*

$$rg(x,r) \geq \theta|r|^{\beta+1} - \tau \qquad a.e.\ in\ \Omega,\ \forall r \in \mathbb{R}, \tag{5.5.6}$$

*respectively for some $\beta > p-1$, $\theta > 0$ and $\tau \geq 0$. Then for any compact $K \subset \Omega$ there exists $c_K = c(N,p,\beta,\theta,\tau,\lambda,K) > 0$, such that any locally upper bounded solution of (5.5.2) satisfies*

$$\sup_{x\in K} \operatorname{ess} u(x) \leq c_K. \tag{5.5.7}$$

*Proof.* Let $B := B_r(a) \subset \overline{B}_r(a) \subset B_{r'}(a) := B' \subset \Omega$ and $\zeta \in C^1_0(B')$, with value 1 on $\overline{B}$, such that $0 \leq \zeta \leq 1$ and $|\nabla\zeta| \leq 2(r'-r)^{-1}\chi_{B'\setminus B}$. We define $m$ by

$$\frac{1}{p'} + \frac{1}{\beta+1} + \frac{1}{m} = 1 \Longleftrightarrow \frac{1}{m} = \frac{\beta+1-p}{p(\beta+1)}.$$

Taking $u_+\zeta^m$ as test function in (5.5.1), we have

$$\int \left( |\nabla u_+|^p \zeta^m - \lambda m \, |\nabla u_+|^{p-1} |\nabla\zeta| \, u_+\zeta^{m-1} + \theta u_+^{\beta+1}\zeta^m - \tau u_+\zeta^m \right) dx \leq 0,$$

and we obtain by Hölder's inequality,

$$\int |\nabla u_+|^{p-1} |\nabla\zeta| \, u_+\zeta^{m-1} dx$$
$$\leq \left(\int_{B'} |\nabla u_+|^p \, \zeta^m dx\right)^{\frac{1}{p'}} \left(\int_{B'} u_+^{\beta+1}\zeta^m dx\right)^{\frac{1}{\beta+1}} \left(\int_{B'} |\nabla\zeta|^r \, dx\right)^{\frac{1}{r}}.$$

We have also

$$\int \left(\theta u_+^{\beta+1}\zeta^m - \tau u_+\zeta^m\right) dx \geq \frac{\theta}{2}\int u_+^{\beta+1}\zeta^m dx - c_2|B'|.$$

Therefore

$$\begin{aligned}\int \left(|\nabla u_+|^p \, \zeta^m + \theta u_+^{\beta+1}\zeta^m\right) dx &\leq 2c_2|B'| + c_3\int_{B'} |\nabla\zeta|^m \, dx \\ &\leq c_4\left(r'^N + \tfrac{r'^{N-1}}{(r'-r)^{m-1}}\right),\end{aligned} \tag{5.5.8}$$

where $c_2 = c_2(q,\theta) > 0$, $c_3 = c_3(p,\beta,\lambda) > 0$, $c_4 = (N,p,\beta,\lambda) > 0$. This is a local energy estimate independent of the solution. It follows by the method of proof of Lemma 4.3.3 that $u$ is bounded from above on any compact subset of $\Omega$ (in this lemma a local bound from below for super solution is proved). □

**Corollary 5.5.2** *Assume $A$ satisfies (4.1.1) and $g : \Omega\times\mathbb{R}\to\mathbb{R}$ is a Caratheodory function which satisfies (5.5.6), is locally bounded in $\Omega\times\mathbb{R}$ and*

$$\begin{aligned}&(i) \qquad r\mapsto g(x,r) \text{ is nondecreasing on } \mathbb{R}^+ \text{ a.e. in } \Omega\\ &(ii) \qquad g(x,0)=0 \qquad \text{a.e. in } \Omega.\end{aligned} \tag{5.5.9}$$

*Then equation (5.5.9) admits a nonnegative maximal solution in $\Omega$.*

*Proof.* We need not suppose that $\Omega$ is bounded. The construction is standard. We consider a sequence of smooth bounded domains $\{\Omega_n\}$ such that $\Omega_n \subset \overline{\Omega}_n \subset \Omega_{n+1}$ and $\cup_n\Omega_n = \Omega$. For each $k > 0$ there exists a solution $u = u_{n,k}$ of

$$\begin{aligned}-\operatorname{div} A(x,\nabla u) + g(x,u) &= 0 \qquad \text{in } \Omega_n\\ u &= k \qquad \text{on } \partial\Omega_n.\end{aligned} \tag{5.5.10}$$

This fact follows easily from Theorem 1.4.4. Furthermore $u_{n,k}$ is unique and $0 \leq u_{n,k} \leq k$. By Proposition 5.5.1, for any compact set $K \subset \Omega_{n_0}$, $u_{n,k}$ is bounded from above in $K$ by a constant independent of $n \geq n_0+1$.

Furthermore $k \mapsto u_{n,k}$ is nondecreasing. When $k \to \infty$, $u_{n,k}$ converges to a solution $u_{\Omega_n}$ of

$$\begin{aligned} -\operatorname{div} A(x, \nabla u) + g(x,u) &= 0 && \text{in } \Omega_n \\ u &= \infty && \text{on } \partial\Omega_n. \end{aligned} \tag{5.5.11}$$

Since $u_{n+1,k} \leq u_{n,k}$, the sequence $\{u_{\Omega_n}\}$ is nonincreasing. Its limit $\overline{u}_\Omega$ is a locally bounded solution of (5.5.1) in $\Omega$. Since $g$ is locally bounded in $\Omega \times \mathbb{R}$, any solution $u$ of the same equation in $\Omega$ is Hölder continuous and thus also locally bounded. As a consequence $u \leq u_{n,k}$ for $n$ fixed and $k$ large enough, and finally $u \leq \overline{u}_\Omega$. □

### 5.5.2 *Potential estimates*

The following space of functions is a little more general than the Marcinkiewicz space $L^{s,\infty}$

**Definition 5.2** Let $\Omega$ be a bounded domain and $1 \leq s \leq \infty$. A function $f \in L^1_{loc}(\Omega)$ belongs to the Morrey class $\mathcal{M}^s(\Omega)$ if

$$\int_{\Omega \cap B_r(x)} |f|\, dy \leq C r^{\frac{N}{s'}} \qquad \forall x \in \Omega,\ \forall r > 0. \tag{5.5.12}$$

The smallest constant such that this inequality holds is denoted by $\|f\|_{\mathcal{M}^s(\Omega)}$ and it is a norm. Note that $L^s(\Omega) \subset L^{s,\infty}(\Omega) \subset \mathcal{M}^s(\Omega) \subset L^1(\Omega)$.

The following variant of (5.2.43) in Theorem 4.2.5 (see [Malý and Ziemer (1997), Th 4.35]) can be verified by a modification of the proof of Theorem 4.3.8.

**Proposition 5.5.3** *Let $\Omega$ be a bounded domain, $1 < p \leq N$ and $0 < \gamma < p$. If $\mu \in \mathfrak{M}_b(\Omega)$ and $f \in \mathcal{M}^{\frac{N}{p-\gamma}}(\Omega)$, any locally renormalized solution of*

$$-\operatorname{div} A(x, \nabla u) + f = \mu \qquad \text{in } \Omega, \tag{5.5.13}$$

*satisfies*

$$u(x) + \|f\|_{\mathcal{M}^{\frac{N}{p-\gamma}}} \geq \inf_\Omega \operatorname{ess} u + c_{11} \mathbf{W}_{1,p}^{\frac{\rho(x)}{4}} [\mu](x) \qquad \text{a.e. in } \Omega. \tag{5.5.14}$$

As in the case $p = 2$ potential estimates lead to sufficient conditions for existence of large solutions. We recall that when $1 < p \leq N$, the

$\mathcal{H}^R_{N-p}$-Hausdorff capacity of a compact set $K$ is defined by

$$\mathcal{H}^R_{N-p}(K) = \inf\left\{\sum_j r_j^{N-p} : K \subset \bigcup_j B_{r_j}(a_j), r_j \leq R\right\}, \tag{5.5.15}$$

(see (1.2.14) for the general definition). Then this set function is extended to general sets using the usual rules developed in Section 1-2-4. In the same way as for the Bessel-Lorentz capacity (see Proposition 1.2.6), it is possible to give a dual definition of the $\mathcal{H}_{N-p}$-capacity linking it to the truncated fractional maximal operator defined in (5.3.2). This result due to [Frostman (1935)] is also proved in [Turesson (2000), Th 3.4.27].

**Proposition 5.5.4** *Let $1 < p < N$ and $R > 0$. There exists $c > 0$ such that for Borel set $E \subset \mathbb{R}^N$ there holds*

$$c\mathcal{H}^R_{N-p}(E) = \sup\left\{\mu(E) : \mu \in \mathfrak{M}_+(E),\ \left\|\mathbf{M}^R_p[\mu]\right\|_{L^\infty} \leq 1\right\} \leq \mathcal{H}^R_{N-p}(E), \tag{5.5.16}$$

*where $\mathfrak{M}_+(E) = \{\mu \in \mathfrak{M}_+(\mathbb{R}^N) : \mu(E^c) = 0\}$.*

The main estimate is the following.

**Theorem 5.5.5** *Let $1 < p < N$. Assume $A$ and $g$ are Caratheodory functions which verify (4.1.1) and (5.5.6) and (5.5.9) respectively with $B_1 = \Omega$. Let $K \subset B_{\frac{1}{4}} \setminus \{0\}$ be a compact subset and $U$ the maximal solution of (5.5.1) in $B_1 \setminus K$.*

*(i) If $g$ satisfies*

$$g(x,r) \leq e^r - 1 \qquad a.e.\ in\ B_1,\ \forall r \geq 0, \tag{5.5.17}$$

*there exist positive constants $c_1, c_2$, depending on $N$, $p$ and $q$, such that*

$$U(0) \geq -c_1 + c_2 \int_0^1 \left(\frac{\mathcal{H}^1_{N-p}(K \cap B_r)}{r^{N-p}}\right)^{\frac{1}{p-1}} \frac{dr}{r}. \tag{5.5.18}$$

*(ii) If $q^* > q_1$, $g$ satisfies*

$$g(x,r) \leq ar^q \qquad a.e.\ in\ B_1,\ \forall r \in \mathbb{R}_+, \tag{5.5.19}$$

*for some* $p-1<q<\frac{pq^*}{N}$, *then*

$$U(0) \geq -c_3 + c_4 \int_0^1 \left( \frac{c^{\mathbb{R}^N}_{p,\frac{q^*}{q^*+1-p}}(K \cap B_r)}{r^{N-p}} \right)^{\frac{1}{p-1}} \frac{dr}{r}. \tag{5.5.20}$$

*Proof.* For $j \in \mathbb{Z}$ we set $r^j = 2^{-j}$, $B_j = B_{r_j}$, $S_j = \overline{B}_{j-1} \setminus B_j$. Let $J \in \mathbb{N}_*$ such that $K \subset B_{2^{-3}} \setminus B_J$. By Proposition 5.5.4, for $j = 3, ..., J$ there exists $\mu_j \in \mathfrak{M}_+(K \cap S_j)$ such that $\left\|\mathbf{M}^1_p[\mu_j]\right\|_{L^\infty(\mathbb{R}^N)} \leq 1$ and

$$c\mathcal{H}^1_{N-p}(K \cap S_j) = \mu(\mathbb{R}^N) \leq \mathcal{H}^R_{N-p}(K \cap S_j) \qquad \forall j, \tag{5.5.21}$$

for $c = c(N,p) > 0$.

*Step 1.* We claim that there exist $\epsilon = \epsilon(N,p) > 0$ small enough and $c_2 > 0$, independent of $J$, such that

$$A := \int_{B_1} \exp\left( \epsilon \mathbf{W}^R_{1,p} \left[ \sum_{k=3}^{J} \mu_k \right](x) \right) dx \leq c_2. \tag{5.5.22}$$

We decompose this sum as follows

$$A = \sum_{j=0}^{\infty} \int_{S_j} \exp\left( \epsilon \mathbf{W}^R_{1,p} \left[ \sum_{k=3}^{J} \mu_k \right](x) \right) dx.$$

For any $j \geq 0$

$$\begin{aligned} \mathbf{W}^R_{1,p}\left[\sum_{k=3}^{J} \mu_k\right] \leq c^* \mathbf{W}^R_{1,p}\left[\sum_{k \geq j+2}^{J} \mu_k\right] + c^* \mathbf{W}^R_{1,p}\left[\sum_{k=3}^{j-2} \mu_k\right] \\ + c^* \mathbf{W}^R_{1,p}\left[\sum_{k=\max\{j-1,3\}}^{j+1} \mu_k\right], \end{aligned}$$

where $c^* = \max\{1, 5^{\frac{2-p}{p-1}}\}$. Since $\exp\left(\sum_{j=1}^{5} a_j\right) \leq \sum_{j=1}^{5} \exp(5a_j)$, we deduce

$$\begin{aligned}\sum_{j=0}^{\infty}\int_{S_j} \exp\left(\mathbf{W}_{1,p}^{R}\left[\sum_{k=3}^{J}\mu_k\right]\right) &\leq \sum_{j=0}^{\infty}\int_{S_j} \exp\left(c_5\mathbf{W}_{1,p}^{R}\left[\sum_{k\geq j+2}^{J}\mu_k\right]\right)\\ &\quad+\sum_{j=0}^{\infty}\int_{S_j} \exp\left(c_5\mathbf{W}_{1,p}^{R}\left[\sum_{k=3}^{j-2}\mu_k\right]\right)\\ &\quad+\sum_{j=0}^{\infty}\int_{S_j} \exp\left(c_5\mathbf{W}_{1,p}^{R}\left[\sum_{k=\max\{j-1,3\}}^{j+1}\mu_k\right]\right)\\ &\leq A_1 + A_2 + A_3.\end{aligned}$$

*Estimate of* $A_3$. By applying Theorem 5.3.3 with $\eta = 0$, $\mu = \mu_k$ and $B = B_{k-1}$, we get

$$\int_{B_{k-2}} \exp\left(c_3\epsilon\mathbf{W}_{1,p}^{R}[\mu_k]\right) dx \leq c_4 r_{k-1}^{N},$$

where $c_3\epsilon \leq \frac{1}{12}$, which is the constant $\delta_0$ in Theorem 5.3.3. For $k = j-1, j, j+1$ this inequality is valid provided the domain of integration $B_{k-2}$ be replaced by $S_j$, thus

$$A_3 \leq c\sum_{j=0}^{\infty} r_j^N = c_5 < \infty. \tag{5.5.23}$$

*Estimate of* $A_1$. If $t \in (0, r_{j+1})$, $x \in S_j$ there holds $\sum_{k\geq j+2} \mu_k(S_j) = 0$, thus

$$\begin{aligned}A_1 &= \sum_{j=1}^{\infty}\int_{S_j} \exp\left(c_3\epsilon\int_{r_{j+1}}^{1}\left(\frac{\sum_{k\geq j+2}\mu_k(B_t(x))}{t^{N-p}}\right)^{\frac{1}{p-1}}\frac{dt}{t}\right)dx\\ &\leq \sum_{j=1}^{\infty}\exp\left(c_3\epsilon\frac{p-1}{N-p}\left(\sum_{k\geq j+2}\mu_k(S_k)\right)^{\frac{1}{p-1}} r_{j+1}^{-\frac{p-1}{N-p}}\right)|S_j|.\end{aligned}$$

Since $\mu_k(S_k) \le \mu_k(B_{r^{k-1}}) \le r_{k-1}^{N-p}$, we are led to

$$\left(\sum_{k=j+2}^{\infty} \mu_k(S_k)\right)^{\frac{1}{p-1}} r_{j+1}^{-\frac{p-1}{N-p}} \le \left(\sum_{k=j+2}^{\infty} r_{k-1}^{N-p}\right)^{\frac{1}{p-1}} r_{j+1}^{-\frac{p-1}{N-p}}$$
$$\le \left(\sum_{k=0}^{\infty} r_{k-1}^{N-p}\right)^{\frac{1}{p-1}}$$
$$\le \left(1-2^{p-N}\right)^{\frac{1}{1-p}}.$$

Therefore

$$A_1 \le \exp\left(c_3\epsilon\frac{p-1}{N-p}\left(1-2^{p-N}\right)^{\frac{1}{1-p}}\right)|B_1| := c_6. \tag{5.5.24}$$

*Estimate of* $A_2$. For $x \in S_j$,

$$\mathbf{W}_{1,p}^{R}\left[\sum_{k=3}^{j-2}\mu_k\right](x) = \int_{r_{j-1}}^{1}\left(\frac{\sum_{k\le j-2}\mu_k(B_t(x)}{t^{N-p}}\right)^{\frac{1}{p-1}}\frac{dt}{t}$$
$$= \sum_{i=1}^{j-1}\int_{r_i}^{r_{i-1}}\left(\frac{\sum_{k\le j-2}\mu_k(B_t(x))}{t^{N-p}}\right)^{\frac{1}{p-1}}\frac{dt}{t}.$$

For all $i = 1, ..., j-1$ and $r_i < t < r_{i-1}$, there holds $\sum_{k\le i-2}\mu_k(B_t(x) = 0$, hence

$$\mathbf{W}_{1,p}^{R}\left[\sum_{k=3}^{j-2}\mu_k\right](x) = \sum_{i=1}^{j-1}\int_{r_i}^{r_{i-1}}\left(\frac{\sum_{i-1\le k\le j-2}\mu_k(B_t(x)}{t^{N-p}}\right)^{\frac{1}{p-1}}\frac{dt}{t}$$
$$\le \sum_{i=1}^{j-1}\int_{r_i}^{r_{i-1}}\left(\frac{\sum_{i-1\le k\le j-2}\mu_k(S_k}{t^{N-p}}\right)^{\frac{1}{p-1}}\frac{dt}{t}$$
$$\le \sum_{i=1}^{j-1}\left(\sum_{i-1}^{j-2} r_{k-1}^{N-p}\right)^{\frac{1}{p-1}} r_i^{\frac{N-p}{p-1}}$$
$$\le c_7 j \quad \text{with } c_7 = \left(\frac{4^{N-p}}{1-2^{p-N}}\right)^{\frac{1}{p-1}}.$$

Therefore

$$A_2 \leq \sum_{j=1}^{\infty} \int_{S_j} e^{c_3 c_7 \epsilon j} dx = |S_1| \sum_{j=1}^{\infty} r_j^N e^{c_3 c_7 \epsilon j} \\ \leq |S_1| \sum_{j=1}^{\infty} r_j^N e^{c_3 c_7 \epsilon j - jN \ln 2} = c_8 < \infty, \tag{5.5.25}$$

provided $\epsilon < \frac{N \ln 2}{2c_3 c_7}$. Combining (5.5.23), (5.5.24) and (5.5.25) we obtain that $A$ is bounded by some constant $c_2 > 0$ depending on $N$ and $p$ provided $\epsilon \leq \epsilon_0(N, p)$.

*Step 2. Proof of (5.5.18).* Using (5.5.12) with the Morrey space $\mathcal{M}^{\frac{2N}{p}}$ we derive from (5.5.22),

$$\left\| \exp\left( \frac{\epsilon p}{2N} \mathbf{W}_{1,p}^R \left[ \sum_{k=3}^{J} \mu_k \right] \right) \right\|_{\mathcal{M}^{\frac{2N}{p}}(B_1)} \\ \leq c_9 \left( \int_{B_1} \exp\left( \epsilon \mathbf{W}_{1,p}^R \left[ \sum_{k=3}^{J} \mu_k \right](x) \right) dx \right)^{\frac{p}{2N}} \\ \leq c_{10}. \tag{5.5.26}$$

Set $B = B_{\frac{1}{4}}$. By Theorem 5.4.2 and (5.4.8), any renormalized solution of

$$\begin{aligned} -\operatorname{div} A(x, \nabla u) + g(x, u) &= \mu && \text{in } B \\ u &= 0 && \text{on } \partial B, \end{aligned} \tag{5.5.27}$$

where $g$ satisfies $rg(.,r) \geq 0$ and $\mu$ is a positive bounded measure satisfies

$$u(x) \leq c_0 \mathbf{W}_{1,p}^{\frac{1}{2}} [\mu](x) \qquad \text{a.e. in } B, \tag{5.5.28}$$

and $c_0$ depends only on $N$ and $p$. We fix $\epsilon \leq \epsilon_0(N, p)$ as above and put $\epsilon_1 = \left( \frac{p\epsilon}{2Nc_0} \right)^{\frac{1}{p-1}}$.

It follows from Theorem 5.4.8 and estimate (5.4.37) in the remark here after that there exists a renormalized solution to

$$\begin{aligned} -\operatorname{div} A(x, \nabla u) + g(x, u) &= \epsilon_2 \sum_{j=3}^{J} \mu_j && \text{in } B \\ u &= 0 && \text{on } \partial B, \end{aligned} \tag{5.5.29}$$

for $\epsilon_2 = \epsilon_2(N, p, c_0) > 0$ independent of $J$ since the $\mu_j$ have disjoint supports. By Proposition 5.5.3 and (5.5.26)

$$u(0) \geq -c_{11} + c_{12}\mathbf{W}^{\frac{1}{2}}_{1,p}\left[\sum_{k=3}^{J}\mu_k\right](0). \tag{5.5.30}$$

Therefore

$$\begin{aligned} u(0) &\geq -c_{11} + c_{12}\sum_{i=2}^{\infty}\int_{r_{i+1}}^{r_i}\left(\frac{\sum\limits_{3\leq k\leq J}\mu_k(B_t(0))}{t^{N-p}}\right)^{\frac{1}{p-1}} \\ &\geq -c_{11} + c_{12}\sum_{i=2}^{J-2}\int_{r_{i+1}}^{r_i}\left(\frac{\mu_{i+2}(B_t(0))}{t^{N-p}}\right)^{\frac{1}{p-1}}\frac{dt}{t} \\ &\geq -c_{11} + c_{12}\sum_{i=2}^{J-2}\int_{r_{i+1}}^{r_i}\left(\frac{\mu_{i+2}(S_{i+2})}{t^{N-p}}\right)^{\frac{1}{p-1}}, \\ &\geq -c_{11} + c_{13}\sum_{i=2}^{J-2}\left(\mathcal{H}^1_{N-p}(K\cap S_{i+2})\right)^{\frac{1}{p-1}} r_i^{-\frac{N-p}{p-1}}, \\ &\geq -c_{11} + c_{13}\sum_{i=4}^{\infty}\left(\mathcal{H}^1_{N-p}(K\cap S_i)\right)^{\frac{1}{p-1}} r_i^{-\frac{N-p}{p-1}}. \end{aligned} \tag{5.5.31}$$

Since $S_i = B_{i-1}\setminus B_i$, there holds

$$\left(\mathcal{H}^1_{N-p}(K\cap S_i)\right)^{\frac{1}{p-1}} \geq \frac{\left(\mathcal{H}^1_{N-p}(K\cap B_{i-1})\right)^{\frac{1}{p-1}}}{\max\left\{1, 2^{\frac{2-p}{p-1}}\right\}} - \left(\mathcal{H}^1_{N-p}(K\cap B_i)\right)^{\frac{1}{p-1}},$$

from which inequality we infer,

$$\begin{aligned} u(0) &\geq -c_{11} + c_{13}\sum_{i=4}^{\infty}\left[\frac{\left(\mathcal{H}^1_{N-p}(K\cap B_{i-1})\right)^{\frac{1}{p-1}}}{\max\left\{1, 2^{\frac{2-p}{p-1}}\right\}} - \left(\mathcal{H}^1_{N-p}(K\cap B_i)\right)^{\frac{1}{p-1}}\right] r_i^{-\frac{N-p}{p-1}} \\ &\geq -c_{11} + c_{13}\left(\frac{2^{\frac{N-p}{p-1}}}{\max\left\{1, 2^{\frac{2-p}{p-1}}\right\}} - 1\right)\sum_{i=4}^{\infty}\left(\mathcal{H}^1_{N-p}(K\cap B_i)\right)^{\frac{1}{p-1}} r_i^{-\frac{N-p}{p-1}} \\ &\geq -c_{14} + c_{15}\int_0^1\left(\frac{\mathcal{H}^1_{N-p}(K\cap B_t)}{t^{N-p}}\right)^{\frac{1}{p-1}}\frac{dt}{t}. \end{aligned}$$

Since the maximal solution $U_1 = U$ of (5.5.1) in $K^c$ satisfies the same equation in $B \setminus K$ and dominates $u$ on $\partial B$, it is larger than $u$. This implies (5.5.18).

*Step 3. Proof of 5.5.20.* By Proposition 1.2.6 there exists $\mu_j \in \mathfrak{M}^+(K \cap S_j)$ such that

$$\mu_j(K \cap S_j) = \int_{\mathbb{R}^N} (\mathbf{G}_p[\mu_j])^{\frac{q^*}{p-1}} dx = c^{\mathbb{R}^N}_{p,\frac{q^*}{q^*+1-p}}(K \cap S_j).$$

For $\theta > 0$ and $k \in \mathbb{N}$ we define $\theta_{k,s}$ by

$$\theta_{k,s} = \begin{cases} 1 & \text{if } 0 < s \leq 1 \\ \left(\frac{1+\theta}{\theta}\right)^{s-1} (\theta+1)^{k(s+1)} & \text{if } s > 1. \end{cases}$$

If $\{a_k\}$ is a sequence of nonnegative real numbers, we have by Jensen inequality,

$$\left(\sum_{k=0}^{\infty} a_k\right)^s \leq \sum_{k=0}^{\infty} \theta_{k,s} a_k^s.$$

Hence, using (5.3.23) and Proposition 1.2.6, we derive

$$\begin{aligned}
\int_{B_1} \left(\mathbf{W}^1_{1,p}\left[\sum_{k=3}^{J} \mu_k\right](x)\right)^{q^*} dx &\leq \int_{B_1} \left(\sum_{k=3}^{J} \theta_{k,\frac{1}{p-1}} \mathbf{W}^1_{1,p}[\mu_k](x)\right)^{q^*} dx \\
&\leq \sum_{k=3}^{J} \theta^{q^*}_{k,\frac{1}{p-1}} \theta_{k,q^*} \int_{B_1} \left(\mathbf{W}^1_{1,p}[\mu_k](x)\right)^{q^*} dx \\
&\leq c_{16} \sum_{k=3}^{J} \theta^{q^*}_{k,\frac{1}{p-1}} \theta_{k,q^*} \int_{\mathbb{R}^N} (\mathbf{G}_p[\mu_k](x))^{\frac{q^*}{p-1}} dx \\
&\leq c_{16} \sum_{k=3}^{J} \theta^{q^*}_{k,\frac{1}{p-1}} \theta_{k,q^*} c^{\mathbb{R}^N}_{p,\frac{q^*}{q^*+1-p}}(K \cap S_j) \\
&\leq c_{17} \sum_{k=3}^{J} \theta^{q^*}_{k,\frac{1}{p-1}} \theta_{k,q^*} 2^{-k\left(N - \frac{pq^*}{q^*+1-p}\right)} \\
&\leq c_{18}.
\end{aligned}$$

Therefore

$$\left\| \left(\mathbf{W}^1_{1,p}\left[\sum_{k=3}^{J} \mu_k\right]\right)^q \right\|_{\mathcal{M}^{\frac{q^*}{q}}} \leq c_{19} \left\| \left(\mathbf{W}^1_{1,p}\left[\sum_{k=3}^{J} \mu_k\right]\right) \right\|^q_{L^{q^*}} \leq c_{20}, \quad (5.5.32)$$

where $c_{20}$ is independent of $J$. Since the measure $\sum_{k=3}^{J}\mu_k$ is absolutely continuous with respect to the $c^{\mathbb{R}^N}_{p,\frac{q}{q+1-p}}$-capacity, it follows from Theorem 5.4.6 that there exists a renormalized solution to

$$\begin{aligned} -\mathrm{div}\, A(x,\nabla u) + g(x,u) &= \sum_{k=3}^{J}\mu_k \qquad &\text{in } B\\ u &= 0 &\text{on } \partial B. \end{aligned} \tag{5.5.33}$$

Thus, using lower estimate Proposition 5.5.3 with $0 \le f = g(x,u) \le au^q$, and (5.5.33), we obtain as in Step 2,

$$\begin{aligned} u(0) &\ge -c_{21} + c_{22}\mathbf{W}^{\frac{1}{4}}_{1,p}\left[\sum_{k=3}^{J}\mu_k\right](0)\\ &\ge -c_{23} + c_{24}\int_0^1\left(\frac{c^{\mathbb{R}^N}_{p,\frac{q^*}{q^*+1-p}}(K\cap B_r)}{r^{N-p}}\right)^{\frac{1}{p-1}}\frac{dr}{r}. \end{aligned} \tag{5.5.34}$$

As in Step 2, the maximal solution $U_2 = U$ of (5.5.1) in $B_1 \setminus K$ dominates $u$. This implies (5.5.20). □

### 5.5.3 *Applications to large solutions*

In order to obtain estimates from below of the maximal solution of (5.5.1) near the boundary, we have to compare it with the restriction to $\Omega$ of solutions defined in a larger domain than $\Omega$. *Therefore, in this section, we assume that $A$ and $g$ are Caratheodory functions respectively defined in $\Omega' \times \mathbb{R}^N$ and $\Omega' \times \mathbb{R}$ where $\Omega'$ is a domain which contains $\overline{\Omega}$ and that the assumptions (4.1.1) on $A$, and the different assumptions on $g$ are verified for $x \in \Omega'$.* For the sake of simplicity we assume that $\{x \in \mathbb{R}^N : \mathrm{dist}\,(x,\Omega) \le 1\} \subset \Omega'$.

**Theorem 5.5.6** *Assume $N \ge 2$, $1 < p < N$ and the function $g$ satisfies (5.5.6) (i), (5.5.9) and (5.5.15). If*

$$\int_0^1\left(\frac{\mathcal{H}^1_{N-p}(\Omega^c\cap B_r(x))}{r^{N-p}}\right)^{\frac{1}{p-1}}\frac{dr}{r} = \infty \quad \forall x \in \partial\Omega, \tag{5.5.35}$$

*then $\overline{u}_\Omega$ is a large solution.*

*Proof.* Let $x_0 \in \partial\Omega$, and for simplicity we assume $x_0 = 0$. Let $0 < \delta < \frac{1}{12}$ and $z_0 \in \Omega \cap \overline{B}_\delta$. We put $K = \Omega^c \cap \overline{B}_{\frac{1}{4}}(z_0)$ and denote by $U_1$ the maximal solution of (5.5.1) in $B_1 \setminus K$. Then $\overline{u}_\Omega \geq U_1$ in $\Omega$. Furthermore

$$\begin{aligned}
U_1(z_0) &\geq -c_1 + c_2 \int_\delta^1 \left( \frac{\mathcal{H}^1_{N-p}(\Omega^c \cap B_r(z_0))}{r^{N-p}} \right)^{\frac{1}{p-1}} \frac{dr}{r} \\
&\geq -c_1 + c_2 \int_\delta^1 \left( \frac{\mathcal{H}^1_{N-p}(\Omega^c \cap B_{r-|z_0|})}{r^{N-p}} \right)^{\frac{1}{p-1}} \frac{dr}{r} \quad \text{as } B_{r-|z_0|} \subset B_r(z_0) \\
&\geq -c_1 + c_2 \int_{2\delta}^1 \left( \frac{\mathcal{H}^1_{N-p}(\Omega^c \cap B_{\frac{r}{2}})}{r^{N-p}} \right)^{\frac{1}{p-1}} \frac{dr}{r} \\
&\geq -c_1 + c_3 \int_\delta^{\frac{1}{2}} \left( \frac{\mathcal{H}^1_{N-p}(\Omega^c \cap B_r)}{r^{N-p}} \right)^{\frac{1}{p-1}} \frac{dr}{r}.
\end{aligned}$$

The proof follows from the next estimate

$$\inf_{B_\delta \cap \Omega} \overline{u}_\Omega \geq \inf_{B_\delta \cap \Omega} \overline{U}_1 \geq -c_1 + c_3 \int_\delta^{\frac{1}{2}} \left( \frac{\mathcal{H}^1_{N-p}(\Omega^c \cap B_r)}{r^{N-p}} \right)^{\frac{1}{p-1}} \frac{dr}{r},$$

and the assumption (5.5.35), by letting $\delta \to 0$. □

*Remark.* In the case $p = N$ the maximal solution is a large solution. This follows from the fact that for any $z_0 \in \partial\Omega$ and any $k > 0$, there holds $\overline{u}_\Omega \geq u_{k\delta_{z_0}}$, where $u_{k\delta_{z_0}}$ is a renormalized solution of

$$\begin{aligned}
-\text{div}\, A(x, \nabla u) + e^u - 1 &= k\delta_{z_0} && \text{in } B \\
u &= 0 && \text{on } \partial B,
\end{aligned} \tag{5.5.36}$$

and $k < M$ defined in (5.4.37). Using Serrin's characterization of isolated singularities Theorem 1.3.7 (ii) we derive that

$$u_{k\delta_{z_0}}(x) \geq \ell \ln(|x|^{-1}) - \ell' \quad \forall x \in B_{\frac{1}{2}}.$$

In the power case, the proof is similar and left to the reader.

**Theorem 5.5.7** *Assume $N \geq 2$, $1 < p < N$, $q^* > q_1$, $g$ satisfies (5.5.6) (i), (5.5.9) and (5.5.19) for some $p - 1 < q < \frac{pq^*}{N}$. If*

$$\int_0^1 \left( \frac{c^{\mathbb{R}^N}_{p, \frac{q^*}{q^*+1-p}}(K \cap B_r(x))}{r^{N-p}} \right)^{\frac{1}{p-1}} \frac{dr}{r} = \infty \quad \forall x \in \partial\Omega, \tag{5.5.37}$$

*then $\overline{u}_\Omega$ is a large solution.*

### 5.5.4 *The case $p = 2$*

When $p = 2$ and $q > 1$ the previous result is not optimal since there exists a necessary and sufficient condition [Labutin (2003)] for the maximal solution of

$$-\Delta u + u^q = 0 \tag{5.5.38}$$

in $\Omega$ to be a large solution. A local version of Labutin's condition [Marcus, Véron (2009)] provides a necessary and sufficient condition in order a positive solution of (5.5.38) tend to infinity at a particular boundary point.

**Theorem 5.5.8** *Assume $N \geq 2$, $q > 1$, $\Omega \subset \mathbb{R}^N$ is a bounded domain and $z \in \partial\Omega$. There exists a positive solution $u$ of (5.5.38) which tends to infinity at $z$ if and only if*

$$\int_0^1 c_{2,q'}^{\mathbb{R}^N}(K \cap B_r(z))\frac{dr}{r^{N-1}} = \infty. \tag{5.5.39}$$

*Furthermore, if the above condition holds for any boundary point $z$, the maximal solution is the unique large solution of (5.5.38) in $\Omega$.*

The proof of a universal capacitary estimate from above of any solution of (5.5.38) in $\Omega$ necessitates some notations and intermediate results.

**Lemma 5.5.9** *Assume $N \geq 3$, $q \geq q_1 := q_{1,2} = \frac{N}{N-2}$, $F \subset \mathbb{R}^N$ is compact and $u$ is nonnegative nontrivial solution of (5.5.38) in $\mathbb{R}^N \setminus F$. Then*

$$\lim_{r\to\infty} \int_{\partial B_r} \frac{\partial u}{\partial \mathbf{n}} dS := \tilde{\ell}_u \leq 0. \tag{5.5.40}$$

*Proof.* We assume and $F \subset B_R$ for some $R > 0$. Since $u$ is nonnegative and nontrivial, it is positive in $F^c$ by the strong maximum principle. By the Keller-Osserman estimate (3.1.7),

$$0 \leq u(x) \leq c_{N,q}\left(|x| - R\right)^{-\frac{2}{q-1}} \qquad \forall x \in B_R^c. \tag{5.5.41}$$

Writing (5.5.38) under the form $-\Delta u + Vu = 0$ where $V = u^{q-1}$ we deduce from Harnack inequality (see Chapter 1) that

$$\sup_{|x|=r} u(x) \leq c_{N,q} \inf_{|x|=r} u(x) \qquad \forall r \geq 2R. \tag{5.5.42}$$

Let $\bar{u}(r)$ be the spherical average of $u(r,\sigma)$ (in spherical coordinates). Hence, by convexity,

$$\bar{u}'' + \frac{N-1}{r}\bar{u}' - \bar{u}^q \geq 0 \quad \text{in } [R,\infty). \tag{5.5.43}$$

Set $y(s) = r^{N-2}\bar{u}(r)$ with $s = \frac{r^{N-2}}{N-2}$, then

$$y'' - a_{N,q}s^{-q-\frac{N-4}{N-2}}y^q \geq 0 \quad \text{in } [S,\infty). \tag{5.5.44}$$

If $y$ is unbounded, then $y(s) \geq cs$ for $s \geq s_0$ by convexity, therefore

$$y'' \geq a_{N,q}c^q s^{-\frac{N-4}{N-2}} \Longrightarrow y(s) \geq c' s^{\frac{N}{N-2}} \quad \text{for } s \geq s_1.$$

Iterating this process, implies that there exist sequences $\{\alpha_n\} \subset \mathbb{R}_+$ tending to $\infty$ and $\{c_n\} \subset \mathbb{R}_+$ such that $y(s) \geq c_n s^{\alpha_n}$ for $s$ large enough which contradicts (5.5.41). Consequently $y$ is bounded, which implies by Harnack inequality (5.5.42),

$$0 \leq u(x) \leq c|x|^{2-N} \qquad \forall x \in B_R^c, \tag{5.5.45}$$

where $c$ may depends on $u$ and $R$. In the case $q = q_1$ this estimate can be improved and (5.5.44) becomes

$$y'' - a_N s^{-2}y^{q_1} \geq 0 \quad \text{in } [S,\infty). \tag{5.5.46}$$

This implies that $y(s) \to 0$ as $s \to \infty$ and $y$ is decreasing. We set $z(t) = y(s)$ with $s = e^t$, then $z(t) \to 0$ as $t \to \infty$ and

$$z'' - z' - a_N z^{q_1} \geq 0 \quad \text{in } [T,\infty). \tag{5.5.47}$$

By the maximum principle, $z$ is dominated on $[T,\infty)$ by any solution of equation $Z'' - Z' - Z^{q_1} = 0$ which tends to 0 at infinity and dominates $z(t)$ for $t = T$. Such a solution $Z$ is easily constructed by approximation on $(T,n)$ and letting $n \to \infty$. It is decreasing and satisfies

$$Z'(t) - a_N e^t \int_t^\infty e^{-\tau} Z^{q_1}(\tau)d\tau = 0 \quad \text{in } [T,\infty). \tag{5.5.48}$$

Therefore $Z'(t) - a_N Z^q(t) \leq 0$. By integration

$$Z(t) \leq \left(\frac{a_N}{t(q-1)}\right)^{\frac{1}{q_1-1}} \Longrightarrow \bar{u}(r) \leq cr^{2-N}(\ln r)^{\frac{2-N}{2}} \quad \text{in } [R,\infty).$$

This implies finally

$$0 \leq u(x) \leq c|x|^{2-N}(\ln(|x|))^{\frac{2-N}{2}} \qquad \forall x \in B_R^c, \tag{5.5.49}$$

with a constant $c$ depending on $N, p, u$ and $R$. For $r > R$ there holds

$$\int_{B_r \setminus B_R} u^q dx - \int_{\partial B_r} \frac{\partial u}{\partial \mathbf{n}} dS = -\int_{\partial B_R} \frac{\partial u}{\partial \mathbf{n}} dS.$$

By estimates (5.5.45) if $q > q_1$ and (5.5.49) if $q = q_1$, $u^q \in L^1(B_R^c)$, therefore there exists

$$\tilde{\ell}_u = \lim_{r \to \infty} \int_{\partial B_r} \frac{\partial u}{\partial \mathbf{n}} dS. \tag{5.5.50}$$

We write $u(r, \sigma) = r^{2-N} v(t, \sigma)$ with $r = e^t$, $t > t_0 := \ln R$. Then $v$ is bounded and satisfies

$$v_{tt} + (2 - N)v_t + \Delta' v - e^{(N-2)(q-q_1)t} v^q = 0 \quad \text{in } (t_0, \infty) \times S^{N-1}. \tag{5.5.51}$$

By linear elliptic equations local regularity theory [Gilbarg, Trudinger], $v_t$, $v_{tt}$, $\nabla' v$, $\nabla' v_t$ and $D'^2 v$ are uniformly bounded in some Hölder norm on $(t_0 + 1, \infty) \times S^{N-1}$. If $\overline{\phi}$ denotes the spherical average of $\phi$, then

$$\overline{v}_{tt} + (2 - N)\overline{v}_t - e^{(N-2)(q_1-q)t} \overline{v^q} = 0,$$

which in turn implies

$$\overline{v}_t(t) = e^{(N-2)t} \int_t^\infty e^{(N-2)(q_1-q-1)\tau} \overline{v^q}(\tau) d\tau. \tag{5.5.52}$$

*Case 1:* $q > q_1$. From the previous estimate, $|\overline{v}_t(t)| \leq c e^{(N-2)(q_1-q)t}$ for some $c > 0$, consequently $\overline{v}(t)$ admits a finite nonnegative limit $\gamma$ when $t \to \infty$. From (5.5.51) there holds

$$\frac{d}{dt} \int_{S^{N-1}} \left( v_t^2 - |\nabla' v|^2 \right) d\sigma = 2 \int_{S^{N-1}} \left( (N-2) v_t^2 + e^{(N-2)(q-q_1)t} v^q v_t \right) d\sigma. \tag{5.5.53}$$

This implies that $v_t \in L^2((t_0, \infty) \times S^{N-1})$. Differentiating the equation with respect to $t$, multiplying by $v_{tt}$ and integrating on $S^{N-1}$ implies that $v_{tt} \in L^2((t_0, \infty) \times S^{N-1})$. Moreover the bounds of $v_t$ and $v_{tt}$ implies that these functions are uniformly continuous in $(t_0, \infty) \times S^{N-1})$, therefore $v_t(t, .)$ and $v_{tt}(t, .)$ converge to 0 in $L^2(S^{N-1})$ as $t \to \infty$. The uniform bounds on $v$ and its covariant derivatives implies that the set of functions $\mathcal{T}[v] := \{v(t, .) : t \geq t_0 + 1\}$ is relatively compact in the $C^2(S^{N-1})$-topology. Thus the $\omega$-limit set $\Omega[\mathcal{T}[v]]$ (see Theorem 3.1.16 in Chapter 3) is a nonempty connected compact subset of the set of solutions of $\Delta' \omega = 0$. Therefore this limit set is composed of constant functions. Because $\overline{v}(t) \to \gamma$ we infer that

$v(t,.) \to \gamma$ in the $C^1(S^{N-1})$ topology. Applying l'Hospital rule to (5.5.52) yields

$$\overline{v}_t(t) = \frac{e^{(N-2)(q_1-q)t}}{N-2}\gamma^q(1+\circ(1)) \quad \text{as } t \to \infty.$$

Since $r^{N-1}u_r(r,\sigma) = (2-N)v(t,\sigma) + v_t(t,\sigma)$ we have

$$\int_{\partial B_r} \frac{\partial u}{\partial \mathbf{n}} dS = r^{N-1}\overline{u}_r(r) = (2-N)\overline{v}(t) + \overline{v}_t(t) \to (2-N)\gamma \leq 0. \quad (5.5.54)$$

*Case 2:* $q = q_1$. Equation (5.5.51) becomes

$$v_{tt} + (2-N)v_t + \Delta' v - v^{q_1} = 0 \qquad \text{in } (t_0,\infty) \times S^{N-1}. \quad (5.5.55)$$

Integrating on $S^{N-1}$ yields $\overline{v}_{tt} + (2-N)\overline{v}_t - \overline{v^{q_1}} = 0$, thus

$$\overline{v}_t(t) = e^{(N-2)t}\int_t^\infty e^{(2-N)\tau}\overline{v^{q_1}}(\tau)d\tau. \quad (5.5.56)$$

Since $v_t(t,.) \leq ct^{\frac{2-N}{2}}$ by (5.5.49), we derive from (5.5.55)

$$\overline{v}_t(t) \leq \frac{c}{N-2}t^{-\frac{N}{2}} \qquad \text{for } t \geq t_1 > t_0.$$

The relation $r^{N-1}u_r(r,\sigma) = (2-N)v(t,\sigma)+v_t(t,\sigma)$ still holds, thus (5.5.54) is also valid which implies

$$\int_{\partial B_r} \frac{\partial u}{\partial \mathbf{n}} dS \to 0 \qquad \text{as } r \to \infty. \quad (5.5.57)$$

□

In order to prove the estimate from above in Theorem 5.5.8, we introduce some notations.

**Definition 5.3** Let $1 < q < \infty$ such that $q > q_1$. If $K$ is a compact subset of $\mathbb{R}^N$ the Riesz capacity $\dot{c}_{2,q'}(K)$ is defined by

$$\dot{c}_{2,q'}(K) = \inf\left\{\left\|D^2\psi\right\|^{q'}_{L^{q'}(\mathbb{R}^N)} : \psi \in C_0^\infty(\mathbb{R}^N), 0 \leq \psi \leq 1, \psi = 1 \text{ on } K\right\}. \quad (5.5.58)$$

The function $\dot{c}_{2,q'}$ is actually a *Riesz capacity* (see Section 1.2.4 and [Adams, Hedberg (1999)] for details). It is extended to general sets by the usual rules of capacities. The main reason for introducing this capacity are the Gagliardo-Nirenberg interpolation inequality (see below) and the homogeneity property (which is not shared by the Bessel capacity),

$$\dot{c}_{2,q'}(aK) = a^{N-2q'}\dot{c}_{2,q'}(K) \qquad \forall a > 0, \tag{5.5.59}$$

and there holds

$$\dot{c}_{2,q'}(K) \leq c_{2,q'}^{\mathbb{R}^N}(K). \tag{5.5.60}$$

Furthermore [Adams, Hedberg (1999), Prop 5.1.4, 5.6.1], for all $m > 0$ there exists $A_m = A(m, N, q) > 0$ such that

$$c_{2,q'}^{\mathbb{R}^N}(K) \leq A_m \dot{c}_{2,q'}(K) \quad \forall E \subset \mathbb{R}^N \text{ s.t. } \operatorname{diam}(E) \leq m, \tag{5.5.61}$$

and there exists $M = M(N, q) > 0$ such that

$$c_{2,q'}^{\mathbb{R}^N}(K) \leq M\left(\dot{c}_{2,q'}(K) + (\dot{c}_{2,q'}(K))^{\frac{N}{N-2q'}}\right) \quad \forall E \subset \mathbb{R}^N. \tag{5.5.62}$$

Notice also this link between the Bessel capacities relative to $\mathbb{R}^N$ and relative to a domain $G \subset \mathbb{R}^N$. Let $\Theta \subset \overline{\Theta} \subset G$ be a bounded domain, then there exists $B = A(G, \Theta, N, q)$ such that

$$c_{2,q'}^{\mathbb{R}^N}(E) \leq c_{2,q'}^{G}(E) \leq B c_{2,q'}^{\mathbb{R}^N}(E) \qquad \forall E \subset \Theta,\ E \text{ Borel}. \tag{5.5.63}$$

**Definition 5.4** If $F \subset \mathbb{R}^N$ is a closed set, $x \in \mathbb{R}^N$, $\delta > 0$ and $m \in \mathbb{Z}$, we denote

$$\begin{aligned}
&T_{m,\delta}(x) = \left\{y \in \mathbb{R}^N : 2^m\delta \leq |x - y| \leq 2^{m+1}\delta\right\}\\
&F_{m,\delta}(x) = F \cap T_{m,\delta}(x) \qquad F^*_{m,\delta}(x) = F \cap \overline{B}_{2^{m+1}\delta}(x)\\
&W_F(x) = \delta^{2-N} \sum_{m=-\infty}^{\infty} 2^{(2-N)m}\dot{c}_{2,q'}(F_{m,\delta}(x))\\
&W^*_F(x) = \delta^{2-N} \sum_{m=-\infty}^{\infty} 2^{(2-N)m}\dot{c}_{2,q'}(F^*_{m,\delta}(x))\\
&\mathcal{W}_F(x) = \int_0^{\infty} \dot{c}_{2,q'}(F \cap B_r(x))\frac{dr}{r^{N-1}}.
\end{aligned} \tag{5.5.64}$$

The function $W_F$ is called the $\dot{c}_{2,q'}$-*Riesz capacitary potential of* $F$.

The following inequalities are easy to verify:

$$W_F(x) \leq W^*_F(x) \leq (1 - 2^{2-N})^{-1}W_F(x) \qquad \forall x \in \mathbb{R}^N, \tag{5.5.65}$$

and since

$$\mathcal{W}_F(x) = \sum_{m=-\infty}^{\infty} \int_{2^m\delta}^{2^{m+1}\delta} \dot{c}_{2,q'}(F \cap B_r(x)) \frac{dr}{r^{N-1}},$$

there holds

$$\frac{1-2^{-N}}{N} W_F^*(x) \leq \mathcal{W}_F(x) \leq \frac{2^N-1}{N} W_F^*(x) \qquad \forall x \in \mathbb{R}^N. \tag{5.5.66}$$

**Lemma 5.5.10** *Assume $F \subset \mathbb{R}^N$ is compact and $u$ satisfies (5.5.38) in $\mathbb{R}^N \setminus F$. Let $\eta \in C_0^\infty(\mathbb{R}^N)$ with value 1 in a neighborhood of $F$ and $0 \leq \eta \leq 1$. Then*

$$\int_{\mathbb{R}^N} u^q (1-\eta)^{2q'} dx \leq c_1 \left\| D^2\eta \right\|_{L^{q'}}^{q'}, \tag{5.5.67}$$

*for some $c_1 = c_1(N, q)$.*

*Proof.* Set $\zeta = (1-\eta)^{2q'}$. Assuming that $\mathrm{supp}(\eta) \subset B_{2R}$, we have for $r > 2R$

$$\begin{aligned}
\int_{B_r} u^q \zeta dx &= \int_{\partial B_r} \frac{\partial u}{\partial \mathbf{n}} dS + \int_{B_r} u \Delta (1-\eta)^{2q'} dx \\
&= \int_{\partial B_r} \frac{\partial u}{\partial \mathbf{n}} dS + \int_{\mathbb{R}^N} u \left( -2q'(1-\eta)^{2q'-1} \Delta\eta \right. \\
&\qquad\qquad \left. + 2q'(2q'-1)(1-\eta)^{2q'-2} |\nabla \eta|^2 \right) dx \\
&\leq \int_{\partial B_r} \frac{\partial u}{\partial \mathbf{n}} dS + c_2 \left( \int_{B_r} u^q \zeta dx \right)^{\frac{1}{q}} \left( \int_{B_r} \left( |\Delta \eta|^{q'} + |\nabla \eta|^{2q'} \right) dx \right)^{\frac{1}{q'}},
\end{aligned}$$

where $c_2 = 2q'(2q'-1)$. By the Gagliardo-Nirenberg interpolation inequality (1.2.34) (using the fact that $\|\eta\|_{L^\infty} = 1$),

$$\left( \int_{\mathbb{R}^N} |\nabla \eta|^{2q'} dx \right)^{\frac{1}{q'}} \leq c_3 \left\| D^2\eta \right\|_{L^{q'}},$$

where $c_3$ depends on $N$ and $q$. Using (5.5.40) we obtain, letting $r \to \infty$,

$$\int_{\mathbb{R}^N} u^q \zeta dx \leq c_4 \left( \int_{\mathbb{R}^N} u^q \zeta dx \right)^{\frac{1}{q}} \left\| D^2\eta \right\|_{L^{q'}},$$

which implies (5.5.67). □

**Corollary 5.5.11** *Assume $F \subset B_R$ is compact and $u$ satisfies (5.5.38) in $\mathbb{R}^N \setminus F$, then*

$$\int_{B_{2R}^c} u^q dx \leq c_1 c_{2,q'}^{B_{2R}}(F) \leq c_1 B c_{2,q'}^{\mathbb{R}^N}(F), \tag{5.5.68}$$

*where $B$ is the constant appearing in (5.5.63).*

*Proof.* If we assume that $\operatorname{supp}(\eta) \subset B_{2R}$ and take the infimum over all $\eta \in C_0^\infty(B_{2R})$ we derive the first inequality in (5.5.68). Using (5.5.63) we obtain the second one. □

*Remark.* The following technical form of the previous estimate will be used in the next lemma

$$\int_{B_{2R}^c} u^q dx - \tilde{\ell}_u \leq c_1 c_{2,q'}^{B_{2R}}(F) \leq c_1 B c_{2,q'}^{\mathbb{R}^N}(F), \quad \forall r \geq 2R. \tag{5.5.69}$$

**Lemma 5.5.12** *Under the assumptions of Corollary 5.5.11 there holds*

$$\int_{\partial B_{2R}} u dS \leq c_2 R c_{2,q'}^{B_{2R}}(F) \leq c_2 B R c_{2,q'}^{\mathbb{R}^N}(F), \tag{5.5.70}$$

*and*

$$u(x) \leq c_4 |x|^{2-N} c_{2,q'}^{B_{2R}}(F) \leq c_4 B |x|^{2-N} c_{2,q'}^{\mathbb{R}^N}(F), \tag{5.5.71}$$

*where the $c_j > 0$, $j = 2, 3, 4$, depend on $N$ and $q$.*

*Proof.* Set

$$\psi_R(x) = \frac{1}{N-2}\left(\left(\frac{1}{2}\right)^{N-2} - \left(\frac{R}{|x|}\right)^{N-2}\right) \quad \forall x \in B_{2R}^c.$$

Then $\psi_R$ is harmonic, vanishes on if $|x| = 2R$, satisfies $0 \leq \psi_R < 1$ in $B_r \setminus B_{2R}$ for any $r > 2R$ and

$$\frac{\partial \psi_R}{\partial \mathbf{n}}\lfloor_{\partial B_{2R}} = \frac{-1}{2^{N-1}R} \quad \text{and} \quad \frac{\partial \psi_R}{\partial \mathbf{n}}\lfloor_{\partial B_r} = \frac{R^{N-1}}{r^{N-1}},$$

if $\mathbf{n}$ stands for the outward unit vector of $B_r \setminus B_{2R}$. Multiplying (5.5.38) by $\psi_R$ and integrating on $B_r \setminus B_{2R}$ yields (since $\psi_R$ is harmonic)

$$\begin{aligned}\int_{B_r \setminus B_{2R}} \psi_R u^q dx &= \int_{B_r \setminus B_{2R}} \psi_R \Delta u dx \\ &= \int_{\partial B_r} \left(\frac{\partial u}{\partial \mathbf{n}}\psi_R - \frac{\partial \psi_R}{\partial \mathbf{n}} u\right) dS - \int_{\partial B_{2R}} \frac{\partial \psi_R}{\partial \mathbf{n}} u dS.\end{aligned}$$

By (5.5.40)

$$\lim_{r\to\infty}\int_{\partial B_r}\left(\frac{\partial u}{\partial \mathbf{n}}\psi_R-\frac{\partial \psi_R}{\partial \mathbf{n}}u\right)dS=\frac{\tilde{\ell}_u}{(N-2)2^{N-2}}\leq 0,$$

and

$$\lim_{r\to\infty}\int_{\partial B_{2R}}\frac{\partial \psi_R}{\partial \mathbf{n}}udS=\frac{-1}{2^{N-1}R}\int_{\partial B_{2R}}udS.$$

Therefore

$$\int_{B^c_{2R}}\psi_R u^q dx-\frac{\tilde{\ell}_u}{(N-2)2^{N-2}}=\frac{1}{2^{N-1}R}\int_{\partial B_{2R}}udS, \tag{5.5.72}$$

which, combined with (5.5.69), implies (5.5.71) since $\psi_R\leq 1$. Let $P^{B^c_{2R}}$ be the Poisson kernel in $B^c_{2R}$. Its expression is easy to obtain by Kelvin transform.

$$P^{B^c_{2R}}(x,y)=c_N(2R)^{N-1}\left|x\right|^N\frac{|x|^2-4R^2}{\left|4R^2x-|x|^2y\right|^N}\quad \forall (x,y)\in B^c_{2R}\times\partial B_{2R}. \tag{5.5.73}$$

Thus, for $|x|\geq 4R$

$$\begin{aligned}u(x)&=c_N(2R)^{N-1}\left|x\right|^N(|x|^2-4R^2)\int_{\partial B_{2R}}\frac{u(y)}{\left|4R^2x-|x|^2y\right|^N}dS(y)\\&\leq c_N\frac{|x|+2R}{R(|x|-2R)^{N-1}}\int_{\partial B_{2R}}u(y)dS(y)\\&\leq\frac{c_N2^{N-1}}{R|x|^{N-2}}\int_{\partial B_{2R}}u(y)dS(y).\end{aligned}$$

Jointly with (5.5.68) we infer (5.5.71). □

*Proof of Theorem 5.5.8: I- Estimate from above.* If $G\subset\mathbb{R}^N$ is a compact set, we denote by $U_G$ the maximal solution of (5.5.38) in $G^c$. If $x\in F^c$, we set $\delta=\operatorname{dist}(x,F)$ and consider the slicing described in (5.5.64). Then $F_m=\emptyset$ if $m<1$ or if $m\geq\ell(x)$ for some integer $\ell(x)$. One can find an integer $K$ such that for each $m\in\mathbb{N}_*$ there exist $K$ points $a_{j,m}$ such that $2^{m-1}\delta\leq|x-a_{j,m}|\leq 2^{m+1}\delta$ and

$$T_{m,\delta}(x)\subset\bigcup_{j=1}^{K}\overline{B}_{2^{m-3}\delta}(a_{j,m}).$$

We set $F_{m,j,\delta}(x) = \overline{B}_{2^{m-3}\delta}(a_{j,m}) \cap F$. Since $\sum_{m=1}^{\ell(x)}\sum_{j=1}^{K} U_{F_{m,j,\delta}(x)}$ is a super solution of (5.5.38) we have that $u \leq \sum_{m=1}^{\ell(x)}\sum_{j=1}^{K} U_{F_{m,j,\delta}(x)}$. In particular

$$u(x) \leq \sum_{m=1}^{\ell(x)}\sum_{j=1}^{K} U_{F_{m,j,\delta}(x)}(x). \tag{5.5.74}$$

Using (5.5.74) we obtain

$$\begin{aligned} u(x) &\leq c_4 \sum_{m=1}^{\ell(x)} (2^m\delta)^{2-N} \sum_{j=1}^{K} c_{2,q'}^{B_{2R}}(F_{m,j,\delta}) \\ &\leq Kc_4 \sum_{m=1}^{\ell(x)} (2^m\delta)^{2-N} c_{2,q'}^{B_{2R}}(F_{m,\delta}) \\ &\leq c_4 K W_F(x). \end{aligned} \tag{5.5.75}$$

Hence, if (5.5.39) does not hold, $u(x)$ remains bounded when $x \to x_0 \in \partial\Omega$.
*Proof. II- Estimate from below.* Thanks to the semilinearity, the estimate of $U_2$ uses, in a sharper way than in Theorem 5.5.5 (ii), the construction of the solution $u$ of

$$\begin{aligned} -\Delta u + |u|^{q-1}u &= \sum_{k=3}^{J} \mu_k && \text{in } B \\ u &= 0 && \text{on } \partial B. \end{aligned} \tag{5.5.76}$$

A solution $u$ of (5.5.38) in $\Omega$ is called *$\sigma$-moderate* if there exists a bounded nonnegative measure $\mu$ in $\mathbb{R}^N$, such that $\mu(\Omega) = 0$ and

$$-\Delta u + |u|^{q-1}u = \mu \qquad \text{in } \mathbb{R}^N. \tag{5.5.77}$$

We recall that the necessary and sufficient condition for a nonnegative measure $\mu$ in order the above problem can be solved is $\mu \ll c_{2,q'}^{\mathbb{R}^N}$. Because of uniqueness we write $u = u_\mu$. The largest $\sigma$-moderate solution of (5.5.38) in $\Omega$ is defined by

$$\underline{U}_\Omega = \sup\{u_\mu : \mu \in \mathfrak{M}^+(\mathbb{R}^N), \mu(\Omega) = 0\}.$$

The estimate from below yields the following estimate

$$\underline{U}_\Omega(x) \geq c_5 W_F(x). \tag{5.5.78}$$

*Proof. III- Uniqueness.* Combining (5.5.75) and (5.5.78) we see that

$$c_8 W_F \leq \underline{U}_\Omega \leq U_\Omega \leq c_6 \underline{U}_\Omega \leq c_7 W_F. \tag{5.5.79}$$

If (5.5.39) holds, $\underline{U}_\Omega$ and $U_\Omega$ are large solutions and $U_\Omega \leq c\underline{U}_\Omega$ for some positive constant $c = c(N, q)$. Then uniqueness follows by the same method as in the proof of Theorem 3.1.27. □

## 5.6 Notes and open problems

*5.6.1.* We conjecture that a compact set $K \subset \Omega$ is removable for equation (5.2.1) in $\Omega \setminus K$ if and only if $c^{\mathbb{R}^N}_{p,\frac{p}{q+1-p}}(K) = 0$. By [Adams, Hedberg (1999), Th 5.5.1] there holds for all $p > 1$, $R < N$ and Borel set $E \subset \mathbb{R}^N$,

$$c^{\mathbb{R}^N}_{1,R}(K) \leq A c^{\mathbb{R}^N}_{p,\frac{R}{p}}(K), \tag{5.6.1}$$

where $A$ depends on $p$, $R$, $N$ and $\operatorname{diam}(E)$. In particular, if $R > pq_p$ and $c^{\mathbb{R}^N}_{p,\frac{R}{p}}(K) = 0$, it follows from Theorem 5.2.4 that $K$ is removable for equation (5.2.1).

*5.6.2.* In Theorem 5.4.2 it is assumed that $\tilde{g}\left(c\mathbf{W}^{2\,diam\,(\Omega)}_{1,p}[\mu_i]\right) \in L^1(\Omega)$, which is a strong requirement which may not satisfied if $\mu_i \in L^1(\Omega)$. In the case $p = 2$ and $\tilde{g}$ satisfies the $\Delta_2$-condition *i.e.*

$$\tilde{g}(a+b) \leq k\left(\tilde{g}(a) + \tilde{g}(b)\right) \qquad \forall a, b \geq 0, \tag{5.6.2}$$

for some $k > 0$ a much sharper assumption is to require that only the singular parts of $\mu_i$ satisfy $\tilde{g}\left(c\mathbf{W}^{2\,diam\,(\Omega)}_{1,p}[\mu_{i\,s}]\right) \in L^1(\Omega)$ (see [Véron (1996), Th 4.2] for a similar problem in the case $p = 2$). We conjecture that this result can be extended to the case $p \neq 2$.

*5.6.3.* Theorem 5.4.6 provides a sufficient condition for the solvability of problem (5.4.2). We conjecture that this condition is also a necessary condition.

*5.6.4.* Theorem 5.2.7 gives only a sufficient condition for the removability of singularities for equation (5.2.1). We conjecture that this condition is necessary. In the same way, we conjecture that the condition $\mu < c^{\mathbb{R}^N}_{1,q_p}$ obtained in Theorem 5.2.8 is also a sufficient condition for solving (5.2.1).

*5.6.5.* If $\Omega \subset \mathbb{R}^N$ is a smooth bounded domain, it could be interesting to study the following nonlinear boundary value problem

$$\begin{aligned} -\,divA(x,\nabla u) &= 0 && \text{in } \Omega \\ \langle A(x,\nabla u),\mathbf{n}\rangle + g(x,u) &= \mu && \text{in } \partial\Omega \end{aligned} \tag{5.6.3}$$

where $g \in C(\partial\Omega \times \mathbb{R})$ and $\mu \in \mathfrak{M}(\partial\Omega)$. Weak solutions verify

$$\int_\Omega \langle A(x,\nabla u,\nabla\zeta\rangle dx + \int_{\partial\Omega} g(.,u)\zeta dS = \int_{\partial\Omega} \zeta d\mu \qquad \forall \zeta \in C^1(\overline{\Omega}). \tag{5.6.4}$$

If $g(x,u) = |u|^{q-1}\,u$, $q > p-1$, a clearly critical exponent is $q_1 = \frac{N(p-1)}{N-p}$. Note 4.4.4 could provide useful tools for this study.

# Chapter 6

# Quasilinear equations with source

In this Chapter we study properties of quasilinear equations of the type

$$-\mathrm{div}A(x,\nabla u) = g(x,u,\nabla u) \tag{6.0.1}$$

in a domain $\Omega$ and, if $\mu$ is a Radon measure in $\Omega$,

$$-\mathrm{div}A(x,\nabla u) = g(x,u,\nabla u) + \mu, \tag{6.0.2}$$

where $A$ satisfies the structural assumptions (4.1.1) and $g : \Omega\times\mathbb{R}\times\mathbb{R}^N \mapsto \mathbb{R}$ is a Caratheodory function called a source type reaction, that is

$$g(x,r,\xi)r \geq 0 \qquad \forall (x,r,\xi) \in \Omega\times\mathbb{R}\times\mathbb{R}^N. \tag{6.0.3}$$

## 6.1 Singularities of quasilinear Lane-Emden equations

The model equation considered here is the following

$$\Delta_p u + |u|^{q-1}u = 0, \tag{6.1.1}$$

in the range of exponents $q > p-1 > 0$. When $p = 2$ it is called a Lane-Emden equation.

### 6.1.1 *Separable solutions in* $\mathbb{R}^N$

With spherical coordinates $(r,\sigma)$, separable solutions of (6.1.1) under the form $u(x) = u(r,\sigma) = r^{-\beta}\omega(\sigma)$ exist if $\beta = \beta_{p,q} = \frac{p}{q+1-p}$ and $\omega$ satisfies

$$\begin{aligned} div'\left(\left(\beta_{p,q}^2\omega^2 + |\nabla'\omega|^2\right)^{\frac{p-2}{2}}\nabla'\omega\right) + |\omega|^{q-1}\omega \\ + \beta_{p,q}\Lambda(\beta_{p,q})\left(\beta_{p,q}^2\omega^2 + |\nabla'\omega|^2\right)^{\frac{p-2}{2}}\omega = 0 \quad \text{in } S^{N-1}, \end{aligned} \tag{6.1.2}$$

where

$$\beta_{p,q}\Lambda(\beta_{p,q}) = \beta_{p,q}((p-1)\beta_{p,q} + p - N) = \frac{p}{q+1-p}\left(N - \frac{pq}{q+1-p}\right). \tag{6.1.3}$$

If $q > q_1 := \frac{N(p-1)}{N-p}$ $(1 < p < N)$ then $\Lambda(\beta_{p,q}) < 0$ and there exist radial singular solutions under the form

$$u(r,\sigma) = r^{-\beta_{p,q}}\tilde{\omega}_c \tag{6.1.4}$$

where

$$\tilde{\omega}_c := \tilde{\omega}_{c,p,q} = \left(-\Lambda(\beta_{p,q})\beta_{p,q}^{p-1}\right)^{\frac{1}{q+1-p}}. \tag{6.1.5}$$

If $p - 1 < q \leq q_1$ then $\Lambda(\beta_{p,q}) \geq 0$; integrating equation (6.1.2) shows that there exists no nontrivial solution to (6.1.2). We will see in next section that there could exist signed solutions provided $q$ is large enough. Existence of positive solutions can be obtained by a bifurcation method. We recall that $S^{N-1}$ can be represented by the angle coordinates $(\theta_1, ..., \theta_{N-1})$, see (2.2.50).

**Proposition 6.1.1** *Assume $1 < p < N - 1$ and set*

$$q_s := \frac{Np - N + 1}{N - 1 - p}. \tag{6.1.6}$$

*Then there exist $\tau_0 > 0$ and a $C^1$ curve $\tau \mapsto (q(\tau), \omega(\tau))$ from $[0, \tau_0)$ into $\mathbb{R} \times C^1(S^{N-1})$ with $(q(0), \omega(0)) = (q_s, \tilde{\omega}_{c,p,q_s})$ such that $\omega(\tau)$ is a nonconstant solution of (6.1.2) satisfying $q = q(\tau)$ and depending only on one spherical coordinates $\theta_j$.*

*Proof.* We first notice that $q_s + 1 = \frac{(N-1)p}{N-1-p}$ is the exponent corresponding to the Gagliardo-Nirenberg inequality for $W^{1,p}(S^{N-1})$. If we linearize the equation at the constant solution $\tilde{\omega}_c$ we obtain

$$-\Delta'\phi + (q+1-p)\beta_{p,q}\Lambda(\beta_{p,q})\phi = 0 \quad \text{in } S^{N-1}. \tag{6.1.7}$$

The bifurcation following the first eigenspace of $-\Delta'$ with functions depending only on one variable occurs when

$$(p-1-q)\beta_{p,q}\Lambda(\beta_{p,q}) = N - 1 \Longleftrightarrow q = q_s. \tag{6.1.8}$$

The fact that the branch of solutions $(q(\tau), \omega(\tau))$ is a $C^1$ curve follows from [Crandall, Rabinowitz (1971)]. □

*Remark.* If $p = 2$, $q_s = \frac{N+1}{N-3}$; it is proved in [Gidas, Spruck (1980)], [Bidaut-Véron, Véron (1991)] that if $1 < q < q_s$ the only positive solution is the constant $\omega_c$. If $q = q_s$ this equation becomes

$$\Delta'\omega - \frac{(N-1)(N-3)}{4}\omega + \omega^{\frac{N+1}{N-3}} = 0. \tag{6.1.9}$$

By a stereographic projection $\pi$ of $S^{N-1} \setminus \{\sigma_S\}$ onto $\mathbb{R}^{N-1}$ where $\sigma_S$ is the south pole, the equation becomes, with $\psi(x) = (\frac{2}{1+|x|^2})^{\frac{N-3}{2}}\omega \circ \pi$

$$\Delta\psi + \psi^{\frac{N+1}{N-3}} = 0 \quad \text{in } \mathbb{R}^{N-1}. \tag{6.1.10}$$

The solutions of (6.1.10) are the extremals of the Gaglardo-Nirenberg inequality in $\mathbb{R}^{N-1}$ with exponent $p = 2$ and they form a N-dimensional family of solutions since they are explicitly given by

$$\psi(x) := \psi_{\epsilon,a}(x) = \left(\frac{\epsilon\sqrt{(N-1)(N-3)}}{\epsilon^2 + |x-a|^2}\right)^{\frac{N-3}{2}}. \tag{6.1.11}$$

This implies that the set of solutions of (6.1.8) has the structure of a $N$-dimensional manifold. When $p \neq 2$, the set of positive solutions of (6.1.1) is at least (N-1)-dimensional. It is a completely open problem to prove that all the positive solutions of (6.1.2) are constant when $p - 1 < q < q_s$.

*Remark.* Equation (6.1.2) admits a variational structure in the case $q = p^* - 1 = \frac{N(p-1)+p}{N-p}$. In that case the solutions are critical points of the functional

$$J(w) = \frac{1}{p}\int_{S^{N-1}} \left(\left(\frac{N-p}{p}\right)^2 w^2 + |\nabla' w|^2\right)^{\frac{p}{2}} d\sigma - \frac{1}{p^*}\int_{S^{N-1}} |w|^{p^*} d\sigma, \tag{6.1.12}$$

which is well defined in $W^{1,2}(S^{N-1})$.

### 6.1.2 *Isolated singularities*

The first result on isolated singularities deals with the subcritical case and the following theorem is an extension of the previous ones due to [Guedda, Véron (1988)] and [Bidaut-Véron (1989)]. We assume that $g : \mathbb{R} \times \mathbb{R}^N \mapsto \mathbb{R}$ is a continuous function satisfying

$$tg(t,\xi) \geq 0 \qquad \forall (t,\xi) \in \mathbb{R} \times \mathbb{R}^N, \tag{6.1.13}$$

and

$$|g(t,\xi)| \leq c_1 |t|^q + c_2 |\xi|^r + c_3 \qquad \forall (t,\xi) \in \mathbb{R} \times \mathbb{R}^N, \tag{6.1.14}$$

for exponents $q > p-1$, $r > 0$ and constants $c_j \geq 0$, $j = 1,2,3$, and we define the critical exponent

$$q_c = \frac{N(p-1)}{N-1}. \tag{6.1.15}$$

**Theorem 6.1.2** *Let $\Omega$ be a domain containing 0, $1 < p \leq N$ and $g$ a continuous function satisfying the assumptions (6.1.13) and (6.1.14) with $q < q_1$ and $r < q_c$. If $u \in C^1(\Omega \setminus \{0\})$ is a nonnegative solution of*

$$-\Delta_p u = g(u, \nabla u), \tag{6.1.16}$$

*in $\Omega \setminus \{0\}$, then $g(u,\nabla u) \in L^1_{loc}(\Omega)$ and there exists $\gamma \geq 0$ such that*

$$-\Delta_p u = g(u,\nabla u) + \gamma\delta_0 \qquad \text{in } \Omega. \tag{6.1.17}$$

*Furthermore*

$$\lim_{|x|\to 0} \frac{u(x)}{\mu_p(x)} = \gamma \quad \text{and} \quad \lim_{r\to 0} r^{\frac{N-1}{N-p}} \nabla u(r,\sigma) = c_{N,p}\gamma\sigma, \tag{6.1.18}$$

*where $\mu_p$ is the fundamental solution of the p-harmonic equation in $\mathbb{R}^N$ defined by (1.5.14).*

*Proof.* We assume $\overline{B}_1 \subset \Omega$. Since $u \geq 0$ and (6.1.13) holds, the right-hand side of (6.1.16) is nonnegative, hence, by Theorem 1.5.14, the following statements hold:
(i) If $1 < p < N$, $u^{p-1} \in L^{\frac{N}{N-p},\infty}_{loc}(\Omega)$, $|\nabla u|^{p-1} \in L^{\frac{N}{N-1},\infty}_{loc}(\Omega)$ and there exist $\gamma \geq 0$ and $h \in L^1_{loc}(\Omega)$ such that

$$-\Delta_p u = h + \gamma\delta_0 \qquad \text{in } \Omega. \tag{6.1.19}$$

(ii) If $p = N$, $u^m \in L^1_{loc}(\Omega)$ for all $m > 0$, $|\nabla u|^s \in L^1_{loc}(\Omega)$ for all $0 < s < N$. Furthermore, if $u(x) \to \infty$ when $x \to 0$, then the conclusion of (i) holds.

In the two cases it implies in particular that $g(u,\nabla u) \in L^1_{loc}(\Omega)$. In order to apply Theorem 1.3.7, we write (6.1.14) under the form

$$|g(u,\nabla u)| \leq c_1 |u|^{q+1-p}|u|^{p-1} + c_2 |\nabla u|^{(r+1-p)_+} |\nabla u|^{p-1} + c_3'.$$

If $p < N$ with $q < q_1$, then $|u|^{q+1-p} \in L^{\frac{N}{p}+\epsilon}_{loc}(\Omega)$ for some $\epsilon > 0$. If $p = N$, $|u|^{q+1-p} \in L^{1+\epsilon}_{loc}(\Omega)$. If $r+1-p > 0$ then $|\nabla u|^{r+1-p} \in L^{\frac{(p-1)(N-\epsilon)}{(r+1-p)(N-1)}}_{loc}(\Omega)$.

Since $r < r_c$, we can find $\epsilon > 0$ such that $\frac{(p-1)(N-\epsilon)}{(r+1-p)(N-1)} > N + \epsilon$. If $r+1-p \leq 0$, then $|\nabla u|^{(r+1-p)_+} = 1$. Therefore assumptions (1.3.6) (i)-(ii) are fulfilled. It follows that either there exists $c > 0$ such that

$$c^{-1}\mu_p(x) \leq u(x) \leq c\mu_p(x) \qquad \forall x \in B_{\frac{1}{2}}, \tag{6.1.20}$$

or $u \in L^\infty_{loc}(\Omega)$.

*Step 1.* If $u \in L^\infty_{loc}(\Omega)$ we claim that 0 is a removable singularity and (6.1.16) is satisfied in whole $\Omega$. Let $\zeta_\epsilon \in C_0^\infty(\mathbb{R}^N)$ with support in $B_1$, satisfying $0 \leq \zeta_\epsilon \leq 1$, $\zeta(x) = 0$ if $|x| \leq \epsilon$ and $|\nabla\zeta_\epsilon| \leq 2\epsilon^{-1}$. Then

$$p\int_{B_1} |\nabla u|^{p-2} \langle \nabla u, \nabla\zeta_\epsilon\rangle \zeta_\epsilon^{p-1} u dx + \int_{B_1} |\nabla u|^p \zeta_\epsilon^p dx = \int_{B_1} g(u, \nabla u)\zeta_\epsilon^p u dx.$$

Hence

$$\int_{B_1} |\nabla u|^p \zeta_\epsilon^p dx - p\left(\int_{B_1} |\nabla u|^p \zeta_\epsilon^p dx\right)^{\frac{1}{p'}} \left(\int_{B_1} |\nabla\zeta_\epsilon|^p u^p dx\right)^{\frac{1}{p}} \leq \int_{B_1} g(u, \nabla u)\zeta_\epsilon^p u dx.$$

Since

$$\left(\int_{B_1} |\nabla\zeta_\epsilon|^p u^p dx\right)^{\frac{1}{p}} \leq c_N \|u\|_{L^\infty(B_1)} \epsilon^{\frac{N}{p}-1},$$

and

$$\int_{B_1} g(u, \nabla u)\zeta_\epsilon^p u dx \leq \|u\|_{L^\infty(B_1)} \int_{B_1} g(u, \nabla u) dx,$$

we obtain that $\nabla u \in L^p_{loc}(\Omega)$. If $\phi \in C_0^\infty(\Omega)$, then

$$\int_\Omega |\nabla u|^{p-2} \langle \nabla u, \nabla\phi\rangle \zeta_\epsilon dx + p\int_\Omega |\nabla u|^{p-2} \langle \nabla u, \nabla\zeta_\epsilon\rangle \phi dx = \int_{B_1} g(u, \nabla u)\zeta_\epsilon \phi dx.$$

By the previous estimate, the second term on the left-hand side tends to 0 when $\epsilon \to 0$, thus

$$\int_\Omega |\nabla u|^{p-2} \langle \nabla u, \nabla\phi\rangle dx = \int_{B_1} g(u, \nabla u)\phi dx. \tag{6.1.21}$$

*Step 2.* Assume (6.1.20) holds for some $c > 0$. We define $\gamma > 0$ by

$$\gamma = \limsup_{|x|\to 0} \frac{u(x)}{\mu_p(x)}. \tag{6.1.22}$$

If $1 < p < N$, we set $u_\lambda(x) = \lambda^{\frac{N-p}{p-1}} u(\lambda x)$ for $0 < \lambda \leq 1$, then

$$\nabla u_\lambda(x) = \lambda^{\frac{N-1}{p-1}} \nabla u(\lambda x).$$

If we define $g_\lambda$ by

$$g_\lambda(v, \xi) = \lambda^N g(\lambda^{-\frac{N-p}{p-1}} v, \lambda^{-\frac{N-1}{p-1}} \xi) \quad \forall (v, \xi) \in \mathbb{R} \times \mathbb{R}^N,$$

there holds

$$\begin{aligned} |g_\lambda(v, \xi)| &\leq c_1 \lambda^{N - q\frac{N-p}{p-1}} |v|^q + c_2 \lambda^{N - r\frac{N-1}{p-1}} |\xi|^r + \lambda^N c_3 \\ &\leq c_1 |v|^q + c_2 |\xi|^r + c_3 \\ &\leq C(|v|)(1 + |\xi|^p), \end{aligned} \tag{6.1.23}$$

for some nondecreasing function $C$, because $0 < \lambda \leq 1$, $q \leq q_1$ and $r \leq q_c$.
Since $u_\lambda$ satisfies

$$-\Delta_p u_\lambda = g_\lambda(u_\lambda, \nabla u_\lambda), \tag{6.1.24}$$

in $\frac{1}{\lambda}\Omega \setminus \{0\}$, we can apply local regularity estimates of Section I-3: for $R > 0$, $\|u_\lambda\|_{L^\infty(B_{2R} \setminus B_{\frac{R}{2}})}$ is bounded by some constant independent of $\lambda$, hence

$$\begin{aligned} &(i) \qquad |\nabla u_\lambda(x)| \leq M && \forall x \in B_{\frac{3R}{2}} \setminus B_{\frac{2R}{3}} \\ &(ii) \qquad |\nabla u_\lambda(x) - \nabla u_\lambda(y)| \leq M\,|x - y|^\alpha && \forall x, y \in B_{\frac{3R}{2}} \setminus B_{\frac{2R}{3}}. \end{aligned} \tag{6.1.25}$$

There exist sequences $\{r_n\} \subset \mathbb{R}_+$ converging to 0 and $\{\sigma_n\} \subset S^{N-1}$ converging to $\sigma$ such that

$$\lim_{r_n \to 0} r_n^{\frac{N-p}{p-1}} u(r_n, \sigma_n) = \gamma = \lim_{r_n \to 0} u_{r_n}(1, \sigma_n). \tag{6.1.26}$$

By (6.1.25) the sequence $\{u_{r_n}\}$ is eventually relatively compact for the $C^1_{loc}$ topology of $\mathbb{R}^N \setminus \{0\}$. Up to a subsequence still denoted by $\{r_n\}$ it converges to some function $U$ which is $p$-harmonic in $\mathbb{R}^N$ because $|g_{r_n}(v, \xi)| \to 0$ by (6.1.23). Moreover $U \leq c\mu_p$ by (6.1.20). Using Corollary 1.5.9, it follows that $U$ is a multiple of $\mu_p$ and finally $U = \gamma\mu_p$ because of (6.1.23). This implies (6.1.18) in the case $1 < p < N$.

If $N = p$, we put $u_\lambda(x) = (-\ln\lambda)^{-1} u(\lambda x)$ for $0 < \lambda \leq \frac{1}{2}$; hence $\nabla u_\lambda(x) = \lambda(-\ln\lambda)^{-1} \nabla u(\lambda x)$. We define $g_\lambda$ by

$$g_\lambda(v, \xi) = \lambda^N g((-\ln\lambda)v, (-\ln\lambda)\lambda^{-1}\xi) \qquad \forall (v, \xi) \in \mathbb{R} \times \mathbb{R}^N.$$

Then

$$|g_\lambda(v,\xi)| \leq (c_1(-\ln\lambda)^q\lambda^N|v|^q + c_2(-\ln\lambda)^r\lambda^{N-r}|\xi|^r + c_3\lambda^N,$$

which implies that $|g_\lambda(v,\xi)| \to 0$ when $\lambda \to 0$, uniformly on compact subsets of $\mathbb{R}\times\mathbb{R}^N$. Since $u_\lambda$ satisfies (6.1.21), we can apply the same method as in the previous case: the set of functions $u_{r_n}$ is eventually relatively compact for the $C^1_{loc}$ topology of $\mathbb{R}^N\setminus\{0\}$, therefore it converges up to a subsequence to $\gamma\mu_N$ (here we use again (6.1.22)). Thus (6.1.18) follows. □

The supercritical case $q \geq q_1$ is much more difficult to handle and the results, although very deep, are only partial. Most of the results have been published for the first time in [Serrin, Zou (2002)].

**Theorem 6.1.3** *Assume $\Omega \subset \mathbb{R}^N$ is a proper domain, $1 < p < N$, $g : \mathbb{R}_+ \mapsto \mathbb{R}_+$ is a contiuous function and $u \in C^1(\Omega)$ is a a weak nonnegative solution of*

$$\Delta_p u + g(u) = 0 \qquad in\ \Omega. \tag{6.1.27}$$

*Then there exists a constant $c = c(N,p,q,\Lambda)$ such that*

$$u(x) \leq c\,(\rho(x))^{-\frac{p}{q+1-p}} \qquad \forall x \in \Omega, \tag{6.1.28}$$

*where $\rho(x) = \operatorname{dist}(x,\partial\Omega)$, if one of the next conditions is verified:*
*(i) There exist $C > 0$ and $p-1 < q < q_1$ such that $C^{-1}u^q \leq g(u) \leq Cu^q$ for all $u > 0$.*
*(ii) $g \in C^1((0,\infty))$ and there exist $C > 0$ and $q \in (p-1, p^*-1)$ such that*

$$C^{-1}u^q \leq g(u) \leq Cu^q \text{ and } qg(u) \geq ug'(u) \qquad \forall u > 0. \tag{6.1.29}$$

If (ii) holds, $g$ is called *subcritical* (for this equation). The first step of the proof of this result is based upon an integral estimate of the infimum of a solution in a ball due to [Bidaut-Véron, Pohozaev (2001), Lemma 2.5].

**Proposition 6.1.4** *Assume $\Omega \subset \mathbb{R}^N$ is a domain, $0 < p-1 < q$ and $u \in W^{1,p}_{loc}(\Omega) \cap L^q_{loc}(\Omega)$ is a nonnegative and satisfies*

$$-\Delta_p u \geq u^q \qquad in\ \Omega. \tag{6.1.30}$$

*Then for any $\gamma \in (0, q-1)$ there exists $c = c(N,p,q,\gamma) > 0$ such that for any $y \in \Omega$ and $R > 0$ such that $B_{2R}(y) \subset \Omega$ there holds*

$$\int_{B_R(y)} u^\gamma dx \leq cR^{N-\frac{\gamma p}{q+1-p}}. \tag{6.1.31}$$

*Furthermore, for any* $\sigma \in (0, p - \frac{p}{q})$,

$$\int_{B_R(y)} |\nabla u|^\sigma \, dx \leq c R^{N - \frac{\sigma(q+1)}{q+1-p}}. \tag{6.1.32}$$

These estimates are obtained by multiplying the inequality by $u^{-\delta}\Phi$ for some suitable $\delta > 0$ and well chosen test function $\Phi$ in a similar way as in the proof of Theorem 6.1.8 below. As a consequence of (6.1.31) we derive,

$$\inf_{x \in B_R(y)} u(x) \leq \left(\frac{c}{\omega_N}\right)^{\frac{1}{\gamma}} R^{-\frac{p}{q+1-p}}. \tag{6.1.33}$$

The second step of the proof is an Harnack inequality which is obtained by using higly technical integral identities.

**Proposition 6.1.5** *Under the structural assumptions (i) or (ii) in Theorem 6.1.3 any nonnegative solution u of (6.1.27) with* $\Omega = B_{2R_0} \setminus \{0\}$ *satisfies*

$$\sup_{x \in B_{\frac{|y|}{2}}(y)} u(x) \leq c \inf_{x \in B_{\frac{|y|}{2}}(y)} u(x) \qquad \forall y \in B_{R_0}, \tag{6.1.34}$$

*for some* $c = c(N, p, q, C) > 0$.

Combining (6.1.33) and (6.1.34) we infer (6.1.28). As a consequence

**Corollary 6.1.6** *If* $1 < p < N$, $\Omega \subset \mathbb{R}^N$ *is a domain containing* 0 *and the structural assumptions (i) or (ii) in Theorem 6.1.3 hold. Then any nonnegative weak solution* $u \in C^1(\Omega \setminus \{0\})$ *of (6.1.27) with* $\Omega$ *replaced by* $\Omega \setminus \{0\}$ *satisfies*

$$u(x) \leq c R^{-\frac{p}{q+1-p}} \qquad \forall x \in \Omega \setminus \{0\}, \tag{6.1.35}$$

*where* $R = \min\{|x|, \rho(x)\}$.

### 6.1.3 *Rigidity theorems*

The following consequence follows

**Corollary 6.1.7** *If* $1 < p < N$ *and* $p - 1 < q < p^* - 1$ *the only weak nonnegative fonction satisfying (6.1.27) in* $\mathbb{R}^N$ *is the zero function.*

*Proof.* It sufficies to apply Theorem 6.1.3 in $B_R$ and to let $R \to \infty$. □

We give below a simple proof of non-existence for a more general class of operators but with a reaction term with a weaker growth. The results are due to [Bidaut-Véron (1989)] and [Bidaut-Véron, Pohozaev (2001)]. They are based upon integral estimates and do not necessitate to differentiate the equation as Theorem 6.1.3 does.

**Theorem 6.1.8** *Assume $A : \mathbb{R}^N \times \mathbb{R}^N \mapsto \mathbb{R}^N$ satisfies the structural assumptions (4.1.1) with $1 < p \leq N$ and $g : \mathbb{R}^N \times \mathbb{R} \times \mathbb{R}^N \mapsto \mathbb{R}$ is a Caratheodory function which satisfies (6.0.3) and*

$$g(x, r, \xi) \geq c_1 r^q \qquad \forall (x, r, \xi) \in \mathbb{R}^N \times \mathbb{R} \times \mathbb{R}^N, \tag{6.1.36}$$

*for some $c_1 > 0$ and $p - 1 < q$.*

*(i) If $1 < p \leq N$ and $p - 1 < q < q_1$, there exists no nonnegative nontrivial weak solution to inequality*

$$-\text{div} A(x, \nabla u) \geq g(x, u, \nabla u), \tag{6.1.37}$$

*in $\mathbb{R}^N$.*

*(ii) If $A(x, \xi) = a(|\xi|)\xi$, $1 < p \leq N$ and $p - 1 < q \leq q_1$ (any $q > p - 1$ in $p = N$), there exists no nonnegative nontrivial weak solution to (6.1.37) in $\mathbb{R}^N \setminus B_\tau$ for $\tau > 0$.*

*Proof.* Assume that the assumptions of (i) hold, let $\alpha \in (1 - p, 0)$, $\ell, \epsilon > 0$ and $\zeta := \zeta_\tau \in C_0^\infty(\mathbb{R}^N)$, $0 \leq \tau \leq 1$, $\zeta = 1$ on $B_\tau$, $\zeta = 0$ on $B_{2\tau}^c$, $|\nabla \zeta| \leq 2\tau^{-1}$ for some $\tau > 0$. We set $u_\epsilon = u + \epsilon$ and take $u_\epsilon^\alpha \zeta^\ell$ as test function, then

$$\begin{aligned} \alpha \int_{\mathbb{R}^N} |\nabla u_\epsilon|^p \, u_\epsilon^{\alpha-1} \zeta^\ell dx + \ell \int_{\mathbb{R}^N} |\nabla u_\epsilon|^{p-2} \, \langle \nabla u_\epsilon, \nabla \zeta \rangle u_\epsilon^\alpha \zeta^{\ell-1} dx \\ \geq \int_{\mathbb{R}^N} g(x, u, \nabla u_\epsilon) u_\epsilon^\alpha \zeta^\ell dx. \end{aligned}$$

Next

$$\begin{aligned} \ell & \left| \int_{\mathbb{R}^N} |\nabla u_\epsilon|^{p-2} \, \langle \nabla u_\epsilon, \nabla \zeta \rangle u_\epsilon^\alpha \zeta^{\ell-1} dx \right| \\ & \leq \left( \int_{\mathbb{R}^N} |\nabla u_\epsilon|^p \, u_\epsilon^{\alpha-1} \zeta^\ell dx \right)^{\frac{1}{p'}} \left( \int_{\mathbb{R}^N} |\nabla \zeta|^p \, u_\epsilon^{\alpha-1+p} \zeta^{\ell-p} dx \right)^{\frac{1}{p}} \\ & \leq \frac{|\alpha|}{2} \int_{\mathbb{R}^N} |\nabla u_\epsilon|^p \, u_\epsilon^{\alpha-1} \zeta^\ell dx + C(\alpha, \ell) \int_{\mathbb{R}^N} |\nabla \zeta|^p \, u_\epsilon^{\alpha-1+p} \zeta^{\ell-p} dx. \end{aligned}$$

Therefore

$$\frac{|\alpha|}{2}\int_{\mathbb{R}^N}|\nabla u_\epsilon|^p u_\epsilon^{\alpha-1}\zeta^\ell dx+\int_{\mathbb{R}^N}g(x,u,\nabla u_\epsilon)u_\epsilon^\alpha\zeta^\ell dx \leq C(\alpha,\ell)\int_{\mathbb{R}^N}|\nabla\zeta|^p u_\epsilon^{\alpha-1+p}\zeta^{\ell-p}dx. \tag{6.1.38}$$

Next

$$\begin{aligned}&\int_{\mathbb{R}^N}|\nabla\zeta|^p u_\epsilon^{\alpha-1+p}\zeta^{\ell-p}dx\\ &\leq\left(\int_{\mathbb{R}^N}u_\epsilon^{\alpha+q}\zeta^\ell dx\right)^{\frac{\alpha+p-1}{\alpha+q}}\left(\int_{\mathbb{R}^N}|\nabla\zeta|^{\frac{p(\alpha+q)}{q+1-p}}\zeta^{\ell-\frac{p(\alpha+q)}{q+1-p}}dx\right)^{\frac{q+1-p}{\alpha+q}}\\ &\leq c_2\rho^{\frac{N(q+1-p)}{\alpha+q}-p}\left(\int_{\mathbb{R}^N}u_\epsilon^{\alpha+q}\zeta^\ell dx\right)^{\frac{\alpha+p-1}{\alpha+q}},\end{aligned} \tag{6.1.39}$$

where $c_2=c_2(N)>0$, provided $\ell\geq\frac{p(\alpha+q)}{q+1-p}$. Since we have assume $p-1<q<q_1$ it is possible to choose $\alpha\in[1-p,0)$ such that $\frac{N(q+1-p)}{\alpha+q}-p<0$. Combining (6.1.35), (6.1.38) and (6.1.39) we obtain

$$\int_{\mathbb{R}^N}u_\epsilon^{\alpha+q}\zeta^\ell dx\leq c_3\rho^{N-\frac{p(\alpha+q)}{q+1-p}}. \tag{6.1.40}$$

Since the exponent of $\rho$ is negative, we let successively $\epsilon\to 0$ and $\rho\to\infty$ and derive

$$\int_{\mathbb{R}^N}u^{\alpha+q}dx=0,$$

which proves (i).

The assumption on $A$ implies $|\xi|^{p-2}\leq a(\xi)\leq\lambda|\xi|^{p-2}$ and that it admits a radial form, namely if $u(x)=u(|x|)=u(r)$, then

$$-\mathrm{div}A(x,\nabla u)=-r^{1-N}\left(r^{N-1}a(|u_r|)u_r\right)_r. \tag{6.1.41}$$

*Step 1.* There is no nontrivial radial solution in $\mathbb{R}^N\setminus B_\tau$. Assuming that such a solution $v$ exists, then

$$-r^{1-N}\left(r^{N-1}a(|v_r|)v_r\right)_r\geq c_1v^q. \tag{6.1.42}$$

Because of (6.1.35) there exists $r_0>\tau$ such that $v_r(r_0)\leq 0$, hence

$$-r^{N-1}\left(a(|v_r|)v_r\right)(r)\geq -r_0^{N-1}\left(a(|v_r|)v_r\right)(r_0)+\int_{r_0}^r v^q(s)s^{N-1}ds.$$

Furthermore, if $v$ is not identically zero for $r \geq r_1 > r_0$, there exists $\gamma > 0$ such that

$$-r^{N-1}\left(a(|v_r|)v_r\right)(r) \geq \gamma \qquad \forall r \geq r_1.$$

Since (4.1.1) holds, this implies $v_r < 0$. Because $v(r) \to 0$ when $r \to \infty$ by (6.1.35), we get

$$v(r) \geq \left(\frac{\gamma}{\lambda}\right)^{\frac{1}{p-1}} \int_r^\infty s^{\frac{1-N}{p-1}} ds = \gamma_1 r^{\frac{p-N}{p-1}},$$

for some $\gamma_1 > 0$ and thus $v^q \geq \gamma_2 r^{\frac{q(p-N)}{p-1}}$. Since $q \leq q_1$,

$$N - 1 + \frac{q(p-N)}{p-1} = N\left(1 - \frac{q}{q_1}\right) - 1 \geq -1.$$

If $q = q_1$ we get

$$-v_r(r) \geq \gamma_3 r^{\frac{1-N}{p-1}} (\ln r)^{\frac{1}{p-1}} \Longrightarrow v(r) \geq \gamma_4 r^{\frac{p-N}{p-1}} (\ln r)^{\frac{1}{p-1}} \quad \text{as } r \to \infty,$$

which contradicts (6.1.35).
If $p - 1 < q < q_1$, we have

$$-v_r(r) \geq \gamma_5 r^{\frac{1-N}{p-1}+\epsilon} \Longrightarrow v(r) \geq \gamma_6 r^m \quad \text{as } r \to \infty$$

where

$$m = \frac{p}{p-1} - \frac{q(N-p)}{(p-1)^2} > \frac{p}{q+1-p},$$

which again contradicts (6.1.35), which is the claim.
*Step 2.* End of the proof. We suppose by contradiction that there exists a nontrivial nonnegative solution $u$. By the strong maximum principle we can assume that $\operatorname{ess\,min} u = \nu > 0$ on $\{|x| = \tau\}$. Let $k > \tau$, we construct a sequence $\{v_{n,k}\}$ of radial functions by

$$\begin{aligned} -\mathrm{div} A(x, \nabla v_{n,k}) &= c_1 v_{n-1,k}^q && \text{in } B_k \setminus B_\tau \\ v_{n,k}(x) &= \nu && \text{on } \partial B_\tau \\ v_{n,k}(x) &= 0 && \text{on } \partial B_k, \end{aligned} \tag{6.1.43}$$

with $v_0 = 0$. The sequence $\{v_{n,k}\}$ is increasing both in $n$ and $k$. Furthermore $v_{n,k} \leq u$ in $B_k \setminus B_\tau$. Letting successively $n \to \infty$ and $k \to \infty$

and using regularity results of Theorem 1.3.9, we infer that there exists a positive radial solutions of

$$\begin{aligned} -\mathrm{div}A(x,\nabla v) &= c_1 v^q && \text{in } \mathbb{R}^N \setminus B_\tau \\ v(x) &= \nu && \text{on } \partial B_\tau \end{aligned} \tag{6.1.44}$$

which satisfies $v \leq u$ in $\mathbb{R}^N \setminus B_\tau$, which contradicts Step 1. □

*Remark.* Corollary 6.1.7 is clearly not valid if $q \geq p^* - 1$. If one consider the differential equation

$$\begin{aligned} &\left(r^{N-1}\left|u_r\right|^{p-2}u_r\right)_r + r^{N-1}u^q = 0 \qquad \text{on } (0, r_{\max}) \\ &u_r(0) = 0\,,\ u(0) = a > 0, \end{aligned} \tag{6.1.45}$$

then $r_{\max} = \infty$, $r \mapsto u(r)$ is decreasing from $(0,\infty)$ to $(a,0)$ and $u(r) \to 0$ when $r \to \infty$. The fact that $u$ cannot vanish follows from a Pohozaev type identity. In the particular case where $q = p^* - 1$ the radial solutions are explicit, depending on two parameters, $\epsilon > 0$ and $x_0 \in \mathbb{R}^N$

$$u(x) = u_{\epsilon,x_0}(x) = \left(\frac{\epsilon^{\frac{1}{p-1}}N^{\frac{1}{p}}\left(\frac{N-p}{p-1}\right)^{\frac{p-1}{p}}}{\epsilon^{\frac{p}{p-1}} + |x-x_0|^{\frac{p}{p-1}}}\right)^{\frac{N-p}{p}} \qquad \forall x \in \mathbb{R}^N. \tag{6.1.46}$$

It is remarkable that this class of solutions describes the whole set of positive solutions of

$$-\Delta_p u = u^{p^*-1} \qquad \text{in } \mathbb{R}^N, \tag{6.1.47}$$

which belong to the space $\mathcal{D}^{1,p}(\mathbb{R}^N)$

$$\mathcal{D}^{1,p}(\mathbb{R}^N) = \left\{v \in L^{p^*}(\mathbb{R}^N) : \int_{\mathbb{R}^N} |\nabla v|^p\,dx < \infty\right\}. \tag{6.1.48}$$

The following result is proved in [Damascelli, Merchán, Montoro, Sciunzi (2014)] when $p \geq 2$ and [Sciunzi (2016)] when $p < 2$.

**Theorem 6.1.9** *Let $N \geq 2$ and $\frac{2N}{N+2} < p < N$. Then any positive solution $u \in \mathcal{D}^{1,p}(\mathbb{R}^N)$ of (6.1.47) is radial and given by (6.1.46).*

Their proofs are based upon a delicate adaptation of the moving plane method developed by [Gidas, Ni, Nirenberg (1979)]. In the case $p = 2$, the assumption $u \in \mathcal{D}^{1,2}(\mathbb{R}^N)$ is unnecessary, see [Gidas, Spruck (1980)].

The following nonexistence result of positive solution in a half-space is proven in [Gidas, Spruck (1980)] in the case $p = 2$ and in the general case $1 < p < N$ in [Zou (2008)].

**Theorem 6.1.10** *Let $1 < p < N$, $p - 1 < q < p^* - 1$ and $g \in C(\mathbb{R}_+) \cap C^1((0,\infty))$ satisfies for some $C > 0$ and $\lambda \in (0, p^* - 1)$,*

$$C^{-1}u^q \leq g(u) \leq Cu^q \text{ and } \lambda g(u) \geq ug'(u) \qquad \forall u > 0. \tag{6.1.49}$$

*Then there exists no nontrivial nonnegative solution to*

$$\begin{aligned} -\Delta_p u &= g(u) && \text{in } \mathbb{R}^N_+ \\ u &= 0 && \text{on } \partial\mathbb{R}^N_+. \end{aligned} \tag{6.1.50}$$

The proof involves several intermediate results which have their own interest. For their proofs we introduce some notations. We set

$$\mathbb{R}^N_+ = \{x = (x', x_N) = (x_1, ..., x_N) : x_N > 0\}, \; \partial\mathbb{R}^N_+ \sim \mathbb{R}^{N-1} = \{x = (x', 0)\},$$

$$B'_r = B_r \cap \partial\mathbb{R}^N_+, \; S'_r = \partial B_r \cap \partial\mathbb{R}^N_+ = \partial' B'_r \text{ and } \Sigma_r = S'_r \times (0, \infty),$$

and $D_{r,s} = \{x = (x', x_N) : |x'| < r, x_N = s\} = B'_r \times \{s\}$.

**Lemma 6.1.11** *Let $u \in C^{1,\alpha}(\overline{\mathbb{R}}^N_+)$ be a solution of (6.1.50), then*

$$\int_{B'_r} |\nabla u(x', 0)|^p \, dx' = \frac{p}{p-1} \int_{\Sigma_r} |\nabla u(x', x_N)|^{p-2} \frac{\partial u}{\partial \mathbf{n}'} \frac{\partial u}{\partial x_N} dS dx_N, \tag{6.1.51}$$

*where $\mathbf{n}'$ is the normal outward unit vector to $\Sigma_r$.*

*Proof.* We first notice that if there exists a nonnegative solution $u$ of (6.1.50), there holds for $x_N > 0$,

$$u(x', x_N) \leq c_1 x_N^{-\frac{p}{q+1-p}}, \tag{6.1.52}$$

by (6.1.28) and

$$|\nabla u(x', x_N)| \leq c_2 x_N^{-\frac{q+1}{q+1-p}}, \tag{6.1.53}$$

by Lemma 3.3.2, for some $c_1, c_2 > 0$ depending on $N, p, q, C$. Furthermore, we write (6.1.50) under the form

$$-\Delta_p u - V(x)u^{p-1} = 0, \tag{6.1.54}$$

where

$$V(x) = \begin{cases} u^{p-1-q}g(u) & \text{if } u > 0 \\ 0 & \text{if } u = 0. \end{cases}$$

Then for any $\xi \in \mathbb{R}^N_+$, $\xi_N > 0$, there holds

$$\max\{|V(x)| : x \in B_{\frac{2|\xi|}{3}}(\xi)\} \leq c_3 |\xi|^{-p},$$

where $c_3 = c_3(N, p, q, C) > 0$. Therefore, by Harnack inequality (see Chapter 1),

$$\max_{x \in B_{\frac{|\xi|}{2}}(\xi)} u(x) \leq c_4 \min_{x \in B_{\frac{|\xi|}{2}}(\xi)} u(x), \tag{6.1.55}$$

where $c_4 = c_4(N, p, q, C) > 0$. For $0 < \epsilon < s$ set $\Theta := \Theta_{r,\epsilon,s} = B'_r \times (\epsilon, s)$, hence

$$\begin{aligned}
\int_\Theta \frac{\partial u}{\partial x_N} \Delta_p u dx &= \int_{\partial\Theta} |\nabla u|^{p-2} \frac{\partial u}{\partial \mathbf{n}'} \frac{\partial u}{\partial x_N} dS - \int_\Theta |\nabla u|^{p-2} \langle \nabla u, \frac{\partial \nabla u}{\partial x_N} \rangle dx \\
&= \int_{\partial\Theta} |\nabla u|^{p-2} \frac{\partial u}{\partial \mathbf{n}'} \frac{\partial u}{\partial x_N} dS - \frac{1}{p} \int_\Theta \frac{\partial |\nabla u|^p}{\partial x_N} dx \\
&= \int_{\partial\Theta} |\nabla u|^{p-2} \frac{\partial u}{\partial \mathbf{n}'} \frac{\partial u}{\partial x_N} dS + \frac{1}{p} \int_{B'_r} (|\nabla u|^p (x', \epsilon) - |\nabla u|^p (x', s)) dx'.
\end{aligned}$$

The previous estimate needs to differentiate $u$ twice, but the computation can be easily justified by replacing $\frac{\partial u}{\partial x_N}$ by $h^{-1}(u(x', x_N + h) - u(x', x_N))$ for $h > 0$ and performing an Abel's transform. By (6.1.53)

$$\lim_{s \to \infty} \frac{1}{p} \int_{B'_r} |\nabla u|^p (x', s) dx' = 0.$$

Moreover

$$\lim_{\epsilon \to 0} |\nabla u|^p (x', \epsilon) = |\nabla u|^p (x', 0), \quad \text{uniformly in } B'_r,$$

$$\lim_{\epsilon \to 0} |\nabla u|^{p-2} \frac{\partial u}{\partial \mathbf{n}'} (x', \epsilon) = |\nabla u|^p (x', 0), \quad \text{uniformly in } B'_r,$$

since $\nabla u = (0, \frac{\partial u}{\partial x_N})$ and $\mathbf{n}' = -\mathbf{e}_N$ on $B'_r \times \{0\}$. In turn this yields

$$\begin{aligned}
\lim_{\epsilon \to 0} \lim_{s \to \infty} \int_{\partial\Theta} |\nabla u|^{p-2} \frac{\partial u}{\partial \mathbf{n}'} \frac{\partial u}{\partial x_N} dS + \frac{1}{p} \int_{B'_r} (|\nabla u|^p (x', \epsilon) - |\nabla u|^p (x', s)) dx' \\
= \frac{1-p}{p} \int_{B'_r} |\nabla u|^p (x', 0) dx' + \int_{\Sigma_r} |\nabla u|^{p-2} \frac{\partial u}{\partial \mathbf{n}'} \frac{\partial u}{\partial x_N} dS dx_N,
\end{aligned}$$

and

$$\begin{aligned}\lim_{\epsilon\to 0}\lim_{s\to\infty}\int_{\Theta}\frac{\partial u}{\partial x_N}\Delta_p u dx \\ = \frac{1-p}{p}\int_{B'_r}|\nabla u|^p(x',0)dx' + \int_{\Sigma_r}|\nabla u|^{p-2}\frac{\partial u}{\partial \mathbf{n}'}\frac{\partial u}{\partial x_N}dS dx_N.\end{aligned} \tag{6.1.56}$$

Set

$$G(u) = \int_0^u g(t)dt,$$

then $\lim_{\epsilon\to 0} G(u(x',\epsilon)) = 0$ uniformly on compact subsets of $\mathbb{R}^{N-1}$. Because

$$\begin{aligned}\int_{\Theta}\frac{\partial u}{\partial x_N}\Delta_p u dx &= -\int_{\Theta}\frac{\partial u}{\partial x_N}g(u)dx \\ &= \int_{B'_r}\left(G(u(x',\epsilon)) - G(u(x',s))\right)dx',\end{aligned}$$

which converges to 0 when $\epsilon\to 0$ and $s\to\infty$, we infer (6.1.51). □

**Lemma 6.1.12** *Under the assumptions of Lemma 6.1.11 there exists* $c = c(N,p,q,C) > 0$ *such that*

$$u(x',x_N) \leq c\left(\frac{\partial u}{\partial x_N}(x',0)\right)^{\frac{p}{q+1}} \qquad \forall x_N > 0, \tag{6.1.57}$$

*and*

$$u(x',x_N) \leq c x_N\frac{\partial u}{\partial x_N}(x',0) \qquad \forall x_N > 0. \tag{6.1.58}$$

*Proof.* Let $x_0 = (x'_0, x_{0,N}) \in \mathbb{R}^N_+$, $x_{0,N} > 0$; put $M = u(x_0) > 0$ and define $w$ by

$$w(x) = M^{-1}u\left((x'_0,0) + M^{\theta}x\right) \quad \forall x \in \mathbb{R}^N_+,$$

for some $\theta$ to be chosen later on. Then

$$\begin{aligned} -\Delta_p w - M^{1-p+\theta p}g(Mw) &= 0 \qquad \text{in } \mathbb{R}^N_+ \\ w &= 0 \qquad \text{in } \partial\mathbb{R}^N_+.\end{aligned}$$

We denote $\tilde{V}(x) = M^{1-p+\theta p}g(Mw)w^{1-p}$, clearly

$$0 \leq \tilde{V}(x) \leq Cc_1^{q+1-p}x_N^{-p}.$$

Set $\xi = (0, M^{-\theta}x_{0,N})$, $B^1 = B_{|\xi|}(\xi)$ and $B^2 = B_{\frac{|\xi|}{2}}(\xi)$, then $w(\xi) = 1$ and by Harnack inequality as in (6.1.55),

$$w(x) \geq c_5 w(\xi) = c_5 \qquad \forall x \in B^2.$$

Define

$$\tilde{w}(x) = 2^{-\ell}c_5 \left|\xi\right|^\ell \left(\left|x-\xi\right|^{-\ell} - \left|\xi\right|^{-\ell}\right) \quad \text{with } \ell = \frac{N-p}{p-1},$$

then $\tilde{w}$ is $p$-harmonic in $\mathbb{R}^N \setminus \{\xi\}$, it vanishes on $\partial B^1$ and

$$\tilde{w}(x) = c_5 \left(1 - 2^{-\ell}) \left|\xi\right|^{-\ell}\right) < c_5 \leq \min_{z\in\partial B^2} w(z) \quad \forall x \in \partial B^2.$$

Since

$$-\Delta_p w + \Delta_p \tilde{w} = M^{1-p+\theta p} g(Mw) > 0 \qquad \text{in } B^1 \setminus B^2,$$

it follows from the comparison principle that $w \geq \tilde{w}$. Therefore

$$\begin{aligned} M^{-1+\theta}\frac{\partial u}{\partial x_N}(x_0',0) &= \frac{\partial w}{\partial x_N}(0) \\ &\geq \frac{\partial \tilde{w}}{\partial x_N}(0) = \ell 2^{-\ell}c_5 \left|\xi\right|^{-1} = c_6 \left|\xi\right|^{-1}, \end{aligned}$$

which yields

$$\frac{\partial u}{\partial x_N}(x_0',0) \geq c_6 M^{1-\theta} \left|\xi\right|^{-1} = c_6 M x_{0,N}^{-1}, \tag{6.1.59}$$

since $|\xi| = M^{-\theta}x_{0,N}$. This implies (6.1.58). Because $x_{0,N}^{-1} \geq c_1^{\frac{q+1-p}{p}} M^{\frac{q+1-p}{p}}$ we derive finally

$$\frac{\partial u}{\partial x_N}(x_0',0) \geq c_6 c_1^{\frac{q+1-p}{p}} M^{\frac{q+1}{p}}, \tag{6.1.60}$$

which is (6.1.57). □

The reverse pointwise estimate of $\nabla u(x)$ with respect to $u(x)$ and the distance of $x$ to $\partial\mathbb{R}_+^N$ is also the consequence of Harnack inequality combined with the local regularity theory Theorem 1.3.9.

**Lemma 6.1.13** *Under the assumptions of Lemma 6.1.11 there holds*

$$|\nabla u(x', x_N)| \leq c x_N^{-1} u(x', x_N) \qquad \forall x_N > 0, \tag{6.1.61}$$

*for some* $c = c(N, p, q, C) > 0$.

*Proof.* Let $\xi = (\xi', \xi_0) \in \mathbb{R}^N_+$, $\xi_0 > 0$. For $r > 0$ we denote by $M$ the supremum of $u$ on $B_{\frac{\xi_0}{2}}(\xi)$. By (6.1.52) and Harnack inequality (6.1.55),

$$M \leq c_1 2^{\frac{p}{q+1-p}} \xi_0^{-\frac{p}{q+1-p}} \text{ and } u(\xi) \geq \inf_{B_{\frac{\xi_0}{2}}(\xi)} u(x) \geq c_4^{-1} M. \tag{6.1.62}$$

If we set $w(x) = M^{-1} u(\xi + \xi_0(x - \xi))$, then $\xi + \xi_0(x - \xi) \in B_{\xi_0}(\xi)$ if and only if $x \in B_1(\xi)$. Furthermore

$$\Delta_p w + M^{1-p} \xi_0^p g(Mw) = 0 \quad \text{in } B_1(\xi).$$

Clearly $0 < w \leq 1$ in $B_{\frac{1}{2}}(\xi)$, where, by (6.1.52), there also holds

$$0 < M^{1-p} \xi_0^p g(Mw(x)) \leq C M^{q+1-p} \xi_0^p w^q(x) \leq c_7 w^q(x) \leq c_7.$$

By the gradient estimates Theorem 1.3.9, it implies

$$\sup_{x \in B_{\frac{\xi_0}{4}}(\xi)} M^{-1} \xi_0 \left| \nabla u(\xi + \xi_0(x - \xi)) \right| = \sup_{x \in B_{\frac{1}{4}}(\xi)} |\nabla w(x)| = c_8. \tag{6.1.63}$$

Combining (6.1.62) and (6.1.63) we derive (6.1.61). □

**Lemma 6.1.14** *Under the assumptions of Lemma 6.1.11, for any $\gamma \in (0,1)$ and $\epsilon \in (0, p-1)$, there exists $c_1 = c_1(N, p, q, C, \gamma, \epsilon)$ such that*

$$\int_{\Sigma_r} |\nabla u|^p \, dS dx_N \leq c_1 \int_{S_r} |\nabla u(x', 0)|^{p_\gamma} \, dS + c_1 \int_{S_r} |\nabla u(x', 0)|^{p_\epsilon} \, dS, \tag{6.1.64}$$

*where*

$$p_\gamma := p - 1 + \gamma + \frac{(1-\gamma)p}{q+1} > p_\epsilon := p - 1 - \epsilon + \frac{(1+\epsilon)p}{q+1}. \tag{6.1.65}$$

*Proof.* By Fubini's theorem

$$\begin{aligned} \int_{\Sigma_r} |\nabla u|^p \, dS dx_N &= \int_0^1 \left( \int_{S_r} |\nabla u|^p \, dS \right) dx_N \\ &\quad + \int_1^\infty \left( \int_{S_r} |\nabla u|^p \, dS \right) dx_N \\ &= I + II. \end{aligned} \tag{6.1.66}$$

By Lemma 6.1.13

$$I \leq c_2^p \int_0^1 x_N^{-p} \int_{S_r} u^p dS dx_N \quad \text{and} \quad II \leq c_2^p \int_1^\infty x_N^{-p} \int_{S_r} u^p dS dx_N.$$

Using (6.1.57) and (6.1.58) for any $\gamma \in (0,1)$,

$$(i) \qquad u^{1-\gamma}(x', x_N) \leq c_3 \left( \frac{\partial u}{\partial x_N}(x', 0) \right)^{\frac{p(1-\gamma)}{q+1}},$$

$$(ii) \qquad u^{p-1+\gamma}(x', x_N) \leq c_4 \left( x_N \frac{\partial u}{\partial x_N}(x', 0) \right)^{p-1+\gamma}.$$

This yields

$$\begin{aligned} x_N^{-p} u^p(x', x_N) &\leq c_5 x_N^{-p} \left( \frac{\partial u}{\partial x_N}(x', 0) \right)^{\frac{p(1-\gamma)}{q+1}} \left( x_N \frac{\partial u}{\partial x_N}(x', 0) \right)^{p-1+\gamma} \\ &\leq c_5 x_N^{\gamma-1} \left( \frac{\partial u}{\partial x_N}(x', 0) \right)^{p_\gamma}, \end{aligned}$$

and

$$I \leq c_6 \int_{S_r} |\nabla u(x', 0)|^{p_\gamma} \, dS. \tag{6.1.67}$$

In the same way

$$II \leq c_7 \int_{S_r} |\nabla u(x', 0)|^{p_\epsilon} \, dS, \tag{6.1.68}$$

which implies (6.1.64). □

*Proof of Theorem 6.1.10.* Since $p - 1 < q < p^* - 1$,

$$p > p - 1 + \frac{p}{q+1} > \frac{p(N-1)}{N} > \frac{p(N-2)}{N-1}.$$

Therefore there exist $\gamma \in (0,1)$ and $\epsilon \in (0, p-1)$ such that

$$\begin{aligned} &(i) \qquad p > p_\gamma := p - 1 + \gamma + \frac{p(1-\gamma)}{q+1} > \frac{p(N-2)}{N-1}, \\ &(ii) \qquad p > p_\epsilon := p - 1 - \epsilon + \frac{p(1+\epsilon)}{q+1} > \frac{p(N-2)}{N-1}. \end{aligned} \tag{6.1.69}$$

We first assume that $N > 2$ and define, for $r > 0$,

$$\Psi(r) = \int_{B'_r} |\nabla u(x', 0)|^p \, dx';$$

then

$$\Psi'(r) = \int_{S'_r} |\nabla u(x',0)|^p \, dS.$$

By Hölder's inequality,

$$\int_{S'_r} |\nabla u(x',0)|^{p_\gamma} \, dS \le c_1 r^{p^\gamma} \left(\Psi'(r)\right)^{\frac{p_\gamma}{p}},$$

and

$$\int_{S'_r} |\nabla u(x',0)|^{p_\epsilon} \, dS \le c_1 r^{p^\epsilon} \left(\Psi'(r)\right)^{\frac{p_\epsilon}{p}},$$

where

$$p^\gamma = \frac{(N-2)(p-p_\gamma)}{p} > 0 \text{ and } p^\epsilon = \frac{(N-2)(p-p_\epsilon)}{p} > 0.$$

By (6.1.64)

$$\int_{\Sigma_r} |\nabla u|^p \, dS dx_N \le c_2 \left( r^{p^\gamma} \left(\Psi'(r)\right)^{\frac{p_\gamma}{p}} + r^{p^\epsilon} \left(\Psi'(r)\right)^{\frac{p_\epsilon}{p}} \right). \tag{6.1.70}$$

Combined with (6.1.51) we get

$$\begin{aligned} \Psi(r) &\le \frac{p}{p-1} \int_{\Sigma_r} |\nabla u|^p \, dS dx_N \\ &\le c_3 \left( r^{p^\gamma} \left(\Psi'(r)\right)^{\frac{p_\gamma}{p}} + r^{p^\epsilon} \left(\Psi'(r)\right)^{\frac{p_\epsilon}{p}} \right). \end{aligned} \tag{6.1.71}$$

Let $r \ge 1$; since $p^\gamma < p^\epsilon$ we get,

$$\begin{aligned} 1 &\le c_3 \left( r^{p^\gamma} \left(\Psi'(r)\right)^{\frac{p_\gamma}{p}} \left(\Psi(r)\right)^{-1} + r^{p^\epsilon} \left(\Psi'(r)\right)^{\frac{p_\epsilon}{p}} \left(\Psi(r)\right)^{-1} \right) \\ &\le \frac{1}{4} + c_4 r^{\frac{pp^\gamma}{p_\gamma}} \Psi'(r) (\Psi(r))^{-\frac{p}{p_\gamma}} + \frac{1}{4} + c_5 r^{\frac{pp^\epsilon}{p_\epsilon}} \Psi'(r) (\Psi(r))^{-\frac{p}{p_\epsilon}} \\ &\le \frac{1}{2} + c_5 r^{\frac{pp^\epsilon}{p_\epsilon}} \left( (\Psi(r))^{-\frac{p}{p_\gamma}} + (\Psi(r))^{-\frac{p}{p_\epsilon}} \right) \Psi'(r), \end{aligned}$$

which yields

$$r^{-\frac{pp^\epsilon}{p_\epsilon}} \le c_6 \left( (\Psi(r))^{-\frac{p}{p_\gamma}} + (\Psi(r))^{-\frac{p}{p_\epsilon}} \right) \Psi'(r).$$

Now

$$\frac{pp^\epsilon}{p_\epsilon} = \frac{(N-2)(p-p_\epsilon)}{p} \frac{p}{p_\epsilon} = (N-2)\left(\frac{p}{p_\epsilon} - 1\right) < 1,$$

by (6.1.69) (ii). Integrating on $(1, r)$ we obtain

$$\frac{p_\epsilon}{p_\epsilon - pp^\epsilon}\left(r^{1-\frac{pp^\epsilon}{p_\epsilon}} - 1\right) \leq c_7\left((\Psi(1))^{1-\frac{p}{p_\gamma}} + (\Psi(1))^{1-\frac{p}{p_\epsilon}}\right),$$

which leads to an impossibility if $r \to \infty$.

In the case $N = 2$, $x = (x_1, x_2)$ $x_2 > 0$, $B'_r = (-r, r)$ and $S'_r = \{-r, r\}$. Then

$$\Psi(r) = \int_{-r}^{r} |\nabla u(x', 0)|^p \, dx' = \int_{-r}^{r} \left|\frac{\partial u}{\partial x_2}(x_1, 0)\right|^p dx_1,$$

and

$$\Psi'(r) = \left|\frac{\partial u}{\partial x_2}(r, 0)\right|^p + \left|\frac{\partial u}{\partial x_2}(-r, 0)\right|^p.$$

Therefore (6.1.64) takes the form

$$\begin{aligned}\int_{\Sigma_r} |\nabla u|^p \, dx_2 &= \int_0^\infty (|\nabla u(r, x_2)|^p + |\nabla u(-r, x_2)|^p) \, dx_2 \\ &\leq c_8\left(\left|\frac{\partial u}{\partial x_2}(r, 0)\right|^{p_\gamma} + \left|\frac{\partial u}{\partial x_2}(-r, 0)\right|^{p_\gamma}\right) \\ &\quad + c_8\left(\left|\frac{\partial u}{\partial x_2}(r, 0)\right|^{p_\epsilon} + \left|\frac{\partial u}{\partial x_2}(-r, 0)\right|^{p_\epsilon}\right) \\ &\leq c_{10}\left((\Psi'(r))^{p_\gamma} + (\Psi'(r))^{p_\epsilon}\right).\end{aligned} \tag{6.1.72}$$

Using Lemma 6.1.11 we arrive to

$$\Psi(r) \leq \frac{pc_{10}}{p-1}\left((\Psi'(r))^{p_\gamma} + (\Psi'(r))^{p_\epsilon}\right). \tag{6.1.73}$$

By Young's inequality we infer that there holds, for $r > 1$,

$$r \leq 1 + c_{11}\left((\Psi(1))^{1-\frac{p}{p_\gamma}} + (\Psi(1))^{1-\frac{p}{p_\epsilon}}\right), \tag{6.1.74}$$

which yields a contradiction. □

The subcriticality assumption on the reaction term can be removed provided some global regularity assumption is made. The next results which point out this fact have been proved in [Farina, Montoro, Sciunzi (2012)], [Farina, Montoro, Sciunzi (2017)], [Farina, Montoro, Riey, Sciunzi (2017)].

**Theorem 6.1.15** *Assume* $1 < p < 2$ *and* $g$ *is a locally Lipschitz continuous. If* $u \in C(\overline{\mathbb{R}^N_+}) \cap W^{1,\infty}(\mathbb{R}^N_+)$ *is a nonnegative solution of (6.1.50), then it is identically zero if one of the following holds:*

*(i)* $N = 2$ *and* $sg(s) > 0$ *if* $s \neq 0$.

*(ii)* $N \geq 3$, $sg(s) > 0$ *if* $s \neq 0$ *and* $g(s) \leq cs^{\frac{(N-1)p}{N-1-p}} + d$ *for some positive constants* $c, d$.

*(iii)* $N \geq 3$, $sg(s) > 0$ *if* $s \neq 0$ *and* $g(s) \geq cs^{\frac{(N-1)(p-1)}{N-1-p}}$ *on* $[0, \delta]$ *for some positive constants* $c, \delta$.

**Theorem 6.1.16** *Assume* $p > 2$. *Any nonnegative function* $u \in C(\overline{\mathbb{R}_+^N})$ *such that* $\nabla u \in L^\infty(\mathbb{R}_+^N)$ *satisfying*

$$\begin{aligned} -\Delta_p u &= u^q \qquad && \text{in } \mathbb{R}_+^N \\ u &= 0 && \text{in } \partial\mathbb{R}_+^N, \end{aligned} \tag{6.1.75}$$

*is identically zero if one of the next conditions holds:*
*(i)* $p - 1 < q < \infty$ *if* $N \leq \frac{p(p+3)}{p-1}$.
*(ii)* $p - 1 < q < q_{N,p}^*$ *if* $N > \frac{p(p+3)}{p-1}$, *where*

$$q_{N,p}^* = \frac{((p-1)N - p)^2 + p^2\left((p-2) - N(p-1) + 2\sqrt{(p-1)(N-1)}\right)}{(N-p)\left((p-1)N - p(p+3)\right)}. \tag{6.1.76}$$

*Furthermore, if we suppose that* $u \in L^\infty(\mathbb{R}_+^N)$, *then the same conclusion holds if we only assume:*
*(i)* $p - 1 < q < \infty$ *if* $N - 1 \leq \frac{p(p+3)}{p-1}$.
*(ii)* $p - 1 < q < q_{N-1,p}^*$ *if* $N - 1 > \frac{p(p+3)}{p-1}$, *where* $q_{N-1,p}^*$ *is defined accordingly.*

There are several results proving monotonicity properties of positive solutions of (6.1.50) under more general conditions on $g$, but with a restriction on the class of solutions $u$.

**Theorem 6.1.17** *Assume* $g$ *is locally Lipschitz continuous on* $\mathbb{R}$ *and satisfies* $sg(s) > 0$ *for* $s \neq 0$. *If* $p > 1$ *any positive solution of (6.1.50) such that* $\nabla u$ *is bounded is monotone increasing in the direction* $x_N$.

*Remark.* The proofs of these results are based upon a delicate use of the moving planes method combined with weak comparison principles in a narrow strip. The assumption that $\nabla u \in L^\infty$ implies that $u$ remains bounded and arbitrarilly small whenever $x_N$ is small enough, which allows to use this comparison method. Notice that the boundedness of $u$ implies the one of $\nabla u$ by extension of $u$ by reflection to $\mathbb{R}_-^N$ and standard regularity results.

### 6.1.4 Separable solutions in $\mathbb{R}_+^N$

The study of boundary isolated singularities of solutions of (6.1.1) in $\mathbb{R}_+^N$, or more generally in a cone $C_S$, generated by a smooth subdomain $S \subset S^{N-1}$, which vanish on $\partial C_S \setminus \{0\}$ leads to the following problem

$$\begin{aligned} div'\left(\left(\beta_{p,q}^2\omega^2 + |\nabla'\omega|^2\right)^{\frac{p-2}{2}}\nabla'\omega\right) + |\omega|^{q-1}\omega \qquad\qquad & \\ + \beta_{p,q}\Lambda(\beta_{p,q})\left(\beta_{p,q}^2\omega^2 + |\nabla'\omega|^2\right)^{\frac{p-2}{2}}\omega = 0 \quad & \text{in } S \\ \omega = 0 \quad & \text{in } \partial S. \end{aligned} \tag{6.1.77}$$

To the problem we associate the spherical $p$-harmonic eigenvalue problem treated in Chapter 2-2

$$\begin{aligned} div'\left(\left(\beta^2\phi^2 + |\nabla'\phi|^2\right)^{\frac{p-2}{2}}\nabla'\phi\right) + \beta\Lambda(\beta)\left(\beta^2\phi^2 + |\nabla'\phi|^2\right)^{\frac{p-2}{2}}\phi = 0 \quad & \text{in } S \\ \phi = 0 \quad & \text{in } \partial S, \end{aligned} \tag{6.1.78}$$

with $\Lambda(\beta) = \beta(p-1) + p - N$. We denote by $(\beta_S, \phi_S)$ the unique couple solution with $\beta_S > 0$ and $\phi_S > 0$. The general nonexistence result holds

**Theorem 6.1.18** *Assume $0 < p - 1$, $S \subset S^{N-1}$ is a $C^2$ domain and $g : \overline{S} \times \mathbb{R} \times TS^{N-1} \mapsto \mathbb{R}$ is a locally Lipschitz continuous function which satisfies*

$$sg(\sigma, s, \xi) > 0 \qquad \forall (\sigma, s, \xi) \in \overline{S} \times \mathbb{R} \times TS^{N-1},\ s \neq 0. \tag{6.1.79}$$

*Then for any $\beta \geq \beta_S$ there exists no nontrivial nonnegative function $\omega \in C^2(S) \cap C(\overline{S})$ satisfying*

$$\begin{aligned} div'\left(\left(\beta^2\omega^2 + |\nabla'\omega|^2\right)^{\frac{p-2}{2}}\nabla'\omega\right) + \beta\Lambda(\beta)\left(\beta^2\omega^2 + |\nabla'\omega|^2\right)^{\frac{p-2}{2}}\omega & \\ + g(.,\omega,\nabla\omega) = 0 \qquad\qquad & \text{in } S \\ \omega = 0 \qquad\qquad & \text{in } \partial S. \end{aligned} \tag{6.1.80}$$

*Proof.* Assume $\omega$ is such a solution. By the strong maximum principle it is positive in $S$ and Hopf boundary lemma holds on the relative boundary $\partial S$. If $\beta = \beta_S$, we denote by $\phi_S$ the largest solution of the spherical $p$-harmonic eigenvalue problem (6.1.78) which is smaller than $\omega$. Then $w := \omega - \phi_S \geq 0$ in $S$. As in the proof of Theorem 3.2.1 we consider the linear elliptic equation satisfied by $w$ near $\sigma_0$. Then (with the notations of the above mentioned theorem)

$$\mathcal{L}w = \frac{-1}{\sqrt{g}}\sum_{\ell,i} a_i^\ell \frac{\partial^2 w}{\partial\sigma_i\partial\sigma_\ell} + \sum_i C_i \frac{\partial w}{\partial\sigma_i} + Cw = g(.,\omega,\nabla\omega) \geq 0. \qquad (6.1.81)$$

Either the graphs of $\omega$ and $\phi_S$ are tangent at some $\sigma_0 \in S$, or $\omega - \phi_S > 0$ in $S$ and there exists some $\sigma_0 \in \partial S$ such that

$$\omega(\sigma_0) = \phi_S(\sigma_0) = 0 \;\text{ and }\; \frac{\partial\omega}{\partial\mathbf{n}'}(\sigma_0) = \frac{\partial\phi_S}{\partial\mathbf{n}'}(\sigma_0) < 0,$$

where $\mathbf{n}'$ is the normal unit outward vector to $\partial S$ in $TS^{N-1}$. In the first case this is a contradiction with Proposition 1.3.11. In the second case it contradicts Proposition 1.3.12.

If $\beta > \beta_S$ we introduce the operator

$$\mathcal{T}(\eta) = div'\left(\left(\beta^2\eta^2 + |\nabla'\eta|^2\right)^{\frac{p-2}{2}}\nabla'\eta\right) + \beta\Lambda(\beta)\left(\beta^2\eta^2 + |\nabla'\eta|^2\right)^{\frac{p-2}{2}}\eta,$$

and set $\theta = \frac{\beta}{\beta_S} > 1$. Then

$$\begin{aligned}\mathcal{T}(\phi_S^\theta) &= \beta\phi_S^\theta\left(\frac{\beta}{\beta_S}\phi_S^{\theta-1}\left(\beta_S^2\phi_S^2 + |\nabla'\phi_S|^2\right)^{\frac{1}{2}}\right)^{p-2}(\Lambda(\beta) - \Lambda(\beta_S))\\ &\quad + \theta(\theta-1)\left(\frac{\beta}{\beta_S}\right)^{p-1}\left(\beta_S^2\phi_S^2 + |\nabla'\phi_S|^2\right)^{\frac{p-2}{2}}|\nabla'\phi_S|^2 > 0.\end{aligned}$$

Again we take the largest $\phi_S$ such that $w := \omega - \phi_S \geq 0$, and since $\theta > 0$ the tangency of the two graphs cannot occur on $\partial S$ and there exists $\sigma_0 \in S$ such that $w(\sigma_0) = 0$. Since $\mathcal{L}w \geq 0$ we infer a contradiction from Proposition 1.3.11 as in the first case. □

The following existence result is proved in [Porretta, Véron (2013)].

**Theorem 6.1.19** *Assume $1 < p < N-1$, $S \subset S^{N-1}$ is a $C^2$ domain and $0 < p-1 < q < q_s$. Then for any $0 < \beta < \beta_S$ there exists a positive function $\omega \in C^2(S) \cap C(\overline{S})$ satisfying (6.1.78).*

The problem is not variational, and since the reaction term is a source, it is difficult to find a supersolution of the equation. In such situations topological and deformations arguments associated to degree theory are often used. Such arguments were thoroughly studied in [Krasnosel'skii (1964)] and the following result is a variant proved in [de Figuereido, Lions, Nussbaum (1982), Prop. 2.1, Remark 2.1].

**Proposition 6.1.20** *Let $X$ be a Banach space with norm $\|.\|_X$ and $K \subset X$ a closed cone with a non-empty interior. Let $F : K \times \mathbb{R}_+ \mapsto K$ be a compact mapping, and let $\Phi(u) = F(u, 0)$ (it is a compact mapping from $K$ to $K$). Assume that there exist $R_1 < R_2$ and $T > 0$ such that*
*(i) $u \neq s\Phi(u)$ for every $s \in [0, 1]$ and every $u$ such that $\|u\|_X = R_1$,*
*(ii) $F(u, t) \neq u$ for every $(u, t)$ such that $\|u\|_X \leq R_2$ and $t \geq T$,*
*(iii) $F(u, t) \neq u$ for every $u$ such that $\|u\|_X = R_2$ and $t \geq 0$.*
*Then the mapping $\Phi$ has a fixed point $u$ such that $R_1 < \|u\|_X < R_2$.*

Another intermediate result is the following one which proof is a mere adaptation of the one of Proposition 2.1.3 ([Porretta, Véron (2013), Appendix]).

**Proposition 6.1.21** *Let $p > 1$, $\beta > 0$, $S \subset S^{N-1}$ and denote*

$$\lambda_{1,\beta} = \inf \left\{ \int_S \left(\beta^2\eta^2 + |\nabla'\eta|^2\right)^{\frac{p}{2}} dS : \eta \in W_0^{1,p}(S), \int_S |\eta|^p \, dS = 1 \right\}. \tag{6.1.82}$$

*Then $\lambda_{1,\beta} > 0$ is achieved by some $\phi \in W_0^{1,p}(S)$ which is positive in $S$ and satisfies*

$$\begin{aligned} -\mathrm{div}'\left( \left(\beta^2\phi^2 + |\nabla'\phi|^2\right)^{\frac{p-2}{2}} \nabla'\phi \right) & \\ + \beta^2 \left(\beta^2\phi^2 + |\nabla'\phi|^2\right)^{\frac{p-2}{2}} \phi = \lambda_{1,\beta} \, |\phi|^{p-2} \phi \qquad & \text{in } S \\ \phi = 0 \qquad & \text{on } \partial S. \end{aligned} \tag{6.1.83}$$

*Moreover $\phi$ is the unique nonnegative nontrivial solution of (6.1.83) and $\lambda_{1,\beta}$ is the smallest of all the $\lambda \in \mathbb{R}$ such that there exists a solution $\eta \in W_0^{1,p}(S)$ to*

$$\begin{aligned} -\mathrm{div}'\left( \left(\beta^2\eta^2 + |\nabla'\eta|^2\right)^{\frac{p-2}{2}} \nabla'\eta \right) & \\ + \beta^2 \left(\beta^2\eta^2 + |\nabla'\eta|^2\right)^{\frac{p-2}{2}} \eta = \lambda \, |\eta|^{p-2} \eta \qquad & \text{in } S \\ \eta = 0 \qquad & \text{on } \partial S, \end{aligned} \tag{6.1.84}$$

*and it is isolated among the set of such $\lambda$.*

*Remark.* In the above statement one can replace the condition $S \subset S^{N-1}$ by $S$ bounded and $S \subset M$, where $(M, g)$ is any Riemannian manifold, in particular $\mathbb{R}^N$.

*Proof of Theorem 6.1.19.* The following functional

$$\mathcal{J}(\eta) = \frac{1}{p}\int_S \left(\beta^2\eta^2 + |\nabla'\eta|^2\right)^{\frac{p}{2}} dS, \tag{6.1.85}$$

defined on $W_0^{1,p}(S)$ is strictly convex and coercive. Its Frechet derivative $\mathcal{A}$ defined by

$$\mathcal{A}(\eta) = -\mathrm{div}'\left(\left(\beta^2\eta^2 + |\nabla'\eta|^2\right)^{\frac{p-2}{2}} \nabla'\eta\right) + \beta^2\left(\beta^2\eta^2 + |\nabla'\eta|^2\right)^{\frac{p-2}{2}} \eta, \tag{6.1.86}$$

is a bijection from $W_0^{1,p}(S)$ into $W^{-1,p'}(S)$ and its inverse $\mathcal{A}^{-1}$ is continuous [Lions (1969)]. We set $X = C_0^1(\bar{S})$ endowed with the norm $C^1(\bar{S})$, a norm for which the imbedding into $W_0^{1,p}(S)$ is continuous. We denote by $K$ the cone of positive function in $S$; it has a non-empty interior. For $t > 0$ we set

$$F(\eta,t) = \mathcal{A}^{-1}\left(\beta(\beta + t + \Lambda(\beta))\left(\beta^2\eta^2 + |\nabla'\eta|^2\right)^{\frac{p-2}{2}}\eta + (\eta+t)^q\right), \tag{6.1.87}$$

and put

$$\Phi(\eta) := F(\eta,0) = \mathcal{A}^{-1}\left(\beta(\beta + \Lambda(\beta))\left(\beta^2\eta^2 + |\nabla'\eta|^2\right)^{\frac{p-2}{2}}\eta + \eta^q\right). \tag{6.1.88}$$

*Step 1: The mapping $F$ is compact.* Actually, if $F(\eta,t) = \phi$, it means that $\phi \in W_0^{1,p}(S)$ satisfies

$$\begin{aligned} &-\mathrm{div}'\left(\left(\beta^2\phi^2 + |\nabla'\phi|^2\right)^{\frac{p-2}{2}} \nabla'\phi\right) + \beta^2\left(\beta^2\phi^2 + |\nabla'\phi|^2\right)^{\frac{p-2}{2}} \phi \\ &\quad = \beta(\beta + t + \Lambda(\beta))\left(\beta^2\eta^2 + |\nabla'\eta|^2\right)^{\frac{p-2}{2}} \eta + (\eta+t)^q \quad \text{in } S \\ &\phi = 0 \text{ on } \partial S. \end{aligned} \tag{6.1.89}$$

Therefore, if we assume that $\eta$ belongs to a bounded subset of $K \cap X$, then the right-hand side of (6.1.89) is bounded in $C(\overline{S})$, hence by Theorem 1.3.10, $\phi$ remains bounded in $C^{1,\gamma}(\overline{S})$ for some $\gamma \in (0,1)$ and therefore relatively compact in $C^1(\overline{S})$.

*Step 2: Condition (i) of Proposition 6.1.20 holds.* Suppose by contradiction that there exists a sequence $\{t_n\}$ such that the problem

$$\begin{aligned} &-\mathrm{div}'\left(\left(\beta^2\eta^2 + |\nabla'\eta|^2\right)^{\frac{p-2}{2}} \nabla'\eta\right) + \beta^2\left(\beta^2\eta^2 + |\nabla'\eta|^2\right)^{\frac{p-2}{2}} \eta \\ &\quad = t_n^{p-1}\beta(\beta + \Lambda(\beta))\left(\beta^2\eta^2 + |\nabla'\eta|^2\right)^{\frac{p-2}{2}} \eta + t_n^{p-1}\eta^q \quad \text{in } S \\ &\eta = 0 \quad \text{on } \partial S, \end{aligned} \tag{6.1.90}$$

admits a positive solutions $\eta_n$ for any $n \in \mathbb{N}$ and that $\|\eta_n\|_X \to 0$ as $n \to \infty$. Then $w_n = \|\eta_n\|_X^{-1} \eta_n$ satisfies

$$
\begin{aligned}
&-\mathrm{div}'\left(\left(\beta^2 w_n^2 + |\nabla' w_n|^2\right)^{\frac{p-2}{2}} \nabla' w_n\right) + \beta^2\left(\beta^2 w_n^2 + |\nabla' w_n|^2\right)^{\frac{p-2}{2}} w_n \\
&\quad = t_n^{p-1}\beta(\beta + \Lambda(\beta))\left(\beta^2 w_n^2 + |\nabla' w_n|^2\right)^{\frac{p-2}{2}} w_n + t_n^{p-1}\|\eta_n\|_X^{q+1-p} w_n^q \quad \text{in } S
\end{aligned}
$$

$w_n = 0$ on $\partial S$.

Up to a subsequence we can assume that $t_n \to t \in [0,1]$, and by compactness, $w_n \to w$ in $C^1(\overline{S})$ where $w$ is positive and $\|w\|_X = 1$. Then $w$ satisfies

$$
\begin{aligned}
&-\mathrm{div}'\left(\left(\beta^2 w^2 + |\nabla' w|^2\right)^{\frac{p-2}{2}} \nabla' w\right) \\
&\qquad = \beta(t^{p-1}\beta - \beta + t^{p-1}\Lambda(\beta))\left(\beta^2 w^2 + |\nabla' w|^2\right)^{\frac{p-2}{2}} w \quad \text{in } S \\
&\qquad w = 0 \text{ on } \partial S.
\end{aligned}
$$

It follows from Theorem 2.2.3 with the notations therein applied with $v = -\frac{1}{\beta}\ln w$ that $t^{p-1}\beta - \beta + t^{p-1}\Lambda(\beta) = \lambda(\beta)$, where $\lambda(\beta)$ is the ergodic constant. Since $\beta < \beta_s$, there holds by Proposition 2.2.5

$$t^{p-1}\beta - \beta + t^{p-1}\Lambda(\beta) > \Lambda_p(\beta_s).$$

Since $t \leq 1$ we get,

$$\Lambda_p(\beta_s) < t^{p-1}\beta - \beta + t^{p-1}\Lambda(\beta) \leq t^{p-1}\Lambda(\beta),$$

which is impossible if we explicit the values of $\Lambda(\beta)$ and $\Lambda_p(\beta_s)$.

*Step 2: Condition (ii) holds.* Since, for $t$ large enough, $\Lambda(\beta) + \beta + t > 0$, using the fact that $q > p-1$, for any $\delta > 0$, one can find $T > 0$ such that

$$
\begin{aligned}
&\beta\left(\Lambda(\beta) + \beta + t\right)\omega\left(\beta^2\omega^2 + |\nabla'\omega|^2\right)^{\frac{p-2}{2}} \\
&\qquad + (\omega + t)^q > (\lambda_{1,\beta} + \delta)\omega^{p-1} \quad \forall t \geq T,\ \forall \omega \geq 0,
\end{aligned}
$$

where $\lambda_{1,\beta}$ has been defined in (6.1.82). Therefore, if $t \geq T$ and $F(\omega, t) = \omega$ for some $\omega \in K \subset X$, $\omega \neq 0$, one obtains

$$
\begin{aligned}
&-\mathrm{div}'\left(\left(\beta^2\omega^2 + |\nabla'\omega|^2\right)^{\frac{p-2}{2}} \nabla'\omega\right) + \beta^2\left(\beta^2\omega^2 + |\nabla'\omega|^2\right)^{\frac{p-2}{2}}\omega \\
&\qquad \geq (\lambda_{1,\beta} + \delta)\omega^{p-1} \quad \text{in } S \\
&\qquad \omega = 0 \quad \text{on } \partial S.
\end{aligned}
\tag{6.1.91}
$$

Since the minimum $\phi$ of (6.1.82) is a subsolution of equation (6.1.84) with $\lambda = \lambda_{1,\beta} + \delta$ and we may adjust it so that it is smaller than $\omega$, it follows from Theorem 1.4.7 that there exists a solution $\psi$ of the same equation such that $\phi \leq \psi \leq \omega$. Since $\delta$ is arbitrary, it would imply that $\lambda_{1,\beta}$ is not isolated, a contradiction. Notice that (ii) holds independently of the norm of $\omega$ for $t \geq T$.

*Step 3: Condition (iii) holds.* Since (ii) holds independently of the norm of $\omega$, it is enough to show that (iii) holds for every $t \leq T$. This is done if one can find an *a priori* estimates, *i.e.* if we can prove the existence of a constant $R_2$ such that for any $t \leq T$ every positive solution of

$$\begin{aligned}
&-\mathrm{div}'\left(\left(\beta^2\omega^2 + |\nabla'\omega|^2\right)^{\frac{p-2}{2}}\nabla'\omega\right) + \beta^2\left(\beta^2\omega^2 + |\nabla'\omega|^2\right)^{\frac{p-2}{2}}\omega \\
&\quad = \beta(\Lambda(\beta) + \beta + t)\left(\beta^2\omega^2 + |\nabla'\omega|^2\right)^{\frac{p-2}{2}}\omega + (\omega + t)^q \quad \text{in } S \qquad (6.1.92)\\
&\quad \omega = 0 \qquad\qquad \text{on } \partial S,
\end{aligned}$$

satisfies $\|\omega\|_X \leq R_2$. By the regularity estimates Theorem 1.3.9, it is sufficient to prove the existence of an *a priori* bound of $\|\omega\|_{W^{1,p}}$. We proceed by contradiction in assuming that there exist sequences $\{t_n\}$ and $\{\omega_n\}$ such that $0 < t_n \leq T$ and $\omega_n$ is a positive solution of (6.1.92) with $t = t_n$ satisfying

$$\|\omega_n\|_{L^\infty} \to \infty \quad \text{as } n \to \infty,$$

and there exists a sequence of points $\sigma_n \in S$ such that $\|\omega_n\|_{L^\infty} = \omega_n(\sigma_n)$. Furthermore, it can be assumed that $\sigma_n \to \sigma_0 \in \overline{S}$. We set $M_n = (\|\omega_n\|_{L^\infty})^{\frac{p-1-q}{p}}$. Using local coordinates in a neighborhood of $\sigma_0$, we set

$$v_n(y) = \frac{\omega_n(\sigma_n + M_n y)}{\|\omega_n\|_{L^\infty}} = M_n^{\frac{p}{q+1-p}}\omega_n(\sigma_n + M_n y). \qquad (6.1.93)$$

The sequence $\{v_n\}$ is uniformly bounded in the ball $B_{\delta M_n^{-1}}(\sigma_0) \subset \mathbb{R}^{N-1}$ and thus it is also eventually locally uniformly bounded in the $C^{1,\alpha}$ topology of $\mathbb{R}^{N-1}$ and hence eventually relatively compact in the $C^1$ topology of $\mathbb{R}^{N-1}$. Up to a subsequence, it converges to some nonnegative $v$ which satisfies (for some $c_0 > 0$)

$$-\Delta_p v = c_0 u^q \qquad (6.1.94)$$

either in $\mathbb{R}^{N-1}$ if $\sigma_0 \in S$, or in $\mathbb{R}_+^{N-1}$ if $\sigma_0 \in \partial S$ and $v$ vanishes on $\partial\mathbb{R}_+^{N-1}$. In both case $v(0) = 1$. Such a solution of (6.1.94) cannot exist, either by Corollary 6.1.7 or by Theorem 6.1.10. □

The exponent $q_s$ is critical for existence of positive solution of problem (6.1.77), provided $S$ verifies some geometric condition: a $C^1$ domain $S \subset S_+^{N-1}$ is *starshaped*, if there exists a nonzero eigenfunction $\phi$ of the Laplacian on $S^{N-1}$ with corresponding eigenvalue $N-1$, positive on $S$, such that

$$\langle \nabla\phi, \mathbf{n}'_a \rangle \leq 0 \qquad \forall a \in \partial S, \tag{6.1.95}$$

where $\mathbf{n}'_a$ is the normal outward unit vector at $a$ in $TS^{N-1}$. The following result is proved in [Porretta, Véron (2013)].

**Theorem 6.1.22** *Assume $1 < p < N-1$, $S \subset S^{N-1}$ is a $C^2$ starshaped domain and $q = q_s$. Then there exists no positive function $\omega \in C^2(S) \cap C(\overline{S})$ satisfying (6.1.77).*

The proof is based upon a series of integral identities dealing with solutions of

$$\begin{aligned} div'\left(\left(\beta^2\omega^2 + |\nabla'\omega|^2\right)^{\frac{p-2}{2}} \nabla'\omega\right) + \beta\Lambda(\beta)\left(\beta^2\omega^2 + |\nabla'\omega|^2\right)^{\frac{p-2}{2}} \omega & \\ + |\omega|^{q-1}\omega = 0 \qquad & \text{in } S \\ \omega = 0 \qquad & \text{in } \partial S. \end{aligned} \tag{6.1.96}$$

**Proposition 6.1.23** *Let $S \subsetneq S_+^{N-1}$ and $\phi_1$ be the first eigenfunction of $-\Delta'$ in $W_0^{1,2}(S_+^{N-1})$. If $\omega \in W_0^{1,p}(S) \cap C(\overline{S})$ is a positive solution of (6.1.96) and if we set $Q = \left(\beta^2\omega^2 + |\nabla'\omega|^2\right)^{\frac{1}{2}}$, then*

$$\begin{aligned} \left(1 - \frac{1}{p}\right) \int_{\partial S} \left|\frac{\partial\omega}{\partial\mathbf{n}'}\right|^p \frac{\partial\phi_1}{\partial\mathbf{n}'} dS = A \int_S \omega^{q+1}\phi_1 d\sigma & \\ + B \int_S Q^{p-2} \left|\nabla'\omega\right|^2 \phi_1 d\sigma + C \int_S Q^{p-2}\omega^2\phi_1 d\sigma, & \end{aligned} \tag{6.1.97}$$

*where*

$$A = A(\beta) = -\frac{N-1}{q+1} - \beta(p\beta + p - N), \tag{6.1.98}$$

$$B = B(\beta) = \frac{N-1-p}{p} - \beta(p\beta + p - N), \tag{6.1.99}$$

$$C = C(\beta) = \beta^2\left(\frac{N-1}{p} - (p\beta + p - N)\Lambda(\beta)\right). \tag{6.1.100}$$

*Proof.* The key step is a general identity valid even for any $S \subset S^{N-1}$. We recall that by Theorem 1.3.9, $\omega \in C^{1,\alpha}(\overline{S})$. Furthermore, since $\left(\beta^2\omega^2 + |\nabla'\omega|^2\right)^{\frac{1}{2}} > 0$ in $S$, $\omega$ is $C^2$ in $S$.

*Step 1:* We claim that for any $\phi \in C^2(\overline{S})$ there holds:

$$\begin{aligned}\left(1-\frac{1}{p}\right)\int_{\partial S}\left|\frac{\partial\omega}{\partial\mathbf{n}'}\right|^p\frac{\partial\phi_1}{\partial\mathbf{n}'}dS &= \int_S\left(\frac{\Delta'\phi}{q+1}-\beta(\beta p+p-N)\phi\right)\omega^{q+1}d\sigma\\ &\quad -\frac{1}{p}\int_S Q^p\Delta'\phi d\sigma + \int_S Q^{p-2}\langle D^2\phi\nabla'\omega,\nabla'\omega\rangle d\sigma\\ &\quad +\beta(\beta p+p-N)\int_S Q^{p-2}\left|\nabla'\omega\right|^2\phi d\sigma\\ &\quad +\beta^2\Lambda(\beta)(\beta p+p-N)\int_S Q^{p-2}\omega^2\phi d\sigma.\end{aligned} \tag{6.1.101}$$

Multiplying (6.1.96) by $\zeta\phi$ where $\zeta \in C_0^1(S)$ and integrating yields

$$\begin{aligned}&\int_S Q^{p-2}\left(\frac{1}{2}\langle\nabla'\left|\nabla'\omega\right|^2,\nabla'\phi\rangle+\langle D^2\phi\nabla'\omega,\nabla'\omega\rangle\right)\zeta d\sigma\\ &\qquad\qquad + \int_S Q^{p-2}\langle\nabla'\omega,\nabla'\zeta\rangle\langle\nabla'\omega,\nabla'\phi\rangle d\sigma\\ &= \beta\Lambda(\beta)\int_S Q^{p-2}\omega\langle\nabla'\omega,\nabla'\phi\rangle\zeta d\sigma + \frac{1}{q+1}\int_S\langle\nabla'\omega^{q+1},\nabla'\phi\rangle\zeta d\sigma.\end{aligned}$$

Since

$$\frac{1}{2}Q^{p-2}\langle\nabla'\left|\nabla'\omega\right|^2,\nabla'\phi\rangle = \frac{1}{p}\langle\nabla' Q^p,\nabla'\phi\rangle - \beta^2\Lambda(\beta)Q^{p-2}\omega\langle\nabla'\omega,\nabla'\phi\rangle,$$

and $\Lambda(\beta) = (p-1)\beta + p - N$, we obtain

$$\begin{aligned}&\frac{1}{p}\int_S\langle\nabla' Q^p,\nabla'\phi\rangle\zeta d\sigma + \int_S Q^{p-2}\langle D^2\phi\nabla'\omega,\nabla'\omega\rangle\zeta d\sigma\\ &\qquad\qquad + \int_S Q^{p-2}\langle\nabla'\omega,\nabla'\zeta\rangle\langle\nabla'\omega,\nabla'\phi\rangle d\sigma\\ &= \beta(\beta p+p-N)\int_S Q^{p-2}\omega\langle\nabla'\omega,\nabla'\phi\rangle\zeta d\sigma + \frac{1}{q+1}\int_S\langle\nabla'\omega^{q+1},\nabla'\phi\rangle\zeta d\sigma.\end{aligned}$$

Integrating by parts the first and the last terms leads to

$$
\begin{aligned}
&-\frac{1}{p}\int_S Q^p\langle\nabla'\zeta,\nabla'\phi\rangle d\sigma+\frac{1}{q+1}\int_S\omega^{q+1}\langle\nabla'\zeta,\nabla'\phi\rangle d\sigma\\
&\quad+\int_S\left(\frac{\omega^{q+1}}{q+1}-\frac{Q^p}{p}\right)\zeta\Delta'\phi d\sigma+\int_S Q^{p-2}\langle D^2\phi\nabla'\omega,\nabla'\omega\rangle\zeta d\sigma\\
&\quad+\int_S Q^{p-2}\langle\nabla'\omega,\nabla'\zeta\rangle\langle\nabla'\omega,\nabla'\phi\rangle d\sigma\\
&\quad=\beta(\beta p+p-N)\int_S Q^{p-2}\omega\langle\nabla'\omega,\nabla'\phi\rangle\zeta d\sigma.
\end{aligned}
\tag{6.1.102}
$$

For $\delta>0$ small enough, we consider $\zeta:=\zeta_\delta$ in the above equalities where the $\{\zeta_\delta\}$ forms a sequence of functions in $C^1_0(S)$ converging to 1 locally uniformly in $S$ and such that $|\nabla'\zeta_\delta|$ remains bounded in $L^1(S)$. Since for any continuous vector field $\mathbf{F}$ on $\overline{S}$ there holds

$$
\lim_{\delta\to 0}\int_S\langle\mathbf{F},\nabla'\zeta_\delta\rangle d\sigma=-\int_{\partial S}\langle\mathbf{F},\mathbf{n}'\rangle dS,
$$

we infer, by letting $\delta\to 0$ and since $\nabla'\omega=-\left|\dfrac{\partial\omega}{\partial\mathbf{n}'}\right|\mathbf{n}'$ on $\partial S$,

$$
\begin{aligned}
\left(1-\frac{1}{p}\right)\int_{\partial S}\left|\frac{\partial\omega}{\partial\mathbf{n}'}\right|^p\frac{\partial\phi}{\partial\mathbf{n}'}dS&=\int_S\left(\frac{\omega^{q+1}}{q+1}-\frac{Q^p}{p}\right)\Delta'\phi d\sigma\\
&+\int_S Q^{p-2}\langle D^2\phi\nabla'\omega,\nabla'\omega\rangle d\sigma+\beta(\beta p+p-N)\int_S Q^{p-2}\omega\langle\nabla'\omega,\nabla'\phi\rangle d\sigma.
\end{aligned}
\tag{6.1.103}
$$

Next we multiply (6.1.96) by $\omega\phi$ and integrate

$$
\begin{aligned}
&\int_S Q^{p-2}\omega\langle\nabla'\omega,\nabla'\phi\rangle d\sigma\\
&\qquad=-\int_S Q^{p-2}\left|\nabla'\omega\right|^2\phi d\sigma+\beta\Lambda(\beta)\int_S Q^{p-2}\omega^2\phi d\sigma+\int_S\omega^{q+1}\phi d\sigma.
\end{aligned}
$$

Replacing the last term in (6.1.102) yields

$$
\begin{aligned}
\left(1-\frac{1}{p}\right)\int_{\partial S}\left|\frac{\partial\omega}{\partial\mathbf{n}'}\right|^p\frac{\partial\phi}{\partial\mathbf{n}'}dS&=\int_S\left(\frac{\omega^{q+1}}{q+1}-\frac{Q^p}{p}\right)\Delta'\phi d\sigma\\
&+\beta(\beta p+p-N)\int_S\left(Q^{p-2}\left(\left|\nabla'\omega\right|^2-\beta\Lambda(\beta)\omega^2\right)-\omega^{q+1}\right)\phi d\sigma\\
&+\int_S Q^{p-2}\langle D^2\phi\nabla'\omega,\nabla'\omega\rangle d\sigma,
\end{aligned}
$$

which is the claim.

*Step 2:* End of the proof. We chose $\phi$ as first eigenfunction of $-\Delta'$ in $W_0^{1,2}(S_+^{N-1})$, thus $-\Delta'\phi = (N-1)\phi$ and $D^2\phi = -\phi g$, hence

$$\begin{aligned}\left(1-\frac{1}{p}\right)\int_{\partial S}\left|\frac{\partial \omega}{\partial \mathbf{n}'}\right|^p \frac{\partial \phi}{\partial \mathbf{n}'} dS &= -\int_S \left(\frac{N-1}{q+1} + \beta(\beta p + p - N)\right)\omega^{q+1}\phi d\sigma \\ &\quad - \int_S Q^{p-2}\left|\nabla'\omega\right|^2 \phi d\sigma + \beta(\beta p + p - N)\int_S Q^{p-2}\left|\nabla'\omega\right|^2 \phi d\sigma \\ &\quad + \frac{N-1}{p}\int_S Q^p \phi d\sigma - \beta^2(\beta p + p - N)\Lambda(\beta)\int_S Q^{p-2}\omega^2\phi d\sigma.\end{aligned} \tag{6.1.104}$$

Replacing $Q$ by its value we obtain (6.1.97) with $A$, $B$ and $C$ given by (6.1.98), (6.1.99) and (6.1.100) respectively. □

*Proof of Theorem 6.1.22.* First we compute the coefficients $A$, $B$ and $C$ in Proposition 6.1.23 with $\beta_{p,q} = \frac{p}{q+1-p}$ and we express it by introducing $\beta_s := \beta_{p,q_s}$ where

$$\beta_s = \frac{p}{q_s + 1 - p} = \frac{N-1-p}{p}.$$

Then

$$\begin{aligned} A &= -\frac{(N-1)\beta_{p,q}}{p(1+\beta_{p,q})} - \beta_{p,q}(p\beta_{p,q} + p - N) \\ &= -\frac{\beta_{p,q}}{\beta_{p,q}+1}\left(\frac{N-1}{\beta_{p,q}} + p(\beta_{p,q}+1)^2 - N(\beta_{p,q}+1)\right) \\ &= -\frac{p\beta_{p,q}}{\beta_{p,q}+1}\left(\beta_{p,q} + 1 - \frac{1}{p}\right)(\beta_{p,q} - \beta_s), \end{aligned}$$

$$B = \beta_s + \beta_{p,q}\left(p(\beta_{p,q} - \beta_s) - 1\right) = \left(\beta_{p,q} - \beta_s\right)\left(p\beta_{p,q} - 1\right),$$

and

$$\begin{aligned} C &= \beta_{p,q}^2\left(\frac{N-1}{p} - (p\beta_{p,q} + p - N)\left((p-1)\beta_{p,q} + p - N\right)\right) \\ &= \beta_{p,q}^2\Big(\beta_s + 1 - \left(p(\beta_{p,q} - \beta_s) - 1\right)\left(p(\beta_{p,q} - \beta_s) - (\beta_{p,q} + 1)\right)\Big) \\ &= (1-p)\beta_{p,q}^2(\beta_{p,q} - \beta_s)\left(p\beta_{p,q} - 1 - \frac{p(N-p)}{p-1}\right). \end{aligned}$$

Therefore $A \geq 0$, $B \geq 0$ and $C \geq 0$ can be obtained only if $q = q_s$ *i.e.* $\beta_{p,q} = \beta_s$, in which case $A = B = C = 0$. Since $\partial_{\mathbf{n}'}\phi \leq 0$ because $S$ is starshaped, we deduce from (6.1.97) that $|\partial_{\mathbf{n}'}\omega|^p \partial_{\mathbf{n}'}\phi = 0$. Unless $\omega$ is identically zero, we have $\partial_{\mathbf{n}'}\omega < 0$ by Hopf boundary lemma. Then $\partial_{\mathbf{n}'}\phi = 0$ on $\partial S$. Using the equation satisfied by $\phi$ and Green formula we derive

$$(N-1)\int_S \phi dS = 0.$$

Thus $\phi = 0$ in $S$, a contradiction with $\phi > 0$. □

*Remark.* It is conjectured that non-existence holds also for any $q > q_s$.

### 6.1.5 *Boundary singularities*

In this section we prove the following universal *a priori* estimate for positive solutions of

$$\begin{aligned} -\Delta_p u &= u^q && \text{in } \Omega \\ u &= 0 && \text{in } \partial\Omega \setminus \{0\}, \end{aligned} \tag{6.1.105}$$

where $\Omega \subset \mathbb{R}^N$ is a $C^2$ bounded domain such that $0 \in \partial\Omega$ and $p-1 < q < p^* - 1$.

**Theorem 6.1.24** *Assume $p-1 < q < p^*-1$, then any positive solution of (6.1.105) satisfies*

$$u(x) \leq c\,|x|^{-\frac{p}{q+1-p}} \qquad \forall x \in \Omega, \tag{6.1.106}$$

*for some $c > 0$ independent of the solution.*

The proof is based upon a method which connects Liouville type results and *a priori* estimate due to [Polàcik, Quittner, Souplet (2007)]. In the case $p = 2$ the above estimate has been obtained in [Bidaut-Véron, Ponce, Véron (2011)]. The main ingredient is a topological result called *the doubling lemma* (see [Polàcik, Quittner, Souplet (2007)]).

**Proposition 6.1.25** *Let $(X, d)$ be a complete metric space $\Gamma \subsetneq X$ and $\gamma : X \setminus \Gamma \mapsto (0, \infty)$. We assume that $\gamma$ is bounded on all compact subset of $X \setminus \Gamma$. Given $k > 0$, let $y \in X \setminus \Gamma$ such that*

$$\gamma(y)\text{dist}\,(y, \Gamma) > 2k.$$

*Then there exists $x \in X \setminus \Gamma$ satisfying*

*(i)* $\gamma(x)\text{dist}\,(x,\Gamma) > 2k$,

*(ii)* $\gamma(x) > \gamma(y)$,

*(iii)* $2\gamma(x) \geq \gamma(z)$, *for all* $z \in B_{\frac{k}{\gamma(x)}}(x)$.

**Proposition 6.1.26** *Assume* $p-1 < q < p^*-1$ *and* $0 < r < \frac{1}{2}\textit{diam}\,(\Omega)$. *Then every nonnegative solution of*

$$\begin{aligned} -\Delta_p u &= u^q && \text{in } \Omega \cap (B_{2r} \setminus \overline{B}_r) \\ u &= 0 && \text{in } \partial\Omega \cap (B_{2r} \setminus \overline{B}_r), \end{aligned} \tag{6.1.107}$$

*satisfies*

$$u(x) \leq c\,(\text{dist}\,(x,\Gamma_r))^{-\frac{p}{q+1-p}} \qquad \forall x \in \Omega \cap (B_{2r} \setminus \overline{B}_r), \tag{6.1.108}$$

*where* $\Gamma_r = \overline{\Omega} \cap (\partial B_{2r} \cup \partial B_r)$ *and* $c > 0$ *is independent of* $u$ *and* $r$.

*Proof.* We proceed by contradiction, assuming that (6.1.108) is false. Then for every $k \in \mathbb{N}^*$, there exists $r_k \in (0, \frac{1}{2}\text{diam}\,(\Omega))$ and a solution $u_k$ of (6.1.107) with $r = r_k$ and $y_k \in \Omega \cap (B_{2r_k} \setminus \overline{B}_{r_k})$ such that

$$u_k(y_k) > (2k)^{\frac{p}{q+1-p}}\,(\text{dist}\,(y_k,\Gamma_{r_k}))^{-\frac{p}{q+1-p}}\,, \tag{6.1.109}$$

where $\Gamma_{r_k} = \overline{\Omega} \cap (\partial B_{2r_k} \cup \partial B_{r_k})$. We apply Proposition 6.1.25 with

$$X = \overline{\Omega} \cap (\overline{B}_{2r_k} \setminus B_{r_k})\,,\ \Gamma = \Gamma_{r_k} \ \text{ and } \ \gamma(x) = (u_k(x))^{\frac{q+1-p}{p}}\,.$$

One can find $x_k \in X \setminus \Gamma_{r_k}$ such that

(i) $u_k(x_k) > (2k)^{\frac{p}{q+1-p}}\,(\text{dist}\,(x_k,\Gamma_{r_k}))^{-\frac{p}{q+1-p}}$,

(ii) $u_k(x_k) > u_k(y_k)$,

(iii) $2^{\frac{p}{q+1-p}}u_k(x_k) \geq u_k(z)$, for all $z \in B_{R_k}(x_k)$ where we denote $R_k = k\,(u_k(x_k))^{-\frac{q+1-p}{p}}$.

By (i) we have $R_k < \frac{1}{2}\text{dist}\,(x_k,\Gamma_{r_k})$ and therefore $B_{R_k}(x_k) \cap \Gamma_{r_k} = \emptyset$. Furthermore $\text{dist}\,(x_k,\Gamma_{r_k}) \leq \frac{1}{2}r_k < \frac{1}{4}\text{diam}\,(\Omega)$, hence, from (i),

$$u_k(x_k) \geq \left(\frac{8k}{\text{diam}\,(\Omega)}\right)^{\frac{p}{q+1-p}}.$$

In particular, $u_k(x_k) \to \infty$ as $k \to \infty$. For every $k \in \mathbb{N}^*$ let

$$\begin{aligned} t_k &= (u_k(x_k))^{-\frac{q+1-p}{p}} \\ D_k &= \left\{\xi \in \mathbb{R}^N : |\xi| \leq 2k \text{ and } x_k + t_k\xi \in \Omega\right\} = \frac{1}{t_k}(\Omega - x_k), \end{aligned}$$

and

$$v_k(\xi) = \frac{u_k(x_k + t_k\xi)}{u_k(x_k)} \qquad \forall \xi \in D_k.$$

Then $v_k$ satisfies

$$\begin{aligned} -\Delta_p v_k &= v_k^q && \text{in } D_k \\ 0 \leq v_k &\leq 2^{\frac{p}{q+1-p}} && \text{in } D_k \\ v_k(0) &= 1. \end{aligned}$$

Up to a subsequence, we encountered the following dichotomy:

(A) either for every $a > 0$ there exists $k_a \leq 1$ such that $B_{at_k}(x_k) \cap \partial\Omega = \emptyset$. Since the sequence $\{v_k\}$ is locally uniformly bounded, it follows from Theorem 1.3.9 and Ascoli's compactness theorem that it is eventually relatively compact in the $C^1_{loc}(\mathbb{R}^N)$ topology and that there exist a subsequence, still denoted by $\{v_k\}$ and a function $v \in C^1(\mathbb{R}^N)$ which satisfies (6.1.1) in $\mathbb{R}^N$, $0 \leq v \leq 2^{\frac{p}{q+1-p}}$ and $v(0) = 1$, which is impossible by Corollary 6.1.7.

(B) or there exists $a_0 > 0$ such that $B_{a_0 t_k}(x_k) \cap \partial\Omega \neq \emptyset$ for all $k \in \mathbb{N}^*$. We set $x'_k = Proj_{\partial\Omega}(x_k)$, then $|x_k - x'_k| \leq a_0 t_k$. Since $v_k$ is bounded in $D_k$, it follows from Theorem 1.3.9 that $\nabla v_k$ remains locally bounded therein, and because $u_k(x'_k) = 0$ it implies that $|x_k - x'_k| \geq a_1 t_k$ for some $a_1 \in (0, a_0)$. Up to a subsequence, we can assume that $t_k^{-1} x_k \to x_0$ and thus $t_k^{-1} x'_k \to x'_0$, that $-\mathbf{e}_N$ is the normal outward unit vector to $\partial\Omega$ at $x'_0$ and that $D_k$ converges to the half-space $H$ passing through $x'_0$ with normal outward unit vector $-\mathbf{e}_N$ and that $x_0 - x'_0 = -a\mathbf{e}_N$ for some $a \in [a_1, a_0]$. We denote by $\tilde{H}$ the union of $H$ and its reflection through $\partial H$. In order to analyze the convergence of $\{v_k\}$, we perform the same reflexion through $\partial(\frac{1}{t_k}\Omega)$ as the one used in Lemma 3.3.2. The reflected function $\tilde{v}_k$ converges, up to a subsequence in the $C^1_{loc}(\tilde{H})$ topology to some $\tilde{v}$ which satisfies (6.1.1) in $\tilde{H}$, vanishes on $\partial H$, is nonnegative in $H$ and such that $v(0) = 1$. By Theorem 6.1.10 such a function cannot exist, which ends to proof. □

*Proof of Theorem 6.1.24.* Let $x \in \Omega$ such that $|x| \leq \frac{3}{4}\text{diam}\,(\Omega)$. We apply Proposition 6.1.26 with $r = \frac{2}{3}|x|$. Since $\text{dist}\,(x, \Gamma_r) = \frac{1}{3}r$ we derive

$$u(x) \leq c\,(\text{dist}\,(x, \Gamma_r))^{-\frac{p}{q+1-p}} = c\left(\frac{r}{3}\right)^{-\frac{p}{q+1-p}} \leq \tilde{c}\,|x|^{-\frac{p}{q+1-p}}. \qquad (6.1.110)$$

□

The following estimate is an immediate consequence of Theorem 6.1.24 and Lemma 3.3.2.

**Corollary 6.1.27** *Under the assumptions of Theorem 6.1.24, any positive solution u of (6.1.105) satisfies*

$$u(x) \leq c\rho(x)\,|x|^{-\frac{q+1}{q+1-p}} \qquad \forall x \in \Omega, \tag{6.1.111}$$

*where* $\rho(x) = \operatorname{dist}(x, \partial\Omega)$, *and*

$$|\nabla u(x)| \leq c\,|x|^{-\frac{q+1}{q+1-p}} \qquad \forall x \in \Omega, \tag{6.1.112}$$

*for some* $c = c(\Omega, p, q) > 0$.

The first critical exponent $q$ for boundary singularities is $q = \tilde{q}_1$ defined by relations (3.2.5), for which value $\beta_{S_+^{N-1}} = \frac{p}{\tilde{q}_1+1-p}$.

**Theorem 6.1.28** *Let* $\Omega$ *be a bounded* $C^2$ *domain such that* $0 \in \partial\Omega$ *and* $\Omega \cap B_{r_0} = B_{r_0}^+ := B_{r_0} \cap \mathbb{R}_+^N$ *for some* $r_0 > 0$. *Assume* $0 < p-1 < q < \tilde{q}_1$. *If u is a positive solution of (6.1.105) there exists* $c > 0$, *depending on u such that*

$$u(x) \leq c\,|x|^{-\beta_{S_+^{N-1}}} \qquad \forall x \in \Omega. \tag{6.1.113}$$

*Proof.* We write the equation under the form

$$-\Delta_p u - V(x)u^{p-1} = 0, \tag{6.1.114}$$

where $V(x) = (u(x))^{q+1-p} \leq c|x|^{-p}$ by Theorem 6.1.24. Using boundary Harnack inequality Corollary 2.3.2, there exists $c_1 > 0$ such that

$$c_1^{-1}\frac{u(x)}{\rho(x)} \leq \frac{u(y)}{\rho(y)} \leq c_1\frac{u(x)}{\rho(x)} \qquad \forall x, y \in B^+_{\frac{r_0}{2}} \text{ s.t. } |x| = |y| \neq 0. \tag{6.1.115}$$

Let us suppose that (6.1.113) does not hold. Then there exists a sequence $\{x_k\} \subset \Omega$ such that

$$u(x_k) \geq k\,|x_k|^{-\beta_{S_+^{N-1}}} \qquad \forall k \in \mathbb{N}^*. \tag{6.1.116}$$

By Theorem 6.1.24 the sequence $\{x_k\}$ converges to 0, and because of (6.1.115) we can assume that $\rho(x_k) = |x_k| = r_k \leq \frac{r_0}{2}$. We recall that

$$V_{\mathbb{R}_+^N}(x) = V_{\mathbb{R}_+^N}(r,\sigma) = r^{-\beta_{S_+^{N-1}}}\phi_{S_+^{N-1}}(\sigma),$$

is the singular spherical $p$-harmonic function defined in $\mathbb{R}_+^N$ constructed in Theorem 2.2.1 and $V_{B_{r_0}^+}$ is the singular $p$-harmonic function defined in

$B_{r_0}^+$ constructed in Proposition 3.3.5 which vanishes on $\partial B_{r_0}^+$ and satisfies (3.3.36), *i.e.*

$$\lim_{x\to 0}\frac{V_{\mathbb{R}_+^N}(x)}{V_{B_{r_0}^+}(x)}=1.$$

By Hopf boundary lemma $\partial_{\mathbf{n}'}\phi_{S_+^{N-1}}\leq\gamma<0$ on $\partial S_+^{N-1}$, hence there exists $c_2>0$ such that

$$c_2^{-1}\frac{\rho(x)}{|x|}\leq\phi_{S_+^{N-1}}\left(\frac{x}{|x|}\right)\leq c_2\frac{\rho(x)}{|x|}.$$

Then

$$\begin{aligned}
u(r_k,\sigma)&=u(r_k,\sigma_k)\frac{u(r_k,\sigma)}{u(r_k,\sigma_k)}\\
&\geq kc_1^{-1}\frac{\rho(r_k,\sigma)}{r_k}r_k^{-\beta_{S_+^{N-1}}}\\
&\geq kc_1^{-1}c_2^{-1}V_{\mathbb{R}_+^N}(r_k,\sigma)\\
&\geq kc_1^{-1}c_2^{-1}V_{B_{r_0}^+}(r_k,\sigma)\qquad\forall\sigma\in S_+^{N-1}.
\end{aligned}$$

By the comparison principle $u\geq kc_1^{-1}c_2^{-1}V_{B_{r_0}^+}$ in $B_{r_0}^+\setminus B_{r_k}^+$. For $j$ fixed and any $k\geq j$, $u\geq kc_1^{-1}c_2^{-1}V_{B_{r_0}^+}$ in $B_{r_0}^+\setminus B_{r_j}^+$, which yields a contradiction when $k\to\infty$. □

**Corollary 6.1.29** *Under the assumptions of Theorem 6.1.28 there exists $\gamma\geq 0$ such that*

$$\begin{aligned}
&\lim_{r\to 0}r^{\beta_{S_+^{N-1}}}u(r,\sigma)=\gamma\phi_{S_+^{N-1}}(\sigma)\\
&\lim_{r\to 0}r^{1+\beta_{S_+^{N-1}}}\nabla'u(r,\sigma)=\gamma\nabla'\phi_{S_+^{N-1}}(\sigma)
\end{aligned}\tag{6.1.117}$$

*uniformly in $S_+^{N-1}$.*

*Proof.* By Theorem 6.1.28 and Lemma 3.3.2 there holds

$$|\nabla u(x)|\leq c\,|x|^{-1-\beta_{S_+^{N-1}}}\qquad\forall x\in B_{r_0}^+,$$

and

$$|\nabla u(x)-\nabla u(y)|\leq c\,|x|^{-1-\alpha-\beta_{S_+^{N-1}}}|x-y|^{\alpha}\quad\forall x,y\in B_{r_0}^+,\ |x|\leq|y|.$$

The remaining of the proof follows the one of Theorem 3.3.14 in the charaterization of weak singularities. □

*Remark.* In the case $p = N$, the assumption that $\Omega \cap B_{r_0} = B^+_{r_0}$ is not needed if we use conformal transformations as in Chapter 3. In that case the critical exponent $\tilde{q}_1$ is $2N - 1$.

## 6.2 Equations with measure data

### 6.2.1 *The case p = 2*

Let $\Omega \subset \mathbb{R}^N$ be a bounded domain and $g : \mathbb{R} \mapsto \mathbb{R}$ a continuous function satisfying

$$rg(r) \geq |r|^{q+1} \qquad \forall r \in \mathbb{R}, \tag{6.2.1}$$

for some $q > 1$. If $u$ is a nonnegative weak solution of the semilinear problem

$$-\Delta u = g(u) + \mu \qquad \text{in } \Omega \qquad u = 0 \qquad \text{on } \partial\Omega, \tag{6.2.2}$$

where $\mu$ is a positive measure in $\Omega$, there holds, by definition of weak solutions,

$$-\int_\Omega u \Delta\psi dx = \int_\Omega g(u)\psi dx + \int_\Omega \psi d\mu,$$

for any $\psi \in C^1_0(\overline{\Omega})$ such that $-\Delta\psi \in L^\infty(\Omega)$. Thus, if we assume that $\psi$ is the first eigenfunction of $-\Delta$ in $W^{1,2}_0(\Omega)$ we have using (6.2.1),

$$-\int_\Omega u\Delta\psi dx = \lambda_\Omega \int_\Omega u\psi dx \leq \lambda_\Omega \left(\int_\Omega g(u)\psi dx\right)^{\frac{1}{q}} \left(\int_\Omega \psi dx\right)^{\frac{1}{q'}},$$

which implies

$$\int_\Omega \psi d\mu \leq \lambda_\Omega^{q'} \left(\frac{1}{q} - \frac{1}{q^q}\right) \int_\Omega \psi dx. \tag{6.2.3}$$

This is an upper estimate of the norm of $\mu$ in $\mathfrak{M}_\rho(\Omega)$. Notice that the following estimate is also satisfied

$$\int_\Omega g(u)\psi dx \leq \lambda_\Omega^{q'} \int_\Omega \psi dx. \tag{6.2.4}$$

If we take $\psi = \varphi^{q'}$ for some $\varphi \in C_0^\infty(\Omega)$, $\varphi \geq 0$, there holds

$$-q'\int_\Omega u\varphi^{q'-1}\Delta\varphi dx \geq \int_\Omega g(u)\varphi^{q'}dx + \int_\Omega \varphi^{q'}d\mu.$$

Because

$$\left|\int_\Omega u\varphi^{q'-1}\Delta\varphi dx\right| \leq \left(\int_\Omega g(u)\varphi^{q'}dx\right)^{\frac{1}{q}}\left(\int_\Omega |\Delta\varphi|^{q'}dx\right)^{\frac{1}{q'}},$$

we infer

$$\int_\Omega g(u)\varphi^{q'}dx \leq q'^{q'}\int_\Omega |\Delta\varphi|^{q'}dx,$$

from which inequality it follows

$$\int_\Omega \varphi^{q'}d\mu \leq q'^{q'}\left(\frac{1}{q} - \frac{1}{q^q}\right)\int_\Omega |\Delta\varphi|^{q'}dx. \tag{6.2.5}$$

Hence, if $E \subset \bar{E} \subset \Omega$ is a Borel set, there holds

$$\mu(E) \leq c(q)cap^\Omega_{2,q'}(E). \tag{6.2.6}$$

This is an estimate of Lipschitz continuity of $\mu$ with respect to $cap^\Omega_{2,q'}$. Furthermore, if $N > 2$ and $1 < q < \frac{N}{N-2}$, $cap^\Omega_{2,q'}(E) > \tau$ for any $E \neq \emptyset$ and some fixed $\tau > 0$. Hence condition (6.2.6) is fulfilled for any measure $\mu$ such that $\mu(\Omega) \leq c(q)\tau$.

Estimates (6.2.3) and (6.2.6) show that in the case $p = 2$ there are two types of necessary conditions linking $\mu$ and $q$, for the existence of positive solutions of (6.2.2). Actually this situation holds for all $p > 1$.

### 6.2.2 *The subcritical case*

In this section $\Omega \subset \mathbb{R}^N$ is a $C^2$ bounded domain, $A : \Omega \times \mathbb{R} \times \mathbb{R}^N \mapsto \mathbb{R}^N$ a Caratheodory vector field satisfying (4.1.1) with $p > 1$ and $g : \Omega \times \mathbb{R} \times \mathbb{R}^N \mapsto \mathbb{R}$ a Caratheodory function satisfying (5.1.18). If $\mu$ is a measure in $\Omega$, we consider the problem

$$\begin{aligned} -\text{div}\, A(x, \nabla u) &= g(x, u, \nabla u) + \mu && \text{in } \Omega \\ u &= 0 && \text{on } \partial\Omega. \end{aligned} \tag{6.2.7}$$

Existence of a renormalized solution to (6.2.7) is based upon Leray-Schauder's theorem.

### 6.2.2.1 *Sufficient condition*

The next result proved in [Bidaut-Véron (2002)] extends to the quasilinear case the result presented in the previous paragraph for the case $p = 2$ and has to be compared with Lemma 5.2.5. We recall that $q_p = \frac{q}{q+1-p}$, has been introduced in (5.2.27).

**Proposition 6.2.1** *Assume $\mu$ is a bounded positive measure. If $g$ satisfies*

$$rg(x, r, \xi) \geq |r|^{q+1} \qquad \forall (x, r, \xi) \in \Omega \times \mathbb{R} \times \mathbb{R}^N, \tag{6.2.8}$$

*for some $q > p - 1$, then for any $R > pq_p$ any nonnegative renormalized solution $u$ of problem (6.2.7) verifies*

$$\int_\Omega \zeta d\mu + \int_\Omega g(x, u, \nabla u)\zeta dx \leq c_1 \left( \int_\Omega \zeta^{1-R} |\nabla \zeta|^R \, dx \right)^{\frac{pq_p}{R}}, \tag{6.2.9}$$

*for any $\zeta \in W_0^{1,p}(\Omega) \cap W^{1,s}(\Omega)$ $(s > N)$ satisfying $0 \leq \zeta \leq 1$, for some $c_1 > 0$ depending on $N$, $p$, $q$, $R$ and $\Omega$. Furthermore, for any $\alpha < 0$, there exists $c_2 = c_2(N, p, q, R, \alpha, \Omega) > 0$ such that*

$$\int_\Omega (1+u)^\alpha |\nabla u|^p \zeta dx \leq c_2 \left( 1 + \int_\Omega g(x, u, \nabla u)\zeta dx \right) \left( \int_\Omega \zeta^{1-R} |\nabla \zeta|^R \, dx \right)^{\frac{q}{R}}. \tag{6.2.10}$$

*Proof.* We decompose $\mu = \mu_0 + \mu_s^+$, $\mu_0 \ll c_{1,p}$ and $\mu_s^+ \perp c_{1,p}$. For $\epsilon, k > 0$ and $\alpha \in (1-p, 0)$ we set $h_{\alpha,k,\epsilon}(r_+) = (T_k(r_+) + \epsilon)^\alpha$, $u_k = T_k(u)$. We choose $\zeta h_{\alpha,k,\epsilon}(u)$ for test function where $\zeta \in W_0^{1,p}(\Omega) \cap W^{1,s}(\Omega)$ $(s > N)$, takes values in $[0, 1]$. We obtain, using (4.1.1),

$$\begin{aligned}
\int_\Omega (u_k + \epsilon)^\alpha \zeta d\mu_0 + (k+\epsilon)^\alpha \int_\Omega \zeta d\mu_s^+ + \int_\Omega (u_k + \epsilon)^\alpha \zeta g(x, u, \nabla u) dx& \\
+ |\alpha| \int_\Omega \langle A(x, \nabla u), \nabla u \rangle (u_k + \epsilon)^{\alpha - 1} \zeta dx& \\
= \int_\Omega \langle A(x, \nabla u), \nabla \zeta \rangle (u_k + \epsilon)^\alpha dx& \\
\leq \lambda \int_\Omega |\nabla u_k|^{p-1} |\nabla \zeta| \, (u_k + \epsilon)^\alpha dx + \lambda (k + \epsilon)^\alpha \int_{\{u>k\}} |\nabla u|^{p-1} |\nabla \zeta| \, dx& \\
\leq \frac{|\alpha|}{2} \int_\Omega |\nabla u|^p \, (u_k + \epsilon)^{\alpha-1} \zeta dx + c_3 \int_\Omega (u_k + \epsilon)^{\alpha+p-1} \zeta^{1-p} |\nabla \zeta|^p \, dx& \\
+ \lambda (k+\epsilon)^\alpha \int_{\{u>k\}} |\nabla u|^{p-1} |\nabla \zeta| \, dx,&
\end{aligned}$$

where $c_3 = c_3(\alpha, \lambda)$. From Hölder's inequality with $\theta = \frac{q}{p-1+\alpha}$,

$$\int_\Omega (u_k+\epsilon)^{\alpha+p-1}\zeta^{1-p}|\nabla\zeta|^p dx \le \left(\int_\Omega (u_k+\epsilon)^q\zeta dx\right)^{\frac{1}{\theta}}\left(\int_\Omega \zeta^{1-p\theta'}|\nabla\zeta|^{p\theta'} dx\right)^{\frac{1}{\theta'}}.$$

If we take $k \geq 1$, we get

$$\frac{|\alpha|}{2}\int_\Omega |\nabla u|^p (u_k+\epsilon)^{\alpha-1}\zeta dx \le c_3\left(\int_\Omega (u_k+\epsilon)^q\zeta dx\right)^{\frac{1}{\theta}}\left(\int_\Omega \zeta^{1-p\theta'}|\nabla\zeta|^{p\theta'} dx\right)^{\frac{1}{\theta'}} + \int_{\{u>k\}} |\nabla u|^{p-1}|\nabla\zeta| dx, \tag{6.2.11}$$

and, letting $\epsilon \to 0$, yields

$$\frac{|\alpha|}{2}\int_\Omega |\nabla u|^p u_k^{\alpha-1}\zeta dx \le c_3\left(\int_\Omega u_k^q\zeta dx\right)^{\frac{1}{\theta}}\left(\int_\Omega \zeta^{1-p\theta'}|\nabla\zeta|^{p\theta'} dx\right)^{\frac{1}{\theta'}} + \int_{\{u>k\}} |\nabla u|^{p-1}|\nabla\zeta| dx. \tag{6.2.12}$$

Since $u$ is a weak solution, we can take $\zeta$ for test function and derive

$$\begin{aligned}\int_\Omega \zeta d\mu_0 + \int_\Omega \zeta d\mu_s^+ + \int_\Omega \zeta g(x,u,\nabla u)dx &= \int_\Omega \langle A(x,\nabla u),\nabla\zeta\rangle dx \\ &\le \lambda\int_\Omega u_k^{\frac{\alpha-1}{p'}}|\nabla u_k|^{p-1}|\nabla\zeta| u_k^{\frac{1-\alpha}{p'}} dx + \lambda\int_{\{u>k\}} |\nabla u|^{p-1}|\nabla\zeta| dx \\ &\le \left(\int_\Omega |\nabla u_k|^p u_k^{\alpha-1}\zeta dx\right)^{\frac{1}{p'}}\left(\int_\Omega u_k^{(1-\alpha)(p-1)}|\nabla\zeta|^p\zeta^{1-p} dx\right)^{\frac{1}{p'}} \\ &\quad + \lambda\int_{\{u>k\}} |\nabla u|^{p-1}|\nabla\zeta| dx.\end{aligned} \tag{6.2.13}$$

Since $q > p-1$, we fix $\alpha \in (1-p, 0)$ such that $\tau = \frac{q}{(1-\alpha)(p-1)} > 1$. Hence,

$$\begin{aligned}
&\int_\Omega \zeta d\mu + \int_\Omega \zeta g(x,u,\nabla u)dx \le \lambda \int_{\{u>k\}} |\nabla u|^{p-1} |\nabla\zeta|\, dx \\
&\qquad + \left(\int_\Omega |\nabla u_k|^p u_k^{\alpha-1}\zeta dx\right)^{\frac{1}{p'}} \left(\int_\Omega u_k^q \zeta dx\right)^{\frac{1}{\tau p}} \left(\int_\Omega \zeta^{1-\tau' p} |\nabla\zeta|^{\tau' p} dx\right)^{\frac{1}{\tau' p}} \\
&\le \lambda \int_{\{u>k\}} |\nabla u|^{p-1} |\nabla\zeta|\, dx \\
&+ \left[c_3 \left(\int_\Omega u_k^q \zeta dx\right)^{\frac{1}{\theta}} \left(\int_\Omega \zeta^{1-p\theta'} |\nabla\zeta|^{p\theta'} dx\right)^{\frac{1}{\theta'}} + \int_{\{u>k\}} |\nabla u|^{p-1} |\nabla\zeta|\, dx\right]^{\frac{1}{p'}} \\
&\times \left(\int_\Omega u_k^q \zeta dx\right)^{\frac{1}{\tau p}} \left(\int_\Omega \zeta^{1-\tau' p} |\nabla\zeta|^{\tau' p} dx.\right)^{\frac{1}{\tau' p}},
\end{aligned}$$

from (6.2.12) and (6.2.13). Because $g(.,u,\nabla u) + |\nabla u|^{p-1} \in L^1(\Omega)$, and using (6.2.8), we can let $k \to \infty$ and obtain

$$\begin{aligned}
\int_\Omega \zeta d\mu + \int_\Omega \zeta g(x,u,\nabla u)dx &\le c_4 \left(\int_\Omega u^q \zeta dx\right)^{\frac{1}{p'\theta} + \frac{1}{\tau p}} \\
&\quad \times \left(\int_\Omega \zeta^{1-p\theta'} |\nabla\zeta|^{p\theta'} dx\right)^{\frac{1}{p'\theta'}} \left(\int_\Omega \zeta^{1-\tau' p} |\nabla\zeta|^{\tau' p} dx\right)^{\frac{1}{\tau' p}}.
\end{aligned} \tag{6.2.14}$$

We notice that $\frac{1}{p'\theta'} + \frac{1}{\tau' p} = \frac{1}{q_p} = 1 - \left(\frac{1}{p'\theta} + \frac{1}{\tau p}\right)$. Using again (6.2.8) we derive in particular

$$\begin{aligned}
&\left(\int_\Omega \zeta g(x,u,\nabla u)dx\right)^{1-\frac{p-1}{q}} \\
&\qquad \le c_4 \left(\int_\Omega \zeta^{1-p\theta'} |\nabla\zeta|^{p\theta'} dx\right)^{\frac{1}{p'\theta'}} \left(\int_\Omega \zeta^{1-\tau' p} |\nabla\zeta|^{\tau' p} dx\right)^{\frac{1}{\tau' p}}.
\end{aligned} \tag{6.2.15}$$

Because $\tau < \frac{p}{q-1} < \theta$, we have by Hölder's inequality

$$\begin{aligned}
\int_\Omega \zeta^{1-p\theta'} |\nabla\zeta|^{p\theta'} dx &\le \left(\int_\Omega \zeta^{1-\tau' p} |\nabla\zeta|^{\tau' p} dx\right)^{\frac{\theta'}{\tau'}} \left(\int_\Omega \zeta dx\right)^{1-\frac{\theta'}{\tau'}} \\
&\le |\Omega|^{1-\frac{\theta'}{\tau'}} \left(\int_\Omega \zeta^{1-\tau' p} |\nabla\zeta|^{\tau' p} dx\right)^{\frac{\theta'}{\tau'}},
\end{aligned}$$

since $0 \leq \zeta \leq 1$. Combined with (6.2.15) it yields

$$\int_\Omega \zeta g(x,u,\nabla u)dx \leq c_5 \left(\int_\Omega \zeta^{1-\tau' p} |\nabla \zeta|^{\tau' p} dx\right)^{\frac{q_p}{\tau'}}. \tag{6.2.16}$$

Jointly with (6.2.14) it implies

$$\int_\Omega \zeta d\mu \leq c_6 \left(\int_\Omega \zeta^{1-\tau' p} |\nabla \zeta|^{\tau' p} dx\right)^{\frac{q_p}{\tau'}}. \tag{6.2.17}$$

We take now $\alpha < 0$ so that

$$pq_p < p\tau' = \frac{q}{q+1-p-|\alpha|(p-1)} \leq R,$$

and we infer (6.2.9) from (6.2.16) and (6.2.17) by Hölder's inequality. If we let $k \to \infty$ in (6.2.11), we also deduce from the assumptions on $g$,

$$\begin{aligned}
&\frac{|\alpha|}{2}\int_\Omega (\epsilon + u)^{\alpha-1} |\nabla u|^p \zeta dx \\
&\leq c_6 \left(\int_\Omega g(x,u+\epsilon,\nabla u)\zeta dx\right)^{\frac{1}{\theta}} \left(\int_\Omega \zeta^{1-p\theta'} |\nabla \zeta|^{p\theta'} dx\right)^{\frac{1}{\theta'}} \qquad (6.2.18)\\
&\leq c_2 \left(\int_\Omega g(x,u+\epsilon,\nabla u)\zeta dx\right) \left(\int_\Omega \zeta^{1-R} |\nabla \zeta|^{R} dx\right)^{\frac{p}{R}},
\end{aligned}$$

for any $0 \leq \epsilon \leq 1$, where $c_2$ depends on $N$, $p$, $q$, $\alpha$, $R$ and $\Omega$. □

It follows from this result a series of necessary conditions in order to solve (6.2.7) with a measure data.

**Corollary 6.2.2** *Assume $g$ satisfies the assumptions of Proposition 6.2.1 and $\mu$ is a bounded nonnegative measure. If there exists a renormalized solution $u \geq 0$ to problem (6.2.7), then for any $R > pq_p$ and any subdomain $G \subset \bar{G} \subset \Omega$ there exist two positive constants $c_1$ depending on $N$, $p$, $q$, $\Omega$, $G$ and $R$ and $c_2$ on $N$, $p$, $q$, $\Omega$, $G$ and $R$ such that*

$$\|\mu\|_{\mathfrak{M}_{\rho^R}} \leq c_1, \tag{6.2.19}$$

*where* $\|\mu\|_{\mathfrak{M}_{\rho^R}} = \int_\Omega \rho^R d|\mu|$ *and* $\rho(x) = \operatorname{dist}(x, \partial\Omega$, *and*

$$\mu(E) \leq c_2 c^\Omega_{1,R}(E) \quad \forall E \subset G,\ E \text{ Borel}. \tag{6.2.20}$$

We will see in Section 5.2.3 that this condition is not a necessary and sufficient condition for solving (6.2.7).

6.2.2.2 *The case* $1 < p < N$

The following general existence result for problem with forcing has to be compared with Theorem 5.1.2. It extends [Grenon (2002), Th 1.1] and [Bidaut-Véron, Hamid (2010), Th 2.18].

**Theorem 6.2.3** *Let* $2 - \frac{1}{N} \le p < N$, $\Omega \subset \mathbb{R}^N$ *be a bounded domain, A satisfies the assumptions (4.1.1) and* $g : \Omega \times \mathbb{R} \times \mathbb{R}^N \mapsto \mathbb{R}$ *verifies (6.0.3), (5.1.18),*

$$\int_1^\infty \tilde{g}\left(s, s^{\frac{N-1}{N-p}}\right) s^{-\frac{p(N-1)}{N-p}} ds < \infty, \tag{6.2.21}$$

*and*

$$\tilde{g}(as, bt) \le \left(\alpha a^{(p-1)\tilde{q}} + \beta b^{(p-1)\tilde{r}}\right) \tilde{g}(s,t), \quad \forall (a,b,s,t) \in \mathbb{R}^4_+, \tag{6.2.22}$$

*for some* $\tilde{q} > 1$, $\tilde{r} > 1$ *and* $\alpha, \beta \ge 0$ *verifying* $\alpha^2 + \beta^2 > 0$. *Then there exists* $m > 0$ *such that for any positive measure* $\mu$ *in* $\Omega$ *with* $\|\mu\|_{\mathfrak{M}} \le m$ *there exists a renormalized solution* $u$ *to (6.2.7).*

*Proof.* We recall that (5.1.18) is

$$|g(x,s,\xi)| \le \tilde{g}(|s|,|\xi|) \qquad \forall (x,s,\xi) \in \Omega \times \mathbb{R} \times \mathbb{R}^N,$$

where $\tilde{g}$ is a continuous function nondecreasing in its two variables. Furthermore (6.2.8) combined with (6.2.21) implies $q < q_1 = \frac{N(p-1)}{N-p}$. Let $a \in C^\infty(\mathbb{R}^*)$ such that $a(r) = 1$ for $0 \le r \le 1$, $a(r) = 0$ for $r \ge 2$ and $a'(r) \le 0$. For $n \in \mathbb{N}^*$ we set $a_n(r) = a(n^{-1}r)$ and

$$g_n(x,r,\xi) = g\left(x, T_n(r), a_n(|\xi|)\xi\right) \qquad \forall (x,r,\xi) \in \Omega \times \mathbb{R} \times \mathbb{R}^N. \tag{6.2.23}$$

Then

$$|g_n(x,r,\xi)| \le \tilde{g}(T_n(|r|), a_n(|\xi|)|\xi|) := \tilde{g}_n(|r|,|\xi|) \le \tilde{g}(|r|,|\xi|), \tag{6.2.24}$$

and we notice that $|g_n(x,r,\xi)| \le \tilde{g}(n,2n)$.

*Step 1: The approximation.* We set

$$X = \left\{ v \in L^{\frac{N(p-1)}{N-p},\infty}(\Omega) \text{ s.t. } \nabla v \in L^{\frac{N(p-1)}{N-1},\infty}(\Omega) \right\}.$$

Since $2 - \frac{1}{N} \le p < N$, $X$ is a Banach space for the norm

$$\|v\|_X = \|v\|_{L^{\frac{N(p-1)}{N-p},\infty}} + \|\nabla v\|_{L^{\frac{N(p-1)}{N-1},\infty}}.$$

We notice that $W_0^{1,p} \subset X$ with continuous imbedding by Gagliardo-Nirenberg inequality. In the same way as in the proof of Theorem 4.3.8 we consider an approximation of $\mu$ by the converging decomposition $\mu_\epsilon = f_\epsilon - \operatorname{div} g_\epsilon + \mu^+_{s,\epsilon}$ (there is no negative singular part). Thus $\mu_\epsilon \in W^{-1,p'}(\Omega)$. If $v \in X$, we denote by $u = \mathcal{T}_n(v)$ the unique solution of

$$\begin{aligned} -\operatorname{div} A(x, \nabla u) &= g_n(x, v, \nabla v) + \mu_\epsilon && \text{in } \Omega \\ u &= 0 && \text{on } \partial\Omega. \end{aligned} \tag{6.2.25}$$

Because the right-hand side of equation (6.2.25) is bounded in $W^{-1,p'}(\Omega)$, $\mathcal{T}_n(v)$ is uniformly bounded in $W_0^{1,p}(\Omega)$ and therefore in $X$. If $\{v_j\} \subset X$ is such that $\|v_j\|_X \leq R$, then up to a subsequence, $g_n(., v_j, \nabla v_j)$ converges in the weak-star topology of $L^\infty$ to some $f \in L^\infty(\Omega)$. By Rellich-Kondrachov theorem the imbedding of $W_0^{1,p}(\Omega)$ in $L^1(\Omega)$ is compact, and by Schauder duality theorem the imbedding of $L^\infty(\Omega)$ into $W^{-1,p'}(\Omega)$ shares this property ([Brezis (2010), Th 6.4]). Thus $g_n(., v_j, \nabla v_j)$ converges to $f$ in $W^{-1,p'}(\Omega)$ which implies that $u_j = \mathcal{T}_n(v_j)$ converges to some $u$ in $W_0^{1,p}(\Omega)$ and therefore in $X$ and there holds

$$\begin{aligned} -\operatorname{div} A(x, \nabla u) &= f + \mu_\epsilon && \text{in } \Omega \\ u &= 0 && \text{on } \partial\Omega, \end{aligned} \tag{6.2.26}$$

by Theorem 4.3.8. If we assume moreover that $v_j \to v$ in $X$, then $f = g_n(., v, \nabla v)$, thus $u = \mathcal{T}_n(v)$ which proves that $\mathcal{T}_n$ is compact.

*Step 2: The a priori bound.* We first show that $g(., v, \nabla v)$ is integrable in $\Omega$ if $v \in X$.

$$A^v_\sigma = \{x \in \Omega : |v(x)| > \sigma\} \quad \text{and} \quad a^v(\sigma) = \operatorname{meas}(A^v_\sigma),$$

$$B^v_\sigma = \{x \in \Omega : |\nabla v| > \sigma^{\frac{N-1}{N-p}}\} \quad \text{and} \quad b^v_\sigma = \operatorname{meas}(B^v_\sigma),$$

and $C^v_\sigma = A^v_\sigma \cap B^v_\sigma$ with $c^v(\sigma) = \operatorname{meas}(C^v_\sigma)$. Then

$$a^v(\sigma) \leq \min\left\{\sigma^{-\frac{N(p-1)}{N-p}} \|v\|^{\frac{N(p-1)}{N-p}}_{L^{\frac{N(p-1)}{N-p},\infty}}, |\Omega|\right\},$$

$$b^v(\sigma) \leq \min\left\{\sigma^{-\frac{N(p-1)}{N-p}} \|\nabla v\|^{\frac{N(p-1)}{N-1}}_{L^{\frac{N(p-1)}{N-1},\infty}}, |\Omega|\right\},$$

and

$$c^v(\sigma) \leq \min\left\{\min\left\{\|v\|^{\frac{N(p-1)}{N-p}}_{L^{\frac{N(p-1)}{N-p},\infty}}, \|\nabla v\|^{\frac{N(p-1)}{N-1}}_{L^{\frac{N(p-1)}{N-1},\infty}}\right\}\sigma^{-\frac{N(p-1)}{N-p}}, |\Omega|\right\}.$$

We set

$$\sigma_1 = \frac{\|v\|_{L^{\frac{N(p-1)}{N-p},\infty}}}{|\Omega|^{\frac{N-p}{N(p-1)}}}, \quad \sigma_2 = \frac{\|\nabla v\|^{\frac{N-p}{N-1}}_{L^{\frac{N(p-1)}{N-1},\infty}}}{|\Omega|^{\frac{N-p}{N(p-1)}}},$$

and $\sigma_3 = \min\{\sigma_1, \sigma_2\}$. Then

$$\begin{aligned}
\int_\Omega |g(x, v, \nabla v)|\, dx &\leq \int_\Omega \tilde{g}(|v|, |\nabla v|) dx \\
&\leq \int_{\Omega\cap C^v_{\sigma_3}} \tilde{g}(|v|, |\nabla v|) dx + \int_{\Omega\cap B^v_{\sigma_3}\cap(A^v_{\sigma_3})^c} \tilde{g}(|v|, |\nabla v|) dx \\
&\quad + \int_{\Omega\cap(A^v_{\sigma_3})^c\cap(B^v_{\sigma_3})^c} \tilde{g}(|v|, |\nabla v|) dx + \int_{\Omega\cap A^v_{\sigma_3}\cap(B^v_{\sigma_3})^c} \tilde{g}(|v|, |\nabla v|) dx.
\end{aligned}$$

Because $\tilde{g}$ is nondecreasing in its two variables and (6.2.21) and (6.2.22) hold, we obtain

$$\begin{aligned}
\int_{\Omega\cap(A^v_{\sigma_3})^c\cap(B^v_{\sigma_3})^c} \tilde{g}(|v|, |\nabla v|) dx &\leq \tilde{g}(\sigma_3, \sigma_3^{\frac{N-1}{N-p}})\, |\Omega| \\
&\leq \left( \alpha\sigma_3^{(p-1)\tilde{q}} + \beta\sigma_3^{\frac{(p-1)(N-1)}{N-p}\tilde{r}} \right) \tilde{g}(1, 1).
\end{aligned}$$

Since $dc^v(\sigma) = 0$ on $[0, \sigma_3)$,

$$\begin{aligned}
\int_{\Omega\cap C^v_{\sigma_3}} \tilde{g}(|v|, |\nabla v|) dx &= -\int_{\sigma_3}^\infty g\left(\sigma, \sigma^{\frac{N-1}{N-p}}\right) dc^v(\sigma) \\
&\leq c_1 \min\left\{ \|v\|^{\frac{N(p-1)}{N-p}}_{L^{\frac{N(p-1)}{N-p},\infty}}, \|\nabla v\|^{\frac{N(p-1)}{N-1}}_{L^{\frac{N(p-1)}{N-1},\infty}} \right\} \int_{\sigma_3}^\infty \tilde{g}(\sigma, \sigma^{\frac{N-1}{N-p}}) \sigma^{-\frac{p(N-1)}{N-p}} d\sigma \\
&\leq c_2 \int_1^\infty \tilde{g}(\sigma_3 t, (\sigma_3 t)^{\frac{N-1}{N-p}}) t^{-\frac{p(N-1)}{N-p}} dt \\
&\leq c_2 \left( \alpha\sigma_3^{(p-1)\tilde{q}} + \beta\sigma_3^{\frac{(p-1)(N-1)\tilde{r}}{N-p}} \right) \int_1^\infty \tilde{g}(t, t^{\frac{N-1}{N-p}}) t^{-\frac{p(N-1)}{N-p}} dt,
\end{aligned}$$

where $c_2 = c_2(N, p, |\Omega|) > 0$. Similarly $da^v(\sigma) = 0$ on $[0, \sigma_1)$, hence

$$\begin{aligned}\int_{\Omega\cap A^v_{\sigma_3}\cap(B^v_{\sigma_3})^c} \tilde{g}(|v|, |\nabla v|)dx &\leq -\int_{\sigma_1}^{\infty} \tilde{g}(\sigma, \sigma_3^{\frac{N-1}{N-p}})da^v(\sigma) \\ &\leq c_1 \|v\|^{\frac{N(p-1)}{N-p}}_{L^{\frac{N(p-1)}{N-p},\infty}} \int_{\sigma_1}^{\infty} \tilde{g}(\sigma, \sigma_3^{\frac{N-1}{N-p}})\sigma^{-\frac{p(N-1)}{N-p}} d\sigma \\ &\leq c_2 \left(\alpha\sigma_1^{(p-1)\tilde{q}} + \beta\sigma_3^{\frac{(p-1)(N-1)\tilde{r}}{N-p}}\right) \int_1^{\infty} \tilde{g}(t, 1)t^{-\frac{p(N-1)}{N-p}} dt,\end{aligned}$$

and, *mutatis mutandis*,

$$\begin{aligned}&\int_{\Omega\cap B^v_{\sigma_3}\cap(A^v_{\sigma_3})^c} \tilde{g}(|v|, |\nabla v|)dx \\ &\qquad\leq c_2 \left(\alpha\sigma_3^{(p-1)\tilde{q}} + \beta\sigma_2^{\frac{(p-1)(N-1)\tilde{r}}{N-p}}\right) \int_1^{\infty} \tilde{g}(1, t^{\frac{N-1}{N-p}})t^{-\frac{p(N-1)}{N-p}} dt.\end{aligned}$$

Combining these inequalities, we infer

$$\begin{aligned}\int_{\Omega} |g(x, v, \nabla v)|\, dx &\leq c_3 \left(\alpha\sigma_1^{(p-1)\tilde{q}} + \beta\sigma_2^{\frac{(p-1)(N-1)\tilde{r}}{N-p}}\right) \\ &\leq c_4 \left(\alpha \|v\|^{(p-1)\tilde{q}}_{L^{\frac{N(p-1)}{N-p},\infty}} + \beta \|\nabla v\|^{(p-1)\tilde{r}}_{L^{\frac{N(p-1)}{N-1},\infty}}\right),\end{aligned} \tag{6.2.27}$$

where

$$c_3 = 4c_2\int_1^{\infty} \tilde{g}\left(s, s^{\frac{N-1}{N-p}}\right) s^{-\frac{p(N-1)}{N-p}} ds \text{ and } c_4 = c_3 \min\left\{|\Omega|^{\frac{(p-N)\tilde{q}}{N}}, |\Omega|^{\frac{(1-N)\tilde{r}}{N}}\right\}.$$

Notice that these inequalities hold uniformly with respect to $n$ if $g$ is replaced by $g_n$ and $\tilde{g}$ by $\tilde{g}_n$. If $v \in X$ and $u = \mathcal{T}_n(v)$, then, by (4.3.26) and (4.3.27) in Proposition 4.3.9,

$$\|u\|^{p-1}_{L^{\frac{N(p-1)}{N-p},\infty}} + \|\nabla u\|^{p-1}_{L^{\frac{N(p-1)}{N-1},\infty}} \leq c_5 \left(\int_{\Omega} |g_n(x, v, \nabla v)|\, dx + \int_{\Omega} d\mu\right), \tag{6.2.28}$$

with $c_5 = c_5(N, p, |\Omega|) > 0$. Combining (6.2.27) and (6.2.28) we derive

$$\begin{aligned}\|u\|_{L^{\frac{N(p-1)}{N-p},\infty}} + \|\nabla u\|_{L^{\frac{N(p-1)}{N-1},\infty}} &\leq c_6 \left(\alpha' \|v\|^{\tilde{q}}_{L^{\frac{N(p-1)}{N-p},\infty}} + \beta' \|\nabla v\|^{\tilde{r}}_{L^{\frac{N(p-1)}{N-1},\infty}}\right) \\ &\quad + c_7 \left(\int_{\Omega} d\mu\right)^{\frac{1}{p-1}}\end{aligned} \tag{6.2.29}$$

where $\alpha' = \alpha^{\frac{1}{p-1}}$ and $\beta' = \beta^{\frac{1}{p-1}}$. For $R > 0$ we define a bounded closed convex subset $\mathcal{K}_R$ by

$$\mathcal{K}_R = \{v \in X : \|v\|_{L^{\frac{N(p-1)}{N-p},\infty}} + \|\nabla v\|_{L^{\frac{N(p-1)}{N-1},\infty}} \leq R\}. \tag{6.2.30}$$

If $v \in \mathcal{K}_R$ and $u = \mathcal{T}_n(v)$, there holds by (6.2.29)

$$\|u\|_{L^{\frac{N(p-1)}{N-p},\infty}} + \|\nabla u\|_{L^{\frac{N(p-1)}{N-1},\infty}} \leq c_6\left(R^{\tilde{q}} + R^{\tilde{r}}\right) + A \tag{6.2.31}$$

where $A = c_7\left(\int_\Omega d\mu\right)^{\frac{1}{p-1}}$. Put $\Phi(R) = c_6\left(R^{\tilde{q}} + R^{\tilde{r}}\right) + A - R$ for $R \geq 0$. Then $\Phi$ is convex, $\Phi(0) = A$ and $\Phi(R) \to \infty$ when $R \to \infty$. Furthermore

$$\Phi'(R) = c_6\left(\tilde{q}R^{\tilde{q}-1} + \tilde{r}R^{\tilde{r}-1}\right) - 1,$$

hence there exists a unique $R_0 > 0$ such that $\Phi'(R_0) = 0$. Consequently there exists a unique $A_0$ such that $\Phi(R_0) = 0$. Therefore

$$\|u\|_{L^{\frac{N(p-1)}{N-p},\infty}} + \|\nabla u\|_{L^{\frac{N(p-1)}{N-1},\infty}} - R_0 \leq \Phi(R_0) = 0, \tag{6.2.32}$$

which implies that $u \in \mathcal{K}_{R_0}$. By the Leray-Schauder theorem, $\mathcal{T}_n$ admits a fixed point $u_n$ in $\mathcal{K}_{R_0}$. Note that the value of $R_0$ depends in $N$ and $p$, but not on $n$, and it is the same with $A_0$.

*Step 3: Convergence.* Let $\{u_n\} \subset \mathcal{K}_{R_0}$ be a sequence of fixed points of $\mathcal{T}_n$. By (6.2.28) the set of functions $\{g_n(., u_n\nabla u_n)\}$ is bounded in $L^1(\Omega)$. Let $E \subset \Omega$ be a Borel set and $\sigma > 0$. Following the method used in the proof of Theorem 5.1.2, we put

$$A_\sigma^{u_n} = \{x \in \Omega : |u_n(x)| > \sigma\} \quad \text{and} \quad a^{u_n}(\sigma) = \text{meas}\,(A_\sigma^{u_n}),$$

$$B_\sigma^{u_n} = \{x \in \Omega : |\nabla u_n(x)| > \sigma^{\frac{N-1+\delta}{N-p}}\} \quad \text{and} \quad b^{u_n}(\sigma) = \text{meas}\,(B_\sigma^{n}),$$

and $C_\sigma^{u_n} = A_\sigma^{u_n} \cap B_\sigma^{u_n}$ with $c^{u_n}(\sigma)(\sigma) = \text{meas}\,(C_\sigma^{n})$. Then

$$\max\{a^{u_n}(\sigma), b^{u_n}(\sigma), c^{u_n}(\sigma)\} \leq \theta_{u_n}(\sigma) \leq \min\left\{c_R s^{-\frac{N(p-1)}{N-p}}, |\Omega|\right\}.$$

Hence we get

$$\begin{aligned}\int_E |g(x,u_n,\nabla u_n)|\,dx &\leq \int_{E\cap C_\sigma^{u_n}} \tilde{g}(|u_n|\,,|\nabla u_n|)dx\\ &+\int_{E\cap B_\sigma^{u_n}\cap(A_\sigma^{u_n})^c} \tilde{g}(|u_n|\,,|\nabla u_n|)dx\\ &+\int_{E\cap A_\sigma^{u_n}\cap(B_\sigma^{u_n})^c} \tilde{g}(|u_n|\,,|\nabla u_n|)dx\\ &+\int_{E\cap(A_\sigma^{u_n})^c\cap(B_\sigma^{u_n})^c} \tilde{g}(|u_n|\,,|\nabla u_n|)dx\\ &\leq c_8\int_\sigma^\infty \tilde{g}(t,t^{\frac{N-1}{N-p}})t^{-\frac{p(N-1)}{N-p}}+\tilde{g}(\sigma,\sigma^{\frac{N-1}{N-p}})\,|E|\,.\end{aligned}$$

This implies that the set of functions $\{g(.,u_n,\nabla u_n)\}$ is uniformly integrable in $\Omega$. Since it is bounded in $L^1(\Omega)$, it is relatively weakly compact by the Dunford-Pettis theorem. Furthermore $T_k(u_n)$ converges to $T_k(u_\epsilon)$ in $W_0^{1,p}(\Omega)$ for any $k>0$ and $\nabla u_n\to\nabla u_\epsilon$ in $(L^r(\Omega))^N)$ for all $r<\frac{N(p-1)}{N-1}$ (see Step 6 in the proof of Theorem 4.3.8). Therefore, up to a subsequence, $g_n(.,u_n,\nabla u_n)$ converges weakly to some $f_\epsilon\in L^1(\Omega)$. By Theorem 4.3.8 $\{T_n(u_n)\}$ converges a.e. in $\Omega$ to a renormalized solution $u_\epsilon$ of

$$\begin{aligned}-\text{div}\,A(x,\nabla u_\epsilon)&=f_\epsilon+\mu_\epsilon &&\text{in }\Omega\\ u_\epsilon&=0 &&\text{on }\partial\Omega.\end{aligned}\tag{6.2.33}$$

Furthermore, $T_k(u_n)$ converges to $T_k(u_\epsilon)$ in $W_0^{1,p}(\Omega)$ for any $k>0$ and $\nabla u_n\to\nabla u_\epsilon$ in $(L^r(\Omega))^N)$ for all $r<\frac{N(p-1)}{N-1}$ (see Step 6 in the proof of Theorem 4.3.8). Up to a subsequence $\nabla u_n\to\nabla u_\epsilon$ a.e. in $\Omega$. By the Vitali convergence theorem we infer that $g_n(.,u_n,\nabla u_n)\to g(.,u_\epsilon,\nabla u_\epsilon)=f_\epsilon$ in $L^1(\Omega)$ which proves that $u_\epsilon$ is a renormalized solution of

$$\begin{aligned}-\text{div}\,A(x,\nabla u_\epsilon)&=g(.,u_\epsilon,\nabla u_\epsilon)+\mu_\epsilon &&\text{in }\Omega\\ u_\epsilon&=0 &&\text{on }\partial\Omega.\end{aligned}\tag{6.2.34}$$

*Step 4: End of the proof.* Since the set $\{g(.,u_\epsilon,\nabla u_\epsilon)\}$ is uniformly bounded in $L^1\Omega$, it is uniformly integrable as in Step 3 thanks to Theorem 4.3.8. We first obtain that $g(.,u_\epsilon,\nabla u_\epsilon)$ converges weakly (up to a subsequence) in $L^1\Omega$ to some $f$, then $u_\epsilon\to u$ a.e. in $\Omega$. Then $u$ is a renormalized solution of

$$\begin{aligned}-\text{div}\,A(x,\nabla u)&=f+\mu &&\text{in }\Omega\\ u&=0 &&\text{on }\partial\Omega.\end{aligned}\tag{6.2.35}$$

Next $T_k(u_\epsilon)$ to $T_k(u)$ in $W_0^{1,p}(\Omega)$ and $\nabla u_\epsilon \to \nabla u$ in $(L^r(\Omega))^N)$ for all $r < \frac{N(p-1)}{N-1}$. We use again Vitali's theorem and derive that $f = g(.,u,\nabla u)$ and $u$ is a renormalized solution of (6.2.7). □

**Applications**

- If $g(x,u,\nabla u) = |u|^q$ then (6.2.21) and (6.2.22) are satisfied provided $p-1 < q < q_1$.
- If $g(x,u,\nabla u) = |\nabla u|^r$ then (6.2.21) and (6.2.22) are satisfied provided $p-1 < r < q_c = \frac{N(p-1)}{N-1}$.
- If $g(x,u,\nabla u) = |u|^q + |\nabla u|^r$, (6.2.21) is satisfied if $p-1 < q < q_1$ and $p-1 < r < q_c$. Since

$$(a|u|)^q + (b|\nabla u|)^r \le (a^q + b^r)(|u|^q + |\nabla u|^r) \quad \forall a, b \ge 0,$$

(6.2.22) is verified.
- If $g(x,u,\nabla u) = |u|^q |\nabla u|^r$ with $qr > 0$, we write

$$(a|u|)^q (b|\nabla u|)^r = a^q b^r |u|^q |\nabla u|^r \le \left(\frac{a^{q\kappa}}{\kappa} + \frac{b^{r\kappa'}}{\kappa'}\right) |u|^q |\nabla u|^r.$$

The condition $p-1 < \min\{\kappa q, \kappa' r\}$ is satisfied for some $\kappa \in (1,\infty)$ if $q + r > p - 1$, and (6.2.21) holds if $0 < q(N-p) + r(N-1) < N(p-1)$.

#### 6.2.2.3 *The case $p = N$*

When $p = N$ the Gagliardo-Nirenberg inequality does not apply since $W_0^{1,N}(\Omega)$ does not imbed in $L^\infty(\Omega)$. The corresponding inequality is due to [Pohozaev (1965)], [Moser (1967)] and [Trudinger (1980)]. The extension of Moser-Pohozaev-Trudinger inequality to Sobolev-Lorentz spaces is essentially obtained in [Brezis, Wainger (1980)] (under a slightly different form) and in [Alvino, Ferone, Trombetti (1996)].

**Proposition 6.2.4** *Let $\Omega$ be a bounded domain in $\mathbb{R}^N$. For any $1 < q \le \infty$ there exists $c_{N,q} > 0$ such that*

$$\sup_{\|\nabla v\|_{L^{N,q}} \le 1} \int_\Omega e^{\theta_{N,q} |u(x)|^{\frac{q}{q-1}}} dx \le c_{N,q} |\Omega|, \qquad (6.2.36)$$

*where $\theta_{N,q} = (N\omega_N^{\frac{1}{N}})^{\frac{q}{q-1}}$ and $\omega_N$ is the volume of the unit ball in $\mathbb{R}^N$.*

In the sequel we set $\theta_{N,\infty} = \theta_N = N\omega_N^{\frac{1}{N}}$ and $c_{N,\infty} = c_N$. A consequence of this estimates is the following result.

**Lemma 6.2.5** *Let $\Omega$ be a bounded domain in $\mathbb{R}^N$. Then for any $v \in C_0^1(\Omega)$ there holds*

$$a^v(\sigma) \leq \min\left\{c_N e^{-\frac{\theta_N \sigma}{\|\nabla v\|_{L^{N,\infty}}}}, 1\right\}|\Omega|, \tag{6.2.37}$$

*where $A_\sigma^v = \{x \in \Omega : |v(x)| > \sigma\}$ and $a^v(\sigma) = \operatorname{meas}(A_\sigma^v)$. Moreover, for any $1 < m < \infty$,*

$$a^v(\sigma) \leq \min\left\{d_{N,m} \|\nabla v\|^m_{L^{N,\infty}} \sigma^{-m}, 1\right\}|\Omega|, \tag{6.2.38}$$

*where $d_{N,m} = c_N \theta_N^{-m} \Gamma(m+1)$.*

*Proof.* Put $\tilde{v} = \frac{v}{\|\nabla v\|_{L^{N,\infty}}}$. Since inequality

$$a^v(\sigma' \|\nabla v\|_{L^{N,\infty}}) e^{\theta_N \sigma'} = a^{\tilde{v}}(\sigma') e^{\theta_N \sigma'} \leq \int_{A_\sigma^{\tilde{v}}} e^{\theta_N |\tilde{v}(x)|} dx \leq c_N |\Omega|,$$

holds by (6.2.36) with $q = \infty$, (6.2.37) follows if we set $\sigma = \sigma' \|\nabla v\|_{L^{N,\infty}}$. Furthermore, for $m > 1$,

$$\frac{\theta_N^m}{\Gamma(m+1)} \int_\Omega |v(x)|^m dx \leq c_N |\Omega| \|\nabla v\|^m_{L^{N,\infty}},$$

($\Gamma(m+1) = r!$ if $m \in \mathbb{N}^*$) which yields

$$\|v\|_{L^m} \leq \theta_N^{-1} \left(c_N \Gamma(m+1) |\Omega|\right)^{\frac{1}{m}} \|\nabla v\|_{L^{N,\infty}}, \tag{6.2.39}$$

and (6.2.38) follows. □

The next result is the mere extension of Theorem 6.2.3 to the case $p = N$.

**Theorem 6.2.6** *Let $\Omega \subset \mathbb{R}^N$ be a bounded domain, $A$ satisfies the assumptions (4.1.1) with $p = N$ and $g : \Omega \times \mathbb{R} \times \mathbb{R}^N \mapsto \mathbb{R}$ verifies (6.0.3), (5.1.18),*

$$\int_1^\infty \tilde{g}\left(s, s^{\frac{q}{N}}\right) s^{-q-1} ds < \infty, \tag{6.2.40}$$

*for $q \geq 1$ and*

$$\tilde{g}(as, bt) \leq \left(\alpha a^{(N-1)\tilde{q}} + \beta b^{(N-1)\tilde{r}}\right) \tilde{g}(s,t), \quad \forall (a,b,s,t) \in \mathbb{R}_+^4, \tag{6.2.41}$$

*for some $\tilde{q} > 1$, $\tilde{r} > 1$ and $\alpha, \beta \geq 0$ verifying $\alpha^2 + \beta^2 > 0$. Then there exists $m > 0$ such that for any positive measure $\mu$ in $\Omega$ with $\|\mu\|_{\mathfrak{M}} \leq m$ there exists a renormalized solution $u$ to (6.2.7).*

*Proof.* Up to some modifications that we indicate the proof follows the one of Theorem 6.2.3 with the same notations. We define $A_\sigma^v$ and $a^v(\sigma)$ as in Theorem 6.2.3 and we have

$$a^v(\sigma) \leq \min\left\{d_{N,q} \|\nabla v\|^q_{L^{N,\infty}} \sigma^{-q}, 1\right\} |\Omega| .$$

We set

$$B_\sigma^v = \left\{x \in \Omega : |\nabla v| > \sigma^{\frac{q}{N}}\right\} \text{ and } b^v(\sigma) = \text{meas}(B_\sigma^v),$$

then

$$b^v(\sigma) \leq \min\left\{\|\nabla v\|^N_{L^{N,\infty}} \sigma^{-q}, |\Omega|\right\} = \min\left\{|\Omega|^{-1} \|\nabla v\|^N_{L^{N,\infty}} \sigma^{-q}, 1\right\} |\Omega| ,$$

and $c^v(\sigma) = \text{meas}(B_\sigma^v \cap A_\sigma^v)$, then

$$c^v(\sigma) \leq \min\left\{\min\left\{d_{N,q} \|\nabla v\|^q_{L^{N,\infty}} , |\Omega|^{-1} \|\nabla v\|^N_{L^{N,\infty}}\right\} \sigma^{-q}, 1\right\} |\Omega| .$$

We put

$$\sigma_1 = d_{N,q}^{\frac{1}{q}} \|\nabla v\|_{L^{N,\infty}} \ , \ \sigma_2 = |\Omega|^{-\frac{1}{q}} \|\nabla v\|^{\frac{N}{q}}_{L^{N,\infty}} \text{ and } \sigma_3 = \min\{\sigma_1, \sigma_2\}.$$

After some computation (6.2.27) is transformed into

$$\int_\Omega |g(x, v, \nabla v)| \, dx \leq c_1 \left(\alpha \sigma_1^{(N-1)\tilde{q}} + \beta \sigma_2^{\frac{(N-1)\tilde{r}}{N}}\right), \tag{6.2.42}$$

where $c_1$ depends on $N$, $q$, $|\Omega|$ and $\int_1^\infty g(t, t^{\frac{q}{N}} t^{-q-1} dt$. If $u = \mathcal{T}_n(v)$, there holds

$$\|u\|_{BMO} + \|\nabla u\|_{L^{N,\infty}} \leq c_2 \left(\|\nabla v\|^{\tilde{q}}_{L^{N,\infty}} + \|\nabla v\|^{\tilde{r}}_{L^{N,\infty}}\right) + \left(\int_\Omega d\mu\right)^{\frac{1}{N-1}}, \tag{6.2.43}$$

where $c_2$ depends on $c_1$. When the integration is not performed on whole $\Omega$ but on a Borel set $E$, the condition (6.2.40) yields the uniform integrability of the sequence $\{g_n(., u_n, \nabla u_n\}$ and $\{g(., u_\epsilon, \nabla u_\epsilon\}$ as in Theorem 6.2.3. The conclusion of the proof follows. □

**Applications**

If $g(x, u, \nabla u) = \ell_1 |u|^q + \ell_2 |u|^{q^*} |\nabla u|^{r^*} + \ell_3 |\nabla u|^r$, then the assumptions of Theorem 6.2.6 are satisfied if $N - 1 < q$ and $N - 1 < r < N$.

**Theorem 6.2.7** *Let $\Omega \subset \mathbb{R}^N$ be a bounded domain, $A$ satisfy the assumptions (4.1.1) with $p = N$ and $g : \Omega \times \mathbb{R} \times \mathbb{R}^N \mapsto \mathbb{R}$ verify (6.0.3),*

*(5.1.18),*

$$\tilde{g}(s,t) \leq \alpha \sinh(\lambda s) + \beta t^{\nu} \qquad \forall (s,t) \in \mathbb{R}_+^2, \tag{6.2.44}$$

*for $0 < \nu < N$, some $\alpha, \beta, \lambda \geq 0$ such that $\alpha^2 + \beta^2 > 0$. Then for any $\lambda_0 > 0$ there exists $m_0 = m_0(\Omega, \nu, N, \lambda_0) > 0$ such that for any $(\alpha, \beta, \lambda) \in \mathbb{R}_+ \times \mathbb{R}_+ \times [0, \lambda_0]$ and any positive measure $\mu$ in $\Omega$, the condition*

$$\alpha\lambda + \beta + \mu(\Omega) \leq m_0, \tag{6.2.45}$$

*implies that there exists a renormalized solution $u$ to (6.2.7).*

*Proof.* We proceed as in Theorem 6.2.3 using the Moser-Trudinger type estimates as in the case $p = N$ in Theorem 5.1.2. We define $g_n$ and $\tilde{g}_n$ by (6.2.23) and denote by $X_N$ the space of functions $v$ in $BMO(\Omega)$ such that $\|\nabla v\|_{L^{N,\infty}}$ It is known that $BMO(\Omega) \subset X_N$ [Tartar (1998)] and (4.3.43) holds, therefore $X_N$ is a Banach space for the norm

$$\|v\|_{X_N} = \|\nabla v\|_{L^{N,\infty}} \sim \|v\|_{BMO} + \|\nabla v\|_{L^{N,\infty}}.$$

*Step 1: Estimate of $\|g(.,v,\nabla v)\|_{L^1}$.* We assume that $v \in X_N$ and define for $\sigma > 0$,

$$A_\sigma^v = \{x \in \Omega : |v(x)| > \sigma\} \text{ and } a^v(\sigma) = \text{meas}\,(A_\sigma^v).$$

From (6.2.37), we get for $\sigma > 0$,

$$a^v(\sigma) \leq \min\left\{2^{-N} c_N \operatorname{csch}^N\left(\tfrac{\theta_N \sigma}{N\|\nabla v\|_{L^{N,\infty}}}\right), 1\right\} |\Omega|. \tag{6.2.46}$$

Next, for some constant $c > 0$ which will be made precise later on,

$$B_\sigma^v = \{x \in \Omega : |\nabla v(x)| > \sinh(c\sigma)\} \text{ and } b^v(\sigma) = \text{meas}\,(B_\sigma^v),$$

then

$$b^v(\sigma) \leq \min\left\{\|\nabla v\|_{L^{N,\infty}}^N \operatorname{csch}^N(c\sigma), |\Omega|\right\}.$$

If we take $c = \frac{\theta_N}{N\|\nabla v\|_{L^{N,\infty}}}$, we obtain

$$b^v(\sigma) \leq \min\left\{\|\nabla v\|_{L^{N,\infty}}^N \operatorname{csch}^N\left(\tfrac{\theta_N \sigma}{N\|\nabla v\|_{L^{N,\infty}}}\right), |\Omega|\right\}. \tag{6.2.47}$$

Finally $C^v_\sigma = A^v_\sigma \cap B^v_\sigma$, $c^v(\sigma) = \operatorname{meas}(C^v_\sigma)$, hence

$$c^v(\sigma) \leq \min\left\{\min\left\{\|\nabla v\|^N_{L^{N,\infty}}, 2^{-N}c_N|\Omega|\right\}\operatorname{csch}^N\left(\tfrac{\theta_N\sigma}{N\|\nabla v\|_{L^{N,\infty}}}\right), |\Omega|\right\}. \tag{6.2.48}$$

Set

$$\sigma_1 = \frac{N\|\nabla v\|_{L^{N,\infty}}}{\theta_N}\operatorname{arsinh}\left(\frac{c_N^{\frac{1}{N}}}{2}\right),\ \sigma_2 = \frac{N\|\nabla v\|_{L^{N,\infty}}}{\theta_N}\operatorname{arsinh}\left(\frac{\|\nabla v\|_{L^{N,\infty}}}{|\Omega|^{\frac{1}{N}}}\right),$$

and $\sigma_3 = \min\{\sigma_1, \sigma_2\}$. As in the proof of Theorem 6.2.3 we write

$$\begin{aligned}\int_\Omega |g(x,v,\nabla v)|\,dx &\leq \int_\Omega \tilde{g}(|v|,|\nabla v|)dx\\ &\leq \int_{\Omega\cap C^v_{\sigma_3}} \tilde{g}(|v|,|\nabla v|)dx + \int_{\Omega\cap B^v_{\sigma_3}\cap(A^v_{\sigma_3})^c}\tilde{g}(|v|,|\nabla v|)dx\\ &+ \int_{\Omega\cap(A^v_{\sigma_3})^c\cap(B^v_{\sigma_3})^c}\tilde{g}(|v|,|\nabla v|)dx + \int_{\Omega\cap A^v_{\sigma_3}\cap(B^v_{\sigma_3})^c}\tilde{g}(|v|,|\nabla v|)dx.\end{aligned}$$

Firstly

$$\begin{aligned}\int_{\Omega\cap(A^v_{\sigma_3})^c\cap(B^v_{\sigma_3})^c}\tilde{g}(|v|,|\nabla v|)dx &\leq \tilde{g}\left(\sigma_3, \sinh(\tfrac{\theta_N\sigma_3}{N\|\nabla v\|_{L^{N,\infty}}})\right)|\Omega|\\ &\leq \tilde{g}\left(\sigma_3, \sinh(\tfrac{c_1\sigma_3}{\sigma_1})\right)|\Omega|,\end{aligned}$$

where $c_1 = \operatorname{arsinh}\left(\frac{c_N^{\frac{1}{N}}}{2}\right)$. Then

$$\begin{aligned}\int_{\Omega\cap A^v_{\sigma_3}\cap(B^v_{\sigma_3})^c}\tilde{g}(|v|,|\nabla v|)dx &\leq -\int_{\sigma_3}^\infty \tilde{g}\left(\sigma, \sinh(\tfrac{\theta_N\sigma_3}{N\|\nabla v\|_{L^{N,\infty}}})\right)da^v(\sigma)\\ &\leq -\int_{\sigma_1}^\infty \tilde{g}\left(\sigma, \sinh(\tfrac{c_1\sigma_3}{\sigma_1})\right)da^v(\sigma)\\ \leq \frac{2^{-N}c_N\theta_N}{\|\nabla v\|_{L^{N,\infty}}}&\int_{\sigma_1}^\infty \tilde{g}\left(\sigma, \sinh(\tfrac{c_1\sigma_3}{\sigma_1})\right)\coth\left(\tfrac{\theta_N\sigma}{N\|\nabla v\|_{L^{N,\infty}}}\right)\operatorname{csch}^N\left(\tfrac{\theta_N\sigma}{N\|\nabla v\|_{L^{N,\infty}}}\right)d\sigma\\ \leq 2^{-N}Nc_Nc_1&\int_1^\infty \tilde{g}\left(\sigma_1 t, \sinh(\tfrac{c_1\sigma_3}{\sigma_1})\right)\coth(c_1t)\operatorname{csch}^N(c_1t)\,dt\\ &\leq c_2\int_1^\infty \tilde{g}\left(\sigma_1 t, \sinh(\tfrac{c_1\sigma_3}{\sigma_1})\right)\operatorname{csch}^N(c_1t)\,dt,\end{aligned}$$

where $c_2 = 2^{-N}Nc_Nc_1\coth c_1$. Next

$$\int_{\Omega\cap B^v_{\sigma_3}\cap(A^v_{\sigma_3})^c}\tilde{g}(|v|,|\nabla v|)dx \le -\int_{\sigma_3}^{\infty}\tilde{g}\left(\sigma_3,\sinh(c\sigma)\right)db^v(\sigma)$$
$$\le \theta_N\int_{\sigma_2}^{\infty}\tilde{g}(\sigma_3,\sinh(c\sigma))\coth(c\sigma)\operatorname{csch}^N(c\sigma)d\sigma$$
$$\le Nc_1\|\nabla v\|^N_{L^{N,\infty}}\int_{\frac{\sigma_2}{\sigma_1}}^{\infty}\tilde{g}\left(\sigma_3,\sinh(c_1t)\right)\coth(c_1t)\operatorname{csch}^N(c_1t)dt.$$

Finally

$$\int_{\Omega\cap C^v_{\sigma_3}}\tilde{g}(|v|,|\nabla v|)dx \le -\int_{\sigma_3}^{\infty}\tilde{g}(\sigma,\sinh(c\sigma))dc^v(\sigma)$$
$$\le Nc_1\min\left\{\|\nabla v\|^N_{L^{N,\infty}},2^{-N}c_N|\Omega|\right\}$$
$$\times\int_{\frac{\sigma_3}{\sigma_1}}^{\infty}\tilde{g}(\sigma_1 t,\sinh(c_1t))\coth(c_1t)\operatorname{csch}^N(c_1t)\,dt.$$

Using (6.2.44) we obtain

$$\int_{\Omega\cap(A^v_{\sigma_3})^c\cap(B^v_{\sigma_3})^c}\tilde{g}(|v|,|\nabla v|)dx \le \left(\alpha\sinh(\lambda\sigma_3)+\beta\sinh^\nu\left(\tfrac{c_1\sigma_3}{\sigma_1}\right)\right)|\Omega|,$$

$$\int_{\Omega\cap A^v_{\sigma_3}\cap(B^v_{\sigma_3})^c}\tilde{g}(|v|,|\nabla v|)dx$$
$$\le c_2\int_1^{\infty}\left(\alpha\sinh(\lambda\sigma_1 t)+\beta\sinh^\nu\left(\tfrac{c_1\sigma_3}{\sigma_1}\right)\right)\operatorname{csch}^N(c_1t)\,dt,$$
$$\le c_3\left(\frac{\alpha\lambda\sigma_1}{N^2c_1^2-\lambda^2\sigma_1^2}+\frac{\beta}{Nc_1}\sinh^\nu\left(\tfrac{c_1\sigma_3}{\sigma_1}\right)\right),$$

where $c_3$ depends only on $N$,

$$\int_{\Omega\cap B^v_{\sigma_3}\cap(A^v_{\sigma_3})^c}\tilde{g}(|v|,|\nabla v|)dx$$
$$\le Nc_1\|\nabla v\|^N_{L^{N,\infty}}\int_{\frac{\sigma_2}{\sigma_1}}^{\infty}(\alpha\sinh(\lambda\sigma_3)+\beta\sinh^\nu(c_1t))\coth(c_1t)\operatorname{csch}^N(c_1t)\,dt$$
$$\le \|\nabla v\|^N_{L^{N,\infty}}\left(\alpha\sinh(\lambda\sigma_3)\operatorname{csch}^N\left(\tfrac{c_1\sigma_2}{\sigma_1}\right)+\frac{N\beta}{N-\nu}\operatorname{csch}^{N-\nu}\left(\tfrac{c_1\sigma_2}{\sigma_1}\right)\right),$$

and

$$\int_{\Omega\cap C^v_{\sigma_3}} \tilde{g}(|v|,|\nabla v|)dx \le Nc_1 \min\left\{\|\nabla v\|^N_{L^{N,\infty}}, 2^{-N}c_N|\Omega|\right\}$$
$$\times \int_{\frac{\sigma_3}{\sigma_1}}^{\infty} (\alpha\sinh(\lambda\sigma_1 t) + \beta\sinh^{\nu}(c_1 t))\coth(c_1 t)\operatorname{csch}^N(c_1 t)\,dt,$$

which is estimated in two steps: Firstly

$$\int_{\frac{\sigma_3}{\sigma_1}}^{1} \alpha\sinh(\lambda\sigma_1 t)\coth(c_1 t)\operatorname{csch}^N(c_1 t)dt = \frac{\alpha c_1}{N}\sinh(\lambda\tfrac{\sigma_3}{\sigma_1})\operatorname{csch}^N\left(c_1\tfrac{\sigma_3}{\sigma_1}\right)$$
$$+\frac{\alpha\lambda c_1}{N}\int_{\frac{\sigma_3}{\sigma_1}}^{1}\cosh(\lambda\sigma_1 t)\operatorname{csch}^N(c_1 t)dt$$
$$\le \frac{\alpha c_1^{1-N}}{N}\sinh(\lambda\tfrac{\sigma_3}{\sigma_1})\left(\frac{\sigma_3}{\sigma_1}\right)^{-N} + \frac{\alpha\lambda\cosh(\lambda\frac{\sigma_3}{\sigma_1})}{N(N-1)c_1^{N-2}}\left(\frac{\sigma_3}{\sigma_1}\right)^{1-N}$$
$$\le \frac{2\alpha\lambda\cosh\lambda}{(N-1)c_1^{N-2}}\left(\frac{\sigma_3}{\sigma_1}\right)^{1-N}.$$

Hence

$$\int_1^{\infty}\alpha\sinh(\lambda\sigma_1 t)\coth(c_1 t)\operatorname{csch}^N(c_1 t)dt \le c_4\alpha\int_1^{\infty}\sinh(\lambda\sigma_1 t)e^{-c_1 N t}dt$$
$$\le \frac{c_4\alpha\lambda\sigma_1}{N^2c_1^2-\lambda^2\sigma_1^2},$$

where $c_4 = c_4(N) > 0$. From these inequalities it follows

$$\int_{\Omega\cap C^v_{\sigma_3}} \tilde{g}(|v|,|\nabla v|)dx \le Nc_1 \min\left\{\|\nabla v\|^N_{L^{N,\infty}}, 2^{-N}c_N|\Omega|\right\}$$
$$\times\left(\frac{N\beta}{N-\nu}\operatorname{csch}^{N-\nu}\left(\tfrac{c_1\sigma_3}{\sigma_1}\right) + \frac{2\alpha\lambda\cosh\lambda}{(N-1)c_1^{N-2}}\left(\frac{\sigma_3}{\sigma_1}\right)^{1-N} + \frac{c_4\alpha\lambda\sigma_1}{N^2c_1^2-\lambda^2\sigma_1^2}\right).$$

*Step 2: The a priori estimates.* For $R > 0$ we set

$$\mathcal{K}_R = \{v \in X_N : \|v\|_{X_N} \le R\},$$

with $R \le R_\kappa$, where

$$R_\kappa = \min\left\{\frac{(c_N|\Omega|)^{\frac{1}{N}}}{2}, \frac{\kappa\theta_N}{\lambda}\right\}, \tag{6.2.49}$$

for $\kappa \in (0,1)$. If $v \in X_N$ there holds $\sigma_1 \leq \frac{Nc_1R}{c_2}$, $\sigma_2 \leq \frac{NR}{c_2}\operatorname{arsinh}\left(\frac{R}{|\Omega|^{\frac{1}{N}}}\right)$, $\sigma_3 = \sigma_2$, $c_1\frac{\sigma_2}{\sigma_1} = \operatorname{arsinh}\left(\frac{\|\nabla v\|_{L^{N,\infty}}}{|\Omega|^{\frac{1}{N}}}\right)$, $\min\left\{\|\nabla v\|^N_{L^{N,\infty}}, 2^{-N}c_N|\Omega|\right\} = \|\nabla v\|^N_{L^{N,\infty}}$ and $\frac{\alpha\lambda\sigma_1}{N^2c_1^2-\lambda^2\sigma_1^2} \leq \frac{\alpha\lambda}{(1-\kappa^2)\theta_N c_1}\|\nabla v\|_{L^{N,\infty}}$. Summarizing the estimates from step 1, we infer

$$\int_\Omega \tilde{g}(|v|,|\nabla v|)dx \leq c_6\alpha\lambda\left(\|\nabla v\|^{N+1}_{L^{N,\infty}} + \|\nabla v\|^2_{L^{N,\infty}} + \cosh\lambda\,\|\nabla v\|_{L^{N,\infty}}\right) + c_7\beta\,\|\nabla v\|^{\nu}_{L^{N,\infty}}, \tag{6.2.50}$$

where $c_6$, depends on $N$, $\kappa$ and $|\Omega|$, and $c_7$ on $N$, $\nu$ and $|\Omega|$. In particular, since $v \in \mathcal{K}_R$,

$$\int_\Omega \tilde{g}(|v|,|\nabla v|)dx \leq c_5\alpha\lambda\left(R^{N+1} + R^2 + \cosh\lambda\,R\right) + c_6\beta R^{\nu}. \tag{6.2.51}$$

*Step 3: End of the proof.* We proceed as in the proof of Theorem 6.2.3-Steps 2,3 in defining the same operator $\mathcal{T}_n$ in $X$ by (6.2.25). For the same reason as here above, $\mathcal{T}_n$ is a compact operator and the only problem is to prove that $\mathcal{T}_n(\mathcal{K}_R) \subset \mathcal{K}_R$. If $u = \mathcal{T}_n(v)$ there holds if $v \in \mathcal{K}_R$,

$$\begin{aligned}
\|\nabla u\|^{N-1}_{L^{N,\infty}} &\leq c_5\left(\int_\Omega |g_n(x,v,\nabla v)|\,dx + \int_\Omega d\mu\right)\\
&\leq c_8\left(\beta R^{\nu} + \alpha\lambda\left(R^{N+1} + R^2 + \cosh\lambda\,R\right)\right) + c_5\mu(\Omega)\\
&\leq c_8\alpha\lambda R^{N+1} + c_5\mu(\Omega) + c_8\left(\beta + \alpha\lambda(1+\cosh\lambda)\right)\\
&\qquad + c_8\left(\beta M(\nu) + \alpha\lambda(\cosh\lambda\,M(1) + M(2))\right)R^{N-1},
\end{aligned} \tag{6.2.52}$$

since, by Young's inequality, [1] there holds, for any $0 < \tau \leq N-1$,

$$R^{\tau} \leq 1 + \frac{\tau}{N-1}\left(\frac{N-1}{N-1-\tau}\right)^{1-\frac{N-1}{\tau}} R^{N-1} := 1 + M(\tau)R^{N-1}.$$

We put

$$\Psi(S) = c_8\alpha\lambda S^{\frac{N+1}{N-1}} - \Big[1 - c_8\left(\beta M(\nu) + \alpha\lambda(\cosh\lambda\,M(1) + M(2))\right)\Big]\,S + A,$$

[1] Here we assume that $N > 3$ in order $M(2)$ be defined. When $N = 3$ only $M(\nu)$ and $M(1)$ need to be introduced.

where $A = c_8\left(\beta + \alpha\lambda(1+\cosh\lambda)\right) + c_5\mu(\Omega)$. Then $\Psi$ is convex on $[0,\infty)$ and

$$\Psi'(S) = c_8\frac{N+1}{N-1}\alpha\lambda S^{\frac{2}{N-1}} - \left[1 - c_8\left(\beta M(\nu) + \alpha\lambda(\cosh\lambda\, M(1) + M(2))\right)\right].$$

If we take $\lambda \leq \lambda_0$, arbitrarily and then $\alpha\lambda$ and $\beta$ small enough so that $1 - c_8\left(\beta M(\nu) + \alpha\lambda(\cosh\lambda_0\, M(1) + M(2))\right) > 0$, there exists a unique $R_0 > 0$ such that $\Psi'(R_0^{N-1}) = 0$. Furthermore $R_0$ can be assumed smaller than $R^*$ by increasing $\alpha\lambda$ or $\beta$. Hence there exists $m_0 > 0$ such that the identity $c_8\left(\beta + \alpha\lambda(1+\cosh\lambda)\right) + c_5\mu(\Omega) = m_0$ implies that

$$c_8\alpha\lambda R_0^{N+1} - \left[1 - c_8\left(\beta M(\nu) + \alpha\lambda(\cosh\lambda\, M(1) + M(2))\right)\right] R_0^{N-1} + m_0 = 0.$$

This yields $\mathcal{T}_n(\mathcal{K}_R) \subset \mathcal{K}_R$. Since $\mathcal{T}_n$ is compact, there exists $u_n \in \mathcal{K}_R$ such that $\mathcal{T}_n(u_n) = u_n$. We write the equation under the form

$$\begin{aligned} -\mathrm{div} A(x,\nabla u_n) &= g_n(x,u_n,\nabla u_n) + \beta_\epsilon := M_{\epsilon,n} && \text{in } \Omega \\ u_n &= 0 && \text{on } \Omega. \end{aligned} \tag{6.2.53}$$

The measure $M_{\epsilon,n}$ is positive and bounded by Step 1. Since $\beta_\epsilon \to \mu$ in the sense of measures, there exists $m > 0$, independent of $\epsilon$ and $n$ such that

$$m^{-1} \leq \|M_{\epsilon,n}\|_{\mathfrak{M}} \leq m. \tag{6.2.54}$$

By Proposition 4.3.11 there exists a constant $c_9 > 0$ such that

$$\begin{aligned} &(i) && \|u_n\|_{BMO(\Omega)} \leq c_9 \|M_{\epsilon,n}\|_{\mathfrak{M}}^{\frac{1}{N-1}} \leq c_9 m^{\frac{1}{N-1}} \\ &(ii) && \mathrm{meas}\left\{|\nabla u_n| > \sigma\right\} \leq c_9\frac{\|M_{\epsilon,n}\|_{\mathfrak{M}}^{\frac{N}{N-1}}}{\sigma^N} \leq c_9\frac{m^{\frac{N}{N-1}}}{\sigma^N}. \end{aligned} \tag{6.2.55}$$

We put

$$\tilde{A}_\sigma^{u_n} = \left\{x \in \Omega : |u_n(x)| \geq N\theta_N^{-1} \|\nabla u_n\|_{L^{N,\infty}} \sigma\right\} \text{ and } \tilde{a}^{u_n}(\sigma) = \mathrm{meas}(\tilde{A}_\sigma^{u_n}),$$

$$\tilde{B}_\sigma^{u_n} = \{x \in \Omega : |\nabla u_n(x)| \geq \sinh\sigma\} \text{ and } \tilde{b}^{u_n}(\sigma) = \mathrm{meas}(\tilde{B}_\sigma^{u_n}),$$

and

$$\tilde{C}_\sigma^{u_n} = \tilde{A}_\sigma^{u_n} \cap \tilde{B}_\sigma^{u_n} \text{ and } \tilde{c}^{u_n}(\sigma) = \mathrm{meas}(\tilde{C}_\sigma^{u_n}).$$

By (6.2.46) and (6.2.55)-(ii) there holds

$$\tilde{a}^{u_n}(\sigma) \leq \min\left\{2^{-N} c_N \operatorname{csch}^N(\sigma), 1\right\} |\Omega|,$$

and

$$\tilde{b}^{u_n}(\sigma) \leq \min\left\{c_9 m^{\frac{N}{N-1}} \operatorname{csch}^N(\sigma), |\Omega|\right\}.$$

The set of functions $\{g_n(.,u_n,\nabla u_n)\}$ (with the notations of the proof of Theorem 6.2.3) is uniformly bounded in $L^1(\Omega)$ since $u_n \in \mathcal{K}_R$. In order to prove its uniform integrability, we consider $\sigma > 0$ and a Borel set $E \subset \Omega$ that we decompose into

$$E = (E\cap\tilde{C}^{u_n}_\sigma)\cup(E\cap\tilde{B}^{u_n}_\sigma\cap(\tilde{A}^{u_n}_\sigma)^c)\cup(E\cap\tilde{A}^{u_n}_\sigma\cap(\tilde{B}^{u_n}_\sigma)^c)\cup(E\cap(\tilde{B}^{u_n}_\sigma)^c\cap(\tilde{A}^{u_n}_\sigma)^c).$$

Then, as in Step 1, and since $\|\nabla u_n\|_{L^{N,\infty}} \leq R$,

$$\begin{aligned}\int_E g_n(.,u_n,\nabla u_n)dx \leq \tilde{g}\left(N\theta_N^{-1}R\sigma, \sinh\sigma\right)|E| - \int_\sigma^\infty \tilde{g}\left(N\theta_N^{-1}Rt, \sinh\sigma\right)d\tilde{a}^{u_n}(t)\\ - \int_\sigma^\infty \tilde{g}\left(N\theta_N^{-1}R\sigma, \sinh t\right)d\tilde{b}^{u_n}(t) - \int_\sigma^\infty \tilde{g}\left(N\theta_N^{-1}Rt, \sinh t\right)d\tilde{c}^{u_n}(t).\end{aligned}$$

Up to changing $m$ we can assume that $2^{-N}c_N|\Omega| \leq c_9 m^{\frac{N}{N-1}}$, therefore

$$\max\{\tilde{a}^{u_n}(t), \tilde{b}^{u_n}(t), \tilde{c}^{u_n}(t)\} \leq \min\left\{c_9 m^{\frac{N}{N-1}} \operatorname{csch}^N(t), |\Omega|\right\}.$$

Using the previous inequalities and some integrations by parts, we derive

$$\begin{aligned}\int_E g_n(.,u_n,\nabla u_n)dx &\leq \tilde{g}\left(N\theta_N^{-1}R\sigma, \sinh\sigma\right)|E|\\ &+ 3Nc_9 m^{\frac{N}{N-1}} \int_\sigma^\infty \tilde{g}\left(N\theta_N^{-1}Rt, \sinh t\right)\coth t \operatorname{csch}^N(t)dt.\end{aligned}$$

Since $\tilde{g}\left(N\theta_N^{-1}Rt, \sinh t\right) \leq \alpha\sinh(\lambda N\theta_N^{-1}Rt) + \beta\sinh^\nu t$, $\lambda N\theta_N^{-1}R \leq N\kappa$ and $\coth t \operatorname{csch}^N(t) \sim 2^N e^{-Nt}$ as $t \to \infty$, the function $t \mapsto \tilde{g}\left(N\theta_N^{-1}Rt, \sinh t\right)\coth t \operatorname{csch}^N(t)$ is integrable on $(0,\infty)$. The same argument as in the proof of Theorem 5.1.2 yields the uniform integrability of $\{g_n(.,u_n,\nabla u_n)\}$. Letting $n \to \infty$ we obtain the existence of a solution of (6.2.34). Since $\{g(.,u_\epsilon,\nabla u_g e)\}$ satisfies the same estimates as $\{g_n(.,u_n,\nabla u_n)\}$, we end the proof as in Theorem 6.2.3-Steps 4. □

### 6.2.3 *The supercritical case*

#### 6.2.3.1 *Necessary and sufficient conditions*

In this section we consider the problem

$$\begin{aligned} -\mathrm{div}A(x,\nabla u) &= u^q + \mu && \text{in } \Omega \\ u &= 0 && \text{on } \partial\Omega, \end{aligned} \tag{6.2.56}$$

where $\Omega \subset \mathbb{R}^N$ is a domain, $A : \Omega \times \mathbb{R}^N \mapsto \mathbb{R}^N$ satisfies (4.1.1) and $\mu$ is a positive measure in $\Omega$. The critical exponent $q$ for this equation is again $q_c$. As for the similar problem (5.4.2) with absorption treated in Theorem 5.4.6, the treatment of the supercritical case $q \geq q_1$ with a source reaction necessitates potential estimates and Bessel or Riesz capacities. The following results have been published in [Phuc, Verbitsky (2006)], [Phuc, Verbitsky (2008)]. The next local estimate is a sharpening of Proposition 6.2.1.

**Theorem 6.2.8** *Assume $1 < p < N$, $p - 1 < q$, $\mu$ is a positive measure in $\Omega$ and $u$ is a nonnegative locally renormalized solution of*

$$-\mathrm{div}A(x,\nabla u) = u^q + \mu \qquad \text{in } \Omega. \tag{6.2.57}$$

*Then for any closed cube $Q \subset \Omega$ and any compact set $E \subset Q$ there holds*

$$\begin{aligned} &(i) && \int_E u^q dx + \mu(E) \leq c_1^* \dot{c}_{p,q_p}^{\mathbb{R}^N}(E) && \text{if } q > q_1, \\ &(ii) && \int_E u^q dx + \mu(E) \leq c_2^* c_{p,q_p}^{\mathbb{R}^N}(E) && \text{if } p-1 < q \leq q_1, \end{aligned} \tag{6.2.58}$$

*with $q_p = \frac{q}{q+1-p}$, $q_1 = \frac{(p-1)N}{N-p}$, where $\dot{c}_{p,q_p}^{\mathbb{R}^N}$ and $c_{p,q_p}^{\mathbb{R}^N}$ are respectively the Riesz and the Bessel capacities and where the constant $c_1^*$ depends only on $N$, $p$ and $q$, while $c_2^*$ depends on $N$, $p$, $q$ and the length of the side of $Q$. Moreover, if $\Omega$ is bounded and $q > q_1$,*

$$\int_E u^q dx + \mu(E) \leq c_3^* C_{p,q_p}^{\Omega}(E) \quad \forall E \subset \Omega,\ E \text{ compact}, \tag{6.2.59}$$

*where $c_3^*$ depends on $N$, $p$, $q$ and $\Omega$ and $C_{p,q_p}^{\Omega}$ denotes the fractional Sobolev capacity associated to the space $W_0^{p,q_p}(\Omega)$.*

The proof uses the local lower estimate of [Kilpeläinen, Malý (1994)] in terms of Wolff potential that every $p$-superharmonic satisfies. One of its main originality lies in the construction of a discrete model of this equation using dyadic decomposition of $\Omega$ [Phuc, Verbitsky (2006)]. The next metric estimate follows easily (see e.g. [Adams, Hedberg (1999), Prop 5.1.2 and 5.1.3])

**Corollary 6.2.9** *Under the assumptions of Theorem 6.2.8 there holds*

$$\begin{aligned} &(i) \qquad \int_{B_r} u^q dx + \mu(B_r) \leq c_4^* r^{N-pq_p} && \text{if } q > q_1, \\ &(ii) \qquad \int_{B_r} u^{q_1} dx + \mu(B_r) \leq c_5^* \left(\ln\left(\tfrac{2R}{r}\right)\right)^{1-\frac{N}{p}} && \text{if } q = q_1, \end{aligned} \tag{6.2.60}$$

*for all $r > 0$ such that $\overline{B}_r \subset \Omega$ in case (i) and $\overline{B}_r \subset B_R \subset B_{2R} \subset \Omega$ in case (ii).*

The following inequality is a sufficient condition for the existence of a renormalized solution to (6.2.56).

**Theorem 6.2.10** *Assume $\Omega \subset \mathbb{R}^N$ is a bounded domain, $1 < p < N$ and $p - 1 < q$ and $\mu$ is a positive bounded measure in $\Omega$. If there holds*

$$\mathbf{W}_{1,p}^{2diam(\Omega)}[\mathbf{W}_{1,p}^{2diam(\Omega)}[\mu]^q] \leq C^* \mathbf{W}_{1,p}^{2diam(\Omega)}[\mu] \qquad a.e. \text{ in } \Omega, \tag{6.2.61}$$

*where*

$$C^* = \left(\frac{q+1-p}{qc\max\{1, 2^{p'-2}\}}\right)^{q(p'-1)} \left(\frac{p-1}{q+1-p}\right), \tag{6.2.62}$$

*and where $c = c(N,p) > 0$ is the constant appearing in estimate (5.4.3) in Theorem 5.4.1, then there exists a nonnegative renormalized solution $u$ to (6.2.56). Furthermore there exists $m > 0$, depending on $N$, $p$, $q$ and the constant $\lambda$ appearing in (4.1.1) such that*

$$u \leq m\mathbf{W}_{1,p}^{2diam(\Omega)}[\mu] \qquad a.e. \text{ in } \Omega. \tag{6.2.63}$$

*Proof.* We consider a sequence $\{u_k\}$ of nonnegative functions defined by the following scheme: $u_0$ is a renormalized solution of

$$\begin{aligned} -\text{div}A(x, \nabla u_0) &= \mu && \text{in } \Omega \\ u_0 &= 0 && \text{on } \partial\Omega, \end{aligned} \tag{6.2.64}$$

and, for $k \in \mathbb{N}^*$, $u_k$ is a renormalized solution of

$$\begin{aligned} -\text{div}A(x, \nabla u_k) &= u_{k-1}^q + \mu && \text{in } \Omega \\ u_k &= 0 && \text{on } \partial\Omega. \end{aligned} \tag{6.2.65}$$

By regularizing $\mu$ as we did in the proof of Theorem 5.4.5 we can assume that the sequence $u_k$ is nondecreasing. By (5.4.3) there holds

$$u_0 \leq c\mathbf{W}_{1,p}^{2diam(\Omega)}[\mu] \text{ and } u_k \leq c\mathbf{W}_{1,p}^{2diam(\Omega)}[u_{k-1}^q + \mu] \quad \forall k \geq 1. \tag{6.2.66}$$

The second identity implies

$$u_k \leq c \max\left\{1, 2^{p'-2}\right\} \left(\mathbf{W}_{1,p}^{2diam(\Omega)}[u_{k-1}^q] + \mathbf{W}_{1,p}^{2diam(\Omega)}[\mu]\right).$$

By induction we construct a sequence $\{c_k\}$ with the property that $u_k \leq c_k\mathbf{W}_{1,p}^{2diam(\Omega)}[\mu]$ such that $c_0 = c$ and

$$c_k = c \max\left\{1, 2^{p'-2}\right\} \left(c_{k-1}^{q(p'-1)}C^* + 1\right) \quad \forall k \geq 1.$$

Since $c_k \leq cq_p \max\left\{1, 2^{p'-2}\right\}$, we conclude that the sequence $\{u_k\}$ is bounded from above by $m\mathbf{W}_{1,p}^{2diam(\Omega)}[\mu]$ for some $m > 0$ and that $u_k^q$ is dominated by a fixed integrable function by (6.2.61). We end the proof by Theorem 4.3.8. $\square$

The next result proved in [Phuc, Verbitsky (2006)] exhibits a necessary and sufficient condition for the solvability of (6.2.56).

**Theorem 6.2.11** *Assume $\Omega \subset \mathbb{R}^N$ is a bounded domain, $1 < p < N$ and $p - 1 < q$ and $\mu$ is a positive measure with compact support in $\Omega$ and set $R = \operatorname{diam}(\Omega)$. Then the following assertions are equivalent.*

*(i) There exists some $\epsilon_0 > 0$ such that for any $0 \leq \epsilon \leq \epsilon_0$ there exists a nonnegative renormalized solution $u$ to*

$$\begin{aligned} -\operatorname{div} A(x, \nabla u) &= u^q + \epsilon\mu && \text{in } \Omega \\ u &= 0 && \text{on } \partial\Omega. \end{aligned} \tag{6.2.67}$$

*(ii) There exists $c_1 > 0$ such that for all compact set $K \subset \Omega$ there holds*

$$\mu(K) \leq c_1 c_{p,q_p}^{\mathbb{R}^N}(K). \tag{6.2.68}$$

*(iii) There exists $c_2 > 0$ such that for any ball $B \subset \Omega$,*

$$\int_B \left(\mathbf{W}_{1,p}^{2diam(\Omega)}[\chi_B\mu](x)\right)^q dx \leq c_2\mu(B). \tag{6.2.69}$$

*(iv) There exists $c_3 > 0$ such that*

$$\mathbf{W}_{1,p}^{2R}[\mathbf{W}_{1,p}^{2R}[\mu]^q] \leq c_3\mathbf{W}_{1,p}^{2R}[\mu] \qquad a.e. \text{ in } \Omega. \tag{6.2.70}$$

6.2.3.2 *Removable singularities*

The previous results allow us to characterize sets which are removable for any positive solution of (6.2.57).

**Theorem 6.2.12** *Assume $\Omega \subset \mathbb{R}^N$ is any domain, $1 < p < N$ and $p - 1 < q$ and $E \subset \Omega$ is compact. Any nonnegative locally renormalized solution $u \in L^q_{loc}(\Omega \setminus E)$ of*

$$-\mathrm{div} A(x, \nabla u) = u^q \tag{6.2.71}$$

*in $\Omega \setminus E$ is a locally renormalized solution in $\Omega$ if and only if $c^{\mathbb{R}^N}_{p,q_p}(E) = 0$.*

*Proof.* Assume that $c^{\mathbb{R}^N}_{p,q_p}(E) = 0$. Then $c^{\mathbb{R}^N}_{1,p}(E) = 0$. Since $u$ is $p$-superharmonic in $\Omega \setminus K$, it follows from Theorem 4.3.6 that it is $p$-superharmonic in $\Omega$ and thus there exists a nonnegative Radon measure $\mu$ in $\Omega$ such that (6.2.57) holds. By Corollary 6.2.9, $u^q \in L^1_{loc}(\Omega)$, and by Theorem 6.2.8, $\mu$ vanishes on sets with zero $c^{\mathbb{R}^N}_{p,q_p}$-capacity, in particular on $E$. Since $\mu(K^c) = 0$ it follows that $\mu = 0$ and (6.2.71) holds in $\Omega$. The reverse assertion is true since, if $c^{\mathbb{R}^N}_{p,q_p}(E) > 0$ there exists a capacitary measure with support in $E$ and this measure satisfies (6.2.68) since it belongs to the dual space $L^{-p,\frac{q}{p-1}}(\mathbb{R}^N)$. □

## 6.3 Quasilinear Hamilton-Jacobi type equations

Let $\Omega \subset \mathbb{R}^N$ be a domain and $p > q > p - 1 > 0$. This section is concerned with properties of the solutions of

$$-\Delta_p u = |\nabla u|^q \tag{6.3.1}$$

in $\Omega$ or $\Omega \setminus \{0\}$. By opposition to the Lane-Emden equation, the *a priori* estimates are easy to obtain and have already been proved in Section 3.4.3.

### 6.3.1 *Separable solutions of quasilinear Hamilton-Jacobi equations*

We denote by $(r, \sigma) \in \mathbb{R}_+ \times S^{N-1}$ the spherical coordinates in $\mathbb{R}^N$. Any separable solution of (6.3.1) has the form

$$u(x) = u(r, \sigma) = r^{-\gamma_{p,q}} \omega(\sigma), \tag{6.3.2}$$

where $\gamma_{p,q} = \frac{p-q}{q+1-p}$ (see Section 5.5) and $\omega$ is a solution of

$$\begin{aligned} -div'\left(\left(\gamma_{p,q}^2\omega^2 + |\nabla'\omega|^2\right)^{\frac{p-2}{2}}\nabla'\omega\right) - \left(\gamma_{p,q}^2\omega^2 + |\nabla'\omega|^2\right)^{\frac{q}{2}} \\ -\gamma_{p,q}\Lambda(\gamma_{p,q})\left(\gamma_{p,q}^2\omega^2 + |\nabla'\omega|^2\right)^{\frac{p-2}{2}}\omega = 0 \quad \text{in } S^{N-1}. \end{aligned} \tag{6.3.3}$$

We recall that $\Lambda(\gamma_{p,q}) = \gamma_{p,q}\left((p-1)\gamma_{p,q} + p - N\right)$. A necessary and sufficient condition for the existence of a positive solution is $\Lambda(\gamma_{p,q}) < 0$, that is

$$\gamma_{p,q} < \frac{N-p}{p-1} \Longleftrightarrow q > q_c := \frac{N(p-1)}{N-1}. \tag{6.3.4}$$

If this is satisfied there exists a singular radial solution of (6.3.1) in $\mathbb{R}^N \setminus \{0\}$,

$$\tilde{U}_s(r) = \tilde{\mu}_{N,p,q} r^{-\gamma_{p,q}}, \tag{6.3.5}$$

where

$$\tilde{\mu}_{N,p,q} = \gamma_{p,q}^{-1}\left(N-p-(p-1)\gamma_{p,q}\right)^{\frac{1}{p+q-1}}. \tag{6.3.6}$$

It is easy to check that there exists no branch of solutions bifurcating from the value $(q, \tilde{\mu}_{N,p,q})$, contrary to the Lane-Emden equation (see Proposition 6.1.1).

Singular solutions in $\mathbb{R}^N_+ = \{(r,\sigma) : r > 0, \sigma \in S^{N-1}_+\}$ vanishing on $\partial\mathbb{R}^N_+ \setminus \{0\}$, under the form (6.3.2) are solutions of

$$\begin{aligned} -div'\left(\left(\gamma_{p,q}^2\omega^2 + |\nabla'\omega|^2\right)^{\frac{p-2}{2}}\nabla'\omega\right) - \left(\gamma_{p,q}^2\omega^2 + |\nabla'\omega|^2\right)^{\frac{q}{2}} \\ -\gamma_{p,q}\Lambda(\gamma_{p,q})\left(\gamma_{p,q}^2\omega^2 + |\nabla'\omega|^2\right)^{\frac{p-2}{2}}\omega = 0 \qquad & \text{in } S^{N-1}_+ \\ \omega = 0 \qquad & \text{on } \partial S^{N-1}_+. \end{aligned} \tag{6.3.7}$$

The critical exponent for existence is defined by (3.5.3):

$$\beta_{S^{N-1}_+} = \gamma_{p,q} \Longleftrightarrow q := \tilde{q}_c = p - \frac{\beta_{S^{N-1}_+}}{\beta_{S^{N-1}_+} + 1}.$$

It is the same as in the absorption case.

**Theorem 6.3.1** *Assume $p > 1$, then the following holds.*

*(i) If $p - 1 < q \leq \tilde{q}_c$ there exists no nontrivial nonnegative solution of (6.3.7).*

*(ii) If $\tilde{q}_c < q < p$ there exists at least one positive solution $\omega := \omega_h$ of (6.3.7).*

The next lemma will be used in the proof.

**Lemma 6.3.2** *Assume $\tilde{q}_c < q < p$, then the set of positive solutions of (6.3.7) is bounded.*

*Proof.* We recall that any solution of (6.2.56) in $\mathbb{R}^N_+$ satisfies (3.4.32), *i.e.*

$$|\nabla u(x)| \leq c_{N,p,q} \left(\rho(x)\right)^{-\frac{1}{q+1-p}} \qquad \forall x \in \mathbb{R}^N_+,$$

where $\rho(x) = \operatorname{dist}(x, \partial\mathbb{R}^N_+) = x_N$. If $\sigma_N$ denotes the north pole of $S^{N-1}_+$, this implies

$$|u(r,\sigma_N) - u(1,\sigma_N)| \leq \frac{q+1-p}{p-q} c_{N,p,q} \left(r^{-\gamma_{p,q}-1}\right) \qquad \forall r \in (0,1]. \tag{6.3.8}$$

If $u(r,\sigma) = r^{-\gamma_{p,q}}\omega(\sigma)$ it yields in particular

$$|\omega(\sigma_N)| \leq \frac{q+1-p}{p-q} c_{N,p,q}. \tag{6.3.9}$$

In order to extend this estimate from the point $\sigma_N$ to whole $S^{N-1}_+$ we use [Pucci, Serrin (2007), Th 8.2.2] which asserts that a positive solution of a nondegenerate quasilinear equation in a ball $B$ which vanishes on $\partial B$ is radially symmetric. Their proof is performed by using the moving plane method which can be easily adapted to the geometric setting on the sphere provided the symmetries with respect to hyperplanes be replaced by symmetries with respect to great *circles*, *i.e.* intersection of $S^{N-1}$ by hyperplanes. As a consequence $\omega$ depends only on the azimuthal angle $\theta_{N-1} := \arccos\left(\frac{x_N}{r}\right) \in (0, \frac{\pi}{2})$ (see (2.2.50)) and is denoted by $\tilde{\omega}$. Furthermore $\theta_{N-1} \mapsto \omega(\theta_{N-1})$ is decreasing. As a consequence

$$\tilde{\omega}(\theta_{N-1}) \leq \tilde{\omega}(0) = \omega(\sigma_N) \leq \frac{q+1-p}{p-q} c_{N,p,q},$$

which ends the proof. □

*Proof of Theorem 6.3.1.* The non-existence is a consequence of Theorem 6.1.18 with $S = S^{N-1}_+$, $\beta = \gamma_{p,q}$ and $g(.,\omega,\nabla\omega) = \left(\gamma^2_{p,q}\omega^2 + |\nabla'\omega|^2\right)^{\frac{q}{2}}$. The existence part follows the method of the proof of Theorem 6.1.19. We define the homeomorphism $\mathcal{A}$ from $W_0^{1,p}(S_+^{N-1})$ onto $W^{-1,p'}(S_+^{N-1})$ by (6.1.86) and put

$$\tilde{F}(\eta,t) = \mathcal{A}^{-1}\Big(\beta(\beta+\Lambda(\beta))\left(\beta^2\eta^2+\nabla'\eta|^2\right)^{\frac{p-2}{2}}\eta \\ +(1+t)|\left(\beta^2\eta^2+|\nabla'\eta|^2\right)^{\frac{q}{2}}\Big). \tag{6.3.10}$$

For the same reason as in Theorem 6.1.19, $F$ is a bounded mapping from $C_0^1(\overline{S}_+^{N-1})$ into $C_0^{1,\gamma}(\overline{S}_+^{N-1})$ for some $\gamma>0$, thus compact in $X := C_0^1(\overline{S}_+^{N-1})$.

We check now condition (i) in Proposition 6.1.20 by contradiction in supposing that there exist sequence $\{s_n\}\subset[0,1]$ and $\{\eta_n\}\subset C_0^{1,\gamma}(\overline{S}_+^{N-1})$ such that $\eta_n\to 0$ in $C_0^1(\overline{S}_+^{N-1})$ and $s_nF(\eta_n,0)=\eta_n$ and $s_n\to s\in[0,1]$. This implies that $\omega := \lim_{n\to\infty}\omega_n = \|\eta_n\|_X^{-1}\eta_n$ is nonnegative and satisfies $\|\omega\|_X = 1$

$$-div'\left(\left(\beta^2\omega^2+|\nabla'\omega|^2\right)^{\frac{p-2}{2}}\nabla'\omega\right) \\ = \beta\Lambda(\beta)\left(\beta^2\omega^2+|\nabla'\omega|^2\right)^{\frac{p-2}{2}}\omega,$$

in $S_+^{N-1}$ and vanishes on $\partial S_+^{N-1}$, which is impossible since $\beta<\beta_{S_+^{N-1}}$.

Next, the argument developed in Step 2 of the proof of Theorem 6.1.19 applies similarly: there exists $T>1$ such that

$$\beta(\beta+\Lambda(\beta))\left(\beta^2\omega^2+|\nabla'\omega|^2\right)^{\frac{p-2}{2}}\omega \\ +(1+t)\left(\beta^2\omega^2+|\nabla'\omega|^2\right)^{\frac{q}{2}} \geq (\lambda_{1,\beta}+\delta)\omega^{p-1} \quad \forall t\geq T\,,\ \forall\omega\geq 0.$$

Thus, if $t\geq T$ and $\omega\geq 0$, $\omega\neq 0$ we infer that it satisfies (6.1.91). The contradiction follows as in the previous proof. Thus condition (ii) in Proposition 6.1.20 holds and it is independent of $\|\omega\|_X$.

In order to prove that condition (iii) in Proposition 6.1.20, we have to show that for $\|\omega\|_X = R_2$ large enough the equation $\tilde{F}(\omega,t)=\omega$ cannot be satisfied. Assuming the contrary, there exists a function $\omega_t$ such that $\|\omega_t\|_X = R_t$. Then $\omega_t$ satisfies

$$-div'\left(\left(\gamma_{p,q}^2\omega_t^2+|\nabla'\omega_t|^2\right)^{\frac{p-2}{2}}\nabla'\omega_t\right)-(1+t)\left(\gamma_{p,q}^2\omega_t^2+|\nabla'\omega_t|^2\right)^{\frac{q}{2}} \\ -\gamma_{p,q}\Lambda(\gamma_{p,q})\left(\gamma_{p,q}^2\omega_t^2+|\nabla'\omega_t|^2\right)^{\frac{p-2}{2}}\omega=0,$$

in $S_+^{N-1}$ and vanishes on $\partial S_+^{N-1}$. Thus $(1+t)^{\frac{1}{q+1-p}}\omega_t$ is a positive solution of (6.3.10). It is bounded by a constant depending only on $N, p, q$ by Lemma 6.3.2. Therefore $\omega_t$, is also uniformly bounded on $S_+^{N-1}$, contradiction. Thus the condition holds. We conclude by Proposition 6.1.20. □

### 6.3.2 *Quasilinear Hamilton-Jacobi equations with measure data*

We give below a result of removability of singularity

**Theorem 6.3.3** *Assume $\Omega \subset \mathbb{R}^N$ is a domain and $p-1<q<p$. Let $K \subset \Omega$ be a compact set such that $c^{\Omega}_{1,q_p}(K)=0$ and $\mu \in \mathfrak{M}_+(\Omega)$ vanishes on $K$. If $u$ is a nonnegative locally renormalized solution of (6.0.2) in $\Omega\setminus K$ where $A$ satisfies (4.1.1) and $g: \Omega\times\mathbb{R}\times\mathbb{R}^N \mapsto \mathbb{R}$ is a Caratheodory function such that*

$$g(x,r,\xi)r \geq |\xi|^q \qquad \forall (x,r,\xi) \in \Omega\times\mathbb{R}\times\mathbb{R}^N, \tag{6.3.11}$$

*then $u$ can be extended to whole $\Omega$ into a locally renormalized solution of (6.0.2) in $\Omega$.*

*Proof.* The function $u$ is $p$-superharmonic in $\Omega\setminus K$. Since $p \geq q$, $q_p \geq p$, thus $c^{\Omega}_{1,p}(K)=0$. From Theorem 4.3.6, $u$ can be extended into a $p$-superharmonic function in $\Omega$ by formula (4.3.18). The $p$-harmonic measure $\mu_u$ coincides with $g(.,u,\nabla u)+\mu$ on $\Omega\setminus K$. In particular $g(.,u,\nabla u) \in L^1_{loc}(\Omega)$, thus $\nabla u \in L^q_{loc}(\Omega)$. Furthermore $u$ satisfies the regularity estimates given in Proposition 4.3.9. If $\zeta \in C^1_0(\Omega)$ and $\{\eta_n\} \subset C^1_0(\Omega)$ is a sequence of nonnegative functions with value 1 in a neighborhood of $K$, $0 \leq \eta_n \leq 1$ and $\|\eta_n\|_{W^{1,q_p}} \to 0$, we put $\tilde{\eta}_n = 1-\eta_n$ and we have

$$\begin{aligned} \int_\Omega \left(\langle A(x,\nabla u),\nabla\zeta\rangle\tilde{\eta}_n - \langle A(x,\nabla u),\nabla\eta_n\rangle\zeta\right)dx \\ = \int_\Omega g(x,u,\nabla u)\zeta\tilde{\eta}_n dx + \int_\Omega \zeta\tilde{\eta}_n d\mu. \end{aligned} \tag{6.3.12}$$

By the dominated convergence theorem

$$\lim_{n\to\infty}\int_\Omega g(x,u,\nabla u)\zeta\tilde{\eta}_n dx + \int_\Omega \zeta\tilde{\eta}_n d\mu = \int_\Omega g(x,u,\nabla u)\zeta dx + \int_\Omega \zeta d\mu,$$

and

$$\lim_{n\to\infty}\int_{\Omega}\langle A(x,\nabla u),\nabla\zeta\rangle\tilde{\eta}_n dx=\int_{\Omega}\langle A(x,\nabla u),\nabla\zeta\rangle dx.$$

Hence, if the support of $\zeta$ is the compact set $\Theta\subset\Omega$,

$$\left|\int_{\Omega}\langle A(x,\nabla u),\nabla\eta_n\rangle\zeta dx\right|\leq\lambda\left\|\chi_{\Theta}\nabla u\right\|_{L^q}^{p-1}\left\|\nabla\eta_n\right\|_{L^{q_p}},$$

and the right-hand side term tends to 0 when $n\to\infty$. Consequently

$$\int_{\Omega}\langle A(x,\nabla u),\nabla\zeta\rangle dx=\int_{\Omega}g(x,u,\nabla u)\zeta dx+\int_{\Omega}\zeta d\mu. \tag{6.3.13}$$

Therefore $u$ is a weak solution of (6.0.2) in $\Omega$. Finally let $u_k=\min\{u,k\}$, $k>0$, and assume $\zeta$ is nonnegative with value in $[0,1]$ and support in $\Theta$, then

$$\int_{\Omega}\left(\langle A(x,\nabla u),\nabla\zeta\rangle u_k\tilde{\eta}_n-\langle A(x,\nabla u),\nabla\eta_n\rangle u_k\zeta+\langle A(x,\nabla u),\nabla u_k\rangle\zeta\tilde{\eta}_n\right)dx$$
$$\leq k\int_{\Theta}g(x,u,\nabla u)dx+k\mu(\Theta). \tag{6.3.14}$$

Since

$$\int_{\Omega}\langle A(x,\nabla u),\nabla u_k\rangle\zeta\tilde{\eta}_n dx\geq\int_{\Omega}|\nabla u_k|^p\,\zeta\tilde{\eta}_n dx,$$

and the two other terms on the left-hand side of (6.3.14) admit a limit when $n\to\infty$, we derive by Fatou's lemma,

$$\int_{\Omega}|\nabla u_k|^p\,\zeta dx\leq\int_{\Omega}\langle A(x,\nabla u),\nabla\zeta\rangle u_k dx+k\int_{\Theta}g(x,u,\nabla u)dx+k\mu(\Theta).$$

We conclude by Proposition 4.2.2 that $u$ is a $p$-superharmonic and therefore it is a renormalized solution of (6.0.2) in $\Omega$. □

The method used in the previous theorem allows to give a necessary condition for solving (6.0.2) [Phuc (2014)].

**Theorem 6.3.4** *Let $A$ and $g$ satisfy the assumptions of Theorem 6.3.3. If $\mu$ is a positive measure in $\Omega$ and there exists a locally renormalized solution of (6.0.2) in $\Omega$, then for any compact subset $Q\subset\Omega$ there exists a positive constant $c_1=c_1(N,p,q,Q)$ such that*

$$\int_E g(x,u,\nabla u)dx+\mu(E)\leq c_1c^{\Omega}_{1,q_p}(E), \tag{6.3.15}$$

*for any compact subset $E$ of $Q$.*

*Proof.* Let $\{\eta_n\} \subset C_0^\infty(\Omega)$ where supp $(\eta_n) \subset Q$ and $0 \leq \eta_n \leq 1$, $\eta_n = 1$ on $E$ and $\|\nabla \eta_n\|_{L^{q_p}}^{q_p} \to c_{1,q_p}^{\Omega}(E)$ as $n \to \infty$. Since

$$\int \langle A(x, \nabla u), \nabla \eta_n^{q_p} \rangle dx = \int_\Omega g(x, u, \nabla u) \eta_n^{q_p} dx + \int_\Omega \eta_n^{q_p} d\mu, \qquad (6.3.16)$$

taking $\eta_n^{q_p}$ as test function, we derive by Hölder's inequality, using (4.1.1),

$$\lambda q_p \left( \int_\Omega |\nabla u|^q \, \eta_n^{q_p} dx \right)^{\frac{p-1}{q}} \left( \int_\Omega |\nabla \eta_n|^{q_p} \, dx \right)^{\frac{1}{q_p}} \geq \int_\Omega |\nabla u|^q \, \eta_n^{q_p} dx + \int_E d\mu.$$

This implies

$$\int_\Omega |\nabla u|^q \, \eta_n^{q_p} dx \leq (\lambda q_p)^{q_p} \int_\Omega |\nabla \eta_n|^{q_p} \, dx,$$

and

$$0 \leq \int \langle A(x, \nabla u), \nabla \eta^{q_p} \rangle dx \leq (\lambda q_p)^{q_p} \int_\Omega |\nabla \eta_n|^{q_p} \, dx. \qquad (6.3.17)$$

Combined with (6.3.16), this yields (6.3.15). □

In order to state existence results we introduce the following definitions (the first one has already been used in Chapter V).

**Definition 6.1** Let $\Omega \subset \mathbb{R}^N$ be a domain.

I- The R-truncated fractional maximal operator of order $\alpha \in [0, N)$ is defined for a bounded measure in $\Omega$ by

$$\mathbf{M}_{\alpha,R}[\mu](x) = \sup_{0<t\leq R} t^{\alpha-N} |\mu|(B_t(x)), \qquad (6.3.18)$$

where $R \leq \operatorname{diam}(\Omega)$, finite or not. For any $q > 1$ and $1 \leq m \leq \infty$, $\left\|\mathbf{M}_\alpha^R[\mu]\right\|_{L^{q,m}}$ is a norm on the set $\mathfrak{M}_{\alpha,q,m}(\Omega)$ of bounded measures such that $\mathbf{M}_\alpha^R[\mu] \in L^{q,m}(\Omega)$.

II- For $\gamma > 0$ and $s > 1$ we denote by $\mathfrak{M}^{\gamma,s}(\Omega)$ the set of bounded measures in $\Omega$ satisfying

$$\|\mu\|_{\mathfrak{M}^{\gamma,s}} = \sup \left\{ \frac{|\mu| \, (K \cap \Omega)}{c_{\gamma,s}^{\mathbb{R}^N}(K \cap \Omega)} : K \text{ compact s.t. } c_{\gamma,s}^{\mathbb{R}^N}(K \cap \Omega) > 0 \right\} < \infty. \qquad (6.3.19)$$

Clearly $\mu \mapsto \|\mu\|_{\mathfrak{M}^{\gamma,s}}$ is a norm on $\mathfrak{M}^{\gamma,s}(\Omega)$.

These definitions are justified by the following regularity results dealing with solutions of (4.2.3) proved in [Phuc (2014), Th 1.4, Th 5.6]. We will not go into their proofs which are delicate and based upon the use of sharp estimates of the measure of the level sets of the differents functions involved in their expression. Furthermore, the Lebesgue measure $dx$ can be replaced by a weighted measure $wdx$ where $w$ belongs to the class of Muckenhoupt-Wheeden weights [Muckenhoupt, Wheeden (1974)], and the $C^1$ condition on the boundary can be replaced by a more general one called the Reifenberg flatness condition [Toro (1997)].

**Proposition 6.3.5** *Assume $\Omega \subset \mathbb{R}^N$ is a bounded domain with a $C^1$ boundary with $R =$diam$(\Omega)$, $2 - \frac{1}{N} < p \leq N$ and $A$ satisfies (4.1.1).*
*I- If $1 \leq q < \infty$, $1 \leq m \leq \infty$ and $\mu \in \mathfrak{M}_{q,m}(\Omega)$, any renormalized solution $u$ of problem (4.2.3) satisfies*

$$\||\nabla u|\|_{L^{q,m}} \leq c_1 \left\|(\mathbf{M}_{1,R}[\mu])^{\frac{1}{p-1}}\right\|_{L^{q,m}} \leq c_1' \|\mathbf{M}_{1,R}[\mu]\|^{\frac{1}{p-1}}_{L^{\frac{q}{p-1},m}}, \tag{6.3.20}$$

*for some $c_1 > 0$ depending on $N$, $m$, $q$, $\Omega$ and $\lambda$.*
*II- If $m > 1$ and $\mu \in \mathfrak{M}^{1,m}(\Omega)$, any renormalized solution $u$ of problem (4.2.3) satisfies*

$$\left\||\nabla u|^{\frac{m(p-1)}{m-1}}\right\|_{\mathfrak{M}^{1,m}} \leq c_2 \|(\mathbf{M}_1[\mu])\|^{\frac{m}{m-1}}_{\mathfrak{M}^{1,m}}, \tag{6.3.21}$$

*for some $c_2 > 0$ depending on $N$, $m$, $\Omega$ and $\lambda$.*

The next theorems provide sufficient conditions for solving the supercritical quasilinear Hamilton-Jacobi equation (6.0.2).

**Theorem 6.3.6** *Let $\Omega \subset \mathbb{R}^N$ be a bounded $C^1$ domain, $R = \mathrm{diam}(\Omega)$ and $2 - \frac{1}{N} < p \leq N$. Assume $A$ satisfies (4.1.1), $g : \Omega \times \mathbb{R} \times \mathbb{R}^N \mapsto \mathbb{R}_+$ is a Caratheodory function such that*

$$g(x,r,\xi) \leq \tilde{g}(|\xi|) \leq a\,|\xi|^{p-1+\frac{q}{N}} + b \qquad \forall (x,r,\xi) \in \Omega \times \mathbb{R}_+ \times \mathbb{R}^N, \tag{6.3.22}$$

*where $a > 0$, $b \geq 0$, $p - 1 < q \leq p$ and $\tilde{g} : \mathbb{R}_+ \mapsto \mathbb{R}_+$ is a continuous nondecreasing function such that*

$$\int_1^\infty \tilde{g}(s)s^{-1-q}ds < \infty. \tag{6.3.23}$$

*There exists $\epsilon_0 > 0$, depending on $p$, $q$, $a$, $N$ and $\Omega$ such that if $\mu \in \mathfrak{M}^b_+(\Omega)$, the inequality*

$$\|\mathbf{M}_{1,R}[\mu]\|_{L^{\frac{q}{p-1},\infty}} + b \leq \epsilon_0, \tag{6.3.24}$$

*implies that the problem*

$$\begin{aligned} -div\,A(x,\nabla u) &= g(x,u,\nabla u)+\mu && in\ \Omega \\ u &= 0 && on\ \partial\Omega, \end{aligned} \tag{6.3.25}$$

*admits a renormalized solution.*

*Proof.* We extend $\mu$ by 0 outside $\Omega$, then, by Theorem 5.3.2, we have

$$c_3^{-1}\left\|\mathbf{G}_1[\mu]\right\|_{L^{\frac{q}{p-1},\infty}} \leq \left\|\mathbf{M}_{1,R}[\mu]\right\|_{L^{\frac{q}{p-1},\infty}} \leq c_3\left\|\mathbf{G}_1[\mu]\right\|_{L^{\frac{q}{p-1},\infty}}. \tag{6.3.26}$$

If $\mu_\epsilon=(\mu_0+\mu_s)*\eta_\epsilon-\operatorname{div}\mathrm{g}$ is an approximation of $\mu$ by the convolution with the smooth regularizing sequence $\{\eta_\epsilon\}$, then $\mathbf{G}_1[\mu_\epsilon]\to\mathbf{G}_1[\mu]$ in $L^{\frac{q}{p-1},\infty}(\Omega)$. Therefore $\mathbf{M}_{1,R}[\mu_\epsilon]\to\mathbf{M}_{1,R}[\mu]$ in $L^{\frac{q}{p-1},\infty}(\Omega)$. We follow the ideas introduced in Theorem 6.2.3. We set $X=\{v\in L^{\frac{N(p-1)}{N-p},\infty}(\Omega):\nabla v\in L^{q,\infty}(\Omega)\}$ with norm

$$\|v\|_X=\|v\|_{L^{\frac{N(p-1)}{N-p},\infty}}+\|\nabla v\|_{L^{q,\infty}},$$

and, for $k>0$, put

$$\mathcal{K}_k=\{v\in X \text{ s.t. } \|v\|_X\leq k\}. \tag{6.3.27}$$

Note that we have assumed that $p<N$ in the definition of $X$, the case $p=N$ needing only a minor adaptation. Then $\mathcal{K}_k$ is a closed convex set of the Banach space $X$. If $v$ in $\mathcal{K}_k$, we have

$$\int_\Omega g(x,v,\nabla v)dx\leq\int_\Omega\tilde g(|\nabla v|)dx\leq-\int_0^\infty\tilde g(t)db^v(t)$$

where

$$B_t^v=\{x\in\Omega:|\nabla v(x)|>t\}\quad\text{and}\quad b^v(t)=\operatorname{meas}(B_t^v).$$

Since $\nabla v\in L^{q,\infty}(\Omega)$, $b^v(t)\leq\max\{\|\nabla v\|_{L^{q,\infty}}t^{-q},|\Omega|\}$, hence

$$\begin{aligned} \int_\Omega g(x,v,\nabla v)dx &\leq q\,\|\nabla v\|_{L^{r,\infty}}\int_{|\Omega|}^\infty\tilde g(t)t^{-1-q}dt \\ &\leq kq\int_{|\Omega|}^\infty\tilde g(t)t^{-1-q}dt. \end{aligned} \tag{6.3.28}$$

*Step 1: Approximation and estimates.* We introduce also the approximation $g_n$ of $g$ by truncation of the variables $r$ and $|\xi|$ used in the proof of

Theorem 6.2.3. If $v \in \mathcal{K}_k$ let $u = \mathcal{T}_n(v)$ be the renormalized solution of

$$\begin{aligned} -divA(x, \nabla u) &= g_n(x, v, \nabla v) + \mu_\epsilon && \text{in } \Omega \\ u &= 0 && \text{on } \partial\Omega. \end{aligned} \tag{6.3.29}$$

By (6.3.22),

$$0 \leq g_n(x, v, \nabla v) \leq a\,|\nabla v|^{p-1+\frac{q}{N}} + b \qquad \text{a.e. in } \Omega.$$

Hence

$$\mathbf{M}_{1,R}[g(., v, \nabla v)] \leq a\mathbf{M}_{1,R}[|\nabla v|^{p-1+\frac{q}{N}}] + c_4 b.$$

By Theorem 5.3.2

$$\left\|\mathbf{M}_{1,R}[|\nabla v|^{p-1+\frac{q}{N}}]\right\|_{L^{\frac{q}{p-1}},\infty} \leq c_4 \left\|\mathbf{G}_1[|\nabla v|^{p-1+\frac{q}{N}}]\right\|_{L^{\frac{q}{p-1}},\infty}.$$

The Riesz kernel satisfies the following Gagliardo-Nirenberg inequality in Lorentz spaces which extends the classical ones valid in $L^p$ spaces:

$$\|\mathbf{I}_1[\Theta]\|_{L^{s^*,\infty}} \leq c_5 \|\Theta\|_{L^{s,\infty}}, \tag{6.3.30}$$

where $s^* = \frac{sN}{N-s}$ and $c_5$ depends on $N$ and $s$. Its proof follows from [Adams, Hedberg (1999), Prop 3.1.4] which asserts that $\mathbf{I}_1$ is a continuous linear map from $L^1(\mathbb{R}^N)$ into $L^{1^*,\infty}(\mathbb{R}^N)$ and from $L^p(\mathbb{R}^N)$ into $L^{p^*}(\mathbb{R}^N)$ (see e.g. [Adams, Hedberg (1999), Prop 3.1.4]). By a classical *real interpolation* theorem [Lions, Peetre (1964), Th 3.1], $\mathbf{I}_1$ extends as a continuous linear maps between all the real interpolations spaces as they are defined in [Lions, Peetre (1964)]. In particular, for any $\theta \in (0,1)$

$$\mathbf{I}_1 \in \mathcal{L}\left(\left[L^1, L^p\right]_{\theta,\infty}, \left[L^{1^*,\infty}, L^{p^*}\right]_{\theta,\infty}\right).$$

It is a classical property of interpolation of Lorentz spaces that

$$\left[L^1, L^p\right]_{\theta,\infty} = L^{s,\infty} \quad \text{with } \frac{1}{s} = 1 - \theta + \frac{\theta}{p},$$

hence

$$\left[L^{1^*,\infty}, L^{p^*}\right]_{\theta,\infty} = L^{\tilde{s},\infty} \quad \text{with } \frac{1}{\tilde{s}} = \frac{1-\theta}{1^*} + \frac{\theta}{p^*} = \frac{1}{s} - \frac{1}{N} = \frac{1}{s^*}.$$

As a consequence, we obtain

$$\left\|\mathbf{I}_1[|\nabla v|^{p-1+\frac{q}{N}}]\right\|_{L^{\frac{q}{p-1}},\infty} \leq c_5 \left\||\nabla v|^{p-1+\frac{q}{N}}\right\|_{L^{\tilde{q},\infty}}, \tag{6.3.31}$$

where

$$\frac{p-1}{q}=\frac{1}{\tilde{q}}-\frac{1}{N}\Longleftrightarrow\tilde{q}=\frac{Nq}{N(p-1)+q},$$

and

$$\left\||\nabla v|^{p-1+\frac{q}{N}}\right\|_{L^{\tilde{q},\infty}}=\|\nabla v\|^{p-1+\frac{q}{N}}_{L^{\tilde{Q},\infty}},$$

with

$$\tilde{Q}=\tilde{q}\left(p-1+\frac{q}{N}\right)=q.$$

Since $G_1\leq I_1$ we finally derive

$$\|\mathbf{M}_{1,R}[g(.,v,\nabla v)]\|_{L^{\frac{q}{p-1},\infty}}\leq ac_6\|\nabla v\|^{p-1+\frac{q}{N}}_{L^{q,\infty}}+c_4b. \tag{6.3.32}$$

Therefore, if $u=\mathcal{T}_n(v)$ and $v\in\mathcal{K}_k$ there holds by Proposition 6.3.5-I

$$\begin{aligned}\|\nabla u\|^{p-1}_{L^{q,\infty}}&\leq c_7\left(\|\mathbf{M}_{1,R}[g_n(.,v,\nabla v)]\|_{L^{\frac{q}{p-1},\infty}}+\|\mathbf{M}_{1,R}[\mu]\|_{L^{\frac{q}{p-1},\infty}}\right)\\&\leq c_8a\|\nabla v\|^{p-1+\frac{q}{N}}_{L^{q,\infty}}+c_9b+c_7\|\mathbf{M}_{1,R}[\mu]\|_{L^{\frac{q}{p-1},\infty}}.\end{aligned} \tag{6.3.33}$$

Combining this estimate with (6.3.28) and Proposition 4.3.9 we finally infer

$$\begin{aligned}\|u\|^{p-1}_X&\leq c_9\|v\|^{p-1+\frac{q}{N}}_X+c_{10}b+c_{11}\|\mathbf{M}_{1,R}[\mu]\|_{L^{\frac{q}{p-1},\infty}}\\&\leq\left(2c_9\|v\|_X^{1+\frac{q}{N(p-1)}}+B\right)^{p-1},\end{aligned} \tag{6.3.34}$$

where

$$B=2\left(c_{10}b+c_{11}\|\mathbf{M}_{1,R}[\mu]\|_{L^{\frac{q}{p-1},\infty}}\right)^{\frac{1}{p-1}}.$$

In particular, since $v\in\mathcal{K}_k$,

$$\|u\|_X\leq 2c_9k^{1+\frac{q}{N(p-1)}}+B.$$

If we set

$$k_0=\left(2c_9\frac{N(p-1)+q}{N(p-1)}\right)^{-\frac{N(p-1)}{q}}\quad\text{and}\quad B_0=\frac{qk_0}{q+N(p-1)},$$

then the function $k\mapsto 2c_9k^{1+\frac{q}{N(p-1)}}+B_0$ is positive and decreasing on $[0,k_0)$ and vanishes at $k=k_0$. Moreover $c_9$ and $B$ are independent of $n$

and $\epsilon$. If we suppose that $v \in \mathcal{K}_{k_0}$ there holds

$$\|u\|_X - k_0 \leq 2c_9 k_0^{1+\frac{r}{N(p-1)}} + B_0 - k_0 = 0,$$

thus $\mathcal{T}_n(\mathcal{K}_{k_0}) \subset \mathcal{K}_{k_0}$. It is interesting to notice that the dependence of $B_0$ with respect to $b$ and $\|\mathbf{M}_{1,R}[\mu]\|_{L^{\frac{r}{p-1}},\infty}$ can be expressed by

$$c_{10}b + c_{11}\|\mathbf{M}_{1,R}[\mu]\|_{L^{\frac{q}{p-1}},\infty} \leq 2^{1-p}B_0^{p-1}. \tag{6.3.35}$$

This is condition (6.3.24).

*Step 2: Convergence.* Because $q \leq p$, there holds $W_0^{1,p}(\Omega) \subset X$. Consequently, the proof of the compactness of the operator $\mathcal{T}_n$ is the same as the one of Theorem 6.2.3, Step 1. Hence $\mathcal{T}_n$ admits a fixed point $u = u_n$ in $\mathcal{K}_{k_0}$. The set of functions $\{g_n(.,u_n,\nabla u_n)\}$ is bounded in $L^1(\Omega)$ independently of $n$ and $\epsilon$. If $E$ is a Borel subset of $\Omega$ and $\sigma > 0$, we put

$$B_\sigma^{u_n} = \{x \in \Omega : |\nabla u_n| > \sigma\} \quad \text{and } b^{u_n}(\sigma) = \text{meas}\,(\mathrm{B}_\sigma^{\mathrm{u_n}}).$$

Since $u_n \in \mathcal{K}_{k_0}$, $b^{u_n}(\sigma) \leq k_0\sigma^{-q}$, thus

$$\begin{aligned}\int_E |g_n(x,u_n,\nabla u_n)|\,dx &\leq \int_E \tilde{g}(|\nabla u_n|)dx \\ &\leq \tilde{g}(\sigma)\,|E| - \int_\sigma^\infty \tilde{g}(\tau)db^{u_n}(\tau) \\ &\leq \tilde{g}(\sigma)\,|E| + qk_0\int_\sigma^\infty \tilde{g}(\tau)\tau^{-q-1}d\tau.\end{aligned}$$

The integrability condition (6.3.23) implies that the set $\{g_n(.,u_n,\nabla u_n)\}$ is uniformly integrable. We conclude as in Theorem 6.2.3-Step 3 that, up to a subsequence, $u_n$ converges a.e. to $u_\epsilon$, $T_k(u_n)$ converges to $T_k(u_\epsilon)$ and $u_\epsilon$ is a renormalized solution of

$$\begin{aligned} -divA(x,\nabla u) &= g(x,v,\nabla v) + \mu_\epsilon \qquad &&\text{in } \Omega \\ u &= 0 &&\text{on } \partial\Omega. \end{aligned} \tag{6.3.36}$$

Similarly as in the proof of Theorem 6.2.3-Step 4, $u_\epsilon$ converges a.e. to a renormalized solution of (6.3.25). □

*Remark.* The condition that $q \leq p$ can be removed if $\mu$ has compact support. In this case we approximate it by $\mu_\epsilon = \mu * \eta_\epsilon$ for a sequence of mollifiers $\{\eta_\epsilon\} \subset C_0^\infty(\mathbb{R}^N)$. If we write $\mu_\epsilon = (\mu_0 + \mu_s - \operatorname{div}\mathrm{g}) * \eta_\epsilon$, then it is a converging decomposition in the sense of Theorem 4.3.8. By regularity the operator $\mathcal{T}_n$ maps $\mathcal{K}_{k_0}$ into a bounded subset of $C^{1,\alpha}(\overline{\Omega})$. Thus compactness follows.

*Remark.* Assumption (6.3.24), jointly with Theorem 5.3.2 and the links between the Bessel potential $\mathbf{G}_1$ and the Riesz potential $\mathbf{I}_1$, implies that $\mu \in \mathfrak{M}_+(\Omega) \cap L^{-1,\frac{r}{p-1},\infty}$. Since $L^{-1,\frac{r}{p-1},\infty}(\mathbb{R}^N)$ is the dual space of $L^{1,\frac{r}{r+1-p},1}(\mathbb{R}^N)$, $\mu$ vanishes on compact sets $K \subset \Omega$ with zero $c^{\mathbb{R}^N}_{1,\frac{r}{r+1-p},1}$-capacity. The following result [Phuc (2014), Th 1.6] is more precise with a more restrictive assumption on $g$.

**Theorem 6.3.7** *Assume $\Omega \subset \mathbb{R}^N$ is a bounded $C^1$ domain, $A$ satisfies (4.1.1) with $2 - \frac{1}{N} < p \leq N$ and $q > p - 1$. Then there exists $c_0$ depending on $N$, $p$, $q$, $\lambda$ and diam$(\Omega)$ such that for all bounded measure $\mu$ in $\Omega$, the assumption $\|\mu\|_{\mathfrak{M}^{1,\frac{q}{q+1-p}}} \leq c_0$, see (6.3.19), implies that the problem*

$$\begin{aligned} -divA(x,\nabla u) &= |\nabla u|^q + \mu && \text{in } \Omega \\ u &= 0 && \text{on } \partial\Omega, \end{aligned} \tag{6.3.37}$$

*admits a solution in $W^{1,q}_0(\Omega)$. Furthermore there exists $c_1$ depending on the same elements as $c_0$ such that*

$$\int_{K\cap\Omega} |\nabla u|^q \, dx \leq c_1 c^{\mathbb{R}^N}_{1,\frac{q}{q+1-p}}(K), \tag{6.3.38}$$

*for all compact set $K \subset \mathbb{R}^N$.*

## 6.4 Notes and open problems

*6.4.1.* Concerning problems on the whole sphere it is not known if all positive solutions of (6.1.2) are constant when $q_1 < q < q_s$. The same question arises with positive solutions of (6.3.3) when $q_c < q < p$.

*6.4.2.* It is an open problem to prove that there is only one positive solution to the problems (6.1.77) with $S = S^{N-1}_+$ and (6.3.7) in the respective range of exponents $\tilde{q}_1 < q < q_s$ and $\tilde{q}_c < q < p$. Since it is easily proved that such solutions depend only on the azymuthal variable $\theta_{N-1}$ the problem is reduced to prove uniqueness of positive solutions of some ODE.

*6.4.3.* It is an open problem to prove that there is only one positive solution to the problems (6.1.77) with $S = S^{N-1}_+$ and (6.3.7). Similarly, is it true that any positive solution of (6.3.1) in $\mathbb{R}_+^N$ which vanishes on $\partial\mathbb{R}_+^N \setminus \{0\}$ is bounded by $c\,|x|^{-\gamma_{p,q}}$ ?

*6.4.4.* The quest of singular separable solutions of

$$-\Delta_p u + \epsilon\,|u|^{q-1}\,u\,|\nabla u|^s = 0, \tag{6.4.1}$$

in $\mathbb{R}^N \setminus \{0\}$ or $\mathbb{R}_+^N$, where $\epsilon = \pm 1$ and $(q+s+1-p)(p-s) > 0$, leads to $u(r,\sigma) = r^{-\tau_{p,q,s}}\omega(\sigma)$ where $\tau := \tau_{p,q,s} = \frac{p-s}{q+s+1-p}$. As for $\omega$ it satisfies

$$\begin{aligned} -div\left(\left(\tau^2\omega^2 + |\nabla\omega|^2\right)^{\frac{p-2}{2}}\nabla\omega\right) &= \tau\Lambda(\tau)\left(\tau^2\omega^2 + |\nabla\omega|^2\right)^{\frac{p-2}{2}}\omega \\ &\quad - \epsilon\,|\omega|^{q-1}\,\omega\left(\tau^2\omega^2 + |\nabla\omega|^2\right)^{\frac{s}{2}}, \end{aligned} \tag{6.4.2}$$

either on $S^{N-1}$ or on $S_+^{N-1}$ with zero boundary conditions on $\partial S_+^{N-1}$. A first critical set of exponents $(q,s)$ is

$$\Lambda(\tau) = 0 \Longleftrightarrow q + s = p - 1 + \frac{1}{N},$$

for the problem in $\mathbb{R}^N$ and

$$\Lambda(\tau) = \Lambda(\beta_{S_+^{N-1}}) \Longleftrightarrow (q+s)\beta_{S_+^{N-1}} + s = (p-1)\beta_{S_+^{N-1}} + p,$$

for the problem in $S_+^{N-1}$. Existence of nontrivial positive solutions of (6.4.2) on $S_+^{N-1}$ vanising on $\partial S_+^{N-1}$ is an interesting question. The proof of an *a priori* estimate for any solution in $\mathbb{R}^N \setminus \{0\}$ is a deeper question.

*6.4.5.* Not much is known for the supercritical problem

$$-\Delta_p u + \epsilon\,|u|^{q-1}\,u\,|\nabla u|^s = \mu. \tag{6.4.3}$$

A natural critical threshold is expressed by $q(N-p) + s(N-1) = N(p-1)$.

*6.4.6.* The study of

$$-\Delta_p u + \epsilon\,|u|^{q-1}\,u + \epsilon'\,|\nabla u|^s = 0, \tag{6.4.4}$$

where $\epsilon, \epsilon' = \pm 1$ is still essentially open, except in the case $\epsilon = \epsilon' = 1$ where one of the two reaction terms can be dominant, the critical case being

$$s = \frac{pq}{q+1}.$$

II $\epsilon = \epsilon' = -1$ no *a priori* estimate is known, even if one term dominates the other. Another interesting question is to find condition (necessary, sufficient) to solve the problem with measure data,

$$-\Delta_p u + \epsilon |u|^{q-1} u + \epsilon' |\nabla u|^s = \mu. \tag{6.4.5}$$

# Bibliography


Adams D. and Hedberg L. (1999). *Function spaces and potential theory*, , Grundleheren der mathematischen Wissenshaften **vol. 314**, Springer-Verlag. Berlin, Heidelberg, New-York, Tokyo.

Alvono A., Ferone V. and Trombetti G. (1996). Moser-Type Inequalities in Lorentz Spaces, *Potential Anal.* **5**, pp. 273–299.

Anane A. (2012). Simplicité et isolation de la première valeur propre du *p*-laplacien avec poids, *C. R. Acad. Sci. Paris Sér. I Math.* **305**, pp. 725–728.

Ancona A. (2012). On positive harmonic functions in cones and cylinders, *Rev. Math. Iberoam* **28**, pp. 201–230.

Arnowitt R., Deser S., Misner C. (1961). Coordinate invariance and energy expressions in general relativity, *Phys. Rev.* **122**, pp. 997–1006.

Aronson G. (1984). On certain singular solutions of the partial differential equation $u_x^2 u_{xx} + 2u_x u_y u_{xy} u_y^2 u_{yy} = 0$, *Manuscripta Math.* **47**, pp. 133–151.

Baras P. and Pierre M. (1984). Singularités éliminables pour des équations semilinéaires, *Ann. Inst. Fourier* **34**, pp. 117–135.

Barles G. (1988). Remarks on uniqueness results of the first eigenvalue of the *p*-Laplacian, *Ann. Fac. Sci. Toulouse Math.* **9**, pp. 65–75.

Belloni A. and Kawhol P. (2002). A direct uniqueness proof for equations involving the *p*-Laplace operator, *Manuscripta Math.* **109**, pp. 219–231.

Bénilan P. and Brezis H. (1975). Nonlinear problems related to the Thomas-Fermi equation. Unpublished paper. After Benilan's death a detailled version appeared in (2003), see the next reference.

Bénilan P. and Brezis H. (2003). Nonlinear problems related to the Thomas-Fermi equation. Dedicated to Philippe Bénilan, *J. Evol. Eq.* **3**, pp. 673–770.

Bénilan P., Boccardo L., Gallouët T., Gariepy R., Pierre M. and Vàzquez J.L. (1995). An $L^1$-theory of existence and uniqueness of solutions of nonlinear elliptic equations, *Ann. Scuola Norm. Sup. Pisa Cl. Sci.* **22**, pp. 241–273.

Bensoussan A., Boccardo L. and Murat F. (1988). On a nonlinear partial differential equation having natural growth terms and unbounded solutions, *Ann. Inst. H. Poincaré-Anal. Non Linéaire* **5**, pp. 347–363.

Berger M., Gauduchon P., Mazet E. (1971). *Le spectre d'une variété riemannienne*, Lecture Notes in Math. **194**, Springer, Berlin-Heidelberg, New-York.

Bhattacharya T. (2004). On the behaviour of $\infty$-harmonic functions near isolated points, *Nonlinear Analysis* **58**, pp. 333–349.

Bhattacharya T. (2005). A note on non-negative singular infinity-harmonic functions in the half-space, *Rev. Mat. Complutense* **18**, pp. 377–385.

Bhattacharya T., Marazzi L. (2013). An eigenvalue problem for the infinity-Laplacian, *Elec. J. Diff. Eq.* **47**, pp. 1–30.

Bidaut-Véron M. F. (1989). Local and global properties of solutions of quasilinear elliptic equations of Emden-Fowler type, *Arch. Rat. Mech. Anal.* **107**, pp. 293–324.

Bidaut-Véron M. F. (2002). Necessary conditions of existence for an elliptic equation with source term and measure data involving $p$-Laplacian, *Electron. J. Differ. Eq. Conf.* **8**, pp. 23–34.

Bidaut-Véron M. F. (2003). Removable singularities and existence for a quasilinear equation with absorption or source term and measure data, *Adv. Nonlinear Stud.* **3**, pp. 25–63.

Bidaut-Véron M. F., Borghol R. and Véron L. (2006). Boundary Harnack inequality and a priori estimates of singular solutions of quasilinear elliptic equations, *Calc. Var. Part. Diff. Eq.* **27**, pp. 159–177.

Bidaut-Véron M. F., Garcia-Huidobro M. and Véron L. (2014). Local and global properties of solutions of quasilinear Hamilton–Jacobi equations, *J. Funct. Anal.* **267**, pp. 3294–3331.

Bidaut-Véron M. F., Garcia-Huidobro M. and Véron L. (2015). Boundary singularities of positive solutions of quasilinear Hamilton-Jacobi equations, *Calc. Var. Part. Diff. Eq.* **54**, pp. 3471–3515.

Bidaut-Véron M. F., Garcia-Huidobro M. and Véron L. (2017). Separable infinite harmonic functions in cones, preprint.

Bidaut-Véron M. F., Hamid H. A. (2010). On the connection between two quasilinear elliptic problems with source terms of order 0 or 1, *Comm. Cont. Math.* **12**, pp. 1–60.

Bidaut-Véron M. F., Nguyen Q. H. and Véron L. (2014). Quasilinear Lane-Emden equations with absorption and measure data, *J. Math. Pures Appl.* **102**, pp. 315–337.

Bidaut-Véron M. F. and Pohozaev S. (2001). Nonexistence results and estimates for some nonlinear elliptic problems, *J. Anal. Math.* **84**, pp. 1–49.

Bidaut-Véron M. F., Ponce A. and Véron L. (2011). Isolated boundary singularities of semilinear elliptic equations, *Calc. Var. Part. Diff. Eq.*, **40**, pp. 183–221.

Bidaut-Véron M. F. and Véron L. (1991). Nonlinear elliptic equations on compact Riemannian manifolds and asymptics of Emden equations, *Invent. Math.* **106**, pp. 489–539.

Boccardo L., Gallouët T. (1989). Nonlinear elliptic and parabolic equations involving measure data, *J. Funct. Anal.* **87**, pp. 149–169.

Boccardo L., Gallouët T. (1992). Nonlinear elliptic equations with right hand side measures, *Comm. Part. Diff. Eq.* **17**, pp. 641–655.

Boccardo L., Gallouët T. and Orsina L. (1996). Existence and uniqueness of entropy solutions for nonlinear elliptic equations with measure data, *Ann.*

*Ist. H. Poincaré Anal. Non Linéaire* **13**, pp. 539–551.

Boccardo L., Murat F. and Puel J. P. (1984). Résultats d'existence pour certains problèmes elliptiques quasilinéaires, *Ann. Scuola Norm. Sup. Pisa, Cl. Scienze* **11**, pp. 213-235.

Borghol R. and Véron L. (2006). Boundary singularities of solutions of N-harmonic equations with absorption, *J. Funct. Anal.* **241**, pp. 611–637.

Borghol R. and Véron L. (2007). Boundary singularities of N-harmonic functions, *Comm. Part. Diff. Eq.* **32**, pp. 1001–1015.

Borsuk M. and Kondratiev V. (2007). *Elliptic Boundary Value Problems of Second Order in Piecewise Smooth Domains*, North-Holland Mathematical Library **69**, Elsevier (2006).

Bourbaki N. (1965). *Éléments de Mathématiques-Intégration* **Livre VI**, Actualités Scientifiques et Industrielles, Hermann, Paris.

Brandolini B., Chiacchio F., Cîrstea F. C. and Trombetti C. (2013). Local behaviour of singular solutions for nonlinear elliptic equations in divergence form, *Calc. Var. Part. Diff. Eq.* **48**, pp. 367–393.

Brelot M. (1967). *Lectures on potential theory*, Notes by K. N. Gowrisankaran and M. K. Venkatesha Murthy. Second edition, revised and enlarged with the help of S. Ramaswamy. Tata Institute of Fundamental Research Lectures on Mathematics **19**, ii+170+iv pp, Bombay.

Brezis H. (1980). Some variational problems of the Thomas-Fermi type (1980). in *Variational inequalities and complementary problems: theory and applications*, (Proc. Internat. School, Erice, 1978) (R. W. Cottle, F. Giannessi and J.-L. Lions ed.), Wiley, pp. 53–73.

Brezis H. (2010). *Functional Analysis, Sobolev Spaces and partial Differential Equations*, Universitext, Springer.

Brezis H. and Wainger S. (1980). A note on limiting cases of Sobolev embeddings, *Comm. in Part. Diff. Eq.* **5**, pp. 773–789.

Chen X. Y., Matano H. and Véron L. (1989). Anisotropic Singularities of Solutions of Nonlinear Elliptic Equations in $\mathbb{R}^2$, *J. Funct. Anal.* **83**, pp. 50–97.

Cîrstea F. and Du Y. (2010). Isolated singularities for weighted quasilinear elliptic equations, *J. Funct. Anal.* **259**, pp. 174–202.

Cîrstea F. and Vétois J. (2015). Fundamental solutions for anisotropic elliptic equations: existence and a priori estimates, *Comm. Partial Differential Equations* **40**, pp. 727–765.

Choquet G. (1953-54). Theory of capacities, *Ann. Inst. Fourier* **5**, pp. 131–295.

Courant R. and Hilbert D. (1953). *Methods of Mathematical Physics*, Vol. 1, Wiley-VCH.

Crandall M. and Rabinowitz P. (1971). Bifurcation from simple eigenvalues, *J. Funct. Anal.* **8**, pp. 321–340.

Dal Maso G. (1983). On the integral representation of certains functional, *Ricerche Mat.* **32**, pp. 85–113.

Dal Maso G., Murat F., Orsina L. and Prignet A. (1999). Renormalized solutions of elliptic equations with general measure data, *Ann. Scuola Norm. Sup. Pisa Cl. Sci.* **28**, pp. 741–808.

Damascelli L., Merchán S., Montoro L. and Sciunzi B.(2014). Radial symmetry and applications for a problem involving the $-\Delta_p$ operator and critical nonlinearity in $\mathbb{R}^N$. *Adv. Math.* **265**, pp. 313–335.

Dat C. T. and Verbitsky I. E. (2015). Finite energy solutions of quasilinear elliptic equations with sub-natural growth terms, *Calc. Var. Part. Diff. Eq.* **52**, pp. 529–546.

de Figueiredo D., Lions, P.-L. and Nussbaum, R (1982). A priori estimates and existence of positive solutions of semilinear elliptic equations, *J. Math. Pures Appl.* **61**, pp. 41–63.

Dhersin J. -S. and Le Gall J.-F. (1997). Wieners test for super-Brownian motion and the Brownian snake, *Probab. Theory Related Fields* **108**, pp. 103–129.

Diaz G. and Letelier A. R. (1993). Explosive solutions of quasilinear elliptic equations: existence and uniqueness, *Nonlinear Anal.* **20**, pp. 97–125.

Dolzmann G., Hungerbühler N. and Müller S (2000). Uniqueness and maximal regularity for nonlinear elliptic systems of n-Laplace type with measure valued right hand side, *J. Reine Angew. Math.* **520**, pp. 1–35.

Dràbek P. and Robinson S. (2004). On the generalization of the Courant nodal domain theorem, *J. Diff. Eq.* **181**, pp. 58–71.

Evans L. (1982). A new proof of local $C^{1,\alpha}$ regularity for solutions of certain degenerate elliptic p.d.e., *J. Diff. Eq.* **45**, pp. 356–373.

Evans L. and Smart K. (2011). Everywhere differentiability of infinity harmonic functions, *Calc. Var. Part. Diff. Eq.* **42**, pp. 289–299.

Faria L.F.O., Miyagaki O.H., Motreanu D. and Tanaka M. (2014). Existence results for nonlinear elliptic equations with Leray-Lions operator and dependence on the gradient, *Nonlinear Anal.* **96**, pp. 154–166.

Farina A., Montoro L. and Sciunzi B. (2012). Monotonicity and one dimensional symmetry for solutions of $-\Delta_p u = f(u)$ in half-spaces, *Calc. Var. Part. Diff. Eq.* **43**, pp. 123–145.

Farina A., Montoro L. and Sciunzi B. (2017). Monotonicity in half-space of positive solutions to $-\Delta_p u = f(u)$ in the case $p > 2$, *ArXiv* 1509.03897 [math. AP], *Ann. Sc. Norm. Sup. Pisa*, to appear.

Farina A., Montoro L., Riey G. and Sciunzi B. (2017). Monotonicity of solutions to quasilinear problems with a first-order term in half-spaces, *ArXiv* 1509.03897 [math. AP], *Ann. Inst. H. Poincaré-Anal. Non Linéaire*, to appear.

Feyel D. and de la Pradelle A. (1977). Topologies fines et compactifications associées certains espaces de Dirichlet, *Ann. Inst. Fourier* **27**, pp. 121–146.

Federer H. (1969). *Geometric Measure Theory*, Springer-Verlag, Heidelberg- New York.

Firenza A. (1998). *A summability condition on the gradient ensuring BMO, Rev. Mat. Complutense* **11**, pp. 313–324.

Friedlander L. (1989). Asymptotic behavior of the eigenvalues of the p-Laplacian, *Comm. Part. Diff. Eq.***14**, pp. 1059–1069.

Friedman A. and Véron L. (1986). Singular solutions of some quasilinear elliptic equations, *Arch. Rat. Mech. Anal.* **96**, pp. 359–387.

Frostman O. (1935). Potentials d'équilibre et capacité des ensembles avec quelques applications à la théorie des fonctions, *Medd. Lunds Univ. Mat. Sem.* **3**,

1-118.

Fukushima M., Sato K. and Taniguchi S. (1991). On the closable part of pre-Dirichlet forms and the fine support of the underlying measures, *Osaka J. Math.* **28**, pp. 517–535.

Gallot S. (1983). A Sobolev inequality and some geometric application, in: *Spectra of Riemannian Manifolds*, Kaigai Publ., pp. 45–55.

Garcia Azorero J. and Peral I. (1988). Comportement asymptotique des valeurs propres du p-laplacien, *C. R. Acad. Sci. Paris Ser. I Math.* **307**, pp. 75–78.

Garciá Melián J. (2009). Large solutions for equations involving the $p$-Laplacian and singular weights, *Z. Angew. Math. Phys* **60**, pp. 594–607.

Gidas B., Ni W. and Nirenberg L. (1979). Symmetries and related properties via the maximum principle, *Comm. Math. Phys.* **68**, pp. 209–243.

Gidas B. and Spruck J. (1980). Local and global behaviour of positive solutions of nonlinear elliptic equations, *Comm. Pure Appl. Math.* **34**, pp. 525–598.

Gilbarg D. and Trudinger N. (1983). *Elliptic Partial Differential Equations of Second Order*, Grundleheren der mathematischen Wissenshaften **vol. 224**, Springer-Verlag. Berlin, Heidelberg, New-York, Tokyo.

Giusti E. (1984). Minimal surfaces and functions of bounded variation, *Monographs in Mathematics* **vol. 80**, Birkhäuser, Boston-Basel-Stuttgart.

Gkikas K. and Véron L. (2017). The spherical p-harmonic eigenvalue problem in general domains, to appear.

Grafakos L. (2008). *Classical Fourier Analysis*, Graduate Texts in Mathematics **vol. 249** (2nd ed.), Springer-Verlag. Berlin, New York.

Grenon N. (2002). Existence results for semilinear elliptic equations with small measure data, *Ann. Inst. H. Poincaré-Anal. Non Linéaire* **19**, pp. 1–11.

Grun-Rehomme M. (1977). Caractérisation du sous-différential d'intégrandes convexes dans les espaces de Sobolev, *J. Math. Pures Appl.* **56**, pp. 149–156.

Guedda M. and Véron L. (1988). Local and global properties of solutions of quasilinear elliptic equations, *J. Diff. Eq.* **76**, pp. 159–189.

Hedberg L. and Wolff T. H. (1983). Thin sets in nonlinear potential theory, *Annales Inst. Fourier* **33**, pp. 161–187.

Heinonen J., Kilpelinen T., Martio O. (2006). *Nonlinear Potential Theory of Degenerate Elliptic Equations*, Dover Publications, Inc., Mineola, NY, (Unabridged republication of the 1993 original).

Honzik P. and Jaye B. (2012). The good $\lambda$-inequality for nonlinear potentials, *Proc. Amer. Math. Soc.* **140**, pp. 4167–4180.

Huentotripay J., Jazar M. and Véron L. (2003). A dynamical system approach to the construction of singular solutions of some degenerate elliptic equations, *J. Diff. Eq.* **195**, 175–193.

John F. and Nirenberg L. (1961). On functions of bounded mean oscillation, *Comm. Pure Appl. Math.* **14**, 415–426.

Juutinen P. (2007). Principal eigenvalue of a very badly degenerate operator and applications, *J. Diff. Eq.* **236**, 532–550.

Juutinen P., Lindqvist P. and Manfredi J. (1999). The $\infty$-eigenvalue problem, *Arch. Rat. Mech. Anal.* **148**, pp. 89–105.

Kichenassamy S. and Véron L. (1988). Singular solutions of the $p$-Laplace equation, *Math. Ann.* **275**, pp. 599–615.

Kilpeläinen T., Kuusi T. and Tuhola-Kujanpää A. (2011). Superharmonic functions are locally renormalized solutions, *Ann. Inst. H. Poincaré-Anal. Nonlinéaire* **28**, pp. 775–795.

Kilpeläinen T. and Li G. (2000). Estimates for $p$-Poisson equations, *Diff. Int. Eq.* **13**, pp. 791–800.

Kilpeläinen T. and Malý J. (1994). The Wiener test and potential estimates for quasilinear elliptic equations, *Acta Math.* **172**, pp. 137–161.

Kilpeläinen T., Shahgholian H. and Zhong X. (2007). Growth estimates through scaling for quasilinear partial differential equations, *Ann. Acad. Sci. Fenn. Math.* **32**, pp. 595–599.

Kilpeläinen T., Shanmugalingam N. and Zhong X. (2008). Maximal regularity via reverse Höder inequalities for elliptic systems of n-Laplace type involving measures, *Ark. Mat.(2008)* **46** , pp. 77–93.

Krasnosel'skii M. A. (1964). *Positive Solutions of Operator Equations*, Noordhoff, Groningen.

Krol I.N. (1973). On the behaviour of the solutions of a quasilinear equation near null salient points of the boundary, *Proc. Steklov Inst. Math.* **125**, pp. 130–136.

Kura T. (1989). The weak supersolution subsolution method for second order quasilinear elliptic equations, *Hiroshima Math. J.* **19**, pp. 1–36.

Labutin D. A. (2003). Wiener regularity for large solutions of nonlinear equations, *Ark. Mat.* **41**, pp. 307–339.

Lasry J. M. and Lions P. L. (1989). Nonlinear elliptic equations with singular boundary conditions and stochastic control with state constraints. I. The model problem, *Math. Ann.* **283**, pp. 583–630.

Leray J. and Lions J. L. (1965). Quelques résultats de Višik sur les problèmes elliptiques non linéaires par les méthodes de Minty-Browder, *Bull. Soc. Math. France* **93**, pp. 97 107.

Lewis J. (1983). Regularity of the derivatives of solutions to certain degenerate elliptic equations, *Indiana Univ. Math. J.* **32**, pp. 849 858.

Lewis J. (2012). Applications of Boundary Harnack Inequalities for p Harmonic Functions and Related Topics, *Lecture Notes in Math.* **2045**, Springer, Heidelberg, pp. 1 72.

Lewis J. and Nyström K. (2010). Boundary behavior and the Martin boundary problem for p harmonic functions in Lipschitz domains, *Ann. Math.* **172**, pp. 1907 1948.

Li P. and Yau S. T. (1980). Estimates of eigenvalues of a compact Riemannian manifold, in *Proc. Sympos. Pure Math.* **XXXVI**, Amer. Math. Soc., pp. 205–239.

Lieberman G. (1988). Boundary regularity for solutions of degenerate elliptic equations, *Nonlinear Anal.* **12**, pp. 1203–1219.

Lieberman G. (1991). The natural generalization of the natural conditions of Ladyzhenskaya and Ural'tseva for elliptic equations, *Comm. in Part. Diff. Eq.* **16**, pp. 311–361.

Lindqvist P. (1990-1992). On the equation $-div\left(|\nabla u|^{p-2}\right)\nabla u+\lambda u^{p-2}u=0$, *Proc.*

*Amer. Math. Soc.* **109**, pp. 157-164. Addendum, ibidem **116**, pp. 583–594.

Lindqvist P. (1993). On nonlinear Rayleigh quotients, *Potential Anal.* **2**, pp. 199–218.

Lindqvist P. (1995). On a nonlinear eigenvalue problem, *Fall School in Analysis (Jyväskylä, 1994)*, Report **68**, Univ. Jyväskylä, Jyväskylä.

Lindqvist P. (2006). Notes on the $p$-Laplace equation, *Report. University of Univ. Jyväskylä Department of Mathematics and Statistics* **102**, Univ. Jyväskylä, Jyväskylä, ii+80 pp. ISBN: 951-39-2586-2

Lindqvist P. and Manfredi J. (1997). Note on $\infty$-superharmonic functions, *Rev. Mat. Univ. Complutense Madrid* **10**, pp. 471–480.

Lions J. L. (1969). *Quelques méthodes de résolution des problèmes aux limites non linéaires*, Études Mathématiques, Dunod, Gauthiers-Villars, Paris.

Lions J. L. and Petree J. (1964). Sur une classe d'espaces d'interpolation, *Pub. Inst. Hautes Études Sci.* **119**, pp. 5–68.

Malý J. and Ziemer W. (1997). *Fine Regularity of Solutions of Elliptic Partial Differential Equations*, Mathematical Surveys and Monographs **51**, Amer. Math. Soc., Providence, RI, xiv+291 pp.

Marcus M. and Nguyen P. T. (2015). Elliptic equations with nonlinear absorption depending on the solution and its gradient., *Proc. Lond. Math. Soc.* **111**, pp. 205–239.

Marcus M. and Véron L. (1997). Uniqueness and asymptotic behaviour of solutions with boundary blow-up for a class of nonlinear elliptic equations, *Ann. Inst. Henri Poincaré. Analyse Non Linéaire* **14**, pp. 237–274.

Marcus M. and Véron L. (2004). Existence and uniqueness results for large solutions of general nonlinear elliptic equations, *J. Evol. Eq.* **3**, pp. 637–652.

Marcus M. and Véron L. (2009). Maximal solutions for $-\Delta u + u^q = 0$ in open and finely open sets, *J. Math. Pures Appl.* **91**, pp. 256–295.

Matano H. (1990). Singular solutions of a nonlinear elliptic equation and an infinite-dimensional dynamical system, *Functional-analytic methods for partial differential equations (Tokyo, 1989)* Lecture Notes in Math., **1450**, Springer, Berlin, 1990, pp. 64–87.

Matei A. M. (2000). First eigenvalue for the p-Laplace, *Nonlinear Analysis* **39**, pp. 1051–1068.

Maz'ya V. G. (1970). The continuity at a boundary point of the solutions of quasilinear elliptic equations (Russian), *Vestnik Leningrad. Univ.* **25**, pp. 42–55.

Mitidieri M. and Pohozaev S. I. (2001). A priori estimates and the absence of solutions of nonlinear partial differential equations and inequalities, *Proc. Steklov Inst. Math.* **3 (234)** pp. 1–362.

Moser J. (1967). A sharp form of an inequality by N. Trudinger, *Indiana Univ. Math. J.* **30**, pp. 473–484.

Muckenhoupt B. and Wheeden R. (1974). Weighted norm inequalities for fractional integrals, *Trans. Amer. Math. Soc.* **192**, pp. 261–274.

Naber A. and Valtorta D. (2014). Sharp estimates on the first eigenvalue of the p-Laplacian with negative Ricci lower bound, *Math. Z* **277**, pp. 867–891.

Nash J. (1954). $C^1$-isometric imbeddings, *Annals Math.* **60**, pp. 383–396.

Nash J. (1956). The imbedding problem for Riemannian manifolds, *Annals Math.*

**63**, pp. 20–63.

Nguyen P. T. and Véron L. (2012). Boundary singularities of solutions to elliptic viscous Hamilton-Jacobi equations, *J. Funct. Anal.* **263**, pp. 1487–1538.

Nguyen Q. H. and Véron L. (2014). Quasilinear and Hessian Type Equations with Exponential Reaction and Measure Data, *Arch. Rat. Mech. Anal.* **214**, pp. 235–267.

Nguyen Q. H. and Véron L. (2015). Wiener Criteria for Existence of Large Solutions of Quasilinear Elliptic Equations with Absorption, *Potential Anal.* **42**, pp. 681–697.

Otani M. (1984). A remark on certain nonlinear elliptic equations. *Proceedings of the Faculty of Science, Tokai University* **19**, pp. 20–28.

Perron O. (1923). Eine neue Behandlung der ersten Randwertaufgabe für $\Delta u = 0$, *Math. Z.* **18**, pp. 42–54.

Phuc N. Q. (2014). Nonlinear Muckenhoupt–Wheeden type bounds on Reifenberg flat domains, with applications to quasilinear Riccati type equations, *Adv. in Math.* **250**, pp. 387–419.

Phuc N. Q. and Verbitsky I. E. (2006). Local integral estimates and removable singularities for quasilinear and Hessian equations with nonlinear source terms, *Comm. Part. Diff. Eq.* **31**, pp. 1779–1791.

Phuc N. Q. and Verbitsky I. E. (2008). Quasilinear and Hessian equations of Lane-Emden type, *Ann. of Math.* **168**, pp. 859–914.

Phuc N. Q. and Verbitsky I. E. (2009). Quasilinear and Hessian equations of Lane-Emden type, *J. Funct. Anal.* **256**, pp. 1875–1906.

Pohozaev S. I. (1965). On the eigenfunctions of the equation $\Delta u = \lambda f(u)$, *Dokl. Akad. Nauk SSSR* **165** (Russian), pp. 36–39.

Pohozaev S. I. (1965). The Sobolev embedding in the case $pl = n$, *Proc. Tech. Sci. Conf. on Adv. Sci., Research 1964–1965*, Mathematics Section (1965), 158–170, Moskov. 'Energet. Inst., Moscow.

Polàcik P., Quittner P. and Souplet Ph. (2007). Singularity and decay estimates in superlinear problems via Liouville- type theorems. I. Elliptic equations and systems, *Duke Math. J.* **139**, pp. 555–579.

Porretta A. and Véron L. (2009). Separable p-harmonic functions in a cone and related quasilinear equations on manifolds, *J. Europ. Math. Soc.* **11**, pp. 1285–1305.

Porretta A. and Véron L. (2013). Separable solutions of quasilinear Lane–Emden equations, *J. Europ. Math. Soc.* **15**, pp. 755–774.

Protter M. and Weinberger H. (1967). *Maximum Principles in Differential Equations*, Prentice-Hall.

Pucci P. and Serrin J. (2007). *The maximum principle*, Progress in Nonlinear Differential Equations and Their Applications **vol. 73**, Bikhäuser. Basel, Boston, Berlin.

Rakotoson J. M. (1994). Uniqueness of renormalized solutions in a $T$-set for the $L^1$-data problem and the link between various formulations, *Indiana Univ. Math. J.* **43**, pp. 685–702.

Ratto A., Rigoli M. and Véron L. (1994). Conformal immersion of complete Riemannian manifolds and extensions of the Schwarz lemma, *Duke Math. J.*

**74**, 223–236.

Reshetniak Y. G. (1966). Mapping with bounded deformation as extremals of Dirichlet type integrals, *Sibirsk. Mat. Zh* **9**, pp. 652–666.

Sakai T. (1997). *Riemannian geometry*, Transl. Math. Monogr. **vol. 149**, Amer. Math. Soc.

Serrin J. (1964). Local behaviour of solutions of quasilinear equations, *Acta Math.* **111**, pp. 247–302.

Serrin J. (1965). Isolated singularities of solutions of quasilinear equations, *Acta Math.* **113**, pp. 219–240.

Serrin J. (1971). A symmetry problem in potential theory, *Arch. Rat. Mech. Anal.* **43**, pp. 304–318.

Serrin J. and Zou H. (2002). Cauchy-Liouville and universal boundedness theorems for quasilinear elliptic equations and inequalities, *Acta Math.* **189**, pp. 79–142.

Sciunzi B. (2015). Monotonicity results in half-spaces for the $p$-Laplacian. *Proc. Roy. Soc. Edinburgh Sect. A* **145**, pp. 597–610.

Sciunzi B. (2012). Classification of positive $\mathcal{D}^{1,p}(\mathbb{R}^N)$-solutions to the critical $p$-Laplace equation in $\mathbb{R}^N$. *Adv. Math.* **291**, pp. 12–23.

Stein E.M. (1971). *Singular Integrals and Differentiability of Functions*, Princeton Univ. Press, Princeton, NJ.

Struwe M. 1996. *Variational methods. Applications to nonlinear partial differential equations and Hamiltonian systems*, 2nd edition, Ergebnisse der Mathematik und ihrer Grenzgebiete (3), **34**, Springer-Verlag, Berlin, xvi+272 pp.

Tartar L. (1998). Imbedding theorems of Sobolev spaces into Lorentz spaces, *Boll. Un. Mat. Ital.* **8**, pp. 479–500.

Toro T. (1997). Doubling and flatness: geometry of measures, *Notices Amer. Math. Soc.* **44**, pp. 1087–1094.

Tolksdorff P. (1983). On the Dirichlet problem for quasilinear equations in domains with conical boundary points, *Comm. Part. Diff. Eq.* **8**, pp. 773–817.

Tolksdorff P. (1984). Regularity for a more general class of quasilinear elliptic equations, *J. Diff. Eq.* **51**, pp. 126–150.

Trudinger N. (1967). On Harnack type inequalities and their applications to quasilinear elliptic equations, *Comm. Pure Appl. Math.* **20**, pp. 721–747.

Trudinger N. (1980). On embeddings into Orlicz spaces and some applications, *J. Math. Mech.* **75**, pp. 59–77.

Trudinger N. and Wang X. J. (2009). Quasilinear elliptic equations with signed measure data, *Discrete and Cont. Dyn. Systems* **23**, pp. 477–494.

Turesson B. (2000). *Nonlinear potential theory and weighted Sobolev spaces*, Lecture Notes in Mathematics **1736**, xiv+173 pp. Springer-Verlag, Berlin.

Uhlenbeck K. (1977). Regularity for a class of non-linear elliptic systems; *Acta Math.* **138**, pp. 219–240.

Valtorta D. (2014). On the p-Laplace operator on Riemannian manifolds, PhD Thesis in Math. Università Degli Studi di Milano. pp. 1–107.

Vazquez J. L. (1981). An a priori interior estimate for the solution of a nonlinear problem representing weak diffusion, *Nonlinear Anal. T., M. & A.* **5**, pp. 119–135.

Vazquez J. L. (1983). On a semilinear equation in $R^2$ involving bounded measures, *Proc. Roy. Soc. Edinburgh* **95A**, pp. 181–202.

Vazquez J. L. (1984). A Strong Maximum Principle for Some Quasilinear Elliptic Equations, *Appl. Math. Opt.* **12**, pp. 191–202.

Vazquez J. L. and Véron L. (1981). Removable singularities of some strongly nonlinear elliptic equations, *Manuscripta Math.* **33**, pp. 129–144.

Verbitsky I. E. (2008). Weighted norm inequalities with positive and indefinite weights, in *Perspectives in partial differential equations, harmonic analysis and applications. Proc. Sympos. Pure Math.* **79** pp. 377–406.

Véron L. (1987). Singularities of some quasilinear equations, *Nonlinear Diffusion Equations and their Equilibrium States*, Math. Sci. Res. Inst. Series **13**, Springer-Verlag Ed., pp. 335–365.

Véron L. (1991). Some existence and uniqueness results for solution of some quasilinear elliptic equations on compact Riemannian manifolds, *Colloq. Math. Soc. Janos Bolyai* **62**, North-Holland, Amsterdam, pp. 317–352.

Véron L. (1996). *Singularities of Solutions of Second Order Quasilinear Equations*, Pitman Research Notes in Math. **353**, Addison Wesley Longman Inc.

Véron L. (2002). Singular p-harmonic functions and related quasilinear equations on manifolds, *Elec. J. Diff. Eq., Conf.* **8**, pp. 133–154.

Véron L. (2004). Elliptic equations involving measures, in *Stationary Partial Differential Equations* **vol. I**, *Handb. Differ. Eq., North-Holland*, Amsterdam, pp. 593–712.

Wiener N. (1924). The Dirichlet problem, *J. Math. and Phys.* **3**, pp. 127–146.

Yang H. C. and Zhong J. Q. (1984). On the estimate of the first eigenvalue of a compact Riemannian manifold, *Sci. Sinica Ser.* **A 27**, pp. 1265–1309.

Zou H. (2008). A priori estimates and existence for quasi-linear elliptic equations, *Calc. Var. Part. Diff. Eq.* **33** pp. 417–437.

# List of symbols

# Glossary